Interdisciplinary Applied Mathematics

Editors
S.S. Antman   P. Holmes
L. Sirovich   K. Sreenivasan

Series Advisors
C.L. Bris   L. Glass
P.S. Krishnaprasad   R.V. Kohn
J.D. Muray   S.S. Sastry

Problems in engineering, computational science, and the physical and biological sciences are using increasingly sophisticated mathematical techniques. Thus, the bridge between the mathematical sciences and other disciplines is heavily traveled. The correspondingly increased dialog between the disciplines has led to the establishment of the series: *Interdisciplinary Applied Mathematics*.

The purpose of this series is to meet the current and future needs for the interaction between various science and technology areas on the one hand and mathematics on the other. This is done, firstly, by encouraging the ways that mathematics may be applied in traditional areas, as well as point towards new and innovative areas of applications; and, secondly, by encouraging other scientific disciplines to engage in a dialog with mathematicians outlining their problems to both access new methods and suggest innovative developments within mathematics itself.

The series will consist of monographs and high-level texts from researchers working on the interplay between mathematics and other fields of science and technology.

Interdisciplinary Applied Mathematics

Andrew Fowler

Mathematical
Geoscience

 Springer

Andrew Fowler
MACSI, Department of Mathematics
& Statistics
University of Limerick
Limerick, Ireland

Series Editors
S.S. Antman
Department of Mathematics
and
Institute for Physical Science
and Technology
University of Maryland
College Park, MD 20742, USA
ssa@math.umd.edu

P. Holmes
Department of Mechanical and Aerospace
Engineering
Princeton University
215 Fine Hall
Princeton, NJ 08544, USA
pholmes@math.princeton.edu

L. Sirovich
Department of Biomathematics
Laboratory of Applied Mathematics
Mt. Sinai School of Medicine
Box 1012
New York, NY 10029, USA
Lawrence.Sirovich@mssm.edu

K. Sreenivasan
Department of Physics
New York University
70 Washington Square South
New York City, NY 10012, USA
katepalli.sreenivasan@nyu.edu

ISSN 0939-6047
ISBN 978-1-4471-6821-8   ISBN 978-0-85729-721-1 (eBook)
DOI 10.1007/978-0-85729-721-1
Springer London Dordrecht Heidelberg New York

British Library Cataloguing in Publication Data
A catalogue record for this book is available from the British Library

Mathematics Subject Classification (2000): 86.02, 76.02, 35.02, 34.02

Cover design: VTeX UAB, Lithuania

Printed on acid-free paper

Springer is part of Springer Science+Business Media (www.springer.com)

This book is dedicated with affection and appreciation to Jim Murray and his wife Sheila

Jim and Sheila Murray in the garden of their home in Connecticut, summer 2010

Preface

The hardest thing to do with this book was to decide what to call it. The original working title was 'Mathematics and the environment', and my aspiration was, and is, to provide a blueprint for the application of mathematical models to problems in the environment which involve the use of differential equations.

The environment is becoming fashionable in applied mathematics, but it often means different things to different people. It may mean oceans and atmospheres, and numerical modelling; it may mean groundwater flow and related pollution problems, for example involving remediation of hydrocarbons or dispersal of phosphates and nitrates in the soil; or it might be the application of statistical methods in the assessment of risk and uncertainty in, for example, hydrological forecasting.

No doubt these subjects concern the environment, but they are particular topics. This book is about general scientific problems concerning phenomena in the world around us. In the sense that 'mathematical biology' is the mathematical study of living things, the logical title for this book would be 'Mathematical Geology', the mathematical study of processes on (or in) the Earth. Unfortunately, Geology is a subject which tends to carry the narrower meaning of the study of rocks, and it is partly to get away from this that university departments have increasingly rechristened themselves as departments of Geology and Geophysics, or of Earth Science, or (most recently) of Earth System Science.

So, this book is not just about mathematical geology: it concerns much more than the study of rocks. Nor is it mathematical geophysics, although it contains a good deal of this also. It is mathematics and the environment, but where the word 'environment' is used in a much wider sense than the narrower uses alluded to above.

The two books which are closest to this in theme and subject matter are Andrew Goudie's 'The Nature of the Environment', and Arthur Holmes's masterful 'Principles of Physical Geology'. The latter book could almost provide the contents list for the present one. The difference of course is that my concern here is in providing mathematical models which can explain some of the physical phenomena which are described in these two books.

Writing about recent theories for subglacial landforms, Clarke (2005) said that 'the work has a daunting mathematical level, uncertain relevance, but potentially interesting implications.' For an applied mathematician working in seriously inter-disciplinary subjects, perhaps this slightly barbed comment is as good as it gets. This book is, I expect, daunting. It is not necessary that hard scientific problems beget hard mathematical problems when they are done properly, but it ought to be what you expect. Decent science does not come cheap.

I personally hope that most of this book *is* relevant, but that is ultimately a matter for the scientific community. Relevance is promoted by a kind of cultural accep-tance, and it needs to be argued through, almost religiously.

This book in its earliest form consisted of written course notes for a sixteen lecture final year undergraduate course at Oxford. I have taught a similar twenty-four lecture course at masters' level at Limerick. For such courses, I select four or five chapters, and selectively teach material from them. For example, the current Oxford and Limerick courses take material from Chaps. 2, 4, 5 and 10. Of course the chapters contain much more material than one could cover in four or six lectures; one could in fact take an entire course from a single chapter. But my purpose here is to allow a freedom for selection, and also to elaborate the material to the point where it becomes of research interest. In writing the book, I have been stimulated to question accepted wisdom, and to explore new ideas, and some of the material has even been written up in the form of research papers after the fact.

There is a danger in trying to write an encompassing book about mathematical geoscience, of which I am only too aware. Most obviously, there are many subjects which have been left out, and for those which are included, there is no space for a comprehensive exposition. A glance at the reference list will show that I have largely followed my own personal view of the subject matter. References are given at the end of each chapter, but do not aim to give a complete review; rather the intention is to provide pointers for those interested, with the hope that others will engage with some of the problems. Geoscience is full of extraordinarily interesting problems.

The audience for this book is largely what is called the GFD community, brought up on fluid mechanics in the oceans and atmosphere, but which has now branched out into many of the subjects dealt with here. It is my hope that applied mathemati-cians may chance on the material, and be stimulated to explore some of the models which are discussed. It is also my hope that geoscientists will find some of the phe-nomena and ideas interesting, even if some of the technical detail becomes at times too threatening.

A large number of people have been of considerable assistance and help in the something like ten years it has taken to finally produce this book. Firstly, I should thank my publishers at Springer, who have been very patient over the years: Karen Borthwick, and more recently, Lauren Stoney. I am grateful to Felix Ng, who rapidly and expertly produced early drafts of some of the figures for Chaps. 2, 4, 7 and 11. Ian Hewitt produced Fig. 10.14 and Fig. 9.15. Christine Butler unearthed a copy of Fig. 11.12 from the vaults of the International Glaciological Society. Bill Shilts, Christian Zdanowicz and Brian Moorman were very helpful concerning the image in Fig. 10.22; Gary Parker, Norm Smith and Terence McCarthy were equally helpful concerning Fig. 5.1. Thanks also to Emanuele Schiavi, Stephen O'Brien, Thomas Vitolo, Dave Cocks, Rachel Zammett, Geoff Evatt, Rob Style, Sarah Mitchell, Chris

Banerji and Sarah McBurnie for their vigilance in spotting errors or providing advice. Neil Balmforth has been very kind in providing photographs and movies of roll waves. Duncan Wingham has been a great help sorting out some of the scaling arguments in Chaps. 10 and 11. Eric Wolff was very kind in providing me with ice core data, and spending time explaining to me how it worked. Torgeir Wiik and Kjartan Rimstad pointed out errors in Sect. 2.5.7.

I solicited comments on individual chapters from many people, and these have been of great use. Firstly, my thanks to Garry Clarke and Chris Clark, who provided images (of Trapridge Glacier and ribbed moraine in Northern Ireland) for the front cover; sadly they could not be used because it took me so long to finish the book that in the meantime Springer changed the series design! Bruce Malamud spent a year in Oxford, and was no end of help in the minutiae of computer technology. I have received useful critical comments from Tom Witelski, Stephen O'Brien, Eric Wolff, Richard Alley, Henry Winstanley, Slava Solomatov, Alison Rust, Ian Hewitt, Garry Clarke, Janet Elliott and Don Drew.

Thanks to Ros Rickaby for discussions on carbon; Andy Ellis and Giles Wiggs for providing images of dunes; Mark McGuinness for Figs. 5.12 and 5.16; Mike Vynnycky for discussion on diapirism, and for providing the computations and the resultant figures in Figs. 8.3, 8.6, 8.10 and 8.11. Thanks also to Sophie Nowicki, for discussions concerning the grounding line; Rich Katz, for his comments on the material on ice streams; Ian Hewitt, for discussions about canals and eskers, together with many other things; my fellow drumliners, Chris Clark, Paul Dunlop, Chris Stokes and Matteo Spagnolo for much information and insight into the geographic setting of drumlins; Peter Howell, for comments on viscous beams; Geoff Evatt, for help in assembling Sect. 11.7.

For a book such as this, it would be remiss not to mention with gratitude the annual GFD summer school at Woods Hole, where I have variously spent long periods of time, most recently in 2010, and where I have benefitted from the experience and wisdom of that excellent community of scholars, in particular Joe Keller, Lou Howard, George Veronis, and Ed Spiegel. Those who have spent time on the porch or in the classroom at Walsh Cottage will know what a privilege it is to be there, in the presence of one of the brightest and wittiest seminar audiences on the planet.

The University of Limerick has supported me through my appointment there as an adjunct Professor and subsequently, through an award by Science Foundation Ireland, as Stokes Professor. The funds they have generously provided have enabled me to maintain a research presence at conferences and workshops, as well as purchasing two of the laptops on which this book was written. They have provided a pleasant and stimulating working environment, not to mention easy access to the best countryside in the world.

This book is dedicated to Jim Murray and his wife Sheila. I first met Jim on a cold, dark December evening in 1970, when I ascended staircase 10 in Corpus Christi College, Oxford, to be interviewed for a place as an undergraduate. We peered at each other in the ancient, wood-panelled room by candlelight (these were the days of miners' strikes and power cuts). Ever since then, Jim has been the torch-bearer for my path in applied mathematics, yielding to no man in his quest for the practical and useful.

My view of science, and the act of doing science, is that at best it is like driving a car on an icy road. You know the car works, the road is flat, but actually, you do not really know what you are doing. You try out a few things and they more or less work. You might hit a slippery bit, but if you are lucky you get there somehow. And if you are not lucky, you end up in the ditch. What you have to avoid is the idea that, if you end up in the ditch, it is the right place to be. Do not get stuck in the ditch. Get out of the car and back on the road.

It was Kolumban Hutter who said: you do not finish a book, you abandon it. He was so right. It is like bringing up a child. You love it, change its nappies, feed it, nurture it, but by the time it is an adult, it is time to go. Be gone!

Limerick, Ireland   A.C. Fowler

Contents

Chapter 1
Mathematical Modelling

This book concerns the application of mathematics to problems in the physical sciences, and particularly to problems which arise in the study of the environment. Much of the environment consists of fluid—the atmosphere, the ocean—and even those parts which are solid may deform in a fluid-like way—ice sheets, glaciers, the Earth's mantle; as a consequence, one way into the study of the environment is through the study of fluid dynamics, although we shall not follow that approach here. Rather, we shall approach the study of environmental problems as applied mathematicians, where the emphasis is on building a suitable mathematical model and solving it, and in this introductory chapter, we set out the stall of techniques and attitudes on which the subsequent chapters are based.

There are two particular points of view which we can bring to bear on the mathematical models which describe the phenomena which concern us: these are the dynamical systems approach, or equivalently the bifurcation theory approach; and the perturbation theory approach. Each has its place in different contexts, and sometimes they overlap.

The bifurcation theory approach is most usually (but not always) brought to bear on problems which have some kind of (perhaps complicated) time-dependent behaviour. The idea is that we seek to understand the observations through the understanding of a number of simpler problems, which arise successively through bifurcations in the mathematical model, as some critical parameter is changed. A classic example of this approach is in the study of the origin of chaos in the Lorenz equations, or the onset of complicated forms of thermal convection in fluids.

In its simplest form (e.g., in weakly nonlinear stability theory) the perturbative approach is similar in method to the bifurcational one; however, the ethos is rather different. Rather than try and approach the desired solution behaviour through a sequence of simpler behaviours, we try and break down the solution by making approximations, which (with luck) are in fact realistic. In real problems, such approximations are readily available, and part of the art of the applied mathematician is having the facility of being able to judge how to make the right approximations. In this book, we follow the perturbative approach. It has the disadvantage of being harder, but it is able to get closer to a description of how realistic systems may actually behave.

A. Fowler, *Mathematical Geoscience*, Interdisciplinary Applied Mathematics 36, DOI 10.1007/978-0-85729-721-1_1, © Springer-Verlag London Limited 2011

1.1 Conservation Laws and Constitutive Laws

The basic building blocks of continuous mathematical models are conservation laws. The continuum assumption adopts the view that the physical medium of concern may be considered continuous, whether it be a porous medium (for example, sand on a beach) or a fluid flow. The continuum hypothesis works whenever the length or time scales of interest are (much) larger than the corresponding microscale. For example, the formation of dunes in a desert (length scale hundreds of metres) can be modelled as a continuous process, since the microscale (sand grain size) is much smaller than the macroscale (dune length). Equally, the modelling of large animal populations or of snow avalanches treats the corresponding media as continuous.

Conservation laws arise as mathematical equations which represent the idea that certain quantities are conserved—for example, mass, momentum (via Newton's law) and energy. More generally, a conservation law refers to an equation which relates the increase or decrease of a quantity to terms representing supply or destruction.

In a continuous medium, the typical form of a conservation law is as follows:

$$\frac{\partial \phi}{\partial t} + \nabla . \mathbf{f} = S. \tag{1.1}$$

In this equation, ϕ is the quantity being 'conserved' (expressed as amount per unit volume of medium, i.e., as a density; $\mathbf{f}$ is the 'flux', representing transport of ϕ within the medium, and S represents source ($S > 0$) or sink ($S < 0$) terms. Derivation of the point form (1.1) follows from the integral statement

$$\frac{d}{dt} \int_V \phi \, dV = - \int_{\partial V} \mathbf{f.n} \, dS + \int_V S \, dV, \tag{1.2}$$

after application of the divergence theorem (which requires $\mathbf{f}$ to be continuously differentiable), and by then equating integrands, on the basis that they are continuous and V is arbitrary. Derivation of (1.1) thus requires ϕ and $\mathbf{f}$ to be continuously differentiable, and S to be continuous.

Two basic types of transport are advection (the medium moves at velocity $\mathbf{u}$, so there is an advective flux $\phi\mathbf{u}$) and diffusion, or other gradient-driven transport (such as chemotaxis). One can thus write

$$\mathbf{f} = \phi\mathbf{u} + \mathbf{J}, \tag{1.3}$$

where $\mathbf{J}$ might represent diffusive transport, for example.

Invariably, conservation laws contain more terms than equations. Here, for example, we have one scalar equation for ϕ, but other quantities $\mathbf{J}$ and S are present as well, and equations for these must be provided. Typically, these take the form of constitutive laws, and are usually based on experimental measurement. For example, diffusive transport is represented by the assumption

$$\mathbf{J} = -D\nabla\phi, \tag{1.4}$$

where D is a diffusion coefficient. In the heat equation, this is known as Fourier's law, and the heat equation itself takes the familiar form

$$\frac{\partial}{\partial t}(\rho c_p T) + \nabla.(\rho c_p T \mathbf{u}) = \nabla.(k \nabla T) + Q, \tag{1.5}$$

where Q represents any internal heat source or sink.

1.2 Non-dimensionalisation

Putting a mathematical model into non-dimensional form is fundamental. It allows us to identify the relative size of terms through the presence of dimensionless parameters. Although technically trivial, there is a certain art to the process of non-dimensionalisation, and the associated concept of scaling. We illustrate some of the precepts by consideration of the heat equation, (1.5). We write it in the form (assuming density ρ and specific heat c_p are constant)

$$\frac{\partial T}{\partial t} + \mathbf{u}.\nabla T = \kappa \nabla^2 T + H, \tag{1.6}$$

where $H = Q/\rho c_p$. We have taken $\nabla.\mathbf{u} = 0$, which follows from the conservation of mass equation

$$\frac{\partial \rho}{\partial t} + \nabla.(\rho \mathbf{u}) = 0, \tag{1.7}$$

together with the supposition of incompressibility in the form $\rho = \text{constant}$.

Suppose we are to solve (1.6) in a domain D of linear magnitude l, on the boundary of which we prescribe

$$T = T_B \quad \text{on } \partial D, \tag{1.8}$$

where T_B is constant. We also have an initial condition

$$T = T_0(\mathbf{x}) \quad \text{in } D, \quad t = 0, \tag{1.9}$$

and we suppose $\mathbf{u}$ is given, of order U.

We can make the variables dimensionless in the following way:

$$\mathbf{x} = l\mathbf{x}^*, \quad \mathbf{u} = U\mathbf{u}^*, \quad t = t_c t^*, \quad T = T_B + (\Delta T)T^*. \tag{1.10}$$

We do this in order that both dependent and independent dimensionless variables be of numerical order one, written $O(1)$. *If* we can do this, then we might suppose *a priori* that derivatives such as $\nabla^* T^*$ ($\nabla = l^{-1}\nabla^*$) will also be of numerical $O(1)$, and the size of various terms will be reflected in certain dimensionless parameters which occur.

In writing (1.10), it is clear that l is a suitable length scale, as it is the size of D. For example, if D was a sphere we might take l as its radius or diameter. We also suppose that the origin is in D; if not, we could write $\mathbf{x} = \mathbf{x}_0 + l\mathbf{x}^*$, where $\mathbf{x}_0 \in D$: evidently $\mathbf{x}^* = O(1)$ in D.

A similar motivation underlies the choice of an 'origin shift' for T. In the absence of a heat source, the temperature will tend to the uniform state $T \equiv T_B$ as $t \to \infty$. If $H \neq 0$, the final state will be raised above T_B (if $H > 0$) by an amount dependent on H. We take ΔT to represent this amount, but we do not know what it is in advance—we will choose it by *scaling*. The subtraction of T_B from T before non-dimensionalisation is because the model for T contains only derivatives of T, so that it is really the variation of T about T_B which we wish to scale.

In a similar way, the time scale t_c is not prescribed in advance, and we will choose it also by scaling, in due course.

With the substitutions in (1.10), the heat equation (1.6) can be written in the form

$$\left(\frac{l^2}{\kappa t_c}\right)\frac{\partial T^*}{\partial t^*} + \left(\frac{Ul}{\kappa}\right)\mathbf{u}^*.\mathbf{\nabla}^* T^* = \nabla^{*2} T^* + \left(\frac{Hl^2}{\kappa \Delta T}\right). \tag{1.11}$$

This equation is dimensionless, and the bracketed parameters are dimensionless. They are somewhat arbitrary, since t_c and ΔT have not yet been chosen: we now do so by scaling.

The solution of the equation can depend only on the dimensionless parameters. It is thus convenient to *choose* t_c and ΔT so that two of these are set to some convenient value. There is no unique way to do this.

The temperature scale ΔT appears only in the source term. Since it is this which determines the temperature rise, it is natural to *choose*

$$\Delta T = \frac{Hl^2}{\kappa}. \tag{1.12}$$

It is also customary to choose the time scale so that the two terms of the advective derivative on the left of (1.11) are the same size, and this gives the convective time scale

$$t_c = \frac{l}{U}. \tag{1.13}$$

It is finally also customary (if sometimes confusing) to remove the asterisks (or whatever equivalent symbol is used). If this is done, the dimensionless equation takes the form

$$Pe\left[\frac{\partial T}{\partial t} + \mathbf{u}.\mathbf{\nabla} T\right] = \nabla^2 T + 1, \tag{1.14}$$

where the Péclet number is

$$Pe = \frac{Ul}{\kappa}, \tag{1.15}$$

and the solution of the model depends only on this parameter (as well as the initial condition). The boundary condition is

$$T = 0 \quad \text{on } \partial D, \tag{1.16}$$

Fig. 1.1 Sub-characteristics and boundary layer for Eq. (1.14) when $Pe \gg 1$. The sub-characteristics are the flow lines $d\mathbf{x}/dt = \mathbf{u}$, and the boundary layer (of thickness $O(1/Pe)$) is on the part of the boundary where the flow lines terminate

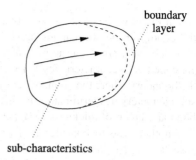

and the initial condition is

$$T = \theta(\mathbf{x}) \quad \text{at } t = 0, \tag{1.17}$$

where

$$\theta(\mathbf{x}) = \frac{T_0(l\mathbf{x}) - T_B}{\Delta T}. \tag{1.18}$$

1.2.1 Scaling

A well-scaled problem generally refers to a model in which the dimensionless parameters are $O(1)$ or less. Evidently, this can be ensured simply by dividing through by the largest parameter in any equation. More importantly, if parameters are numerically small, then (as we discuss below) approximate solutions can be obtained by neglecting them. The problem is well-scaled if the resulting approximation makes sense. For example, (1.14) is well-scaled for any value of Pe. However, the problem $\varepsilon T_t = \varepsilon \nabla^2 T + 1$, with $\varepsilon \ll 1$, is not well scaled. One makes a problem well-scaled in this situation by *rescaling* the variables, and we will see examples in our subsequent discussion.

1.2.2 Approximations

Let us consider (1.14) with (1.16) and (1.17), and suppose that $\theta \le O(1)$. If $Pe \ll 1$, we obtain an approximation by putting $Pe = 0$: $\nabla^2 T + 1 \approx 0$. Evidently, we cannot satisfy the initial condition, and this suggests that we rescale t: put $t = Pe\,\tau$, so that (approximately)

$$\frac{\partial T}{\partial \tau} = \nabla^2 T + 1; \tag{1.19}$$

now we can satisfy the initial condition (at $\tau = 0$) too. Often one abbreviates the rescaling by simply saying, 'rescale $t \sim Pe$, so that $T_t \approx \nabla^2 T + 1$'.

On the other hand, if $Pe \gg 1$, then $T_t + \mathbf{u}.\nabla T \approx 0$, and we can satisfy the initial condition; but we cannot satisfy the boundary condition on the whole of the boundary ∂D, since the approximating equation is hyperbolic (its characteristics are called 'sub-characteristics'). To remedy this, one has to rescale $\mathbf{x}$ near the part of the boundary where the boundary condition cannot be satisfied, and this is where the sub-characteristics terminate. This gives a spatially thin region, called (evidently) a boundary layer, of thickness $1/Pe$ (see Fig. 1.1).

Another case to consider is if $\theta \gg 1$, say $\theta \sim \Lambda \gg 1$. We discuss only the case $Pe \gg 1$ (see also Question 1.6). Since $T \sim \Lambda$ initially, we need to rescale T, say $T = \Lambda \tilde{T}$. Then $Pe[\tilde{T}_t + \mathbf{u}.\nabla \tilde{T}] = \nabla^2 \tilde{T} + \frac{1}{\Lambda}$, and with $\tilde{T} = O(1)$, we have $\tilde{T}_t + \mathbf{u}.\nabla \tilde{T} \approx 0$ for $Pe \gg 1$. The initial function is simply advected along the flow lines (sub-characteristics), and the boundary condition $\tilde{T} = 0$ is advected across D. In a time of $O(1)$, the initial condition is 'washed out' of the domain. Following this, we revert to T, thus $T_t + \mathbf{u}.\nabla T = \frac{1}{Pe}(\nabla^2 T + 1)$. Evidently T will remain ≈ 0 in most of D, and in fact $T \sim O(\frac{1}{Pe})$. Putting $T = \frac{\chi}{Pe}$, χ satisfies $\chi_t + \mathbf{u}.\nabla \chi = \frac{1}{Pe}\nabla^2 \chi + 1$, and there is a boundary layer near the boundary as shown in Fig. 1.1. If n is the coordinate normal to ∂D in this layer, then $n \sim \frac{1}{Pe}$ in the boundary layer. The final steady state has $T \sim \frac{1}{Pe}$, and this applies also for $\theta \lesssim O(1)$.

These ideas of perturbation methods are very powerful, but a full exposition is beyond the scope of this book. Nevertheless, they will relentlessly inform our discussion. While it is possible to use formal perturbation expansions, it is sufficient in many cases to give more heuristic forms of argument, and this will typically be the style we choose.

1.3 Qualitative Methods for Differential Equations

The language of the description of continuous processes is the language of differential equations, and these will form the instrument of our discussion. The simplest differential equation is the ordinary differential equation, and the simplest ordinary differential equation (or ODE) is the first order autonomous equation

$$\dot{x} = f(x), \tag{1.20}$$

where the notation $\dot{x} \equiv \frac{dx}{dt}$ indicates the first derivative, and the use of an overdot is normally associated with the use of time t as the independent variable, i.e., $\dot{x} = dx/dt$.

The solution of (1.20) with initial condition $x(t_0) = x_0$ can be written as the quadrature

$$t = t_0 + \int_{x_0}^{x} \frac{d\xi}{f(\xi)}, \tag{1.21}$$

and, depending on the function f, this may be inverted to find x explicitly. So, for example, the solution of $\dot{x} = 1 - x^2$ is $x = \tanh(t + c)$ (if $|x(t_0)| < 1$).

Fig. 1.2 The evolution of the solutions of $\dot{x} = f(x)$ (here $f = 1 - x^2$) depends only on the sign of x

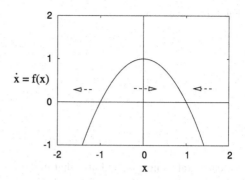

Going on with this latter example, we see that $x \to 1$ as $t \to \infty$ (and $x \to -1$ as $t \to -\infty$), and in practice, this may be all we want to know. If a population is subject to constant immigration and removal by mutual pair destruction, so that $\dot{x} = 1 - x^2$, then after a transient (a period of time dependence), the population will equilibrate stably to $x = 1$. But to ascertain this, all we need to know is the shape of the curve $f(x) = 1 - x^2$. Simply by finding the zeros of $1 - x^2$ and the slope of the graph there, we can immediately infer that for all initial values $x(0) > -1$, $x \to 1$ as $t \to \infty$, while if $x(0) < -1$, then $x \to -\infty$ as $t \to -\infty$: see Fig. 1.2. And this can be done for *any* function $f(x)$ in the equation $\dot{x} = f(x)$.

This simple example carries an important message. Approximate or qualitative methods may be just as useful, or more useful, than the ability to obtain exact results. An extension of this insight suggests that it may often be the case that approximate analytic insights can provide more information than precise, computational results.

1.3.1 Oscillations

If we move from first order systems to second order systems of the form

$$\begin{aligned}
\dot{x} &= f(x, y), \\
\dot{y} &= g(x, y),
\end{aligned} \tag{1.22}$$

more interesting phenomena can occur. This is the subject of phase plane analysis, and the fundamental distinction between first and second order systems is that periodic oscillations can occur. An illuminating example is illustrated in Fig. 1.3, and is typified by (but is not restricted to) the equations

$$\begin{aligned}
\dot{x} &= y - g(x), \\
\dot{y} &= h(x) - y,
\end{aligned} \tag{1.23}$$

where the functions g and h are as shown in the figure: g is unimodal (e.g., like $g = xe^{-x}$) and h is monotonic decreasing (e.g., like $h = 1/(x - c)$). The graphs of $g(x)$ and $h(x)$ (and more generally, the curves where $\dot{x} = 0$ and $\dot{y} = 0$) are called the nullclines of x and y, and it is simple to see that where they intersect, there is

Fig. 1.3 Nullclines for (1.23)

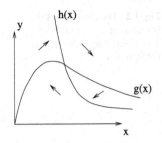

a steady state solution, and also that in the four regions separated by the nullclines, the trajectories wind round the fixed point in a clockwise manner.

The next issue is whether the fixed point is unstable. If we denote it as (x^*, y^*), write $x = x^* + X$, $y = y^* + Y$, and linearise for small X and Y, then

$$\dot{\mathbf{U}} \approx \begin{pmatrix} -g' & 1 \\ h' & -1 \end{pmatrix} \mathbf{U}, \tag{1.24}$$

where $\mathbf{U} = \begin{pmatrix} X \\ Y \end{pmatrix}$, and the derivatives are evaluated at the fixed point. The stability of such a two by two system with community matrix $A = \begin{pmatrix} -g' & 1 \\ h' & -1 \end{pmatrix}$ is governed by the trace and determinant of A. Solutions of (1.24) proportional to $e^{\sigma t}$ exist if $\sigma^2 - \sigma \operatorname{tr} A + \det A = 0$, and this delineates the stability regions in the $(\operatorname{tr} A, \det A)$ space as indicated in Fig. 1.4. In the present case, $\operatorname{tr} A = -g' - 1$, $\det A = g' - h'$, so that for the situation shown in Fig. 1.3, where $h' < g' < 0$, $\det A > 0$, and the fixed point is an unstable spiral (or node) if $g' < -1$. When $g' = -1$, there is a Hopf bifurcation, and if the system has bounded trajectories (as is normal for a model of a physical process) then one expects a stable periodic solution to exist. Figure 1.5 illustrates a possible example.

1.3.2 Relaxation Oscillations

It is a general precept of the applied mathematician that there are three kinds of numbers: small, large, and of order one. And the chances of a number being $O(1)$

Fig. 1.4 Characterisation of fixed point stability in terms of trace and determinant of the community matrix A. The curve separating spirals from nodes is given by $\det A = \frac{1}{4}(\operatorname{tr} A)^2$

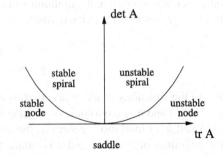

Fig. 1.5 Typical form of the limit cycle for a system with nullclines as in Fig. 1.3

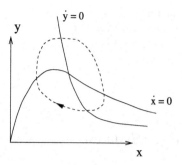

Fig. 1.6 Typical form of relaxation oscillation in phase plane for (1.25)

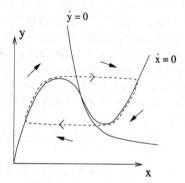

are not great. Thus for systems of the form (1.22), it is often the case in practice that the time scales for each equation are different, so that in suitable dimensionless units, the second order system (1.23) might take the form

$$\varepsilon \dot{x} = y - g(x),$$
$$\dot{y} = h(x) - y,$$

$$(1.25)$$

where the parameter ε is small. Now suppose that the nullclines $y = g(x)$ and $y = h(x)$ for the system (1.25) are as shown in Fig. 1.6, i.e., g has a cubic shape. Trajectories rotate clockwise, and linearisation about the fixed point yields a community matrix A with $\operatorname{tr} A = -(g'/\varepsilon) - 1$, $\det A = (g' - h')/\varepsilon$, thus with $g' > h'$, the fixed point is a spiral or node, and with $\varepsilon \ll 1$, $\operatorname{tr} A \approx -g'/\varepsilon > 0$, so it is unstable. Thus we expect a limit cycle, and because $\varepsilon \ll 1$, this takes the form of a *relaxation oscillation* in which the trajectory jumps rapidly backwards and forwards between branches of the x nullcline. For $\varepsilon \ll 1$, x rapidly jumps to its quasi-equilibrium $y \approx g(x)$, and then y migrates slowly ($\dot{x} \approx [h(x) - g(x)]/g'(x)$) until $g' = 0$ and x jumps rapidly to the other branch of g. Figure 1.7 shows the time series of the resulting oscillation. The motion is called 'relaxational' because the fast variable x 'relaxes' rapidly to a quasi-stationary state after each transient excursion.

Fig. 1.7 Time series for x
corresponding to Fig. 1.6

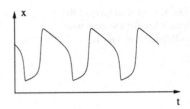

1.3.3 Hysteresis

Lighting a match is an everyday experience, but an understanding of why it occurs is less obvious. As the match is lit, a reaction starts to occur which is exothermic, i.e., it releases heat. The amount of heat released is proportional to the rate of reaction, and this itself increases with temperature (coal burns when hot, but not at room temperature). The heat released is given by the Arrhenius expression $A \exp(-E/RT)$, where E is the activation energy, R is the gas constant, T is the absolute temperature, and we take A as constant (it actually depends on reactant concentration). A simple model for the match temperature is then

$$c\frac{dT}{dt} = -k(T - T_0) + A \exp(-E/RT), \qquad (1.26)$$

where c is a suitable specific heat capacity, k is a cooling rate coefficient, and T_0 is ambient (e.g., room) temperature. The terms on the right represent the source term due to the reactive heat release, and a Newtonian cooling term (cooling rate proportional to temperature excess over the surroundings).

We can solve (1.26) as a quadrature, but it is much simpler to look at the problem graphically. Bearing in mind that T is absolute temperature, the source and sink terms typically have the form shown in Fig. 1.8, and we can see that there are three equilibria, and the lowest and highest ones are stable. Of course, one could have only the low equilibrium (for example, if k is large or T_0 is low) or the high equilibrium (if k is small or T_0 is high). The low equilibrium corresponds to the quiescent state—the match in the matchbox; the high one is the match alight. If we vary T_0, then the equilibrium excess temperature $\Delta\ (= T - T_0)$ varies as shown in Fig. 1.9: the upper and lower branches are stable.

We can model lighting a match as a local perturbation to Δ; the heat of friction in striking a match raises the temperature excess from near zero to a value above the unstable equilibrium on the middle branch, and Δ then migrates to the stable upper branch, where the reaction (like that of a coal fire) is self-perpetuating. Figure 1.9 also explains why it is difficult to light a wet match, but a match will spontaneously light if held at some distance above a lighted candle.

Figure 1.9 exhibits a form of hysteresis, meaning non-reversibility. Suppose we place a (very large, so it will not burn out) match in an oven, and we slowly raise the ambient temperature from a very low value to a very high value, and then lower it once again. Because the variation is slow, the excess temperature will follow the equilibrium curve in Fig. 1.9. At the value T_+, Δ suddenly jumps (spontaneous

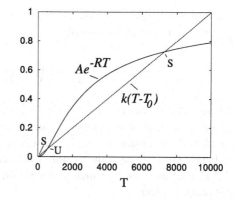

Fig. 1.8 Plots of the functions $A \exp[-E/R(T + T_m)]$ and $k(T - T_0)$ using values $T_m = 273$ (so T is measured in centigrade), with values $A = 1$, $E = 20,000$, $R = 8.3$, $k = 10^{-4}$, $T_0 = 15°C$

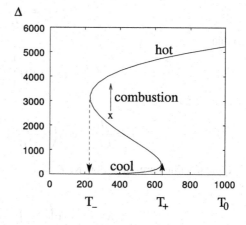

Fig. 1.9 Equilibrium curve for $\Delta = T - T_0$ as a function of T_0, parameters as for Fig. 1.8, but $E = 35,000$. An initial condition above the unstable middle branch leads to combustion

combustion) to the hot branch, and remains on this if T_0 is increased further. Now if T_0 is decreased, Δ remains on the hot branch until $T_0 = T_-$, below which it suddenly drops to the cool branch again (extinction).[1] The path traced out in the (T_0, Δ) plane is not reversible (it is not an arc but a closed curve).

The reason the multiple equilibria exist (at least for matches) is that for many reactions, E/R is very large and also A is very large. This just says that it is possible that $Ae^{-E/RT}$ is very small near T_0 but jumps rapidly at higher T to a large asymptote. To be more specific, we non-dimensionalise (1.26) by putting

$$T = T_0 + (\Delta T)\theta, \qquad t = [t]t^*, \tag{1.27}$$

---

[1] We can understand why T follows the equilibrium curve as follows. We can write (1.26) in terms of suitable dimensionless variables as $\dot{\Delta} = T_0 - g(\Delta)$, where $g(\Delta)$ is a cubic-like curve similar to the function $T_0(\Delta)$ depicted in Fig. 1.9. if T_0 is slowly varying, then $T_0 = T_0(\delta t)$ where $\delta \ll 1$, and putting $\tau = \delta t$, we have $\delta d\Delta/d\tau = T_0(\tau) - g(\Delta)$; thus on the slow time scale τ, Δ will tend rapidly to a (quasi-equilibrium) zero of the right hand side.

and in fact we choose the cooling time scale $[t] = c/k$. Then we have, dropping the asterisk, and after some simplification,

$$\dot{\theta} = -\theta + \frac{A}{k\Delta T}\exp\left(-\frac{E}{RT_0}\right)\exp\left[\frac{E\Delta T}{RT_0^2}\frac{\theta}{1+\varepsilon\theta}\right], \qquad (1.28)$$

where $\varepsilon = \Delta T/T_0$. The temperature rise scale ΔT has to be chosen, and there are two natural choices: to set the exponent coefficient $E\Delta T/RT_0^2$ to one, or the pre-multiplicative constant to one. In one way, the latter seems the better choice: it seems to balance the source with the sink. But because E/R is large, we might then find $E\Delta T/RT_0^2$ to be large, which would ruin the intention. So we choose (but it does not really matter)

$$\Delta T = \frac{RT_0^2}{E}, \qquad (1.29)$$

so that

$$\dot{\theta} = -\theta + \lambda\exp\left[\frac{\theta}{1+\varepsilon\theta}\right], \qquad (1.30)$$

where

$$\lambda = \frac{EA}{kRT_0^2}\exp\left(-\frac{E}{RT_0}\right), \qquad \varepsilon = \frac{RT_0}{E}. \qquad (1.31)$$

If typical values are $T_0 = 300$ K, $E/R = 10,000$ K, we see that $\varepsilon \ll 1$, and also, since

$$\lambda = \frac{\lambda_0}{\varepsilon^2}\exp\left(-\frac{1}{\varepsilon}\right), \qquad \lambda_0 = \frac{AR}{kE}, \qquad (1.32)$$

λ is extremely sensitive to ε and thus T_0.

So long as $\theta = O(1)$, or at least $\theta \ll 1/\varepsilon$ (i.e. $T - T_0 \ll T_0$), we can neglect the $\varepsilon\theta$ term, so that

$$\dot{\theta} \approx -\theta + \lambda e^{\theta}. \qquad (1.33)$$

This gives the lower part of the S-shaped curve in Fig. 1.9, and the equilibria are given by $\theta e^{-\theta} = \lambda$, the roots of which coalesce and disappear if $\lambda > e^{-1}$. This corresponds to the value of $T_0 = T_+$ in Fig. 1.9, and implies

$$\frac{E}{RT_+} \approx 1 + \ln\lambda_0 + 2\ln\left(\frac{E}{RT_+}\right). \qquad (1.34)$$

There are two roots to this, but only one has $E/RT_+ \gg 1$. Further, since $x \gg 2\ln x$ if $x \gg 1$, we have, approximately,

$$T_+ \approx \frac{E}{R[1 + \ln\lambda_0 + 2\ln\{1 + \ln\lambda_0\}]}. \qquad (1.35)$$

If $E/R \gg T_0$, then the fact that one can light matches at room temperature suggests that λ_0 is large, and specifically $\ln \lambda_0 \sim E/RT_0$. (Note that this does not imply $\lambda = O(1)$.)

Carrying on in this vein, let us suppose that we define a temperature T_c by

$$\lambda_0 = \exp\left[\frac{E}{RT_c}\right], \tag{1.36}$$

and we suppose $T_c \sim T_0$. It follows that $T_+ \approx T_c$, or more precisely,

$$T_+ \approx \frac{T_c}{1 + \varepsilon_c\{1 + 2\ln(1 + \varepsilon_c^{-1})\}}, \tag{1.37}$$

where $\varepsilon_c = RT_c/E$. The stable cool branch and unstable middle branch are then the roots of

$$\theta e^{-\theta} \approx \lambda = \frac{1}{\varepsilon^2} \exp\left[-\frac{1}{\varepsilon}\left(1 - \frac{T_0}{T_c}\right)\right], \tag{1.38}$$

and in general $\lambda \ll 1$ (if $T_0 < T_c$), so that we find the stable cool branch (when $\theta \ll 1$)

$$\theta \approx \lambda \approx \left(\frac{E}{RT_0}\right)^2 \exp\left[\frac{E}{R}\left(\frac{1}{T_c} - \frac{1}{T_0}\right)\right], \tag{1.39}$$

and the unstable middle branch (where $\theta \gg 1$),

$$\theta \approx \frac{1}{\varepsilon}\left(1 - \frac{T_0}{T_c}\right) + O(|\ln \varepsilon|) \approx \frac{E}{R}\left(\frac{1}{T_0} - \frac{1}{T_c}\right). \tag{1.40}$$

Evidently θ becomes $O(1/\varepsilon)$ on the middle branch, and to allow for this, we put

$$\theta = \Theta/\varepsilon, \tag{1.41}$$

and (1.30) becomes

$$\dot{\Theta} = -\Theta + \frac{1}{\varepsilon} \exp\left[\frac{1}{\varepsilon}\left\{\frac{\Theta}{1+\Theta} - \left(1 - \frac{T_0}{T_c}\right)\right\}\right]. \tag{1.42}$$

Equating the right hand side to zero gives an equilibrium which can be written approximately as[2]

$$\Theta \approx \frac{T_c - T_0}{T_0} + O(\varepsilon|\ln \varepsilon|), \tag{1.43}$$

and Θ tends to infinity as $T_0 \to 0$. The hot branch is recovered for even higher values of Θ, so that $\Theta \gg 1$, in which case the equilibrium of (1.42) is given by

$$\Theta \approx \frac{1}{\varepsilon} \exp\left[\frac{T_0}{\varepsilon T_c}\right], \tag{1.44}$$

and increases again with T_0.

---

[2]Note that as $T_0 \to T_c$, (1.43) matches with (1.40).

At a fixed value of T_0 (and thus λ), the critical value of T for ignition is that on the unstable middle branch, as this gives the necessary temperature which must be generated in order for combustion to occur. From (1.43) (ignoring terms in ε), this can be written dimensionally in the simple approximate form

$$T \approx T_c, \tag{1.45}$$

which is approximately the critical temperature at the nose of the curve in Fig. 1.9. The fact that T is approximately constant on the unstable branch is due to the steepness of the exponential curve in Fig. 1.8, which is in turn due to the large value of E/R. In terms of the parameters of the problem, the critical (ignition) temperature is thus

$$T_c \approx \frac{E}{R \ln\left(\frac{AR}{kE}\right)}. \tag{1.46}$$

Hysteresis and multiplicity of solutions is a theme which will recur again and again in this book.

1.3.4 Resonance

Swinging a pendulum is an everyday experience, and one which students learn about in a first year mechanics course. If the point of suspension itself oscillates, then one has a forced pendulum, and an interesting phenomenon occurs. At low forcing frequencies, the pendulum oscillates in phase with the oscillating point of support. At high forcing frequencies, it oscillates out of phase with the support. Moreover, this change in phase appears to occur abruptly, at a particular value of the forcing frequency. At the same time, there is also a sudden rise in amplitude of the motion, although it is less easy to see this in a casual experiment. This observation is associated with the phenomenon of resonance, and can be easily experienced by jumping on a springboard.

To illustrate the phenomenon of resonance mathematically, we solve the equation of a forced oscillator, and an example of such a system is the forced pendulum. To be specific, we take as a model equation

$$\ddot{u} + \beta \dot{u} + \Omega_0^2 \sin u = \varepsilon \sin \omega t. \tag{1.47}$$

This represents the motion of a damped, non-linear pendulum, with a forcing on the right hand side which mimics (it is not a precise model) the effect on the pendulum of an oscillating support. We suppose that the model is dimensionless, and that ε is small, so that the response amplitude of u will be also. We also suppose that the damping term β is small.

The simplest approximation of (1.47) neglects β altogether, and linearises $\sin u$, so that

$$\ddot{u} + \Omega_0^2 u \approx \varepsilon \sin \omega t, \tag{1.48}$$

Fig. 1.10 Resonant
amplitude response

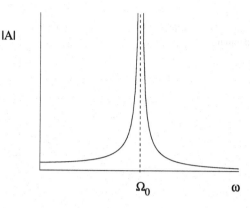

to which the forced solution is

$$u = A \sin \omega t, \qquad (1.49)$$

where the response amplitude A is given by

$$A = \frac{\varepsilon}{\Omega_0^2 - \omega^2}. \qquad (1.50)$$

Plotting $|A|$ versus ω gives the familiar resonant response diagram of Fig. 1.10, in which the amplitude tends to infinity as $\omega \to \Omega_0$. (If one actually solves (1.48) at $\omega = \Omega_0$, one obtains a solution whose amplitude grows linearly in time.)

The two effects we have neglected, damping and non-linearity, have two separate effects on this diagram. If we include only damping, so that

$$\ddot{u} + \beta \dot{u} + \Omega_0^2 u = \varepsilon \operatorname{Im} e^{i\omega t}, \qquad (1.51)$$

then the forced solution is again

$$u = \operatorname{Im}\big[A e^{i\omega t}\big], \qquad (1.52)$$

where now

$$A = \frac{\varepsilon}{\Omega_0^2 + i\beta\omega - \omega^2}, \qquad (1.53)$$

and the presence of the damping term causes a phase shift which caps the response amplitude, as shown in Fig. 1.11, since

$$|A| = \frac{\varepsilon}{[(\Omega_0^2 - \omega^2)^2 + \beta^2\omega^2]^{1/2}}; \qquad (1.54)$$

the peak amplitude at resonance is $|A| = \varepsilon/\beta\omega$.

The other effect is non-linearity, which is less easy to deal with. In fact, one can use perturbation methods to assess its effect in a formal manner, but our present purpose is more rough and ready. Our idea is this: resonance occurs when the forcing

Fig. 1.11 Resonant
amplitude response with
damping

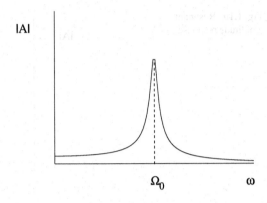

frequency ω equals the frequency of the underlying oscillator. The difference which
occurs for a non-linear pendulum is that this frequency (call it Ω) now depends on
the amplitude of the oscillation A: $\Omega = \Omega(A)$.

To be specific, we again put $\beta = 0$, and consider simply the unforced pendulum:

$$\ddot{u} + \Omega_0^2 \sin u = 0. \tag{1.55}$$

A first (energy) integral is

$$\frac{1}{2}\dot{u}^2 + \Omega_0^2(1 - \cos u) = E, \tag{1.56}$$

where E is constant (and depends on amplitude, with $E(A)$ increasing with A). The
phase plane is shown in Fig. 1.12 and is symmetric about both u and $\dot{u}$ axes. Thus a
quadrature of (1.56) implies the period P is given by

$$P = \frac{2\sqrt{2}}{\Omega_0} \int_0^A \frac{du}{[\cos u - \cos A]^{1/2}}, \tag{1.57}$$

Fig. 1.12 Phase plane for the
simple pendulum

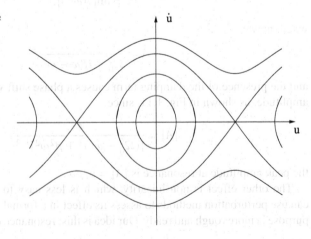

Fig. 1.13 Non-linearity
bends the resonant response
curve, producing hysteresis

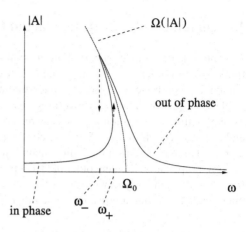

where we have used the fact that the amplitude A is given by

$$E = \Omega_0^2 (1 - \cos A). \tag{1.58}$$

From (1.57), we find that the frequency $\Omega = 2\pi/P$ is given by

$$\Omega(A) = \frac{\pi \Omega_0}{\sqrt{2} \int_0^A \frac{du}{[\cos u - \cos A]^{1/2}}}. \tag{1.59}$$

Ω is a monotonically decreasing function of A in $(0, \pi)$, with $\Omega(0) = \Omega_0$ and $\Omega(\pi) = 0$, and this is represented as the dotted curve in Fig. 1.13.

Without now actually solving the forced, damped, non-linear equation, we can guess intelligently what happens. For small amplitude oscillations, $|A|$ starts to increase as ω approaches Ω_0; but as $|A|$ increases, the natural frequency Ω decreases, and as it is the approach of ω to the natural frequency which is the instrument of resonance, so the amplitude response curve bends round, as shown in Fig. 1.13, to try and approach the dotted $\Omega(A)$ curve. Finally, the effect of damping can be expected to be as in the linear case, to put a cap on the two asymptotes to $\Omega(A)$. Thus, we infer the response diagram shown in Fig. 1.13, and this is in fact correct. Moreover, (1.50) suggests $A \gtrless 0$ for $\omega \lessgtr \Omega_0$, i.e., the solution is in phase with the forcing for $\omega < \Omega_0$, and out of phase for $\omega > \Omega_0$. Extending this to the non-linear case, we infer that at low frequencies, the response is in phase, but that it is out of phase at high frequencies (as observed).

The response also involves hysteresis (if damping is small enough). If ω is increased gradually, then at a value $\omega_+ < \Omega_0$, there is a sudden jump to an out of phase oscillation with higher amplitude. Equivalently, as ω is reduced for this high frequency response there is a sudden jump down in amplitude to an in-phase oscillation at a value $\omega_- < \omega_+$. This response diagram explains what one sees in the simple experiment and illustrates the important effects of non-linearity.

1.4 Qualitative Methods for Partial Differential Equations

Any introductory course on partial differential equations will provide the classification of second order partial differential equations into the three categories: elliptic, parabolic, hyperbolic; and one also finds the three simple representatives of these: Laplace's equation $\nabla^2 u = 0$, governing steady state temperature distribution (for example); the heat equation $u_t = \nabla^2 u$, which describes diffusion of heat (or solute); and the wave equation $u_{tt} = \nabla^2 u$, which describes the oscillations of a string or of a drum. These equations are of fundamental importance, as they describe diffusion or wave propagation in many other physical processes, but they are also linear equations; however, the way in which they behave carries across to non-linear equations, but of course non-linear equations have other behaviours as well.

1.4.1 Waves

In the linear wave equation (in one dimension, describing waves on strings) $u_{tt} = c^2 u_{xx}$, the general solution is $u = f(x + ct) + g(x - ct)$, and represents the superposition of two travelling waves of speed c moving in opposite directions. In more than one space dimension, the equivalent model is $u_{tt} = c^2 \nabla^2 u$, and the solutions are functions of $(\mathbf{k}.\mathbf{x} \pm \omega t)$, where ω is frequency and $\mathbf{k}$ is the wave vector; the waves move in the direction of the vector $\mathbf{k}$, while the wave speed is then $c = \omega/|\mathbf{k}|$.

Even simpler to discuss is the first order wave equation

$$u_t + c u_x = 0, \tag{1.60}$$

which is trivially solved by the method of characteristics to give

$$u = f(x - ct), \tag{1.61}$$

representing a wave of speed c. The idea of finding characteristics generalises to systems of the form

$$A\mathbf{u}_t + B\mathbf{u}_x = \mathbf{0}, \tag{1.62}$$

where $\mathbf{u} \in \mathbf{R}^n$ and A and B are constant $n \times n$ matrices. We can solve this system as follows. The eigenvalue problem

$$\lambda A\mathbf{w} = B\mathbf{w} \tag{1.63}$$

will in general have n solution pairs $(\mathbf{w}, \lambda)$, where each value of λ is one of the roots of the nth order polynomial

$$\det(\lambda A - B) = 0. \tag{1.64}$$

Suppose the n eigenvalues λ_i, $i = 1, \ldots, n$, are distinct (which is the general case); then the corresponding $\mathbf{w}_i$ are independent, and the matrix P formed by the eigenvectors as columns (i.e., $P = (\mathbf{w}_1, \ldots, \mathbf{w}_n)$) satisfies $BP = APD$, where D is the diagonal matrix $\mathrm{diag}(\lambda_1, \ldots, \lambda_n)$. P is invertible, and if we write $\mathbf{v} = P^{-1}\mathbf{u}$, then $AP\mathbf{v}_t + BP\mathbf{v}_x = 0$, whence $\mathbf{v}_t + D\mathbf{v}_x = 0$, and the general solution is

$$\mathbf{u} = P\mathbf{v} = \sum_{i,j} P_{ij} f_j(x - \lambda_j t)\mathbf{e}_i, \tag{1.65}$$

where $\mathbf{e}_i$ is the ith unit vector, and the functions f_j are arbitrary; this represents the superposition of n travelling waves with speeds λ_i. This procedure works providing A is invertible, and also (practically) if all the λ_i are real, in which case we say the system is hyperbolic.

More generally, we can use the above prescription to solve the non-linear equation

$$A\mathbf{u}_t + B\mathbf{u}_x = \mathbf{r}(x, t, \mathbf{u}), \tag{1.66}$$

where we allow A and B to depend on x and t also. The diagonalisation procedure works exactly as before, leading to

$$A\frac{\partial}{\partial t}(P\mathbf{v}) + B\frac{\partial}{\partial x}(P\mathbf{v}) = \mathbf{r}[x, t, P\mathbf{v}]; \tag{1.67}$$

now, however, λ, $\mathbf{w}$ and therefore also P will depend on x and t. Thus we find

$$\mathbf{v}_t + D\mathbf{v}_x = P^{-1}A^{-1}\mathbf{r} - [P^{-1}P_t + DP^{-1}P_x]\mathbf{v}, \tag{1.68}$$

and the components of $\mathbf{v}$ can be solved as a set of coupled ordinary differential equations along the characteristics $dx/dt = \lambda_i$.

If A and B depend also on $\mathbf{u}$ (the *quasi-linear* case), the procedure is less simple for systems. The characterisation of the system as hyperbolic based on the reality of the eigenvalues of (1.63) is still appropriate, but the diagonalisation and reduction to the equivalent of (1.68) are less clear. In the particular case where P depends only on $\mathbf{u}$ (and not on x and t), and if P^{-1} is a Jacobian matrix (i.e., $(P^{-1})_{ij} = \frac{\partial v_i}{\partial u_j}$ for some vector $\mathbf{v}(\mathbf{u})$), then the function $\mathbf{v}$ is given by the (well-defined) line integral

$$\mathbf{v} = \int P^{-1}\, d\mathbf{u}, \tag{1.69}$$

and $\mathbf{v}_t = P^{-1}\mathbf{u}_t$, $\mathbf{v}_x = P^{-1}\mathbf{u}_x$; hence we can derive the diagonalised form

$$\mathbf{v}_t + D\mathbf{v}_x = P^{-1}A^{-1}\mathbf{r}. \tag{1.70}$$

This shows how the characteristic equations can be derived, but in general the equations cannot be solved, since the elements of D will depend on all the components of $\mathbf{v}$. An example of this type occurs in river flow, and will be discussed in Chap. 4.

However, the method of characteristics always works in one dimension, so we now return our attention to this case. Consider as an example the non-linear evolution equation

$$u_t + uu_x = 0, \tag{1.71}$$

Fig. 1.14 Non-linearity
causes wave steepening

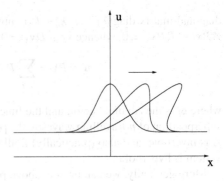

Fig. 1.15 Intersection of
characteristics leads to shock
formation

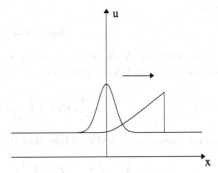

to be solved on the whole real axis. The method of characteristics leads to the implicitly defined general solution

$$u = f(x - ut),\qquad\qquad (1.72)$$

which is analogous to (1.61), and represents a wave whose *speed depends on its amplitude*. Thus higher values of u propagate more rapidly, and this leads to the wave steepening depicted in Fig. 1.14.

In fact, it can be seen that eventually u becomes multi-valued, and this signifies a break down of the solution. The usual way in which this multi-valuedness is avoided is to allow for the formation of a *shock*, which consists of a point of discontinuity of u. The characteristic solution applies in front of and behind the shock, and the characteristics intersect at the shock, whose propagation forwards is described by an appropriate *jump condition*: see Fig. 1.15.

This seemingly arbitrary escape route is motivated by the fact that evolution equations such as (1.71) are generally derived from a conservation law, here of the form

$$\frac{d}{dt}\int_A^B u\,dx = -\left[\frac{1}{2}u^2\right]_A^B,\qquad\qquad (1.73)$$

where the square-bracketed term represents the jump in $\frac{1}{2}u^2$ between A and B. The deduction of the point form (1.71) from (1.73) required the additional assumption that u was continuously differentiable; however, it is possible to satisfy (1.73) at a point of discontinuity of u. Suppose u is discontinuous at $x = x_S(t)$, and denote the jump in a quantity q across the shock by $[q]_-^+ = q(x_{S+}, t) - q(x_{S-}, t)$. Then by letting $B \rightarrow x_{S+}$, $A \rightarrow x_{S-}$, we find that (1.73) implies the jump condition

$$\dot{x}_S = \frac{[\frac{1}{2}u^2]_-^+}{[u]_-^+} = \frac{1}{2}(u_+ + u_-). \tag{1.74}$$

An Example

We illustrate how to solve a problem of this type by considering the initial function for u

$$u = u_0(x) = \frac{1}{1+x^2} \quad \text{at } t = 0. \tag{1.75}$$

The implicitly defined solution is then

$$u = \frac{1}{1 + (x - ut)^2}, \tag{1.76}$$

or, in characteristic form,

$$u = u_0(\xi) = \frac{1}{1+\xi^2}, \quad x = \xi + ut. \tag{1.77}$$

This defines a single-valued function so long as u_x is finite everywhere. Differentiating (1.77) leads to

$$u_x = \frac{u_0'(\xi)}{1 + tu_0'(\xi)}, \tag{1.78}$$

and this shows that $u_x \rightarrow -\infty$ as $t \rightarrow t_c = \min_{\xi\,:\,u_0' < 0}[-\frac{1}{u_0'(\xi)}]$. Since $-u_0' = 2\xi/(1+\xi^2)^2$, we find the relevant value of ξ is $1/\sqrt{3}$, and thus $t_c = \frac{8}{3\sqrt{3}}$ and the corresponding value of x is $x_c = \sqrt{3}$. Thus (1.76) applies while $t < t_c$, and thereafter the solution also applies in $x < x_S(t)$ and $x > x_S(t)$, where

$$\dot{x}_S = \frac{1}{2}[u(x_S+) + u(x_S-)], \tag{1.79}$$

with

$$x_S = \sqrt{3} \quad \text{at } t = \frac{8}{3\sqrt{3}}. \tag{1.80}$$

As indicated in Fig. 1.16, the characteristics intersect at the shock, and it is geometrically clear from Fig. 1.14, for example, that u_+ and u_- are the largest and smallest roots of the cubic (1.76). An explicit solution for x_S is not readily available, but it is of interest to establish the long term behaviour, and for this we need approximations to the roots of (1.76) when $t \gg 1$.

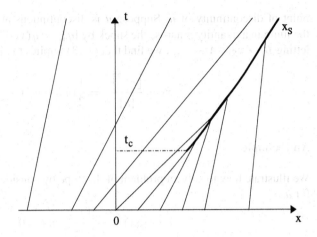

Fig. 1.16 Characteristic diagram indicating shock formation

We write the cubic (1.76) in the form

$$u = \frac{x}{t} \pm \frac{1}{t}\left(\frac{1-u}{u}\right)^{1/2}. \tag{1.81}$$

We know that $u \leq 1$, and we expect x_S to tend to infinity as $t \to \infty$, so that we suppose $x \gg 1$. In that case $u \approx x/t$ if $u = O(1)$, and the next corrective term gives

$$u \approx \frac{x}{t} \pm \frac{1}{t}\left(\frac{t-x}{x}\right)^{1/2}. \tag{1.82}$$

This evidently gives the upper two roots for $x < t$ (since they coalesce at $u = 1$ when $x = t$). For large x, the other root must have $u \ll 1$, and in fact

$$u \approx \frac{1}{x^2}, \tag{1.83}$$

in order that (1.81) not imply (1.82).[3] Alternatively, (1.83) follows from consideration of (1.76) in the form

$$t^2 u^3 - 2xtu^2 + (x^2 + 1)u - 1 = 0, \tag{1.84}$$

providing $x \gg t^{1/3}$.

To find the location of the 'noses' of the solution, we note that the approximation that $u \approx x/t$ breaks down (see (1.82)) when $x \sim t^{1/3}$, which is also where (1.83) becomes invalid. This suggests writing

$$u = \frac{x}{t} W(X), \quad X = \frac{x}{t^{1/3}}, \tag{1.85}$$

---

[3]We need $u \lesssim O(\frac{1}{x^2})$ in order that the second term in (1.81) be significant (otherwise we regain (1.82)), and in fact we need the two terms to be approximately equal, so that $0 < u < 1$: hence (1.83).

Fig. 1.17 Determination of $W(X)$

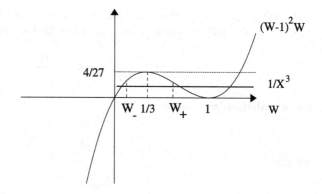

Fig. 1.18 Large time solution of the characteristic solution

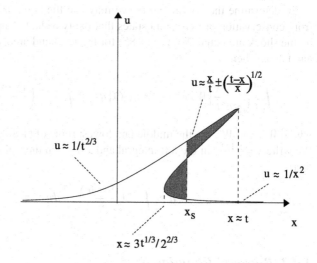

and then $W(X)$ is given approximately, for large t, by

$$W(W-1)^2 = \frac{1}{X^3}, \tag{1.86}$$

and for $X = O(1)$ there are three roots providing $X > 3/2^{2/3}$; at $X = 3/2^{2/3}$, the two lower roots coalesce at $W = \frac{1}{3}$: this describes the left nose of the curve.

As X becomes large, the upper two roots approach $W = 1$, thus $u \approx x/t$, while the lower approaches zero, specifically $W \approx 1/X^3$, and hence $u \approx 1/x^2$: see Fig. 1.17. Thus these roots match to the approximations in (1.82) and (1.83). As X becomes small, the remaining root is given by $W \approx 1/X$, so that $u \approx 1/t^{2/3}$, and (1.84) shows that this is the correct approximation as long as $|x| \ll t^{1/3}$. The situation is shown in Fig. 1.18.

In order to determine the shock location x_S, we make the ansatz that $t^{1/3} \ll x_S \ll t$, i.e., that the shock is far from both noses. In that case

$$u_+ \approx \frac{1}{x_S^2}, \qquad u_- \approx \frac{x_S}{t}, \tag{1.87}$$

and at leading order we have

$$\dot{x}_S \approx \frac{x_S}{2t}, \tag{1.88}$$

whence

$$x_S \approx at^{1/2}, \tag{1.89}$$

confirming our assumption that $t^{1/3} \ll x_S \ll t$.

To determine the coefficient a, we may use the equal area rule, which follows from conservation of mass, and states that the two shaded areas in Fig. 1.18 cut off by the shock are equal. We use (1.85) for the left hand area, and (1.82) for the right hand area. Then

$$\int_{3t^{1/3}/2^{2/3}}^{at^{1/2}} \frac{x}{t}\left[W_+(X) - W_-(X)\right] dx \approx \int_{at^{1/2}}^{t} \frac{2}{t}\left(\frac{t - x}{x}\right)^{1/2} dx, \tag{1.90}$$

where W_+ and W_- are the middle and lowest roots of (1.86), as shown in Fig. 1.17. We write $x = t^{1/2}\xi$ in the left integral and $x = t\eta$ in the right, and hence we deduce that

$$a \approx \int_0^1 2\left(\frac{1 - \eta}{\eta}\right)^{1/2} d\eta = \pi. \tag{1.91}$$

1.4.2 Burgers' Equation

Although the presence of a shock for (1.71) is entirely consistent with the derivation of the equation from an integral conservation law, nature appears generally to avoid discontinuities and singularities, and it is usually the case that in writing an equation such as (1.71), we have neglected some term which acts to smooth the shock, so that the change of u is rapid but not abrupt.

The most common type of neglected term which provides the necessary smoothing is a diffusion term, which is manifested in the adjusted equation as a second derivative term. The resulting equation is known as Burgers' equation:

$$u_t + uu_x = \kappa u_{xx}. \tag{1.92}$$

Sometimes, as for example in the smoothing effect of heat conduction or viscosity on sonic shock waves, such a term genuinely represents a physically diffusive process (e.g., diffusion of heat or momentum); sometimes it arises for more subtle reasons, as for example in the smoothing of waves on rivers (see, for example, the derivation of Eq. (4.57) in the discussion of the monoclinal flood wave in Chap. 4).

More generally, even-order derivative terms of the form $(-1)^{n-1}\kappa \frac{\partial^{2n} u}{\partial x^{2n}}$ are smoothing. (This can be seen by the fact that solutions of the resulting linearised equation $u_t = (-1)^{n-1}\kappa \frac{\partial^{2n} u}{\partial x^{2n}}$ have damped solutions $\exp(ikx + \sigma t)$ in which $\sigma = -\kappa k^{2n}$.) A fourth order smoothing term occurs in the smoothing of capillary waves by surface tension, for example.

How does the presence of a diffusive term modify the structure of the solutions? If κ is small, we should suppose that it has little effect, so that shocks would start to form. However, the neglect of the diffusion term becomes invalid when the derivatives of u become large. In fact, the diffusion term is trying to do the opposite of the advective term. The latter is trying to fold the initial profile together like an accordion, while the former is trying to spread everything apart. We might guess that a balanced position is possible, in which the non-linear advective term keeps the profile steep, but the diffusion prevents it actually folding over (and hence causing a discontinuity), and this will turn out to be the case.

Shock Structure

We suppose $\kappa \ll 1$, so that $u_t + u u_x \approx 0$, and a shock forms at $x = x_S(t)$. Our aim is to show that (1.92) supports a *shock structure*, i.e., a region of rapid change for u near x_S from u_- to u_+.

To focus on the shock, we need to rescale x near x_S, and we do this by writing

$$x = x_S(t) + \kappa X. \tag{1.93}$$

Burgers' equation becomes

$$\kappa u_t - \dot{x}_S u_X + u u_X = u_{XX}. \tag{1.94}$$

We expect the characteristic solution (with $\kappa = 0$) to be approximately valid far from x_S, and so appropriate conditions (technically, these are *matching conditions*) are

$$u \to u_\pm \quad \text{as } X \to \pm\infty, \tag{1.95}$$

and we take these values as prescribed from the *outer* solution (i.e., the solution of $u_t + u u_x = 0$ as $x \to x_S\pm$).

Since $\kappa \ll 1$, (1.94) suggests that u relaxes rapidly (on a time scale $t \sim \kappa \ll 1$) to a quasi-steady state (quasi-steady, because u_+ and u_- will vary with t) in which

$$-\dot{x}_S u_X + u u_X \approx u_{XX}, \tag{1.96}$$

whence

$$K - \dot{x}_S u + \frac{1}{2} u^2 \approx u_X, \tag{1.97}$$

and prescription of the boundary conditions implies

$$K = \dot{x}_S u_+ - \frac{1}{2} u_+^2 = \dot{x}_S u_- - \frac{1}{2} u_-^2, \qquad (1.98)$$

whence

$$\dot{x}_S = \frac{[\frac{1}{2} u^2]_-^+}{[u]_-^+}, \qquad (1.99)$$

which is precisely the jump condition we obtained in (1.74). The solution for u of (1.97) is then

$$u = c - (u_- - c) \tanh \left[\frac{1}{2} (u_- - c) X \right], \qquad (1.100)$$

where $c = \dot{x}_S$.

1.4.3 The Fisher Equation

In Burgers' equation, a wave arises as a balance between non-linear advection and diffusion. In Fisher's equation,

$$u_t = u(1 - u) + u_{xx}, \qquad (1.101)$$

a wave arises as a mechanism for transferring a variable from an unstable steady state ($u = 0$) to a stable one ($u = 1$). Whereas Burgers' equation balances two transport terms, Fisher's equation balances diffusive transport with an algebraic source term. It originally arose as a model for the dispersal of an advantageous gene within a population, and has taken a plenary rôle as a pedagogical example in mathematical biology of how reaction (source terms) and diffusion can combine to produce travelling waves.

We pose (1.101) with boundary conditions

$$\begin{aligned} u \to 1, \quad x \to -\infty, \\ u \to 0, \quad x \to +\infty. \end{aligned} \qquad (1.102)$$

It is found (and can be proved) that any initial condition leads to a solution which evolves into a travelling wave of the form

$$u = f(\xi), \quad \xi = x - ct, \qquad (1.103)$$

where

$$f'' + cf' + f(1 - f) = 0, \qquad (1.104)$$

and

$$f(\infty) = 0, \qquad f(-\infty) = 1. \qquad (1.105)$$

Fig. 1.19 Phase portrait of Fisher equation, (1.106), for $c = 2$. Note how close the connecting trajectory (*thick line*) is to the g nullcline. This is why the large c approximation is accurate *for this trajectory*

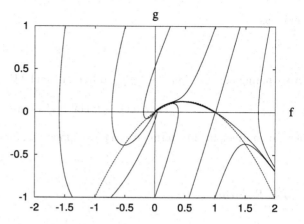

In the (f, g) phase plane, where $g = -f'$, we have

$$\begin{aligned} f' &= -g, \\ g' &= f(1 - f) - cg, \end{aligned} \tag{1.106}$$

and a travelling wave corresponds to a trajectory which moves from $(1, 0)$ to $(0, 0)$.

Linearisation of (1.106) near the fixed point $(f^*, 0)$ via $f = f^* + F$ leads to

$$\begin{pmatrix} F \\ g \end{pmatrix}' = \begin{pmatrix} 0 & -1 \\ 1 - 2f^* & -c \end{pmatrix} \begin{pmatrix} F \\ g \end{pmatrix}, \tag{1.107}$$

with solutions $e^{\lambda \xi}$, where $\lambda^2 + c\lambda + (1 - 2f^*) = 0$. We anticipate $c > 0$; then $(1, 0)$ is a saddle point, while $(0, 0)$ is a stable node if $c \geq 2$ (and a spiral if $c < 2$). For $c \geq 2$, a connecting trajectory exists as shown in Fig. 1.19: in practice the minimum wave speed $c = 2$ is selected. (Connecting trajectories also exist if $c < 2$, but because $(0, 0)$ is a spiral, these have oscillating tails as $u \to 0$, which are unstable and also (for example, if u represents a population) unphysical.)

Explicit solutions for (1.104) are not available, but an excellent approximation is easily available. We put

$$\xi = c\Xi, \tag{1.108}$$

so

$$vf'' + f' + f(1 - f) = 0, \tag{1.109}$$

with $v = 1/c^2 = 1/4$ for $c = 2$. Taking $v \ll 1$ and writing $f = f_0 + vf_1 + \cdots$, we have

$$\begin{aligned} f_0' + f_0(1 - f_0) &= 0, \\ f_1' + (1 - 2f_0)f_1 &= -f_0'', \end{aligned} \tag{1.110}$$

and thus

$$f_0 = \frac{e^{-\Xi}}{1 + e^{-\Xi}}. \tag{1.111}$$

Also, noting that $1 - 2f_0 = -f_0''/f_0'$ (differentiate $(1.110)_1$),

$$f_1 = f_0(1 - f_0) \ln[f_0(1 - f_0)], \tag{1.112}$$

and so on. Even the first term gives a good approximation, and even for $c = 2$.

1.4.4 Solitons

The Fisher wave is an example of a solitary travelling wave. Another type of solitary wave is the soliton, as exemplified by solutions of the Korteweg–de Vries equation

$$u_t + u u_x + u_{xxx} = 0. \tag{1.113}$$

This has travelling wave solutions $u = f(\xi)$, $\xi = x - ct$, where

$$f''' + ff' - cf' = 0, \tag{1.114}$$

and solitary waves with $f \to 0$ at $\pm\infty$ satisfy the first integral

$$f'' + \frac{1}{2}f^2 - cf = 0, \tag{1.115}$$

and thus

$$\frac{1}{2}f'^2 + \frac{1}{6}f^3 - \frac{1}{2}cf^2 = 0, \tag{1.116}$$

with solution

$$f = \frac{3}{2}c \operatorname{sech}^2\left(\frac{\sqrt{c}\xi}{2}\right). \tag{1.117}$$

Thus there is a one-parameter family of these solitary waves, and they are called solitons, because they have the remarkable particle-like ability to 'pass through' each other without damage, except for a change of relative phase. Despite the non-linearity, they obey a kind of superposition principle. Soliton equations (of which there are many) have many other remarkable properties, beyond the scope of the present discussion.

Some understanding of the solitary wave arises through an understanding of the balance between non-linearity (uu_x) and dispersion (u_{xxx}). The dispersive part of the equation, $u_t + u_{xxx} = 0$, is so called because waves $\exp[ik(x - ct)]$ have wave speed $c = -k^2$ which depends on wave number k; waves of different wavelengths ($2\pi/k$) move at different speeds and thus disperse. On the other hand, the non-linear advection equation $u_t + uu_x$ has a focussing effect, which (from a spectral point of view) concentrates high wave numbers near shocks (rapid change means large derivatives means high wave number). So the non-linearity tries to move high wave number modes in from the left, while the dispersion tries to move them to the left: again a balance is struck, and a travelling wave is the result.

1.4.5 Non-linear Diffusion: Similarity Solutions

Like travelling wave solutions, similarity solutions are important indicators of solution behaviour. A particularly illuminating illustration of this behaviour is provided by the general non-linear diffusion equation

$$u_t = \left(u^m u_x\right)_x, \tag{1.118}$$

which arises in many contexts. We shall illustrate the derivation of this equation for a fluid droplet below. Typically, (1.118) represents the conservation of the density of some quantity u with a diffusive flux $-u^m u_x$. A standard kind of problem to consider is then the release of a concentrated amount at $x = 0$ at $t = 0$. We can idealise this by supposing that at $t = 0$ (in suitable units),

$$u = 0 \quad \text{for } x \neq 0, \qquad \int_{-\infty}^{\infty} u(x)\,dx = 1. \tag{1.119}$$

This apparently contradictory prescription idealises the concept of a very concentrated local injection of u. For example, (1.118) with (1.119) could represent the diffusion of sugar in hot (one-dimensional) tea from an initially emplaced sugar grain. (1.119) defines the delta function $\delta(x)$, an example of a generalised function. One can think of generalised functions as being (defined by) the equivalence classes of well-behaved functions u_n with appropriate limiting behaviour. For example, the delta function is defined by the class of well-behaved functions u_n for which

$$\int_{-\infty}^{\infty} u_n(x) f(x)\,dx \to f(0) \tag{1.120}$$

as $n \to \infty$ for all well-behaved $f(x)$. As a shorthand, then,

$$\int_{-\infty}^{\infty} \delta(x) f(x)\,dx = f(0) \tag{1.121}$$

for any f, but the ulterior definition is really in (1.120). In practice, however, we think of a delta function as a 'function' of x, zero everywhere except for a (very) sharp spike at $x = 0$.

In solving (1.118), we also apply boundary conditions

$$u \to 0 \quad \text{as } x \to \pm\infty, \tag{1.122}$$

and these, together with the equation and initial condition, imply that

$$\int_{-\infty}^{\infty} u\,dx = 1 \qquad (1.123)$$

for all time.

A similarity solution is appropriate because there are no intrinsic space or time scales for the problem. It is in this context that one can expect the solution to look the same at different times on different scales. In general, as t varies, then the length scale might vary as $\xi(t)$ and the amplitude of the solution u might vary as $U(t)$. That is, if we look at u/U as a function of x/ξ, it will look the same for all t. This in turn implies that the solution takes the form

$$u = U(t)f\left[\frac{x}{\xi(t)}\right], \qquad (1.124)$$

and this is one of the forms of a similarity solution.

It is often the case that U and ξ are powers of t, and the exponents are to be chosen so that the problem has such a solution. This is best seen by example. If we denote $\eta = x/\xi(t)$, and substitute the form (1.124) into (1.118), (1.122) and (1.123), we find

$$\frac{U'}{U}f - \frac{\xi'}{\xi}\eta f' = \frac{U^m}{\xi^2}[f^m f']', \qquad (1.125)$$

where $U' = dU/dt$, $\xi' = d\xi/dt$, but $f' = df/d\eta$. The initial/boundary conditions become

$$f(\pm\infty) = 0, \qquad (1.126)$$

and the normalisation condition (1.123) is

$$U\xi \int_{-\infty}^{\infty} f\,d\eta = 1. \qquad (1.127)$$

A solution can be found provided the t dependence vanishes from the model, and this requires $U\xi = 1$ (the constant can be taken as one without loss of generality), whence (1.125) becomes

$$[f^m f']' + \xi^{m+1}\xi'(\eta f)' = 0, \qquad (1.128)$$

and $\xi^{m+1}\xi'$ must be constant. It is algebraically convenient to choose $\xi^{m+1}\xi' = 2/m$, thus

$$\eta = x\left[\frac{m}{2(m+2)t}\right]^{\frac{1}{m+2}}, \qquad (1.129)$$

and a first integral of (1.128) is

$$f^m f' + \frac{2}{m}\eta f = 0, \qquad (1.130)$$

Fig. 1.20 $f(\eta)$ given by (1.131)

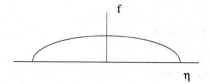

with the constant of integration being zero (because $f \to 0$ as $\eta \to \pm\infty$). Thus either $f = 0$, or

$$f = \left[\eta_0^2 - \eta^2\right]^{1/m}, \tag{1.131}$$

so that the solution has the form of a cap of finite extent, given by (1.131) (for $|\eta| < \eta_0$, and $f = 0$ for $|\eta| > \eta_0$. The value of η_0 is determined from $\int_{-\infty}^{\infty} f \, d\eta = 1$, and is

$$\eta_0 = \frac{1}{\left[2 \int_0^{\pi/2} \cos^{\frac{m+2}{m}} \theta \, d\theta\right]^{\frac{m}{m+2}}}. \tag{1.132}$$

The finite extent of the profile is due to the degeneracy of the equation when $m > 0$. (The limit $m \to 0$ regains the Gaussian solution of the heat equation by first putting $\eta = \sqrt{m}\eta_0\zeta$, $f = F/\sqrt{m}$, and noting that $\eta_0 \approx (\pi m)^{-m/2}$ as $m \to 0$ (this last following by application of Laplace's method to (1.132)).) The graph of $f(\eta)$ is shown in Fig. 1.20.

1.4.6 The Viscous Droplet

An example of where the non-linear diffusion equation can arise is in the dynamics of a drop of viscous fluid on a level surface. If the fluid occupies the region $0 < z < h(x, y, t)$ and is shallow, then lubrication theory gives the approximation

$$\nabla p = \mu \frac{\partial^2 \mathbf{u}}{\partial z^2}, \tag{1.133}$$

$$p_z = -\rho g,$$

in which $\mathbf{u} = (u, v, 0)$ is the horizontal component of velocity, and ∇ is the horizontal gradient $(\partial/\partial x, \partial/\partial y, 0)$. With $p = 0$ at $z = h$, we have the hydrostatic pressure $p = \rho g(h - z)$, so that $\nabla p = \rho g \nabla h$, and three vertical integrations of $(1.133)_1$ (with zero shear stress $\partial \mathbf{u}/\partial z = 0$ at $z = h$ and no slip $\mathbf{u} = 0$ at $z = 0$) yield the horizontal fluid flux

$$\mathbf{q} = \int_0^h \mathbf{u} \, dz = -\frac{\rho g}{3\mu} h^2 \nabla h. \tag{1.134}$$

Conservation of fluid volume for an incompressible fluid is $h_t + \nabla \cdot \mathbf{q} = 0$, and thus

$$h_t = \frac{\rho g}{3\mu} \nabla \cdot \left[h^3 \nabla h\right], \tag{1.135}$$

Fig. 1.21 The surface shown has positive curvature when the radius of curvature is measured from below the surface; in this case equilibrium requires $p > p_a$

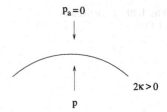

corresponding to (1.118) (in two space dimensions) with $m = 3$. A drop of fluid placed on a table will spread out at a finite rate.

That this does not continue indefinitely is due to surface tension. Rather than having $p = 0$ at $z = h$ (where the atmospheric pressure above is taken as zero), the effect of surface tension is to prescribe

$$p = 2\gamma\kappa, \qquad (1.136)$$

where γ is the surface tension, and κ is the mean curvature relative to the fluid droplet (i.e., $\kappa > 0$ if the interface is concave[4], as illustrated in Fig. 1.21). The curvature is defined as $2\kappa = \nabla.\mathbf{n}$, where $\mathbf{n}$ is the unit normal pointing away from the fluid (i.e., upwards). At least this shorthand definition works if we define

$$\mathbf{n} = \frac{(-h_x, -h_y, 1)}{[1 + |\nabla h|^2]^{1/2}}; \qquad (1.137)$$

thus

$$2\kappa = -\nabla.\left[\frac{\nabla h}{\{1 + |\nabla h|^2\}^{1/2}}\right]. \qquad (1.138)$$

It is less obvious that it will work more generally, since there are many ways of defining the interface in the form $\phi(x, y, z) = 0$ and thus $\mathbf{n} = \nabla\phi/|\nabla\phi|$ (that in (1.137) uses $\phi = z - h$); but in fact it does not matter, since we may generally take $\phi = (z - h)P$ for some arbitrary smooth function P, so that $\nabla\phi = (-h_x, -h_y, 1)P$ on $z = h$, and $\nabla\phi/|\nabla\phi|$ is the same expression as in (1.137).

For shallow flows, we replace $p = 0$ on $z = h$ by $p = -\gamma\nabla^2 h$ there, and thus

$$p \approx \rho g(h - z) - \gamma\nabla^2 h, \qquad (1.139)$$

and (via (1.134)), (1.135) is modified to

$$h_t = \nabla.\left[\frac{h^3}{3\mu}\nabla\{\rho g h - \gamma\nabla^2 h\}\right]. \qquad (1.140)$$

The fourth order term is also 'diffusive', insofar as it is a smoothing term, as already mentioned: high wave number (high gradient) modes are rapidly damped. The effect of surface tension relative to the diffusional gravity term is given by the Bond

---

[4]Geomorphologists would call this surface convex; see Chap. 6.

number

$$Bo = \frac{\rho g l^2}{\gamma},$$ (1.141)

where l is the lateral length scale of the drop. This is the (only) dimensionless parameter which occurs when (1.140) is written dimensionlessly.

1.4.7 Advance and Retreat: Waiting Times

The similarity solution (1.131) predicts an infinite slope at the margin (where $f = 0$) if $m > 1$ (and a zero slope if $m < 1$). If one releases a finite quantity at $t = 0$, then one expects the long time solution to be this similarity solution. The question then arises as to how this similarity solution is approached, in particular if the initial droplet has finite slope at the margin.

This question can be addressed in a more general way by studying the behaviour near the margin $x = x_S(t)$ of a solution $h(x, t)$ of (1.118),

$$h_t = \left(h^m h_x\right)_x.$$ (1.142)

Suppose that $h \sim c(x_S - x)^\nu$ for x near x_S. Then satisfaction of (1.142) requires

$$\dot{x}_S \approx c^m \left[\nu(m + 1) - 1\right](x_S - x)^{\nu m - 1}.$$ (1.143)

Note that the similarity solution (1.131) has $\dot{x}_S$ finite when $\nu = 1/m$, consistent with (1.143), and more generally we see that the margin will advance at a rate $\dot{x}_S \approx c^m/m$ if $h \sim c(x_S - x)^{1/m}$.

Suppose now that $m > 1$, and we emplace a droplet with finite slope, $\nu = 1$. Then the right hand side of (1.143) is zero at $x = x_S$, and thus $\dot{x}_S = 0$: the front does not move. What happens in this case is that the drop flattens out: there is transport of h towards the margin, which steepens the slope at x_S until it becomes infinite, at which point it will move. This pause while the solution fattens itself prior to margin movement is called a *waiting time*.

Conversely, if $m < 1$, then the front moves (forward) if the slope is zero there, and $\nu = 1/m$. If the slope is finite, $\nu = 1$, then (1.143) would imply infinite speed. An initial drop of finite margin slope will instantly develop zero front slope as the margin advances.

(1.143) does not allow for the possibility of retreat, because it describes a purely diffusive process. The possibility of both advance and retreat is afforded by a model of a viscous droplet with accretion, one example of which is the mathematical model of an ice sheet.[5] Essentially, an ice sheet, such as that covering Antarctica or Greenland, can be thought of as a (large) viscous droplet which is nourished by an accumulation rate (of ice formed from snow). A general model for such a nourished

---

[5]Ice sheets and their marginal movement are discussed further in Chap. 10.

droplet is

$$h_t = \left(h^m h_x\right)_x + a,\tag{1.144}$$

where a represents the accumulation rate. Unlike the pure diffusion process, (1.144) has a steady state

$$h = \left[\frac{1}{2}(m+1)a\left(x_0^2 - x^2\right)\right]^{1/(m+1)},\tag{1.145}$$

where x_0 must be prescribed. (In the case of an ice sheet, we might take x_0 to be at the continental margin.) (1.145) is slightly artificial, as it requires $a = 0$ for $x > x_0$, and allows for a finite flux $-h^m h_x = ax_0$ where $h = 0$. More generally, we might allow for accumulation and ablation (snowfall and melting), and thus $a = a(x)$, with $a < 0$ for large $|x|$. In that case the steady state is

$$h = \left[(m+1)\int_x^{x_0} B\,dx\right]^{1/(m+1)},\tag{1.146}$$

where the balance function s is

$$B = \int_0^x a\,dx,\tag{1.147}$$

and x_0 is defined to be where accumulation balances ablation,

$$\int_0^{x_0} a\,dx = 0.\tag{1.148}$$

This steady state is actually stable, and both advance and retreat can occur. Suppose the margin is at x_S, where $a = a_S = -|a_S|$ ($a_S < 0$, representing ablation). If we put $h \approx c(x_S - x)^\nu$, then (1.144) implies

$$v c\dot{x}_S(x_S - x)^{\nu-1} \approx \nu c^{m+1}\left[\nu(m+1) - 1\right](x_S - x)^{[\nu(m+1)-2]} - |a_S|,\tag{1.149}$$

and there are three possible balances of leading order terms.
    The first is as before,

$$\dot{x}_S \approx c^m\left[\nu(m+1) - 1\right](x_S - x)^{\nu m - 1},\tag{1.150}$$

and applies generally if $\nu < 1$. Supposing $m > 1$, then we have advance, $\dot{x}_S \approx c^m/m$ if $\nu = 1/m$, but if $\nu > 1/m$, this cannot occur, and the margin is stationary if $1/m < \nu < 1$. If $\nu = 1$, then $\nu(m+1) - 2 = m - 1 > 0$, so that

$$\dot{x}_S \approx -|a_S|/c,\tag{1.151}$$

and the margin retreats; if $\nu > 1$, then instantaneous adjustment to finite slope and retreat occurs.
    The ice sheet exhibits the same sort of waiting time behaviour as the viscous droplet without accretion. For $1/m < \nu < 1$, the margin is stationary, and if $x_S < x_0$

Fig. 1.22 Maximum value of steady solutions u of (1.152), $u(0)$, as a function of the parameter λ. Blow-up occurs if $\lambda \gtrsim 0.878$

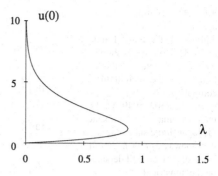

then the margin slope will steepen until $\nu = 1/m$, and advance occurs. On the other hand, if $x_S > x_0$, then the slope will decrease until $\nu = 1$, and retreat occurs. In the steady state, a balance is achieved (from (1.146)) when $\nu = 2/(m+1)$.

1.4.8 Blow-up

Further intriguing possibilities arise when the source term is non-linear. An example is afforded by the non-linear (reaction–diffusion) equation

$$u_t = u_{xx} + \lambda e^u, \tag{1.152}$$

which arises in the theory of combustion. Indeed, as we saw earlier, combustion occurs through the fact that multiple steady states can exist for a model such as (1.30), and the same is true for (1.152), which can have two steady solutions. In fact, if we solve $u'' + \lambda e^u = 0$ with boundary conditions $u = 0$ on $x = \pm 1$, then the solutions are

$$u = 2\ln\left[A \, \mathrm{sech}\left\{ \sqrt{\frac{\lambda}{2}} Ax \right\} \right], \tag{1.153}$$

where $A = \exp[u(0)/2]$, and A satisfies

$$A = \cosh\left[\sqrt{\frac{\lambda}{2}} A \right], \tag{1.154}$$

which has two solutions if $\lambda < 0.878$, and none if $\lambda > 0.878$: the situation is depicted in Fig. 1.22. If we replace e^u by $\exp[u/(1 + \varepsilon u)]$, $\varepsilon > 0$, we regain the top (hot) branch also, as in Fig. 1.9.

One wonders what the absence of a steady state for (1.152) if $\lambda > \lambda_c$ implies. The time-dependent problem certainly has a solution, and an idea of its behaviour can be deduced from the spatially independent problem, $u_t = \lambda e^u$, with solution $u = \ln[1/\{\lambda(t_0 - t)\}]$: u reaches infinity in a finite time. This phenomenon is known as *thermal runaway*, and more generally the creation of a singularity of the solution in finite time is called *blow-up*. Numerical solutions of Eq. (1.152) including the

Fig. 1.23 Solution of $u_t = u_{xx} + e^u$ on $[-1, 1]$, with $u = 0$ at $x = -1, 1$ and $t = 0$. The solution is shown for four times close to the blow-up time, which in this computation is $t_c = 3.56384027594971$. The many decimal places indicate the logarithmic suddenness of the runaway as $t \to t_c$, but the value of t_c itself will depend on the numerical approximation used

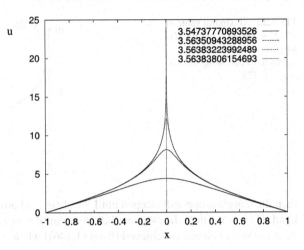

diffusion term show that blow-up still occurs, but at an isolated point; Fig. 1.23 shows the approach to blow-up as t approaches a critical blow-up time t_c.

In fact, one can prove generally that no steady solutions exist for λ greater than some critical value, and also that in that case, blow-up will occur in finite time. To do this, we use some slightly more sophisticated mathematics.

Suppose we want to solve the more general problem

$$u_t = \nabla^2 u + \lambda e^u \quad \text{in } \Omega, \tag{1.155}$$

with $u = 0$ in the boundary $\partial \Omega$, and $u = 0$ at $t = 0$ (these conditions are for convenience rather than necessity). We will be able to prove results for (1.155) which are comparable to those for the ordinary differential equation version (cf. (1.33))

$$\dot{w} = -\mu_1 w + \lambda e^w, \tag{1.156}$$

because, in some loose sense, the Laplacian operator ∇^2 resembles a loss term.

More specifically, we recall some pertinent facts about the (Helmholtz) eigenvalue problem

$$\nabla^2 \phi + \mu \phi = 0 \quad \text{in } \Omega, \tag{1.157}$$

with $\phi = 0$ on $\partial \Omega$. There exists a denumerable sequence of real eigenvalues $0 < \mu_1 \leq \mu_2 \ldots$, with $\mu_n \to \infty$ as $n \to \infty$, and corresponding (real) eigenfunctions $\phi_1, \phi_2, \ldots$ which form an orthonormal set (using the L^2 norm), thus

$$(\phi_i, \phi_j) \equiv \int_\Omega \phi_i \phi_j \, dV = \delta_{ij}, \tag{1.158}$$

where δ_{ij} is the Kronecker delta ($= 1$ if $i = j$, 0 if $i \neq j$). These eigenvalues satisfy a variational principle of the form

$$\mu_i = \min \int_\Omega |\nabla \phi|^2 \, dV, \tag{1.159}$$

where ϕ ranges over functions of unit norm, $\|\phi\|_2 = \{\int \phi^2 \, dV\}^{1/2} = 1$, which are orthogonal to ϕ_j for $j < i$; (more generally $\mu_i = \min\{\int |\nabla\phi|^2 \, dV / \int \phi^2 \, dV\}$ if ϕ is not normalised on to the unit sphere $\|\phi\|_2 = 1$). In particular

$$\mu_1 = \min_{\|\phi\|_2 = 1} \int_\Omega |\nabla\phi|^2 \, dV, \tag{1.160}$$

and the corresponding ϕ_1 is of one sign, let us say positive.

We take the inner product of Eq. (1.155) with ϕ_1 and divide by $\int \phi_1 \, dV$; defining

$$v(t) = \frac{\int_\Omega u\phi_1 \, dV}{\int_\Omega \phi_1 \, dV} = \int_\Omega u \, d\omega, \tag{1.161}$$

where $d\omega = \phi_1 \, dV / \int_\Omega \phi_1 \, dV$ is a measure on Ω (with $\int_\Omega d\omega = 1$), and using Green's theorem, we find

$$\dot{v} = \lambda \int_\Omega e^u \, d\omega - \mu_1 v, \tag{1.162}$$

and the equation for v is close to the ordinary differential equation (1.156).

Now we use Jensen's inequality. This says that if we have an integrable function $g(\mathbf{x})$ on Ω and a convex function $f(s)$ on $\mathbf{R}$ (i.e., one that bends upwards, $f'' > 0$), then

$$f\left[\int_\Omega g(\mathbf{x}) \, d\omega\right] \le \int_\Omega f[g(\mathbf{x})] \, d\omega \tag{1.163}$$

for any measure ω on Ω such that $\int_\Omega d\omega = 1$. We have chosen ω to be so normalised, and e^u is convex: thus

$$\int_\Omega \exp(u) \, d\omega \ge \exp\left[\int_\Omega u \, d\omega\right] = e^v, \tag{1.164}$$

so that

$$\dot{v} \ge \lambda e^v - \mu_1 v. \tag{1.165}$$

It is now easy to prove non-existence of steady states and blow-up for λ greater than some critical value λ_c. Firstly, u must be positive, and hence also v. (For suppose $u < 0$: since $u = 0$ at $t = 0$ and on $\partial\Omega$, then u attains its minimum in Ω at some $t > 0$, at which point $u_t \le 0$, $u_{xx} \ge 0$, which is impossible, since then $u_t - u_{xx} = \lambda e^u \le 0$.) For any v, $e^v \ge ev$, thus $\dot{v} \ge (\lambda e - \mu_1)v$. In a steady state we must have $\dot{v} = 0$, and also $v > 0$ (since clearly $u = 0$ is not a steady solution), and this pair of conditions is impossible if

$$\lambda > \mu_1/e. \tag{1.166}$$

This implies non-existence of a steady solution for $\lambda > \lambda_c$, where $\lambda_c \le \mu_1/e$.

In a similar vein, if $\lambda > \mu_1/e$, then

$$\dot{v} > \mu_1[e^{v-1} - v], \tag{1.167}$$

and $v > w$, where

$$\dot{w} = \mu_1(e^{w-1} - w), \qquad w(0) = 0. \tag{1.168}$$

(This is a standard comparison argument: $v = w$ at $t = 0$, and $\dot{v} > \dot{w}$ there, so $v - w$ is initially positive. It remains so unless at some future time $v - w$ reaches zero again, when necessarily $\dot{v} - \dot{w} \leq 0$—which is impossible, since $\dot{v} > \dot{w}$ whenever $v = w$.) But $w \to \infty$ in finite time ($\dot{w} > 0$ so that $w \to \infty$ as t increases, and as $w \to \infty$, $e^{-w}\dot{w} \approx \mu e^{-1}$, so e^{-w} reaches zero in finite time); therefore also v reaches infinity in finite time. Finally

$$v = \int_\Omega u\, d\omega \leq \sup_\Omega u, \tag{1.169}$$

since $\int_\Omega d\omega = 1$: hence $u \to \infty$ in finite time.

In fact $u \to \infty$ at isolated points, and usually at one isolated point. As blow-up is approached, one might suppose that the nature of the solution in the vicinity of the blow-up point would become independent of the initial (or boundary) conditions, and thus that some form of local similarity solution might be appropriate.

This is indeed the case, although the precise structure is rather complicated. We examine blow-up in one spatial dimension, x. As a first guess, the logarithmic nature of blow-up in the spatially independent case, together with the usual square-root behaviour of the space variable in similarity solutions for the diffusion equation, suggests that we define

$$\tau = -\ln(t_0 - t), \qquad \eta = \frac{x - x_0}{(t_0 - t)^{1/2}}, \qquad u = -\ln[\lambda(t_0 - t)] + g(\eta, \tau), \tag{1.170}$$

where blow-up occurs at $x = x_0$ at $t = t_0$; hence g satisfies

$$g_\tau = g_{\eta\eta} - \frac{1}{2}\eta g_\eta + e^g - 1. \tag{1.171}$$

The natural candidate for a similarity solution is then a steady solution $g(\eta)$ of (1.171), satisfying

$$g'' - \frac{1}{2}\eta g' + e^g - 1 = 0, \tag{1.172}$$

and matching to a far field solution $u(x, t_0)$ would suggest

$$g \sim -2\ln|\eta| \quad \text{as } \eta \to \pm\infty. \tag{1.173}$$

Solutions of (1.172) with this asymptotic structure do exist as either $\eta \to \infty$ or $\eta \to -\infty$—but not at both ends simultaneously. (1.172) admits even solutions, and if we restrict ourselves to these, then we may take

$$g'(0) = 0, \qquad g(0) \neq 0. \tag{1.174}$$

(If $g(0) = 0$, then $g \equiv 0$ is the solution.) However, it is found that such solutions have a different asymptotic behaviour as $\eta \to \infty$, namely

$$g \sim -\frac{A}{|\eta|} \exp\left[\frac{1}{4}\eta^2\right], \tag{1.175}$$

and $A = A[g(0)] > 0$ for $g(0) \neq 0$ (and $A(0) = 0$), and these cannot match to the outer solution. If one alternately prescribes (1.173) as $\eta \to +\infty$, for example, then the solution is asymmetric, and has the exponential behaviour (1.175) as $\eta \to -\infty$. Thus the appealingly simple similarity structure implied by steady solutions of (1.171) is wrong (and actually, the solution of the initial value problem (1.171) satisfying (1.173) tends to zero as $\tau \to \infty$).

However, (1.171) itself develops a local similarity structure as $\tau \to \infty$, using a further similarity variable

$$z = \frac{\eta}{\tau^{1/2}} = \frac{x - x_0}{(t_0 - t)^{1/2}[-\ln(t_0 - t)]^{1/2}}. \tag{1.176}$$

Rewriting (1.171) in terms of z and τ yields

$$g_\tau + \frac{1}{2}zg_z + 1 - e^g = \frac{1}{\tau}\left[g_{zz} + \frac{1}{2}zg_z\right]. \tag{1.177}$$

At leading order in τ^{-1} this has a solution

$$g = -\ln\left[1 + \frac{1}{4}cz^2\right], \tag{1.178}$$

where c is indeterminate, and this forms the basis for a formal expansion. It is algebraically convenient to use (1.178) to define c as a new variable, and also to write

$$s = \ln \tau. \tag{1.179}$$

Then (1.177) becomes

$$c_z = \frac{2}{\tau z^3}\left[2c + 4zc_z + z^2c_{zz} + z^2\left\{-\frac{[c + \frac{1}{2}zc_z]^2}{1 + \frac{1}{4}cz^2} + c + \frac{1}{2}zc_z - c_s\right\}\right]. \tag{1.180}$$

We seek a solution for (1.180) in the form

$$c \sim c_0(z, s) + \frac{1}{\tau}c_1(z, s) + \frac{1}{\tau^2}c_2(z, s) + \cdots, \tag{1.181}$$

and then, since $\tau d/d\tau = d/ds$, we have

$$c_s \sim \dot{c}_0 + \frac{1}{\tau}(\dot{c}_1 - c_1) + \frac{1}{\tau^2}(\dot{c}_2 - 2c_2) + \cdots, \tag{1.182}$$

where $\dot{c}_i \equiv \partial c_i / \partial s$. Substituting this into (1.180) and equating powers of τ, we find

$$c_0 = C_0(s), \tag{1.183}$$

where C_0 is arbitrary, and

$$c_{1z} = \frac{2}{z^3}\left[2C_0 + z^2\left\{-\frac{C_0^2}{1 + \frac{1}{4}C_0 z^2} + C_0 - \dot{C}_0\right\}\right]. \tag{1.184}$$

The arbitrary function C_0 arises because the order of the approximate equation is reduced. In order to specify it, and other arbitrary functions of s which arise at each order, we require that the solutions c_i be smooth, and this requires that there be no term on the right hand side of (1.184) proportional to $1/z$ as $z \to 0$, in order that logarithmic singularities not be introduced. Specifically, we require at each stage of the approximation that

$$\frac{\partial c_i}{\partial z} = \frac{2}{z^3}[a_{0i} + a_{1i}z + a_{3i}z^3 + \cdots]; \tag{1.185}$$

so that $z^2 c_i$ is smooth. Applying this to (1.184) requires that

$$\dot{C}_0 = C_0(1 - C_0), \tag{1.186}$$

so that $C_0 \to 1$ as $s \to \infty$, and then

$$c_1 = -\frac{2C_0}{z^2} + C_1(s) + C_0^2 \ln\left[1 + \frac{1}{4}C_0 z^2\right]. \tag{1.187}$$

At $O(1/\tau^2)$, we then have

$$c_{2z} = \frac{2}{z^3}\left[2c_1 + 4zc_{1z} + z^2 c_{1zz} + z^2\left\{-(\dot{c}_1 - c_1) + \left(c_1 + \frac{1}{2}zc_{1z}\right)\right.\right.$$
$$\left.\left. -\frac{2c_0(c_1 + \frac{1}{2}zc_{1z})}{1 + \frac{1}{4}c_0 z^2} + \frac{1}{4}c_0^2 c_{1z}^2\left(1 + \frac{1}{4}c_0 z^2\right)^2\right\}\right], \tag{1.188}$$

and applying the regularity condition (1.185), we find, after some algebra,

$$\dot{C}_1 = 2(1 - C_0)C_1 + \frac{5}{2}C_0^3, \tag{1.189}$$

so that $C_1 \to C_{10} + \frac{5}{2}s$ as $s \to \infty$. Thus, finally we obtain the local similarity solution

$$u \approx -\ln\left[\lambda\left\{t_0 - t + \frac{c(x - x_0)^2}{4[-\ln(t_0 - t)]}\right\}\right], \tag{1.190}$$

where $c \approx C_0(s)$, $s = \ln \tau = \ln[-\ln(t_0 - t)]$.

1.4.9 Reaction–Diffusion Equations

The development of mathematical biology in the last thirty years has led to one particular pedagogical example of wave and pattern formation, and that is in the coupled sets of equations known as reaction–diffusion equations. The general type is

$$\frac{\partial u_i}{\partial t} = f_i(\mathbf{u}) + \nabla.[D_{ij}\nabla u_j], \tag{1.191}$$

for n reactants $u_1, \ldots, u_n$, where the summation convention (sum over repeated suffixes, here j) is implied, but much of what is known about the behaviour of such systems can be illustrated with the two-species equations

$$u_t = f(u, v) + D_1\nabla^2 u,$$
$$v_t = g(u, v) + D_2\nabla^2 v. \tag{1.192}$$

The phenomena which we find are closely allied to the behaviour of the underlying dynamical system

$$\dot{u} = f(u, v),$$
$$\dot{v} = g(u, v), \tag{1.193}$$

and we will discuss three types of behaviour: wave trains, solitary waves, and stationary patterns.

Wave Trains

One way in which periodic travelling waves, or wave trains, can arise is when the underlying kinetics described by (1.193) is oscillatory. Diffusion causes the oscillations to propagate in space, and a periodic travelling wave results. It suffices to consider components which diffuse equally rapidly, so that we may consider the suitably scaled equation

$$\mathbf{w}_t = \mathbf{f}(\mathbf{w}) + \nabla^2\mathbf{w}, \tag{1.194}$$

where $\mathbf{w} \in \mathbf{R}^n$.

Suppose that the reaction kinetics admit an attractive limit cycle for the underlying system $\mathbf{w}_t = \mathbf{f}(\mathbf{w})$, and denote this as $\mathbf{W}_0(t)$, i.e.

$$\mathbf{W}_0' = \mathbf{f}(\mathbf{W}_0). \tag{1.195}$$

Suppose further that we look for solutions which are slowly varying in space. We define slow time and space scales τ and $\mathbf{X}$ as

$$\tau = \varepsilon t, \qquad \mathbf{X} = \sqrt{\varepsilon}\mathbf{x} \tag{1.196}$$

and seek formal solutions of (1.194) in the form $\mathbf{w}(\mathbf{X}, t, \tau)$, where

$$\mathbf{w}_t + \varepsilon \mathbf{w}_\tau = \mathbf{f}(\mathbf{w}) + \varepsilon \nabla^2 \mathbf{w}, \tag{1.197}$$

and $\nabla = \nabla_\mathbf{X}$ now. Expanding $\mathbf{w}$ as

$$\mathbf{w} \sim \mathbf{w}_0 + \varepsilon \mathbf{w}_1 + \cdots \tag{1.198}$$

leads to

$$\begin{aligned} \mathbf{w}_{0t} &= \mathbf{f}(\mathbf{w}_0), \\ \mathbf{w}_{1t} - J \mathbf{w}_1 &= -\mathbf{w}_{0\tau} + \nabla^2 \mathbf{w}_0, \end{aligned} \tag{1.199}$$

and so on; here $J = D\mathbf{f}(\mathbf{w}_0)$ is the Jacobian of $\mathbf{f}$ at $\mathbf{w}_0$. After an initial transient, we may take

$$\mathbf{w}_0 = \mathbf{W}_0(t + \psi), \tag{1.200}$$

where $\psi(\tau, \mathbf{X})$ is the slowly varying phase, and $J = D\mathbf{f}(\mathbf{W}_0)$ is a time-periodic matrix. Thus we find that $\mathbf{w}_1$ satisfies

$$\mathbf{w}_{1t} - J \mathbf{w}_1 = -\left(\psi_\tau - \nabla^2 \psi \right) \mathbf{W}_0' + |\nabla \psi|^2 \mathbf{W}_0''. \tag{1.201}$$

Note that $\mathbf{s} = \mathbf{W}_0'$ satisfies the homogeneous equation $\mathbf{s}_t - J\mathbf{s} = \mathbf{0}$. It follows that the solution of (1.201) is

$$\mathbf{w}_1 = -t \left(\psi_\tau - \nabla^2 \psi \right) \mathbf{s} + |\nabla \psi|^2 \mathbf{u}, \tag{1.202}$$

where

$$\mathbf{u} = M(t) \int_0^t M^{-1}(\theta) J(\theta) \mathbf{s}(\theta) \, d\theta + M(t), \tag{1.203}$$

and M is a fundamental matrix for the homogeneous equation, i.e., $M' = JM$, $M(0) = I$. Floquet's theorem implies that

$$M = P e^{t\Lambda}, \tag{1.204}$$

where P is a periodic matrix of period T (the same as that of the limit cycle $\mathbf{W}_0$). We can take the matrix Λ to be diagonal if the characteristic multipliers are distinct, and since we assume $\mathbf{W}_0$ is attracting, the eigenvalues of Λ will all have negative real part, except one of zero corresponding to $\mathbf{s}$. With a suitable choice of basis, we then have

$$\left(e^{t\Lambda} \right)_{ij} \to \delta_{i1} \delta_{j1} \quad \text{as } t \to \infty, \tag{1.205}$$

i.e., a matrix with the single non-zero element being unity in the first element. In this case the first column of P is $\mathbf{s}$, i.e., $P_{i1} = s_i$.

From (1.203), we have

$$\mathbf{u} = P(t) \int_0^t e^{\eta \Lambda} P^{-1}(t-\eta) J(t-\eta) \mathbf{s}(t-\eta) \, d\eta + M\mathbf{c}. \qquad (1.206)$$

The effect of the transient dies away as $t \to \infty$, and if we ignore it, then we can take $M_{ij} = s_i \delta_{j1}$, whence $M\mathbf{c} = c_1 \mathbf{s}$, and thus

$$\mathbf{u} = \mathbf{s} \left[\int_0^t \alpha(\eta) \, d\eta + c_1 \right], \qquad (1.207)$$

where the periodic function α is given by[6]

$$\alpha = \left(P^{-1} \right)_{1m} J_{mj} s_j. \qquad (1.208)$$

We define the mean of α to be

$$\bar{\alpha} = \frac{1}{T} \int_0^T \alpha(\eta) \, d\eta, \qquad (1.209)$$

so that

$$\beta = \int_0^t (\alpha - \bar{\alpha}) \, d\eta \qquad (1.210)$$

is periodic with period T. Then (1.202) is

$$\mathbf{w}_1 = \left[t\left\{ -\psi_\tau + \nabla^2 \psi + \bar{\alpha} |\nabla \psi|^2 \right\} + c_1 + \beta \right] \mathbf{s}, \qquad (1.211)$$

and in order to suppress secular terms (those which grow in t), we require the phase ψ to satisfy the evolution equation

$$\psi_\tau = \nabla^2 \psi + \bar{\alpha} |\nabla \psi|^2. \qquad (1.212)$$

This is an integrated form of Burgers' equation; in one dimension, $u = -\psi_X / 2\bar{\alpha}$ satisfies $u_\tau + u u_X = u_{XX}$. Disturbances will form shocks, which are jumps of phase gradient. More generally, if $\mathbf{u} = -\nabla \psi / 2\bar{\alpha}$, then (bearing in mind that $\nabla \times \mathbf{u} = \mathbf{0}$) we find

$$\mathbf{u}_\tau + (\mathbf{u}.\nabla)\mathbf{u} = \nabla^2 \mathbf{u}, \qquad (1.213)$$

which is the Navier–Stokes equation with no pressure term. Phase gradients move down phase gradients, and form defects where the (sub-)characteristics intersect.[7]

Solutions of (1.212) which vary with X correspond to travelling wave trains. For example, in one dimension, waves travel locally at speed $dX/dt \approx -(\partial \psi / \partial X)^{-1}$.

---

[6]We use the summation convention, which implies summation over repeated suffixes.

[7]Physicists call (1.212) the KPZ equation (after Kardar et al. 1986). The substitution $u = \exp(\bar{\alpha}\psi)$ reduces it to the diffusion equation for u; this is the Hopf–Cole transformation (see Whitham 1974).

In general, however, the phase of the oscillation becomes constant at long times if zero flux boundary conditions $\partial \psi / \partial n = 0$ are prescribed at container boundaries, and wave trains die away. However, this takes a long time (if ε is small), and while spatial gradients are present, the solutions have the form of waves. For example, *target patterns* are created when an impurity creates a local inhomogeneity in the medium.

Suppose the effect of such an impurity is to decrease the natural oscillation period by a small amount (of $O(\varepsilon)$) near a point, which we take to be the origin. To be specific, suppose that the impurity is circular, of radius a; then it is appropriate to specify

$$\psi = \tau + c \quad \text{at } R = a, \tag{1.214}$$

where R is the polar radius and c is an arbitrary constant (it merely fixes the time origin), and we expect ψ to tend towards the solution $\psi = \tau - f(R)$ as $t \to \infty$, where f satisfies

$$f'' + \frac{1}{R} f' - \bar{\alpha} f'^2 + 1 = 0, \tag{1.215}$$

together with $f(a) = c$ and an appropriate no flux condition at large R; such a condition can always be implemented by consideration of a small boundary layer near the boundary. Alternatively, we can restrict attention to a target pattern centred at the impurity by suppressing incoming waves (this is known as a radiation condition). The relevant solution if $\bar{\alpha} > 0$ is

$$f(R) = \frac{1}{\bar{\alpha}} \ln K_0(\sqrt{\bar{\alpha}} R), \tag{1.216}$$

where K_0 is the modified Bessel function of the second kind of order zero. The other Bessel function I_0 is suppressed because of the radiation condition (it produces incoming waves). At large R, $\psi \sim -R / \sqrt{\bar{\alpha}}$, which represents an outward travelling wave of speed $dR/dt \approx \sqrt{\bar{\alpha}}$. If, on the other hand, $\bar{\alpha} < 0$, then K_0 is replaced by a combination of the Bessel functions J_0 and Y_0, and the solution blows up at finite R, and travelling wave solutions of this type do not exist. More generally, if $\psi = \beta \tau$ on $R = a$, then target patterns exist if $\bar{\alpha} \beta > 0$.

Activator–Inhibitor System

An example of a system supporting travelling wave solutions is the activator–inhibitor system

$$u_t = f(u, v) + \nabla^2 u,$$
$$v_t = g(u, v) + \nabla^2 v, \tag{1.217}$$

where the nullclines of the kinetics are as shown in Fig. 1.24 (cf. Fig. 1.6). This system is called an activator–inhibitor system because $\partial f / \partial v > 0$, thus increased

Fig. 1.24 Phase diagram for
kinetics of (1.217)

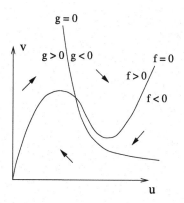

Fig. 1.25 Phase plane for
excitable kinetics

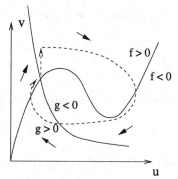

v activates u, while $\partial g/\partial u < 0$, so increased u inhibits v. When the intersection
is on the decreasing part of $f = 0$, as shown, then $\partial f/\partial u > 0$, $\partial g/\partial v < 0$, and
$-f_u/f_v > -g_u/g_v$, whence the determinant D of the Jacobian of $(u, v)^T$ at the
fixed point is positive. Hence the fixed point is unstable if $f_u + g_v > 0$, and a limit
cycle exists in this case if trajectories are bounded. For example, if $f = F/\varepsilon$, $\varepsilon \ll 1$,
this is the case, and the limit cycle takes the relaxational form shown in Fig. 1.6.
The addition of diffusion allows travelling wave trains to exist, as described above.

Solitary Waves in Excitable Media

Suppose now the intersection point of the nullclines $f = 0$ and $g = 0$ is as shown
in Fig. 1.25. The fixed point of the underlying dynamical system is now stable,
but relatively small perturbations to v can cause large excursions in u, as shown.
When diffusion is included, these large excursions can travel as solitary waves. The
simplest way to understand how this comes about is if we allow u to have fast
reaction kinetics and take v as having zero diffusion coefficient.

Fig. 1.26 Phase plane for solitary wave trajectory

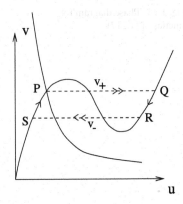

In one dimension, a suitably scaled model is then

$$\varepsilon u_t = f(u, v) + \varepsilon^2 u_{xx},$$
$$v_t = g(u, v), \tag{1.218}$$

and we look for a travelling wave solution of the form

$$u = u(\xi), \qquad v = v(\xi), \quad \xi = ct - x, \tag{1.219}$$

where c (assumed positive) is to be found. Then

$$\varepsilon c u' = f + \varepsilon^2 u'',$$
$$c v' = g, \tag{1.220}$$

and the idea is to seek a trajectory for which $(u, v) \to (u^*, v^*)$ as $\xi \to \pm\infty$ (here (u^*, v^*) is the fixed point of the system). The form of this trajectory is shown in Fig. 1.26. On the slow parts of the wave, $f \approx 0$ and $cv' \approx g$. On the fast parts, we put $\xi = \varepsilon \Xi$; then $v \approx$ constant, and we denote $v_+ \, (= v^*)$ and v_- as the corresponding values of v; v_- is unknown (as is c).

On the fast parts of the wave, we define $u' = w$ (where now $u' = du/d\Xi$), so that

$$u' = w,$$
$$w' = cw - f_\pm(u), \tag{1.221}$$

where $f_\pm(u) = f(u, v_\pm)$. The graphs of f_+ and f_- are similar, and are shown in Fig. 1.27, where we see that construction of the connecting branches PQ and RS requires that the fixed points P and Q, or R and S, of (1.221) have a connecting trajectory. In general, this will not be the case, but we can choose c to connect P to Q (since v_+ is known), and then we choose v_- to connect R to S (with this same value of c). The form of the resulting travelling wave is shown in Fig. 1.28.

Fig. 1.27 Phase plane
connection for the fast parts
of the travelling wave

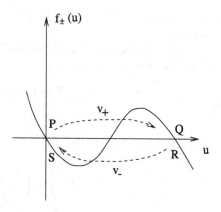

Pattern Formation

We have seen that an activator (v)-inhibitor (u) system

$$\dot{u} = f(u, v),$$
$$\dot{v} = g(u, v),$$

(1.222)

admits periodic travelling waves when the uniform state is unstable, and solitary
waves when it is stable (and the activator diffuses slowly). Stationary patterns can
occur when a stable steady state of (1.222) is rendered spatially unstable by different
component diffusivities. Suppose that

$$u_t = f(u, v) + u_{xx},$$
$$v_t = g(u, v) + dv_{xx},$$

(1.223)

is an activator–inhibitor system with $f_v > 0$, $g_u < 0$; the restriction to one spatial
dimension is inconsequential. The parameter d here represents the ratio of activator
to inhibitor diffusivities. Note that when $d \to 0$, we expect solitary wave propaga-

Fig. 1.28 Spatial form of the
travelling wave

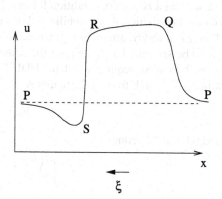

tion, at least for the phase diagram of Fig. 1.25, where also $f_u < 0$, $g_v < 0$ at the fixed point.

With the stationary state denoted as (u^*, v^*), we assume it is stable in the absence of diffusion; thus assume

$$T = f_u + g_v < 0,$$
$$\Delta = f_u g_v - f_v g_u > 0,$$

(1.224)

both evaluated at (u^*, v^*). We put

$$\begin{pmatrix} u \\ v \end{pmatrix} = \begin{pmatrix} u^* \\ v^* \end{pmatrix} + \mathbf{w} e^{\sigma t + ikx};$$

(1.225)

linearisation of (1.223) then yields

$$\left(M - k^2 D - \sigma\right)\mathbf{w} = \mathbf{0},$$

(1.226)

where

$$M = \begin{pmatrix} f_u & f_v \\ g_u & g_v \end{pmatrix}, \qquad D = \begin{pmatrix} 1 & 0 \\ 0 & d \end{pmatrix}.$$

(1.227)

The eigenvalues σ are the roots of

$$\sigma^2 - T_d \sigma + \Delta_d = 0,$$

(1.228)

where

$$T_d = T - (1 + d)k^2,$$
$$\Delta_d = \Delta - k^2(d f_u + g_v) + dk^4.$$

(1.229)

The steady state is stable if and only if $T_d < 0$ and $\Delta_d > 0$ (cf. Fig. 1.4). Now $T < 0$ and $\Delta > 0$ by assumption: hence $T_d < 0$, and thus instability occurs if and only if $\Delta_d < 0$. Since $\Delta > 0$, we see from (1.227) that this can only occur if $d f_u + g_v > 0$. Thus either $f_u > 0$ or $g_v > 0$, and the system cannot be excitable. Since $f_u + g_v < 0$, we see that a necessary condition for instability is that $d \neq 1$. Because d is the ratio of two diffusivities, this instability is known as *diffusion-driven instability* (DDI), or *Turing instability*, after the originator of the theory.

To be specific, let us suppose the situation to be that of Fig. 1.24, i.e., $f_u > 0$, $g_v < 0$: then we require $d > 1$ for DDI. The precise criterion for instability is that $\min \Delta_d < 0$, and, from (1.229), this is

$$d f_u + g_v > 2[\Delta d]^{1/2},$$

(1.230)

and this can be reduced to

$$d > \left[\frac{\Delta^{1/2} + \{f_v |g_u|\}^{1/2}}{f_u}\right]^2.$$

(1.231)

The resulting instability is direct and not oscillatory (in time), though it is oscillatory in space. We can therefore expect stationary finite amplitude patterns to emerge as the stable solutions, and this is indeed what often occurs.

The form of these putative steady solutions as d becomes large can be studied by seeking (spatially) periodic solutions of

$$u_{xx} + f(u, v) = 0,$$
$$v_{xx} + \varepsilon^2 g(u, v) = 0, \tag{1.232}$$

where we define $\varepsilon^2 = 1/d \ll 1$.

We begin by seeking solutions with period of $O(1)$. As u varies over distances of $x = O(1)$, $v = \bar{v}$ is approximately constant, and thus the equation for u can be integrated to give the first integral

$$\frac{1}{2} u_x^2 + V(u, \bar{v}) = E, \tag{1.233}$$

where

$$V(u, v) = \int_0^u f(u, v)\, du, \tag{1.234}$$

and E is constant.

The forms of the curves $f(u, v) = 0$ (defining v as a function of u), $f(u, v)$ as a function of u for various fixed v, and $V(u, v)$ as a function of u are shown in Fig. 1.29. For constant v, solutions for u will be periodic if they lie in the potential well of V. Given $\bar{v}$ and E, these periodic solutions are fully determined, and in particular their period P is a function of $\bar{v}$ and E, thus $P = P(\bar{v}, E)$. The choice of $\bar{v}$ and E must then be made so that v is periodic. We can choose the origin of x so that u is maximum there; then in fact u is even, and hence so is $g[u(x; \bar{v}, E), \bar{v}]$. Integration of $(1.232)_2$ then yields

$$v = \bar{v} - \varepsilon^2 \int_{-P/2}^x (x - \xi) g[u(\xi; \bar{v}, E), \bar{v}]\, d\xi, \tag{1.235}$$

where periodicity of v requires that

$$\int_{-P/2}^{P/2} g[u(\xi; \bar{v}, E), \bar{v}]\, d\xi = 0. \tag{1.236}$$

(We also require that

$$\int_{-P/2}^{P/2} \xi g[u(\xi; \bar{v}, E), \bar{v}]\, d\xi = 0, \tag{1.237}$$

but this is satisfied automatically since the integrand is odd.)

Given $\bar{v}$, (1.236) appears to determine E, and thus provide a one-parameter family of periodic solutions. However, it is unlikely that (1.236) can generally be satisfied for a given function g. Consideration of Fig. 1.29 suggests that it is more likely

Fig. 1.29 Definition of the values $v_{\pm}$ defined by the function $f(u, v)$. The *upper graph* shows the curve defined implicitly by $f(u, v) = 0$ (compare Fig. 1.24). The *middle graph* shows the function $f(u, v)$ as a function of u for $v = v_{+}, \bar{v}, v_{-}$, and the *lowest graph* is the potential $V(u, v) = \int_0^u f(u, v)\, du$ for the value of $v = \bar{v}$ corresponding to the middle of these three curves. The choice of $\bar{v}$ in the figure is that for which the two maxima of V are equal. The particular function used in the illustrations is $f(u, v) = v - [u^3 - 8u^2 + 17u]$, for which the value of $\bar{v}$ where the maxima are equal is $\bar{v} \approx 7.407$; the values of v_{+} and v_{-} are $v_{+} \approx 10.879$ and $v_{-} \approx 3.935$

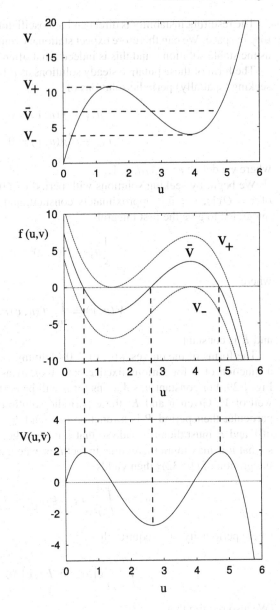

that, given $\bar{v}$ and E, satisfaction of (1.236) will depend on the precise location of the curve $g = 0$. For a function $g(u, v; \alpha)$ dependent on a single parameter α, such as $g = \alpha - u^3 v$, this suggests that (1.236) may be satisfied (if at all) for a unique value of $\alpha(\bar{v}, E)$. Since also $P = P(\bar{v}, E)$, this suggests a one-parameter family of spatially periodic solutions in which $P = P(\alpha)$.

The other possibility for periodic solutions involves the existence of regions in which u is constant, separated by boundary layers in which u changes rapidly. In

Fig. 1.30 The nullclines
$f(u, v) = 0$ and $g(u, v) = 0$.
The f nullcline defines
locally two functions $u_\pm(v)$.
During the oscillation, v
moves from A to B and back
to A, while $g > 0$, and
similarly from C to D and
back to C while $g > 0$. When
v reaches $\bar{v}$, a boundary layer
in u switches the solution
between its two branches

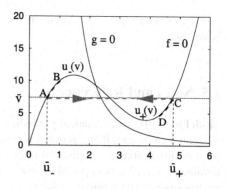

this case, the longer space scale $X = \varepsilon x$ comes into play, and the resultant form of
Eq. (1.232),

$$\varepsilon^2 u_{XX} + f = 0,$$

$$v_{XX} + g = 0,$$

(1.238)

is clearly suggestive of a boundary layer structure.

The boundary layers themselves are still described by (1.233), but now we require that u tends to constants $\bar{u}_+$ and $\bar{u}_-$ as $x \to \pm\infty$; this requires $\bar{v}$ to have the particular value where the local maxima of $V(u, \bar{v})$ are the same (and these occur at $\bar{u}_-$ and $\bar{u}_+$). For the value of E equal to this maximum, there are then boundary layer solutions in which either u goes from $\bar{u}_-$ to $\bar{u}_+$ as x increases, or from $\bar{u}_+$ to $\bar{u}_-$.

The periodic solutions are filled out by solving

$$v_{XX} + g(u, v) = 0,$$

(1.239)

in which u is determined by $f(u, v) = 0$. There are two branches of the resultant function $u(v)$, which we denote by $u_-(v)$ and $u_+(v)$ (and $u_\pm(\bar{v}) = \bar{u}_\pm$), as indicated in Fig. 1.30; if we define

$$W(v) = \begin{cases} \int_{\bar{v}}^{v} g[u_-(v'), v']\,dv' & \text{for } \bar{v} < v, v_+, \\ \int_{\bar{v}}^{v} g[u_+(v'), v']\,dv' & \text{for } v_- < v < \bar{v}, \end{cases}$$

(1.240)

then W is a V-shaped function defined in $[v_-, v_+]$, with a minimum at $v = \bar{v}$. Solutions for v are determined from

$$\frac{1}{2}v_X^2 + W(v) = F,$$

(1.241)

for constant F, and determine a one-parameter family of periodic solutions. Note that this family occurs for a fixed choice of f and g, and the parameter can be taken to be the period. This family is then naturally interpreted as the continuation to large

d of the bifurcating family dependent on wave number which arises when (1.231) is satisfied.

1.5 Notes and References

Modelling   By mathematical modelling, I mean the formulation of a problem in mathematical terms. If the process is continuous, usually the model will take the form of differential equations, and in this book we further confine ourselves to deterministic models, as opposed to stochastic models. Stochastic models are of increasing popularity, aiming as they do to represent the noisiness of a system, but they can also be something of an excuse to sweep things we do not understand under the carpet.

The original classic book which set out the applied mathematician's stall is that by Lin and Segel (1974). It contains the ethos of applied mathematics, but retained a somewhat austere choice of applications. Another classic book which dealt much more with practical (mostly industrial) applications is that by Tayler (1986). My own book (Fowler 1997) is in a similar spirit.

These books, certainly the latter two, are aimed at graduate level. There are a number of books which deal more gently, but still genuinely, with modelling. The classic of this type is perhaps that by Haberman (1998), a reprinted edition of his 1977 text. More recent books in this direction are those by Fowkes and Mahony (1994), Howison (2005), and Holmes (2009).

Asymptotics and Perturbation Theory   Like modelling, there are many books on perturbation methods. To my mind, the pre-eminent ones are those by Kevorkian and Cole (1981), Bender and Orszag (1978) and Hinch (1991). Van Dyke's (1975) book is also a classic. Other well known books are those by Nayfeh (1973) and Holmes (1995).

The flavours of these books are subtly different. Bender and Orszag's blockbuster, taught at M.I.T. in a one-semester course (the whole book), has as its central part the asymptotic study of boundary layers. The book has the novelty of giving many numerical illustrations of how good (or bad) the approximations are, and when they appear to break down.

Kevorkian and Cole's book (an expanded edition of Cole's original 1968 monograph) focusses more on multiple scale methods, and takes these to levels of sophistication a good deal beyond more elementary texts, and there are expositions of some classic problems: the derivation of the Korteweg–de Vries equation describing long waves in shallow water, and the relaxational van der Pol oscillator, for example.

Van Dyke's book is slightly more formal in nature, and mostly concerned with fluid mechanics. It is one of the few places where one can learn the method of strained coordinates, a method which is particularly useful in dealing with the motion of margins and fronts. Hinch's and Nayfeh's books include a chapter on strained coordinates also, as well as the other staple contents. Hinch's is short, to the point, succinct. Holmes's book includes a chapter on homogenisation.

Combustion, Non-linear Diffusion and Blow-up   Two early accounts of combustion and exothermic reactions are those by Aris (1975) and Buckmaster and Ludford (1982). The first of these largely deals with reaction in (solid) permeable catalysts, while combustion theory of the second tends to deal with gaseous combustion, where the theory has all the complication of compressible gas dynamics together with the species reaction kinetics. A more mathematical book is that by Bebernes and Eberly (1989). Other books on this subject include those of Williams (1985), Barnard and Bradley (1985) and Glassman (1987), the latter two more descriptive than Williams's voluminous work. A similar analytic approach is that by Liñán and Williams (1993), but this book is more concise than that of Williams. Combustion really applies to any reaction, but by convention refers specifically to reactions where there is a large change of temperature. If this is such that the reactants become luminous, we have a flame. If the change of temperature is rapid, we have a thermal explosion. Since in gases, increase of temperature is associated with increase of pressure, explosions tend to be associated with shock waves, or detonation waves, and this is the explosive 'blast'.

The classical treatment of thermal explosions (in solids) is much as described in Sect. 1.4.8, and involves the positive feedback associated with exothermic reactive heating, which causes the runaway. Explosive runaway can also be caused by autocatalytic feedback in the reaction scheme, much as in a nuclear explosion; this is the 'chain' reaction. Systems with autocatalysis are also prone to oscillatory bifurcations and waves, and are dealt with in the book by Gray and Scott (1990). Ignition of explosions may be caused by impact or friction (as in striking a match). Both events cause a localised hotspot to occur, that of impact being due to the sudden compression of small gas bubbles, see Bowden and Yoffe (1985).

Reactions in a diffusive flame (i.e., one where fuel and oxidant are not pre-mixed) can be analysed using large activation energy asymptotics; the reactions occur in a narrow front which spreads as a deflagration wave, whose speed is less than the sound speed, and is rate-limited by the supply of reactant to the front. The detonation wave is a reactive shock wave, in which the reaction is triggered not by supply of reactant, but by gas compression and consequent heating within the shock.

The book by Samarskii et al. (1995) provides a wealth of information about non-linear diffusion equations, and their associated solution properties of compact support and blow-up. The asymptotic description given here of the local similarity structure for the blow-up of solutions of $u_t = u_{xx} + \lambda e^u$ is based on that of Dold (1985).

Burgers' Equation   Burgers' equation relates to a model introduced by Burgers (1948) to describe turbulence in fluid flow in a pipe. In its original form, his model is given by the pair of equations

$$b\frac{dU}{dt} = P - \frac{\nu U}{b} - \frac{1}{b}\int_0^b v^2 \, dy,$$

$$\frac{\partial v}{\partial t} + 2v\frac{\partial v}{\partial y} = \frac{Uv}{b} + \nu\frac{\partial^2 u}{\partial y^2}.$$

(1.242)

This is a toy model which aims to mimic the classical procedure of Reynolds averaging, leading to an evolution equation for the mean flow $U(t)$, and another for the fluctuating velocity field $v(y, t)$. The cross stream variable is y, and the width of the 'pipe' is b. Burgers' equation follows from the assumption that $U = 0$, and arises in the original paper as an approximation to describe the transition region near shocks; Burgers gives the travelling wave front solution for this case. A thorough discussion of Burgers' equation is given by Whitham (1974).

Fisher's Equation   The geneticist R.A. Fisher wrote down his famous equation (Fisher 1937) to describe the propagation of an advantageous gene in a population situated in a one-dimensional continuum—Fisher had in mind a shore line as an example. The genes (or more properly *alleles*, i.e., variants of genes), reside in the members of a population, and the proportion of different alleles of any particular gene is described by Hardy–Weinberg kinetics. If one allele has a slight evolutionary advantage, then its proportion p will vary slowly from generation to generation, and its rate of change is given in certain circumstances by the logistic equation $\dot{p} = kp(1 - p)$. The effect of diffusion allows the genes to migrate through the migration of the carrier population. See Hoppensteadt (1975) for a succinct description. Fisher did not bother with all this background, but simply wrote his equation down directly. As well as this paper, he authored or co-authored eight other papers in the same volume, as well as being the journal editor!

Solitons   There are many books on solitons. An accessible introduction is the book by Drazin and Johnson (1989), and a more advanced treatment is that of Newell (1985). The subject is rich and fascinating, as is also the curious discovery of the 'first' soliton, or 'great wave of translation' by John Scott Russell in 1834, as he followed it on horseback along the Edinburgh to Glasgow canal. The Korteweg–de Vries equation which appears successfully to describe such waves was introduced by them much later (Korteweg and de Vries 1895), by which time they are referred to as solitary waves. Korteweg and de Vries also wrote down the periodic (but unstable) *cnoidal* wave solutions.

There are many other equations which are now known to possess soliton solutions, and their folklore has crept into many subjects. Under the guise of 'magmons', for example, they have appeared in the subject of magma transport, which we discuss in Chap. 9.

Reaction–Diffusion Equations   Any book on mathematical biology (and there are a good number of these) will discuss reaction–diffusion equations. The gold standard of the type is the book (now in two volumes) by Murray (2002), which also contains much other subject matter. A more concise book just on reaction–diffusion equations is that by Grindrod (1991). These books span the undergraduate/graduate transition. The book by Edelstein-Keshet (2005) is gentler, and aimed at a lower level.

Kopell and Howard (1973) and Howard and Kopell (1977) studied waves in reaction–diffusion equations using the ideas of bifurcation theory and multiple scales. Keener (1980, 1986) studied spiral wave formation in excitable media, using as a template a singularly perturbed pair of equations, essentially of Fitzhugh–Nagumo type.

Meinhardt (1982) studied pattern formation in reaction–diffusion systems, and later (Meinhardt 1995) studied the relation between a suite of mathematical models and actual observed patterns on sea shells. The comparison is striking as well as pictorially sumptuous.

1.6 Exercises

1.1 Suppose

$$Pe\left[\frac{\partial T}{\partial t} + \mathbf{u}.\nabla T\right] = \nabla^2 T + 1 \quad \text{in } D,$$

with

$$T = 0 \quad \text{on } \partial D,$$

$$T = \Lambda\Theta(\mathbf{x}) \quad \text{in } D \text{ at } t = 0,$$

and $\Theta = O(1)$, $\Lambda \gg 1$, $Pe \ll 1$. Discuss appropriate scales for the various phases of the solution.

1.2 The differential equation

$$\dot{x} = a - xe^{-x}, \quad x > 0, \quad a > 0,$$

may have 0, 1 or 2 steady states. Determine how these depend on a, and describe how solutions behave for $a > e^{-1}$ and $a < e^{-1}$, depending on the value of $x(0)$.

1.3 Each of the equations

$$z^5 - \varepsilon z - 1 = 0,$$

$$\varepsilon z^5 - z - 1 = 0,$$

has five (possibly complex) roots. Find leading order approximations to these if $\varepsilon \ll 1$. Can you refine the approximations?

1.4 u and v satisfy the ordinary differential equations

$$\dot{u} = k_1 - k_2 u + k_3 u^2 v,$$

$$\dot{v} = k_4 - k_3 u^2 v,$$

where $k_i > 0$. By suitably scaling the equations, show that these can be written in the dimensionless form

$$\dot{u} = a - u + u^2 v,$$

$$\dot{v} = b - u^2 v,$$

where a and b should be defined. Show that if u, v are initially positive, they remain so. Draw the nullclines in the positive quadrant, show that there is a unique steady state and examine its stability. Are periodic solutions likely to exist?

1.5 The relaxational form of the van der Pol oscillator is

$$\varepsilon \ddot{x} + (x^2 - 1)\dot{x} + x = 0, \quad \varepsilon \ll 1.$$

A suitable phase plane is spanned by (x, y), where $y = \varepsilon \dot{x} + \frac{1}{3}x^3 - x$. Describe the motion in this phase plane, and find, approximately, the period of the relaxation oscillation. What happens if $\varepsilon < 0$?

1.6 Find a scaling of the combustion equation

$$c\frac{dT}{dt} = -k(T - T_0) + A \exp\left(-\frac{E}{RT}\right),$$

so that it can be written in the form

$$\dot{\theta} = \theta_0 - g(\theta),$$

where $\theta_0 = RT_0/E$ and $g = \theta - \alpha e^{-1/\theta}$. Give the definition of α. Hence show that the steady state θ is a multiple-valued function of θ_0 if $\alpha > \frac{1}{4}e^2$.

Find approximations to the smaller and larger positive roots of $x^2 e^{-x} = \varepsilon$, where ε is small and positive. Hence find the approximate range (θ_-, θ_+) of θ_0 for which there are three steady solutions.

Suppose that $\alpha > \frac{1}{4}e^2$, and θ_0 varies slowly according to

$$\dot{\theta}_0 = \varepsilon(\theta^* - \theta),$$

where $\varepsilon \ll 1$. Show that there are three possible outcomes, depending on the value of θ^*, and describe them.

1.7 A forced pendulum is modelled by the (dimensional) equation

$$l\ddot{\theta} + k\dot{\theta} + g \sin \theta = \alpha \sin \lambda t.$$

By non-dimensionalising the equation, show how to obtain (1.47), and identify the parameters $\varepsilon, \beta, \Omega_0$ and ω.

1.8 It is asserted after (1.59) that $\Omega(A)$ is a decreasing function of A for $0 < A < \pi$, or equivalently, that the function

$$p(A) = \frac{1}{\sqrt{2}} \int_0^A \frac{du}{[\cos u - \cos A]^{1/2}}$$

is increasing. Show that this is true by writing p in the form

$$p = \int_0^1 \left(\frac{\theta}{\sin \theta}\right)^{1/2} \left(\frac{\phi}{\sin \phi}\right)^{1/2} \frac{dw}{(1 - w^2)^{1/2}}$$

for some functions $\theta(w, A)$ and $\phi(w, A)$, and using the fact that $\theta/\sin\theta$ is an increasing function of θ in $(0, \pi)$.

[Hint: $\cos u - \cos A = 2\sin(\frac{A-u}{2})\sin(\frac{A+u}{2})$.]

1.9 A simple model for the two-phase flow of two fluids along a tube is

$$\alpha_t + (\alpha v)_z = 0,$$

$$-\alpha_t + \left[(1 - \alpha)u\right]_z = 0,$$

$$\rho_g\left[(\alpha v)_t + (\alpha v^2)_z\right] = -\alpha p_z,$$

$$\rho_l\left[\{(1 - \alpha)u\}_t + \{D_l(1 - \alpha)u^2\}_z\right] = -(1 - \alpha)p_z,$$

where p is pressure, u and v are the two fluid velocities, α is the volume fraction of the fluid with speed v, ρ_g is its density, and ρ_l is the density of the other fluid. Show that there are two characteristic speeds $dz/dt = \lambda$, satisfying

$$(\lambda - u)^2 = (D_l - 1)\left[u^2 + 2u(\lambda - u)\right] - s^2(\lambda - v)^2,$$

where

$$s = \left[\frac{\rho_g(1 - \alpha)}{\rho_l\alpha}\right]^{1/2}.$$

Deduce that the characteristic speeds are real if, when $D_l - 1 \ll 1$, $s \ll 1$,

$$D_l \gtrsim 1 + \left\{\frac{s(u - v)}{u}\right\}^2.$$

In particular, show that the roots are complex if $D_l = 1$ and $u \neq v$. What does this suggest concerning the well-posedness of the model?

1.10 The function $u(x, t)$ satisfies

$$u_t + uu_x = \alpha\left(1 - u^2\right)$$

for $-\infty < x < \infty$, with $u = u_0(x)$ at $t = 0$, and $0 < u_0 < 1$ everywhere. Show that the characteristic solution can be written parametrically in the form

$$u = \frac{u_0(s) + \tanh\alpha t}{1 + u_0(s)\tanh\alpha t}, \qquad \exp\left[\alpha(x - s)\right] = \frac{\operatorname{sech}\alpha t}{1 - u\tanh\alpha t}.$$

Sketch the form of the characteristics for an initial function such as $u_0(s) = a/(1 + s^2)$. Show that, in terms of s and t, u_x is given by

$$u_x = \frac{[\alpha \operatorname{sech}^2\alpha t]u_0'(s)}{[1 + u_0(s)\tanh\alpha t][\alpha + \{u_0'(s) + \alpha u_0(s)\}\tanh\alpha t]},$$

and deduce that a shock will form if $u_0' + \alpha(1 + u_0)$ becomes negative for some s. Show that if $u_0 = a/(1 + s^2)$ and a is small, this occurs if

$$\alpha \lesssim \frac{3a\sqrt{3}}{8}.$$

1.11 Discuss the formation of shocks and the resulting shock structure for the equation

$$u_t + u^\alpha u_x = \varepsilon\left[u^\beta u_x\right]_x,$$

where $\alpha, \beta > 0$, and $\varepsilon \ll 1$. (Assume $u > 0$, and $u \to 0$ at $\pm\infty$.)
   Show that the equation

$$u_t + u u_x = \varepsilon u u_{xx}$$

admits a shock structure when $\varepsilon \ll 1$, but that the shock speed is *not* given by $\dot{x}_S = \frac{1}{2}(u_+ + u_-)$ (cf. (1.74)). Why should this be so?

1.12 Use phase plane methods to study the existence of travelling wave solutions to the equation

$$u_t = u^p\left(1 - u^q\right) + \left[u^r u_x\right]_x,$$

when (i) $p = 1, q = 2, r = 0$; (ii) $p = 1, q = 1, r = 1$.

1.13 Two examples of integrable partial differential equations which admit soliton solutions are the non-linear Schrödinger (NLS) equation

$$i u_t = |u|^2 u + u_{xx},$$

and the sine-Gordon equation

$$u_{tt} - u_{xx} = \sin u.$$

Show that these equations admit solitary wave solutions (which are in fact solitons).

1.14 Write down the equation satisfied by a similarity solution of the form $u = t^\beta f(\eta)$, $\eta = x/t^\alpha$, for the equation

$$u_t = \left(u^m u_x\right)_x \quad \text{in } 0 < x < \infty,$$

where $m > 0$, with $u^m u_x = -1$ at $x = 0$, $u \to 0$ as $x \to \infty$, $u = 0$ at $t = 0$. Show that $\int_0^\infty f \, d\eta = 1$, and hence show that in fact f reaches zero at a finite value η_0. Is the requirement that $m > 0$ necessary?

1.15 u satisfies the equation

$$u_t = \left[D(u)u_x\right]_x \quad \text{in } 0 < x < \infty,$$

with $u = 0$ at $x \to \infty$ and $t = 0$. For a general function D (not a power of u), for what kind of boundary condition at $x = 0$ does a similarity solution exist? What if, instead, $D = D(u_x)$? Write down suitable equations and boundary conditions for the similarity function in each case.

1.16 The depth of a small droplet, h, satisfies the surface-tension controlled equation

$$h_t = -\frac{\gamma}{3\mu} \nabla \cdot [h^3 \nabla \nabla^2 h].$$

Suppose that a small quantity $\int h \, dA = M$ is released at time zero at the origin. Find a suitable similarity solution in one and two horizontal spatial dimensions.

1.17 A gravity-driven droplet of fluid spreads out on a flat surface. Its viscosity μ is a function of shear rate, so that a lubrication approximation leads to the model for its depth h, shear stress τ and velocity $\mathbf{u}$:

$$\rho g \nabla h = \frac{\partial \tau}{\partial z},$$

$$\frac{\partial \mathbf{u}}{\partial z} = A |\tau|^{n-1} \tau.$$

(A constant viscosity fluid has $n = 1$.) Show that the horizontal fluid flux is

$$\mathbf{q} = -\frac{A(\rho g)^n}{n+2} |\nabla h|^{n-1} h^{n+2} \nabla h,$$

and deduce that

$$\frac{\partial h}{\partial t} = \frac{A(\rho g)^n}{n+2} \nabla \cdot [h^{n+2} |\nabla h|^{n-1} \nabla h].$$

Non-dimensionalise the model, assuming initial emplacement of a finite volume M at the origin, and find similarity solutions in one and two dimensions for the depth. What happens as $n \to \infty$ or $n \to 0$?

1.18 The depth h of a symmetric (two-dimensional) droplet under the influence of gravity and surface tension is described by the dimensionless equation

$$\frac{\partial h}{\partial t} = \frac{\partial}{\partial x} \left[h^3 \frac{\partial}{\partial x} \left\{ Bh - \frac{\partial^2 h}{\partial x^2} \right\} \right],$$

subject to the conditions

$$h = 0, \qquad h_x = \mp \tan \theta \quad \text{at } x = \pm x_0, \qquad \int_{-x_0}^{x_0} h \, dx = A,$$

where θ is the contact angle. Show that there is a steady state solution $h = h_0 u(x)$, in which

$$\int_u^1 \frac{du}{[(1-u)(\rho - u)]^{1/2}} = \sqrt{B}|x|,$$

the coefficient ρ is defined by

$$\rho = \frac{\tan^2 \theta}{B h_0^2},$$

and the maximum depth h_0 is given by

$$\frac{A\sqrt{B}}{2h_0} = I(\rho),$$

where

$$I(\rho) = \int_0^1 \frac{u\,du}{[(1-u)(\rho-u)]^{1/2}}.$$

By considering (graphically) both sides of the equation for h_0 as functions of ρ, show that there is a unique value of h_0 satisfying this equation, and thus a unique solution for h.

By evaluating the integrals explicitly, show that

$$u = 1 - (\rho - 1)\sinh^2\left(\frac{\sqrt{B}x}{2}\right),$$

and that ρ is determined by

$$\frac{AB}{2\tan\theta} = -1 + \frac{\rho+1}{2\sqrt{\rho}}\ln\left(\frac{\sqrt{\rho}+1}{\sqrt{\rho}-1}\right).$$

Find explicit approximations for h when $AB \gg \tan\theta$ and $AB \ll \tan\theta$, and hence show that the margin positions are approximately given by

$$x_0 = \begin{cases} \frac{A\sqrt{B}}{\tan\theta}, & AB \gg \tan\theta, \\ (\frac{3A}{2\tan\theta})^{1/2}, & AB \ll \tan\theta. \end{cases}$$

[Note that if θ is the actual contact angle, then implicitly the depth scale and lateral length scale have been taken equal, and the derivation of the equation for h via lubrication theory is only self-consistent if $h_0 \ll 1$ or $x_0 \gg 1$. Since a length scale can be prescribed from the initial droplet size, we can choose $A = 1$ without loss of generality. We can then find conditions on B and $\tan\theta$ which ensure self-consistency.]

1.19 Let u satisfy

$$u_t = \lambda u^p + u_{xx},$$

with $u = 1$ on $x = \pm 1$ and $t = 0$. Prove that if λ is large enough, u must blow up in finite time if $p > 1$. Supposing this happens at time t_0 at $x = 0$, show that a possible local similarity structure is of the form

$$u = \frac{f(\xi)}{(t_0 - t)^\beta}, \quad \xi = \frac{x}{(t_0 - t)^{1/2}},$$

and prove that $\beta = 1/(p-1)$. Show that in this case, f would satisfy

$$f'' - \frac{1}{2}\xi f' + \lambda f^p - \beta f = 0,$$

and explain why appropriate boundary conditions would be

$$f \sim |\xi|^{-2\beta} \quad \text{as } \xi \to \pm\infty,$$

and show that such solutions might be possible. Are any other limiting behaviours possible?

1.20 When an oscillatory reaction–diffusion system has an imperfection of size comparable to, or larger than, the wave length, then spiral waves can occur. This is because the wave trains need not be in phase round the boundary of the obstacles. For example, consider a slowly varying system (1.194) with solutions $\mathbf{w} \approx \mathbf{W}_0(t + \psi)$, where ψ satisfies the equation

$$\psi_\tau = \nabla^2 \psi + \bar{\alpha} |\nabla \psi|^2.$$

Suppose that the imperfection is of radius a, and that the effect of the surface is to alter the period, so that we take $\psi = \beta\tau + m\theta + c$ on $r = a$, where m is an integer (so that $\mathbf{w}$ is single valued, if we suppose the period of $\mathbf{W}_0$ is normalised to be 2π); c is an arbitrary constant, which we can choose for convenience.

Put $\psi = \beta\tau + m\theta - \phi(r)$, and show that ϕ satisfies the equation

$$\phi'' + \frac{1}{r}\phi' - \bar{\alpha}\left[\phi'^2 + \frac{m^2}{r^2}\right] + \beta = 0.$$

Hence show that

$$\phi = -\frac{1}{\bar{\alpha}} \ln w(\lambda r),$$

where $w(z)$ satisfies Bessel's equation in the form

$$w'' + \frac{1}{z}w' + \left[s - \frac{v^2}{z^2}\right]w = 0, \qquad (*)$$

providing we choose

$$\lambda = |\bar{\alpha}\beta|^{1/2}, \qquad v = i|\bar{\alpha}m|, \qquad s = -\operatorname{sgn}(\bar{\alpha}\beta).$$

The solutions of $(*)$ when $s = 1$, i.e., $\bar{\alpha}\beta < 0$, are the Hankel functions

$$H_v^{(1,2)}(z) = J_v(z) \pm i Y_v(z) \sim \left(\frac{2}{\pi z}\right)^{1/2} \exp\left[\pm i\left(z - \frac{1}{2}v\pi - \frac{1}{4}\pi\right)\right]$$

as $z \to \infty$. If $\bar{\alpha}\beta > 0$, so that $s = -1$, then the solutions are the modified
Bessel functions $I_\nu(z)$ and $K_\nu(z)$, and we have

$$I_\nu \sim \frac{1}{\sqrt{2\pi z}} e^z, \qquad K_\nu \sim \left(\frac{\pi}{2z}\right)^{1/2} e^{-z}$$

as $z \to \infty$.[8]

Deduce that solutions of this type exist if $\bar{\alpha}\beta > 0$, and that in this case the
presumption of outward travelling waves (the radiation condition) requires us
to choose $w = K_\nu(z)$ if $\bar{\alpha} > 0$. Show that as $r \to \infty$ in this case,

$$\mathbf{w} \approx \mathbf{W}_0\left[t + \beta\tau + m\theta - \left(\frac{\beta}{\bar{\alpha}}\right)^{1/2} r + O(\ln r) \right].$$

This solution represents a spiral wave. Note that the integer m is unconstrained. Its specification would require a model for the reaction on the surface
of the impurity at $r = a$. It is plausible to imagine that such angle dependent
phases arise through bifurcation of the surface reaction model as the impurity
size increases.

1.21 The Fitzhugh–Nagumo equations are

$$\varepsilon u_t = u(a - u)(u - 1) - v + \varepsilon^2 u_{xx},$$

$$v_t = bu - v,$$

where $0 < a < 1$, $\varepsilon \ll 1$, and b is positive and large enough that $u = v = 0$ is
the only steady state. Show that the system is excitable, and show, by means
of a phase plane analysis, that solitary travelling waves of the form $u(\xi)$, $v(\xi)$,
$\xi = ct - x$, are possible with $c > 0$ and $u, v \to 0$ as $\xi \to \pm\infty$.

1.22 u and v satisfy the equations

$$\delta u_t = \varepsilon^2 u_{xx} + f(u, v),$$

$$v_t = v_{xx} + g(u, v),$$

where

$$f(u, v) = u\big[F(u) - v\big], \qquad g(u, v) = v\big[u - G(v)\big],$$

and $F(u)$ is a unimodal function ($F'' < 0$) with $F(0) = 0$, while $G(v)$
is monotone increasing ($G' > 0$) and $G(0) > 0$, and there is a unique
point (u_0, v_0) in the positive quadrant where $f(u_0, v_0) = g(u_0, v_0) = 0$, and
$F'(u_0) < 0$. (For example $F = u(1 - u)$, $G = 0.5 + v$.)

Examine the conditions on δ and ε^2 which ensure that diffusion-driven instability of (u_0, v_0) occurs.

---

[8] See Watson (1944, pp. 199 f.) for these results.

If the upper and lower branches of F^{-1} are denoted as $u_+(v) > u_-(v)$, explain why u_- is unstable when $\varepsilon \ll 1$. By constructing phase portraits for v when $u = 0$ and when $u = u_+(v)$, and 'gluing' them together at a fixed value $v = v^*$, show that spatially periodic solutions exist which are 'patchy', in the sense that u alternates rapidly between $u_+(v)$ and 0.

Chapter 2
Climate Dynamics

The most noticeable facets of the weather are those which directly impinge on us: wind, rain, sun, snow. It is hotter at the equator than at the poles simply because the local intensity of incoming solar radiation is greater there, and this differential heating drives (or tries to), through its effect on the density of air, a poleward convective motion of the atmosphere: rising in the tropics, poleward in the upper atmosphere, down at the poles and towards the equator at the sea surface. The buoyancy-induced drift is whipped by the rapid rotation of the Earth into a predominantly *zonal flow*, from west to east in mid-latitudes. In turn, these zonal flows are *baroclinically* unstable, and form waves (Rossby waves) whose form is indicated by the isobar patterns in weather charts.[1]

All this frenetic activity obscures the fact that the weather is a rather small detail in the determination of the basic climate of the planet. The mean temperature of the planetary atmosphere and of the Earth's surface is determined by a balance between the radiation received by the Earth from the Sun (the incoming solar radiation), and that re-emitted into space by the Earth.

2.1 Radiation Budget

We denote the incoming solar radiation by Q; it has a value $Q = 1370 \text{ W m}^{-2}$ (watts per square metre). A fraction a of this (the albedo) is reflected back into space, while the rest is absorbed by the Earth; for the Earth, $a \approx 0.3$. In physics we learn that a perfect radiative emitter (a *black body*) at absolute surface temperature T emits energy at a rate

$$E_b = \sigma T^4, \tag{2.1}$$

---

[1]This overly simple description is inaccurate in one main respect, which is that the hemispheric polewards circulation actually consists of three cells, not one: a tropical cell, a mid-latitude cell and a polar cell. The prevailing winds are westerly (from the west) only in the mid-latitude cells; tropical winds (the *trade* winds), for example, are easterlies (from the east).

A. Fowler, *Mathematical Geoscience*, Interdisciplinary Applied Mathematics 36, DOI 10.1007/978-0-85729-721-1_2, © Springer-Verlag London Limited 2011

where σ is the Stefan–Boltzmann constant, given by $\sigma = 5.67 \times 10^{-8}$ W m^{-2} K^{-4}. If we assume that the Earth acts as a black body of radius R with *effective* (radiative) temperature T_e, and that it is in radiative *equilibrium*, then

$$4\pi R^2 \sigma T_e^4 = \pi R^2 (1 - a) Q,$$

whence

$$T_e = \left[\frac{(1 - a)Q}{4\sigma} \right]^{1/4}. \tag{2.2}$$

Computing this value for the Earth using the parameters above yields $T_e \approx 255$ K. A bit chilly, but not in fact all that bad!

Actually, if the average effective temperature is *measured* (T_m) via the black body law from direct measurements of emitted radiation, one finds $T_m \approx 250$ K, which compares well with T_e. On the other hand, the Earth's (average) surface temperature is $T_s \approx 288$ K. The fact that $T_s > T_e$ is due to the *greenhouse* effect, to which we will return later. First we must deal in some more detail with the basic mechanisms of radiative heat transfer.

2.2 Radiative Heat Transfer

We are familiar with the idea of conductive heat flux, a vector with magnitude and direction, which depends on position $\mathbf{r}$. Radiant energy transfer is a more subtle concept. A point in a medium will emit radiation of different frequencies ν (or different wavelengths λ: they are conventionally related by $\lambda = c/\nu$, where c is the speed of light), and the intensity of emitted radiation will depend not only on position $\mathbf{r}$, but also on *direction*, denoted by $\mathbf{s}$, where $\mathbf{s}$ is a *unit* vector. Also, like heat flux, emitted radiation is an area-specific quantity (i.e., it denotes energy emitted per unit area of emitting surface), and because it depends on orientation, this causes also a dependence on angle between emitting surface and direction: the intensity you receive from a torch depends on whether it is shone at you or not.

So, the *radiation intensity* $I_\nu(\mathbf{r}, \mathbf{s})$ is defined via the relation

$$dE_\nu = I_\nu \cos\theta \, d\nu \, dS \, d\omega \, dt, \tag{2.3}$$

where dE_ν is the energy transmitted in time dt through an area dS in the frequency range $(\nu, \nu + d\nu)$ over a *pencil* of rays of *solid angle* $d\omega$ in the direction $\mathbf{s}$; see Fig. 2.1. θ is the angle between $\mathbf{s}$ and dS.

The solid angle (element) $d\omega$ is the three-dimensional generalisation of the ordinary concept of angle, and is defined in an analogous way. The solid angle $d\omega$ subtended at a point O by an element of surface area dS located at $\mathbf{r}$ is simply

$$d\omega = \frac{\mathbf{r} \cdot d\mathbf{S}}{r^3}. \tag{2.4}$$

Fig. 2.1 A pencil of rays
emitted from a point **r** in the
direction of **s**

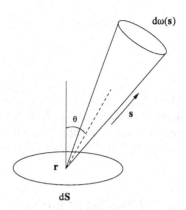

The solid angle subtended at O by a surface Σ is just $\omega = \int_\Sigma \frac{\mathbf{r} \cdot d\mathbf{S}}{r^3}$, and for example $\int_{O} d\omega = 4\pi$, representing the solid angle over all directions from a point, and $\int_{\frown} d\omega = 2\pi$, representing the solid angle subtended over all upward directions.

Three processes control how the intensity of radiation varies in a medium.

- *Absorption* occurs when a ray is absorbed by a molecule, e.g. of H_2O or CO_2 in the atmosphere, or by water droplets or particles. The rate of absorption is proportional to the density of the medium ρ and the radiation intensity I_ν, and is thus given by $\rho \kappa_\nu I_\nu$, where κ_ν is the *absorption coefficient*.
- *Emission* occurs (in all directions) when molecules or particles emit radiation; this occurs at a rate proportional to the density ρ, and is thus ρj_ν, where j_ν is the *emission coefficient*.
- *Scattering* can be thought of as a combination of absorption and emission, or alternatively as a local reflection. An incident ray on a molecule or particle—a *scatterer*—is re-directed (not necessarily uniformly) by its interaction with the scatterer. The process is equivalent to instantaneous absorption and re-emission. Reflection at a surface is simply the integrated response of a distribution of scatterers. Scattering leads to an effective scattering emission coefficient $j_\nu^{(s)}$, and is discussed further below in Sect. 2.2.6.

2.2.1 Local Thermodynamic Equilibrium

In order to prescribe j_ν, we will make the assumption of local thermodynamic equilibrium. More or less, this means that the medium is sufficiently dense that a local (absolute) temperature T can be defined, and *Kirchhoff's law* then defines j_ν as

$$j_\nu = \kappa_\nu B_\nu(T), \tag{2.5}$$

where $B_\nu(T)$ is the Planck function given by

$$B_\nu(T) = \frac{2h\nu^3}{c^2[e^{h\nu/kT} - 1]}, \tag{2.6}$$

where $h = 6.6 \times 10^{-34}$ J s is Planck's constant, $k = 1.38 \times 10^{-23}$ J K^{-1} is Boltzmann's constant, and $j_\nu \, d\nu$ represents the emitted energy per unit mass per unit time per unit solid angle in the frequency range $(\nu, \nu + d\nu)$. The formula (2.6) can be used to derive the Stefan–Boltzmann law (2.1) (see Question 2.2).

2.2.2 Equation of Radiative Heat Transfer

Considering Fig. 2.1, the rate of change of the radiation intensity I_ν in the direction **s** is given by

$$\frac{\partial I_\nu}{\partial s} = -\rho \kappa_\nu I_\nu + \rho \kappa_\nu B_\nu, \tag{2.7}$$

and this is the equation of radiative heat transfer. Note that the meaning of $\partial I_\nu / \partial s$ in (2.7) is that it is equal to $\mathbf{s} . \nabla I_\nu$, where ∇ is the gradient with respect to **r**. (2.7) is easily derived from first principles, given the definition of absorption and emission coefficients.

2.2.3 Radiation Budget of the Earth

We will use (2.7) to derive a model for the vertical variation of the intensity of radiation in the Earth's atmosphere. We need to do this in order to explain the discrepancy between the effective black body temperature of the Earth (250 K) and the observed surface temperature (290 K). The discrepancy is due to the greenhouse effect of the atmosphere, which acts both as an absorber and emitter of radiation. Importantly, the absorptive capacity of the atmosphere as a function of wavelength λ is very variable. Figure 2.2 shows the variation of κ_ν (or, we might write κ_λ) as a function of λ, or more specifically, $\log_{10} \lambda$. Above it we have also the black body radiation curves for two temperatures corresponding to those of the effective Earth emission temperature, and to that at the surface of the Sun. (To obtain these, we write the Planck function as a density B_λ in wavelength λ, using the fact that $\nu = \frac{c}{\lambda}$, where c is the speed of light, thus $d\nu = -\frac{c \, d\lambda}{\lambda^2}$, and therefore we define

$$B_\lambda = \frac{c B_\nu}{\lambda^2} = \frac{2hc^2}{\lambda^5[e^{hc/k\lambda T} - 1]}.) \tag{2.8}$$

From the graphs in Fig. 2.2, we see that solar radiation is concentrated at short wavelengths, including the band of visible light ($\lambda = 0.4$–0.7 μm), whereas the emitted

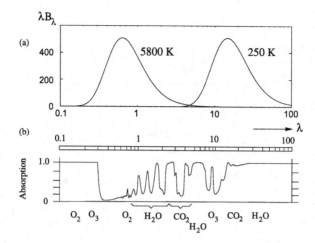

Fig. 2.2 Absorption spectrum of the Earth's atmosphere. The upper graphs indicate the different wavelength dependence of the radiation emitted by the Earth and the Sun. λ is measured in μm, and the solar output (from (2.8)) is scaled by 3.45×10^{-6} so that it overlays the Earth's output, if additionally λ in (2.8) is scaled by 0.043. In this case the areas under the two curves (note that $\int B_\lambda \, d\lambda = \ln 10 \int \lambda B_\lambda \, d \log_{10} \lambda$) are equal, as they should be in radiative balance. The factor 3.45×10^{-6} represents the product of $\frac{1}{4}(1-a)$ (cf. (2.2)) with the square of the ratio of the Sun's radius to the distance from the Earth to the Sun. The radius of the Sun is 6.96×10^8 m and the distance from the Earth to the Sun is 1.5×10^{11} m, so that the value of the square ratio is about 21.53×10^{-6}. Multiplying this by the discount factor $\frac{1}{4}(1-a)$ gives 3.45×10^{-6} if the albedo $a = 0.36$. The curves can be made to overlap for the measured albedo of $a = 0.3$ by, for example, taking Earth and Sun radiative temperatures to be 255 K and 5780 K, but this is largely a cosmetic exercise. The lower curve represents the absorption by atmospheric gases over a clear vertical column of atmosphere (i.e., it does not represent the absorption coefficient); we see that there is a long-wave window for wavelengths between about 8 and 15 μm. This figure is redrawn from Fig. 2.1 of Houghton (2002), by permission of Cambridge University Press

radiation is all infra-red (IR). Furthermore, the absorption coefficient variation with λ is such that the atmosphere is essentially *transparent* ($\kappa \approx 0$) to solar radiation (in the absence of clouds), but (mostly) *opaque* to the emitted long-wave radiation, with the exception of an IR window between 8 and 14 μm. It is this concept of transparency to solar radiation in the presence of only a small emission window, which leads to the analogy of a greenhouse.[2] The outgoing radiation is trapped by the atmosphere, and it is this which causes the elevated surface temperature.

The actual radiation budget of the Earth's atmosphere is shown in Fig. 2.3, which indicates the complexity of the transfer processes acting between the Earth's surface, the atmosphere and cloud cover, and which also shows the rôle played by *sensible* heat loss (i.e., due to convective or conductive cooling) and *latent* heat loss (due to evaporation from the oceans, for instance).

---

[2]The analogy is probably rather loose, since it is more the absence of convective (rather than radiative) cooling of the greenhouse which causes its elevated temperature.

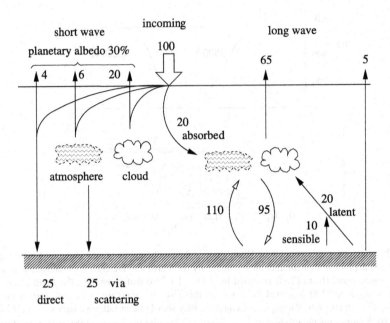

Fig. 2.3 Radiation budget of the Earth. Versions of this figure, differing slightly in the numerical values, can be found in many books. See, for example, Gill (1982), Fig. 1.6

As indicated in this figure, and as can be seen also from Fig. 2.2, one can essentially think of the short-wave budget and long-wave budget as separate systems. We shall be concerned here with the variation of IR radiation intensity, by solving (2.7). If κ_ν varies with ν, the problem requires computational solution. However, we can gain significant insight by introducing the idea of a *grey atmosphere*. This is one for which $\kappa_\nu = \kappa$ is independent of ν (and as mentioned, we will restrict this assumption to the long-wave budget).

We then define the radiation intensity I and emission density B as

$$I = \int_0^\infty I_\nu \, d\nu, \qquad B = \int_0^\infty B_\nu \, d\nu. \qquad (2.9)$$

Note that we have

$$B = \frac{\sigma T^4}{\pi}, \qquad (2.10)$$

where σ is the Stefan–Boltzmann constant; thus $B = E_b/\pi$. The factor of π arises because E_b represents the radiation per unit surface area emitted normally to the surface, while B represents emission per unit area per unit solid angle in any direction. It is important to understand the distinction between the two.

From (2.7), we have for a grey atmosphere

$$\frac{\partial I}{\partial s} = -\kappa \rho (I - B). \qquad (2.11)$$

We now consider the important case of a one-dimensional atmosphere. Let z be the direction in the upward vertical, and let θ be the (polar) angle to the z-axis. We also define the optical depth

$$\tau = \int_z^\infty \kappa\rho\,dz, \tag{2.12}$$

and put $\mu = \cos\theta$. For a one-dimensional atmosphere, we have $I = I(\tau,\mu)$, where τ represents the vertical position, and μ represents the direction of the ray pencil in Fig. 2.1. Note also that $ds = dz/\mu$ (some care is needed here: z and s are independent, but this relation correctly interprets $\partial/\partial s \equiv \mathbf{s}.\nabla_\mathbf{r}$ for the one-dimensional case), so that (2.11) is

$$\mu\frac{\partial I}{\partial \tau} = I - B, \tag{2.13}$$

for a *one-dimensional, grey* atmosphere.

This seems simple enough, but note that B depends on T, which is as yet unconstrained. In order to constitute B, we define the *average intensity*

$$J = \frac{1}{4\pi}\int_0 I\,d\omega = \frac{1}{2}\int_{-1}^1 I(\tau,\mu)\,d\mu, \tag{2.14}$$

and we make the assumption of *local radiative equilibrium*[3] that $J = B$, i.e., that the total absorbed radiation at a point is equal to that determined by black body emission (note that this does not necessarily imply $I = B$ for all θ, however). The radiative intensity equation for a one-dimensional, grey atmosphere is thus

$$\mu\frac{\partial I}{\partial \tau} = I - \frac{1}{2}\int_{-1}^1 I(\tau,\mu)\,d\mu, \tag{2.15}$$

and is in fact an integro-differential equation.

We require two further pieces of information to determine I completely. In view of our previous discussion, we take I as referring to long-wave radiation, and therefore it is appropriate to specify

$$I = 0 \quad \text{for } \mu < 0 \text{ at } \tau = 0, \tag{2.16}$$

i.e., no incoming long-wave radiation at the top of the atmosphere. Furthermore, we can see from Eq. (2.15) that the net upward flux

$$\int_0 I\cos\theta\,d\omega = 2\pi\int_{-1}^1 \mu I\,d\mu = \Phi \tag{2.17}$$

is conserved (i.e., is independent of depth). (The factor 2π is due to integration with respect to the azimuthal angle ϕ.) Since this is $2\pi[\int_0^1 \mu I\,d\mu - \int_{-1}^0(-\mu I)\,d\mu] =$

---

[3]This now specifically assumes that no other energy transport processes occur.

outgoing IR radiation minus incoming IR radiation, it is in fact equal to the net emission of IR radiation. By the assumption of global radiative balance, Φ is equal to the net received short-wave radiation, thus

$$\Phi = \frac{(1-a)Q}{4} = \sigma T_e^4, \tag{2.18}$$

where the factor 4 allows for the variation of received solar radiation per unit area with latitude. (Strictly, the assumption of a one-dimensional atmosphere assumes horizontal variations due to latitude are rapidly removed, e.g. by mixing, but in fact the horizontal variation is small anyway, because the atmosphere is geometrically *thin*.) In fact, even if there is global imbalance, as in climatic energy-balance models (see Sect. 2.4), we still have $\Phi = \sigma T_e^4$.

2.2.4 The Schuster–Schwarzschild Approximation

The solution of (2.15) with (2.16) and (2.17) is possible but technically difficult, and is described in Appendix A. A simple approximate result can be obtained by defining the outward and inward flux integrals

$$\begin{aligned} I_+ &= \int_0^1 I \, d\mu, \\ I_- &= \int_{-1}^0 I \, d\mu, \end{aligned} \tag{2.19}$$

and then approximating $\int_0^1 \mu I \, d\mu \approx \frac{1}{2} I_+$, $\int_{-1}^0 \mu I \, d\mu = -\frac{1}{2} I_-$, based on the idea that $\int_0^1 \mu \, d\mu = \frac{1}{2}$. This causes (2.15) to be replaced by

$$\begin{aligned} I'_+ &= I_+ - I_-, \\ I'_- &= I_+ - I_- \end{aligned} \tag{2.20}$$

so that $I_+ - I_- = \Phi/\pi$ is the conservation law (2.17), and thus (with $I_- = 0$ at $\tau = 0$)

$$\begin{aligned} I_- &= \Phi \tau / \pi, \\ I_+ &= \frac{\Phi}{\pi}(1+\tau). \end{aligned} \tag{2.21}$$

It follows that the average intensity

$$J = \frac{1}{2}(I_+ + I_-) = \frac{\Phi}{2\pi}(1+2\tau) = B, \tag{2.22}$$

and using (2.10) and (2.18), we thus find the atmospheric temperature T in terms of the emission temperature T_e:

$$T = T_e \left[\frac{(1 + 2\tau)}{2} \right]^{1/4} . \tag{2.23}$$

The surface temperature is determined by the black body emission temperature corresponding to I_+ at the surface, where $\tau = \tau_s$, that is, $I_+ = B = \sigma T_s^4 / \pi$, so that the ground surface temperature is

$$T_s = T_e (1 + \tau_s)^{1/4} , \tag{2.24}$$

whereas the surface air temperature T_{as} is, from (2.23),

$$T_{as} = T_e \left(\frac{1}{2} + \tau_s \right)^{1/4} . \tag{2.25}$$

Note that there is a discontinuity in temperature at the surface, specifically

$$T_s^4 - T_{as}^4 = 0.5 T_e^4 ; \tag{2.26}$$

molecular heat transport (conduction) will in fact remove such a discontinuity. If we use $T_s = 290$ K and $T_e = 255$ K, then (2.24) implies that the optical depth of the Earth's atmosphere is $\tau_s = 0.67$.

2.2.5 Radiative Heat Flux

Although radiative heat transfer is the most important process in the atmosphere, other mechanisms of heat transport are essential to the thermal structure which is actually observed, notably conduction and convection. In order to incorporate radiative heat transfer into a more general heat transfer equation, we need to define the *radiative heat flux*. This is a vector, analogous to the conductive heat flux vector, and is defined (for a grey medium) by

$$\mathbf{q}_R = \int_\bigcirc I(\mathbf{r}, \mathbf{s}) \, \mathbf{s} \, d\omega(\mathbf{s}). \tag{2.27}$$

Note that $\mathbf{q}_R . \mathbf{n} = \int_\bigcirc I \cos\theta \, d\omega$ (see Fig. 2.1) is the energy flux density through a surface element dS with normal $\mathbf{n}$. Determination of $\mathbf{q}_R$ requires the solution of the radiative heat transfer equation for I, but a simplification occurs in the optically dense limit, when $\tau \gg 1$ (i.e., $\kappa\rho$ is small). We write

$$I = B - \frac{1}{\rho\kappa} \mathbf{s} . \nabla I, \tag{2.28}$$

Fig. 2.4 Scattering from
direction s′ to s

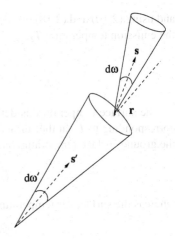

and solve for I using a perturbation expansion in powers of $1/\rho\kappa$. One thus obtains

$$I = B - \frac{1}{\rho\kappa}\, \mathbf{s}.\nabla B + \cdots, \tag{2.29}$$

and substitution into (2.27) leads to the expression

$$\mathbf{q}_R \approx -\frac{4\pi}{3\kappa\rho}\nabla B = -\frac{4\sigma}{3\kappa\rho}\nabla T^4, \tag{2.30}$$

so that for an optically dense atmosphere, the radiative heat flux is akin to a conductive heat flux, with a nonlinear temperature-dependent (radiative) conductivity. Because of its simplicity, we will often use this expression for the radiative flux despite its apparent inappropriateness for the Earth.

2.2.6 Scattering

In a scattering atmosphere, a beam of radiation is scattered as it is transmitted, as indicated in Fig. 2.4. At any position $\mathbf{r}$, an incident beam of frequency v in the direction $\mathbf{s}'$ will be deflected to a new direction $\mathbf{s}$ with a probability distribution which we define to be $p_v(\mathbf{s}, \mathbf{s}')/4\pi$; thus the integral of p_v over all directions is one, i.e.,

$$\int_O p_v(\mathbf{s}, \mathbf{s}')\frac{d\omega(\mathbf{s}')}{4\pi} = 1. \tag{2.31}$$

If all the incident radiation is scattered, then we have perfect scattering: no radiation is lost. More generally, we may suppose that a fraction a_v is scattered (and the rest is absorbed), and a_v is called the albedo for single scattering. Thus we define $\frac{a_v p_v\, d\omega}{4\pi}$ to be the probability that incident radiation from the direction $\mathbf{s}'$ will be scattered in the direction **s** over a solid angle increment $d\omega$. In general, p_ν depends on frequency, and we also suppose it depends only on the angle between **s'** and **s**, thus $p_\nu = p_\nu(\mathbf{s}.\mathbf{s'})$.

Integrating this probability over all directions **s'**, we obtain the emission coefficient for scattering as

$$j_\nu^{(s)} = a_\nu \kappa_\nu \int_O p_\nu(\mathbf{s}, \mathbf{s'}) I_\nu(\mathbf{r}, \mathbf{s'}) \frac{d\omega(\mathbf{s'})}{4\pi}, \qquad (2.32)$$

where κ_ν is the emission coefficient. The equation of radiative transfer is modified from (2.7) to

$$\frac{\partial I_\nu}{\partial s} = -\rho \kappa_\nu \left[I_\nu - (1 - a_\nu) B_\nu - a_\nu \int_O p_\nu(\mathbf{s}, \mathbf{s'}) I_\nu(\mathbf{r}, \mathbf{s'}) \frac{d\omega(\mathbf{s'})}{4\pi} \right]. \qquad (2.33)$$

Scattering in the atmosphere is most closely associated with Rayleigh's explanation for the blue colour of the sky. For the visible spectrum we can ignore short-wave emission, $B_\nu = 0$. Rayleigh derived an expression for the scattering distribution of sunlight by air molecules. Importantly, the intensity of scattered radiation is proportional to ν^4 (or $1/\lambda^4$), and thus is much larger for high frequency, or short-wavelength, radiation. In terms of the visible spectrum, this is the blue end. The wavelength of blue light is about 0.425 μm, while that of red light is 0.65 μm, so that blue light is scattered about five times more than red light. Hence the blue sky.

Rayleigh scattering applies to scattering by entities which are much smaller than the radiation wavelength, and in particular, molecules. Scattering by objects much larger than the wavelength (dust particles, water droplets, etc.) is called Mie scattering and is determined by WKB theory applied to the electromagnetic wave equation.

2.2.7 Troposphere and Stratosphere

Thus far, we have not considered the vertical structure of the atmosphere. The principal feature of the atmosphere is that it is stratified: the density decreases, more or less exponentially, with height. This is why it becomes difficult to breathe at high altitude. The reason for this decrease is simply that the atmospheric pressure at a point depends on the weight of the overlying air, which obviously decreases with height. Since density is proportional to pressure, it also decreases with height.

To quantify this, we use the fact that for a *shallow* atmosphere (whose depth d is much less than a relevant horizontal length scale l), the pressure p is nearly hydrostatic, that is,

$$\frac{dp}{dz} = -\rho g, \qquad (2.34)$$

where z is height, ρ is air density, and g is gravitational acceleration (approximately constant). If we assume (reasonably) that air behaves as a perfect gas, then

$$\rho = \frac{M_a p}{RT},\tag{2.35}$$

where M_a is the *molecular weight*[4] of air, R is the perfect gas constant, and T is absolute temperature. For a perfect gas, the thermal expansion coefficient $-\frac{1}{\rho}\frac{\partial \rho}{\partial T}$ is simply $1/T$.

In terms of the temperature, the pressure and density are then found to be

$$p = p_0 \exp\left[-\int_0^z \frac{dz}{H}\right], \qquad \rho = \rho_0 \exp\left[-\int_0^z \frac{dz}{H}\right],\tag{2.36}$$

where the *scale height* is

$$H = \frac{RT}{M_a g},\tag{2.37}$$

having a value in the range 6–8 km. The temperature varies by less than a factor of two over most of the atmosphere, and an exponential relation between pressure or density and height is a good approximation.

We mentioned earlier, in deriving (2.15), that we assumed local radiative equilibrium, that is to say, radiative transport dominates the other transport mechanisms of convection and heat conduction. As we discuss further below, this is a reasonable assumption if the atmospheric density is small. As a consequence of the decrease in density with height, the atmosphere can therefore be divided into two layers. The lower layer is the *troposphere*, of depth about 10 km, and is where convective heat transport is dominant, and the temperature is *adiabatic*, and decreases with height: this is described in Sect. 2.3 below. The troposphere is separated from the *stratosphere* above it by the *tropopause*; atmospheric motion is less relevant in the stratosphere, and the temperature is essentially governed by radiative equilibrium. In fact the adiabatic decrease in temperature in the troposphere stops around the tropopause, and the temperature increases again in the stratosphere to about 270 K at 50 km height (the *stratopause*), before decreasing again (in the *mesosphere*) and then finally rising at large distances (in the *thermosphere*, >80 km).

The temperature structure of the atmosphere can thus be represented as in Fig. 2.5: the convection in the troposphere mixes the otherwise radiative temperature field to produce the adiabatic gradient which is observed.

---

[4]The molecular weight is effectively the weight of a molecule of a substance. Equivalently, it is determined by the weight of a fixed number of molecules, known as a *mole*, and equal to *Avogadro's number* 6×10^{23} molecules. For air, a mixture predominantly of nitrogen (78%), oxygen (21%) and argon (0.9%), the molecular weight is given by the equivalent quantity for the mixture. It has the value $M_a = 28.8 \times 10^{-3}$ kg mole^{-1}. Useful references for such quantities and their units are Kaye and Laby (1960) and Massey (1986).

Fig. 2.5 Atmospheric temperature profile. Below the tropopause, convection stirs the temperature field into an adiabatic gradient. Above it, radiative balance is dominant

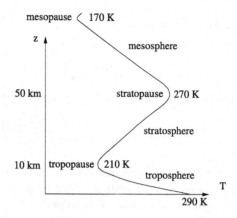

2.2.8   The Ozone Layer

The elevated vertical temperature profile in the stratosphere is basically due to a radiative balance between ultraviolet absorption by ozone and long wave emission by carbon dioxide. As is indicated in Fig. 2.2 (and as is well known), ozone (O_3) in the stratosphere is responsible for removing ultraviolet radiation, which would otherwise be lethal to life on Earth. Ozone is produced in the stratosphere through the photodissociation of oxygen. The basic sequence of reactions describing this process is due to Sydney Chapman:

$$O_2 + h\nu \overset{j_2}{\to} 2O,$$

$$O + O_2 + M \overset{k_2}{\to} O_3 + M,$$

$$O_3 + h\nu \overset{j_3}{\to} O + O_2,$$

$$O + O_3 \overset{k_3}{\to} 2O_2. \tag{2.38}$$

The first of these reactions represents the breakdown of oxygen by absorption of ultraviolet radiation of wavelength less than 0.24 µm ($h\nu$ is Planck's quantum of energy). The next two reactions are fast. The arbitrary air molecule M catalyses the first of these. The final reaction represents the removal of ozone. Overall, the reaction can be written as

$$3O_2 \underset{r_-}{\overset{r_+}{\rightleftharpoons}} 2O_3, \tag{2.39}$$

with the first two reactions of (2.38) providing the forward reaction, and the last two the backward reaction. If we assume (as is the case) that j_3 and k_2 are sufficiently large that

$$\varepsilon = \frac{j_3}{k_2[O_2][M]} \ll 1, \qquad \delta = \left[\frac{j_2 k_3}{j_3 k_2[M]} \right]^{1/2} \ll 1, \tag{2.40}$$

then one can show (see Question 2.8) that the forward and backward rates for (2.39) are

$$r_+ = \frac{2}{3}j_2, \qquad r_- = \frac{j_3 k_3}{k_2[O_2][M]},$$ (2.41)

and the (stable) equilibrium ozone concentration is given by

$$[O_3] = \left[\frac{j_2 k_2[M]}{j_3 k_3} \right]^{1/2} [O_2].$$ (2.42)

Ozone occurs principally in the *ozone layer*, at heights between 15 and 50 km (i.e., in the stratosphere), where it attains concentrations of about 10 ppmv (parts per million by volume). It is formed here because the reactions in (2.38) require UV radiation to be absorbed, which in itself requires the presence of oxygen. So at the top of the stratosphere, where the pressure and thus also density are both small, absorption is small and little ozone is formed. Deeper in the stratosphere, density increases, which allows increased production of ozone, but also less UV radiation can penetrate to deeper levels, and so the source for the ozone forming reaction disappears at the base of the stratosphere. The ozone which is produced itself enhances the absorption of UV radiation, of course.

A simple model for the formation of this structure, which is called a *Chapman layer*, assumes a constant volume concentration, or mixing ratio, for ozone. The radiative transfer equation for incoming shortwave radiation of intensity I can be written

$$\frac{\partial I}{\partial z} = \kappa \rho I;$$ (2.43)

there is no radiative source term, and the incoming beam is unidirectional, and here taken to be vertical (the Sun is overhead). We suppose a constant pressure scale height so that

$$\rho = \rho_0 \exp(-z/H).$$ (2.44)

With I negative, and $I \to -I_\infty$ as $z \to \infty$, the solution to this is

$$I = -I_\infty \exp\left[-\kappa \rho_0 H e^{-z/H}\right],$$ (2.45)

and the consequent heating rate $Q = -\frac{\partial I}{\partial z}$ is given by

$$Q = \frac{\tau_0 I_\infty}{H} \exp\left[-\frac{z}{H} - \tau_0 e^{-z/H} \right],$$ (2.46)

where

$$\tau_0 = \kappa \rho_0 H$$ (2.47)

is a measure of the opacity of the stratospheric ozone layer.

If τ_0 is sufficiently high, the heating rate exhibits an internal maximum, as seen in Fig. 2.6. This is the distinguishing feature of the Chapman layer. Since Q is also

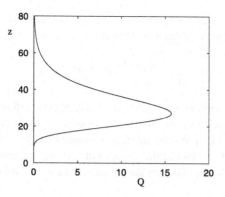

Fig. 2.6 Variation of heating rate Q given by (2.46) with $I_\infty = 342$ W m^{-2}, $H = 8$ km, and $\tau_0 = 30$, this somewhat arbitrary value being chosen to show a maximum heating rate at 30 km altitude. Units of z are km, and of Q W m^{-3}. The choice of $I_\infty = 342$ W m^{-2} refers to all incoming short wave radiation, whereas it is only a small fraction of this in the ultraviolet range which is absorbed in the stratosphere

volumetric absorption rate of radiation, it indicates maximal production of ozone in the stratosphere, as is found to be the case. This structure additionally explains why the temperature rises with height through the stratosphere, because of the increased heating rate.

In the stratosphere much of the short-wave absorption is due to ozone. There is very little water vapour. The resultant heating is almost exactly balanced by long-wave radiation, mostly from carbon dioxide, the remnant being from ozone again. While the resulting radiation balance controls the temperature, there is very little radiant energy lost. As can be seen from Fig. 2.2, the UV tail is taken off by ozone and oxygen, but the visible and infra-red spectrum passes through the stratosphere relatively unscathed.

In the troposphere, the water vapour concentration is much higher than that of ozone, which is virtually absent, and also of carbon dioxide. Although discussions of global warming are fixated by the greenhouse gases—carbon dioxide, methane, and so on, it needs to be borne in mind that water vapour is also a greenhouse gas, and is in fact the most important one. Adding to that the dominating influence of clouds and their somewhat mysterious influence on climate, one sees that an understanding of moisture is of principal concern in determining radiative processes in the troposphere.

2.3 Convection

We have seen that for a purely radiative atmosphere, a discontinuity in temperature occurs at the Earth's surface. Such a discontinuity does not occur in reality, because of molecular conduction. In fact, atmospheric motion causes heat transport in the troposphere to be more importantly due to convection rather than conduction—the transport of heat is primarily due to the motion of the atmosphere itself. The temperature of the atmosphere is described by the heat equation

$$\rho c_p \frac{dT}{dt} - \beta T \frac{dp}{dt} = k\nabla^2 T - \mathbf{\nabla}.\mathbf{q}_R, \qquad (2.48)$$

where the terms represent, respectively, advection of heat, adiabatic (compression) heating, thermal conduction and radiative heat transfer. The time derivative d/dt is a *material derivative*, and represents the rate of change of a property following a fluid element. Thus, $dT/dt = 0$ means the temperature of a fluid element is conserved as it moves. It is related to the ordinary partial derivative by the relation

$$\frac{d}{dt} = \frac{\partial}{\partial t} + \mathbf{u}.\mathbf{\nabla}, \qquad (2.49)$$

where $\mathbf{u}$ is the fluid velocity, c_p is the specific heat, $\beta = -\rho^{-1}\partial\rho/\partial T$ is the thermal expansion coefficient, k is the thermal conductivity, and $\mathbf{q}_R$ was defined above in (2.27).

If we use the optically dense approximation (2.30), then

$$\mathbf{q}_R \approx -k_R \mathbf{\nabla} T, \qquad (2.50)$$

where

$$k_R = \frac{16\sigma T^3}{3\kappa\rho}. \qquad (2.51)$$

We will use (2.50) as a pedagogic tool rather than as an accurate model. To estimate k_R, we use values $\kappa\rho d = 0.7$, $d = 10$ km; then we find $k_R \sim 10^5$ W m^{-1} K^{-1}. This compares to a molecular thermal conductivity of order 10^{-2} W m^{-1} K^{-1}, which is therefore negligible. In fact, atmospheric flows are turbulent, and a better measure of the effective heat conduction is the eddy thermal conductivity, of order $\rho c_p U d$ times a small dimensionless drag coefficient, where d is depth scale and U is wind speed scale. This is discussed further in Chap. 3; we find that eddy conductivity is found to be comparable to the nominal radiative value deduced above.

A measure of the importance of the advective terms is provided by the *Péclet number*, which represents the size of the ratio $(\rho c_p \, dT/dt)/\mathbf{\nabla}.\mathbf{q}_R$, and is given by

$$Pe = \frac{\rho c_p U d^2}{k_R l}, \qquad (2.52)$$

where U and d are velocity and depth scales as mentioned above, and l is a relevant horizontal length scale. Using values $\rho \sim 1$ kg m^{-3}, $c_p \sim 10^3$ J kg^{-1} K^{-1}, $U \sim 20$ m s^{-1}, $d \sim 10$ km, $l \sim 10^3$ km (representing the length scale of planetary waves in the atmosphere), we find $Pe \sim 20$, so that in fact atmospheric motion plays a significant rôle in the redistribution of heat. Since Pe is large, we can obtain an approximation to the vertical thermal structure of the atmosphere by neglecting the radiative and conductive transport terms altogether. This leads to the *adiabatic lapse rate*, which is determined by putting the left hand side of (2.48) to zero, and thus

$$\frac{dT}{dp} = \frac{\beta T}{\rho c_p}. \tag{2.53}$$

To obtain the variation of T with height z, we use the fact that the pressure is hydrostatic, given by (2.34), and assume a perfect gas law (2.35); then we find the (dry) adiabatic lapse rate

$$\frac{dT}{dz} = -\Gamma_d = -\frac{g}{c_p}, \tag{2.54}$$

having a value of about $10\ \mathrm{K\,km}^{-1}$. In practice, the observed temperature gradient is nearer $6\ \mathrm{K\,km}^{-1}$, a value which is due to the presence of water vapour in the atmosphere, the effect of which is considered below.

One of the basic reasons for the presence of convection in the troposphere is the presence of an unstable thermal gradient. The higher temperature at the ground causes the air there to be lighter; convection occurs as the warm air starts to rise, and it is the resultant overturning which causes the mixing which creates the adiabatic gradient. On a larger scale, and as we discuss further in Chap. 3, the unstable thermal gradient which drives large scale atmospheric motion is due to the energy imbalance between the equator and the poles.

Perturbations to the adiabatic gradient occur; for example, temperature inversions can occur under clear skies at night when IR radiation from the Earth is larger. The resultant temperature structure is convectively stable (the inversion is cold and therefore heavy), and its removal by solar irradiation can be hampered by the presence of smog caused by airborne dust particles. Moreover, the cool inversion causes fog (condensed water vapour), and the condensation is also facilitated by airborne pollutant particles, which act as nucleation sites. Hence the infamous smogs in London in the 1950s, and the consequent widespread ban of open coal fires in cities.

While temperature inversions are convectively stable and thus persistent, super-adiabatic temperatures are convectively unstable, and cannot be maintained.

2.3.1 The Wet Adiabat

For a parcel of air of density ρ_a containing water vapour of density ρ_v, the *mixing ratio* is defined as

$$m = \frac{\rho_v}{\rho_a}. \tag{2.55}$$

A typical value in the troposphere is $m \approx 0.02$, so that we can practically take the density of moist air as constant. As m increases, the air can become *saturated* and thus the water vapour will condense. This happens when the *partial pressure* p_v

Fig. 2.7 Phase diagram for
water substance (not to scale)

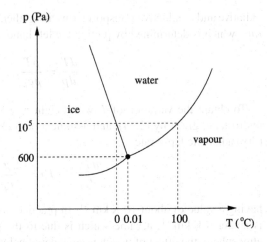

reaches the *saturation vapour pressure* p_{sv}, which depends on temperature via the
Clausius–Clapeyron equation

$$\frac{dp_{sv}}{dT} = \frac{\rho_v L}{T},$$ (2.56)

where L is the latent heat and T in (2.56) is the saturation value T_{sat}. Figure 2.7
shows the phase diagram for water, delineating the curves in (T, p) space at which
freezing, condensation and sublimation occur. (2.56) describes the water/vapour
curve in this figure. The ratio p_v / p_{sv} (normally measured as a percentage) is called
the *relative humidity*. It is an anthropocentric measure of discomfort, since when the
(relative) humidity is high, very little exertion will cause one to sweat.

Let us now suppose that the atmosphere is (just) saturated. The existence of
clouds actually negates this proposition, but not too badly, in the sense that we suppose rainfall removes condensed water droplets. As a moist parcel of air moves
about, the increment of heat content per unit volume due to changes in T, p and ρ_v
is then $\rho_a c_p\, dT - dp + L\, d\rho_v$ (using $\beta T = 1$), and thus (2.48) is modified to

$$\rho_a c_p \frac{dT}{dt} - \frac{dp}{dt} + \rho_a L \frac{dm}{dt} = k\nabla^2 T - \nabla.\mathbf{q}_R.$$ (2.57)

Using the definition of m in (2.55), and the perfect gas laws

$$p = \frac{\rho_a R T}{M_a}, \qquad p_{sv} = \frac{\rho_v R T}{M_v},$$ (2.58)

where M_v is the molecular weight of water vapour, we find that the temperature
gradient is given approximately (by ignoring the right hand side of (2.57)) by

$$\frac{dT}{dz} = -\Gamma_w = -\Gamma_d \left[\frac{1 + \frac{\rho_v L}{p}}{1 + \frac{\rho_v L}{p}\left(\frac{M_v}{M_a}\frac{L}{c_p T}\right)} \right],$$ (2.59)

which is the wet adiabat. Using values $\rho_v = 0.01$ kg m^{-3}, $M_v = 18 \times 10^{-3}$ kg mole^{-1}, $M_a = 28.8 \times 10^{-3}$ kg mole^{-1}, $L = 2.5 \times 10^6$ J kg^{-1}, $c_p = 10^3$ J kg^{-1} K^{-1}, $p \approx 10^5$ Pa, $T \approx 300$ K, we find a typical value $\Gamma_w \approx 5.4$ K km^{-1}, close to that which is observed in practice.

2.4 Energy Balance Models

Although convective transport is the dominant mechanism of energy transfer *within* the atmosphere, the rôle of radiative transport is fundamental to the determination of the average temperature. Moreover, this is equally true if we do *not* assume radiation balance, and this allows us to study long term variations in climate which are of relevance to the evolution of paleoclimatic temperatures, quaternary ice age climates, and more recently, the effect of CO_2 levels on global temperature. All of these phenomena can be roughly understood on the basis of *energy-balance models*.

Since most of the mass of the atmosphere is contained in the troposphere, we define the mean temperature $\bar{T}$ of the atmosphere to be the vertically averaged temperature of the troposphere. Suppose the temperature is adiabatic, with constant lapse rate Γ, and of depth d. The surface temperature T is thus

$$T = \bar{T} + \frac{1}{2}\Gamma d. \tag{2.60}$$

In a purely radiative atmosphere, we found earlier that the greenhouse effect causes the surface temperature to be warmer than the planetary long-wave emission temperature (cf. (2.24)). Let us define a greenhouse factor

$$\gamma = \left(\frac{T_e}{T}\right)^4, \tag{2.61}$$

where for (2.24), this would be

$$\gamma = \frac{1}{1 + \tau_s}; \tag{2.62}$$

this enables us to write the emitted long-wave radiation in terms of the mean surface temperature, and a quantity γ which depends on atmospheric radiative properties.

We can still define a greenhouse factor by (2.61) for a radiative-convective atmosphere, but consultation of Fig. 2.3 shows that its theoretical determination in terms of atmospheric properties is likely to be non-trivial. Nevertheless, we shall suppose γ can be defined; for the Earth $\gamma \approx 0.61$ at present (based on $T_e = 255$ K, $T = 288$ K).

The incoming solar radiation per unit area is $(1 - a)Q$ (a is the albedo, the fraction of short-wave radiation which is reflected back to space), while the emitted IR radiation per unit area is σT_e^4 (units are W m^{-2}). It follows that the net received radiation over the planetary surface is $\pi R^2(1 - a)Q - 4\pi R^2 \sigma T_e^4$, with units of

W, and we can equate this to the rate of change of the atmospheric heat content,[5] $4\pi R^2 d\rho_a c_p \frac{d\bar{T}}{dt}$, where d is the depth of the troposphere; c_p is the specific heat, and R is the planetary radius. Since $\rho_a R^2 d$ has units kg, c_p has units $J\,kg^{-1}\,K^{-1}$, and dT/dt has units $K\,s^{-1}$, this also has units W, and thus (adopting (2.61))

$$\rho_a c_p d \frac{dT}{dt} = \frac{1}{4}(1-a)Q - \sigma\gamma T^4, \tag{2.63}$$

in view of (2.60), since we take $\frac{1}{2}\Gamma d$ to be constant.

For constant Q, (2.63) is a simple first order differential equation with stable positive steady state, the radiative equilibrium state

$$T = T_0 = \left[\frac{(1-a)Q}{4\sigma\gamma}\right]^{1/4}. \tag{2.64}$$

The response time for small deviations from T_0 is then determined by the linearised equation, where we put $T = T_0 + \theta$, whence

$$\frac{\rho_a c_p d T_0}{(1-a)Q}\dot{\theta} \approx -\theta, \tag{2.65}$$

and the response time is

$$t_R \sim \frac{\rho_a c_p T_0 d}{(1-a)Q}. \tag{2.66}$$

With a density $\rho_a = 1\ kg\,m^{-3}$, $c_p = 10^3\ J\,kg^{-1}\,K^{-1}$, $T_0 = 288$ K, $d = 10^4$ m, $a = 0.3$, $Q = 1370\ W\,m^{-2}$, we have $t_R \sim 35$ days, so that climatic response is relatively rapid.

2.4.1 Zonally Averaged Energy-Balance Models

Energy balance models are obviously crude, but attractive nonetheless because they portray the essential truth about atmospheric energy balance. One of the more obvious features of the planetary climate is the temperature difference between equator and poles, due to the latitudinal variation of received solar variation. Indeed, it is this imbalance which drives the atmospheric weather systems, as we shall see in Chap. 3. A simple modification to the 'zero-dimensional' energy-balance model (2.63) is to allow a latitudinal variation in temperature. We denote latitude (angle north of the equator) by λ and we define

$$\xi = \sin\lambda, \tag{2.67}$$

---

[5]This is something of a simplification. Net addition of radiant energy to the atmosphere can cause changes in sensible heat (via temperature), latent heat (via moisture) or gravitational potential energy (via thermal expansion); we thus implicitly neglect the latter two; see also Eq. (3.25) in Sect. 3.2.3, and the next footnote.

thus $-1 < \xi < 1$, and $\xi = 0$ at the equator, $\xi = 1$ at the north pole. We suppose $T(\xi, t)$ is the *zonally averaged* (i.e., integrated over longitude) temperature, and we pose the zonally averaged energy-balance equation

$$C\frac{\partial T}{\partial t} = D\frac{\partial}{\partial \xi}\left[(1 - \xi^2)\frac{\partial T}{\partial \xi}\right] + \frac{1}{4}Q(1 - a)S(\xi) - I(T). \qquad (2.68)$$

In this equation, C is a heat capacity coefficient. For a dry atmosphere,[6] (2.63) indicates $C = \rho_a c_p d$. D is an effective thermal conduction coefficient, scaled with d/R^2, and thus having units of $\mathrm{W\,m^{-2}\,K^{-1}}$; it represents the poleward transport of energy through the eddy diffusive effect of large weather systems in mid-latitudes, which will be discussed further in the following chapter. $I(T)$ represents the outgoing long-wave radiation, supposed to depend only on mean surface temperature. Finally, $S(x)$ represents the latitudinal variation of received solar radiation, normalised so that $\int_0^1 S(\xi)\,d\xi = 1$. If the albedo a is constant, then we regain (2.63) by integrating from $\xi = -1$ to $\xi = 1$, assuming T is regular at the poles. If $S \equiv 1$, then $T = T(t)$, and we also regain the earlier model.

In the formulation of (2.68), we again interpret T as the mean surface temperature, in view of (2.60), and this is what is conventionally done, though without explicit mention. It is also conventional, in view of the limited range of T, to take a linear dependence of I on T, thus

$$I = A + BT, \qquad (2.69)$$

with values of A and B from measurements. Typical such values[7] are $A = 200\ \mathrm{W\,m^{-2}}$ and $B = 2\ \mathrm{W\,m^{-2}\,K^{-1}}$.

The resulting linear equation for T can then be solved as a Fourier–Legendre expansion if the albedo is known. For example, let us suppose that a as well as S is an even function of ξ (thus exhibiting north-south symmetry). It is convenient to write Eq. (2.68) in terms of I, thus

$$C^*\frac{\partial I}{\partial t} = D^*\frac{\partial}{\partial \xi}\left[(1 - \xi^2)\frac{\partial I}{\partial \xi}\right] + \frac{1}{4}Q(1 - a)S(\xi) - I, \qquad (2.70)$$

where $D^* = D/B$, $C^* = C/B$. We solve this in the steady state by writing

$$I = \sum_{n \text{ even}} i_n P_n(\xi), \qquad (2.71)$$

---

[6]For a moist, saturated atmosphere, we may take the moisture mixing ratio m to be a function of T, and in this case the latent heat $\rho_a Lm$ (L being latent heat) simply modifies the heat capacity coefficient. Question 2.11 shows how to calculate $m(T)$. See also Sect. 3.2.7.

[7]The value of A assumes T is measured in degrees Celsius.

where P_n is the n-th Legendre polynomial (and is an even function of ξ for n even). If we expand

$$\frac{1}{4}Q(1-a)S(\xi) = \sum_{n \text{ even}} q_n P_n(\xi), \tag{2.72}$$

where

$$q_n = \frac{(2n+1)Q}{4} \int_0^1 (1-a)S(\xi)P_n(\xi)\,d\xi, \tag{2.73}$$

then the coefficients i_n are given by

$$i_n = \frac{q_n}{1+n(n+1)D^*}. \tag{2.74}$$

For example, if we take a to be constant and the realistic approximation $S = 1 - \alpha P_2(\xi)$, $\alpha \approx 0.48$, then

$$I = \frac{1}{4}Q(1-a)\left[1 - \frac{\alpha P_2(\xi)}{1+6D^*}\right]. \tag{2.75}$$

A better approximation uses $a = a_0 + a_2 P_2(\xi)$, where $a = 0.68$ and $a_2 = -0.2$; this represents to some extent the higher albedo (due to ice cover) in the polar regions. The resultant two term approximation for the temperature, $T = (i_0 - A + i_2 P_2(\xi))/B$, then yields a good approximation to the observed mean surface temperature if we take $D = 0.65$ W m^{-2} K^{-1}.

2.4.2 Carbon Dioxide and Global Warming

If we are interested in the gradual evolution of climate over long time scales, then in practice we can neglect the time derivative term in (2.63), and suppose that T is in a quasi-equilibrium state. Figure 2.8 shows the rising concentration of CO_2 in the atmosphere over the last two hundred years. Essentially, the secular rise is due to the increased industrial output since the industrial revolution.

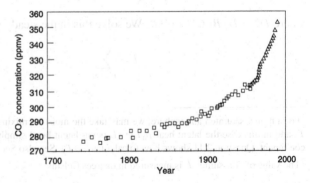

Fig. 2.8 Rise in atmospheric concentration of CO_2 since 1750. The *squares* indicate measurements from Antarctic ice cores, and the *triangles* represent direct measurements from Mauna Loa observatory in Hawaii

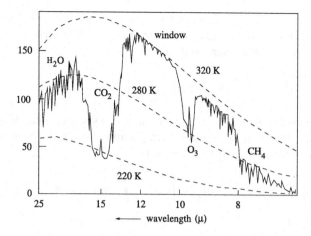

Fig. 2.9 Vertical thermal emission from the Earth measured over the Sahara. The *horizontal axis* is linear in wave number, hence the irregular intervals for the wavelength in microns. The units of radiation are $mW\,m^{-2}\,sr^{-1}\,(cm^{-1})^{-1}$, the last two indicating inverse steradian (the unit of solid angle) and wave number. The *dashed lines* are the black body radiation curves at the indicated temperatures. Redrawn from Fig. 12.7 of Houghton (2002), by permission of Cambridge University Press

Although CO_2 is only present in small quantities, it is an important absorber for the long-wave emitted IR radiation. The effect of increasing its concentration is to increase the optical density, and thus to decrease γ. Let us suppose then that the change in CO_2 leads to a change in the greenhouse coefficient γ given by

$$\gamma = \gamma_0 - \tilde{\gamma}; \qquad (2.76)$$

γ_0 is the pre-industrial reference state, and $\tilde{\gamma}$ represents the (positive) secular change due to CO_2. With $\tilde{\gamma} \ll 1$, we thus have the quasi-equilibrium given by (2.64), which leads to

$$T \approx T_0 + \frac{\tilde{\gamma} T_0}{4\gamma_0}. \qquad (2.77)$$

Of course the difficulty lies in evaluating an effective dependence of $\tilde{\gamma}$ on CO_2 levels, and in reality, the problem is made more difficult by the non-greyness of the atmosphere. To understand this, let us consider the long-wave thermal emission as a function of wavelength. This is shown in Fig. 2.9, together with black body irradiance curves at various temperatures. The emission curve divides quite neatly into a number of distinct wavelength intervals, in each of which the emission quite closely follows the black body radiation corresponding to distinct temperatures. We see a window between 10 and 13 μ, where there is little absorption, and the effective emission temperature is that at ground level. At higher wavelength, (14–16 μ), there is a CO_2 absorption band, and the radiation appears to emanate from the lower stratosphere.

In order to understand how this can be, we revisit the concept of the Chapman layer discussed above in Sect. 2.2.8. We write the radiation intensity equation (2.7) for a one-dimensional atmosphere in the form

$$\mu \frac{\partial I_\nu}{\partial z} = -\kappa_\nu \rho_0 e^{-z/H}[I_\nu - B_\nu], \tag{2.78}$$

where H is the scale height, taken as constant. For local thermodynamic equilibrium, $B_\nu(T)$ is an increasing function of temperature given by (2.6). When $\mu = 1$, the solution for upwards travelling radiation is

$$I_\nu = I_\nu^0 \exp[-\tau_\nu(1 - e^{-\zeta})] + \tau_\nu \exp[\tau_\nu e^{-\zeta}] \int_0^\zeta B_\nu(T) \exp[-\zeta - \tau_\nu e^{-\zeta}] d\zeta, \tag{2.79}$$

where

$$\zeta = z/H, \qquad \tau_\nu = \kappa_\nu \rho_0 H. \tag{2.80}$$

When τ_ν is small, as for the window between 10 and 13 μ, then

$$I_\nu \approx I_\nu^0. \tag{2.81}$$

When $\tau_\nu > 1$ the kernel of the integrand has an internal maximum at $\zeta = \ln \tau_\nu$, and by putting

$$\zeta = \ln \tau_\nu + Z, \tag{2.82}$$

we have for large τ_ν the approximation

$$I_\nu \approx \exp[e^{-Z}] \int_{-\ln \tau_\nu}^Z B_\nu(T) \exp[-Z' - e^{-Z'}] dZ'. \tag{2.83}$$

The kernel $\exp[-Z' - e^{-Z'}]$ of the integrand is a peaked function with a maximum at $Z' = 0$. It thus filters out the values of B in the vicinity of $\zeta = \ln \tau_\nu$. If we idealise the kernel as a delta function centred on $\zeta = \ln \tau_\nu$, then we have

$$I_\nu|_{Z \to \infty} \approx B_\nu(T)|_{\zeta = \ln \tau_\nu}, \tag{2.84}$$

and it is in this sense that the thermal emission picks out black body radiation at the level corresponding to the opacity at that frequency.

We denote the effective emission altitude for a particular frequency as z_ν, thus

$$z_\nu = H \ln[\kappa_\nu \rho_0 H]. \tag{2.85}$$

Inspection of Fig. 2.9 then suggests that the variation of z_ν with frequency (or wavelength) in the 15 μ CO_2 absorption band is as indicated in Fig. 2.10, this variation being due to the variation of absorption coefficient with ν.

We can now infer the effect of increasing CO_2 density. Increasing ρ_0 has the effect of shifting the emission altitude upwards. In the stratosphere, this increases the

Fig. 2.10 Schematic
variation of the effective
emission height with
wavelength in the CO_2
absorption band

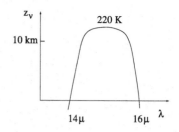

temperature and therefore also the emission rate. Because of this, the stratosphere
will cool under increased CO_2. On the other hand, the upwards shift of emission
height at the fringes of the absorption band causes a cooling in the adiabatic troposphere and thus decreased emission. It is this shift of the emission height which is
the cause of tropospheric heating under raised CO_2 levels.

Estimates of the consequential effect of increasing CO_2 levels is rendered uncertain because of various feedback effects which will occur in association. In particular, water vapour is also a major greenhouse gas (as can be seen from Fig. 2.9), and
increased temperature causes increased evaporation and thus enhances the greenhouse effect. Perhaps more importantly, change of cloud cover can have a strong
effect on temperature, because of its multiple influences: short-wave albedo, as well
as long-wave absorption and emission (see Fig. 2.3). It is partly because of the uncertainty in parameterising cloud formation and structure that there is so much uncertainty associated with forecasts of global warming. Current estimates suggest
that doubling CO_2 leads to a global increase of surface temperature in the region of
2–4 K. It has become popular to relate recent anomalous weather patterns (hurricane
frequency, floods and heat waves, for example) to the effects of CO_2, but although
this may indeed be the cause, nevertheless the natural variability of climate on short
time scales does not allow us to make this deduction with any real justification.
An alternative viewpoint is that since we know that CO_2 causes warming, it is a
likely consequence that weather patterns will tend to become more variable, and it
would then be of little surprise if this is actually happening. Indeed, the retreat of
the glaciers since the nineteenth century is consistent with (but does not prove) the
idea that global warming is not a recent phenomenon.

2.4.3 The Runaway Greenhouse Effect

If the blanketing effect of the greenhouse gases is the cause of the Earth's relatively
temperate climate, what of Venus? Its surface temperature has been measured to
be in the region of 700 K, despite (see Question 2.1) an effective emission temperature of 230 K. That the discrepancy is due to the greenhouse effect is not in
itself surprising; the atmosphere is mostly CO_2 and deep clouds of sulphuric acid
completely cover the planet. What is less obvious is why the Venusian atmosphere
should have evolved in this way, since in other respects, Venus and Earth are quite
similar planets.

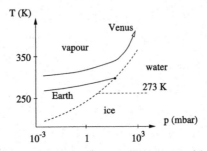

Fig. 2.11 A schematic representation of the evolution of temperatures on Venus and Earth. As the atmospheric water vapour increases on Earth, condensation occurs, leading to clouds, rainfall and ocean formation. On Venus this does not occur, and the water vapour is ultimately lost through dissociation, hydrogen escape and surface oxidation reactions

A possible explanation can be framed in terms of the simple energy-balance model proposed above, together with a consideration of the evolution of the amount of water vapour in the atmosphere. Initially, primitive terrestrial planets have no atmosphere (and no oceans or land ice). The internal heat generated by planetary accretion and by radioactive heat release is, however, substantial, and causes a huge amount of volcanism. In the eruption of magma, dissolved gases including H_2O and CO_2 are exsolved (for example, by pressure release, in much the same way bubbles form when a champagne bottle is opened). On the Earth, the increasing atmospheric density causes a slow rise in the temperature, while simultaneously the increasing partial pressure p_v of water vapour brings the atmosphere closer to saturation. On the Earth, it is supposed (see Fig. 2.11) that p_v reaches the saturation vapour pressure p_{sv} when $p_v > 600$ Pa (the triple point pressure). Clouds form of water droplets, and the ensuing rain forms the oceans and rivers. Most of the CO_2 is then removed from the atmosphere to form carbonate rocks.

On Venus, on the other hand, the slightly higher received solar radiation causes the (T, p_v) path which is traced to be higher. As p_v increases, so does T, and we suppose (see Fig. 2.11) that saturation never occurs. The water vapour continues to increase, leading to ever higher greenhouse temperatures.

A subsidiary question is then, what happens to the H_2O on Venus? The atmosphere is essentially devoid of H_2O. Here the idea is that UV radiation in the upper atmosphere dissociates the hydrogen from oxygen, the hydrogen then escapes to space, while the oxygen is used up in oxidising reactions with surface rocks.

The mechanism above is attractively simple, and can be understood using the concept of radiation balance in a grey atmosphere. The equilibrium temperature from (2.64) is

$$T = \left[\frac{(1-a)Q}{4\sigma\gamma} \right]^{1/4}, \tag{2.86}$$

and we can expect both the albedo a and greyness γ to depend on the density of water vapour ρ_v. For simplicity, take $a = 0$ and for γ we use a formula suggested by (2.62) in the optically dense limit ($\tau_s \gg 1$):

$$\gamma = \frac{1}{\tau_s}. \tag{2.87}$$

Taking $\tau_s = \kappa \rho_v d$, we then have

$$T \approx \left[\frac{Q \kappa d}{4\sigma} \rho_v \right]^{1/4}, \tag{2.88}$$

and using the perfect gas law (2.58) in the form $p_v = \rho_v R T / M_v$ gives

$$T \approx \left[\frac{Q \kappa d M_v}{4\sigma R} p_{sv} \right]^{1/5}, \tag{2.89}$$

and we write this in the form

$$T = \left[\frac{Q \kappa d M_v p_{sv}^0}{4\sigma R} \right]^{1/5} e^\xi, \tag{2.90}$$

where

$$\xi = \frac{1}{5} \ln(p_v / p_{sv}^0), \tag{2.91}$$

and p_{sv}^0 is a reference value of the saturation vapour pressure, which we will take to be the triple point pressure, 6 mbar. On the other hand, the saturation temperature is determined by solving the Clausius–Clapeyron equation (2.56). The exact solution of this is

$$p_{sv} = p_{sv}^0 \exp\left[a \left\{ 1 - \frac{T_{sat}^0}{T_{sat}} \right\} \right], \tag{2.92}$$

where T_{sat} is the saturation temperature, $a = M_v L / R T_{sat}^0$, and for $T_{sat} - T_{sat}^0 \ll T_{sat}^0$, this is

$$T_{sat} \approx T_{sat}^0 [1 + \nu \xi], \tag{2.93}$$

where

$$\nu = \frac{5}{a} = \frac{5 R T_{sat}^0}{M_v L}, \tag{2.94}$$

with approximate value $\nu \approx 1/4$. T_{sat}^0 is the saturation temperature at the triple point, $T_{sat}^0 \approx 273$ K. If we write $T_{sat} = T_{sat}^0 \theta_{sat}$, $T = T_{sat}^0 \theta$, then the planetary and saturation temperature curves are given, respectively, by

$$\theta = r e^\xi,$$

$$\theta_{sat} = 1 + \nu \xi, \tag{2.95}$$

where

$$r = \frac{1}{T_{sat}^0} \left[\frac{Q \kappa d M_v p_{sv}^0}{4\sigma R} \right]^{1/5}. \tag{2.96}$$

The definition of r here should not be taken too seriously, as we implicitly assumed that absorption was entirely due to water vapour. However, the intersection of the curves in (2.95) makes the point that the runaway effect can be expected if r is large enough, specifically if

$$r > r_c = v \exp\left(\frac{1-v}{v}\right) \approx 5 \tag{2.97}$$

for $v = 1/4$, and this corresponds to a sufficiently large value of Q. Hence, the distinction between Earth and Venus, for which the value of Q is twice that of Earth. The situation is illustrated in Fig. 2.11.

2.5 Ice Ages

Most people are probably aware that we live in glacial times. During the last two million years, a series of ice ages has occurred, during which large ice sheets have grown, principally on the northern hemisphere land masses. The Laurentide ice sheet grows to cover North America down to the latitude of New York, while the Fennoscandian ice sheet grows in Scandinavia, reaching into the lowlands of Germany, and possibly connecting across the north sea to a British ice sheet which covers much of Britain and Ireland down to Kerry in the west and Norfolk in the east. The global ice volume which grows in these ice ages is sufficient to lower sea level by some 120 metres, thus exposing vast areas of continental shelf.

These Pleistocene ice ages occur with some regularity, with a period of 100,000 years (although prior to the last 900,000 years, a periodicity of 40,000 years appears more appropriate). The great ice sheets grow slowly over some 90,000 years, and there is then a fairly sudden deglaciation. This is illustrated in Fig. 2.12, which shows a proxy measurement of temperature over the last 740,000 years, obtained from an Antarctic ice core. Five sharp rises in temperature can be seen separating the last four ice ages, which show a characteristic slow decline in temperature. As we shall see, the mechanism which causes this sequence of pseudo-periodic oscillations in the climate is not very well understood.

The present glacial climate may be a result of a gradual cooling initiated by the collision of India with Asia starting some 50 million years ago, and causing the rise of the Himalayas. Although these mountains affect weather systems directly, their effect on climate may be due to the increasing precipitation and thus weathering which they induce, which leads to a removal of carbon from the atmosphere and a consequent cooling of the atmosphere. It is certainly the case that CO_2 has faithfully followed climatic temperatures through the recent ice ages, and it is difficult not to suppose that it has been a major causative factor in their explanation.

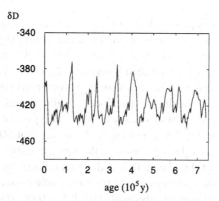

Fig. 2.12 A proxy measurement from deuterium isotope data of the climate of the last 740,000 years. The measurements come from an Antarctic ice core (see the EPICA community members' paper, Agustin et al. (2004), and were provided by Eric Wolff. Each data point is the measurement of deuterium isotope ratios in a column of ice representing 3,000 years accumulation (i.e., the data represent 3,000 year averages). Time moves from right to left along the abscissa, and the deuterium isotope ratio is a proxy measurement of prevailing climatic temperature

If we go further back in time, we encounter much warmer climates. The time of the dinosaurs, extending back to the Triassic, some 200 million years ago, saw a very warm climate and some very large creatures. There were no ice sheets: the Antarctic ice sheet only began to grow some 34 million years ago after the India–Asia collision.

Further back, however, we find evidence of major glaciated periods of Earth history, for example in the Carboniferous period some 300 million years ago. The glacial deposits which indicate this are located in India, South Africa, Australia and South America. But at the time of the glaciation, these continental masses were all sutured together in the great palaeo-continent of Gondwanaland, and they resided at the south pole. The break up of Gondwanaland to form the continents as we now see them only began some 200 million years ago, and is more or less coincident with the global rise in temperature and the flourishing of the dinosauria.

Even earlier in time, we have evidence of further massive glaciation on the super-continent of Rhodinia during Proterozoic times, some 600 million years ago. The fact that these glaciations occur at then equatorial positions has led to the challenging concept of the 'snowball Earth', the idea that the whole planet was glaciated. Like most outrageous ideas, this is both enticing and controversial; we shall say more about it in Sect. 2.6 below.

2.5.1 Ice-Albedo Feedback

The simplest type of model to explain why ice ages may occur in a sequential fashion is the energy-balance model of Sect. 2.4. On its own, it predicts a stable climatic response to solar radiative input, but when the feedback effect of ice is included, this

alters dramatically. Although simple in concept, the energy-balance model provides
the platform for more recent models of 'intermediate complexity'.

The mechanism of the ice-albedo feedback is this. In winter, Antarctica is surrounded by sea ice, and the Arctic ocean is permanently covered by sea ice. Land ice
is also present on the Earth near the poles, or in mountainous regions. The presence
of ice has a dramatic effect on the surface albedo. While the reflectivity (the fraction
of radiation which is reflected) of oceans or forest is typically 0.1, that of sea ice
or snow is in the range 0.6–0.8. From Fig. 2.3, we see that 50% of the incoming
solar radiation Q (i.e., $0.5Q$) is received at the surface, either directly or through
scattering. The albedo of the planet, 0.3, is due to a reflectivity of 0.26 from cloud
and atmosphere, and a reflectivity of 0.04 from the surface: since $0.5Q$ reaches the
surface, this represents a surface albedo of $0.04/0.5 = 0.08$. However, if the planet
were covered in ice, the surface albedo might be 0.7, so that $0.7 \times 0.5 = 0.35$ of
the solar radiation would be reflected. Consequently, the planetary albedo would be
doubled, from $0.26 + 0.04 = 0.3$, to $0.26 + 0.35 = 0.61$, from this effect alone.

It is thus of interest to examine the effect on the energy-balance equation of
including this effect of ice and thus temperature on albedo, since the occurrence of
precipitation as snow or rain is essentially related to the atmospheric temperature.
We write (2.63) in the form

$$c\dot{T} = R_i - R_o, \tag{2.98}$$

where $c = \rho_a c_p d$ is the specific heat capacity of the atmosphere, and

$$R_i = \frac{1}{4}(1 - a)Q, \qquad R_o = \sigma \gamma T^4 \tag{2.99}$$

are, respectively, the incoming short-wave radiation and the emitted IR radiation.

The effect of decreasing temperature on the albedo is to increase the extent of
land and sea ice, so that a will increase. It is convenient to define a family of equilibrium albedo functions

$$a_{eq}(T) = a_1 - \frac{1}{2}a_2\left[1 + \tanh\left(\frac{T - T^*}{\Delta T}\right)\right], \tag{2.100}$$

one example of which is shown in Fig. 2.13. The epithet 'equilibrium' refers to
the assumption that the land ice cover is in dynamical equilibrium with the ground
surface temperature: more on this below. The effect of the albedo variation on the
emitted radiation is shown in Fig. 2.14: R_o is an increasing function of T, but the
sigmoidal nature of R_i can lead, for a range of Q and suitable choices of the albedo
function, to the existence of multiple steady states. If this is the case, then the equilibrium response diagram for steady states T in terms of Q is as shown in Fig. 2.15.

The parameters used in Fig. 2.14 are chosen to illustrate the multiple intersection
of R_i with R_o, but do not correspond to modern climate (for which $T = 288$ K
and $R_i = R_o = 235$ W m^{-2}). The reason is that with more appropriate parameters,
such as those used in Fig. 2.15, the two curves become very close, and the range of
Q over which multiplicity occurs is very small. Insofar as these parameterisations

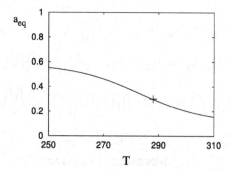

Fig. 2.13 A representation of the possible variation of equilibrium surface albedo $a_{eq}(T)$ due to variations in ice cover due to climatic temperature. The function plotted is $a_{eq}(T)$ given by (2.100), with $a_1 = 0.58$, $a_2 = 0.47$, $T^* = 283$ K, $\Delta T = 24$ K,which tends to 0.58 for small T, and equals 0.3 at about $T = 288$ K (the point marked $+$)

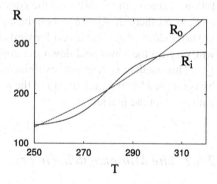

Fig. 2.14 Variation of R_i and R_o with temperature T. Parameters used are $a_1 = 0.6$, $a_2 = 0.45$, $\gamma = 0.6$, $T^* = 280$ K, $\Delta T = 15$ K

apply to the Earth, it does suggest that the current climate is close to a switching point, as seen in Fig. 2.15, corroborating this explanation for ice age formation.

We have seen this kind of S-shaped diagram before in Sect. 1.3.3, where it was used in describing combustion. It is easy to see from Fig. 2.14 that the upper and lower branches in Fig. 2.15 are stable, while the middle branch is unstable; this

Fig. 2.15 Multivalued response curve for T in terms of Q. Parameters used are $a_1 = 0.58$, $a_2 = 0.47$, $\gamma = 0.6175$, $T^* = 283$ K, $\Delta T = 24$ K. Also shown is the point $(+)$ corresponding to current climate $(Q = 1370 \text{ W m}^{-2}$, $T = 288$ K)

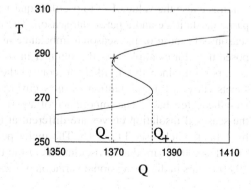

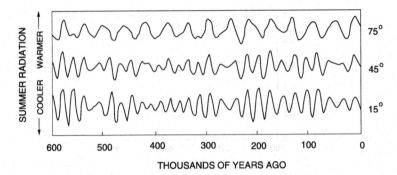

Fig. 2.16 Milankovitch radiation curves for 15° N, 45° N, and 75° N (see Bolshakov 2003). The *lower two curves* indicate the ~22,000 year precession cycle, while the upper one shows more clearly the 41,000 year tilt cycle

follows because the stability of the equilibria of (2.96) is determined by the slope of $R_i - R_o$ there: if $R_i' < R_o'$, then the equilibrium is stable, and vice versa. Thus if Q varies slowly backwards and forwards beyond Q_- and Q_+, then the temperature will vary up the lower branch and down the upper branch, with sudden jumps at Q_+ and Q_-. This oscillatory response exhibits *hysteresis*, it is *irreversible*, and it forms the basis for the Milankovitch theory of the ice ages, since the lower branch is associated with widespread glaciation.

2.5.2 The Milankovitch Theory

The solar radiation received seasonally on the Earth is not in fact constant. Due to variations in the Earth's orbit, the value of Q at a point varies by about ±5% either side of its mean. Nor are these variations periodic. Because the solar system has planets (and moons) other than the Earth, and because also the planets do not act exactly as point masses, the orbit of the Earth is not precisely a Keplerian ellipse. The Earth's axis of rotation precesses, its angle of tilt (from the plane of the ecliptic) oscillates, and the eccentricity of the orbit itself oscillates. All of these astronomical features cause the value of Q to oscillate quasi-periodically, when considered for a particular latitude and a particular season. The reason for focussing on a particular season is because of the seasonal imbalance in snowfall, whence it might be supposed that, for example, it is the summer insolation received at (say) 65° N which is important, since this is likely to control the inception of northern latitude ice sheets via year round retention of snow cover and the consequent operation of the ice-albedo feedback. The importance of a particular latitude is due to the fact that the seasonal insolation curves are different at different latitudes, as indeed found by Milankovitch—see Fig. 2.16. The major periodicities in the signals consist of one of 41,000 years due to oscillations in the tilt axis, and periods of 23,000 and 19,000 years in the precessional variation of the rotation axis. The third component, eccentricity, causes a variation over a period of 100,000 years, though its amplitude is much smaller.

The test of the Milankovitch theory that variations in climate (and thus ice ages) are associated with the variation in Q, can then be made by computing the Fourier power spectrum of a record of past climatic temperature. Oxygen isotope ratios in deep-sea sediment cores (or in ice cores) provide a proxy measurement of temperature, and Fig. 2.12 showed just such a record. When a spectral analysis of records of this type is made, it is indeed found that the principal frequencies are (in order of decreasing amplitude) 100 ka (100,000 years), 41, 23 and 19 ka. This seems to serve as dramatic confirmation of the Milankovitch theory. In our simple energy-balance model, the concept is enunciated by the hysteretic oscillations exhibited by the system as Q varies.

2.5.3 Nonlinear Oscillations

There is currently a consensus that the Milankovitch orbital variation indeed acts as pacemaker for the Quaternary ice ages, but it is as well to point out that there is an essential problem with the Milankovitch theory, even if the basis of the concept is valid. The spectral insolation frequencies do match those of the proxy climate record, with one essential discrepancy: the largest climatic signal is the 100 ka period. Ice ages essentially last 90 ka, with an interval of 10 ka between (and since the last ice age terminated about 10 ka ago, as the Scottish ice sheet withdrew from the lochs in the Highlands, and the North American Laurentide ice sheet shrank from the Great Lakes, we might be on the verge of starting the next). But the 100 ka astronomical signal is very weak, and it is unrealistic to imagine that the forcing can directly drive the strong response which is observed. What may happen is that the weak 100 ka forcing *resonates* with the climatic system, suggesting that the climate is essentially a (nonlinear) oscillator, with a natural period close to 100 ka, which is tuned by the astronomical forcing. A mathematical paradigm would be the forced Van der Pol oscillator

$$\ddot{x} + \varepsilon(x^2 - 1)\dot{x} + \omega^2 x = f(t), \tag{2.101}$$

where $f(t)$ would represent the astronomical forcing. If ε is small, the oscillator has a natural frequency close to $2\pi/\omega$, and if forced by a frequency close to this, tuning can occur. If the oscillator is nonlinear, other exotic effects can occur: subharmonics, chaos; no doubt these effects are present in the forced climate system too.

The simplest kind of model which can behave in an oscillatory manner is the energy-balance model (2.98) subjected to an oscillating radiation input which can drive the climate back and forth between the cold and warm branches, presumably representing the glacial and interglacial periods. Two questions then arise; where does the 100,000 year time scale come from, and why is the climatic evolution through an ice age (slow development, rapid termination) so nonlinear?

There are three principal components of the climate system which change over very different time scales. These are the atmosphere, the oceans and the ice sheets. The time scale of response of ice sheets is the longest of these, and is measured by l/u, where l is a horizontal length scale and u is a horizontal velocity scale. Estimates of $l \sim 1000$ km and $u \sim 100$ m y^{-1} suggest a time scale of 10^4 y, which is within range of the value we seek. Since ice sheet extent is directly associated with albedo. it suggests that a first realistic modification of the energy-balance model is to allow the ice sheets, and therefore also the albedo, to respond to a changing temperature over the slow ice sheet time scale t_i. A simple model to do this is to write

$$c\dot{T} = \frac{1}{4}Q(1 - a) - \sigma\gamma T^4,$$

$$t_i\dot{a} = a_{\mathrm{eq}}(T) - a, \tag{2.102}$$

and $a_{\mathrm{eq}}(T)$ is the equilibrium albedo represented in Fig. 2.13. Since the thermal response time scale is so rapid (months), we may take the temperature to be the equilibrium temperature,

$$T = T(a, Q) = \left[\frac{Q(1 - a)}{4\sigma\gamma}\right]^{1/4}. \tag{2.103}$$

The energy-balance model thus reduces to the first order albedo evolution equation

$$t_i\dot{a} = I(a, Q) - a, \tag{2.104}$$

where

$$I(a, Q) = a_{\mathrm{eq}}\left[T(a, Q)\right]. \tag{2.105}$$

As Q varies backwards and forwards about the critical switching values in Fig. 2.15, the ice extent (as indicated by a) changes on the slow time scale t_i, aiming to follow the hysteretically switching equilibria. Oscillatory inputs Q do indeed cause oscillations; if t_i is sufficiently small, these are large scale, going from warm branch to cold branch and back, whereas at larger t_i, two different oscillatory climates are possible, a cold one and a warm one (see Question 2.14). None of these solutions bears much resemblance to real Quaternary ice ages, for which a more sophisticated physical model is necessary.

2.5.4 Heinrich Events

The study of climate is going through some exciting times. The pulse of the ice ages can be seen in Fig. 2.12, but the signal appears noisy, with numerous irregular jumps. Twenty or thirty years ago, one might have been happy to ascribe these to the influence of different spectral components of the Milankovitch radiation curves on a nonlinear climatic oscillator, together with a vague reference to 'noisy' data.

Increasingly in such circumstances, however, one can adopt a different view: what you see is what you get. In other words, sharp fluctuations in apparently noisy data are actually signals of real events. Put another way, noise simply refers to the parts of the signal one does not understand.

It has become clear that there are significant climate components which cause short term variations, and that these events are written in the data which are exhumed from ocean sediment cores and ice cores. Perhaps the most dramatic of these are *Heinrich events*. Sediment cores retrieved from the ocean floor of the North Atlantic reveal, among the common ocean sediments and muds, a series of layers (seven in all have been identified) in which there is a high proportion of lithic fragments. These fragments represent ice rafted debris, and are composed of carbonate mudstones, whose origin has been identified as Hudson Bay. The spacing between the layers is such that the periods between the Heinrich events are 5,000–10,000 years.

What the Heinrich events are telling us is that every 10,000 years or so (more or less periodically) during the last ice age, there were episodes of dramatically increased iceberg production, and that the ice in these icebergs originated from the Hudson Bay underlying the central part of the Laurentide ice sheet. Ice from this region drained through an ice stream some 200 km wide which flowed along the Hudson Strait and into the Labrador sea west of Greenland.

The generally accepted cause of these events is also the most obvious, but equally the most exciting. The time scale of 10,000 years is that associated with the growth of ice sheets (for example, by accumulation of $0.2 \, \mathrm{m \, y^{-1}}$ and depth of 2000 m), and so the suggestion is that Heinrich events occur through a periodic surging of the ice in the Hudson Strait, which then draws down the Hudson Bay ice dome. This would sound like a capricious explanation, were it not for the fact that many glaciers are known to surge in a similar fashion; we shall discuss the mechanism for surging in Chap. 10.

Another feature of Heinrich events is that they appear to be followed by sudden dramatic warmings of the Earth's climate, which occur several hundred years after the Heinrich event. Dating of these can be difficult, because dating of ice cores and also of sediment cores sometimes requires an assumption of accumulation or sedimentation rates, so that precise association of timings in different such cores can be risky.

How would Heinrich events affect climate? There are two obvious ways. A sudden change in an ice sheet elevation might be expected to alter storm tracks and precipitation patterns. Perhaps more importantly, the blanketing of the North Atlantic with icebergs is likely to affect oceanic circulation. Just like the atmosphere, the ocean circulation is driven by horizontal buoyancy induced by the difference between equatorial and polar heating rates. This large scale flow is called the global thermohaline circulation, and its presence in the North Atlantic is the cause of the gulf stream (see also Sect. 3.9), which promotes the temperate climate of Northern Europe, because of the poleward energy flux it carries. If this circulation is disrupted, there is liable to be an immediate effect on climate.

If the North Atlantic is covered by ice, one immediate effect is a surface cooling, because of the increased albedo. This is liable to cause an increase in the thermohaline circulation, but would not cause atmospheric warming until the sea ice melted.

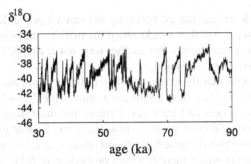

Fig. 2.17 Oxygen isotope ratio ($\delta^{18}O$) measurements from the GRIP ice core on Greenland, as a function of age in ka (1 ka = 1000 years). This is a proxy for surface temperature (with four units corresponding to about 10 K on the vertical axis). The data represent averages from segments of 55 cm length, and the age scale is determined from a model of ice burial rate. Near the surface, the separate measurements are two-to-three yearly, but the compression of ice with burial causes the ice segments to encompass longer and longer time periods. At the age of 90,000 years ago (at a depth of some 2685 m), each segment is a time average of some 120 years. Thus the data are increasingly sparsely resolved further into the past

On the other hand, the melting itself releases fresh water, which is buoyant in a saline ocean, suggesting a shutdown of ocean circulation. As we discuss further below, this can lead, following a delay, to a massive restart of ocean circulation and thus sudden warming.

2.5.5 Dansgaard–Oeschger Events

There are other rapid changes in the climate which are seen during the last ice age. Figure 2.17 shows a segment of oxygen isotope measurements (a proxy for surface temperature) from the GRIP ice core on Greenland. Time marches from right to left on this diagram. There are numerous sudden rises in temperature that can be seen, followed by a more gentle sinking of temperature. These sharp rises are called *Dansgaard–Oeschger events*. Between 30,000 and 45,000 years B. P., for example there are seven of these events, thus, like Heinrich events, they occur at reasonably regular intervals, with a typical repetition period being in the region of about 1,500 years. The association of the D–O events with oceanic salt oscillations is described by Schmidt et al. (2006), for example.

Let us examine one of these events in greater detail: that of the D–O event between 44,000 and 45,000 years B. P. In the GRIP core, this ice lies between 2,316 and 2,330 metres depth. A higher resolution data set is that of Sigfus Johnsen, and is shown in Fig. 2.18. This shows that the climatic temperature changes abruptly, over a time scale of about a century. Other such inspections show that the transitions can be even shorter.

What is the cause of these warming events? Why are they so rapid, and why do they have a regular period of some 1,500 years? The idea here is that the climate

Fig. 2.18 GRIP core data between 2,316 m and 2,330 m. The sharp jump near 2,324 m occurs over a range of about 1.3 metres, corresponding to a time interval of some 90 years

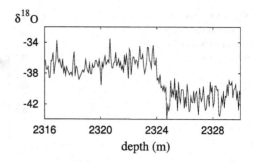

in the northern hemisphere is essentially controlled by the oceanic conveyor circulation, and so the change in climate occurs because of a sudden disruption to this. Model studies have shown that an injection of a massive pulse of fresh water into the North Atlantic can cause just such a disruption.[8]

The mechanism is, however, counter-intuitive. A warm climate is associated with a vigorous circulation, and a cold one with the circulation off, but a freshwater pulse has the initial effect, being buoyant, of switching the (relatively weak) circulation off. This causes a climatic cooling. However, the cooling is temporary, because a situation with no circulation is unstable. When convection begins again, it does so dramatically, with deep water formation occurring further north (as it does in interglacial times), causing a sudden shift to a warmer climate. The same model studies have shown that even larger meltwater pulses, such as would occur following the melting induced by large scale iceberg production, can lead, after the initial cooling of the north Atlantic, to a subsequent extreme warming comparable to that seen following Heinrich events.

If freshwater pulses are the cause of the sudden climate shifts, what is their origin? For Heinrich events, the ice rafted debris gives the clue; for Dansgaard–Oeschger events, there is apparently no such clue. However, it is pertinent to note that these events are associated with the presence of large ice sheets. If we seek an explanation by means of freshwater pulses, then the most obvious (and really, the only) candidate for the source of the pulses is that they come from meltwater from the ice, and one way in which meltwater drainage is known to occur episodically is in the large sub-glacial floods known as jökulhlaups. As with surges, these are well documented from beneath glaciers. It has been less common to imagine that they could occur from beneath modern day ice sheets, but in fact such floods are being increasingly observed to happen below the Antarctic Ice Sheet. As a hypothesis it seems sensible to suggest that Dansgaard–Oeschger events might arise as a consequence of semi-regular Laurentide jökulhlaups which occur with a rough periodicity of one to two thousand years. The question then arises as to whether such floods are dynamically possible, and whether they could produce the necessary fresh water at the required frequency to do the job. We shall re-examine this question in Chap. 11.

---

[8]This is not the only possible mechanism. Another is the North Atlantic salt oscillator, discussed in Sect. 2.5.7.

Fig. 2.19 Oxygen isotope data from the GRIP core at the transition to the Holocene interglacial

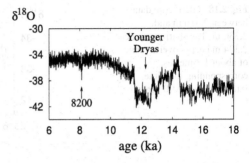

Fig. 2.20 High resolution data set from 20 metres of ice near the 8,200 year event

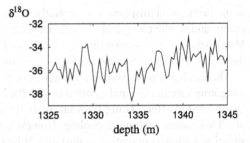

2.5.6 *The 8,200 Year Cooling Event*

One climatic event which is thought to have been caused by a sub-glacial flood is a sudden cooling event dated to 8,200 years B. P. This is shown in context in Fig. 2.19. which also shows the termination of the ice age after the Younger Dryas readvance of the ice sheets between 13,000 and 11,600 years B. P. Two Heinrich events precede the two warmings at about 15,000 and 12,000 years B. P. Following the Younger Dryas, there is a gradual return to an interglacial climate by 9,500 years B. P., and the onset of the current (Holocene) period. A cursory glance might suggest that the 8,200 year dip is just a noisy outlier, but this is not the case. It represents a genuine climatic cooling of some 4 K.

Figure 2.19 shows a high resolution record of this event. Inspection of the coarse (55 cm samples) GRIP data shows that the 8,200 year event is actually (on the age scale used) at 8,126 years B. P., and occurs in a single 55 cm segment at a depth of 1334 metres; blink, and you miss it. At this depth, six metres of ice (1,331–1,337) is considered to represent 65 years of accumulation. Plotting the data using Johnsen's higher resolution data set over a more restricted range, we can see (Fig. 2.20) various features.

One is that the event occupies three data points from an ice depth between 1,334 and 1,335 metres depth. These three samples are 27.5 cm long, as are the two immediately above and below. This suggests that the 'duration' of the event is between 82 and 137 cm, which corresponds to a period of between 9 and 15 years. This is incredibly fast.

The other thing to notice from Fig. 2.20 is that there are a good number of other large spikes and oscillations. Since, more or less, each data point represents a three

Fig. 2.21 The cooling trend
of the 8,200 year event

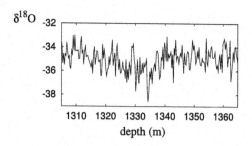

year average, these are not seasonal fluctuations. Do they represent real events, or
simply the natural fluctuation of the climate from year to year? If one looks at a
slightly larger slice of the time series, from 1305 m to 1365 m, it is apparent (see
Fig. 2.21) that these short term fluctuations sit on top of a broader cooling trend from
about 1340 to 1315 m, with rapid decrease in the first 5 m (~54 years), and slower
recovery over the following 20 m (~217 years). It is perhaps easier to imagine that
this slower average trend represents the underlying event.

The explanation which is currently thought to apply to this event is that it is
caused by a sub-glacial jökulhlaup which drains the massive proglacial Lake Agassiz into the Hudson Strait, whence it pours into the Labrador Sea and the North
Atlantic. As the remnant of the Laurentide ice sheet dwindles, it builds up a massive
proglacial lake on its southern margin. The topography is such that this lake is pre-
vented from outflow to the south, and at some point it drains catastrophically, either
over or more probably under the ice sheet to the north. The resulting fresh water
efflux to the North Atlantic causes the cooling event.

One might wonder, if glacial meltwater pulses cause convective shutdown, cooling, and then subsequent warming, why would an interglacial one produce only the
cooling? The putative answer to this lies in our idea of what a meltwater pulse will
actually do. In an interglacial climate, the ocean circulation is strong, and meltwater
weakens it temporarily: a cooling. In a glacial climate, the circulation is weaker,
and deep water formation occurs further south, say near Iceland, than it does currently. Then a meltwater pulse may shut down the circulation entirely, which would
indeed cause further cooling, but the resultant overshoot when circulation resumes
causes the warming. Since Dansgaard–Oeschger events occur at the end of cooling
cycles, the initial cooling is swamped by the trend. It is interesting to note that the
D–O warming events in Fig. 2.19 are initiated at 14,500 B. P. and 11,600 B. P., the
interval between these being 2,900 years. The interval between the Younger Dryas
and the 8,200 event is about 3,500 years. If the D–O events are due to sub-glacial
floods, then possibly the 8,200 event is simply the last of these. It is then tempting
to look further on for similar, smaller events. There is one at 5,930 B. P., for example, and another at 5,770 B. P.; these are about another 2,400 years further on. It is
a natural consequence of the hypothesis that jökulhlaups occurred from below the
Laurentide ice sheet to suppose that they will occur also from beneath Greenland
and Antarctica, and that this may continue to the present day. It has been suggested,
for instance, that the cool period in Europe between 1550 A. D. and 1900 was due
to a similar upset of the oceanic circulation.

Fig. 2.22 Stommel's box
model of the North Atlantic
circulation

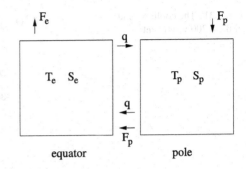

2.5.7 *North Atlantic Salt Oscillator*

Deeply embroiled in this whole saga of Quaternary climate and the ice ages is the rôle of the North Atlantic ocean circulation. For the descriptions we have given of Heinrich events and Dansgaard–Oeschger events to work, the ocean needs to be able to circulate in different ways. That this is indeed the case has been found in a number of model studies, and the resultant flip-flop circulation is sometimes known as the 'bipolar seesaw'. In its original form, the idea is due to Henry Stommel, and can be described with a simple 'box' model, as illustrated in Fig. 2.22.

In this model, we parameterise the thermohaline circulation in the North Atlantic by considering it to be partitioned between two compartments, an equatorial and a polar one. We label the temperature T, salinity (mass fraction of salt) S, density ρ and volume V of each box by a suffix 'e' or 'p', and we write conservation laws of mass, energy and solute, and an equation of state, for each box. Transports in and out of each box are considered to be a freshwater flux F_p to the polar cell, an evaporative flux F_e from the equatorial cell, and a convective flux q due to buoyancy difference from equatorial to polar cell. (The reverse flux is then $q + F_p$ in order to allow conservation of the sizes of both cells.)

Suitable equations to describe the convective flow are then

$$\frac{d}{dt}(\rho_e c V_e T_e) = A_e H_e + \rho_p c T_p (F_p + |q|) - \rho_e c T_e |q|,$$

$$\frac{d}{dt}(\rho_p c V_p T_p) = A_p H_p - \rho_p c T_p (F_p + |q|) + \rho_e c T_e |q|,$$

$$\frac{d}{dt}(\rho_e V_e S_e) = -\rho_e |q| S_e + \rho_p S_p (F_p + |q|),$$

$$\frac{d}{dt}(\rho_p V_p S_p) = \rho_e |q| S_e - \rho_p S_p (F_p + |q|), \qquad (2.106)$$

$$\frac{d}{dt}[\rho_e V_e (1 - S_e)] = \rho_p^0 F_p - \rho_e F_e,$$

$$\frac{d}{dt}[\rho_p V_p (1 - S_p)] = 0,$$

$$\rho_e = \rho_0\big[1 - \alpha(T_e - T_0) + \beta(S_e - S_0)\big],$$

$$\rho_p = \rho_0\big[1 - \alpha(T_p - T_0) + \beta(S_p - S_0)\big],$$

$$q = k(\rho_e - \rho_p).$$

The terms in these equations are fairly self-explanatory. α and β are coefficients of thermal and saline expansion, ρ_p^0 is the freshwater density, A_e and A_p are the equatorial and polar ocean surface areas, and for simplicity we take $A_p = A_e = A$. The heating terms H represent the heat flux to the deep ocean from the surface waters. The sea surface temperature is determined by a radiative balance, which determines equatorial and polar surface temperatures T_e^0 and T_p^0, say. We then suppose that heat transfer to the deep ocean can be parameterised by a suitable heat transfer coefficient h_T, thus we put

$$H_e = h_T(T_e^0 - T_e), \qquad H_p = h_T(T_p^0 - T_p). \tag{2.107}$$

If we add the two energy equations, we have

$$\frac{d}{dt}(\rho_e c V_e T_e + \rho_p c V_p T_p) = h_T A(T_e^0 + T_p^0 - T_e - T_p). \tag{2.108}$$

We use the facts that $\rho_{e,p} \approx \rho_0$ and we will suppose that V_e and V_p, both approximately constant, are also approximately equal, $V_{e,p} \approx V_0$. It then follows from (2.108) that, after an initial transient,

$$T_e + T_p \approx T_e^0 + T_p^0, \tag{2.109}$$

and we suppose this generally to be the case. We define V_0 by

$$\rho_e V_e T_e + \rho_p V_p T_p = 2\rho_0 V_0 T_0, \tag{2.110}$$

where

$$T_0 = \frac{1}{2}(T_e^0 + T_p^0), \tag{2.111}$$

and we then define the temperature excess T via

$$T_e = T_0 + T, \qquad T_p = T_0 - T. \tag{2.112}$$

If we now make the assumptions that $\rho_e \approx \rho_0$, $V_e \approx V_0$ in (2.106)$_1$, then we have the approximate equation for T:

$$\rho_0 c V_0 \dot{T} = \frac{1}{2}\big[h_T A(T_e^0 - T_p^0) + \rho_0 c F_p(T_e^0 + T_p^0)\big] - (h_T A + \rho_0 c F_p + 2\rho_0 c|q|)T. \tag{2.113}$$

In a similar way, we have

$$\rho_e V_e S_e + \rho_p V_p S_p = 2\rho_0 V_0 S_0, \tag{2.114}$$

where S_0 is constant, and we define

$$S_e = S_0 + S, \qquad S_p = S_0 - S. \tag{2.115}$$

With the same Boussinesq type assumption, that $\rho_e \approx \rho_0$ and $V_e \approx V_0$, we obtain

$$V_0 \dot{S} = F_p S_0 - (F_p + 2|q|)S. \tag{2.116}$$

Equations (2.113) and (2.116) are essentially Stommel's box model. Their validity relies on the use of the other equations to show that it is indeed realistic to take ρ and V as constant, though these assumptions appear fairly reasonable ones. Note that with the definitions of the variables, we have

$$q = 2k\rho_0(-\alpha T + \beta S). \tag{2.117}$$

To parameterise the heat transfer coefficient h_T, we use the ideas of Reynolds averaging for turbulent flow (see Appendix B). This suggests choosing

$$h_T = \frac{\varepsilon_T \rho_0 c q_0}{A}, \tag{2.118}$$

where the number ε_T is typically chosen to be in the range 0.001–0.01, and q_0 is a suitable scale for q, defined below in (2.124).

We non-dimensionalise the box model by writing

$$T = \Delta T \, \theta, \qquad S = \Delta S \, s, \qquad t \sim t_0, \qquad q \sim q_0, \tag{2.119}$$

where we choose

$$\Delta T = \frac{1}{4}\left[\varepsilon_T(T_e^0 - T_p^0) + \frac{F_p}{q_0}(T_e^0 + T_p^0)\right], \qquad \Delta S = \frac{F_p S_0}{2q_0}, \qquad t_0 = \frac{V_0}{2q_0}. \tag{2.120}$$

Using the values in Table 2.1, we find $\Delta T \approx 1$ K, $\Delta S \approx 1.1 \times 10^{-4}$, $t_0 \approx 150$ y. We use a value of q_0 as observed, rather than k, which we would in any case choose in order that q was the right size, some 16 Sv (Sverdrups: 1 Sv $= 10^6$ m^3 s^{-1}).

The observed surface temperature variation is of order 30 K, and the observed surface salinity variation is of order 30×10^{-4}. However, these values represent the concentrative effect of surface evaporation and heating; at depth (as is more relevant) the variations are much smaller, of order 2 K for temperature and 4×10^{-4} for salinity at 1000 m depth. The time scale is comparable to the time scales over which Dansgaard–Oeschger events occur. These features suggest that this simple model has the ring of truth.

We can write the model in dimensionless form as

$$\dot{\theta} = 1 - (\mu + |q|)\theta,$$
$$\dot{s} = 1 - (\varepsilon + |q|)s, \tag{2.121}$$
$$q = \kappa(-\theta + Rs),$$

Table 2.1 Typical parameter values for the Stommel box model

Parameter	Value
ρ_0	10^3 kg m^{-3}
c	$4.2 \times 10^3 \text{ J kg}^{-1} \text{ K}^{-1}$
V_0	$1.6 \times 10^{17} \text{ m}^3$
A	$0.4 \times 10^{14} \text{ m}^2$
ε_T	0.01
T_e^0	300 K
T_p^0	270 K
T_0	285 K
F_p	$10^5 \text{ m}^3 \text{ s}^{-1}$
q_0	$1.6 \times 10^7 \text{ m}^3 \text{ s}^{-1}$
α	$1.8 \times 10^{-4} \text{ K}^{-1}$
β	0.8
S_0	0.035

where the parameters are given by

$$\varepsilon = \frac{F_p}{2q_0}, \qquad \mu = \frac{1}{2}\left(\varepsilon_T + \frac{F_p}{q_0}\right),$$

$$R = \frac{\beta \Delta S}{\alpha \Delta T}, \qquad \kappa = \frac{2k\rho_0 \alpha \Delta T}{q_0}. \tag{2.122}$$

Typical values of these are, from Table 2.1,

$$\varepsilon \sim 0.003, \qquad \mu \sim 0.005, \qquad R \sim 0.5, \tag{2.123}$$

and we can assume without loss of generality that $\kappa = 1$, which fixes the value of q_0 (given k):

$$q_0 = 2k\rho_0 \alpha \Delta T. \tag{2.124}$$

Both ε and μ are small, and we will take advantage of this below.

It is straightforward to analyse (2.121) in the phase plane. Figure 2.23 shows the steady states of q as a function of R when $\mu = 0.005$, $\varepsilon = 0.003$. Neglecting ε and taking μ to be small, we deduce that the steady states are given by

$$q \approx (R-1)^{1/2}, \quad R > 1,$$
$$q \approx -(1-R)^{1/2}, \quad R < 1, \tag{2.125}$$

if $q = O(1)$, and

$$q \approx \pm \frac{\mu R}{1-R}, \quad R < 1, \tag{2.126}$$

Fig. 2.23 Steady states of
(2.125) as a function of R

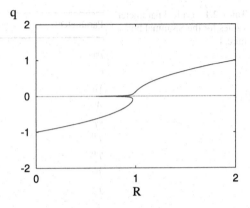

if $q = O(\mu)$. As we might expect, the upper and lower branches are stable, and the middle one is unstable. The upper branch corresponds to present climate, with a northwards circulation at the surface. The stable lower branch corresponds to a reversed haline circulation (thermal buoyancy dominates saline buoyancy because $R < 1$).

Stommel's box model is not an oscillator as such, but it does point out the possibility of multiple convective states of the North Atlantic, and this feature has been found to be robust in other models. What appears to distinguish more realistic models from the Stommel box model is that they allow North Atlantic deep water formation to occur at different latitudes. Thus rather than simply switching from a northerly flow to a southerly one, adjustments can occur between strong northerly flows with deep water formation in the Norwegian sea, and weaker flows with deep water formation further south. It seems that these switches are instrumental in causing the rapid climatic changes during ice ages.

2.6 Snowball Earth

The story of climate on the Earth becomes perhaps more fascinating if we shift our gaze from the relatively recent geologic past to that of more ancient times. There is evidence of glaciation on Earth throughout geologic time, and on all continents. Since the continents move, through the process of plate tectonics, on time scales of hundreds of millions of years, and since their positions and configuration are instrumental in determining ocean circulations and carbon budget (as described below), it seems that plate tectonics is implicated in the long term control of climate.

Recently, one of these periods of glaciation has been at the centre of a scientific controversy concerning what has been picturesquely termed the 'snowball Earth'. In the Neoproterozoic era, between about 750 and 550 million years ago, there was a sequence of glacial episodes. At that time, the Earth's land masses were assembled into a supercontinent called Rodinia, which broke up in a similar way to that in which Gondwanaland fragmented some 200 million years ago.[9] The glaciation of Rodinia would not in itself be surprising, except for the fact that it seems that the supercontinent was located near the equator. It is not impossible for glaciers to exist in equatorial regions at high altitudes (there is an ice cap today on Mount Kilimanjaro in Tanzania), but the suggestion for the Neoproterozoic is that there were widespread ice sheets, and that in fact the land masses were covered with ice. If we suppose also that the oceans were largely ice covered, we see how the concept of a snowball Earth arises.

Although the concept of an ice-covered Earth is entirely consistent with a simple energy-balance model, it is less easy to explain in detail. At that period, the Sun was 6% fainter than today. Model simulations appear able to produce equatorial glaciation providing there is very little CO_2 in the atmosphere, but it is not obvious how to produce such low levels. Nor is it easy to see how to terminate a snowball glaciation.

An interesting idea to explain this latter conundrum is the widespread occurrence of cap carbonate rocks overlying the glacigenic tillites formed from the sub-glacial basal sediments. The idea is that with widespread glaciation and very low temperatures, there would be no water vapour in the atmosphere. Subglacial volcanic eruptions would continue to produce CO_2 however, and with no clouds or water vapour to dissolve it (and rain it out), it would simply build up in the atmosphere. Eventually, the consequent greenhouse effect would cause a rise in temperature, followed by massive deglaciation, moistening of the atmosphere, and thus widespread acid rain. The resulting weathering processes produce the cap carbonate rocks which are seen overlying the glacial tillites.

If this end part of the story is enticing, it is not easy to initiate an equatorial glaciation. One possible way is to allow increased weathering of an equatorial supercontinent (because of tropical climate) which causes reduction of atmospheric CO_2; this then causes the cooling which initiates the glaciation. Once under way, the ice-albedo feedback effect leads to the snowball. Evidently, the whole account relies strongly on the interaction of the carbon cycle with climate. This idea is attractive, because it is widely thought that the onset of the current ice age climate originated with the collision of India and Asia some fifty million years ago. The resulting (and continuing) uplift of the Himalayas resulted in massively increased weathering rates, and therefore reduction of atmospheric CO_2 and consequent planetary cooling. It is thought that the initial growth of the Antarctic Ice Sheet some 34 million years ago is a consequence of this cooling.

2.6.1 The Carbon Cycle

Just as living organisms have a variety of cycles (sleep-wake cycle, menstrual cycle, cell renewal and so on), so the Earth has a number of cycles. Water, rock, topography

---

[9]The problems of plate tectonics are discussed in Chap. 8.

all go through cycles, which we will describe later in this book. There is also a carbon cycle, which we now describe, which is central to plant and animal life, and is also central to the long term control of the Earth's temperature. We have only to look at what has happened on Mars and Venus to see how delicate the control of climate is.

Carbon dioxide is produced as a by-product of volcanism. When mantle rocks melt, some CO_2 is dissolved in the melt, and depressurisation of the ascending magma causes exsolution. This eruptive production adds about 3×10^{11} kg y^{-1} to the atmosphere. On the Earth, water in the atmosphere dissolves the CO_2, forming a weak carbonic acid, and thus when rain falls, it slowly dissolves the silicate rocks of the continental crust. This process is called weathering. One typical reaction describing this dissolution is

$$CaSiO_3 + 2CO_2 + H_2O \longrightarrow Ca^{2+} + 2HCO_3^- + SiO_2 : \qquad (2.127)$$

water dissolves calcium silicate (wollastonite) in the presence of carbon dioxide to form calcium ions, bicarbonate ions and silica. A similar reaction produces magnesium ions.

The ionic species thus produced run off in streams and rivers to the oceans, where the further reaction

$$Ca^{2+} + 2HCO_3^- \longrightarrow CaCO_3 + CO_2 + H_2O \qquad (2.128)$$

creates carbonate sediments. These sink to the ocean floor where they are eventually subducted back into the Earth's mantle. Overall, the pair of reactions (2.127) and (2.128) can be summed to represent

$$CaSiO_3 + CO_2 \longrightarrow CaCO_3 + SiO_2. \qquad (2.129)$$

A very simple model to describe the evolution of the atmospheric CO_2 concentration is then

$$\dot{m}_{CO_2} = -A_L W + v_{CO_2}, \qquad (2.130)$$

where m_{CO_2} is the mass of CO_2 in the atmosphere, A_L is the available land surface for weathering, W is the rate of weathering, and v_{CO_2} is the eruptive production rate of CO_2. It is common practice in discussing CO_2 levels to measure the amount of CO_2 as a pressure, i.e., in bars. The conversion is done by defining the partial pressure of CO_2 as

$$p_{CO_2} = \frac{m_{CO_2} g M_a}{A M_{CO_2}}, \qquad (2.131)$$

where g is gravity, A is total planetary surface area, and M_a and M_{CO_2} are the molecular weights of air and CO_2, respectively. The argument for this is the following. If m_a is the atmospheric mass (of air), then m_a/M_a is the number of moles of air in the atmosphere, while m_{CO_2}/M_{CO_2} is the number of moles of CO_2 in the atmosphere. Then Dalton's law of partial pressures states that

$$\frac{p_{CO_2}}{p_a} = \frac{m_{CO_2} M_a}{M_{CO_2} m_a},$$

(2.132)

and also the atmospheric air pressure p_a is given by

$$p_a = \frac{m_a g}{A}.$$

(2.133)

The current atmospheric carbon mass is around 750 Gt (gigatonnes, 10^{12} kg). Multiplying by the ratio 44/12 of the molecular weights of carbon dioxide and carbon yields the current value of $m_{CO_2} \approx 2.75 \times 10^{15}$ kg. Using $g = 9.81$ m s^{-2}, $A = 5.1 \times 10^{14}$ m^2, $M_a = 28.8 \times 10^{-3}$ kg mole^{-1}, $M_{CO_2} = 44 \times 10^{-3}$ kg mole^{-1}, this converts to a value of $p_{CO_2} = p_0 = 0.35 \times 10^{-3}$ bars, or 35 Pa. In fact, the actual partial pressure of CO_2 in the atmosphere in 2000 was about 36 Pa, or 360 µatm, or 370 ppmv (parts per million by volume) of dry air. It is the latter figure which is commonly reported, and it continues to rise relentlessly.

Weathering Rate

In general we may suppose that $W = W(p_{CO_2}, T, r)$, where T is temperature and r is runoff rate of water to the oceans. This dependence encapsulates the reaction rate of (2.129), and the rate of product removal by runoff. Weathering rates have been measured and range from 0.25×10^{-3} kg m^{-2} y^{-1} in arid regions to 16×10^{-3} kg m^{-2} y^{-1} in the tropics.[10] If we suppose that (2.130) applies in equilibrium, then the consequent current average value would be $W_0 \approx 2.2 \times 10^{-3}$ kg(CO_2) m^{-2} y^{-1}, which appears reasonable. This uses values of $A_L = 1.5 \times 10^{14}$ m^2 and volcanic production rate $v_{CO_2} = 3.3 \times 10^{11}$ kg y^{-1}.[11]

One relation which has been used to represent weathering data is

$$W = W_0 \left(\frac{p_{CO_2}}{p_0} \right)^{\mu} \exp\left[\frac{T - T_0}{\Delta T_c} \right],$$

(2.134)

where $\mu = 0.3$, and the subscript zero represents present day values: thus $T_0 \approx 288$ K, as well as the values of p_0 and W_0 given above. The current value of the

---

[10]The units here are in terms of silica, SiO_2. If we suppose that weathering is described by the reaction (2.129), then one mole of CO_2 (of weight 44 grams) is used to produce one mole of SiO_2 (of weight 60 grams). So to convert units of kg(SiO_2) m^{-2} y^{-1} to units of kg(CO_2) m^{-2} y^{-1}, multiply by $44/60 \approx 0.73$.

[11]The current *net* annual addition of CO_2 to the atmosphere because of fossil fuel consumption and deforestation is about 3.5 Gt carbon, or 1.3×10^{13} kgCO_2 y^{-1}; this is forty times larger than the volcanic production rate. (The actual rate of addition is more than twice as large again, but is compensated by net absorption by the oceans and in photosynthesis.)

Earth's runoff is $r_0 \approx 4 \times 10^{13}$ m^3 y^{-1}, and in general runoff will depend on temperature (by equating runoff to precipitation to evaporation). This dependence is subsumed into the exponential in (2.134). In general, $\partial W/\partial p_{CO_2} > 0$, so that with constant production rate, CO_2 will reach a stable steady state. An inference would be that dramatic variations of climate and CO_2 levels in the past have been due to varying degrees of volcanism or precipitation on altered continental configurations, associated with long time scale plate tectonic processes.

Energy Balance

In seeking to describe how climate may depend on the carbon cycle, we use an energy-balance model. Thus, we combine the ice sheet/energy-balance model (2.102) with (2.130), to find the coupled system for T, a and p_{CO_2}:

$$c\dot{T} = \frac{1}{4}Q(1-a) - \sigma\gamma T^4,$$

$$t_i\dot{a} = a_{eq}(T) - a, \qquad (2.135)$$

$$\frac{M_{CO_2}A}{M_a g}\dot{p}_{CO_2} = -A_L W + v_{CO_2}.$$

We take $a_{eq}(T)$ to be given by (2.100), and W to be given by (2.134).

We model the climatic effect of the greenhouse gases CO_2 and H_2O by supposing that γ depends on p_{CO_2}:

$$\gamma = \gamma_0 - \gamma_1 p_{CO_2}; \qquad (2.136)$$

the value $\gamma_0 < 1$ represents the H_2O dependence, while the small corrective coefficient γ_1 represents the CO_2 dependence.

We have already seen that the response time of T is rapid, about a month, whereas the time scale for albedo adjustment is slower, with the time scale of growth of continental ice sheets being of order 10^4 years. An estimate for the time scale of adjustment of the atmospheric CO_2, based on this model, is

$$t_c = \frac{M_{CO_2}Ap_0}{M_a g v_{CO_2}}. \qquad (2.137)$$

Using values $A = 5.1 \times 10^{14}$ m^2, $p_0 = 36$ Pa, $g = 9.81$ m s^{-2} and $v_{CO_2} = 3.3 \times 10^{11}$ kg y^{-1}, this is $t_c \sim 0.9 \times 10^4$ y, comparable to the ice sheet growth time.

Although (2.135) is a third order system, it is clear that T relaxes rapidly to a well-defined 'slow manifold'

$$T \approx T(a, p_{CO_2}) = \left[\frac{Q(1-a)}{4\sigma\gamma(p_{CO_2})}\right]^{1/4}, \qquad (2.138)$$

on which the dynamics are governed by the slower a and p equations. The nullclines in the (a, p) phase plane are shown in Fig. 2.24. The a nullcline is multivalued for

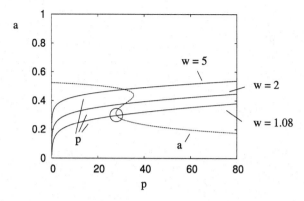

Fig. 2.24 a and p nullclines for (2.135) assuming that T has rapidly equilibrated to $T(a, p_{CO_2})$. The three curves occur for the values of the weathering coefficient $w = 1.08$, 2 and 5, corresponding to pre-industrial climate, oscillatory ice ages and snowball Earth. The parameters for a_{eq} are the same as those in Fig. 2.13, and other values used are $p_0 = 36$ Pa, $\mu = 0.3$, $\Delta T_c = 13$ K, $T_0 = 288$ K, $\sigma = 5.67 \times 10^{-8}$ W m^{-2} K^{-4}, $Q = 1370$ W m^{-2}, $\gamma_0 = 0.64$, $\gamma_1 = 0.8 \times 10^{-3}$ Pa^{-1}. The *circle* marks the value $p = 28$ Pa, $a = 0.3$, which corresponds to pre-industrial climate

the same reason that Fig. 2.15 indicates multiplicity, since both graphs are described by the same equations, the only difference being that p_{CO_2} (and thus γ) is used rather than Q. The horizontal axis of Fig. 2.15 could equally be taken to be Q/γ and thus (for fixed Q) p_{CO_2}.

The analysis of this model is indicated in Question 2.16. The solutions depend on the two critical dimensionless parameters

$$w = \frac{A_L W_0}{v_{CO_2}}, \qquad \delta = \frac{v_{CO_2} g t_i M_a}{A p_0 M_{CO_2}}, \tag{2.139}$$

which are measures of weathering rate and volcanic production. These can vary depending on current tectonic style. The three indicated intersection points in Fig. 2.24 correspond to steady states at low (current), intermediate and high weathering rates (relative to volcanic output). The solution on the upper branch indicates a snowball at enhanced weathering rates, as might be expected when the continental land masses are clustered at the equator, promoting tropical climate. Upper and lower branch solutions are stable, but the intermediate solution is oscillatorily unstable if δ is sufficiently small. If δ is very small, then the motion becomes relaxational.

Figure 2.25 shows an oscillatory solution illustrating this discussion. The corresponding time series is shown in Fig. 2.26. It does not look much like the sawtooth oscillation of the Pleistocene ice ages, and the period is too long, some half million years. No doubt one can find something more persuasive by fiddling with parameters, but it is probably not worth the effort, given the enormous simplicity of the model. The main use of the model is to illustrate the point that the carbon cycle contains a feedback effect which is capable of generating self-sustaining oscillations.

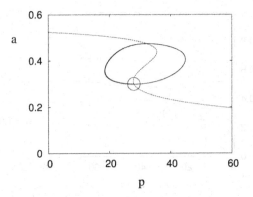

Fig. 2.25 Limit cycle oscillation which passes through pre-industrial climate (indicated by the *circle*) ($T = 288$ K, $a = 0.3$, $p_{CO_2} = 28$ Pa). Also shown is the a nullcline of (2.135). The parameters are those of Fig. 2.24, with $t_i = 10^4$ y, $A = 5.1 \times 10^{14}$ m^2, $A_L = 1.5 \times 10^{14}$ m^2, $g = 9.81$ m s^{-2}. The temperature is taken to be the quasi-equilibrium value of (2.138), and the weathering and eruption rates are taken to be $W_0 = 0.211 \times 10^{-3}$ kg m^{-2} y^{-1}, $v_{CO_2} = 0.15 \times 10^{11}$ kg y^{-1}. With these values, the parameters in (2.139) are $w = 2.11$ and $\delta = 0.0525$. In solving the equations, we take $c = 10^{11}$ J m^{-2} K^{-1}, and thus $\varepsilon = 0.012$, in order to avoid the necessity for impossibly small time steps

Fig. 2.26 Time series of temperature for the periodic oscillation of Fig. 2.25

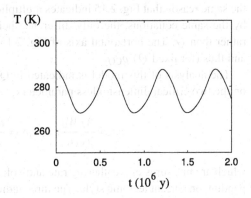

2.6.2 *The Rôle of the Oceans*

The major simplification which has been made in the above discussion is that we have ignored the part played by the oceans. The oceans hold a good deal more carbon than the atmosphere, although the concentration (as volume fraction) is comparable. They thus act as a buffering mechanism for alterations to atmospheric CO$_2$.

The oceans also play an important part in another dramatic feature of the Earth's climate, which is indicated by a comparison of the proxy temperature in Fig. 2.12 with a similar graph of atmospheric CO$_2$ content (see Fig. 2.27). Apart from variations due to noise, the two graphs are essentially the same, which would seem to indicate that ice ages are caused by oscillations in CO$_2$, since we know that global temperature responds promptly to changes in CO$_2$. This would be consistent with

Fig. 2.27 Variation of CO_2 (*upper graph*, units ppmv) with proxy temperature as in Fig. 2.12; horizontal time scale in thousands of years. The deuterium isotope values have been scaled and shifted vertically to point out their resemblance to the CO_2 values

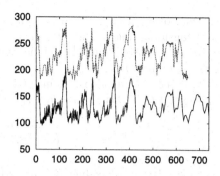

the discussion above. However, temperature also has a direct effect on CO_2, because of respiration of CO_2 by plankton in the oceans, which depends on temperature, and also since the solubility of CO_2 in ocean surface waters depends on temperature; these two facts could then be the mechanism whereby CO_2 conforms with temperature.

A very simple representation of the ocean buffering effect is to add an ocean carbon compartment to the model equations (2.135). Let us denote the concentration of (dissolved inorganic) carbon (not carbon dioxide) in the bulk ocean by C, with units of $mol\,kg^{-1}$. This will be different from the surface value, which we denote by C_s, and this difference induces a transport of CO_2 into the ocean proportional to $C_s - C$. To relate this flux to p_{CO_2}, we need to understand the ideas of solubility and of acid–base buffering.

Acid–base buffering, which we analyse further below, describes the partitioning of carbon between dissolved carbon dioxide, bicarbonate, and carbonate ions. Most of the carbon in the oceans resides in the bicarbonate ion reservoir, which is partially maintained by the reaction

$$H_2O + CO_2 \rightleftharpoons HCO_3^- + H^+, \tag{2.140}$$

and from this we find that the CO_2 concentration in the ocean is related to the total dissolved inorganic carbon (DIC) concentration by a partition equation of the form

$$[CO_2] \approx \frac{C}{\left(1 + \frac{K_1}{[H^+]}\right)}, \tag{2.141}$$

where K_1 is the equilibrium constant of (2.140), and [A] denotes the concentration of A.

Next, thermodynamic equilibrium between the CO_2 in the surface waters and the atmosphere is determined by Henry's law, which relates oceanic concentration of CO_2 to the CO_2 partial pressure in the atmosphere. Henry's law takes the form

$$[CO_2]_s = K_H p_{CO_2}, \tag{2.142}$$

where the subscript s denotes the ocean surface value. The solubility K_H decreases with increasing temperature, and has units of $mol\,kg^{-1}\,atm^{-1}$ (1 atm $= 10^5$ Pa).

Because of the temperature dependence, polar surface waters contain more CO_2 than equatorial waters. The value of K_H in saline oceanic water at 20°C is around 3.5×10^{-2} mol kg^{-1} atm^{-1} (and the value decreases by a factor of about 2.5 between 0°C and 30°C).

With these assumptions, we may take the flux of CO_2 from the atmosphere (in units of kgCO$_2$ y^{-1}) to be $q = h(p_{CO_2} - \frac{C}{K})$, where

$$K = K_H \left[1 + \frac{K_1}{[H^+]} \right], \qquad (2.143)$$

and then the corresponding flux of carbon to the ocean DIC compartment (in mol y^{-1}) is $\frac{q}{M_{CO_2}}$.

In addition, the ocean loses carbon due to the biological pumping effect of carbon uptake by phytoplankton, and its subsequent deposition as organic carbon particles. We take this rate to be bC. The coefficient b will also depend on temperature, and increases with T, due to increased metabolic rates at higher temperatures. A single compartment model for the ocean CO_2 fraction is then

$$\rho_{H_2O} V_{oc} \dot{C} = \frac{1}{M_{CO_2}} \left\{ h \left(p_{CO_2} - \frac{C}{K} \right) + A_L W \right\} - bC, \qquad (2.144)$$

where V_{oc} is ocean volume and ρ_{H_2O} is water density. The dynamics of this extended model are the subject of Question 2.17. The essential result is that the atmospheric CO_2 partial pressure follows the ocean DIC concentration, which itself changes on a longer time scale of about 160,000 years. Bistability and even oscillations are still possible, in particular if the biopump coefficient b decreases with increasing ice volume (and thus a). The mechanism whereby this can occur is an interesting one. As ice sheets grow, the continental shelves are exposed, and the biopump effect due to shallow biomass such as coral reefs is removed. It is a noteworthy fact that in the Pleistocene, the ice sheet maximum extent is such that sea level is lower by some 120 m, thus exposing a significant portion of the continental shelves. It is also noteworthy that the snowball Earth class of ice age is associated with pre-Cambrian periods, when hard-shelled creatures did not exist, and the biopump was thus largely absent.

2.6.3 Ocean Acidity

There are a number of chemical constituents of the ocean (most obviously, salt) which affect the acidity of the ocean. But by far the most important of these is the acid–base buffering system of carbon, which is also instrumental in determining the magnitudes of the different carbon reservoirs. In order to determine the partition coefficient in (2.141), it is thus necessary to consider acid–base buffering.

The two principal reactions involve the dissolution of carbon dioxide gas in water to form carbonic acid, which then dissociates to hydrogen ion (acid) H$^+$ and bicarbonate ion HCO_3^-. The bicarbonate further dissociates to carbonate ion CO_3^{2-} and acid. The reactions involved are

$$H_2O + CO_2 \underset{k_{-1}}{\overset{k_1}{\rightleftharpoons}} HCO_3^- + H^+,$$

$$HCO_3^- \underset{k_{-2}}{\overset{k_2}{\rightleftharpoons}} CO_3^{2-} + H^+,$$

$$(2.145)$$

and the corresponding rate equations, based on the law of mass action, are

$$[\dot{CO_2}] = -R_1,$$

$$[\dot{HCO_3^-}] = R_1 - R_2,$$

$$[\dot{H^+}] = R_1 + R_2,$$

$$[\dot{CO_3^{2-}}] = R_2$$

$$(2.146)$$

(we subsume the essentially constant water fraction into the forward rate k_1); square brackets denote concentrations, with units of mol kg^{-1}. The reaction rates are given by

$$R_1 = k_1[CO_2] - k_{-1}[HCO_3^-][H^+],$$

$$R_2 = k_2[HCO_3^-] - k_{-2}[CO_3^{2-}][H^+].$$

$$(2.147)$$

These reactions are fast, and equilibrate in a few minutes. Thus we may take $R_1 = R_2 = 0$, whence we find the buffering relationships

$$[HCO_3^-][H^+] = K_1[CO_2],$$

$$[CO_3^{2-}][H^+] = K_2[HCO_3^-],$$

$$(2.148)$$

where

$$K_1 = \frac{k_1}{k_{-1}}, \qquad K_2 = \frac{k_2}{k_{-2}}. \qquad (2.149)$$

Two further relationships are necessary to determine all four concentrations, and these arise from the two independent conservation laws which can be formed from summation of constituents of (2.146). One such is for the total dissolved inorganic carbon (DIC):

$$[CO_2] + [HCO_3^-] + [CO_3^{2-}] = C, \qquad (2.150)$$

which we obtain by adding the first, second and fourth of (2.146), and which represents conservation of carbon. In generally, we would also have conservation of oxygen and conservation of hydrogen, but these are not available here since we have not included H_2O as an independent substance. Instead, we can appeal to conservation of charge, which implies (assuming a zero constant of integration)

$$[HCO_3^-] + 2[CO_3^{2-}] = [H^+], \tag{2.151}$$

and which is evidently an independent conservation law from (2.146).

The conservation of charge equation (2.151) can be used to determine $[H^+]$, wrongly as it will turn out, but the other three equations retain their practical validity when more ionic species are included. In terms of $[H^+]$ we find, using (2.148) and (2.150),

$$[CO_2] = \frac{C}{1 + \frac{K_1}{[H^+]} + \frac{K_1 K_2}{[H^+]^2}}, \tag{2.152}$$

and this is where we obtain (2.143), since the term in K_2 is reasonably small.

We mainly leave it as an exercise to carry through the calculation of $[H^+]$ based on the charge equation (2.151). From (2.148), (2.150) and (2.151), we can derive the cubic equation

$$\xi^3 + \lambda_1 \xi^2 - \lambda_1 (1 - \lambda_2)\xi - 2\lambda_1 \lambda_2 = 0, \tag{2.153}$$

where

$$\xi = \frac{[H^+]}{C}, \qquad \lambda_1 = \frac{K_1}{C}, \qquad \lambda_2 = \frac{K_2}{C}. \tag{2.154}$$

We use values $C = 2 \times 10^{-3}$ mol kg^{-1}, $K_1 = 1.4 \times 10^{-6}$ mol kg^{-1}, $K_2 = 1.07 \times 10^{-9}$ mol kg^{-1}, and from these we find

$$\lambda_1 \approx 0.7 \times 10^{-3}, \qquad \lambda_2 \approx 0.53 \times 10^{-6}. \tag{2.155}$$

It is easy to show that for positive λ_1 and λ_2, there is precisely one positive solution, and taking $\lambda_2 \sim \lambda_1^2 \ll 1$, this is given approximately by $\xi = \sqrt{\lambda_1}$, and thus

$$[H^+] \approx \sqrt{K_1 C}. \tag{2.156}$$

Using the values given for K_1 and C above, we would compute the pH of seawater to be

$$\text{pH} = -\log_{10}[H^+] \approx 4.28. \tag{2.157}$$

The actual pH of seawater is 8.2, slightly alkaline, whereas this calculation suggests a strongly acid ocean! Thus in using (2.152), we use the actual value of $[H^+] = 0.63 \times 10^{-8}$ mol kg^{-1}. Note that since K_2 is an order of magnitude smaller than this, it is reasonable to neglect the K_2 term in (2.152).

The dynamics of the acid–base buffering system are discussed further in Question 2.18, where the alkalinity of the oceans is indicated as being due to the rôle of calcium carbonate in the carbon buffering system.

2.7 Notes and References

A good, recent book which addresses most of the issues of concern in this chapter is the book by Ruddiman (2001), which provides an expert's view. The book is aimed at undergraduates, and is very accessible. It is only marred by an addiction to design and graphics, which makes the book expensive and rather over the top—it is a book where there is a production team. Despite this, it is very up to date and informative.

Radiative Heat Transfer The classic treatise on radiative heat transfer is the book by Chandrasekhar (1960), although it is dated and not so easy to follow. A more recent book aimed at engineers is that by Sparrow and Cess (1978). Most books on atmospheric physics will have some material on radiative heat transfer, for example those by Houghton (2002) and Andrews (2000). Other books are more specialised, such as those by Liou (2002) and Thomas and Stamnes (1999), but are not necessarily any easier to follow.

Rayleigh scattering is described by Strutt (1871), J. W. Strutt being Lord Rayleigh's given name.

The Ozone Layer The description of the ozone layer dynamics essentially follows Chapman (1930). An elegant exposition is in the book by Andrews (2000). Reality is of course more complicated than the version presented here, and many more reactions can be included, in particular involving catalytic cycles, in which various chemical species catalyse the conversion of ozone to oxygen.

Chlorine species created by man-made chlorofluorocarbons have been implicated in the destruction of stratospheric ozone in the Antarctic, with the formation of the well known 'ozone hole' (Solomon 1999).

Energy Balance Models The original energy-balance models are due to Budyko (1969) and Sellers (1969). They differ essentially only in the choice of parameterisation of emitted long-wave radiation, and consider only the global balance of energy. North (1975a) allows latitude dependent albedo, and additionally allows for a parameterisation of poleward heat transport by oceans and atmosphere through a diffusive term, as in (2.68). North (1975b) added the time derivative. These meridionally averaged energy-balance models do a rather good job of simulating the mean latitude dependent temperature profile, and have formed the basis for the atmospheric component of the more recent models of 'intermediate complexity'. A later review is given by North et al. (1983).

The Greenhouse Effect The first person who is generally credited with discussing the greenhouse effect is Arrhenius (1896), but Arrhenius himself refers to an earlier discussion by Fourier in 1827, where he refers to the atmosphere acting like the glass of a hothouse. Arrhenius's assessments of the effect of CO_2 are rather more severe than today's considered opinion. For a more recent discussion, see Houghton et al. (1996).

Ice Ages   The data shown in Figs. 2.17–2.21 are taken from the GRIP (Greenland ice core project) ice core, drilled through the central part of the Greenland ice sheet. These data are provided by the National Snow and Ice Data Centre of the University of Colorado at Boulder, Colorado, and the World Data Centre–A for Paleoclimatology at the National Geophysical Data Centre, also in Boulder, Colorado. This and other such data are publicly available at http://www.ngdc.noaa.gov/paleo/icecore/greenland/summit/index.html and have been reported in a number of publications, for example Johnsen et al. (1992), who, in particular, describe Dansgaard–Oeschger events. The higher resolution data sets in Figs. 2.20 and 2.21 were provided by Sigfus Johnsen, through the agency of Eric Wolff.

Abrupt climate change is documented by Severinghaus and Brook (1999) and Taylor et al. (1997). Taylor et al. (1993) find evidence of rapid ice age climate change in measurements of dust content in ice cores.

The cooling event at 8,200 years B. P. is described by Alley et al. (1997); Leuenberger et al. (1999) calibrate the temperature scale indicated by oxygen isotope variation by studying nitrogen isotope variations, suggesting that the cooling at 8,200 B. P. was of the order of 7 K; see also Lang et al. (1999).

Heinrich Events   Heinrich events were first described in North Atlantic deep-sea sediment cores by Heinrich (1988). MacAyeal (1993) introduced his 'binge-purge' model to explain them as a consequence of ice sheet oscillations induced by thermal instability, but assumed that a melting base would automatically cause large ice velocities. Fowler and Schiavi (1998) proposed a more physically realistic model which introduced the concept of hydraulic runaway, and Calov et al. (2002) showed that large scale ice sheets could oscillate somewhat as these earlier studies suggested, using a climate model of 'intermediate complexity' (essentially resolved oceans and ice sheets, and an averaged energy-balance model of atmospheric fluxes).

Dansgaard–Oeschger Events   Rahmstorf (2002) gives a nice review of the interplay of oceans and ice sheets in causing climatic oscillations during the last ice age. Ganopolski and Rahmstorf (2001) show how fluctuating freshwater delivery to the North Atlantic can cause abrupt alterations in circulation. The idea that the freshwater pulses might be due to sub-Laurentide jökulhlaups was voiced by Evatt et al. (2006), while similar floods beneath the Antarctic have been described by Goodwin (1988), Wingham et al. (2006) and Fricker et al. (2007).

Oceans and Climate   Stommel (1961) introduced the idea of different possible North Atlantic circulations. His model is not too realistic, but nevertheless simple and compelling. Rahmstorf (1995) uses a model of intermediate complexity to examine multiple circulation patterns in the North Atlantic. Depending on the freshwater flux to the North Atlantic, he finds hysteretic switches between different possible flows. Stocker and Johnsen (2003) provide a more recent addition to the subject. Ganopolski and Rahmstorf (2001) provide a convincing picture of how switches of ocean circulation can cause rapid climate change. Their intermediate complexity model indicates hysteretic switches in ocean circulation due to changes in freshwater flux to the North Atlantic of unknown origin; we have suggested that the origin could be periodic sub-glacial floods. Broecker et al. (1990) and Manabe and Stouffer (1995) provide a similar thesis.

A great advocate of the 1000–2000 year rhythm in climate was Gerard Bond; for example Bond et al. (1999) describe this rhythm, and also suggest that it has continued beyond the end of the ice age (into the Holocene), its most recent manifestation being the little ice age of 1500–1900. See also Bond et al. (1997).

Snowball Earth   The idea of a snowball Earth is discussed by Hoffman et al. (1998), for example, although the idea of ancient glaciations had been extant for a long time before that (Harland 1964, 2007). Various modelling efforts have been made to assess the snowball's viability, for example, see Crowley and Baum (1993), Hyde et al. (2000), Chandler and Sohl (2000), and Pierrehumbert (2004).

The Carbon Cycle   Our (too) simple model of the interaction of the carbon cycle with ice sheet growth and climate change is based on the discussion of Walker et al. (1981), although their emphasis was on the rôle of CO_2 as a buffer in stabilising climate over geological time, despite the increasing solar luminosity. These ideas are elaborated by Kasting and Ackermann (1986) and Kasting (1989), who consider the effects of very large atmospheric CO_2 concentrations in early Earth history. Kasting (1989) suggests that because of the buffering effect of CO_2, a terrestrial (i.e., with liquid water) planet could be viable out as far as the orbit of Mars. In view of the plentiful evidence of water on Mars in *its* early history, this raises the intriguing prospect of a hysteretic switch from early temperate Mars to present cold Mars.

The buffering effect of CO_2 on climate and the rôle of continental location is discussed by Marshall et al. (1988). Berner et al. (1983) and Lasaga et al. (1985) discuss more complicated chemical models of weathering, and their effect on atmospheric CO_2 levels.

Petit et al. (1999) document the close relation between CO_2 levels and atmospheric temperature over the past 400,000 years, and Agustin et al. (2004) extend this further back in time, as shown in Fig. 2.27. Unlike the result in Fig. 2.25, which together with (2.138), indicates that temperature and CO_2 will vary independently, the data show that there is an excellent match. The model could be made more consistent with this observation if the relaxation time t_i were to be reduced. And indeed, this would not be unreasonable, since the change of albedo due to sea ice coverage will be very fast, and this will shift the effective albedo time scale downwards.

However, it is currently thought that it is the buffering rôle of the oceans which is principally involved in explaining the short term correlation of CO_2 with temperature, and its variation through the ice ages (Toggweiler et al. 2006; Köhler and Fischer 2006). The single compartment model we propose in (2.144) may be the simplest additional complication to add to the basic lumped energy-balance models, but it falls well short of currently fashionable models, which include separate compartments for shallow and deep waters, as well as different compartments for the different oceans (Munhoven and François 1996; Köhler et al.

2005). The apparent necessity for such complexity is to allow a description of the solubility pump (hence the effect of latitude) and the biological pump (hence the effect of depth). The convective interchange between the different compartments occurs on a time scale of several hundred years, consistent with the observation that changes in temperature (at glacial terminations) actually lead the changes of CO_2 by a similar time scale. The surface layers take up CO_2 from the atmosphere, and this is transported to the deep ocean via the global oceanic circulation in the North Atlantic and the Southern Oceans, in particular. Because of the CO_2 solubility dependence on temperature, decreasing temperatures cause an increased flux to the ocean, reducing atmospheric CO_2, and thus providing a positive feedback.

Takahashi et al. (2002) describe the surface flux of CO_2 to the ocean, currently estimated as about 2 PgC y^{-1} (i.e., petagrams, 10^{15} g $= 10^{12}$ kg; 1 kgC corresponds to $10^3/12$ mole C, and if this resides in $10^3/12$ mole CO_2, then this is $44/12 = 3.67$ kg CO_2.

The carbon cycle, and the interchange of CO_2 between ocean and atmosphere, are described in the books by Bigg (2003), Emerson and Hedges (2008) and Krauskopf and Bird (1995). A description together with data on equilibrium constants and solubility is given by Millero (1995). A very useful source is the book by Zeebe and Wolf-Gladrow (2001). While the basic reaction scheme of the carbon buffering system (2.145) is easy enough to understand, it is less easy to get a clear understanding of how the two extra conserved quantities (as for instance (2.150) and (2.151)) should be prescribed. Standard practice (Millero 1995) seems to be simply to measure for example the dissolved inorganic carbon

$$C = [HCO_3^-] + [CO_2] + [CO_3^{2-}], \qquad (2.158)$$

and the carbonate alkalinity

$$A = [HCO_3^-] + 2[CO_3^{2-}], \qquad (2.159)$$

and from these one can calculate the other concentrations in the system, and thus also the pH. This is the strategy adopted by Emerson and Hedges (2008), and also in Question 2.18. An important addition to the carbon buffering system (2.145) is the dissolution of calcium carbonate, and the total carbon is then determined by this dissolution, as well as by transport from the atmosphere and loss via biological pumping. Charge neutrality does not apply, because of the many other ionic species present. Quite how charge should be determined is not very clear from a theoretical perspective, but the prescription of the (measurable) alkalinity circumvents the necessity of doing this. Krauskopf and Bird (1995, p. 68) do provide a calculation of pH of seawater (they obtain a value of 8.4) based essentially on carbon/calcium conservation (their Eq. (3-11); $P = 0$ in Question 2.18) and on charge neutrality (their Eq. (3-12); $Q = 0$ in Question 2.18). But they also use other observed values, and the calculation is hard to follow. Question 2.18 indicates that these assumptions are not correct, however.[12]

---

[12]In fact the corrected approach is to assume charge neutrality, but allowing for the net negative charge of the conservative ions: chloride, sodium, etc.

2.8 Exercises

2.1 The planetary albedos of Venus, Mars and Jupiter are 0.77, 0.15, 0.58, respectively, and their distances from the Sun are 0.72, 1.52, 5.20 *astronomical units* (1 a.u. = distance from Earth to the Sun). Calculate the equilibrium temperature of these planets, and compare them with the measured effective black body temperatures, $T_m = 230$ K, 220 K, 130 K. Which, if any, planets appear not to be in equilibrium; can you think why this might be so?

2.2 Show that $\int_{\Omega} \cos\theta \, d\omega = \pi$, and deduce that $E_{b\nu} = \pi B_{\nu}$, where $E_{b\nu}$ is the black body radiation emitted normally from a surface, per unit area.

Use Planck's law

$$B_{\nu}(T) = \frac{2h\nu^3}{c^2[e^{h\nu/kT} - 1]}$$

to derive the Stefan–Boltzmann law in the form

$$E = \int_0^{\infty} E_{b\lambda} \, d\lambda = \sigma T^4,$$

where

$$\sigma = \frac{2\pi k^4}{c^2 h^3} \int_0^{\infty} \frac{z^3 \, dz}{e^z - 1}.$$

By evaluating the integral and using the values $c = 2.998 \times 10^8$ m s^{-1}, $k = 1.381 \times 10^{-23}$ J K^{-1}, $h = 6.626 \times 10^{-34}$ J s, evaluate the Stefan–Boltzmann constant σ.

Hint:

$$\sum_1^{\infty} \frac{1}{n^4} = \frac{\pi^4}{90}.$$

2.3 In a one-dimensional atmosphere, show that the average intensity is given by

$$J = \frac{1}{2} \int_{-1}^{1} I(\tau, \mu') \, d\mu',$$

and show also that if the energy flux vector is

$$\mathbf{q}_R = \int_O I(\mathbf{r}, \mathbf{s}) \mathbf{s} \, d\omega(\mathbf{s}),$$

then for a grey atmosphere

$$\nabla \cdot \mathbf{q}_R = -4\pi \kappa \rho [J - B].$$

Deduce that in radiative equilibrium, $J = B$.

2.4 For a purely absorptive atmosphere, show, by interpreting the radiation intensity along a ray path as a probability distribution function for the photon free path length (before absorption), that the mean free path is $1/\rho\kappa_v$. Deduce that an optically thin layer is one for which the photon mean free path is larger than the layer thickness.

2.5 In a purely scattering atmosphere, emission occurs by the scattering of radiation in all directions. Suppose that for a beam of intensity I_v, the loss in intensity in a distance ds due to scattering is $\kappa_v I_v\,ds$, of which a fraction $P_v(\mathbf{s}, \mathbf{s}')\,d\omega(\mathbf{s}')/4\pi$ is along a pencil of solid angle $d\omega(\mathbf{s}')$ in the direction $\mathbf{s}'$. Explain why it is reasonable to suppose that the scattering function P_v should depend only on $\mathbf{s}.\mathbf{s}'$, and show that the equation of radiative transfer can be written (assuming a grey atmosphere)

$$\frac{\partial I}{\partial s} = \rho\kappa\left[-I + \frac{1}{4\pi}\int_O P(\mathbf{s}, \mathbf{s}')I(\mathbf{r}, \mathbf{s}')\,d\omega(\mathbf{s}')\right].$$

Deduce that for isotropic scattering, where $P \equiv 1$, the radiative flux $\mathbf{q}_R$ (see Question 2.3) is divergence free.

For a plane parallel atmosphere in which $I = I(\tau, \mu)$, show that

$$\mu\frac{\partial I}{\partial \tau} = I - \frac{1}{4\pi}\int_{-1}^{1}\int_{0}^{2\pi} P(\mathbf{s}.\mathbf{s}')I(\tau, \mu')\,d\mu'\,d\phi',$$

where ϕ' is the azimuthal angle associated with $\mathbf{s}'$. Use spherical polar coordinates to show that

$$\mathbf{s}.\mathbf{s}' = \mu\mu' + \left(1 - \mu^2\right)^{1/2}\left(1 - \mu'^2\right)^{1/2}\cos(\phi - \phi'),$$

and deduce that for Rayleigh scattering, where $P(\cos\Theta) = \frac{3}{4}(1 + \cos^2\Theta)$, I satisfies

$$\mu\frac{\partial I}{\partial \tau} = I - \frac{3}{16}\left[3I_0 - I_2 - \mu^2(I_0 - 3I_2)\right],$$

where $I_0 = \int_{-1}^{1} I\,d\mu$, $I_2 = \int_{-1}^{1}\mu^2 I\,d\mu$.

2.6 By non-dimensionalising the radiative heat transfer equation for a grey atmosphere using a length scale d (atmospheric depth) and an appropriate radiation intensity scale, show that in the optically thick limit, the equation takes the dimensionless form

$$I = B - \varepsilon\,\mathbf{s}.\nabla I,$$

where $\varepsilon \ll 1$ and should be specified. Find an approximate solution to this equation, and hence show that the (dimensional) radiative energy flux vector $\mathbf{q}_R$ is given approximately by

$$\mathbf{q}_R = -\frac{4\sigma}{3\kappa\rho}\nabla T^4.$$

2.7 The equation of radiative transfer in a grey, one-dimensional atmosphere is given by

$$\mu \frac{\partial I}{\partial \tau} = I - B,$$

with $I = 0$ at $\tau = 0$, $\mu < 0$, and $I = B_S \equiv B(\tau_S)$ at $\tau = \tau_S$, $\mu > 0$. Write down the formal solution assuming B is known, and hence show that the radiative flux $q_R = 2\pi \int_{-1}^{1} \mu I \, d\mu$ is given by

$$q_R = 2\pi \left[-\int_0^\tau B(\tau') E_2(\tau - \tau') \, d\tau' + B_S E_3(\tau_S - \tau) \right.$$

$$\left. + \int_\tau^{\tau_S} B(\tau') E_2(\tau' - \tau) \, d\tau' \right],$$

where the exponential integrals are defined by

$$E_n(z) = \int_1^\infty \frac{e^{-zt} \, dt}{t^n},$$

and $B_S \equiv B(\tau_S)$.

Show that $E_n' = -E_{n-1}$, $E_n(0) = \frac{1}{n-1}$, and deduce that

$$\frac{\partial q_R}{\partial \tau} = 2\pi \left[-2B + B_S E_2(\tau_S - \tau) + \int_0^{\tau_S} B(\tau') E_1(|\tau - \tau'|) \, d\tau' \right].$$

Show also that the intensity $J = \frac{1}{2} \int_{-1}^{1} I \, d\mu$ is given by

$$J = \frac{1}{2} \left[\int_0^{\tau_S} B(\tau') E_1(|\tau - \tau'|) \, d\tau' + B_S E_2(\tau_S - \tau) \right].$$

By integrating the expression for q_R by parts, show that

$$q_R = 2\pi \left[B_0 E_3(\tau) + \int_0^{\tau_S} B'(\tau') E_3(|\tau - \tau'|) \, d\tau' \right]. \qquad (*)$$

If τ_S is large, so that B varies slowly with τ, show that when τ is large,

$$q_R \approx \frac{4\pi}{3} B'(\tau)$$

(essentially, this uses Laplace's method for the asymptotic evaluation of integrals).

Use the integral expression $(*)$ for q_R to show that if $q_R = \pi B_0$ at $\tau = 0$, then

$$\int_0^{\tau_S} B'(\tau') E_3(\tau') \, d\tau' = 0,$$

and deduce that the temperature gradient cannot be monotonic for such an atmosphere.

2.8 Chapman's model for the production of ozone in the stratosphere is

$$O_2 + h\nu \xrightarrow{j_2} 2O,$$

$$O + O_2 + M \xrightarrow{k_2} O_3 + M,$$

$$O_3 + h\nu \xrightarrow{j_3} O + O_2,$$

$$O + O_3 \xrightarrow{k_3} 2O_2.$$

Write down the rate equations for the concentrations X, Y and Z of oxygen atoms O, oxygen O_2 and ozone O_3, and show that

$$X + 2Y + 3Z = 2[O_2],$$

where $[O_2]$ is constant.

Suppose, as is observed, that $X \ll \frac{j_3}{k_3}$ and $Y \gg \frac{j_3}{k_2[M]}$, where $[M]$ is the concentration of M. Use these observations to scale the equations to the form

$$\varepsilon \frac{dx}{dt} = z - xy + 2\delta y - \delta xz,$$

$$\frac{dz}{dt} = xy - z - \delta xz,$$

$$y + \frac{1}{2}\lambda(3z + \varepsilon x) = 1,$$

where

$$\varepsilon = \frac{j_3}{k_2[O_2][M]}, \qquad \delta = \left[\frac{j_2 k_3}{j_3 k_2[M]}\right]^{1/2}, \qquad \lambda = \left[\frac{j_2 k_2[M]}{j_3 k_3}\right]^{1/2}.$$

Assuming $\varepsilon, \delta, \lambda \ll 1$, show that the model can be partially solved to produce the approximate equation

$$\frac{dz}{d\tau} = 2(1 - z^2),$$

where $t = \tau/\delta$.

Hence show that $[O_3] \to \lambda[O_2]$ on a time scale $t \sim (\frac{k_2[M]}{j_2 j_3 k_3})^{1/2}$, and that the reaction scheme can be represented by the overall reaction

$$3O_2 \underset{r_-}{\overset{r_+}{\rightleftharpoons}} 2O_3.$$

where

$$r_+ = \frac{2}{3} j_2, \qquad r_- = \frac{j_3 k_3}{k_2 [O_2][M]}.$$

2.9 Suppose that stratospheric heating by absorption of ultraviolet radiation is given by

$$Q = -\frac{\partial I}{\partial z},$$

where

$$I = -I_\infty \exp\left[-\tau_0 e^{-z/H}\right],$$

and

$$\tau_0 = \kappa \rho_0 H, \qquad \tau_c = \kappa \rho_c H.$$

Suppose also that the (upwards) long-wave radiative flux is given by

$$q_R = -k_R \frac{\partial T}{\partial z},$$

where the radiative conductivity is given by

$$k_R = \frac{16 \sigma T^3 e^{z/H}}{3 \kappa \rho_c}.$$

Write down the energy equation describing radiant energy transport, and show that the temperature T is given by

$$T = T_0 \left[A - \phi e^{-\zeta} - \theta \exp\left(-\tau_0 e^{-\zeta}\right) \right]^{1/4},$$

where A and ϕ are constants, and

$$T_0 = \left(\frac{3 I_\infty}{4\sigma}\right)^{1/4}, \qquad \theta = \frac{\tau_c}{\tau_0}, \qquad \zeta = \frac{z}{H}.$$

Suppose that $\phi, \theta, A \sim O(1)$, and that $\tau_0 \gg 1$. Find approximations for T for $\zeta < \ln \tau_0$ and $\zeta \sim \ln \tau_0$, and deduce that T has a maximum at $z \sim H \ln \tau_0$. How is this discussion related to estimation of the temperature in the stratosphere?

2.10 Using values $d = 10$ km, $\kappa \rho d = 0.67$, show that a representative value of the radiative conductivity k_R defined by $q_R = -k_R \nabla T$ for an opaque atmosphere is $k_R \simeq 1.08 \times 10^5$ W m^{-1} K^{-1}. Hence show that a typical value for the effective Péclet number

$$Pe = \frac{\rho c_p U d^2}{k_R l}$$

is about 20, if $U \approx 20$ m s^{-1}, $l \approx 1000$ km. Explain the implication of this in terms of the heat equation

$$\rho c_p \frac{dT}{dt} = \nabla.[k_R \nabla T].$$

2.11 A wet adiabat is calculated from the isentropic equation

$$\rho_a c_p \frac{dT}{dz} - \frac{dp}{dz} + \rho_a L \frac{dm}{dz} = 0,$$

where

$$m = \frac{\rho_v}{\rho_a}, \qquad p = \frac{\rho_a RT}{M_a}, \qquad p_{SV} = \frac{\rho_v RT}{M_v},$$

and

$$\frac{dp_{SV}}{dT} = \frac{\rho_v L}{T}, \qquad \frac{dp}{dz} = -\rho_a g.$$

Deduce that T and p_{SV} can be calculated from the equations

$$\frac{dT}{dz} = -\Gamma_w(\rho_v, p, T),$$

$$\frac{dp_{SV}}{dz} = -\frac{\rho_v L}{T} \Gamma_w,$$

where $\rho_v = \rho_v(p_{SV}, T)$, and Γ_w should be determined. Using values $M_v/M_a = 0.62$, $L = 2.5 \times 10^6$ J kg^{-1}, $T = 290$ K, $c_p = 10^3$ J kg^{-1} K^{-1}, $\rho_v = 0.01$ kg m^{-3}, $p = 10^5$ Pa, $g = 10$ m s^{-2}, $\rho_a = 1$ kg m^{-3}, show that a typical value of Γ_w is 6 K km^{-1}.

By assuming that $T \approx$ constant (why?), derive a differential equation for p_{SV} as a function of z in terms of two dimensionless coefficients

$$a = \frac{M_v L}{RT}, \qquad \beta = \frac{M_v}{M_a} \frac{L}{c_p T},$$

and estimate their values (you will need also the values $M_v = 18 \times 10^{-3}$ kg mole^{-1}, $R = 8.3$ J mole^{-1} K^{-1}). Derive from this an autonomous differential equation for the *molar specific humidity* $h = p_{SV}/p$. Assuming a surface value of $h \approx 0.02$, show that $H = \beta ah \sim O(1)$, and by writing $z = RTZ/M_a g$ (cf. (2.37)), show that

$$\frac{dH}{dZ} = -\frac{(\beta - 1)H}{1 + H}.$$

Deduce that for $Z \sim O(1)$,

$$H \approx H_0 \exp[H_0 - (\beta - 1)Z]:$$

humidity decreases rapidly with altitude.

2.12 Show that the solution of the Clausius–Clapeyron equation for saturation vapour pressure p_{SV} as a function of temperature T is

$$p_{SV} = p_{SV}^0 \exp\left[a\left\{1 - \frac{T_0}{T}\right\}\right],$$

where for water vapour, we may take $T_0 = 273$ K at $p_{SV}^0 = 6$ mbar ($= 600$ Pa), the *triple point*, and $a = M_v L/R T_0$. Show that if T is close to T_0, then

$$p_{SV} \approx p_{SV}^0 \exp\left[a\left(\frac{T - T_0}{T_0}\right)\right].$$

If the long-wave radiation from a planet is $\sigma \gamma T^4$, where T is the mean surface temperature, if the solar flux is Q (and planetary albedo is zero), and the greyness factor is taken to be given by

$$\gamma^{-1/4} = 1 + b\left(p_v/p_{SV}^0\right)^c,$$

where p_v is the H_2O vapour pressure, show that the occurrence of a runaway greenhouse effect is controlled by the intersection of the two curves

$$\theta = 1 + \lambda\xi, \qquad \theta = \rho\left(1 + be^{c\xi}\right),$$

where $\lambda = 1/a$, $\rho = (Q/4\sigma T_0^4)^{1/4}$. Show that runaway occurs if $\rho > \rho_c$, where

$$\rho_c + \delta = 1 + \delta \ln[\delta/b\rho_c]$$

with $\delta = \lambda/c$. Show that this determines a unique value of ρ_c, and that if δ is small,

$$\rho_c \approx 1 + \delta \ln(\delta/b) - \delta.$$

Estimate values of ρ and λ appropriate to the present Earth, and comment on the implications of these values for climatic evolution if we choose $b = 0.06$, $c = 1/4$. What are the implications for Venus, if the solar flux is twice as great? What if solar radiation were 30% lower when the planetary atmospheres were being formed?

2.13 For the energy-balance model

$$c\dot{T} = R_i - R_o,$$

where $R_i = \frac{1}{4}Q(1 - a)$, $R_o = \sigma \gamma T^4$, and $a = a_+$ for $T < T_i$, $a = a_-$ for $T > T_w$ ($> T_i$), $a_+ > a_-$, with $a(T)$ linear between these two ranges, show that possible steady state values of T are $T = T_i$ when $Q = Q_+$ and $T = T_w$ when $Q = Q_-$, where

$$Q_- = \frac{4\sigma \gamma T_w^4}{1 - a_-}, \qquad Q_+ = \frac{4\sigma \gamma T_i^4}{1 - a_+}.$$

By considering the graphs of R_o and R_i, and the slope of $R_o(T)$ at T_i, show that for Q just less than Q_+, multiple steady states will occur if

$$\frac{T_w - T_i}{T_i} < \frac{a_+ - a_-}{4(1 - a_+)},$$

and in this case show that they will exist in a range $Q_c < Q < Q_+$, and prove that the upper and lower branches are stable, but the intermediate one is unstable.

By considering the slope of $R_o(T)$ at T_w, show that if

$$\frac{T_w - T_i}{T_w} < \frac{a_+ - a_-}{4(1 - a_-)},$$

then $Q_c = Q_-$.

By normalising Q and T with respect to present day values Q_0, T_0 satisfying $Q_0(1 - a_-) = 4\sigma\gamma T_0^4$, show that the corresponding dimensionless solar fluxes and mean atmospheric temperatures, q and θ, satisfy

$$q_- = \theta_w^4,$$

$$q_+ = \theta_i^4\left(\frac{1 - a_-}{1 - a_+}\right),$$

and that multiple steady states will occur providing

$$\frac{\theta_w - \theta_i}{\theta_w} < \frac{a_+ - a_-}{4(1 - a_-)}.$$

If $\theta_w = 1$ (we are starting an ice age *now*) show that if $\theta_i = 1 - \delta$, $a_+ = a_- + v$, where $\delta, v \ll 1$, then regular ice ages will occur providing

$$\delta < \frac{v}{4(1 - a_-)},$$

and providing the solar flux q oscillates beyond the limits

$$q_+ \approx 1 + \frac{v}{1 - a_-} - 4\delta$$

and $q_- = 1$.

2.14 Suppose that the planetary albedo a is given by the ordinary differential equation

$$t_i \dot{a} = I(a, Q) - a,$$

where

$$I(a, Q) = a_{eq}[T(a, Q)],$$

$$a_{eq}(T) = a_1 - \frac{1}{2}a_2\left[1 + \tanh\left(\frac{T - T^*}{\Delta T}\right)\right],$$

$$T(a, Q) = \left[\frac{Q(1-a)}{4\sigma\gamma}\right]^{1/4}.$$

Determine the graphical dependence of I as a function of a, and how this varies with Q, and hence describe the form of oscillations if Q is periodic, and t_i is sufficiently small.

For large t_i, show that the equation can be written in the dimensionless form

$$\dot{a} = \varepsilon\big[I\{a, Q(t)\} - a\big], \qquad (*)$$

where $\varepsilon \ll 1$. The *method of averaging* implies that a varies slowly, and thus can be written approximately as the series

$$a \sim A_0(\tau) + \varepsilon A_1(t, \tau) + \cdots,$$

where $\tau = \varepsilon t$, and

$$\dot{A} = \overline{I(A_0, Q)} - A_0,$$

in which $\overline{I(A_0, Q)}$ denotes the time average of I over a period of Q. Deduce that for a range of values of $\bar{Q}$, two periodic solutions can exist, and comment on their climatic interpretation.

Give explicit approximate solutions of $(*)$ for the cases $\varepsilon \ll 1$ and $\varepsilon \gg 1$ when ΔT is very small.

2.15 Ocean temperature θ and salinity s are described by Stommel's box model

$$\dot{\theta} = 1 - \big(\mu + |\theta - Rs|\big)\theta,$$

$$\dot{s} = \big(1 - |\theta - Rs|\big)s,$$

where μ and R are positive. By analysing the equations in the phase plane, show that up to three steady states can exist, and assess their stability.

By drawing the phase portrait, discuss the nature of the solutions when there is one steady state, and when there are three.

2.16 The temperature T, CO_2 pressure p, and planetary albedo a satisfy the ordinary differential equations

$$c\dot{T} = \frac{1}{4}Q(1-a) - \sigma\gamma T^4,$$

$$t_i\dot{a} = a_{eq}(T) - a,$$

$$\frac{M_{CO_2}A}{M_a g}\dot{p} = -A_L W + v,$$

where

$$a_{eq}(T) = a_1 - \frac{1}{2}a_2\left[1 + \tanh\left(\frac{T - T^*}{\Delta T}\right)\right],$$

where $a_1 = 0.58$, $a_2 = 0.47$, $T^* = 283$ K, $\Delta T = 24$ K,

$$W = W_0 \left(\frac{p}{p_0}\right)^\mu \exp\left[\frac{T - T_0}{\Delta T_c}\right],$$

and

$$\gamma(p) = \gamma_0 - \gamma_1 p.$$

Show how to non-dimensionalise the system to the dimensionless form

$$\varepsilon \dot\theta = 1 - a - (1 - a_0)\left(1 + \frac{1}{4}v\theta\right)^4 (1 - v\lambda p),$$

$$\dot a = B(\theta) - a,$$

$$\dot p = \alpha\left[1 - wp^\mu e^\theta\right],$$

and show that

$$\alpha = \frac{vgt_i M_a}{A p_0 M_{CO_2}}, \qquad w = \frac{A_L W_0}{v}, \qquad \varepsilon = \frac{4c\Delta T_c}{t_i Q},$$

$$v = \frac{4\Delta T_c}{T_0}, \qquad \lambda = \frac{\gamma_1 p_0}{v\gamma_0}.$$

What is the function $B(\theta)$? What is the definition of a_0?

Using the values $v = 3 \times 10^{11}$ kg y^{-1}, $g = 9.81$ m s^{-2}, $M_a = 28.8 \times 10^{-3}$ kg mole^{-1}, $M_{CO_2} = 44 \times 10^{-3}$ kg mole^{-1}, $t_i = 10^4$ y, $A = 5.1 \times 10^{14}$ m^2, $p_0 = 36$ Pa, $A_L = 1.5 \times 10^{14}$ m^2, $W_0 = 2 \times 10^{-3}$ kg m^{-2} y^{-1}, $c = 10^7$ J m^{-2} K^{-1}, $Q = 1370$ W m^{-2}, $\Delta T_c = 13$ K, $T_0 = 288$ K, $\gamma_0 = 0.64$, $\gamma_1 = 0.8 \times 10^{-3}$ Pa^{-1}, $\mu = 0.3$, show that

$$\alpha \approx 1.05, \qquad w \approx 1, \qquad \varepsilon \approx 1.2 \times 10^{-6}, \qquad v \approx 0.18, \qquad \lambda \approx 0.25,$$

and find the value of a_0, assuming $\sigma = 5.67 \times 10^{-8}$ W m^{-2} K^{-4}.

Hence show that θ rapidly approaches a quasi-steady state given by

$$\theta \approx \Theta(a, p) = \kappa(a_0 - a) + \lambda p,$$

where

$$\kappa = \frac{1}{v(1 - a_0)}.$$

In the phase plane of a and p satisfying

$$\begin{aligned} \dot a &= B(\Theta) - a, \\ \dot p &= \alpha\left[1 - wp^\mu e^\Theta\right], \end{aligned} \qquad (*)$$

show that the p nullcline is a monotonically increasing function $a_p(p)$ of p, and that the a nullcline is a monotonically decreasing function $a_a(p)$ of p, providing $-B'(\theta) < v(1 - a_0)$ for all θ. Show conversely that if there is a range of θ for which $-B'(\theta) > v(1 - a_0)$, then the a nullcline is multivalued.

Suppose that the a nullcline is indeed multivalued, and that there is always a unique steady state. Show that at low, intermediate and high values of w, this equilibrium can lie on the lower, intermediate or upper branch of the a nullcline.

By consideration from the phase plane of the signs of the partial derivatives of the right hand sides of $(*)$ (and without detailed calculation), show that when they exist, the upper and lower branch steady states are stable, but that the intermediate steady state will be oscillatorily unstable if α is small enough.

How would you expect the solutions to behave if $\alpha \ll 1$?

2.17 Suppose now that Question 2.16 is augmented by the addition of a compartment representing ocean carbon storage. Thus we consider the set of equations

$$c\dot{T} = \frac{1}{4}Q(1 - a) - \sigma\gamma T^4,$$

$$t_i \dot{a} = a_{eq}(T) - a,$$

$$\frac{M_{CO_2} A}{M_a g}\dot{p} = -A_L W + v - h(p - p_s),$$

$$\rho_{H_2O} V_{oc}\dot{C} = \frac{h(p - p_s) + A_L W}{M_{CO_2}} - bC,$$

where in addition to the variables in Question 2.16, we define the atmospheric partial pressure of CO_2 at the ocean surface to be p_s, and the dissolved inorganic carbon to be C; it is related to the dissolved carbon dioxide $[CO_2]$ by the approximate partitioning relationship

$$C \approx \frac{K_1[CO_2]}{[H^+]},$$

where $[H^+] \approx 0.63 \times 10^{-8}$ mol kg^{-1} is the hydrogen ion concentration, and $K_1 \approx 1.4 \times 10^{-6}$ mol kg^{-1} is the equilibrium constant for the dissociation of carbonic acid to bicarbonate and hydrogen ions. ρ_{H_2O} is the density of seawater, V_{oc} is the volume of the oceans, h is a transport coefficient from atmosphere to the ocean surface, and b is an oceanic biological pump rate coefficient.

Dissolved CO_2 in the ocean is related to the atmospheric surface CO_2 partial pressure p_s by Henry's law,

$$[CO_2] = K_H p_s.$$

Show how to derive a scaled model in the form

$$\varepsilon\dot{\theta} = 1 - a - (1 - a_0)\left(1 + \frac{1}{4}v\theta\right)^4 (1 - v\lambda p),$$

$$\dot{a} = B(\theta) - a,$$

$$\dot{p} = \alpha\left[1 - wp^{\mu}e^{\theta} - \Lambda\left(p - \frac{C}{s}\right)\right],$$

$$\frac{\beta}{\delta}\dot{C} = p - \frac{C}{s} + \frac{w}{\Lambda}p^{\mu}e^{\theta} - \beta C,$$

where

$$s = \frac{K_H}{K_H^0},$$

and show that the additional parameters (to those in Question 2.16) are defined by

$$\Lambda = \frac{hp_0}{v}, \qquad \beta = \frac{bK_1K_H^0 M_{CO_2}}{h[H^+]}, \qquad \delta = \frac{bt_i}{\rho_{H_2O}V_{oc}},$$

and that the scale for DIC is

$$C_0 = K_0p_0,$$

where

$$K_0 = \frac{K_1K_H^0}{[H^+]}.$$

Using the values of Question 2.16, together with a reference value $K_H^0 = 3.465 \times 10^{-2}$ mol kg^{-1} atm^{-1} (and 1 atm $= 10^5$ Pa), $\rho_{H_2O} = 1.025 \times 10^3$ kg m^{-3}, $V_{oc} = 1.35 \times 10^{18}$ m^3, $h = 0.73 \times 10^{17}$ kg y^{-1} atm^{-1} and $b = 0.83 \times 10^{16}$ kg y^{-1}, show that

$$\Lambda \approx 88, \qquad \delta \approx 0.06, \qquad \beta \approx 3.9 \times 10^{-2}.$$

(The values of h and b, and the precise choice of K_H^0, are determined by assuming a biopump flux of 0.2 GtC y^{-1}, current net CO_2 flux to the ocean of 2 GtC y^{-1}, and zero net flux in pre-industrial times; we take $C = 2 \times 10^{-3}$ mol kg^{-1}, $p = 36$ Pa now, and $p = 27$ Pa pre-industrially. These fluxes are those given by Bigg (2003, p. 98); note that 1 GtC $= 10^{12}$ kgC $= 3.67 \times 10^{12}$ kgCO$_2$, the factor of 3.67 being the ratio of the molecular weights of CO_2 and carbon. The expression K_H given by Emerson and Hedges (2008, p. 98) varies from 7.8×10^{-2} mol kg^{-1} atm^{-1} at 0°C to 3×10^{-2} mol kg^{-1} atm^{-1} at 30°C; the value chosen corresponds to a temperature of 24.25°C. (Note that Emerson and Hedges state that $K_H = 3.24 \times 10^{-2}$ mol kg^{-1} atm^{-1} at 20°C and 35 ppt salinity, whereas the value according to their own tabulated expression would be 3.9×10^{-2}.)
Show that we can take

$$p - \frac{C}{s} + \frac{w}{\Lambda}p^{\mu}e^{\theta} \approx \frac{1}{\Lambda},$$

$$\theta \approx \Theta(a, C) = \frac{\lambda C}{s} + \kappa(a_0 - a)$$

(as in Question 2.16), and deduce that a and C satisfy approximately

$$\dot{a} = B(\Theta) - a,$$

$$\dot{C} = \delta(C^* - C),$$

where $B(\Theta)$ is the same monotonically decreasing function as in Question 2.16, and

$$C^* = \frac{1}{\beta \Lambda} \approx 0.3.$$

Deduce that the ocean carbon relaxes to an equilibrium value over the biopump throughput time scale of $\frac{\rho_{H_2O} V_{oc}}{b} \approx 160$ ky.

Suppose that the biopump transport coefficient varies with a and Θ, thus

$$b = b_0 b^*(a, \Theta).$$

Using b_0 as the scale for b, write down the corresponding model for a and C. How are the dynamics of C affected by the solubility pump and biological pump dependence on Θ (s decreases with Θ and b^* increases with Θ)? If, instead, $s = 1$ and b^* are independent of Θ, what is the effect of b decreasing with a? Can you think of a mechanism why b should have such a dependence?

2.18 Calcium carbonate, $CaCO_3$, in the form of calcite or aragonite, dissolves in acid to form calcium and bicarbonate ions according to the reaction

$$(CaCO_3 +) H^+ \underset{k_{-3}}{\overset{k_3}{\rightleftharpoons}} Ca^{2+} + HCO_3^-.$$

In addition, the bicarbonate buffering system is described by the reactions

$$(H_2O +) CO_2 \underset{k_{-1}}{\overset{k_1}{\rightleftharpoons}} HCO_3^- + H^+,$$

$$HCO_3^- \underset{k_{-2}}{\overset{k_2}{\rightleftharpoons}} CO_3^{2-} + H^+,$$

where the brackets on H_2O and $CaCO_3$ indicate that these substances are present in unlimited supply, and are thus ignored in writing the rate equations.

Write down the rate equations for the reactant concentrations $[H^+]$, $[Ca^{2+}]$, $[HCO_3^-]$, $[CO_2]$ and $[CO_3^{2-}]$, and by assuming equilibrium, derive three equations for the concentrations in terms of the equilibrium constants

$$K_1 = \frac{k_1}{k_{-1}}, \qquad K_2 = \frac{k_2}{k_{-2}}, \qquad K_3 = \frac{k_3}{k_{-3}},$$

and by suitable summation of the equations, derive the additional relations

$$[HCO_3^-] - [Ca^{2+}] + [CO_2] + [CO_3^{2-}] = P,$$

$$[H^+] + 2[Ca^{2+}] - 2[CO_3^{2-}] - [HCO_3^-] = Q,$$

where P and Q are constants.

Define the dissolved inorganic carbon C to be

$$C = [HCO_3^-] + [CO_2] + [CO_3^{2-}],$$

and the alkalinity to be

$$A = [HCO_3^-] + 2[CO_3^{2-}].$$

By writing

$$\xi = \frac{[H^+]}{C}, \qquad \eta = \frac{[CO_2]}{C}, \qquad p = \frac{[Ca^{2+}]}{C}, \qquad \lambda_i = \frac{K_i}{C},$$

show that

$$\xi + 2p = q + \alpha,$$

$$\frac{\lambda_1 \eta}{\xi^2}(\xi + 2\lambda_2) = \alpha,$$

$$\frac{\lambda_3 \xi^2}{\lambda_1 \eta} = p,$$

$$\eta = \frac{1}{1 + \frac{\lambda_1}{\xi} + \frac{\lambda_1 \lambda_2}{\xi^2}},$$

where

$$P = C(1 - p), \qquad Q = Cq, \qquad A = C\alpha.$$

If all the dissolved carbon is formed from calcium carbonate, we may suppose $P = 0$, and if the system is charge neutral, we may take $Q = 0$. Show in this case that ξ satisfies the two equations

$$\xi + 2 = \frac{\lambda_1(\xi + 2\lambda_2)}{\xi^2 + \lambda_1 \xi + \lambda_1 \lambda_2} = \lambda_3(\xi + 2\lambda_2).$$

(*The extra equation occurs because C is not known.*) Show that an exact solution of this pair of equations occurs for $\xi = 0$, $\lambda_2 \lambda_3 = 1$, and deduce that the dissolved carbon concentration is

$$C = \sqrt{K_2 K_3}.$$

Using the values $K_1 = 1.4 \times 10^{-6}$ mol kg^{-1}, $K_2 = 1.07 \times 10^{-9}$ mol kg^{-1} (Emerson and Hedges 2008, p. 105), $K_2 K_3 = 1.6 \times 10^{-8}$ mol^2 kg^{-2} (Krauskopf and Bird 1995, p. 76), show that this implies that $C \approx 0.126 \times 10^{-3}$ mol kg^{-1}, which is about sixteen times lower than the observed value. The discrepancy may be ascribed to the presence of many other ionic species, and the presence of other carbonate reactions, so that the assumptions $P = 0$, $Q = 0$ are invalid. Instead we will take the observed values for DIC of $C = 2 \times 10^{-3}$ mol kg^{-1}, and for carbonate alkalinity $A = 2.3 \times 10^{-3}$ mol kg^{-1}. Show in this case that $\alpha = 1.15$, and that

$$\lambda_2 \ll \lambda_1 \ll 1 \ll \lambda_3.$$

By anticipating that $\lambda_2 \lesssim \xi \ll \lambda_1$, show that

$$\xi \approx \left(\frac{2 - \alpha}{\alpha - 1} \right) \lambda_2, \qquad \eta \approx \frac{(2 - \alpha)^2 \lambda_2}{(\alpha - 1)\lambda_1},$$

and deduce that $\xi \approx 0.3 \times 10^{-5}$, $\eta \approx 0.36 \times 10^{-2}$, and that pH $= -\log_{10}[H^+] \approx 8.2$, as observed.

Show that the observed concentration of $[Ca^{2+}] \approx 10^{-2}$ mol kg^{-1} implies that $p \approx 5$, and that then $q \approx 8.85$. Show also that this value of p requires that $\lambda_3 \approx 2.48 \times 10^6$, and thus that $K_2 K_3 = 5.3 \times 10^{-6}$ mol^2 kg^{-2}, as opposed to the value quoted above.

2.19 A simple model of the Earth's climate is described by the equations

$$c\dot{T} = \frac{1}{4} Q(1 - a) - \sigma \gamma(p) T^4,$$

$$t_i \dot{a} = a_{eq}(T) - a,$$

$$\frac{M_{CO_2} A}{M_a g} \dot{p} = -A_L W + v - h(p - p_s),$$

$$\rho_{H_2O} V_{oc} \dot{C} = \frac{h(p - p_s) + A_L W}{M_{CO_2}} - bC,$$

in which T is the absolute temperature of the atmosphere, a is the planetary albedo, p is the mean atmospheric CO_2 partial pressure, p_s is the value of the atmospheric CO_2 partial pressure just above the ocean surface, and C is the dissolved inorganic carbon in the ocean. Explain the meaning of the terms in the equations.

Derive expressions for the relaxation times t_T, t_p and t_C of the T, p and C equations, using the values $a = 0.3$, $T = 288$ K, $Q = 1370$ W m^{-2}, $c = 10^7$ J m^{-2} K^{-1}, $M_{CO_2} = 44 \times 10^{-3}$ kg mole^{-1}, $A = 5.1 \times 10^{14}$ m^2, $M_a = 28.8 \times 10^{-3}$ kg mole^{-1}, $g = 9.81$ m s^{-2}, $h = 0.73 \times 10^{17}$ kg y^{-1} atm^{-1}, $\rho_{H_2O} = 1.025 \times 10^3$ kg m^{-3}, $V_{oc} = 1.35 \times 10^{18}$ m^3 and $b = 0.83 \times 10^{16}$ kg y^{-1}, where also 1 atm $= 10^5$ Pa. Assuming that $t_i = 10^4$ y, show that both T and p rapidly relax to a quasi-steady state, and deduce that a and

C satisfy the approximate pair of equations

$$t_T \dot{a} = B(a, p) - a,$$

$$t_C \dot{C} = C_0 - C.$$

Define C_0, and estimate its value, given that (pre-industrially) $v = 3 \times 10^{11}$ kg y^{-1}. How does this compare with the present day value of $C = 2 \times 10^{-3}$ mol kg^{-1}?

Assuming $p_s = \frac{C}{K}$, where $K = 7.1$ mol kg^{-1} atm^{-1}, and also $A_L = 1.5 \times 10^{14}$ m^2, $W = 2 \times 10^{-3}$ kg m^{-2} y^{-1}, find the pre-industrial value of p (using the present day value of C).

Assuming that current net industrial production of $v_i = 10^{13}$ kg y^{-1} is maintained indefinitely, show that on a time scale of centuries, p will reach an approximate equilibrium, and find its value. Show also that thereafter p will continue to increase more slowly, and sketch the evolution of p with time. What is the eventual value of p?

Chapter 3
Oceans and Atmospheres

If we had to define what the subject of mathematics and the environment was about, we might be tempted to limit ourselves to physical oceanography and numerical weather prediction. The wind and the sea are the most obvious examples of fluids in motion around us, and the nightly weather forecast is a commonplace in our perception of our surroundings. Certainly, groundwater levels and river flood forecasting are other environmental fluid flows of concern, but they are more often associated with stochastic behaviour and uncertainty, whereas we all know that ocean currents and weather systems are described, however inexactly, by partial differential equations. The general idea (which may or may not be correct) is that we know, at least in principle, the governing equations. The difficulty with weather prediction is then that the solutions are chaotic.[1]

Oceanography and atmospheric sciences, together tagged with the epithet of geophysical fluid dynamics (GFD), are huge and related subjects which each can and do have whole books devoted to them. This (thus, rather ambitious) chapter aims to describe some of the principal stories of GFD with a view to making sense of how the Earth's oceans and winds operate. The advantage of brevity is succinctness; the evident disadvantage is oversimplification.

3.1 Atmospheric and Oceanic Circulation

The atmosphere is a layer of thin fluid draped around the Earth. The Earth has a radius of some 6,370 kilometres, but the bulk of the atmosphere lies in a film only 10 kilometres deep. This layer is called the *troposphere*. The atmosphere extends above this, into the stratosphere and then the mesosphere, but the fluid density is very small in these upper layers (though not inconsequential), and we will simplify the discussion by conceiving of atmospheric fluid motion as being (largely) confined to the troposphere.

---

[1]This paradigm, that we know the model but cannot solve it well enough, is one which is a matter of current concern in weather forecasting circles.

A. Fowler, *Mathematical Geoscience*, Interdisciplinary Applied Mathematics 36, DOI 10.1007/978-0-85729-721-1_3, © Springer-Verlag London Limited 2011

Atmospheric winds (and thus weather) are driven by heating from the Sun. The Sun heats the Earth non-uniformly, because of the curvature of the Earth's surface, but the outgoing long wave radiation is much more uniform. Consequently, there is an energy imbalance between the equator and the poles. The equator is differentially heated, and the poles are differentially cooled. It is important to realise that the primary climatic energy balance (which, as we saw in Chap. 2, determines the mean temperature of the Earth) is between net incoming short wave radiation and outgoing long wave radiation; the Earth's weather systems and general circulation arise as a consequence of spatial variation in this balance, and as such are a perturbation to the basic energy balance. Weather is a *detail*.

The oceans are similar. The fluid is water and not air, but the oceans also lie in a thin layer on the Earth. For various reasons, their motion is more complicated and less well understood. For a start, their motion is baulked by continents. The great oceans lie in basins, and their global circulation is dictated to some extent by the topography of these basins. The atmosphere may have to flow over mountains, but it can do so: oceans have to flow round continents.

In addition, the oceans are driven not only by the same differential heating which drives the atmosphere, but also by the atmospheric winds themselves; this is the *wind-driven circulation*. It is not even clear whether this is the primary driving force. A final complication is that the density of ocean water depends on salinity as well as temperature, so that oceanic convection is double-diffusive in nature. (One might say in compensation that cloud formation in the atmosphere means that atmospheric convection is multi-phase convection, but this is not conceived of as being fundamental to the nature of atmospheric motion.)

The basic nature of the atmospheric general circulation is thus that it is a convecting fluid. Hot air rises, and so the equatorial air will rise at the expense of the cold polar air. In the simplest situation, the Earth's differential heating would drive a convection cell with warm air rising in the tropics and sinking at the poles; this circulation is called the *Hadley circulation*.

In reality, the hemispheric circulation consists of three cells rather than one. The tropical cell (terminating at about 30° latitude) is still called the Hadley cell, then there is a mid-latitude cell and a polar cell. This basic circulation is strongly distorted by the rotation of the Earth, which as we shall see is rapid, so that the north/south Hadley type circulations are flung to the east (at mid-latitudes): hence the prevailing westerly winds of common European experience.[2]

This eastwards wind is called the *zonal wind*. And it is unstable: a phenomenon called *baroclinic instability* causes the uniform zonal wind to form north to south waves, and these meandering waves form the weather systems which can be seen on television weather forecast charts. At a smaller scale, such instabilities lead to weather fronts, essentially like shocks, and in the tropics these lead to cyclones and hurricanes. In order to begin to understand how this all works, we need a mathematical model, and this is essentially a model of shallow water theory (or shallow air theory) on a rapidly rotating sphere.

---

[2]A westerly wind is one coming from the west. It will be less confusing to call such a wind eastwards, and *vice versa* for easterlies, i.e., westwards winds.

3.2 The Geostrophic Circulation

The basic equations describing atmospheric (or indeed, oceanic) motion are those of mass, momentum and energy in a rotating frame, and can be written in the form

$$\frac{d\rho}{dt} + \rho \mathbf{V}.\mathbf{u} = 0,$$

$$\rho \left\{ \frac{d\mathbf{u}}{dt} + 2\mathbf{\Omega} \times \mathbf{u} \right\} = -\nabla p - \rho \nabla \Phi + \mathbf{F}, \tag{3.1}$$

$$\rho c_p \frac{dT}{dt} - \beta T \frac{dp}{dt} = \nabla.\mathbf{q} + Q.$$

In these equations, ρ is the density, $\mathbf{u}$ is the velocity, p is the pressure, T is the temperature. d/dt is the material derivative following a fluid element, i.e., $d/dt = \partial/\partial t + \mathbf{u}.\nabla$. $\mathbf{\Omega}$ is the angular velocity of the Earth, and the equations have been written with respect to a set of coordinates fixed in the (rotating) Earth.[3]

Φ is called the geopotential; it is the gravitational potential corrected for the effect of centrifugal force, and is defined by

$$\Phi = \Phi_g - \frac{1}{2}|\mathbf{\Omega} \times \mathbf{r}|^2, \tag{3.2}$$

where Φ_g is the gravitational potential. The surface $\Phi = 0$ is called sea level; the surface of the oceans would be this geopotential surface in the absence of motion. We take z to be the coordinate normal to $\Phi = 0$; essentially it is in the radial direction, and to a good approximation we can take $\Phi = gz$, where g is called the gravitational acceleration (although in fact it includes a small component due to centrifugal force).

(3.1) must be supplemented by an equation of state. In the atmosphere, we take the perfect gas law

$$p = \frac{\rho RT}{M_a} \tag{3.3}$$

(cf. (2.35)), where R is the gas constant and M_a is the molecular weight of dry air.

3.2.1 Eddy Viscosity

The force $\mathbf{F}$ represents the effects of friction. Molecular viscosity is insignificant in the atmosphere and oceans, but the flows are turbulent, and the result of this is that

---

[3]The effect of the rotating coordinate system is that time derivatives of vectors $\mathbf{a}$ are transformed as $\frac{d\mathbf{a}}{dt}|_{\text{fix}} = \frac{d\mathbf{a}}{dt}|_{\text{rot}} + \mathbf{\Omega} \times \mathbf{a}$, because in differentiating $\mathbf{a} = a_i \mathbf{e}_i$, both the components a_i and the unit vectors $\mathbf{e}_i$ change with time, and $\dot{\mathbf{e}}_i = \mathbf{\Omega} \times \mathbf{e}_i$.

momentum transport by small scale eddying motion is often modelled by a diffusive frictional term of the form $\rho\varepsilon_T\nabla^2\mathbf{u}$, where ε_T is an 'eddy' (kinematic) viscosity. More generally, ε_T varies with distance from rough boundaries.[4] Some discussion of eddy viscosity is given in Appendix B. A complication in the atmosphere is that the vertical motion is much smaller than the horizontal, and this leads to the idea that different eddy viscosities are appropriate for horizontal and vertical momentum transport. We denote these coefficients as ε_H and ε_V, and take them as constants. To be precise, we then represent the frictional terms in the form

$$\mathbf{F} = \rho\varepsilon_H\nabla_H^2\mathbf{u} + \rho\varepsilon_V\frac{\partial^2\mathbf{u}}{\partial z^2}, \qquad (3.4)$$

where $\nabla_H^2 = \frac{\partial^2}{\partial x^2} + \frac{\partial^2}{\partial y^2}$, and x, y are 'horizontal' coordinates, z is the vertical coordinate. Later we discuss a more precise definition of the relation of these local Cartesian coordinates to the appropriate spherical coordinates of the system. Since the friction terms will only be important in boundary layers where the sphericity is unimportant, we need not concern ourselves with such niceties in defining $\mathbf{F}$.

Estimates of the eddy coefficients are given later. Frictional effects are generally relatively small. In the atmosphere, they are confined to a 'boundary layer' adjoining the surface, having a typical depth of 1000 metres, and bulk motion above this layer is effectively inviscid.

3.2.2 Energy Transport

The catch-all term Q in the energy equation represents internal heating due both to absorption of short wave radiation and to latent heat release by condensation, while $\mathbf{q}$ includes both sensible and radiative heat flux. As for viscosity, molecular thermal conductivity is negligible, but there is an eddy diffusive transport which is larger. In addition, there is a radiative transport term, which was defined in (2.27). For an opaque medium, this also takes the form of a diffusive flux, as given by (2.30). The principal internal heating terms are due to absorption of solar short wave radiation, and to condensation and cloud formation. In a saturated atmosphere, this was represented in Chap. 2 by the term $-\rho_a L\frac{dm}{dt}$ in (2.57). As we shall see, more care is required if we wish to provide boundary conditions for the temperature. The term Q is generally small but not insignificant in the troposphere; it is dominant in the stratosphere.

Because our main concern is with the dynamics generated by the momentum equation, it is tempting simply to prescribe Q (for example, from measured atmospheric temperature profiles). We wish to avoid doing this, because it is precisely the local imbalance in incoming and outgoing radiative transport which drives the

---

[4]And then, we write $\mathbf{F} = \nabla.\boldsymbol{\tau}_T$, $\boldsymbol{\tau}_T = \frac{1}{2}\varepsilon_T(\nabla\mathbf{u} + \nabla\mathbf{u}^T)$.

atmospheric motions we are describing. In particular, we need to understand how these imbalances are manifested in the boundary conditions for $(3.1)_3$.

To be specific, we will write the energy equation for a saturated, grey, opaque atmosphere. Certainly the greyness and opacity are inaccurate, and the atmosphere is by no means always saturated. Nevertheless, our discussion serves an important pedagogic function. The energy equation is similar to (2.57) (including a source term due to absorption of short wave radiation, see Fig. 2.3), but we deal more specifically with the source term associated with water vapour. We write the energy equation as

$$\rho c_p \frac{dT}{dt} - \frac{dp}{dt} = \nabla.(k_T \nabla T) - \nabla.\mathbf{q}_R + Q_a + \rho LC, \qquad (3.5)$$

where Q_a is the short wave absorption term, and C is the condensation rate of water vapour, with the units being measured as (minus) the rate of change of mixing ratio (ρ_v/ρ) per unit time. We have used the fact that, for the perfect gas law, the thermal expansion coefficient $\beta = 1/T$. The long wave radiative flux is given, from (2.27) and (2.28), by

$$\mathbf{q}_R = \int_O I(\mathbf{r}, \mathbf{s}) \, \mathbf{s} \, d\omega(\mathbf{s}), \qquad (3.6)$$

and (using (2.10))

$$I = \frac{\sigma T^4}{\pi} - \frac{1}{\rho\kappa} \mathbf{s}.\nabla I. \qquad (3.7)$$

I satisfies the boundary condition at the top of the atmosphere (taken to be $z = \infty$) that

$$I = 0 \quad \text{as } z \to \infty, \qquad \mathbf{s}.\mathbf{k} < 0, \qquad (3.8)$$

i.e., there is no incoming long wave radiation ($\mathbf{k}$ is a unit vector pointing upwards). At the ground surface, we assume that temperature is continuous, so that

$$I = B = \frac{\sigma T^4}{\pi} \quad \text{on } z = 0, \qquad \mathbf{s}.\mathbf{k} > 0. \qquad (3.9)$$

The effective thermal conductivity k_T represents the eddy conductivity associated with turbulent flow; generally, the eddy thermal *diffusivity* will be similar or equal to the eddy *kinematic* viscosity (and thus also will have different values in the horizontal and the vertical).

The opaque limit treats the derivative in (3.7) as a small perturbation, and a regular expansion then leads to (2.29), i.e., $I = \sigma T^4/\pi + \cdots$, from which we derive (2.30):

$$\mathbf{q}_R \approx -k_R \nabla T, \qquad (3.10)$$

where the effective radiative conductivity (see (2.51)) is

$$k_R = \frac{16\sigma T^3}{3\kappa\rho}.$$ (3.11)

Because this is a singular perturbation, the approximate solution does not necessarily satisfy the boundary conditions (3.8) and (3.9) (though in fact it does approximately satisfy the latter), and thus the effective flux given by (3.10) may not be accurate near the top or bottom of the atmosphere. This should not seriously affect its use in the energy equation (3.5).

In fact, the opaque limit can only apply, if at all, in the troposphere, where the density is reasonably high. So the issue of its use in the stratosphere, and the associated boundary condition (3.8), is somewhat irrelevant. What in fact we will wish to do is to apply an effective thermal boundary condition for (3.5) at the tropopause. We return to the consideration of this below.

The term Q_a represents the volumetric absorption within the troposphere of the short wave radiation which is received at the top of the atmosphere. Denoting this latter term (a flux) as q_i, we see from Fig. 2.4 that, over the whole atmosphere, the absorbed short wave radiation is some $0.2q_i$. Only a small amount of this is absorbed in the stratosphere, so that the quantity absorbed within the troposphere, $hQ_a \sim 0.2q_i$. The reflected short wave radiation (via the planetary albedo) $q_r \approx 0.3q_i$, and the amount absorbed at the ground $q_s \approx 0.5q_i$.

Evaporation provides another important energy flux at the surface. If we denote the rate of evaporation (as a velocity, metres of water per second) as E, then the latent heat flux due to evaporation is $\rho_w L E$, where ρ_w is the density of water, and this is comparable to the short wave absorption, $\approx 0.2q_i$, as shown in Fig. 2.3. Identifying evaporative mass transport as a latent heat flux is confusing even if convenient, and we now discuss its interpretation in detail. To do this, we have to write a moisture mass balance equation, and associated mass and energy balance boundary conditions.

Let q_- denote the heat flux delivered to the ocean/atmosphere surface from the ocean. (A similar discussion can be made for the continents, but evaporation is then negligible.) q_- consists of both a radiative part and a convective part, which can be prescribed following a prescription of energy transfer in the ocean. Let q_0 denote the radiative and convective heat flux at the surface to the atmosphere. We can write

$$q_0 = -\bar{k}\frac{\partial T}{\partial z},$$ (3.12)

where

$$\bar{k} = k_T + k_R,$$ (3.13)

(and there would be a similar expression for q_-, in terms of the oceanic temperature gradient). We can then write a boundary condition of Stefan type for (3.5) at the surface as

$$\rho_w L E = q_s - q_0 + q_-;$$ (3.14)

this represents the net evaporation E, measured as a velocity, at the ocean surface due to the net energy received at the surface. (On continents, $E = 0$ and this condition determines q_0.)

The evaporation term is a surface source term for atmospheric moisture content (measured as mixing ratio m) which thus satisfies a conservation equation of the form

$$\frac{dm}{dt} = \nabla \cdot (\varepsilon_T \nabla m) - C, \tag{3.15}$$

where ε_T is the eddy diffusivity (again, anisotropic as discussed previously), and C is the same condensation rate which appears in the energy equation (3.5). The interpretation of this equation in relation to the discussion in Chap. 2, where the assumption of saturation leads, via (2.55), (2.56) and (2.58), to an expression for m as a function of T and p, $m = m_s(T, p)$, is that when (3.15) applies for unsaturated air, i.e., while $m < m_s(T, p)$, then $C = 0$. For saturated air, we have $m = m_s(T, p)$, and (3.15) determines the condensation rate C.[5] The boundary condition for (3.15) at the surface is

$$-\rho \varepsilon_V \frac{\partial m}{\partial z} = \rho_w E \tag{3.16}$$

(and that at the tropopause can be taken to be that the evaporative flux is zero).

Lastly, we attempt to write a boundary condition for (3.5) at the tropopause, $z = h$. We denote the upwards radiative and convective heat flux there as $(\mathbf{q}_R - k_T \nabla T) \cdot \mathbf{k} = q_h$. At the tropopause, if we suppose the radiant heat flux is given by Stefan's law, the radiative boundary condition would be

$$q_h = -\bar{k}\frac{\partial T}{\partial z} = \sigma T^4, \tag{3.17}$$

presuming a vacuum beyond.

This condition would be suitable if the atmosphere really was all contained in the troposphere, and then the temperature at the tropopause would be the effective long wave emission temperature of the Earth, 255 K. However, as discussed in Chap. 2 (see Sects. 2.2.7 and 2.2.8) some ten per cent of the atmosphere lies above the troposphere, most of it in the stratosphere. Ozone and other gases absorb solar radiation, particularly ultra-violet radiation, in the stratosphere, and the absorption of this radiation in the stratosphere raises its temperature. In fact, the temperature at the tropopause dips to about 220 K, and then rises in the stratosphere to about 270 K at a height of about 50 km (the stratopause). According to Stefan's fourth power law, if 70% of incoming short wave radiation is emitted as long wave radiation at 255 K,

---

[5]The discussion could be elaborated to include a conservation equation for cloud density, i.e., for water density conservation, including a source term due to condensation and a sink term due to precipitation. It is not necessary, at least as regards energy conservation, since there is little energy transport associated with precipitation.

then 114% is emitted at 288 K (consistent with Fig. 2.3), and only 39% at 220 K. A realistic tropopause thermal boundary condition requires some description of the stratosphere.

The condition we need to apply is that the radiative and convective heat flux is continuous at the tropopause. If the opaque approximation (3.10) could be applied throughout the atmosphere, then this would simply imply continuity of temperature gradient. The temperature structure shown in Fig. 2.5 would then imply accumulation of heat at the tropopause, which makes no sense. In fact the approximation (3.10) can only apply where the atmosphere is optically deep, and if at all, this is in the troposphere. More generally, the radiative part of the flux is given by an integral of the emission density B. For example, in a one-dimensional, grey atmosphere, the radiative flux is given by (see Question 2.7)

$$q_R = 2\pi \left[B_0 E_3(\tau) + \int_0^{\tau_S} B'(\tau') E_3\big(|\tau' - \tau|\big) d\tau' \right], \tag{3.18}$$

where the exponential integrals are given by

$$E_n(z) = \int_1^\infty \frac{e^{-zt} \, dt}{t^n}, \tag{3.19}$$

and $B_0 = B(0)$. From this we can derive the opaque approximation (see Question 2.7), and if we suppose vacuum beyond the stratopause, so that $q_R = \pi B_0$ at $\tau = 0$, then (since $E_3(0) = \frac{1}{2}$),

$$\int_0^{\tau_S} B'(\tau') E_3(\tau') \, d\tau' = 0, \tag{3.20}$$

which shows that the temperature structure must have a structure like that in Fig. 2.5. Despite this, the radiative heat flux at the tropopause is upwards. We apply the boundary condition

$$-\bar{k} \frac{\partial T}{\partial z} = q_h \quad \text{at } z = h, \tag{3.21}$$

and will suppose that the radiative flux above the tropopause is known, although in fact we should have to solve an integral equation (such as (3.18), if q_R is known in the stratosphere) in order to find it.

The discussion above appears to be consistent with the concept of the troposphere as the dense, well-mixed fluid layer in which radiative and sensible heat transport terms are small, and thus the temperature is adiabatic, as described in Chap. 2; above this lies the much less dense stratosphere in which radiative transport dominates, as a consequence of which $\partial q_R / \partial z \approx Q_a$, and the net radiative flux rises from its value at the tropopause. Effectively, the tropopause is like an interface between well-mixed, vigorously convecting fluid, and quiescent, stably stratified fluid.

3.2.3 Global Energy Balance

Integration of (3.5) over a vertical tropospheric column $0 < z < h$ yields (approximately) the energy balance equation[6]

$$\frac{d}{dt}(I + P) = q_0 - q_h + \int_0^h Q_a\, dz + \int_0^h \rho LC\, dz, \qquad (3.22)$$

where q_0 and q_h are the combined radiative and sensible heat fluxes upwards at sea level and the tropopause, respectively, $I = \int_0^h \rho c_p T\, dz$ is the internal enthalpy and $P = \int_0^h \rho \Phi\, dz$ is the potential energy. From (3.15) and (3.16), we have

$$\int_0^h \rho LC\, dz = \rho_w LE - \frac{d}{dt}\int_0^h \rho Lm\, dz. \qquad (3.23)$$

Defining the latent heat of moisture as

$$M = \int_0^h \rho Lm\, dz, \qquad (3.24)$$

we thus have

$$\frac{d}{dt}(I + P + M) = q_0 - q_h + \int_0^h Q_a\, dz + \rho_w LE. \qquad (3.25)$$

The four terms on the right are delineated in Fig. 2.3; q_0 is the combined radiative and sensible surface heat flux, q_h is the long wave radiative heat lost from the atmosphere, $\int_0^h Q_a\, dz$ is the internal heating due to short wave absorption, and $\rho_w LE$ is the latent heat flux. Although latent heat appears as a source term in the global equation, it occurs in the boundary conditions of the point forms of the equations. Approximate (globally averaged) values of these quantities (see Fig. 2.3) are $q_0 \approx 0.3q_i$, $q_h \approx 0.7q_i$, $\int_0^h Q_a\, dz \approx 0.2q_i$, $\rho_w LE \approx 0.2q_i$, where q_i is the received short wave radiation, having a globally averaged value of $342\ \mathrm{W\,m^{-2}}$.

In summarising this discussion, we see that (3.5) is a convective diffusion equation for the temperature, with terms on the right hand side representing both sources and transport. The temperature field is driven by heating from below (q_0 in (3.12)) and cooling from above (q_h in (3.17)). Estimated values of these are $q_h \approx 0.7q_i$, $q_0 \approx 0.3q_i$, while the internal condensation term is $\rho LCh \approx 0.2q_i$, and the absorbed radiation term is $Q_a h \approx 0.2q_i$. Because the troposphere is being heated both from below and within, it is convectively unstable. Because the heating is differential (q_i and thus q_s decreases from equator to poles), there is a secondary circulation (Hadley, mid-latitude and polar cells) directed polewards; and because of the rapid rotation of the Earth, this secondary circulation is diverted zonally, along lines of latitude. To see all this, we need to non-dimensionalise the equations.

---

[6]The approximation assumes an approximate hydrostatic balance, and an approximately one-dimensional atmosphere.

3.2.4 *Choosing Coordinates*

We now write the equations in terms of spherical polar coordinates. We take r to be the radius measured from the Earth's centre, λ to be the angle of latitude, and ϕ to be the angle of longitude. In terms of the more usual definition of spherical polar coordinates (r, θ, ϕ), r and ϕ are the same, and $\lambda = \frac{\pi}{2} - \theta$. We denote velocity components in ϕ, λ, r directions as u, v, w (because we are setting up ϕ, λ, r, i.e., east, north, upwards, as future x, y, z Cartesian variables), and we denote the vector velocity $\mathbf{u} = (u, v, w)$.[7] Then the material derivative takes the form

$$\frac{d}{dt} = \frac{\partial}{\partial t} + \mathbf{u}.\nabla, \tag{3.26}$$

and conservation of mass can be written in the form

$$\frac{d\rho}{dt} + \rho \nabla . \mathbf{u} = 0, \tag{3.27}$$

where the definitions of the vector derivatives are

$$\nabla . \mathbf{u} = \frac{1}{r \cos \lambda} \frac{\partial u}{\partial \phi} + \frac{1}{r \cos \lambda} \frac{\partial (v \cos \lambda)}{\partial \lambda} + \frac{1}{r^2} \frac{\partial}{\partial r} (r^2 w),$$

$$\nabla = \left(\frac{1}{r \cos \lambda} \frac{\partial}{\partial \phi}, \frac{1}{r} \frac{\partial}{\partial \lambda}, \frac{\partial}{\partial r} \right). \tag{3.28}$$

The momentum equations have the form

$$\frac{du}{dt} + \frac{uw}{r} - \frac{uv}{r} \tan \lambda - 2\Omega v \sin \lambda + 2\Omega w \cos \lambda = -\frac{1}{\rho r \cos \lambda} \frac{\partial p}{\partial \phi} + \frac{F_\phi}{\rho},$$

$$\frac{dv}{dt} + \frac{vw}{r} + \frac{u^2}{r} \tan \lambda + 2\Omega u \sin \lambda = -\frac{1}{\rho r} \frac{\partial p}{\partial \lambda} + \frac{F_\lambda}{\rho}, \tag{3.29}$$

$$\frac{dw}{dt} - \frac{(u^2 + v^2)}{r} - 2\Omega u \cos \lambda = -\frac{1}{\rho} \frac{\partial p}{\partial r} - g + \frac{F_r}{\rho}.$$

The energy equation in (3.1) is

$$\rho c_p \frac{dT}{dt} - \frac{dp}{dt} = \nabla . (\bar{k} \nabla T) + Q_a + \rho LC, \tag{3.30}$$

---

[7]It should be pointed out that the Earth deviates noticeably from being a sphere; it is more nearly an oblate spheroid, whose radius varies by some 20 km between pole and equator. This is of some conceptual importance, since gravity is the most important force, and the use of a purely spherical coordinate system would yield large 'horizontal' forces in the momentum equations. The correct procedure is to define the level 'horizontal' surfaces to be geopotential surfaces, so that there are no horizontal gravitational forces. But the geometric deviation from sphericity is so small that in effect we regain the form of the equations in spherical polars, as presented here.

where we assume that $\bar{k}$ represents a combined effective radiative and sensible thermal conductivity.

These are awkward equations, but they can be simplified by scaling and approximation. One of the features of the Earth's weather systems is that they have a horizontal length scale which, though large, is not global in extent. The description of such systems is facilitated by using a local, near Cartesian coordinate system.

However, there is a difficulty in doing this. It is necessary to choose a particular latitude on which to put the Cartesian origin, and this then limits the applicability of the resulting approximate model to phenomena appropriate to this latitude. Luckily, as we have seen, there is a natural division of the global circulation into three bands (one in each hemisphere): tropical, mid-latitude and polar. We associate these three latitudes with values of λ near zero, of $O(1)$, and near $\pm\frac{\pi}{2}$. Particularly, the polar régime is an awkward one, because of the degeneracy of the equations near $\lambda = \frac{\pi}{2}$. We will concentrate our discussion on mid-latitude phenomena, and take $\lambda = \lambda_0$ to define the x–z plane.

Specifically, we define east, north and vertical coordinates x, y and z by the relations

$$x = \phi r \cos \lambda_0, \qquad y = (\lambda - \lambda_0)r, \qquad z = r - r_0, \qquad (3.31)$$

where r_0 is the radius at sea level. We then have

$$\frac{1}{r \cos \lambda} \frac{\partial}{\partial \phi} = \mu \frac{\partial}{\partial x}, \qquad \frac{1}{r} \frac{\partial}{\partial \lambda} = \frac{\partial}{\partial y}, \qquad \frac{\partial}{\partial r} = \frac{\partial}{\partial z} + \frac{1}{r}\left(x \frac{\partial}{\partial x} + y \frac{\partial}{\partial y}\right), \qquad (3.32)$$

where

$$\mu = \frac{\cos \lambda_0}{\cos \lambda}, \qquad (3.33)$$

so that

$$\mathbf{\nabla} = \left(\mu \frac{\partial}{\partial x}, \frac{\partial}{\partial y}, \frac{\partial}{\partial z}\right) + \frac{\mathbf{k}}{r}\left(x \frac{\partial}{\partial x} + y \frac{\partial}{\partial y}\right),$$

$$\mathbf{\nabla} \cdot \mathbf{u} = \mu \frac{\partial u}{\partial x} + \mu \frac{\partial(v/\mu)}{\partial y} + \frac{\partial w}{\partial z} + \frac{1}{r}\left\{x \frac{\partial w}{\partial x} + y \frac{\partial w}{\partial y} + 2w\right\}. \qquad (3.34)$$

The mass and energy equations are still (3.27) and (3.30), and the momentum equations are then

$$\frac{du}{dt} - 2\Omega v \sin \lambda + 2\Omega w \cos \lambda + \frac{1}{r}\left[uw - uv \tan \lambda\right] = -\frac{\mu}{\rho} \frac{\partial p}{\partial x} + f_x,$$

$$\frac{dv}{dt} + 2\Omega u \sin \lambda + \frac{1}{r}\left[vw + u^2 \tan \lambda\right] = -\frac{1}{\rho} \frac{\partial p}{\partial y} + f_y, \qquad (3.35)$$

$$\frac{dw}{dt} - 2\Omega u \cos \lambda - \frac{(u^2 + v^2)}{r} = -\frac{1}{\rho}\left[\frac{\partial p}{\partial z} + \frac{1}{r}\left(x \frac{\partial p}{\partial x} + y \frac{\partial p}{\partial y}\right)\right] - g + f_z,$$

where

$$f_x = \frac{F_\phi}{\rho}, \qquad f_y = \frac{F_\lambda}{\rho}, \qquad f_z = \frac{F_r}{\rho}. \tag{3.36}$$

Following (3.4), we take the vector $\mathbf{f} = (f_x, f_y, f_z)$ as

$$\mathbf{f} = \mathcal{F}\mathbf{u}, \tag{3.37}$$

where

$$\mathcal{F} = \varepsilon_H \nabla_H^2 + \varepsilon_V \frac{\partial^2}{\partial z^2}, \tag{3.38}$$

and

$$\nabla_H = \left(\frac{\partial}{\partial x}, \frac{\partial}{\partial y} \right). \tag{3.39}$$

3.2.5 Non-dimensionalisation

There are three obvious length scales of immediate relevance. These are the depth h of the troposphere, the radius r_0 of the Earth, and the length scale l of horizontal atmospheric motions. We have $h = 10$ km, $r_0 = 6370$ km, and the largest (*synoptic*) scales of mid-latitude weather systems are observed to be $l = 1000$ km. These lengths combine to form two dimensionless parameters,

$$\delta = \frac{h}{l}, \qquad \Sigma = \frac{l}{r_0} \tag{3.40}$$

both of which are small: $\delta \approx 0.01$, $\Sigma \approx 0.16$. The ideas of lubrication theory, using the fact that $\delta \ll 1$, suggest that in the vertical momentum equation, $\frac{\partial p}{\partial z} \approx -\rho g$, i.e., the pressure is approximately hydrostatic, as in our basic state. Lubrication theory also suggests that if U is a suitable horizontal velocity scale, then the appropriate vertical velocity scale is hU/l, in order that the material derivative retains vertical acceleration.

Sphericity in the equations is manifested by the terms in $1/r$ and the trigonometric terms in λ. The terms in $1/r$ are generally small, of order Σ or less, and serve as a regular perturbation to the Cartesian derivative terms, except near the poles, where $\tan \lambda \to \infty$ and a different discussion is necessary.

We scale the variables as follows:

$$x, y \sim l, \qquad z \sim h, \qquad u, v \sim U, \qquad w \sim \delta U,$$
$$t \sim \frac{l}{U}, \qquad \rho \sim \rho_0, \qquad p \sim p_0, \qquad T \sim T_0, \tag{3.41}$$

where we choose

$$p_0 = \frac{\rho_0 R T_0}{M_a} = \rho_0 g h \tag{3.42}$$

(which actually defines h as the (dry) atmospheric scale height, cf. Question 3.2). The length scales l and r_0 are those we have described, the horizontal wind speed U is typically about 20 m s^{-1}, and the density and temperature scales ρ_0 and T_0 are their values at sea level. (These are determined by the mass of the atmosphere and the effective radiative temperature.) For the moment we assume they are constant. This is a reasonable approximation for p_0 but less so for temperature.

We then have dimensionless expressions

$$\nabla \cdot \mathbf{u} = \mu \frac{\partial u}{\partial x} + \frac{1}{\cos \lambda} \frac{\partial (v \cos \lambda)}{\partial y} + \frac{\partial w}{\partial z} + \delta \Sigma \left\{ x \frac{\partial w}{\partial x} + y \frac{\partial w}{\partial y} + 2w \right\},$$

$$\frac{d}{dt} = \frac{\partial}{\partial t} + \mathbf{u} \cdot \nabla = \frac{\partial}{\partial t} + \mu u \frac{\partial}{\partial x} + v \frac{\partial}{\partial y} + w \frac{\partial}{\partial z} + \delta \Sigma w \left(x \frac{\partial}{\partial x} + y \frac{\partial}{\partial y} \right),$$

$$(3.43)$$

and the momentum equations take the dimensionless form

$$Ro \frac{du}{dt} - v \left(\sin \lambda + \frac{1}{2} Ro\, \Sigma v \tan \lambda \right) + \delta w \left(\cos \lambda + \frac{1}{2} Ro\, \Sigma u \right) = -\frac{Ro}{F^2} \frac{\mu}{\rho} \frac{\partial p}{\partial x} + f_x^*,$$

$$Ro \frac{dv}{dt} + u \left(\sin \lambda + \frac{1}{2} Ro\, \Sigma u \tan \lambda \right) + \frac{1}{2} \delta Ro\, \Sigma vw = -\frac{Ro}{F^2} \frac{1}{\rho} \frac{\partial p}{\partial y} + f_y^*,$$

$$\delta \left[\delta Ro \frac{dw}{dt} - u \cos \lambda - Ro\, \Sigma (u^2 + v^2) \right]$$

$$= -\frac{Ro}{F^2} \left[\frac{1}{\rho} \left\{ \frac{\partial p}{\partial z} + \delta \Sigma \left(x \frac{\partial p}{\partial x} + y \frac{\partial p}{\partial y} \right) \right\} + 1 \right] + \delta f_z^*,$$

$$(3.44)$$

in which

$$\lambda = \lambda_0 + \Sigma y, \tag{3.45}$$

$$f_k^* = \frac{f_k}{2 \Omega U} \quad \text{for } k = x, y, z, \tag{3.46}$$

and the extra parameters are a form of the Rossby number,

$$Ro = \frac{U}{2 \Omega l}, \tag{3.47}$$

and the Froude number

$$F = \frac{U}{\sqrt{gh}}. \tag{3.48}$$

For $U = 20$ m s^{-1}, $\Omega = 0.7 \times 10^{-4}$ s^{-1}, $l = 10^3$ km, $g = 10$ m s^{-2}, $h = 10$ km, we have $Ro \approx 0.14$, $F \approx 0.06$, and thus $F^2/Ro \approx 0.03$. Evidently the pressure is essentially hydrostatic, as we expect for a shallow flow.

The energy equation is commonly written in terms of the *potential temperature*, defined as

$$\theta = T \left(\frac{p_0}{p} \right)^{R/M_a c_p};$$

(3.49)

the usefulness of this variable lies in the fact that

$$\rho c_p T \frac{d\theta}{\theta} = \rho c_p \, dT - dp,$$

(3.50)

so that θ is constant for the dry adiabatic basic state of Question 3.2.[8] If we scale θ as well as T with T_0, then the dimensionless definition of θ is

$$\theta = \frac{T}{p^\alpha},$$

(3.51)

in which

$$\alpha = \frac{R}{M_a c_p}.$$

(3.52)

The equation of state is simply

$$\rho = \frac{p}{T}.$$

(3.53)

The dimensionless energy equation takes the form

$$\frac{p}{\theta} \frac{d\theta}{dt} = \frac{1}{Pe} \left[\frac{\partial}{\partial z} \left(k^* \frac{\partial T}{\partial z} \right) + O\left(\delta^2 \right) \right] + Q_a^* + C^*;$$

(3.54)

the reduced Péclet number, internal heating rate and condensation rate are given by

$$Pe = \frac{U h^2}{\kappa_0 l}, \qquad Q_a^* = \frac{Q_a l}{\rho_0 c_p T_0 U}, \qquad C^* = \frac{L C l \rho}{c_p T_0 U},$$

(3.55)

where we have written

$$\bar{k} = k_0 k^* \quad (k^* = O(1)),$$

(3.56)

and

$$\kappa_0 = \frac{k_0}{\rho_0 c_p}.$$

(3.57)

---

[8] Thus $s = c_p \ln \theta$, where s is entropy.

3.2.6 Day and Night, Land and Ocean

The thermal boundary condition at the ground is a flux condition as given in (3.12) and (3.14). In dimensionless terms, the heat flux is

$$-k^* \frac{\partial T}{\partial z} = q_0^* \quad \text{at } z = 0, \tag{3.58}$$

where

$$q_0^* = \frac{q_0 h}{k_0 T_0}. \tag{3.59}$$

The heat flux scale $k_0 T_0 / h \approx 10^4$ W m^{-2} (see Sect. 3.2.7 below), while the combined radiative and sensible heat flux from the ground is $q_0 \approx 102$ W m^{-2}. Thus $q_0^* \approx 0.01$, and is very small. Equally, the heat flux through the tropopause is very small. The point is that the time scale of response of the energy balance of the atmosphere is much longer, $O(10^7)$ seconds, than the shorter response time of atmospheric dynamics, $l/U \sim 10^5$ seconds. In this sense, the energy of the atmosphere is like the water in a bath, being filled by a tap and emptied through the plug hole. The source and sink are small, and control the amount of water in the bath over a long time scale, while the dynamics of the motion have a much faster time scale.

There is a fundamental distinction between land and ocean, and between day and night. In the ocean, the temperature must remain at or below the saturation temperature and above the freezing temperature. At saturation, the thermal boundary condition (3.14) determines the rate of evaporation; the thermal boundary condition is that $T = T_{sat}$, the saturation temperature. If $T < T_{sat}$, then $E = 0$ and the sea surface temperature is set by the incoming radiation, as we must have

$$\sigma T^4 = q_0 = q_s + q_-. \tag{3.60}$$

The same is true on land, except that since evaporation is essentially absent, the surface temperature is always determined by incoming short wave radiation.

Evidently, it is cold at night and warm in the day. At sea, evaporation switches on in the daytime. As the moist air is brought by the circulation over the warm land, it rises and thus forms clouds through condensation at higher (thus cooler) altitudes. The clouds we see scudding across the sky are the tops of convective plumes weaving their way across the countryside. This is why it always rains in Seattle, for example.[9]

I live in the Thames valley, say 100 km east of Bristol, perhaps 200 km from the sea. At that distance a wind of 20 m s^{-1} takes 10^4 s, about three hours, to make its way from the sea. And indeed, it is commonly the case on a Sunday morning that the skies are clear in the early morning, but by mid-morning it has clouded over. This is why.

---

[9]The effect is worsened by the topographic effect of the coastal mountain range.

3.2.7 Parameter Estimates

We have already estimated typical values $\delta \approx 0.01$, $Ro \approx 0.14$, $F \approx 0.06$, $\Sigma \approx 0.16$, and we need further to estimate values of f_k^*, Pe, Q_a^* and C^*. We estimate the internal radiative heating $Q_a h \sim 0.2 q_i \sim 68$ W m^{-2}; using values $\rho \approx 1$ kg m^{-3}, $h \approx 10$ km, $c_p \approx 10^3$ J kg^{-1} K^{-1}, $l = 10^3$ km, $T_0 = 288$ K, $U = 20$ m s^{-1}, we obtain $Q_a^* \sim 1.2 \times 10^{-3}$. Internal radiative heating is therefore very small: this is consistent with the discussion concerning thermal boundary conditions above.

In order to estimate Pe and f_k^*, we need estimates of eddy viscosities. A typical estimate in the horizontal is $\varepsilon_H \sim 0.1 U h \sim 10^4$ m^2 s^{-1}, and a typical estimate in the vertical is $\varepsilon_V \sim 0.1 \delta U h \sim 10^2$ m^2 s^{-1}. Therefore $\varepsilon_H \nabla_H^2 \sim 10^{-8}$ s^{-1}, $\varepsilon_V \partial^2 / \partial z^2 \sim 10^{-6}$ s^{-1}, so that the vertical diffusivity is dominant. Then $f_{x,y}^* \sim \varepsilon_V / 2\Omega h^2 \sim 10^{-2}$, while $f_z^* \sim \delta f_{x,y}^*$.

We already estimated $Pe \sim 20$ in Question 2.10, based on a radiative effective thermal conductivity of $k_R \approx 10^5$ W m^{-1} K^{-1}. A corresponding estimate for the (vertical) eddy thermal conductivity is $k_T \approx \rho c_p \varepsilon_V \approx 10^5$ W m^{-1} K^{-1}, comparable to the radiative value. This suggests that $\bar{k} \sim k_0 \approx 2 \times 10^5$ W m^{-1} K^{-1} is a reasonable estimate, which would then suggest that $Pe \sim 10$.

In order to estimate the dimensionless condensation rate C^*, we use (3.15). The eddy diffusive term is small relative to the advective term (the ratio is of order $\varepsilon_V l / U h^2 \sim 0.05$), and is only of concern within the planetary boundary layer, so we can take $C \approx -dm/dt$, assuming saturation. We use the formula in Question 2.12 for p_{SV},

$$p_{SV} = p_{SV}^0 \exp\left[a\left(1 - \frac{T_0}{T} \right) \right], \qquad (3.61)$$

where

$$a = \frac{M_v L}{R T_0}. \qquad (3.62)$$

Appropriate values are $a \approx 18.8$ and $p_{SV}^0 \approx 1{,}688$ Pa.[10] From these we find

$$m = \frac{M_v p_{SV}^0}{M_a p} \exp\left[a\left(1 - \frac{T_0}{T} \right) \right], \qquad (3.63)$$

and in terms of the dimensionless temperature and pressure,

$$m = \nu M(T, p), \qquad (3.64)$$

---

[10]This is different from the triple point value of 600 Pa because we use 288 K as the reference temperature, not 273 K.

where[11]

$$M(T, p) = \frac{1}{p} \exp\left[a\left(1 - \frac{1}{T}\right)\right], \tag{3.65}$$

and

$$\nu = \frac{M_v p_{SV}^0}{M_a p_0}. \tag{3.66}$$

Approximately, $\nu \approx 0.01$. In dimensionless terms, we thus have

$$C^* = \nu St\left(-\rho \frac{dM}{dt}\right), \tag{3.67}$$

where the Stefan number is

$$St = \frac{L}{c_p T_0}. \tag{3.68}$$

The value of St is 8.7, so that $\nu St \approx 0.087$. Because a is large and M is $O(1)$, $dM/dt \sim aM$, and thus $C^* \sim O(1)$ (the value of $\nu St a$ is ≈ 1.6).

3.2.8 Basic Reference State

Using the definitions of M in (3.65), and of ρ and T in (3.53) and (3.51), we can write the energy equation (3.54) in the form

$$\frac{p}{\theta}\left[1 + \frac{\nu St\, a M}{T^2}\right]\frac{d\theta}{dt} = -\frac{\nu St\, M(\alpha a - T)}{T^2}\frac{dp}{dt} + \frac{1}{Pe}\left[\frac{\partial}{\partial z}\left(k^* \frac{\partial T}{\partial z}\right)\right], \tag{3.69}$$

in which we neglect Q_a^* and $O(\delta^2 Pe)$.

If we further neglect the conductive term of $O(1/Pe)$, then to leading order, (3.44) and (3.69) can be written as

$$\frac{\partial p}{\partial z} = -\rho, \qquad \frac{p}{\theta}\left[1 + \frac{\nu St\, a M}{T^2}\right]\frac{d\theta}{dt} = -\frac{\nu St\, M(\alpha a - T)}{T^2}\frac{dp}{dt}, \tag{3.70}$$

representing a wet adiabatic hydrostatically balanced atmosphere. This tells us that in such an atmosphere, θ is a well-defined function of p, and hence (because of hydrostatic balance) also of z. We define this basic wet potential temperature function as $\theta_w(p)$, and the corresponding pressure and density profiles as p_w and ρ_w.

---

[11] This definition of M should not be confused with its use as the tropospheric latent heat term in (3.24), which we no longer have use for.

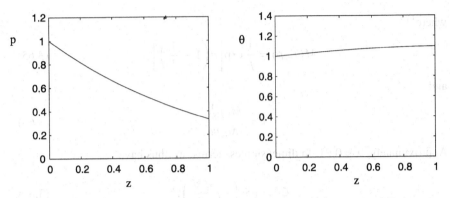

Fig. 3.1 Solution of (3.71). The pressure is excellently approximated by $p \approx e^{-1.08z}$, and the potential temperature is excellently approximated by $\theta \approx 1 + 0.15z - 0.05z^2$

Thus θ_w and p_w are determined by solving the simultaneous differential equations (noting that $\rho = p^{1-\alpha}/\theta$ and $T = \theta p^{\alpha}$)

$$
\frac{dp_w}{dz} = -\frac{p_w^{1-\alpha}}{\theta_w},
$$

$$
\frac{d\theta_w}{dz} = \frac{\nu St\,(a\alpha - \theta_w p_w^{\alpha})M}{[\theta_w^2 p_w^{2\alpha} + \nu St\,aM]p_w^{\alpha}},
$$

(3.71)

with $p_w = \theta_w = 1$ at $z = 0$. We have $\alpha \approx 0.29$, $a \approx 18.8$, and thus $\alpha a \approx 5.45$, and so $d\theta_w/dz > 0$. Also $\nu \approx 0.01$, $St \approx 8.7$, so that $\nu St \approx 0.087$, and the potential temperature gradient appears on this basis to be small, of $O(Ro)$. Figure 3.1 shows a numerical solution of (3.71), which shows that pressure decreases approximately exponentially (with scale height of about 10 km) and θ_w increases approximately linearly, in this model. The numerical solution indicates that the potential temperature gradient is indeed small, of order 0.1. We associate this with the fact that $\nu St \approx 0.087$ is small. Below, we define a parameter ε (the Rossby number) which is of the same order as the wet potential temperature gradient; then $\theta_w = 1 + O(\varepsilon)$ defines a wet adiabat, whereas a dry reference state in which the moisture term is absent is simply $\theta = 1$. Reality is somewhere between the two, though nearer the wet state.

3.2.9 A Reduced Model

In order to approximate the model, we note that

$$
\delta \sim Ro\, \Sigma \sim \frac{F^2}{Ro} \sim f_{x,y}^* \sim 10^{-2},
$$

$$
\frac{1}{Pe} \sim \nu St \sim Ro \sim \Sigma \sim 10^{-1},
$$

(3.72)

and the other parameters Q_a^* and f_z^* are much smaller. These suggest that we should think of $1/Pe$, νSt, Ro and Σ as small, but an order of magnitude larger than δ and F^2/Ro. In fact, $\partial T/\partial z \sim \alpha$, and $\alpha/Pe \approx 0.04$; therefore we shall consider the conductive term in (3.69) to be of $O(\delta)$ (see also Question 3.3). In fact, to be specific, we now define the length scale l and velocity scale U by requiring that

$$\frac{F^2 \sin \lambda_0}{Ro} = \frac{\alpha}{Pe} = \varepsilon^2, \tag{3.73}$$

where it is conventional to define the *Rossby number* as

$$\varepsilon = \frac{Ro}{\sin \lambda_0} = \frac{U}{fl}, \tag{3.74}$$

in which the Coriolis parameter f is defined as

$$f = 2\Omega \sin \lambda_0. \tag{3.75}$$

This leads to definitions

$$U = \left(\frac{\alpha \kappa_0 g}{fh}\right)^{1/2}, \qquad l = U \left(\frac{h^2}{\alpha \kappa_0 f^2}\right)^{1/3}, \tag{3.76}$$

and calculation of these using values used previously leads to $U \approx 26 \text{ m s}^{-1}$, $l \approx 1290$ km.

Next, we adopt the formal asymptotic limits

$$\nu St \sim Ro \sim \Sigma \sim \varepsilon,$$
$$\delta \sim Ro \, \Sigma \sim f_{x,y}^* \sim \varepsilon^2. \tag{3.77}$$

Expanding the equations in powers of ε, the vertical momentum equation is

$$\frac{\partial p}{\partial z} \approx -\rho + O(\varepsilon^3), \tag{3.78}$$

where

$$\rho = \frac{p}{T}, \qquad \theta = \frac{T}{p^\alpha}. \tag{3.79}$$

Also,

$$\nabla \cdot \mathbf{u} \approx \mu \frac{\partial u}{\partial x} + \mu \frac{\partial (v/\mu)}{\partial y} + \frac{\partial w}{\partial z} + O(\varepsilon^3),$$
$$\frac{d}{dt} = \frac{\partial}{\partial t} + \mathbf{u} \cdot \nabla \approx \frac{\partial}{\partial t} + \mu u \frac{\partial}{\partial x} + v \frac{\partial}{\partial y} + w \frac{\partial}{\partial z} + O(\varepsilon^3), \tag{3.80}$$

the horizontal momentum equations are approximately

$$\varepsilon \sin \lambda_0 \frac{du}{dt} - v \sin \lambda = -\frac{\sin \lambda_0}{\varepsilon^2} \frac{\mu}{\rho} \frac{\partial p}{\partial x} + O(\varepsilon^2),$$

$$\varepsilon \sin \lambda_0 \frac{dv}{dt} + u \sin \lambda = -\frac{\sin \lambda_0}{\varepsilon^2} \frac{1}{\rho} \frac{\partial p}{\partial y} + O(\varepsilon^2),$$

(3.81)

and the energy equation is approximately

$$\frac{p}{\theta}\left[1 + \frac{\nu St a M}{T^2}\right]\frac{d\theta}{dt} = -\frac{\varepsilon s M(\alpha a - T)}{T^2}\frac{dp}{dt} + \varepsilon^2 \left[\frac{\partial}{\partial z}\left(\frac{k^*}{\alpha}\frac{\partial T}{\partial z}\right)\right],$$

(3.82)

where we have written

$$\nu St = \varepsilon s$$

(3.83)

to delineate the smallness of νSt (but noting that $\nu St a \approx 1.64$ is $O(1)$). Together with the conservation of mass equation

$$\frac{d\rho}{dt} + \rho \nabla . \mathbf{u} = 0,$$

(3.84)

this completes the basic approximate model, valid locally everywhere except near the poles (where μ and $\tan \lambda \to \infty$). There are seven equations in (3.78), (3.79), (3.81), (3.82) and (3.84) for the seven variables θ, ρ, T, p, u, v and w. The frictional terms $f^*_{x,y}$ can be neglected in the main flow, but they are important in the planetary boundary layer, as we discuss later.

3.2.10 Geostrophic Balance

Geostrophic flow is described by the leading order approximation which considers both curvature and inertial effects to be small, that is, $\varepsilon \ll 1$. At leading order, the pressure is hydrostatic, and (3.81) indicates that the correction is of $O(\varepsilon^2)$. This is consistent with (3.82), which indicates that $\theta = \bar{\theta}(z) + O(\varepsilon^2)$. We do not yet assume that $\bar{\theta}$ is equal to the reference state θ_w defined in (3.71); this would have to be deduced. But we do anticipate that $\bar{\theta}'(z) = O(\varepsilon)$ (since also $\theta'_w = O(\varepsilon)$). We put

$$p = \bar{p}(z) + \varepsilon^2 P,$$

(3.85)

where $\bar{p}$ is the hydrostatic pressure corresponding to $\bar{\theta}$, and we denote the corresponding density and temperature as $\bar{\rho}$ and $\bar{T}$. Then, since $\mu \approx 1, \lambda \approx \lambda_0$ and $\rho \approx \bar{\rho}$, the momentum equations become

$$\bar{\rho}v \approx \frac{\partial P}{\partial x},$$

$$\bar{\rho}u \approx -\frac{\partial P}{\partial y},$$

(3.86)

and mass conservation reduces to

$$\frac{\partial(\bar{\rho}u)}{\partial x} + \frac{\partial(\bar{\rho}v)}{\partial y} + \frac{\partial(\bar{\rho}w)}{\partial z} \approx 0. \tag{3.87}$$

Together (3.86) and (3.87) imply

$$\frac{\partial(\bar{\rho}w)}{\partial z} = 0, \tag{3.88}$$

and thus w is determined by its value on the surface, where it is prescribed by the no flow-through boundary condition. In the absence of topography, we have $w = 0$ at $z = 0$, so that $w = 0$ everywhere. The flow is purely two-dimensional, and the horizontal velocity vector $\mathbf{u}_H = (u, v)$ is given by

$$\bar{\rho}\mathbf{u}_H = \mathbf{k} \times \nabla_H P, \tag{3.89}$$

where $\nabla_H = (\frac{\partial}{\partial x}, \frac{\partial}{\partial y})$. (3.89) defines the *geostrophic wind*, and shows that $\mathbf{u}_H \cdot \nabla_H p = 0$, i.e., wind velocities are along isobars. In the northern hemisphere, the wind moves anti-clockwise about regions of low pressure (depressions, or cyclones). The closer the isobars, the higher the wind speed.

3.3 The Planetary Boundary Layer

By neglecting the frictional terms in (3.29), we are unable to satisfy the condition of no slip at the Earth's surface. We now reconsider these terms in order to see how this condition can be met. Although the frictional terms are small, they become important in the *planetary boundary layer*, a layer with a depth of about a kilometre adjoining the surface.

Following (3.37) and (3.46), we add the dimensionless friction terms to (3.81), to obtain

$$-v = -\frac{1}{\bar{\rho}}\frac{\partial P}{\partial x} + E\frac{\partial^2 u}{\partial z^2} + O(\varepsilon),$$
$$u = -\frac{1}{\bar{\rho}}\frac{\partial P}{\partial y} + E\frac{\partial^2 v}{\partial z^2} + O(\varepsilon), \tag{3.90}$$

where the Ekman number is given by

$$E = \frac{\varepsilon_V}{fh^2}. \tag{3.91}$$

With $f \sim 10^{-4}\,\mathrm{s}^{-1}$, $\varepsilon_V \sim 10^2\,\mathrm{m\,s}^{-1}$, $h \sim 10^4$ m, we have $E \sim 10^{-2} \sim \varepsilon^2$ as previously stated.

The frictional terms are indeed negligible, except in a boundary layer of thickness ε, in which we rescale

$$z = \varepsilon Z, \qquad w = \varepsilon W, \tag{3.92}$$

so that to leading order, mass and momentum conservation equations are

$$-v = -\frac{1}{\bar{\rho}}\frac{\partial P}{\partial x} + \frac{E}{\varepsilon^2}\frac{\partial^2 u}{\partial Z^2},$$

$$u = -\frac{1}{\bar{\rho}}\frac{\partial P}{\partial y} + \frac{E}{\varepsilon^2}\frac{\partial^2 v}{\partial Z^2},$$

$$\frac{\partial u}{\partial x} + \frac{\partial v}{\partial y} + \frac{1}{\bar{\rho}}\frac{\partial(\bar{\rho}W)}{\partial Z} = 0;$$

(3.93)

also $\partial P/\partial Z = O(\varepsilon)$, so that in common with other viscous boundary layers, we can take $P = P(x, y)$, and equal to the free stream value at the surface.

Using (3.86), and denoting the surface values of the free stream velocity as u_0 and v_0, (3.93) can be elegantly solved subject to no slip on the boundary and attainment of the free stream velocities as $Z \to \infty$ in the form

$$u + iv = (u_0 + iv_0)\left[1 - \exp\left\{-\frac{(1+i)\varepsilon Z}{\sqrt{2E}}\right\}\right].$$

(3.94)

This solution is known as the Ekman spiral, as the horizontal velocities spiral round as they approach the free stream velocity.

Of later importance will be the change of W across the Ekman layer. Integration of $(3.93)_3$ from $Z = 0$ to $Z = \infty$ (bearing in mind that $\bar{\rho} = 1 + O(\varepsilon)$) yields the value of W_0, the value of W at the edge of the boundary layer:

$$W_0 = \sqrt{\frac{E}{2\varepsilon^2}}\left(\frac{\partial v_0}{\partial x} - \frac{\partial u_0}{\partial y}\right).$$

(3.95)

This generation of a vertical velocity by the free stream vorticity is known as *Ekman pumping*.

3.4 Poincaré and Kelvin Waves

The geostrophic wind given by Eq. (3.89) is an approximate solution to the governing equations which is *quasi-static*, in the sense that the acceleration terms in the momentum equation are ignored; implicitly, any more rapid transients have died out. Before we proceed to the higher order approximation which it is necessary to take in order to determine the perturbed pressure in (3.89), we consider various classes of wave motion which arise in the model on this shorter transient time scale.

Atmospheric motions are dominated by various kinds of waves. Two particular sorts of waves which are familiar in fluid mechanics are *sound waves* and *gravity waves*. Sound waves are associated with compressibility; they travel at a speed (the speed of sound) which depends on density but is independent of wave number: they are *monochromatic*. At sea level this speed is about 330 m s^{-1}: much faster than typical wind speeds; as a consequence, we might expect sound waves to be high frequency phenomena which are not relevant to common atmospheric motions. If we denote the sound wave speed as c_s, then the dispersion relation relating frequency ω to wave speed and wave number k is just $\omega = kc_s$. When this is written in dimensionless units, as above, we have

$$c_s^2 = gh\bar{c}^2, \quad \bar{c} = \left(\frac{d\bar{p}}{d\bar{\rho}}\right)^{1/2}, \tag{3.96}$$

and the corresponding dimensionless dispersion relation is just

$$\frac{\omega}{k} = \frac{\bar{c}}{F}, \tag{3.97}$$

where F is the Froude number defined by (3.48). Note from (3.73) that $F = \varepsilon^{3/2}$.

Gravity waves are familiar as the waves which propagate on the surface of the sea. The ingredients of the theory which describes them are mass conservation (where horizontal divergence is accommodated by vertical contraction and expansion), acceleration, gravity, pressure gradient, and a vertical stratification which, in the simplest form of the theory, is manifested by the interface between dense underlying fluid (e.g., water) and a lighter overlying fluid (e.g., air). Gravity waves can be seen propagating at the interface between two incompressible liquids such as oil and water, and gravity waves will similarly propagate in a continuously stratified fluid contained in a vertically confined channel; in this case the waves are less easily visualised, and they are often called internal waves, or internal gravity waves.

In the sense that the atmosphere consists of a dense troposphere beneath a light stratosphere, we can expect gravity waves to propagate as undulations in the tropopause altitude. More generally, gravity waves will propagate as internal waves in the stratified atmosphere. Gravity waves can be seen commonly in the atmosphere, because the vertical undulations of the air causes periodic cloud formation as air rises (and thus cools). Figure 3.2 shows a particular striking example from Lapland of low lying periodic gravity waves.

For the simple case of an incompressible fluid of depth h, the dispersion relation between frequency and wave number is $\omega^2 = gk \tanh kh$. In the case of a shallow fluid (such as the atmosphere), the long wave limit $kh \ll 1$ may be appropriate, and then the wave speed is constant, and $\omega \approx k\sqrt{gh}$. This applies to waves of wavelength larger than 10 km (the waves in Fig. 3.2 are of smaller wavelength). In dimensionless terms, the dispersion relation becomes

$$\frac{\omega}{k} = \frac{1}{F}. \tag{3.98}$$

Comparing (3.98) with (3.97), we see that long gravity waves in the atmosphere are essentially the same as sound waves. In an incompressible fluid, density is manifested as fluid column depth, and the pressure is proportional to this, so that the dimensionless 'sound' speed is equal to one. For internal waves, the height of the column need not change, but the common factor is that the height of geopotential surfaces propagates in both types of wave.

Fig. 3.2 Periodic gravity waves in Lapland, Northern Finland, October 2004

We can recover gravity waves from the scaled atmospheric model by focussing on long waves of wave number $k \sim O(\sqrt{\varepsilon})$, and time scales of $O(\varepsilon)$ (i.e., frequencies $\omega \sim O(1/\varepsilon)$). (Note that then $\omega/k \sim 1/\varepsilon^{3/2} = 1/F$, from (3.73) and (3.74), consistent with (3.97) and (3.98).) We write

$$t = \varepsilon\tau, \qquad (x, y) = (X, Y)/\sqrt{\varepsilon}, \qquad P = \frac{\Pi}{\sqrt{\varepsilon}} \qquad (3.99)$$

(note that P is defined in (3.85)), and retain leading order terms in Eqs. (3.81) and (3.84), assuming that $w \sim \varepsilon$. Note that $\rho = p^{1-\alpha}/\theta$, and that $\partial\theta/\partial t \approx 0$, so that

$$\frac{1}{\rho}\frac{\partial\rho}{\partial t} \approx \left(\frac{1-\alpha}{p}\right)\frac{\partial p}{\partial t}. \qquad (3.100)$$

At leading order, mass conservation takes the form

$$\left(\frac{1-\alpha}{\bar{p}}\right)\frac{\partial\Pi}{\partial\tau} + \frac{\partial u}{\partial X} + \frac{\partial v}{\partial Y} = 0; \qquad (3.101)$$

compressibility and stratification are manifested by the first term in this equation.

At leading order, the momentum equations take the form

$$\frac{\partial u}{\partial \tau} - v \approx -\frac{1}{\bar{\rho}} \frac{\partial \Pi}{\partial X},$$

$$\frac{\partial v}{\partial \tau} + u \approx -\frac{1}{\bar{\rho}} \frac{\partial \Pi}{\partial Y}. \tag{3.102}$$

We can write these equations in terms of the horizontal divergence $\Delta = u_X + v_Y$, the vorticity $\zeta = v_X - u_Y$, and the pressure perturbation Π. We obtain

$$\frac{\partial \Delta}{\partial \tau} - \zeta = -\frac{1}{\bar{\rho}} \nabla^2 \Pi,$$

$$\frac{\partial \zeta}{\partial \tau} + \Delta = 0, \tag{3.103}$$

$$\frac{\partial \Pi}{\partial \tau} + \bar{\rho} \bar{c}^2 \Delta = 0,$$

where

$$\bar{c} = \left[\frac{\bar{p}}{(1 - \alpha)\bar{\rho}} \right]^{1/2} \tag{3.104}$$

is the dimensionless isentropic sound speed.

These are linear equations, and solutions exist of the form

$$\begin{pmatrix} \Delta \\ \zeta \\ \Pi \end{pmatrix} = \mathbf{w} \exp \{ i (kX + lY + \omega \tau) \}, \tag{3.105}$$

provided

$$\begin{pmatrix} 0 & 1 & \frac{(k^2+l^2)}{\bar{\rho}} \\ -1 & 0 & 0 \\ -\bar{\rho}\bar{c}^2 & 0 & 0 \end{pmatrix} \mathbf{w} = i \omega \mathbf{w}. \tag{3.106}$$

Solutions to this exist provided either $\omega = 0$, or

$$\omega^2 = 1 + \left(k^2 + l^2 \right) \bar{c}^2, \tag{3.107}$$

and this latter equation is the dispersion relation for gravity waves in a rotating stratified atmosphere. These waves are called *Poincaré waves*.

Another kind of wave can be found by seeking solutions in which $v = 0$. Such waves are particularly relevant to propagation in a confined zonal channel (for example in the ocean), where the condition $v = 0$ at the north and south boundaries forces $v = 0$ everywhere. This requires $\partial \Delta / \partial Y = -\partial \zeta / \partial X$, and substitution into (3.106) then shows that we must have $l = -ik/\omega$, and thus solutions are exponential in y, and

$$\omega = k\bar{c}; \tag{3.108}$$

these waves are called *Kelvin waves*. They are *edge waves*, because they decay exponentially away from one or other boundary. Together with the geostrophic mode $\omega = 0$, Poincaré and Kelvin waves form the complete spectrum of waves for the flow. The mode $\omega = 0$ is associated with low frequency waves which emerge in the higher order quasi-geostrophic approximation (which is derived in the next section); these slow waves are called *Rossby waves*, or *planetary waves*.

The constant term in (3.107) arises from rotation and the Coriolis force. In the high frequency limit, we see that $\omega \approx k\bar{c}$ (for unidirectional waves), and this is consistent with the long wave limit of gravity wave theory, and the acoustic wave speed given in (3.97). Gravity waves are essentially long wavelength sound waves, and Poincaré waves are their modification by the effects of rotation. The critical length scale $l/\sqrt{\varepsilon}$ above which rotation becomes important is known as the Rossby radius of deformation. Using (3.76), it is found to be equal to $\sqrt{gh}/f$. For atmospheric motion, it is of order 3000 km, so that rotation is unimportant for smaller scale gravity waves, such as those in Fig. 3.2.

3.5 The Quasi-geostrophic Approximation

We now return to the problem of finding the pressure for the geostrophic approximation in which (3.89) applies. To do this, we need to carry the approximation to next order in ε, and this will allow us to deduce the *quasi-geostrophic potential vorticity equation*. The equation of mass conservation (3.84) can be written in the form

$$\frac{\partial \rho}{\partial t} + \mu \frac{\partial (\rho u)}{\partial x} + \mu \frac{\partial (\rho v/\mu)}{\partial y} + \frac{\partial (\rho w)}{\partial z} = O(\varepsilon^3). \tag{3.109}$$

Since $w = 0$ at leading order, we put

$$w = \varepsilon W. \tag{3.110}$$

We also define the perturbed potential temperature Θ by

$$\theta = \bar{\theta}(z) + \varepsilon^2 \Theta; \tag{3.111}$$

evidently $\bar{\theta}(z)$ is the time and space-horizontal average of θ correct to $O(\varepsilon^2)$, and we can in fact define it to be the exact such average of θ, without loss of generality. More generally, we might take $\bar{\theta} = \bar{\theta}(z, t)$, but the energy equation then simply implies that $\bar{\theta}_t = 0$. We might have expected $\bar{\theta}$ to be equal to the wet adiabatic potential temperature θ_w, defined in (3.71), but as we shall see, there is a subtle distinction, and it is necessary to delineate the difference in the equations. Because the hydrostatic correction in (3.78) is $O(\varepsilon^3)$, expansion of that equation to $O(\varepsilon^2)$ yields the hydrostatic approximation for the perturbation pressure P, defined in (3.85):

$$\Theta = \bar{\theta}^2 \frac{\partial}{\partial z}\left[\frac{P}{\bar{p}^{1-\alpha}}\right]. \tag{3.112}$$

The geostrophic wind approximation (3.86) suggests that we write

$$P = \bar{\rho}\psi, \tag{3.113}$$

where ψ is the geostrophic stream function, thus

$$u = -\frac{\partial \psi}{\partial y}, \qquad v = \frac{\partial \psi}{\partial x}. \tag{3.114}$$

Bearing in mind that $\bar{\rho} = \bar{p}^{1-\alpha}/\bar{\theta}$, it follows that

$$\Theta = \bar{\theta}^2 \frac{\partial}{\partial z}\left[\frac{\psi}{\bar{\theta}}\right] = \frac{\partial \psi}{\partial z} + O(\varepsilon), \tag{3.115}$$

on the assumption that $\bar{\theta}'(z) = O(\varepsilon)$. This relation, together with the geostrophic wind approximation, gives us the *thermal wind equations*:

$$\frac{\partial u}{\partial z} = -\frac{\partial \Theta}{\partial y}, \qquad \frac{\partial v}{\partial z} = \frac{\partial \Theta}{\partial x}. \tag{3.116}$$

Next we form an equation for the (vertical) vorticity

$$\zeta = \frac{\partial v}{\partial x} - \frac{\partial u}{\partial y} = \nabla^2 \psi \tag{3.117}$$

by cross differentiating (3.81) (with some care) to eliminate the pressure derivatives. Using the conservation of mass equation, together with (3.110) and the fact that $\rho = \bar{\rho}(z) + O(\varepsilon^2)$, we derive the vorticity equation

$$\frac{D\zeta}{Dt} + \beta \frac{\partial \psi}{\partial x} = \frac{1}{\bar{\rho}}\frac{\partial(\bar{\rho}W)}{\partial z}, \tag{3.118}$$

where D/Dt denotes the horizontal material derivative, and the term in β arises from the variation of $\sin \lambda$ with latitude; β is defined by

$$\beta = \frac{\Sigma \cot \lambda_0}{\varepsilon}, \tag{3.119}$$

and the horizontal material derivative is defined by

$$\frac{D}{Dt} = \frac{\partial}{\partial t} + u\frac{\partial}{\partial x} + v\frac{\partial}{\partial y} = \frac{\partial}{\partial t} - \frac{\partial \psi}{\partial y}\frac{\partial}{\partial x} + \frac{\partial \psi}{\partial x}\frac{\partial}{\partial y}. \tag{3.120}$$

Next, we consider the energy equation (3.82). Expanding in powers of ε, this can be written in the form, correct to terms of $O(\varepsilon^2)$,

$$\varepsilon W \frac{d\bar{\theta}}{dz} + \varepsilon^2 \frac{D\Theta}{Dt} = \varepsilon W \frac{d\theta_w}{dz} + \varepsilon^2 H, \tag{3.121}$$

where

$$H = \frac{\frac{\partial}{\partial z}\left(\frac{k^*}{\alpha}\frac{\partial \bar{T}}{\partial z}\right)}{\frac{\bar{p}}{\bar{\theta}}\left[1 + \frac{vSt\alpha M(\bar{T},\bar{p})}{\bar{T}^2}\right]} \tag{3.122}$$

is the heating term.

Now we can see the nature of the assumption about the average potential temperature. Bearing in mind that $d\theta_w/dz = O(\varepsilon)$, we see that the *ansatz* that $d\bar{\theta}/dz = O(\varepsilon)$ is indeed correct. However, it is generally not the case that $\bar{\theta} = \theta_w$. The question then arises how to determine it.

Let us denote the stratification function $S(z)$ by

$$S(z) = \frac{1}{\varepsilon}\left[\frac{d\bar{\theta}}{dz} - \frac{d\theta_w}{dz}\right], \tag{3.123}$$

and note that by observation (and assumption) it is positive and $O(1)$. It is related to the Brunt–Väisälä frequency N, which is the frequency of small vertical oscillations in the atmosphere; in fact $S \propto N^2$. Positive S (and thus real N) indicates a stably stratified atmosphere. If S were to become negative, the atmosphere would become unstably stratified and it would overturn. The energy equation is thus

$$\frac{D\Theta}{Dt} = H - WS. \tag{3.124}$$

In summary, we have the vorticity ζ and potential temperature Θ defined in terms of the stream function ψ by (3.117) and (3.115). Two separate equations for ζ and Θ are then (3.118) and (3.124), from which W and $S(z)$ must also be determined, the latter by averaging the equations.

By an application of Green's theorem in the plane, we have

$$\iint_A \frac{D\Gamma}{Dt}\,dS = \frac{\partial}{\partial t}\iint_A \Gamma\,dS - \oint_{\partial A} \Gamma\,d\psi, \tag{3.125}$$

where A is any horizontal area at fixed z. In particular, if A is a closed region on the boundaries of which ψ is constant in space, i.e., there is no flow through ∂A, then the boundary integral is zero.[12] Let an overbar denote a space horizontal average over A. Putting $\Gamma = \Theta$, it follows that

$$\frac{\partial\bar{\Theta}}{\partial t} = H - \overline{W}S, \tag{3.126}$$

---

[12] We have in mind that A is the region of zonal mid-latitude flow, bounded to the north by the polar front, and to the south by the tropical front. We can allow A to be a periodic strip on the sphere also.

where $\overline{W}(z)$ is the horizontal average of W. Applying the same procedure to (3.118), we have

$$\frac{\partial \bar{\zeta}}{\partial t} = \frac{1}{\bar{\rho}} \frac{\partial}{\partial z} [\bar{\rho} \overline{W}]. \tag{3.127}$$

According to the Ekman pumping boundary condition (3.95), the value of $\overline{W}$ at $z = 0$ is

$$\overline{W}_0 = E^* \bar{\zeta}_0, \tag{3.128}$$

where $\bar{\zeta}_0$ is the space-averaged vorticity at the surface, and

$$E^* = \sqrt{\frac{E}{2\varepsilon^2}}. \tag{3.129}$$

Integrating (3.127), we have (using $\bar{\rho} = 1$ at $z = 0$)

$$\bar{\rho} \overline{W} = \int_0^z \bar{\rho} \bar{\zeta}_t \, dz + E^* \bar{\zeta}_0, \tag{3.130}$$

and it follows from this that the stratification parameter is defined by the relation

$$\frac{\bar{\rho}}{S} = \frac{\int_0^z \bar{\rho} \bar{\zeta}_t \, dz + E^* \bar{\zeta}_0}{H - \bar{\Theta}_t}. \tag{3.131}$$

We can go further if we assume that the solutions are stationary (not necessarily steady), i.e., a well-defined time average exists.[13] The time averages of the time derivative terms are zero, and thus it simply follows (since H, S and $\bar{\rho}$ are functions only of z) that

$$H = \widehat{W} S, \qquad \bar{\rho} \widehat{W} = \widehat{W}_0, \tag{3.132}$$

where $\widehat{W}$ is the time average of $\overline{W}$, and the constant $\widehat{W}_0$ is the value of the surface boundary value of $\widehat{W}$ at $z = 0$.[14] The Ekman pumping boundary condition (3.128) implies that

$$\widehat{W}_0 = E^* \bar{\zeta}_0, \tag{3.133}$$

where $\bar{\zeta}_0$ is the space-averaged vorticity at the surface.

---

[13] This is what we would generally expect. Unbounded drift of ψ would indicate breakdown of the perturbation expansion because of the presence of secular terms.

[14] The question arises at this point, why can we not take $\overline{W}_0$, and thus $\widehat{W}_0$, equal to zero? $\overline{W}_0$ is the average (scaled) vertical velocity above the planetary boundary layer. If it is not zero, then apparently there would be a non-zero mass flux into or out of this layer. While that is feasible, the time average should apparently be zero, unless there is secular growth or decline of the layer thickness. This would follow from (3.109) were we dealing with the exact horizontal velocities (u, v). However, the geostrophic stream function prescription in (3.114) is only accurate to $O(\varepsilon)$, and thus if we were to use (3.109) to calculate $W = w/\varepsilon$, there would be an (unknown) $O(1)$ contribution from the corrections to the horizontal velocities. The point is that we cannot actually use mass conservation to determine W (and thus we cannot use a natural inference such as $\widehat{W}_0 = 0$).

The two equations in (3.132) define S and $\widehat{W}$, and in particular we find that

$$\frac{\bar{\rho}}{S} = \frac{E^* \hat{\zeta}_0}{H}. \tag{3.134}$$

This equation thus defines the stratification function $S(z)$ for a stationary (but not necessarily steady) atmosphere.[15] Evidently, the wet adiabatic profile ($S = 0$) is obtained (in stationary conditions) only if the heating rate H is zero.

We can now use the identity

$$\frac{\partial}{\partial z}\left[K(z)\frac{D\Theta}{Dt}\right] = \frac{D}{Dt}\left[\frac{\partial}{\partial z}\left(K(z)\frac{\partial\psi}{\partial z}\right)\right] \tag{3.135}$$

to show, using (3.124), that

$$\frac{1}{\bar{\rho}}\frac{\partial}{\partial z}[\bar{\rho}W] = \frac{1}{\bar{\rho}}\frac{\partial}{\partial z}\left[\frac{\bar{\rho}H}{S}\right] - \frac{D}{Dt}\left[\frac{1}{\bar{\rho}}\frac{\partial}{\partial z}\left(\frac{\bar{\rho}}{S}\frac{\partial\psi}{\partial z}\right)\right], \tag{3.136}$$

and therefore (3.118) can be written

$$\frac{D}{Dt}\left[\nabla^2\psi + \beta y + \frac{1}{\bar{\rho}}\frac{\partial}{\partial z}\left(\frac{\bar{\rho}}{S}\frac{\partial\psi}{\partial z}\right)\right] = \frac{1}{\bar{\rho}}\frac{\partial}{\partial z}\left[\frac{\bar{\rho}H}{S}\right]. \tag{3.137}$$

This is one form of the *quasi-geostrophic potential vorticity equation*. It is a single equation for the geostrophic stream function ψ, providing the stratification S is known. In most treatments of its solutions, the stratification parameter S is assumed known (from measurements), and then the Eq. (3.137) can be considered on its own.

3.5.1 Boundary Conditions

We wish to solve the quasi-geostrophic equation for ψ in a geometric domain consisting of a rectangular channel, representing roughly the mid-latitude cell. It is simplest to think of fixed boundaries at $y = \pm 1$, for example, although moving boundaries (adjoining the Hadley and polar cells) are more appropriate. We suppose the flow is unbounded in the x-direction (the circumference is of $O(1/\varepsilon)$, and thus large). Finally the flow is bounded by an interface at the tropopause, across which pressure and density are continuous, but temperature gradient is effectively discontinuous, as a consequence of the different stratospheric thermal régime.[16]

---

[15]This derivation is somewhat similar to that of Pedlosky (1987); however, he did not provide an explicit recipe for $S(z)$. See also Question 3.8.

[16]The temperature gradient is in fact continuous; indeed the temperature condition at the tropopause is a suitably dimensionless version of the flux condition (3.21); but heat conduction is provided by a singular highest derivative term, so that the energy equation is essentially conductionless. It is a consequence of this that we may consider the temperature gradient to be discontinuous across the tropopause.

The basic model, (3.1), is one of inviscid flow in a shallow layer with a free boundary, driven by an imposed poleward temperature gradient due to solar insolation. Consequently, we expect to provide velocity conditions of no flow through the base; but, as discussed in Sect. 3.3, the planetary viscous boundary layer induces a non-zero Ekman velocity above it given by (3.95), so that the boundary condition for (3.118) is in fact

$$W = E^* \nabla^2 \psi \quad \text{on } z = 0. \tag{3.138}$$

Other conditions of this type are no-flow-through conditions at the side walls $y = \pm 1$, and boundedness or periodicity conditions in the x-direction. An initial condition for the quasi-geostrophic potential vorticity

$$q = \nabla^2 \psi + \beta y + \frac{1}{\bar{\rho}} \frac{\partial}{\partial z} \left(\frac{\bar{\rho}}{S} \frac{\partial \psi}{\partial z} \right) \tag{3.139}$$

is sufficient for (3.137), and if q is known, then the periodicity or no-flow-through conditions in x and y will provide the necessary horizontal boundary conditions to solve the elliptic (3.139) for ψ. However, we also need to specify two vertical conditions for ψ at the tropopause and surface.

At the tropopause, we expect a kinematic condition and a pressure condition. We define the tropopause to be at $z = 1 + \varepsilon \eta(x, y, t)$, noting that such a variation is consistent with observation (the tropopause slopes from perhaps 15 km at the equator to perhaps 10 km at the poles). The kinematic condition stating that the tropopause is a material interface then takes the scaled form

$$\frac{D\eta}{Dt} = W \quad \text{at } z \approx 1, \tag{3.140}$$

confirming the suggestion that $\eta = O(1)$. However, just as the planetary boundary layer induces an Ekman pumping term which modifies the boundary condition on W at $z = 0$, so also a (less severe) boundary layer at the troposphere will modify (3.140) at $z = 1$. We conceive of the stratosphere as a blanket of less dense air which acts as a brake on the troposphere, and we pose the scaled boundary condition

$$\frac{\partial \mathbf{u}_H}{\partial z} = -\gamma \mathbf{u}_H \tag{3.141}$$

to represent this, where $\mathbf{u}_H = (u, v)$. In Question 3.4, it is shown that the appropriate modification of (3.140) is then

$$\frac{D\eta}{Dt} = W + \Gamma \nabla^2 \psi \quad \text{at } z \approx 1, \tag{3.142}$$

where

$$\Gamma = \frac{\gamma \sqrt{\frac{E}{2\varepsilon^2}} \left(\gamma + \sqrt{\frac{2}{E}} \right)}{\left[\left(\gamma + \frac{1}{\sqrt{2E}} \right)^2 + \frac{1}{2E} \right]}. \tag{3.143}$$

We do not offer much assessment of the likely size of Γ. We might expect γ not to be large (otherwise there is a significant shear layer at the troposphere), and that E would be less than its value at the surface. In this case, $\Gamma \approx \frac{\gamma E}{\varepsilon}$, and may be small.

We must also prescribe continuity of pressure; expanding hydrostatic tropospheric and stratospheric pressures in Taylor series about $z = 1$, and using continuity of density and pressure at the tropopause, but allowing for a jump in the vertical temperature gradient, we find the pressure condition becomes

$$\psi = c\eta^2 \quad \text{at } z = 1, \tag{3.144}$$

where

$$c = \frac{1}{2\bar{T}} \left[\frac{\partial \bar{T}}{\partial z} \right]_-^+. \tag{3.145}$$

Together, (3.142) and (3.144) give a condition on ψ at $z = 1$, in which W is given from the energy equation (3.124). We need one further boundary condition for ψ at $z = 0$.

The quantity ψ represents both pressure ($P = \bar{\rho}\psi$) and potential temperature ($\Theta = \partial\psi/\partial z$). We have already specified our one pressure condition, so any boundary condition on ψ must come from the thermal boundary condition at the surface. We need to be careful about this, however.

A suitable sort of boundary condition on Θ would appear to be to prescribe $\Theta = \Theta^{S,L}$, where for the sea (Θ^S) we would take the saturation potential temperature derived from (2.92), and on land, $\Theta^L(y, t)$ would represent the daily oscillation about a mean temperature which varies with latitude. Latitudinal variation of average surface (and thus potential) temperature from equator to pole is about 60 K over a distance of 10^4 km,[17] or 6 K per 10^3 km, a dimensionless amplitude of about $0.02 \sim \varepsilon^2$, consistent with a secular variation of mean Θ with latitude y.

However, the energy equation (3.124) is hyperbolic for Θ, and the loss of the conductive terms means that the prescription of surface potential temperature has no (short term) effect on tropospheric potential temperature beyond the planetary boundary layer. For example, inclusion of an appropriate eddy diffusive term $\varepsilon^2\nabla^2\Theta$ leads to the conclusion that on the relevant daily time scale (note that $l/U \sim 14$ hours) a surface fluctuating potential temperature only penetrates a distance of $O(\varepsilon)$ into the troposphere.

In obtaining the quasi-geostrophic equation (3.137), we took a z derivative of the energy equation. Therefore the undifferentiated form (3.124) provides the extra boundary condition for $\Theta = \partial\psi/\partial z$ at $z = 0$: specifically,

$$\frac{D}{Dt}\left(\frac{\partial\psi}{\partial z}\right) = H - SE^*\nabla^2\psi \quad \text{on } z = 0. \tag{3.146}$$

---

[17] The distance is very close to this exact value because that is how the French Academy of Sciences intended to define the metre in 1791.

To see how these conditions determine a solution (and how a numerical scheme might be implemented), suppose that S and W are known. (3.140) and (3.144) then give a boundary condition for ψ at $z = 1$; (3.146) gives a boundary condition for $\partial \psi / \partial z$ at $z = 0$. Together with appropriate x and y flow conditions, we can solve the potential vorticity equation for ψ. The unknowns W and S are then determined by (3.118) with the boundary condition (3.138), and the consistency condition (3.126).

Surface Temperature and Planetary Vorticity

The energy equation (3.124) is thus applied in the boundary conditions at both $z = 0$ and $z = 1$. The question arises as to what a suitable choice of initial condition for Θ is. This question is associated with the only remaining closure of the model to be made, which is the choice of the surface temperature scale T_0. Although the boundary fluxes are small, it is they which determine the mean temperature over long time scales.[18] Thus (3.25) states that the total energy $I + P + M$ is constant on short time scales, but over long time scales its value is determined by the fluxes on the right hand side. Thus the correct choice for T_0 follows from a global energy balance, as we earlier surmised.

A more detailed consideration of energy balance considers the variation of solar variation with latitude. If we simply time-average (but not space-average) (3.25), then we find a latitude-dependent surface temperature which varies slowly with y. Taking account of angle of solar insolation, Question 3.6 suggests an appropriate choice is

$$\Theta \approx 1 - s_1 y - s_2 y^2 \quad \text{at } z = 0, \tag{3.147}$$

where $s_1 \approx 1, s_2 \approx 0.14$. We emphasise that it is not necessary to apply (3.147) as the initial condition for the potential temperature perturbation Θ, but it is a physically sensible choice which reflects the long time-average surface temperature.

In a similar way, there are small horizontal diffusive terms missing from the potential vorticity equation. Over long time scales, we might expect these to render q in (3.139) uniform in y, even though this is not required by (3.148) (below).

3.5.2 The Day After Tomorrow

Although it is common to suppose that the stratification parameter S is prescribed, we have seen that in fact it is determined (for the normal case of stationary solutions)

---

[18]This argument is similar to that of the Prandtl–Batchelor theorem, which says that in a steady two-dimensional high Reynolds number flow, vorticity is constant inside any closed streamline. The argument is that $\mathbf{u}.\nabla \omega = Re^{-1} \nabla^2 \omega \approx 0$, so that $\omega \approx \omega(\psi)$, but exact integration of the vorticity equation round the closed streamline using the divergence theorem implies that $\omega'(\psi) \approx 0$, thus $\omega \approx$ constant. The same kind of argument yields the isothermal core of convection cells at high Rayleigh number (see Chap. 8).

from (3.134). In stationary conditions, since $\widehat{\zeta}_0$ is constant, we see that the right hand side of (3.137) vanishes, and the quasi-geostrophic equation takes the form

$$\frac{D}{Dt}\left[\nabla^2\psi + \beta y + \frac{E^*\widehat{\nabla^2\psi_0}}{\bar{\rho}}\frac{\partial}{\partial z}\left(\frac{1}{H}\frac{\partial\psi}{\partial z}\right)\right] = 0. \qquad (3.148)$$

The single equation (3.148) is thus a (nonlinear) integro-differential equation for ψ, and is to be solved subject to the boundary conditions (3.142), (3.144) and (3.146). Using (3.134), (3.124) and (3.115), the boundary conditions on surface and tropopause can be written in the form

$$\frac{D}{Dt}\left(\frac{\partial\psi}{\partial z}\right) = H\left[1 - \frac{\nabla^2\psi}{\widehat{\nabla^2\psi_0}}\right] \quad \text{on } z = 0 \qquad (3.149)$$

and

$$\frac{1}{\sqrt{c}}\frac{D\sqrt{\psi}}{Dt} = \left(H - \frac{D\psi_z}{Dt}\right)\frac{E^*\widehat{\nabla^2\psi_0}}{\bar{\rho}H} + \Gamma\nabla^2\psi \quad \text{on } z = 1. \qquad (3.150)$$

It is presently observed in the atmosphere that S is positive, or equivalently that $\widehat{\nabla^2\psi_0}$ is positive (if we suppose $H > 0$ for a near adiabatic temperature gradient), and this is necessary for basic static stability; but it is not mathematically obvious that the solution of (3.148) will always give a positive mean surface vorticity $\widehat{\zeta}_0$; nor is it necessary that $H > 0$. So long as the average surface vorticity remains positive, the weather remains fluctuating but stable. There are storms, sometimes violent, but they die away in time. In fact, mid-latitude depressions have a cyclonic (anti-clockwise) rotation,[19] and thus have positive vorticity. The rôle of storms may be thought of as a means of generating the positive vorticity necessary for well-posedness.[20]

There are two ways in which the quasi-geostrophic model can break down. One is if H becomes negative. In Question 3.5, it is shown that H can be approximately represented as

$$H = \frac{(4\alpha - 1)T^5}{p^2(T^2 + \nu StaM(T, p))}, \qquad (3.151)$$

in which we use the adiabatic approximation that T and p are given by

$$T(z) = 1 - \alpha z, \qquad (3.152)$$
$$p(z) = (1 - \alpha z)^{1/\alpha},$$

---

[19] In the northern hemisphere: clockwise in the southern hemisphere.

[20] There is a potentially interesting analogy with two-phase flow models here; see the discussion following Eq. (3.155).

Fig. 3.3 $H(z)$ computed
from (3.151) using values
$\alpha = 0.29, a = 18.8, St = 8.7,$
$\nu = 0.01$

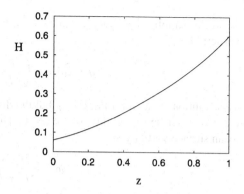

and $M(T, p)$ is given by (3.65). Figure 3.3 shows the variation of the dimensionless heating according to this expression for parameters appropriate to the Earth at present. We note that $H > 0$, and is an increasing function. The size of the increase is exaggerated, since it is observed that S decreases with height (a factor of two between surface and tropopause), while density decreases by a factor of about three, and thus we would expect (given (3.134)) an increase of H of about 1.5. Obvious quantitative reasons for the discrepancy lie in the assumptions of a grey, opaque, saturated atmosphere. The value of H given by (3.151) is very sensitive to variations in the parameter α. If α is reduced from 0.29 to 0.25, then $H = 0$, and for $\alpha < 0.25$, $H < 0$. Variation of α is largely due to atmospheric composition. For a CO_2 atmosphere, such as we have on Venus, $\alpha = 0.19$,[21] and then we would have $H < 0$.[22]

Zonal Flow

The other way of breaking down is if $\widehat{\nabla^2 \psi_0} < 0$. In order to see whether this could occur, it is necessary to solve the quasi-geostrophic potential vorticity equation. In general this is a numerical task, but some insight can be gained from the consideration of simple solutions. The simplest kind of flow is a zonal flow, in which there is no x dependence and a purely westerly flow, $\psi = \psi(y, z)$. There is no unique such solution, but one possibility, which satisfies the equations and boundary conditions (3.148), (3.149) and (3.150), is

$$\psi = (1 - s_1 y)z - \frac{\bar{\rho} H \beta}{2E^* \widehat{\zeta_0}} y z^2 + \frac{1}{2} \widehat{\zeta_0} y^2 \left[1 - \left(\frac{E^*}{\bar{\rho} \Gamma} + 1 \right) z \right], \qquad (3.153)$$

---

[21] The molecular weight of CO_2 is 44.

[22] And indeed, weather on Venus is very different, but whether this has anything to do with the present discussion is unclear; the planetary rotation rate is extremely small, so that the Rossby number is likely to be large, and the lower atmosphere is very dense, and thus more like an ocean. Only the upper atmosphere is more active, and there is some evidence for lightning storms there.

where for simplicity we take H and $\bar{\rho}$ to be constant. The corresponding potential vorticity is

$$q = \widehat{\zeta_0}\left[1 - \left(\frac{E^*}{\bar{\rho}\Gamma} + 1\right)z\right]. \qquad (3.154)$$

This solution exists for arbitrary $\widehat{\zeta_0}$, but the discussion concerning (3.147) suggests that we prescribe Θ and thus ψ_z on $z = 0$. The choice (3.147) then determines the mean surface vorticity as

$$\widehat{\zeta_0} = \frac{2\bar{\rho}\Gamma s_2}{\bar{\rho}\Gamma + E^*}, \qquad (3.155)$$

and thus positive, which is encouraging.

Suppose for the sake of argument that a choice such as (3.155) applies even for unsteady flows. The stratification parameter $S = \frac{\bar{\rho}H}{E^*\zeta_0}$ is thus positive for $H > 0$ and negative for $H < 0$. For increasing CO_2 composition of the atmosphere (for example), S will become negative, and the quasi-geostrophic potential vorticity equation becomes ill-posed. What then?

In Sect. 3.6.2 below, we show that a steady zonal flow becomes unstable for sufficiently small (positive) S. There is an interesting analogy here concerning this instability of a zonal flow as S is reduced, and the ill-posedness which occurs if S becomes negative. Bubbly two-phase flows are thought to become unstable to kinematic waves as the bubble volume (void) fraction increases, and this heralds the breakdown of the bubbly flow régime as the bubbles coalesce, forming a slug flow. It is also the case that appropriate two-phase flow models become ill-posed as the void fraction is increased further above the wave instability threshold. Apparently, the flow régime selects itself in such a way as to avoid ill-posedness in the corresponding model.

The analogy lies in the idea that the potential vorticity equation is ill-posed if $S < 0$; but an instability occurs before this can happen, if $S < S_c$ for some positive S_c. This instability is a herald of the ill-posedness, and the consequent breakdown of the quasi-geostrophic régime. On the Earth, this instability is already in place. It is a herald of the breakdown of our weather systems.

We are used to the atmosphere behaving in one (quasi-geostrophic) way, but there is little to say that alternative behaviours are not possible. We are now used to the idea that oceanic circulation and ice sheet extent can occur in different states: ice sheets in the last ice age surged over time scales of hundreds of years, oceanic circulation switched off and on over time scales of decades. We have not countenanced the more frightening idea that atmospheric circulation might also change rapidly.

If the mean surface vorticity becomes negative, then the surface temperature starts to rise on the convective time scale (14 hours, with $l = 1300$ km, $U = 26$ m s^{-1}), and the negative stratification causes global storms and massive planet-wide atmospheric overturn. The gentle, quasi-geostrophic régime is lost, and the weather becomes relentlessy stormy. The resultant massive cloud cover causes an abrupt increase in the greenhouse effect, with the consequent rise in temperature giving a positive feedback effect on surface evaporation.

Some of these effects are graphically portrayed in the film, *The Day After Tomorrow*, in which collapsing ice shelves and ocean currents magically combine to cause a meteorological upheaval in a matter of days. One might suppose that this is the stuff of fantasy, but it may be that a régime change on the Earth is possible, leading to weather changes far worse than that imagined in the film. If a transition were imminent, then one might expect to see increasing signs of instability: worse storms, more extreme events. This, of course, is thought by some to be exactly what has been happening over the last several decades.

3.6 Rossby Waves

We now seek a wave motion corresponding to the zero frequency geostrophic gravity wave mode satisfying (3.106) with $\omega = 0$. This is the Rossby wave, and it is most simply examined by studying (3.137) in the absence of heating, and assuming that the stratification parameter S is prescribed. (Such simplifications are in fact commonly made in studying the properties of (3.137).) We define a vertical eigenfunction $\Psi(z)$ satisfying the ordinary differential equation

$$\frac{1}{\bar{\rho}}\left[\frac{\bar{\rho}}{S}\Psi'\right]' = -m^2\Psi, \tag{3.156}$$

where for suitable homogeneous boundary conditions on Ψ, m^2 will be positive. With $H = 0$, $\psi = 0$ is a solution of (3.137), and small amplitude solutions of the equation will satisfy the linearised equation

$$\frac{\partial}{\partial t}\left[\nabla^2\psi + \frac{1}{\bar{\rho}}\frac{\partial}{\partial z}\left(\frac{\bar{\rho}}{S}\frac{\partial\psi}{\partial z}\right)\right] + \beta\frac{\partial\psi}{\partial x} = 0. \tag{3.157}$$

This has solutions

$$\psi = \Psi(z)\exp\left[i(kx + ly + \omega t)\right], \tag{3.158}$$

providing

$$\omega = \frac{k\beta}{k^2 + l^2 + m^2}. \tag{3.159}$$

These are Rossby waves. The wave speed $-\omega/k$ is negative, so that the waves move westwards. The sphericity of the Earth (i.e., $\beta > 0$) is essential in causing the waves to move. If there is a constant zonal flow U, then a similar analysis shows that the wave speed is

$$-\frac{\omega}{k} = U - \frac{\beta}{k^2 + l^2 + m^2}, \tag{3.160}$$

so that the westward drift is relative to the mean flow.

3.6.1 Baroclinic Instability

Gravity waves are the sound of the atmosphere. Like a bell which reverberates when struck, gravity waves are excited externally. For example, when the atmosphere flows over mountains, the waves are visualised by the periodic rows of clouds which form in the lee. However, they do not play a prominent part in large scale weather flows, because they are damped fairly rapidly by friction, and they are generated by external effects such as topographic forcing, not by internal dynamics.

Rossby waves, on the other hand, do play an important part in the day to day weather, and this is because they are continually generated by an instability in the underlying basic zonal flow. This instability is called *baroclinic instability*, and it is responsible for the basic wave-like nature of the circulation in mid-latitudes.

We consider the stability of a basic state which is taken to be a purely zonal flow. Because the quasi-geostrophic model is essentially inviscid (and conductionless), there is no unique such state. In the absence of the heating term H on the right hand side of (3.137), *any* zonal stream function $\psi(y, z)$ satisfies the QG equation (3.137). However, we would expect that over sufficiently long time scales, the potential temperature Θ of a zonal flow would become equal to the underlying surface temperature $\Theta_0(y)$, which ultimately is what drives the flow. A local expansion on the mid-latitude length scale of the global $O(\varepsilon)$ variation in θ suggests the prescription of $\Theta_0 = -y$ at $z = 0$. The choice $\Theta = -y$ implies the zonal flow

$$\psi = k - yz; \qquad (3.161)$$

generally, $k = k(z)$ but we will take it as constant. We will use (3.161) as the basic state whose stability we wish to study. However, we have to be careful to ensure that the model we study is consistent with this basic state! In terms of the full quasi-geostrophic model, (3.161) satisfies (3.124) and (3.118) with $W = 0$ from (3.138), $H = 0$ and S arbitrary. The conditions (3.140) and (3.144) then allow

$$\eta = \left(\frac{k - y}{c}\right)^{1/2}, \qquad (3.162)$$

where

$$c = \frac{1}{2T}\left[\frac{\partial T}{\partial z}\right]_-^+ > 0. \qquad (3.163)$$

3.6.2 The Eady Model

The simplest model in which baroclinic instability is manifested is the Eady model. In this model, the tropopause is considered to be a rigid lid, so that we impose

$$W = 0 \quad \text{at } z = 1. \qquad (3.164)$$

This follows from (3.140) and (3.144) in the limit that $c \to \infty$. Basal friction is ignored, corresponding to $E^* \to 0$ in (3.138), so that

$$W = 0 \quad \text{at } z = 0. \tag{3.165}$$

The Earth's sphericity is ignored by putting $\beta = 0$, we take the heating term $H = 0$ (consistent with the basic state (3.161)), and both the density $\bar{\rho}$ and the stratification S are taken as constant. The equation to be solved is thus the QG equation in the form

$$\frac{D}{Dt}\left[\nabla^2 \psi + \frac{1}{S}\frac{\partial^2 \psi}{\partial z^2} \right] = 0, \tag{3.166}$$

with boundary conditions which derive from (3.124):

$$\frac{D}{Dt}\left(\frac{\partial \psi}{\partial z} \right) = 0 \quad \text{at } z = 0, 1, \tag{3.167}$$

together with the no flow conditions $\partial \psi / \partial x = 0$ on $y = \pm 1$. In addition, (3.118) implies that

$$\frac{D}{Dt}\int_0^1 \zeta \, dz = 0. \tag{3.168}$$

This is automatically satisfied when ψ satisfies (3.166) and (3.167).

We write (taking $k = 1$ without loss of generality)

$$\psi = 1 - yz + \Psi, \tag{3.169}$$

and linearise for small Ψ to find

$$\left(\frac{\partial}{\partial t} + z\frac{\partial}{\partial x} \right)\left[\nabla^2 \Psi + \frac{1}{S}\frac{\partial^2 \Psi}{\partial z^2} \right] = 0, \tag{3.170}$$

subject to

$$\left(\frac{\partial}{\partial t} + z\frac{\partial}{\partial x} \right)\frac{\partial \Psi}{\partial z} - \frac{\partial \Psi}{\partial x} = 0 \quad \text{on } z = 0, 1,$$
$$\Psi = 0 \quad \text{on } y = \pm 1. \tag{3.171}$$

We seek solutions as linear combinations of the form

$$\Psi = A(z)e^{\sigma t + ikx + il_n y}, \tag{3.172}$$

where $l_n = n\pi/2$, and n is an integer. The appropriate linear combination of the y-dependent part is $\sin l_n y$ for n even, and $\cos l_n y$ for n odd. Then

$$(ikz + \sigma)\left[A'' - \mu^2 A \right] = 0, \tag{3.173}$$

Fig. 3.4 Wave speed of
perturbations in the Eady
model. Instability occurs
where the wave speeds are
complex conjugates, for
$\mu \lesssim 2.4$

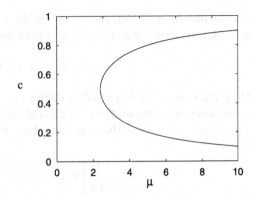

where

$$\mu^2 = (k^2 + l^2)S, \tag{3.174}$$

and

$$(ikz + \sigma)A' - ikA = 0 \quad \text{on } z = 0, 1. \tag{3.175}$$

Smooth solutions of (3.173) are linear combinations of $\cosh \mu z$ and $\sinh \mu z$, and
the dispersion relation which results from satisfaction of the boundary conditions in
(3.173) is

$$c = -\frac{\sigma}{ik} = \frac{1}{2} \pm \frac{1}{\mu}\left[\left(\frac{\mu}{2} - \coth\frac{\mu}{2}\right)\left(\frac{\mu}{2} - \tanh\frac{\mu}{2}\right)\right]^{1/2}, \tag{3.176}$$

where c is the wave speed. Figure 3.4 shows the (real) value of c as a function of
(positive) μ. Since $\mu/2 > \tanh(\mu/2)$, it is clear that c is complex for $\mu < \mu_c$, where

$$\frac{\mu_c}{2} = \coth\frac{\mu_c}{2}, \quad \mu_c \approx 2.399. \tag{3.177}$$

Complex conjugate values of c indicate instability, and this occurs for $\mu < \mu_c$.
Instability occurs if $k^2 + l^2 < \mu_c^2/S$, and thus is effected by the minimum values
$k = 0, l = \pi/2$, and the Eady instability criterion is

$$S < \frac{4\mu_c^2}{\pi^2} \approx 2.218; \tag{3.178}$$

this is readily satisfied in the Earth's atmosphere.

Evidently, the waves (stable or unstable) move to the east in the northern hemisphere, as is observed. The wave speed of unstable waves is 0.5, and the growth rate
is

$$\sigma_R = \frac{k}{\mu}\left[\left(\coth\frac{\mu}{2} - \frac{\mu}{2}\right)\left(\frac{\mu}{2} - \tanh\frac{\mu}{2}\right)\right]^{1/2}. \tag{3.179}$$

The growth rate goes to zero as $k \to 0$, and also as $\mu \to \mu_c$. Since for the fundamental mode $n = 1$, $\mu^2 = (k^2 + \frac{\pi^2}{2})S$ increases with k, the growth rate is maximum

Fig. 3.5 Growth rate σ_R of perturbations in the Eady model as a function of wave number k when the stratification $S = 0.25$. The growth rate is well approximated by $\sigma_R \approx 0.145 k (k_c - k)^{1/2}$, where $k_c = \sqrt{\frac{\mu_c^2}{S} - \frac{\pi^2}{4}}$ is the maximum wave number for instability

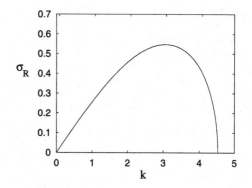

for an intermediate value of k. Indeed, Fig. 3.5 shows a typical graph of the growth rate plotted as a function of wave number k. Although linear stability gives us no information about the eventual form of the growing waves, it is plausible that the maximum growth rate at wave number k_m selects the preferred wavelength of disturbances as $2\pi / k_m$. This appears to be consistent with actual synoptic scale waves in mid-latitudes.

3.7 Frontogenesis

What has all this to do with the weather? If we look at a weather map, or listen to a weather forecaster on a mid-latitude television station, we will hear about fronts and depressions, low pressure systems, cyclones and anti-cyclones. These are indeed the standard bearers of the atmosphere, bringing their associated good and bad weather, storms, rainfall and snow. We are now in a position at least to describe how these features occur.

The weather is described, at least in essence, by some form of the geostrophic or quasi-geostrophic equations. Dissipative effects due to eddy viscosity and eddy thermal conductivity have a short term (days) effect in the planetary boundary layer within a kilometre or so of the surface, but only control the mean temperature of the troposphere over much longer time scales. As a consequence, weather is effectively described by a conservative system, indeed certain approximate models can be written as a Hamiltonian system, and as a consequence it is subject to the same sort of large amplitude fluctuations as those which characterise instability in such systems.

The basic poleward gradient of surface temperature attempts to drive a zonal flow, which is linearly unstable in the presence of a sufficiently small stratification parameter S. The very simplest representation of this instability is found in the Eady model (3.166) and (3.167), which is a nonlinear hyperbolic equation for the potential vorticity q. The consequence of the instability is that the steady, parallel characteristics of the zonal flow are distorted and intersect, forming a shock, as illustrated in Fig. 3.6. This is a front. It consists of a tongue of cold air intruded under warmer air, and the width of the front is typically of order 100 km.

Fig. 3.6 Contours of
temperature (*dashed lines*)
and potential temperature
(*solid lines*) in a forming front

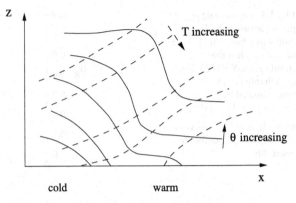

As the front develops, the baroclinic instability also distorts the flow in a wave-like pattern. The effect of this is to bend the front round, as illustrated in Fig. 3.7, forming a series of vortex-like rings. In the atmosphere, these are the cyclonic disturbances which form the mid-latitude low pressure storm systems, with typical dimensions of 2000 km. They also occur in the ocean, forming coherent rings of some 50 km diameter.

The description above is a little idealistic. On the Earth, fronts are an intrinsic consequence of the difference in properties between different air masses. The mid-latitude cells, for example, are bounded north and south by fronts across which the wind direction and the temperature changes. The warm mid-latitude westerlies are bounded polewards by the cold polar easterlies. The situation is complicated by continents and oceans. Continental air is dry, whereas oceanic air is moist. As a consequence of these geographic variations there are a number of different types of air masses, and the boundaries between these provide the seeds for frontal development. The fronts move and distort as shown in Fig. 3.7, but it is more sensible to think of the roll-up of a planar front and the formation of storm systems as a result of (Kelvin–Helmholtz like) instability of a linear vortex sheet, rather than as a consequence of shock formation in the nonlinear wave evolution of the quasi-geostrophic potential vorticity (QGPV) equation. In fact, the QGPV equation does not do a very good job of numerical weather front prediction.[23]

3.7.1 Depressions and Hurricanes

The storm systems which develop as shown in Fig. 3.7 are called cyclones. They are like vortices which rotate anti-clockwise, and are associated with low pressure at their centres (thus they are also called depressions). Conversely, a high pressure vortex rotating clockwise is called an anti-cyclone. A severe storm with central pressure of 960 millibars represents a dimensionless amplitude of $0.04 \sim \varepsilon^2$, and is thus within the remit of the quasi-geostrophic scaling.

---

[23]This comment is due to Peter Lynch.

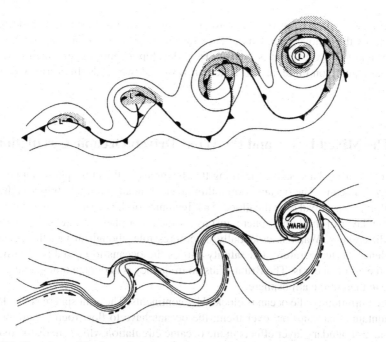

Fig. 3.7 Two views of the formation of cyclonic depressions from a baroclinically unstable front. The illustration resembles the Kármán vortex street which forms at moderate Reynolds number in the flow past a cylinder. The *upper diagram* shows isobars, the front, and cloud cover (*stippled*); the *lower diagram* shows isotherms, and flow of cold air (*solid arrows*) and warm air (*dashed arrows*). From Barry and Chorley (1998), page 162; their image is derived from a figure in Shapiro and Keyser (1990), and is reproduced with permission of the American Meteorological Society

In the tropics, tropical cyclones occur, and the most severe of these is the hurricane, or typhoon. In essence, the hurricane is very similar to the mid-latitude depression, consisting of an anti-clockwise rotating vortex, with wind convergence at the surface, and divergence at the tropopause. It is, however, fuelled by convection, and can be thought of as the result of a strong convective plume interacting with the Coriolis force, which causes the rotation, and in fact organises it into a spiral wave structure, as can be seen in satellite images by the spiral cloud formations.

The hurricane is distinguished by its high winds, high rainfall and relatively small size (hundreds rather than the thousands of kilometres of a mid-latitude cyclone). The strongest hurricane on record was hurricane Gilbert in 1988, where the central pressure fell to 888 mbar, and maximum windspeeds were in excess of 55 m s^{-1} (200 km hr^{-1}). The strong convection is a consequence of evaporation from a warm ocean, and it is generally thought that hurricane formation requires a sea surface temperature above 27° centigrade, or 300 K. Relative to a mean surface temperature of 288 K, this is an amplitude of 12 K, and dimensionlessly $12/288 \approx 0.04$, of $O(\varepsilon^2)$. In the tropics, the Rossby number ε is higher, and near the equator the quasi-geostrophic approximation breaks down, but hurricanes do not form in a band near the equator.

Hurricanes typically move westwards in the prevailing tropospheric winds, and dissipate as they move over land, where the fuelling warm oceanic water is not present, and surface friction is greater. They develop a central eye, which is relatively calm and cloud free, and in which air flow is downwards. In hurricanes, this eye is warm.

3.8 The Mixed Layer and the Wind-Driven Oceanic Circulation

Much of what we have said concerning the dynamics of the atmosphere applies to the world's oceans. The oceans form a thin layer of mean depth (slightly less than) four kilometres, spread over the globe. The dynamics of the oceans are thus those of a shallow layer of fluid on a sphere, just as for the atmosphere. There are, however, some differences. Water is essentially incompressible, though in fact the density dependence on temperature and salinity causes the oceans to be stably stratified, just as the atmosphere is. The Brunt–Väisälä frequency is about ten times smaller in the ocean than in the atmosphere.

More importantly, the ocean is blocked by continents. The atmosphere is blocked by mountains, but can flow over them; the oceans have to flow round continents. This causes boundary layer effects in the oceanic circulation which are distinctive.

The other major difference between oceans and atmosphere is in the driving mechanism for the flow. Differential heating between equator and pole drives the atmospheric flow, and this also drives the global thermohaline circulation (see Sect. 3.10) of the ocean, but the atmospheric circulation itself drives a circulation by means of wind stress at the surface. The global convective circulation due to differential heating and the wind-driven circulation interfere with each other, and it is not even clear which, if either, is dominant in determining the flow. In this sense, oceanic flow is much less well understood than atmospheric flow.

The vertical structure of the oceans is as follows. Near the surface there is a *mixed layer*, of typical thickness of the order of 50–100 metres, in which the density is uniform. This layer exists by virtue of the atmospheric wind stress, which mixes the surface waters. Below the mixed layer, the density begins to increase, and there is a thermocline over which the temperature changes from its warm surface value to the cooler deep ocean value. The thermocline has a thickness of the order of a kilometre, and the temperature contrast (warm at the surface, cool at depth) exists throughout temperate latitudes. It does not exist at the poles, but here the thermal structure is determined by the presence of sea ice. At the poles, salinity is of greater importance in determining the density profile. The thermal structure of the oceans is consistent with the concept of a thermally driven convective flow, which we describe later. First, we describe the wind-driven circulation.

The principal feature of the near-surface circulation of the oceans is the presence of circulatory flows with strong western boundary currents. In the North Atlantic, there is a clockwise circulation, with a strong current running up the Eastern seaboard of the United States. This Gulf Stream separates and flows across towards Europe, and is instrumental in providing Northern Europe with its anomalously warm climate. A similar current (the Kuroshio) occurs in the Western Pacific. These strong currents are due to the effects of the wind-driven circulation. In order to understand them, we need to formulate a model for ocean circulation in response to surface forcing.

Our starting point is with the dimensionless shallow water equations of (3.43) and (3.44). While the dimensionless variables are defined the same way as for the atmosphere, the scales are somewhat different. Typical velocities in the ocean are of order $U \sim 0.1$ m s^{-1}, a typical ocean horizontal length scale is $l \sim 3{,}000$ km, while the ocean depth is taken as $h \sim 4$ km. Assuming these values, we find $\delta \sim 10^{-3}$, $\Sigma \sim 0.5$, $Ro \sim 0.25 \times 10^{-3}$ and $F \sim 0.5 \times 10^{-3}$. We immediately assume that δ and Ro are negligible. We also assume that sea water is (approximately) incompressible, so that $\nabla \cdot \mathbf{u} = 0$, and we suppose that the ocean depth is uniform, of dimensionless depth one. Our coordinate system assumes that $z = 0$ defines mean sea level. From (3.43), we thus have

$$\frac{\cos \lambda_0}{\cos \lambda} \frac{\partial u}{\partial x} + \frac{1}{\cos \lambda} \frac{\partial (v \cos \lambda)}{\partial y} + \frac{\partial w}{\partial z} = 0. \qquad (3.180)$$

The third component of the momentum equation implies hydrostatic equilibrium, and in view of the smallness of F^2/Ro, we write the solution as

$$p = -z + \frac{F^2}{Ro} \eta(x, y). \qquad (3.181)$$

η represents sea surface elevation, and for $F^2/Ro \sim 10^{-3}$, values of $\eta \sim O(1)$ correspond to elevations of order four metres. The two horizontal components of the momentum equations thus take the form

$$-v \sin \lambda = -\frac{\cos \lambda_0}{\cos \lambda} \eta_x + f_x^*,$$
$$u \sin \lambda = -\eta_y + f_y^*, \qquad (3.182)$$

and the friction terms are prescribed as in (3.46), (3.36), (3.37) and (3.38), which yields

$$\mathbf{f}^* = E_H \nabla^2 \mathbf{u} + E_V \frac{\partial^2 \mathbf{u}}{\partial z^2}, \qquad (3.183)$$

where ∇^2 denotes the horizontal Laplacian, and

$$E_H = \frac{\varepsilon_H}{2 \Omega l^2}, \qquad E_V = \frac{\varepsilon_V}{2 \Omega h^2} \qquad (3.184)$$

define horizontal and vertical Ekman numbers, respectively. If we take $\varepsilon_V \sim 10^{-2}$ m^2 s^{-1}, $\varepsilon_H \sim 10^2$ m^2 s^{-1}, then we find $E_V \sim 0.4 \times 10^{-5}$, $E_H \sim 0.4 \times 10^{-7}$.

(3.180), (3.182) and (3.183) constitute the system we want to solve, subject to the conditions of no flow through the surface or base,

$$w = 0 \quad \text{at } z = 0 \text{ and } z = -1, \qquad (3.185)$$

and subject to an applied (dimensional) surface wind stress $\boldsymbol{\tau}_w$, which implies

$$\frac{\partial \mathbf{u}}{\partial z} = \frac{h \boldsymbol{\tau}_w}{\rho \varepsilon_V U} \quad \text{at } z = 0. \tag{3.186}$$

In addition we apply a no-slip condition at the base, thus

$$\mathbf{u} = \mathbf{0} \quad \text{at } z = -1. \tag{3.187}$$

The velocity scale U must be chosen by a suitable balance of the driving boundary condition (3.186). It is in fact not quite obvious how to do this. To do so, we need to anticipate the nature of the solution.

The Ekman numbers E_H and E_V are very small. Therefore they can be neglected except in boundary layers, and the flow is approximately geostrophic. If we integrate the mass conservation equation upwards from the bottom, we find that as $z \to 0-$, the vertical velocity w will be non-zero, and of $O(1)$, if we suppose (by choice of U) that $u, v \sim O(1)$ in the bulk flow. Therefore in order that w decrease to zero, we require a boundary (Ekman) layer near the surface, where the vertical Ekman viscous term becomes important. Evidently, this layer is of thickness $O(E_V^{1/2})$, and in order for w to decrease by $O(1)$ in the boundary layer, we need $u \sim v \sim E_V^{-1/2}$ in the Ekman layer. Hence we must have $\frac{\partial \mathbf{u}}{\partial z} \sim \frac{1}{E_V}$ in the surface Ekman layer, and this allows us to define the velocity scale. In view of (3.184), this suggests that if τ_0 is a scale for the wind stress, and we define

$$\boldsymbol{\tau}_w = \tau_0 \boldsymbol{\tau}, \tag{3.188}$$

then we should choose

$$\tau_0 = 2\rho U \Omega h. \tag{3.189}$$

The boundary condition (3.186) then becomes

$$E_V \frac{\partial \mathbf{u}}{\partial z} = \boldsymbol{\tau} \quad \text{at } z = 0. \tag{3.190}$$

We may now proceed to a solution. Away from all boundaries, we have the outer geostrophic solution $u = u_0(x, y)$, $v = v_0(x, y)$, where

$$-v_0 \sin \lambda = -\frac{\cos \lambda_0}{\cos \lambda} \eta_x, \tag{3.191}$$

$$u_0 \sin \lambda = -\eta_y.$$

By eliminating η and using the fact that $\frac{\partial}{\partial y} = \Sigma \frac{\partial}{\partial \lambda}$, we can show that the mass conservation equation takes the form

$$\frac{\partial w}{\partial z} = \Sigma v_0 \cot \lambda, \tag{3.192}$$

and thus

$$w_{\text{outer}} \to w_B + \Sigma v_0 \cot \lambda \quad \text{as } z \to 0-, \tag{3.193}$$

where w_B is an apparent surface vertical velocity due to Ekman pumping, which we now calculate.

In the surface Ekman layer, we put

$$z = -E_V^{1/2} \zeta, \qquad \mathbf{u} = E_V^{-1/2} \mathbf{U}, \tag{3.194}$$

so that, if we define the complex velocity and stress

$$S = U + iV, \qquad \tau = \tau_1 + i\tau_2, \tag{3.195}$$

where $\boldsymbol{\tau} = (\tau_1, \tau_2)$ and $\mathbf{U} = (U, V)$, then approximately

$$S_{\zeta\zeta} = iS \sin \lambda, \tag{3.196}$$

together with

$$-\frac{\partial S}{\partial \zeta} = \tau \quad \text{on } \zeta = 0, \qquad S \to 0 \quad \text{as } \zeta \to \infty. \tag{3.197}$$

The solution is

$$S = \frac{\tau}{B} e^{-B\zeta}, \tag{3.198}$$

where

$$B = \frac{(1 \pm i)}{\sqrt{2}} (\pm \sin \lambda)^{1/2}, \tag{3.199}$$

where we select the upper or lower sign depending on whether $\lambda > 0$ or $\lambda < 0$, respectively.

The continuity equation in the Ekman layer is

$$\frac{\partial w}{\partial \zeta} = \frac{\cos \lambda_0}{\cos \lambda} \frac{\partial U}{\partial x} + \frac{1}{\cos \lambda} \frac{\partial (V \cos \lambda)}{\partial y}; \tag{3.200}$$

integrating from $\zeta = 0$ to $\zeta = \infty$ and matching to the outer solution (3.193) then requires

$$w_B + \Sigma v_0 \cot \lambda = \frac{\cos \lambda_0}{\cos \lambda} \frac{\partial}{\partial x} \int_0^\infty U \, d\zeta + \frac{1}{\cos \lambda} \frac{\partial}{\partial y} \left[\cos \lambda \int_0^\infty V \, d\zeta \right]. \tag{3.201}$$

Simple calculation gives

$$\int_0^\infty S \, d\zeta = -\frac{i\tau}{\sin \lambda}, \tag{3.202}$$

and therefore (3.201) becomes

$$w_B + \Sigma v_0 \cot \lambda = \mathbf{k}.\nabla \times \left(\frac{\tau}{\sin \lambda} \right), \qquad (3.203)$$

where the vertical ($\mathbf{k}$) component of $\nabla \times \mathbf{F}$ in the present pseudo-spherical coordinates is defined as

$$\mathbf{k}.\nabla \times \mathbf{F} = \frac{\cos \lambda_0}{\cos \lambda} \frac{\partial F_2}{\partial x} - \frac{1}{\cos \lambda} \frac{\partial}{\partial y}(F_1 \cos \lambda). \qquad (3.204)$$

If we ignore the small Ekman pumping velocity w_B, then

$$\Sigma v_0 \cot \lambda = \mathbf{k}.\nabla \times \left(\frac{\tau}{\sin \lambda} \right) \qquad (3.205)$$

describes the so-called *Sverdrup flow* of the oceans due to the applied wind stress. Being purely algebraic, it pays no attention to continents. Therefore, the no-slip condition that we would like to apply at a continental margin cannot be applied; to do this we need to bring back the horizontal friction terms involving the horizontal Ekman number E_H. In fact, the basal Ekman pumping term also allows a regularisation, and we consider its form first.

Near the base, we write

$$z = -1 + E_V^{1/2} \zeta, \qquad (3.206)$$

and with $s = u + iv$, s satisfies

$$s_{\zeta\zeta} = i(s - s_0) \sin \lambda, \qquad (3.207)$$

where $s_0 = u_0 + iv_0$. The solution satisfying $s = 0$ at $\zeta = 0$ and with $s \to s_0$ as $\zeta \to \infty$ is

$$s = s_0 \left[1 - e^{-B\zeta} \right], \qquad (3.208)$$

where B is given by (3.199). Mass conservation then implies

$$\frac{\partial w}{\partial \zeta} = -E_V^{1/2} \left[\frac{\cos \lambda_0}{\cos \lambda} \frac{\partial u}{\partial x} + \frac{1}{\cos \lambda} \frac{\partial (v \cos \lambda)}{\partial y} \right], \qquad (3.209)$$

and in turn this implies that

$$w \sim w_B + \Sigma E_V^{1/2} v_0 \zeta \cot \lambda \quad \text{as } \zeta \to \infty, \qquad (3.210)$$

where

$$w_B = -E_V^{1/2} \left[\frac{\cos \lambda_0}{\cos \lambda} \frac{\partial}{\partial x} \int_0^\infty (u - u_0) \, d\zeta + \frac{1}{\cos \lambda} \frac{\partial}{\partial y} \left\{ \cos \lambda \int_0^\infty (v - v_0) \, d\zeta \right\} \right]. \qquad (3.211)$$

From (3.208),

$$\int_0^\infty (s - s_0)\, d\zeta = -\frac{s_0}{B},$$ (3.212)

and therefore we find

$$w_B = \sqrt{\frac{E_V}{2}} \left[\frac{\cos \lambda_0}{\cos \lambda} \frac{\partial}{\partial x} \left\{ \frac{1}{\sqrt{\sin \lambda}} (u_0 + v_0) \right\} + \frac{1}{\cos \lambda} \frac{\partial}{\partial y} \left\{ \frac{\cos \lambda}{\sqrt{\sin \lambda}} (v_0 - u_0) \right\} \right].$$ (3.213)

This expression is written for $\lambda > 0$ in the northern hemisphere. We give the corresponding recipe for the southern hemisphere below.

Next, we consider how to include the horizontal friction terms. To see how to do this, we reconsider (3.182), which we write, using (3.183), in the form (away from the vertical Ekman layers)

$$-v_0 \sin \lambda = -\frac{\cos \lambda_0}{\cos \lambda} \eta_x + E_H \nabla^2 u_0,$$

$$u_0 \sin \lambda = -\eta_y + E_H \nabla^2 v_0.$$ (3.214)

The solution of (3.200), integrated through the surface boundary layer, is still

$$w|_{\zeta=\infty} = \mathbf{k}.\nabla \times \left(\frac{\boldsymbol{\tau}}{\sin \lambda} \right),$$ (3.215)

and the solution of (3.180) still implies that this must be equal to

$$w|_{z=0-} = w_B - \left[\frac{\cos \lambda_0}{\cos \lambda} \frac{\partial u_0}{\partial x} + \frac{1}{\cos \lambda} \frac{\partial (v_0 \cos \lambda)}{\partial y} \right].$$ (3.216)

Equating these two results, and eliminating η in (3.214), we find after some algebra that (3.203) generalises to

$$w_B + \Sigma v_0 \cot \lambda = \mathbf{k}.\nabla \times \left(\frac{\boldsymbol{\tau}}{\sin \lambda} \right) + \frac{E_H \cos \lambda_0}{\cos \lambda \sin \lambda} \left[\frac{\partial}{\partial x} \nabla^2 v_0 - \frac{\partial}{\partial y} \nabla^2 u_0 \right].$$ (3.217)

In most derivations of this model, the fixation with latitude λ has long since disappeared, and when we look at the form of the Ekman terms in E_H and E_V, it is easy to see why. There are two things that help us. Both are based on the fact that the Ekman term will be completely negligible, except in boundary layers. Therefore the geostrophic approximation (3.191) is appropriate outside boundary layers. However, if we only require the velocity field to satisfy the no-flow-through condition at continents (and not the no-slip condition), then only the gradient of the velocity field will change in the continental margin boundary layer, and to leading order (3.191) will still apply. This enables us to use the geostrophic approximation in the friction term. We cannot use this argument if we wish to apply the no-slip boundary conditions, but we will ignore this subtlety here, and suppose that in both friction terms, u_0 and v_0 are given by (3.191).

The other assistance comes from the fact that because the viscous term is only relevant in thin boundary layers, then since λ will be approximately constant in such boundary layers, it is valid to ignore the derivatives of λ which arise in the Laplacian. Specifically, the definition (ignoring terms of $O(\delta)$ and the like) of the Laplacian is

$$\nabla^2 = \frac{\partial^2}{\partial x^2} + \frac{1}{\cos\lambda}\frac{\partial}{\partial y}\left[\cos\lambda\frac{\partial}{\partial y}\right], \tag{3.218}$$

and can in boundary layers be taken to be $\nabla^2 = \frac{\partial^2}{\partial x^2} + \frac{\partial^2}{\partial y^2}$. Adopting the geostrophic approximation (3.191), we then finally obtain an equation for a 'stream function' ψ, which we define by

$$\psi = \frac{\eta}{\sin\lambda}, \tag{3.219}$$

which is

$$\beta\psi_x = \sin\lambda\,\mathbf{k}.\nabla \times \left(\frac{\tau}{\sin\lambda}\right) - \sqrt{\frac{E_V}{2\sin\lambda}}\frac{\cos\lambda_0}{\cos\lambda}\left(\frac{\cos\lambda_0}{\cos\lambda}\psi_{xx} + \psi_{yy}\right)$$

$$+ \frac{E_H\cos\lambda_0}{\cos\lambda}\left[\frac{\cos\lambda_0}{\cos\lambda}\nabla^2\psi_{xx} + \nabla^2\psi_{yy}\right]. \tag{3.220}$$

In the southern hemisphere, the corresponding equation can be shown to have the same form, providing we write $\sqrt{\sin\lambda}$ as $\sqrt{|\sin\lambda|}$ and take y as pointing polewards (though evidently this is redundant in (3.220) since only second derivatives in y appear). The parameter β is defined here[24] by

$$\beta = \Sigma\cos\lambda_0. \tag{3.221}$$

We will study the boundary layer structure of this equation, and the formation of the western boundary currents, in the following section.

3.9 Western Boundary Currents: The Gulf Stream

Now at last we will ignore the largely irrelevant latitude terms in (3.220), and we will consider the case of an ocean in a box $B: 0 < x < 1, 0 < y < 1$, and we will require no flow through each side of the box; we may also require no slip if we consider the sides as representing continents. We have in mind a representation of the North Atlantic, with $x = 0$ representing the North American coastline, and $x = 1$ representing Africa and Europe.

We assume that the wind stress is purely meridional, but varying linearly with latitude, thus $\tau = (\frac{1}{2} + y, 0)$, and we inconsequentially ignore the trigonometric

---

[24] The parameter β appears as the same coefficient in other derivations, but is usually dimensional.

terms in the definition of $\nabla \times \boldsymbol{\tau}$. This wind field provides a representation of prevailing westerlies in mid-latitudes, and the easterly trade winds near the equator. The version of Eq. (3.220) we aim to solve is thus

$$\beta \psi_x = -1 - \varepsilon \nabla^2 \psi + E_H \nabla^4 \psi, \tag{3.222}$$

where we define

$$\varepsilon = \sqrt{\frac{E_V}{2 \sin \lambda}}, \tag{3.223}$$

and the boundary conditions of no flow through imply that the stream function is constant, i.e.,

$$\psi = 0 \quad \text{on } \partial B. \tag{3.224}$$

If, in addition, we prescribe no slip at the boundary, then also

$$\frac{\partial \psi}{\partial n} = 0 \quad \text{on } \partial B. \tag{3.225}$$

We are interested in the boundary layer structure of the solution for small E_H and small ε.

3.9.1 Effects of Basal Drag

Both of the small terms, in ε and E_H, represent singular perturbations to the basic Sverdrup flow, and we will consider their regularising effects separately. First, we suppose $\varepsilon \ll 1$ and neglect E_H. The equation to be solved is thus

$$\beta \psi_x = -1 - \varepsilon \nabla^2 \psi. \tag{3.226}$$

We will be able with this model to satisfy only the no-flow-through condition $\psi = 0$ on ∂B, since (3.226) is second order and elliptic.

The sub-characteristics go to the left, and therefore any boundary layer will exist at the left of the domain; this is the western boundary current, and the cause of the Gulf Stream. The outer solution is the Sverdrup flow, and is given by

$$\psi = \frac{1 - x}{\beta}, \tag{3.227}$$

which represents a southerly flow $v = -1/\beta$ (since $u \approx -\psi_y$, $v \approx \psi_x$). There is a boundary layer of thickness $O(\varepsilon)$ adjoining the western boundary $x = 0$, in which we put

$$x = \varepsilon X, \tag{3.228}$$

so that

$$\beta \psi_X \approx -\psi_{XX}, \tag{3.229}$$

with boundary conditions

$$\psi = 0 \quad \text{at } X = 0, \qquad \psi \to \frac{1}{\beta} \quad \text{as } X \to \infty; \tag{3.230}$$

the solution is

$$\psi = \frac{1 - e^{\beta X}}{\beta}, \tag{3.231}$$

and represents a northwards current of magnitude $v \sim \frac{1}{\varepsilon}$ at the western boundary. Thus the circulation is highly skewed.

There are also boundary layers adjoining the upper and lower boundaries, and these are similar to each other. For example, near the lower boundary, we put

$$y = \varepsilon^{1/2} Y, \tag{3.232}$$

so that

$$-\beta \psi_x \approx 1 + \psi_{YY}, \tag{3.233}$$

with boundary conditions

$$\psi = 0 \quad \text{at } Y = 0, \qquad \psi \to \frac{1 - x}{\beta} \quad \text{as } X \to \infty; \tag{3.234}$$

the appropriate 'initial' condition for the parabolic equation (3.233) is that

$$\psi = 0 \quad \text{at } x = 1. \tag{3.235}$$

(3.233) has a similarity solution, given by

$$\psi = \frac{1 - x}{\beta} [1 - f(\eta)], \qquad \eta = \frac{Y}{2} \left(\frac{\beta}{1 - x} \right)^{1/2}, \tag{3.236}$$

where f satisfies the differential equation

$$f'' + 2\eta f' - 4f = 0, \tag{3.237}$$

with boundary conditions

$$f(0) = 1, \qquad f(\infty) = 0. \tag{3.238}$$

The solution is the error function integral[25]

$$f(\eta) = i^2 \operatorname{erfc} \eta. \tag{3.239}$$

---

[25] See Abramowitz and Stegun (1964). The error function integrals are defined iteratively by $i^n \operatorname{erfc} \eta = i^{n-1} \operatorname{erfc} \eta$, $i^0 \operatorname{erfc} \eta = \operatorname{erfc} \eta$, and satisfy the equations $f_n'' + 2\eta f_n' - 2n f_n = 0$, where $f_n(\eta) = i^n \operatorname{erfc} \eta$; this is easily shown inductively by differentiating the equation, which shows that $f_n' = f_{n-1}$.

The assumption of a square box is irrelevant to the method of solution. An arbitrary domain will have a solution structure of the same form, with an attached western boundary layer of thickness $O(\varepsilon)$, in which $\psi_x \sim 1/\varepsilon$. We can now assess the neglect of the lateral drag term in E_H. The size of this term in the western boundary layer is $O(E_H/\varepsilon^4)$, and therefore the boundary layer structure above is valid, providing $E_H \ll \varepsilon^3$, i.e.,

$$E_H \ll \left(\frac{E_V}{2\sin\lambda}\right)^{3/2}. \tag{3.240}$$

If we use our estimates, then we have $E_H \sim 0.4 \times 10^{-7}$, $(\frac{E_V}{2\sin\lambda})^{3/2} \sim \frac{0.8 \times 10^{-8}}{(2\sin\lambda)^{3/2}}$, and (3.240) is barely feasible. This suggests that it may be more realistic to suppose that the lateral drag term in E_H controls the western boundary layer structure, and we now consider its effect. In any case, the basal drag term can only allow the no-flow-through condition, and the lateral term is necessary to bring the velocity to zero.

3.9.2 Effects of Lateral Drag

For simplicity, we neglect the basal drag term, so that the model for the stream function is

$$\beta\psi_x = -1 + E_H\nabla^4\psi, \tag{3.241}$$

together with the conditions

$$\psi = \frac{\partial\psi}{\partial n} = 0 \quad \text{on } \partial B. \tag{3.242}$$

The outer solution $\psi \sim \frac{1-x}{\beta}$ is as before, and the appropriate rescaling in the western boundary layer is

$$x = \frac{X}{E_H^{1/3}}, \tag{3.243}$$

and then the boundary layer equation is

$$\beta\psi_X = \psi_{XXXX}, \tag{3.244}$$

together with the boundary conditions

$$\psi = \psi_X = 0 \quad \text{at } X = 0, \qquad \psi \to \frac{1}{\beta} \quad \text{as } X \to \infty. \tag{3.245}$$

The solution of this is

$$\psi = \frac{1}{\beta}\left[1 - \exp\left(-\frac{1}{2}\beta^{1/3}X\right)\left\{\cos\frac{\beta^{1/3}\sqrt{3}\,X}{2} + \frac{1}{\sqrt{3}}\sin\frac{\beta^{1/3}\sqrt{3}\,X}{2}\right\}\right]; \tag{3.246}$$

note the oscillatory decay away from the boundary layer. We leave the solution in the horizontal boundary layers as an exercise (see Question 3.13).

3.10 Global Thermohaline Circulation

While the surface winds drive an oceanic circulation which is confined to the relatively near surface, there is a deeper circulation which is driven ultimately by the same source as that which drives the weather systems, that is to say, the radiatively induced poleward temperature gradient. While the atmospheric circulation can be viewed as a form of thermal convection mediated by the effects of a strong rotation, the deep oceanic circulation can be viewed as a form of thermal convection mediated by the strong effects of salinity. As such, this large scale convection is called the global thermohaline circulation, and it is often, slightly misleadingly, described as a conveyor belt, with descending water in the North Atlantic travelling southwards as North Atlantic Deep Water (NADW) to the Antarctic, where the conveyor sends it to the Indian and Pacific Oceans. There it rises, and eventually returns to the North Atlantic as surface water.

The poleward convection in the oceans is not affected by rotation in the same way as it is in the atmosphere, because of the presence of continents. In particular, convection in the Atlantic is channelled by the confining continents of the Americas to the west, and Europe and Africa to the east, and so it runs north to south. However, the oceans are saline, and this has a significant effect on the convection, because of the large contribution of salt to the density. While there is no source or sink of salt, salinity gradients are generated either by (stabilising) freshwater inputs via continental river outflow, or by (destabilising) evaporation, which provides a freshwater vapour flux to the atmosphere and a consequent salinification of the ocean surface.

If we remove the wind-driven circulation from the picture entirely, we think of competing forms of thermal and saline convection, for example in the North Atlantic. A purely thermal convection is produced by the equator to pole temperature gradient, and will cause a convective circulation in the form of a large scale roll. The Rayleigh number is so enormous that the steady roll may be unstable, with intermittent plumes developing out of the surface boundary layer, but one would expect the convective style to be essentially circulatory.

If, on the other hand, one removes the thermal buoyancy entirely, then the evaporation of the surface waters near the equator will lead to a destabilising surface salinity, but the consequent convection will be more finger-like, and localised, since there is no large scale imposed salinity gradient.

Superimposing these two notions, we might suppose a circulatory thermal convection, with the unstable saline surface boundary layer providing a series of localised downwelling plumes. In practice, such deep water formation regions do indeed exist, but there are not many of them. The two principal ones are in the North Atlantic, which forms the North Atlantic Deep Water, and in the Weddell Sea in the Antarctic, which forms the Antarctic Bottom Water (ABW). Enormous mixing takes place at the interface between these two water masses, and the Antarctic circumpolar current, which rotates west to east round Antarctica, acts as a kind of mixer, spraying out the NADW into the Pacific and Indian oceans, where it eventually wells up and returns to the North Atlantic surface water by various routes: through the Drake Passage between South America and Antarctica, from the Arctic via the Bering Strait, through Indonesia and round South Africa.

Although the origin of the thermohaline circulation may reside in the poleward thermal gradient, its nature may be largely salinity driven. The Atlantic surface waters are more saline than those of the Pacific, there being a net freshwater vapour flux from the Atlantic basin towards the Pacific. As was discussed in Sect. 2.5.7, it is thought that the rapid climate changes indicated by Dansgaard–Oeschger events may be associated with switches in the strength of the North Atlantic circulation—the so-called North Atlantic salt oscillator. The idea of this is that when the circulation is strong, it is warmer in the north, so that ice sheet melting is increased. The increased freshwater flux to the North Atlantic reduces the salinity of the surface ocean, thus reducing the air temperature, until eventually the circulation may even switch off. As the air temperature is reduced, however, melting on the ice sheets decreases and may cease entirely, allowing the ice sheets to regrow. The consequent decreased freshwater flux can then allow the oceanic circulation to restart.

3.11 Tides and Tsunamis

We go to the beach, and if we are paying attention, we notice that the tide comes in twice a day. Most of us know that tides are due to the gravitational attraction of the Sun and the Moon, and this seems to make sense. The Moon (which has the dominant effect) exerts an attraction on the water envelope of the oceans, pulling the water towards the Moon. Since the Earth rotates once a day, the high water remains stationary with respect to the Moon, and so we get the diurnal tide, apparently. But why are there then two tides a day? Worse, why is there only one tide a day in some places, and worst of all why is there sometimes almost no tide at all in certain locations, for example in the Mediterranean?

The answer to the most obvious of these problems, that of the semi-diurnal tide, is indicated in Fig. 3.8. Intuitively, we think that the pull of the Moon will cause a bulge in the oceans only on the side nearest to the Moon. This is because we are thinking at laboratory scale, and are forgetting the variation of gravity with distance. The Moon pulls the centre of the Earth with a certain force. On side N of the Earth in Fig. 3.8, this force is greater, because N is nearer to the Moon; consequently the ocean surface is pulled towards the Moon. So also is the Earth's surface, but: the

Fig. 3.8 The attractive effect of the Moon on the Earth's oceans

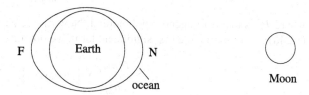

Fig. 3.9 Tide-generating
force diagram

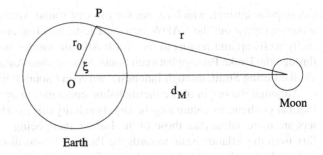

Earth is essentially rigid, and this deformation is inconsequential. On the far side of
the Earth, the force of attraction is correspondingly weaker, and relative to the force
on the Earth, the oceans experience a repulsion. Hence the bulge is as shown, and
thus as the Earth rotates, there are two tides a day.

3.11.1 The Tidal Equations

Suppose at a point P on the Earth, the centre of the Moon is at distance r, as shown
in Fig. 3.9. The distance of the centre of the Earth from the centre of the Moon
is denoted d_M, and the radius of the Earth is r_0. The fluid envelope of the Earth
experiences the gravitational force due to the Earth, but in addition there is a force
towards the Moon. However, to compute the tide-generating force, we must subtract
from this the attractive force of the Moon on the Earth. Thus the tide-generating
force per unit mass at P is

$$\mathbf{f}_{TG} = \nabla \left(\frac{GM}{r} \right) - \left(\frac{GM}{d_M^2} \right) \mathbf{i}, \qquad (3.247)$$

where $\mathbf{i}$ is the unit vector from the centre of the Earth to the centre of the Moon, G
is the gravitational constant, and M is the mass of the Moon. We can equivalently
write this force as the gradient of a potential,

$$\mathbf{f}_{TG} = GM \nabla \left[\frac{1}{r} - \frac{r_0 \cos \xi}{d_M^2} \right]. \qquad (3.248)$$

We can simplify this by using the expansion

$$\frac{1}{r} = \frac{1}{d_M} \left[1 - \frac{2 r_0 \cos \xi}{d_M} + \frac{r_0^2}{d_M^2} \right]^{-1/2} = \frac{1}{d_M} \sum_{n=0}^{\infty} \left(\frac{r_0}{d_M} \right)^n P_n(\cos \xi), \qquad (3.249)$$

where P_n is the nth Legendre polynomial. Now $r_0 \ll d_M$; substituting (3.249) into
(3.248) and retaining the first significant term, we obtain

$$\mathbf{f}_{TG} \approx \frac{GM r_0^2}{d_M^3} \nabla \left[P_2(\cos \xi) \right]. \qquad (3.250)$$

Fig. 3.10 Spherical
trigonometry relating the
angle ξ to the hour angle H,
the declination δ, and the
latitude λ. M indicates the
position of the
tide-generating body (e.g., the
Moon), and P is the local
position on the Earth

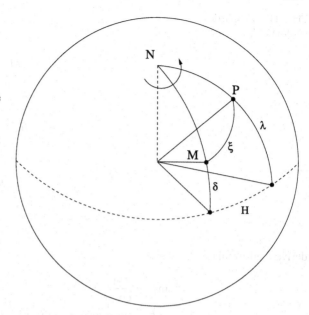

The second Legendre polynomial is defined by

$$P_2(\cos\xi) = \frac{1}{2}\left(3\cos^2\xi - 1\right). \tag{3.251}$$

Next we need to identify the angle ξ in terms of the normal angles of spherical polar coordinates. To do this we need a little spherical trigonometry. The geometry of the situation is indicated in Fig. 3.10, where we take the sphere radius to be one, without loss of generality. We want to relate the angle ξ to the declination of the Moon δ, the latitude λ, and the so-called hour angle H. This is simply longitude, except that the rotation of the Earth causes it to increase with time, specifically

$$H = \omega t + \phi, \tag{3.252}$$

where ω is the angular speed of rotation of the Earth. The bare bones of Fig. 3.10 are shown in Fig. 3.11. To relate ξ to the other variables, we consider triangles on the unit sphere, such as that shown in Fig. 3.12. If the lengths of the sides are a, b, c, and the corresponding opposite angles are α, β and γ, then we have the following formulae, which are, respectively, the first cosine rule and the sine rule:

$$\cos a = \cos b \cos c + \sin b \sin c \cos \alpha,$$

$$\frac{\sin\alpha}{\sin a} = \frac{\sin\beta}{\sin b} = \frac{\sin\gamma}{\sin c}. \tag{3.253}$$

Applying these formulae to the two triangles in Fig. 3.11 which constitute the quadrilateral, and bearing in mind that the two basal angles are right angles, we

Fig. 3.11 The spherical
quadrilateral

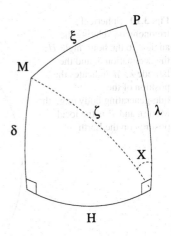

derive the formulae

$$\sin \zeta = \frac{\sin \delta}{\cos X},$$
$$\cos \xi = \cos \zeta \cos \lambda + \sin \delta \sin \lambda, \tag{3.254}$$
$$\cos \zeta = \cos \delta \cos H,$$

and from these we find

$$\cos \xi = \sin \lambda \sin \delta + \cos \lambda \cos \delta \cos H. \tag{3.255}$$

Finally, the tide-generating force can be written as

$$\mathbf{f}_{TG} = D \nabla \chi, \tag{3.256}$$

where

$$D = \frac{3GMr_0^2}{4d_M^3} \tag{3.257}$$

Fig. 3.12 Sides and angles of
a spherical triangle

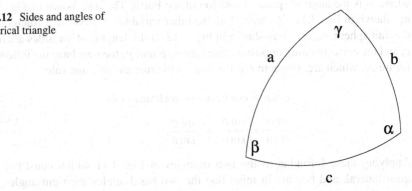

is known as the Doodson number (although it has dimensions), and $\chi = 2\cos^2 \xi$, whence

$$\chi = \cos^2 \lambda \cos^2 \delta \cos 2H + \sin 2\lambda \sin 2\delta \cos H + \left(\cos^2 \lambda \cos^2 \delta + 2\sin^2 \lambda \sin^2 \delta\right). \tag{3.258}$$

The time dependence of the forcing is expressed in the hour angle H, and we see that the three components represent, respectively, a semi-diurnal forcing ($\propto \cos 2H$), a diurnal forcing ($\propto \cos H$), and a 'long period' forcing, independent of Earth's rotation, but dependent on longer term orbital variations. Evidently the comparable but smaller effect of the Sun can be considered in the same way, and will add further ingredients to the tide-generating force.

Our model for tides is based on the Eqs. (3.1), except applied to an incompressible ocean. We use a depth integrated shallow water theory with a free upper boundary, but it is convenient to write the shallow water equations in vector form, delaying the intricacies of spherical polar coordinates until later. We define a vertical coordinate

$$z = r - r_0, \tag{3.259}$$

and we denote the ocean surface as $z = \eta$, and the ocean floor as $z = b$. The ocean depth is thus $h = \eta - b$, and we suppose that the depth-averaged horizontal velocity field is $\mathbf{u}$. From first principles, mass conservation yields the equation

$$\frac{\partial h}{\partial t} + \nabla \cdot (h\mathbf{u}) = 0, \tag{3.260}$$

where ∇ denotes the horizontal gradient vector. The (horizontal) momentum equation is obtained from (3.1), and is

$$\rho \left\{ \frac{d\mathbf{u}}{dt} + 2\mathbf{\Omega}_3 \times \mathbf{u} \right\} = -\nabla p + \rho D \nabla \chi. \tag{3.261}$$

In deriving this (note that $\mathbf{u}$ and ∇ are horizontal) we have integrated over the depth and then used the mass conservation equation. The term $2\mathbf{\Omega}_3 \times \mathbf{u}$ is the horizontal component of the Coriolis force, which is obtained by defining $\mathbf{\Omega}_3$ to be the vertical (in the z direction) component of the Earth's angular velocity. In addition, shallow water theory implies that

$$p \approx \rho g(\eta - z), \tag{3.262}$$

and thus

$$\nabla p \approx \rho g \nabla \eta. \tag{3.263}$$

Next we scale the equations. We denote the horizontal distance vector on the sphere as $\mathbf{x}$, and we define a dimensionless parameter ε as

$$\varepsilon = \frac{D}{gd}, \tag{3.264}$$

where d is mean ocean depth. Values of D/g are 0.27 m for the Moon, and 0.12 m for the Sun, while $d \approx 3,800$ m, so the parameter ε is very small, having a typical value of order 10^{-4}. We scale the variables as follows:

$$\mathbf{x} \sim r_0, \qquad \eta \sim \varepsilon d, \qquad b, h \sim d, \qquad \mathbf{u} \sim \varepsilon \sqrt{gd}, \qquad t \sim \frac{r_0}{\sqrt{gd}}, \qquad (3.265)$$

and this yields the non-dimensional system

$$\eta_t + \nabla \cdot (h\mathbf{u}) = 0,$$

$$h = -b + \varepsilon \eta, \qquad (3.266)$$

$$\mathbf{u}_t + \varepsilon (\mathbf{u} \cdot \nabla)\mathbf{u} + 2S \sin \lambda \, \mathbf{k} \times \mathbf{u} = -\nabla \eta + \nabla \chi,$$

where $\mathbf{k}$ is the unit vector in the vertical,

$$S = \frac{\omega r_0}{\sqrt{gd}} \qquad (3.267)$$

is a Strouhal number, and χ is given by (3.258), with now

$$H = \phi + St. \qquad (3.268)$$

With values $\omega = 7.27 \times 10^{-5}$ s^{-1}, $r_0 = 6.37 \times 10^6$ m, $g = 9.8$ m s^{-2}, $d = 3.8 \times 10^3$ m, we find $S \approx 2.4$. Neglecting terms of $O(\varepsilon)$ in (2.19), we have the tidal model

$$\eta_t + \nabla \cdot (h\mathbf{u}) = 0,$$
$$\mathbf{u}_t + 2S \sin \lambda \, \mathbf{k} \times \mathbf{u} = -\nabla \eta + \nabla \chi, \qquad (3.269)$$

in which we can take $h(\mathbf{x})$ independent of time.

3.11.2 Ocean Tides

We begin by taking uniform depth $h = 1$ and ignoring the Coriolis force, thus we put $S = 0$ in (3.269)$_2$ (but not in the definition of χ). From this there follows

$$\eta_t + \nabla \cdot \mathbf{u} = 0,$$
$$\mathbf{u}_t = -\nabla \eta + \nabla \chi, \qquad (3.270)$$

whence

$$\eta_{tt} = \nabla^2 \eta - \nabla^2 \chi. \qquad (3.271)$$

In the spherical polar coordinates ϕ and λ,

$$\nabla = \left(\frac{1}{\cos \lambda} \frac{\partial}{\partial \phi}, \frac{\partial}{\partial \lambda} \right), \qquad \cos^2 \lambda \, \nabla^2 = \frac{\partial^2}{\partial \phi^2} + \frac{\partial^2}{\partial \nu^2}, \qquad (3.272)$$

(cf. (3.28)), where we define

$$v = \ln\left[\frac{1 + \tan(\lambda/2)}{1 - \tan(\lambda/2)}\right], \qquad \frac{\partial}{\partial v} = \cos\lambda\frac{\partial}{\partial\lambda}. \qquad (3.273)$$

The tide-generating potential given by (3.258) contains separate components due to semi-diurnal, diurnal, and long period variations. The combined effect of these (and of the tidal effects of the Sun) can be obtained by linear superposition. For simplicity we will consider only the semi-diurnal lunar tide, denoted M_2, and suppose that the tide-generating potential is just

$$\chi = \cos^2\delta\cos^2\lambda\cos 2(\phi + St). \qquad (3.274)$$

Newton's equilibrium theory (illustrated in Fig. 3.8) assumes that $\eta = \chi$, but evidently this could only be approximately valid for slowly varying χ, i.e., $S \ll 1$. This is not the case on the Earth, and consequently the times of high tides lag the times of maximum attractive force.

The simplest case to consider is that of a narrow canal at a fixed latitude, which circumtraverses the globe. To obtain a solution in this case, we write (3.271) in the form (using (3.268) and (3.272))

$$S^2\cos^2\lambda\,\eta_{HH} = \left(\frac{\partial^2}{\partial H^2} + \frac{\partial^2}{\partial v^2}\right)(\eta - \chi), \qquad (3.275)$$

where we assume that η depends only on the combination $H = \phi + St$. Supposing the variation of v is small, we write $\eta = \eta^{(0)} + \eta^{(1)} + \cdots$, etc.,[26] and then we have to leading order

$$\eta^{(0)} - \chi^{(0)} \approx f(H), \qquad (3.276)$$

and at the next order

$$\frac{\partial^2(\eta^1 - \chi^{(1)})}{\partial v^2} = \left(S^2\cos^2\lambda - 1\right)f'' + S^2\cos^2\lambda\,\chi_{HH}. \qquad (3.277)$$

The boundary conditions of no flow through the side walls require $\eta_v - \chi_v = 0$, and therefore integration of (3.277) between the walls gives an integrability condition for its solution; this determines f and thus η (omitting the superscript zero), and the result is

$$\eta = \frac{\chi}{1 - S^2\cos^2\lambda}. \qquad (3.278)$$

This represents a westward travelling wave of speed $-S$ (since $\chi \propto \phi + St$), whose amplitude is modulated by latitude.

At the equator, $\lambda = 0$ and (since $S > 1$) the canal tides are out of phase with the tide-generating potential (the lag time is one quarter of a lunar day, slightly over

---

[26]To be more formal, we would write $v = v_0 + \varepsilon\tilde{v}$, take ε (here denoting the dimensionless canal width) to be small, expand as $\eta = \eta^{(0)} + \varepsilon^2\eta^{(1)} + \cdots$, and so on, but the end point is the same.

six hours). At the poles the tides are in phase. In-phase tides are called direct, and out-of-phase tides are called indirect. At a latitude of $\cos^{-1}(1/S) \approx 65°$, resonance occurs and tides can be significantly higher than the peak equilibrium tide of about 0.4 m.

Ocean Basins

In reality, there are continents, and these act as boundaries to the oceanic tidal flow. The free canal tidal wave given by (3.278) is a particular solution of the forced wave equation (3.271), but in an ocean basin, or in a canal with ends, free waves of the system are excited in order to prevent overflow at the ends. A solution for the case of a narrow canal at latitude λ with ends at longitude $\phi = \pm\frac{L}{2}$ is easily found with the same method as above, except that we allow the solution to depend on t and ϕ independently. We find $\eta = \chi + f(\phi, t)$, and the integrability condition for f takes the form

$$\cos^2\lambda \, f_{tt} - f_{\phi\phi} = 4S^2 \cos^2\lambda \, \chi, \tag{3.279}$$

and solving this subject to $f_\phi = 0$ at $\phi = \pm\frac{L}{2}$, we finally obtain the solution (see also Question 3.11)

$$\eta = \frac{\cos^2\delta \cos^2\lambda}{1 - S^2 \cos^2\lambda}[A \cos 2St + B \sin 2St], \tag{3.280}$$

where

$$A = \cos 2\phi - \frac{S \cos \lambda \sin L \cos(2S\phi \cos \lambda)}{\sin(LS \cos \lambda)},$$

$$B = -\sin 2\phi + \frac{S \cos \lambda \cos L \sin(2S\phi \cos \lambda)}{\cos(LS \cos \lambda)}. \tag{3.281}$$

The first terms in the expressions for A and B represent the forced tidal wave, and the second terms represent the free oscillations in the basin at the same frequency. We see that resonance occurs if $S \cos \lambda = 1$, as before, or if

$$2LS \cos \lambda = n\pi, \tag{3.282}$$

for integral n.

3.11.3 Seiches

A particular example of the tidal oscillations which occur in closed basins is afforded by the waves observed in lakes, termed *seiches*. For these, rotational effects are indeed small, and the relevant value of L is also. In addition, the depth of lakes is much less than that of the ocean, so that the tidal forcing coefficient S is larger.

For example, for a lake of length 80 km and depth 100 m, we have $L \sim 0.02$ and $S \sim 15$. Supposing the lake to be narrow and aligned with ϕ, we write

$$\phi = \phi_0 + \frac{\xi}{L}, \tag{3.283}$$

and then (3.281) is approximately (for $L \ll 1$)

$$A = \cos 2\phi - \frac{\Lambda \cos 2\Lambda\xi}{\sin \Lambda},$$
$$B = -\sin 2\phi + \frac{1}{L}\left\{\frac{\Lambda \sin 2\Lambda\xi}{\cos \Lambda}\right\}, \tag{3.284}$$

where

$$\Lambda = LS \cos \lambda. \tag{3.285}$$

The dominant coefficient is B, since Λ can be a good deal larger than L. With the lake dimensions above, at $\lambda = \frac{\pi}{4}$, $\Lambda \approx 0.2$. After some algebra and approximation, we find the tide given by (3.280) to be approximately

$$\eta \approx \frac{\cos^2 \delta}{S} \sin 2\Lambda\xi \sin 2St, \tag{3.286}$$

giving a standing wave of fairly small amplitude. The decrease of depth at the lake margins can enhance the amplitude considerably.

3.11.4 Amphidromic Points

The solution above in (3.280) for a canal represents the superposition of the forced westward travelling wave with two waves (bound by the basin) having the natural speed of the oceans (which is one, in these dimensionless units). When resonance occurs, this canal solution takes on the form of a standing wave.

Suppose, for example, that

$$LS \cos \lambda \approx \frac{\pi}{2}; \tag{3.287}$$

then the dominant part of the solution in (3.280) can be written as

$$\eta \approx -\left(\frac{\pi \cos L}{L(2LS \cos \lambda - \pi)}\right) \sin \frac{\pi\phi}{L} \sin 2St, \tag{3.288}$$

and is a standing wave; in particular, there is a nodal point at $\phi = 0$ where $\eta \approx 0$: at that point there is approximately no tide. If we consider the dominant part of the coefficient A in (3.281), then we find that it is generally non-zero at the resonant canal length; however, if we choose

$$2L \approx \pi \sin L \qquad (3.289)$$

as well as (3.282), then there is a genuine nodal point.[27]

In our discussion so far, we have neglected Coriolis force and the sphericity of the Earth. It is really not sensible to do this, since tidal forcing gives fundamentally two-dimensional motion, but the simpler analysis does illuminate two ideas, which turn out to be central in understanding how the tides work. The first is the idea of the tide as a wave, and the second is the idea that there can be nodal points. These two features essentially describe the real tide. The nodal points where the tidal amplitude vanishes are called *amphidromic points*.

We go back to the linearised model (3.269), again taking $h = 1$. It is clear that tidally forced solutions will be periodic, and we therefore write

$$\chi = \mathrm{Re}\big[\cos^2\delta\, G e^{2iSt}\big],$$
$$\eta = \mathrm{Re}\big[\cos^2\delta\, N e^{2iSt}\big], \qquad (3.290)$$
$$\mathbf{u} = \mathrm{Re}\big[\cos^2\delta\, U e^{2iSt}\big],$$

where we define

$$G = e^{2i\phi}\cos^2\lambda. \qquad (3.291)$$

We substitute these into (3.269), and can then determine U by taking the cross product of (3.269)$_2$ with $\mathbf{k}$, and this yields

$$\mathbf{U} = \frac{i\nabla\psi - \sin\lambda\,\mathbf{k}\times\nabla\psi}{2S\cos^2\lambda}, \qquad (3.292)$$

where we define

$$\psi = N - G. \qquad (3.293)$$

It follows from this that ψ satisfies

$$\nabla.\left\{\frac{\nabla\psi + i\sin\lambda\,\mathbf{k}\times\nabla\psi}{\cos^2\lambda}\right\} + 4S^2\psi = -4S^2G. \qquad (3.294)$$

The boundary conditions on the ocean-continent boundary are taken to be $\mathbf{U}.\mathbf{n} = 0$, and if we let $\mathbf{n}$ denote the outward normal away from continents, and $\mathbf{t}$ the tangent vector at the continent when its boundary is traversed counter-clockwise, then $\mathbf{k}\times\mathbf{n} = \mathbf{t}$, and the no-flow-through boundary condition can be written

$$\frac{\partial\psi}{\partial n} - i\sin\lambda\,\frac{\partial\psi}{\partial t} = 0 \quad \text{on } \partial C, \qquad (3.295)$$

---

[27] Thus $L = \frac{\pi}{2}$, and using (3.287), $S\cos\lambda = 1$.

where ∂C denotes the continental boundary. In spherical coordinate form, the Helmholtz type equation (3.294) can be written, after some manipulation, in the form

$$\frac{1}{(1-\mu^2)^2}\frac{\partial^2\psi}{\partial\phi^2}+\frac{\partial^2\psi}{\partial\mu^2}+i\left(\frac{1+\mu^2}{(1-\mu^2)^2}\right)\frac{\partial\psi}{\partial\phi}+4S^2\psi=-4S^2G, \qquad (3.296)$$

where we define

$$\mu=\sin\lambda. \qquad (3.297)$$

(3.296) looks a little strange with the imaginary term, but in fact the equation behaves essentially as a Helmholtz equation. Both the tidally forced (particular) solution and the free (homogeneous) solutions have separable forms

$$\psi=\Psi(\mu)\,e^{i\alpha\phi}, \qquad (3.298)$$

where

$$\Psi''+\left[4S^2-\frac{\alpha(\alpha+1+\mu^2)}{(1-\mu^2)^2}\right]\Psi=0, \qquad (3.299)$$

though such solutions have limited applicability in a realistic geometry.

They do, however, point the way to understanding the solution behaviour which is actually observed. For sufficiently small longitudinal wave number α, Ψ will also be oscillatory, and the resulting waves are essentially Poincaré or gravity waves, which we have seen before in the atmosphere (in Sect. 3.4). For large α, Ψ is exponential, and the resultant waves are edge waves, attached to coastal boundaries, and are Kelvin waves. In the open ocean, an oscillatory solution of the form

$$N\propto\exp\{-i(\alpha\phi+\beta\lambda)\} \qquad (3.300)$$

corresponds to a tidal wave moving in the direction (α,β), and this is the form of the solution locally providing $N\neq0$.

More generally, suppose the solution ψ of (3.296) with (3.295) is computed, and thus N is determined. We define the amplitude $R_{CR}(\phi,\lambda)$ and the phase $t_{CT}(\phi,\lambda)$ by

$$N=R_{CR}\exp(-2i\,St_{CT}). \qquad (3.301)$$

Then the surface elevation η is given by

$$\eta=\cos^2\delta\,R_{CR}(\phi,\lambda)\cos\{2S[t-t_{CT}(\phi,\lambda)]\}. \qquad (3.302)$$

The lines $t_{CT}(\phi,\lambda)=$ constant are called *co-tidal lines*; they represent the crest of the tidal wave as it circulates round the world's oceans. The quantity R_{CR} is called the *co-range*. It is a measure of the tidal amplitude at a point.[28] If N was an ana-

---

[28]In practice, the co-tidal phase and co-range amplitude are dimensional quantities; the phase is measured in hours (of the lunar day), while the co-range is the elevation distance between high and low water, and thus twice the amplitude of the underlying sine wave.

lytic function (i.e., $\nabla^2 N = 0$), then the co-tidal phase lines and co-range amplitude lines would be orthogonal. This is not the case in practice, but they do intersect transversely, and thus retain much of the same topology.

The image of a global tidal wave washing round the oceans breaks down at points where $N = 0$, since then the solution can no longer be approximately exponential as in (3.300). Because N is complex, the condition $N = 0$ requires both Re $N(\phi, \lambda) = 0$ and Im $N(\phi, \lambda) = 0$, and thus occurs at isolated points: these are the amphidromic points. In their vicinity, N varies linearly with ϕ and λ, and the local structure may be recovered by consideration of (3.296) when $N \approx 0$, i.e., $\psi \approx -G$: if $N = 0$ at (ϕ_0, μ_0), then we put

$$\phi = \phi_0 + X, \qquad \mu = \mu_0 + aY, \tag{3.303}$$

where we will choose $a > 0$ later for convenience. For small X and Y, (3.296) implies, approximately,

$$\frac{1}{(1 - \mu_0^2)^2} \frac{\partial^2 \psi}{\partial X^2} + \frac{1}{a^2} \frac{\partial^2 \psi}{\partial Y^2} + i \left(\frac{1 + \mu_0^2}{(1 - \mu_0^2)^2} \right) \frac{\partial \psi}{\partial X} \approx 0, \tag{3.304}$$

and this has local solutions of the form

$$\psi = -G_0 \exp[i\alpha X + \beta Y], \tag{3.305}$$

where $G = G_0$ at the amphidromic point, and

$$\beta = \pm \frac{a}{(1 - \mu_0^2)} [\alpha^2 + (1 + \mu_0^2)\alpha]^{1/2}. \tag{3.306}$$

Note that if solutions are oscillatory in the ϕ direction, then they are locally exponential in the λ direction, providing $\alpha > 0$ or $\alpha < -(1 + \mu_0^2)$.

In the vicinity of the amphidromic point,

$$G \approx G_0 \exp\left\{ 2iX - \frac{2\mu_0 aY}{1 - \mu_0^2} \right\}, \tag{3.307}$$

and thus

$$N \approx G_0 \left[\exp\left\{ 2iX - \frac{2\mu_0 aY}{1 - \mu_0^2} \right\} - \exp\{i\alpha X + \beta Y\} \right]$$

$$\approx i(2 - \alpha)G_0[X + i\gamma Y], \tag{3.308}$$

where

$$\gamma = \frac{a[2\mu_0 \pm \{\alpha^2 + (1 + \mu_0^2)\alpha\}^{1/2}]}{(1 - \mu_0^2)(2 - \alpha)}. \tag{3.309}$$

By choosing a appropriately, we therefore have the local structure

$$N \sim (X \pm iY) \tag{3.310}$$

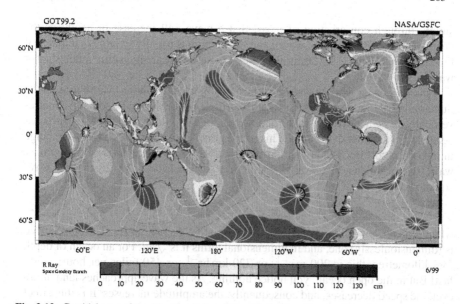

Fig. 3.13 Co-tidal map of the M_2 tide, deduced from Topex/Poseidon satellite altimeter measurements. The *colour scale* indicates the amplitude of this component of the tide. *White phase lines* are shown every 30°, with the *heavier lines* representing 0° (corresponding to when the mean moon passes the 0° or 180° meridians). Figure courtesy of Richard Ray, NASA Goddard Space Flight Center, and kindly provided by Helen Johnson

if β is real. In local polar coordinates $X + iY = Re^{i\theta}$, we then have

$$R_{CR} \sim R, \qquad 2St_{CT} = \mp\theta, \qquad (3.311)$$

and this gives the local structure of the co-tidal lines and co-range lines in the vicinity of an amphidromic point. The co-tidal lines fan out from the point, forming a web whose circular strands are the co-range lines. At the amphidromic point the tidal range is zero, and in its vicinity the tidal wave spins round as if glued to the amphidromic point. The rotation is anti-clockwise if the plus sign is selected in (3.311), and clockwise for the minus sign. Generally rotation is anti-clockwise in the northern hemisphere and clockwise in the southern hemisphere, but not always. Figure 3.13 shows a map of the co-tidal lines for the world oceans. The rôle of the amphidromic points in organising the global tidal wave is clear.

If we supposed that $-(1 + \mu_0^2) < \alpha < 0$, then β would be imaginary, and the value of γ is complex, but as long as the real part is non-zero, the conclusion is essentially unaltered.

3.11.5 Tsunamis

We have not discussed the effect of shallowing of the ocean on the continental slope which joins the continental shelf to the abyssal ocean bottom. It is fairly evident, simply through conservation of mass, that a wave will increase in height as the depth decreases, but for tidal waves, the wavelength is usually so large that this is of little relevance. This is not so for tsunamis, because they are by nature much shorter wavelength waves.

Tsunamis occur when the ocean surface is subjected to sudden disturbance. In the Sumatran earthquake of 2004, for example, the sea floor shifted by some five metres vertically along a horizontal fault running hundreds of kilometres, the whole process occurring in a matter of minutes. This sudden displacement causes a localised bulge in the ocean surface, which then propagates away from its centre as a free gravity wave. In our dimensionless variables, the free surface is described by the free wave equation

$$\eta_{tt} = \mathbf{V} \cdot (h \mathbf{V} \eta), \tag{3.312}$$

where it is reasonable to ignore rotation on relatively small scales. The wave speed is (dimensionlessly) one, and in dimensional terms it is $\sqrt{gd}$. For an ocean of depth four kilometres, this is 200 m s^{-1}, or 720 km hr^{-1}, or 450 miles per hour: pretty fast! But in the open ocean, you notice nothing. As the wave approaches land, however, the speed decreases, and consequently, the amplitude increases. It is this effect which causes the anthropocentrically massive tidal waves with amplitudes of tens of metres, which are the dramatic shoreline expression of a tsunami. From the point of view of the ocean, such massive waves are a tiny dribble at the margin.

In the Open Ocean

To describe the result of an initial localised disturbance to the ocean surface, we write (3.312) in cylindrical coordinates, assuming constant depth $h = 1$. Thus

$$\eta_{tt} = \eta_{rr} + \frac{1}{r}\eta_r, \tag{3.313}$$

and we will suppose that

$$\eta = \eta_0(r), \qquad \eta_t = 0 \quad \text{at } t = 0. \tag{3.314}$$

The initial value problem must be solved numerically, but some information is available if we limit attention to the behaviour at large r. If we define

$$\eta = \frac{\phi}{\sqrt{r}}, \tag{3.315}$$

then ϕ satisfies

$$\phi_{tt} = \phi_{rr} + \frac{\phi}{4r^2}, \tag{3.316}$$

with leading order behaviour $\phi \sim \phi_0(r - t)$, thus

$$\eta \sim \frac{\phi_0(r - t)}{\sqrt{r}}. \tag{3.317}$$

To extend this result to higher order, we write $r - t = \xi$, and change to variables ξ and t. A solution can then be found in the form

$$\eta \sim \frac{\phi_0(r-t)}{\sqrt{r}} + \frac{\frac{1}{8}\int_0^{r-t} \phi_0(\xi)\,d\xi}{\sqrt{rt}} + \cdots \tag{3.318}$$

(see also Question 3.14).

An alternative approach is to solve the problem (3.313) and (3.314) directly using an appropriate transform, which is here the Hankel transform defined, together with its inverse, by

$$\hat{g}(\rho) = \int_0^\infty r J_0(\rho r) g(r)\,dr,$$
$$g(r) = \int_0^\infty \rho J_0(\rho r) \hat{g}(\rho)\,d\rho. \tag{3.319}$$

(A generalisation of this to Bessel functions of order ν is also available, and is appropriate for the solution of the wave equation in cylindrical coordinates when there is dependence on angle as well.) The point of using this transform is that

$$\widehat{\eta_{rr} + \frac{1}{r}\eta_r} = -\rho^2 \hat{\eta}, \tag{3.320}$$

so that the solution for η satisfying the boundary conditions is

$$\hat{\eta} = \hat{\eta}_0(\rho)\cos\rho t, \tag{3.321}$$

and thus

$$\eta = \int_0^\infty \rho J_0(\rho r)\cos\rho t \int_0^\infty s J_0(\rho s)\eta_0(s)\,ds\,d\rho. \tag{3.322}$$

We can now obtain asymptotic limiting behaviour for large t directly from this form of the solution.

Most obviously, it seems we should write

$$\eta = \int_0^\infty s\eta_0(s)K(r,t,s)\,ds, \tag{3.323}$$

where

$$K(r,t,s) = \int_0^\infty \rho J_0(\rho r) J_0(\rho s)\cos\rho t\,d\rho, \tag{3.324}$$

and then evaluate K asymptotically for large t. This is not possible: integration by parts does not work. Inspection of tables of integrals or of transforms also narrowly fails to reveal an explicit expression for K. With some thought, this is hardly surprising, since the integral defined in (3.324) does not generally exist.

In writing (3.323), we essentially assumed Fubini's theorem, which allows replacement of the order of integration when the integrand is Lebesgue integrable, and in particular absolutely integrable. Evidently this is not the case. We can get around this by defining

$$K(r, z, s) = \int_0^\infty \rho J_0(\rho r) J_0(\rho s) e^{-\rho z} \, d\rho, \tag{3.325}$$

where $\operatorname{Re} z > 0$, and then

$$\eta(r, z) = \int_0^\infty s \eta_0(s) K(r, z, s) \, ds. \tag{3.326}$$

The solution is then obtained by letting $z \to it + 0+$. The asymptotics at large t are now straightforward. Using the limit of the Bessel function at large values of its argument, we find, as before, that

$$\eta \sim \frac{\phi_0(r - t)}{\sqrt{r}}, \tag{3.327}$$

where

$$\phi_0(\xi) = \frac{1}{\sqrt{2\pi}} \int_0^\infty \sqrt{\rho} \cos\left(\rho\xi - \frac{1}{4}\pi\right) \int_0^\infty s J_0(\rho s) \eta_0(s) \, ds \, d\rho. \tag{3.328}$$

When $r = O(1)$, then integration by parts of (3.325) as $z \to \infty$ followed by putting $z = it$ yields

$$\eta \sim -\frac{1}{t^2} \int_0^\infty s \eta_0(s) \, ds \tag{3.329}$$

as $t \to \infty$.

At the Coast

When a tsunami arrives at the coast, it slows down and bulks up. The slow down is because the open ocean wave speed $\sqrt{gd}$ decreases, and the growth of the wave amplitude is then a consequence of conservation of mass. A simple model which describes this follows from taking the depth to increase linearly from the shoreline, so that in two dimensions the surface elevation satisfies

$$\eta_t = \frac{\partial}{\partial x}[x\eta_x], \tag{3.330}$$

where x measures seawards distance from the shore. This has separable solutions of the form $\eta = e^{i\omega t} f(x)$, where f satisfies

$$(xf')' + \omega^2 f = 0, \tag{3.331}$$

of which the solutions are Bessel functions $J_0(2\omega\sqrt{x})$ and $Y_0(2\omega\sqrt{x})$, or equivalently the Hankel functions $H_0^{(1)}(2\omega\sqrt{x}) = J_0(2\omega\sqrt{x}) + iY_0(2\omega\sqrt{x})$ and

$H_0^{(2)}(2\omega\sqrt{x}) = J_0(2\omega\sqrt{x}) - iY_0(2\omega\sqrt{x})$, and the asymptotic behaviour at large x of the corresponding solutions $\eta^{(1)}$ and $\eta^{(2)}$ is given by

$$\eta^{(1)} \sim \frac{(1-i)}{\sqrt{2\pi\omega}} x^{-1/4} \exp[i\omega(t + 2\sqrt{x})],$$

$$\eta^{(2)} \sim \frac{(1+i)}{\sqrt{2\pi\omega}} x^{-1/4} \exp[i\omega(t - 2\sqrt{x})]. \tag{3.332}$$

The function $\eta^{(1)}$ represents the incoming wave, while $\eta^{(2)}$ represents the reflected wave; the general solution will be

$$\eta = \int_{-\infty}^{\infty} \left[I(\omega)e^{i\omega t} H_0^{(1)}(2\omega\sqrt{x}) + R(\omega)e^{i\omega t} H_0^{(2)}(2\omega\sqrt{x}) \right] d\omega; \tag{3.333}$$

$I(\omega)$ indicates the incoming wave amplitude, and $R(\omega)$ the reflected wave amplitude.

The asymptotic form of the solutions at large x explains the basic amplification of the wave as it approaches the shore. The wave speed is $\sqrt{x}$ (thus the wave slows down) and the wave amplitude is $1/x^{1/4}$. Of apparent concern in this solution is that the Hankel functions are singular as $x \to 0$, specifically $H_0^{(1,2)}(2\omega\sqrt{x}) \sim \pm\frac{i}{\pi}\ln x + O(1)$ as $x \to 0$, suggesting that η becomes infinite as the wave reaches the shore. This is an artefact of the implicit assumption that the wave is bounded by the fixed shoreline, and that the depth is independent of surface elevation in the shallows.

Consulting (3.266), we can see that a better model would be

$$\eta_{tt} = \frac{\partial}{\partial x}\left[(x + \varepsilon\eta)\frac{\partial\eta}{\partial x} \right]. \tag{3.334}$$

This is the sort of problem which can be treated by the method of strained coordinates, when ε is small, as here. To do this, we consider η to be a function of a strained coordinate s and time t, and we write

$$x = s + \varepsilon X(s, \tau) + \cdots,$$

$$t = \tau, \tag{3.335}$$

and then expand η as an asymptotic expansion $\eta \sim \eta^{(0)} + \varepsilon\eta^{(1)} + \cdots$, whence we find that

$$\eta_{tt}^{(0)} - \frac{\partial}{\partial s}\left[s\frac{\partial\eta^{(0)}}{\partial s} \right] = 0,$$

$$\eta_{tt}^{(1)} - \frac{\partial}{\partial s}\left[s\frac{\partial\eta^{(1)}}{\partial s} \right] = 2X_t\eta_{st}^{(0)} + X_{tt}\eta_s^{(0)} - X_s\eta_{tt}^{(0)} \tag{3.336}$$

$$+ \frac{\partial}{\partial s}\left[(X + \eta^{(0)} - sX_s)\eta_s^{(0)} \right],$$

and so on. (We retain t as the time variable since it is the same as τ; note only that the time derivatives in (3.336) are with respect to constant s.) We suppose that the leading order solution is the monochromatic wave given by the integrand of (3.333), thus

$$\eta^{(0)} = e^{i\omega t}\left[(I + R)J_0(2\omega\sqrt{s}) + i(I - R)Y_0(2\omega\sqrt{s})\right] + \text{(cc)}, \qquad (3.337)$$

where (cc) denotes the complex conjugate. As $s \to 0$, the Bessel functions have behaviours

$$J_0(2\omega\sqrt{s}) \sim 1 - \omega^2 s, \qquad Y_0(2\omega\sqrt{s}) \sim \frac{2}{\pi}\left[\ln\omega + \frac{1}{2}\ln s + \gamma\right], \qquad (3.338)$$

where γ is Euler's constant, $\gamma \approx 0.5772$. As a consequence, the right hand side of $(3.336)_2$ is singular as $s \to 0$, and the method of stationary phase dictates that we choose the straining X so that the solution for $\eta^{(1)}$ is no more singular than $\eta^{(0)}$ at $s = 0$. The inhomogeneity causing possible singular terms is in the last term on the right hand side of $(3.336)_2$, and so this dictates our initial choice for X,

$$X = U(t) + V(t)\ln s + \cdots, \qquad (3.339)$$

where

$$V = -\frac{ie^{i\omega t}(I - R)}{\pi} + \text{(cc)},$$

$$U = V - \left\{e^{i\omega t}\left[I + R + \frac{2i}{\pi}(I - R)(\ln\omega + \gamma)\right] + \text{(cc)}\right\}. \qquad (3.340)$$

Note that U and V are real (as indeed they must be).

This determines the straining to leading order, but provides no information on the amplitude R of the reflected wave in terms of the incident amplitude I. In addition, the choice of X in (3.339) produces further (weaker) singular terms on the right hand side of $(3.336)_2$, particularly in the first and second terms, and these can be removed by correcting (3.339) to

$$X = U(t) + V(t)\ln s + W(t)s\ln^2 s + Y(t)s\ln^3 s + \cdots, \qquad (3.341)$$

and after some algebra, the suppression of terms proportional to $\frac{1}{s}$ and $\frac{\ln s}{s}$ on the right hand side of $(3.336)_2$ leads to the choice

$$Y = \frac{1}{6}\left[\ddot{V} + \frac{2\dot{V}^2}{V}\right], \qquad W = \frac{1}{2}\left[-\ddot{V} + \ddot{U} + \frac{2\dot{V}\dot{U}}{V}\right], \qquad (3.342)$$

which dictates that we must choose $V = 0$ in order that Y and W be bounded. This implies that

$$R = I; \qquad (3.343)$$

the incoming wave is thus perfectly reflected, and (taking I to be real) we have

$$V = Y = 0, \qquad U = -4I \cos \omega t, \qquad W = 2\omega^2 I \cos \omega t. \qquad (3.344)$$

The straining is thus given by

$$x \sim s + \varepsilon I \cos \omega t \left[-4 + 2\omega^2 s \ln^2 s \right] + \cdots, \qquad (3.345)$$

while for small s the surface is given by

$$\eta^{(0)} \sim 4I \cos \omega t. \qquad (3.346)$$

The position x_s of the shoreline is given by $x_s + \varepsilon \eta = 0$, thus $s = 0$, and hence

$$x_s \approx -4\varepsilon I \cos \omega t. \qquad (3.347)$$

A more elaborate theory is necessary to describe the nonlinear amplification of the tsunami wave which occurs in some cases.

3.12 Notes and References

Of the books on geophysical fluid dynamics, that by Pedlosky (1987) is perhaps the most mathematical, and the present chapter is perhaps most influenced by his approach. Another influential book is that by Gill (1982), which is similar in scope but less detailed in the mathematical niceties. The books by Houghton (2002) and particularly Barry and Chorley (1998) are more concerned with weather. Other books on general meteorology are those by Holton (2004) and Vallis (2006), both comprehensive texts, and Andrews (2000), shorter and more like Houghton, and including chapters on radiation and stratospheric chemistry. The book edited by Colling (2001) is a useful primer on ocean circulation. It is an Open University course text. Ghil and Childress (1987) treat the subject from a dynamical systems perspective.

A more recent book which relates the primitive equations of atmospheric flow to the problems of numerical weather prediction is that by Kalnay (2003), and a corresponding book dealing with issues of numerical ocean modelling is that by Miller (2007). The review paper by Olbers (2001) describes, as it says, a gallery of mathematical models relevant to climate physics, meteorology and oceanography.

Eddy Viscosity Apart from our discussion in Appendix B, Pedlosky (1987, pp. 181 ff.) gives an account of Reynolds stresses, and discusses the merits of the use of eddy viscosity as a way of parameterising these. He also discusses the anisotropy of the eddy viscosity in the atmosphere, and gives estimates for the coefficients ε_V and ε_H in (3.4) (denoted A_V and A_H by him).

Tides   A very nice little book on tides is that by Defant (1958), which is short and to the point. Lamb (1945) has a whole hundred page chapter on tides, unfortunately rather dated now.

Geopotential Surfaces   The choice of a correct coordinate system using geopotential surfaces as the horizontal plane is lucidly described by Gill (1982), although be careful; his conservation of mass equation 4.12.11 is not a correct deduction from 4.12.9 and 4.12.10.

Quasi-Geostrophic Potential Vorticity Equation   The derivation of the quasi-geostrophic potential vorticity equation provided here largely follows Pedlosky (1987) in its exposition, up till the point where the stratification parameter S is discussed. At that point in his discussion, Pedlosky declines the challenge of deriving it, and simply takes it as a prescribed or measured quantity. Other authors follow suit, without noting that the stratification of the atmosphere must itself be determined by the solution of the model. The presentation here is not perhaps the most lucid, but it suggests that from the point of view of perturbation theory, the determination of S follows from an integrability condition from a multiple time scale expansion of the governing primitive equations; but this is a topic which is worthy of further investigation.

Two-Phase Flow   The discussion following (3.155) on two-phase flow relates to the well-known ill-posedness of the simplest averaged models; see Fowler (1997), for instance. Two phase flows exist in a number of different régimes—bubbly, slug, churn, annular—but it is not known what causes the transition between them. One suggestion for the bubbly to slug transition is that bubbly flow becomes unstable to kinematic waves as the bubble volume (void) fraction increases (Matuszkiewicz et al. 1987). The onset of instability is a harbinger for ill-posedness, but instability occurs before ill-posedness (Prosperetti and Satrape 1990).

The Global Thermohaline Circulation   The idea of the deep ocean circulation as a conveyor belt is associated with its chief proponent Wally Broecker, see for example Broecker (1991). Somewhat unfortunately, the phrase 'conveyor belt', together with the commonly produced cartoon of this, suggests a one-dimensionality of the motion which is misleading in detail. Broecker's article paints a more sophisticated picture, although the basic concept is still very useful.

It is also Broecker's idea that during ice ages, the circulation can oscillate because of the interplay of the North Atlantic climate and the quantity of ice sheet ablation. This idea is attractive, because the response time of the North Atlantic is of the right magnitude, decades to centuries, for the sudden warmings to occur. Less clear is what might control the millennial recurrence times.

L. F. Richardson and Weather Prediction   If there is an unsung hero of the present chapter, it would be the appealing figure of Lewis Fry Richardson, author of the well-known verse describing the essence of the turbulent energy cascade,[29] proponent of the mathematical theory of war, and author of an astonishingly precocious effort at numerical weather prediction published in 1922. Richardson calculated a weather forecast by hand, some thirty years before the first computer weather forecast, and was only thwarted in this endeavour by the inevitable parasitism of gravity waves in the solution, which wrecks the prediction. Indeed, filtering of gravity waves is one of the keys to successful modern weather forecasting. Richardson's attempt is described in the meticulous book by Lynch (2006).

3.13 Exercises

3.1 The energy equation in the atmosphere is taken to be

$$\rho c_p \frac{dT}{dt} - \frac{dp}{dt} = \nabla \cdot \mathbf{q},$$

where $\mathbf{q}$ is the combined radiative and sensible heat flux. Show how to derive the equation of global energy balance

$$\frac{d}{dt}(I + P) = q_0 - q_h,$$

where q_0 and q_h are the combined radiative and sensible heat fluxes upwards at sea level and the tropopause, respectively, $I = \int_0^h \rho c_p T \, dz$ is the internal enthalpy and $P = \int_0^h \rho \Phi \, dz$ is the potential energy, with Φ being the gravitational potential. You should assume a one-dimensional atmosphere, that the mass conservation equation

$$\frac{d\rho}{dt} + \rho \nabla \cdot \mathbf{u} = 0$$

implies $\frac{d}{dt} \rho \, dV = 0$ for material volume elements dV, and that the pressure p is related to Φ by

$$p_z = -\rho \Phi_z,$$

where p, ρ and Φ may be taken to be functions of z.

29

Big whorls have little whorls
That feed on their velocity,
And little whorls have lesser whorls
And so on to viscosity.

3.2 Derive a reference state for a dry atmosphere (no condensation) by using the equation of state

$$p = \frac{\rho RT}{M_a},$$

the hydrostatic pressure

$$\frac{\partial p}{\partial z} = -\rho g,$$

and the dry adiabatic temperature equation

$$\rho c_p \frac{dT}{dt} - \frac{dp}{dt} = 0.$$

Show that

$$\bar{T} = T_0 - \frac{gz}{c_p}, \qquad \bar{p} = p_0 p^*(z),$$

where

$$p^*(z) = \left(1 - \frac{gz}{c_p T_0}\right)^{M_a c_p / R}.$$

Use the typical values (see Question 2.11) $c_p T_0 / g \approx 29$ km, $M_a c_p / R \approx 3.4$, to show that the pressure can be adequately represented by

$$\bar{p} = p_0 \exp(-z/H),$$

where here the scale height is defined as

$$H = \frac{RT_0}{M_a g} \approx 8.4 \text{ km}.$$

(A slightly better numerical approximation near the tropopause is obtained if the scale height is chosen as 7 km.)

3.3 Use the hydrostatic pressure equation

$$\frac{dp}{dz} = -\frac{p^{1-\alpha}}{\theta}$$

to show that, for $\theta = 1 + O(\varepsilon)$ and α relatively small, $p \approx e^{-z}$. Use this to show that the conductive heating term

$$\frac{1}{Pe}\left[\frac{\partial}{\partial z}\left(k^* \frac{\partial T}{\partial z}\right)\right] \approx \frac{\alpha(5\alpha - 1)}{Pe} e^{-(5\alpha - 1)z},$$

assuming that the radiative conductivity is $k^* = \frac{T^3}{\rho}$, and that

$$\theta = \frac{T}{p^\alpha}, \qquad \rho = \frac{p}{T}.$$

Hence show that for $Pe = 7$ and $\alpha = 0.29$, the heating term is less than 0.02 in magnitude.

3.4 The Ekman boundary layer equations for the horizontal velocity (u, v) in the atmospheric boundary layer can be written in the form

$$-v = -v^* + E u_{zz},$$

$$u = u^* + E v_{zz},$$

where (u^*, v^*) denotes the limiting value of the troposphere velocity as the Earth's surface is approached. The vertical (scaled) velocity W satisfies the mass conservation equation

$$u_x + v_y + \varepsilon W_z \approx 0,$$

and $u_x^* + v_y^* \approx 0$.

Show that $U = u + iv$ satisfies

$$U_{zz} = \frac{i}{E}(U - U^*),$$

and deduce that

$$U = U^* + A \exp\left[-\frac{(1+i)z}{\sqrt{2E}}\right],$$

where $A = A(x, y)$ is to be chosen.

Show that if $U = 0$ and $W = 0$ on $z = 0$, then the value W^* of W outside the boundary layer (i.e., as $z/\sqrt{E} \to \infty$) is given by

$$W^* = (v_x^* - u_y^*)\sqrt{\frac{E}{2\varepsilon^2}}.$$

Now suppose there is an Ekman boundary layer at $z = 1$, where we pose the condition $U_z = -\gamma U$. Solve the problem in this case, and show that the corresponding Ekman pumping term is

$$W|_{z=1} - W|_{(z-1)/\sqrt{E} \to -\infty} = \Gamma(v_x^* - u_y^*),$$

where

$$\Gamma = \frac{\gamma\sqrt{\frac{E}{2\varepsilon^2}}\left(\gamma + \sqrt{\frac{2}{E}}\right)}{\left[\left(\gamma + \frac{1}{\sqrt{2E}}\right)^2 + \frac{1}{2E}\right]}.$$

3.5 Show that an explicit expression for the atmospheric heating term

$$H = \frac{\frac{\partial}{\partial z}\left(\frac{k^*}{\alpha}\frac{\partial T}{\partial z}\right)}{\frac{p}{\theta}\left[1 + \frac{\nu St\, aM(T, p)}{T^2}\right]}$$

is given by

$$H = \frac{(4\alpha - 1)T^5}{p^2(T^2 + vStaM(T, p))},$$

in which you should use the adiabatic approximations that

$$T(z) = 1 - \alpha z,$$

and

$$\theta = \frac{T}{p^\alpha} = 1, \qquad \rho = \frac{p}{T}.$$

3.6 Consider a planet whose polar axis is at right angles to the direction of the Sun. If the dimensionless surface temperature T_0 is proportional to the $\frac{1}{4}$-power of the incident solar radiation, show that

$$T_0 \propto \cos^{1/4} \lambda,$$

where λ is the angle of latitude. Hence show, with $\lambda = \lambda_0 + \Sigma y$, $\Sigma \ll 1$, and $T_0 = 1 + \varepsilon^2 \Theta_0$, that

$$\Theta_0 \approx 1 - s_1 y - s_2 y^2,$$

where

$$s_1 = \frac{\Sigma \tan \lambda_0}{4\varepsilon^2}, \qquad s_2 = \frac{(4 + 3\tan^2 \lambda_0)\Sigma^2}{32\varepsilon^2}.$$

Find typical values of s_1 and s_2 for $\Sigma = 0.16$, $\lambda_0 = \frac{\pi}{4}$, $\varepsilon = 0.2$.

3.7 What is wrong with the following argument? By Green's theorem in the plane, we have

$$\iint_A \frac{D\Theta}{Dt} \, dS = \frac{\partial}{\partial t} \iint_A \Theta \, dS - \oint_{\partial A} \Theta \, d\psi,$$

where A is any horizontal area. Since $\Theta \approx \partial \psi / \partial z$, we have

$$\oint_{\partial A} \Theta \, d\psi = \frac{\partial}{\partial z} \oint_{\partial A} \psi \, d\psi = \frac{\partial}{\partial z} \left[\frac{1}{2} \psi^2 \right]_{\partial A} = 0,$$

and therefore

$$\iint_A \frac{D\Theta}{Dt} \, dS = \frac{\partial}{\partial t} \iint_A \Theta \, dS.$$

This is true for *any* horizontal closed region A, and therefore by shrinking A to a point, we must have

$$\frac{D\Theta}{Dt} = \frac{\partial \Theta}{\partial t}$$

everywhere. This then implies that $\Theta = f(\psi)$.

3.8 Suppose that θ satisfies the equation

$$\frac{D\theta}{Dt} + \varepsilon W \frac{\partial \theta}{\partial z} = \varepsilon^2 \Gamma W + \varepsilon^2 H, \qquad (*)$$

where Γ and H are constants, $W = W(x, y)$ and the horizontal material derivative is given by

$$\frac{D}{Dt} = \frac{\partial}{\partial t} - \frac{\partial \psi}{\partial y} \frac{\partial}{\partial x} + \frac{\partial \psi}{\partial x} \frac{\partial}{\partial y},$$

where ψ is the geostrophic stream function.

The equation is to be solved in the region $V: -L < x < L, -1 < y < 1, 0 < z < 1$, with the boundary condition $\theta = 1 + \varepsilon^2 \Theta_0(y)$ on $z = 0$, and an initial condition for θ. We can assume without loss of generality that the average of Θ_0 over y is zero. (Why?) Assume that $\psi = \pm 1$ on $y = \pm 1$, and that it is periodic in x (with period $2L$). Comment on the suitability of the initial and boundary conditions. Does it matter whether W is positive or negative?

If A is any horizontal section of V, show that

$$\int_A \frac{D\theta}{Dt} dS = \frac{\partial}{\partial t} \int_A \theta \, dS,$$

and deduce that the equation

$$\frac{D\theta}{Dt} = g$$

only has a bounded solution if $\bar{g}(z) = 0$, where $\bar{g}$ is the time average of $\int_A g \, dS$.

By expanding θ as $\theta_0 + \varepsilon \theta_1 + \varepsilon^2 \theta_2 + \cdots$ and assuming that the solution remains regular, find the equations satisfied by θ_i, $i = 1, 2, 3$, and show that a solution exists in which $\theta_0 = \theta_0(z)$; whence also

$$\theta_0 = 1$$

and $\theta_1 = \theta_1(z)$, and θ_1 is given by

$$\theta_1 = \left(\Gamma + \frac{H}{W} \right) z;$$

whence

$$\frac{D\theta_2}{Dt} = H \left(1 - \frac{W}{W} \right). \qquad (†)$$

Suppose now that $\theta_2 = \frac{\partial \psi}{\partial z}$; show that $\frac{D}{Dt}[\frac{\partial \theta_2}{\partial z}] = \frac{\partial}{\partial z}(\frac{D\theta_2}{Dt})$, and deduce that a solution for θ_2 can be found in the form $\theta_2 = \bar{\theta}_2(z) + \Theta(x, y)$, where $\Theta(x, y)$ is a particular solution of (†), and show that the secularity constraint at $O(\varepsilon^3)$ implies that we can take $\bar{\theta}_2 = 0$. Deduce that $\psi = z\Theta(x, y)$.

Suppose now that a diffusion term $\varepsilon^2 \frac{\partial^2 \theta}{\partial z^2}$ is added to the right hand side of
(∗). Show that the preceding discussion still applies, but now Θ represents an
outer solution for θ_2 away from the boundary $z = 0$. By writing $\theta_2 = \Theta + \chi$
and $z = \varepsilon Z$, show that χ satisfies the approximate boundary layer equation

$$\frac{D\chi}{Dt} + W\frac{\partial \chi}{\partial Z} = \frac{\partial^2 \chi}{\partial Z^2},$$

with boundary conditions

$$\chi \to 0 \quad \text{as } Z \to \infty,$$

$$\chi = \chi_0(x, y) = \Theta_0 - \Theta \quad \text{on } Z = 0.$$

For the particular case of a steady zonal flow in which $\frac{D}{Dt} = u\frac{\partial}{\partial x}, u = u(y)$,
$W = W(y)$ and $\chi_0 = \sum_k \hat{\chi}_k(y)e^{ikx}$, show that

$$\chi = \sum_k \hat{\chi}_k(y)e^{ikx - \alpha Z},$$

where

$$\alpha = \left(\frac{W^2}{4} + iku\right)^{1/2} - \frac{W}{2}. \tag{‡}$$

By writing $\frac{W^2}{4} + iku = (p + iq)^2$, $p > 0$, and defining the square root in (‡)
as having $p > 0$, show that $\mathrm{Re}\,\alpha > 0$ irrespective of the sign of W. How would
you expect Θ to behave over long time scales in this case?

3.9 The quasi-geostrophic potential vorticity equation is given by

$$\frac{D}{Dt}\left[\nabla^2\psi + \beta y + \frac{1}{\bar{\rho}}\frac{\partial}{\partial z}\left(\frac{\bar{\rho}}{S}\frac{\partial \psi}{\partial z}\right)\right] = \frac{1}{\bar{\rho}}\frac{\partial}{\partial z}\left[\frac{\bar{\rho}H}{S}\right], \tag{∗}$$

and the stratification parameter S is determined by

$$\frac{\bar{\rho}}{S}[H - \bar{\psi}_{zt}] = \int_0^z \bar{\rho}\overline{\nabla^2\psi}_t\, dz + E^*\overline{\nabla^2\psi}_0, \tag{∗∗}$$

where the overbars denote a horizontal space average.

In deriving the expression

$$\frac{\bar{\rho}H}{S} = E^*\widehat{\nabla^2\psi}_0 \tag{†}$$

for S, where the hat denotes a time and space average for stationary solutions,
we have supposed that $S = S(z)$ is independent of t, although this does not
appear necessary from (∗∗). Show that in fact this assumption is consistent
(i.e., that (†) implies (∗∗)) by using the averaging result

$$\overline{\frac{D\Gamma}{Dt}} = \frac{\partial\bar{\Gamma}}{\partial t}$$

to show that

$$\overline{\nabla^2 \psi}_t + \frac{1}{\bar{\rho}} \frac{\partial}{\partial z} \left[\frac{\bar{\rho}}{S} \bar{\psi}_{zt} \right] = 0,$$

and that the boundary condition

$$\frac{D}{Dt} \left(\frac{\partial \psi}{\partial z} \right) = H - SE^* \nabla^2 \psi \quad \text{on } z = 0$$

implies

$$\bar{\psi}_{zt} = H - SE^* \overline{\nabla^2 \psi} \quad \text{on } z = 0.$$

Deduce that

$$\int_0^z \bar{\rho} \overline{\nabla^2 \psi}_t \, dz + \left[\frac{\bar{\rho}}{S} \bar{\psi}_{zt} \right]_0^z = 0,$$

and hence show that, given (∗) and (†), (∗∗) is true if and only if $\frac{\partial H}{S}$ is constant (which is indeed the case for (†)).

3.10 Show that the solution $A(z)$ of the Eady model equations

$$(ikz + \sigma) \left[A'' - \mu^2 A \right] = 0,$$

where

$$(ikz + \sigma) A' - ikA = 0 \quad \text{on } z = 0, 1,$$

can be written in the form

$$A = \alpha \cosh \mu z + \beta \cosh \left[\mu (1 - z) \right],$$

providing $c = -\sigma / ik$ satisfies

$$\begin{vmatrix} 1 & \cosh \mu - \mu c \sinh \mu \\ \cosh \mu - \mu (1 - c) \sinh \mu & 1 \end{vmatrix} = 0,$$

whence

$$c^2 - c + \frac{\coth \mu}{\mu} - \frac{1}{\mu^2} = 0.$$

Deduce that

$$c = -\frac{\sigma}{ik} = \frac{1}{2} \pm \frac{1}{\mu} \left[\left(\frac{\mu}{2} - \coth \frac{\mu}{2} \right) \left(\frac{\mu}{2} - \tanh \frac{\mu}{2} \right) \right]^{1/2}.$$

[*The identity* $\coth \mu = \frac{1}{2} (\tanh \frac{\mu}{2} + \coth \frac{\mu}{2})$ *may be useful.*]

3.11 The semi-diurnal M_2 tides on the Earth are described, neglecting Coriolis force, by the dimensionless equation

$$\cos^2 \lambda \, \eta_{tt} = \nabla^2 (\eta - \chi),$$

where

$$\nabla^2 = \frac{\partial^2}{\partial \phi^2} + \frac{\partial^2}{\partial v^2},$$

$$v = \ln\left[\frac{1 + \tan\frac{1}{2}\lambda}{1 - \tan\frac{1}{2}\lambda}\right],$$

and ϕ and λ are longitude and latitude angles.

Show that if solutions are sought in a narrow canal at latitude λ with ends at $\phi = \pm\frac{L}{2}$, and boundary conditions of no flow through the boundaries, i.e.,

$$\frac{\partial}{\partial n}(\eta - \chi) = 0,$$

then

$$\eta = \frac{\cos^2\delta \cos^2\lambda}{1 - S^2 \cos^2\lambda}[A \cos 2St + B \sin 2St],$$

where

$$A = \cos 2\phi - \frac{S \cos\lambda \sin L \cos(2S\phi \cos\lambda)}{\sin(LS \cos\lambda)},$$

$$B = -\sin 2\phi + \frac{S \cos\lambda \cos L \sin(2S\phi \cos\lambda)}{\cos(LS \cos\lambda)}.$$

3.12 A tsunami is modelled by the wave equation

$$\frac{\partial^2\eta}{\partial t^2} = \frac{1}{r}\frac{\partial}{\partial r}\left(r\frac{\partial\eta}{\partial r}\right),$$

subject to the conditions that

$$\eta = 0 \quad \text{at } t = 0, \ r \neq 0,$$

$$\eta \to 0 \quad \text{as } r \to \infty,$$

$$2\pi \int_0^\infty r\eta \, dr \equiv V.$$

Show that a similarity solution of this problem can be found in which

$$\eta = \frac{1}{t^2}f(\xi), \quad \xi = \frac{r}{t},$$

and that f satisfies the equation

$$(\xi^3 f)'' = (\xi f')';$$

write down the boundary condition and integral constraint for f.
   Show that there exists a solution of the form

$$f = \frac{A}{(1-\xi^2)^{3/2}} \int_\xi^1 \frac{(1-s^2)^{1/2}\,ds}{s}, \quad 0 < s < 1,$$

$$f = \frac{A}{(\xi^2-1)^{3/2}} \int_1^\xi \frac{(s^2-1)^{1/2}\,ds}{s}, \quad s > 1,$$

if we assume f is finite at $\xi = 1$.
   Show that this solution is continuous at $\xi = 1$ but has discontinuous derivative there. Show also that the solution in $\xi > 1$ cannot satisfy the integral constraint on f.
   Suppose instead that the solution above in $\xi < 1$ is correct, but $f = 0$ in $\xi > 1$. Use the integral constraint to show that

$$A = \frac{V}{2\pi[1 - \ln 2]},$$

and show that the wave front at $r = t$ is of height $\frac{1}{3}A$.
   Does this solution make sense? Is the position of the wave front uniquely defined? Show that $f \sim -A\ln\xi$ as $\xi \to 0$, and deduce (explaining why) that no solution of this type is appropriate.

3.13 The stream function ψ of the ocean circulation satisfies the equation

$$\beta\psi_x = -1 + E_H\nabla^4\psi$$

in the closed domain B, with boundary conditions

$$\psi = \frac{\partial\psi}{\partial n} = 0 \quad \text{on } \partial B.$$

If B is the box $0 < x < 1, 0 < y < 1$, find a suitable scaling for the boundary layer near $y = 0$, and hence show that in terms of the rescaled boundary layer coordinates x and Y, ψ satisfies the boundary layer equation

$$\beta\psi_x = -1 + \psi_{YYYY},$$

together with the boundary conditions

$$\psi = \psi_Y = 0 \quad \text{on } Y = 0, \qquad \psi \to \infty \quad \text{as } Y \to \infty,$$

and the 'initial' condition

$$\psi = 0 \quad \text{on } x = 1.$$

Show that a similarity solution for this equation exists in the form

$$\psi = \left(\frac{1-x}{\beta}\right)[1 - f(\eta)], \qquad \eta = Y\left(\frac{\beta}{1-x}\right)^{1/4},$$

and that f satisfies the equation

$$f^{iv} - \eta f' + 4f = 0,$$

with

$$f(0) = 1, \qquad f'(0) = 0, \qquad f(\infty) = 0.$$

Find four independent possible asymptotic behaviours for f as $\eta \to \infty$, and show that only two tend to zero. Hence deduce that the given boundary conditions should be sufficient to determine the solution uniquely.

3.14 The normalised amplitude of a tsunami wave satisfies the equation

$$\phi_{tt} = \phi_{rr} + \frac{\phi}{4r^2}.$$

Seek solutions valid for large r and t by changing variables to $\xi = r - t$ and $\tau = t$, and show that an asymptotic solution for large t can be found with the assumption that

$$\phi \sim \sum_0^\infty \frac{\phi_n(\xi)}{t^n}.$$

Show that $\phi_1' = \frac{1}{8}\phi_0$, $\phi_2' = -\frac{1}{8}\xi\phi_0 - \frac{7}{16}\phi_1$, and hence find the expansion for ϕ correct to terms of $O(\frac{1}{t^2})$, assuming that ϕ_0 is known.

Chapter 4
River Flow

Much of the environment consists of fluids, and much of this book is therefore concerned with fluid mechanics. Oceans and atmosphere consist of fluids in large scale motion, and even later, when we deal with more esoteric subjects: the flow of glaciers, convection in the Earth's mantle, it is within the context of fluid mechanics that we formulate relevant models. This chapter concerns one of the most obvious common examples of a fluid in motion, that of the mechanics of rivers.

Fluid mechanics in the environment is, however, altogether different to the subject we study in an undergraduate course on viscous flow, and the principal reason for this is that for most of the common environmental fluid flows with which we are familiar, the flow is *turbulent*. (Where it is not, for example in glacier flow, other physical complications obtrude.) As a consequence, the models which we use to describe the flow are different to (and in fact, simpler than) the Navier–Stokes equations.

4.1 The Hydrological Cycle

Rainwater which falls in a catchment area of a particular river basin makes its way back to the ocean (or sometimes to an inland lake) by seepage into the ground, and then through groundwater flow to outlet streams and rivers. In severe storm conditions, or where the soil is relatively impermeable, the rainfall intensity may exceed the soil infiltration capacity, and then direct runoff to discharge streams can occur as overland flow. Depending on local topography, soil cover, vegetation, one or other transport process may be the norm. Overland flow can also occur if the soil becomes saturated. The hydrological cycle is completed when the water, now back in the ocean, is evaporated by solar radiation, forming atmospheric clouds which are the instrument of precipitation.

River flow itself occurs on river beds that are typically quasi-one-dimensional, sinuous channels with variable and rough cross section. Moreover, if the channel

A. Fowler, *Mathematical Geoscience*, Interdisciplinary Applied Mathematics 36, DOI 10.1007/978-0-85729-721-1_4, © Springer-Verlag London Limited 2011

discharge is Q (m^3 s^{-1}), and the wetted perimeter length of the cross section is l (m), then an appropriate Reynolds number for the flow is

$$Re = \frac{Q}{\nu l}, \tag{4.1}$$

where $\nu = \mu/\rho$ is the kinematic viscosity (and μ is the dynamic viscosity). If $l = 20$ m, $\nu = 10^{-6}$ m^2 s^{-1}, $Q = 10$ m^3 s^{-1}, then $Re \sim 0.5 \times 10^6$. Inevitably, river flow is turbulent for all but the smallest rivulets. A different measure of the Reynolds number is

$$Re = \frac{uh}{\nu}, \tag{4.2}$$

where u is mean velocity and h is mean depth. In a wide channel, we find that the width is approximately l, so that $Q \approx ulh$, and this gives the same definition as (4.1). Thus, to model river flow, and to explain the response of river discharge to storm conditions, as measured on flood hydrographs, for instance, one must model a flow which is essentially turbulent, and which exists in a rough, irregular channel.

The classical way in which this is done is by applying a time average to the Navier–Stokes equations, which leads to Reynolds' equation, which is essentially like the Navier–Stokes equation, but with the stress tensor being augmented by a *Reynolds stress tensor*. The procedure is described in Appendix B.

For a flow $\mathbf{u} = (u, v, w)$ which is locally unidirectional *on average*, such as that in a river, we may take the mean velocity $\bar{\mathbf{u}} = (\bar{u}, 0, 0)$, and then the x component of the momentum equation becomes

$$\rho \frac{\partial}{\partial z}(\overline{u'w'}) \approx -\frac{\partial \bar{p}}{\partial x} + \mu \frac{\partial^2 \bar{u}}{\partial z^2}, \tag{4.3}$$

because in a shallow flow, the other Reynolds stress terms are smaller. Integration over the depth shows that the resistance to motion is provided by the wall stress τ, and this is

$$\tau = \mu \frac{\partial \bar{u}}{\partial z} + \{-\rho \overline{u'w'}\}, \tag{4.4}$$

evaluated at the wetted perimeter of the flow. Strictly, the Reynolds stress vanishes at the boundary (because the fluid velocity is zero there), and the molecular stress changes rapidly to compensate, in a very thin laminar wall layer. Normally one evaluates (4.4) just outside this layer, close to but not at the boundary, where the molecular stress is negligible and the Reynolds stress is parameterised in some way. A common choice is to use a friction factor, thus

$$\tau = f\rho \bar{u}^2, \tag{4.5}$$

where the dimensionless number f (called the friction factor) is found to depend rather weakly on the Reynolds number.[1] A crude but effective assumption is simply that f is constant, with a typical value for f of 0.01.

---

[1]More precisely, the stress should be $\tau = f\rho|\bar{u}|\bar{u}$, since the friction acts in the opposite direction to the flow. For unidirectional flows, this reduces to (4.5). Later (in Sect. 4.5.3), we will have need for this more precise formula.

4.2 Chézy's and Manning's Laws

Our starting point is that the flow is essentially one-dimensional: or at least, we focus on this aspect of it. As well as the cross-sectional area (of the *flow*) A and discharge Q, we introduce a longitudinal, curvilinear distance coordinate s, and we assume that the river axis changes direction slowly with s. Then conservation of mass is, in its simplest form,

$$\frac{\partial A}{\partial t} + \frac{\partial Q}{\partial s} = M. \tag{4.6}$$

This source term M represents the supply to the river due to infiltration seepage and overland flow from the catchment.

(4.6) must be supplemented by an equation for Q as a function of A, and this arises through consideration of momentum conservation. There are three levels at which one may do this: by exact specification, as in the Navier–Stokes momentum equation; by ignoring inertia and averaging, as in Darcy's law; and most simply, by ignoring inertia and applying a force balance using a semi-empirical friction factor. We begin by opting for this last choice, which should apply for sufficiently 'slow' (in some sense) flow. Later we will consider more complicated models.

We have already defined the Reynolds number Re in terms of Q and A, or equivalently a mean velocity $u = Q/A$ and a channel depth $d \sim A^{1/2}$. 'Slow' here means a small *Froude number*, defined by

$$Fr = \frac{u}{(gd)^{1/2}} = \frac{Q}{g^{1/2}A^{5/4}}. \tag{4.7}$$

If $Fr < 1$, the flow is *tranquil*; if $Fr > 1$, it is *rapid*. Gravity is of relevance, since the flow is ultimately due to gravity.

Now let l be the wetted perimeter of a cross section, and let τ be the mean shear stress exerted at the bed (longitudinally) by the flow. If the downstream angle of slope is α, then a force balance gives

$$l\tau = \rho g A \sin\alpha, \tag{4.8}$$

where ρ is density. For turbulent flow, the shear stress is given by the friction law

$$\tau = f\rho u^2, \tag{4.9}$$

where the friction factor f may depend on the Reynolds number. Since

$$u = Q/A, \tag{4.10}$$

and defining the hydraulic radius

$$R = A/l, \tag{4.11}$$

we derive the relations

$$u = (g/f)^{1/2} R^{1/2} S^{1/2}, \tag{4.12}$$

where

$$S = \sin\alpha, \tag{4.13}$$

and

$$Q = \left(\frac{g}{fl}\right)^{1/2} A^{3/2} S^{1/2}. \tag{4.14}$$

For wide, shallow rivers, l is essentially the width. For a more circular cross section, $l \sim A^{1/2}$, and

$$Q = (g/f)^{1/2} A^{5/4} S^{1/2}. \tag{4.15}$$

The relation (4.12) is the Chézy velocity formula, and $C = (g/f)^{1/2}$ is the Chézy roughness coefficient. Notice that the Froude number, in terms of the hydraulic radius, is

$$Fr = \frac{u}{(gR)^{1/2}} = (S/f)^{1/2}, \tag{4.16}$$

and tranquillity (at least in uniform flow) is basically due to slope.

Alternative friction correlations exist. That due to Manning is an empirical formula to fit measured stream velocities, and is of the form

$$u = R^{2/3} S^{1/2} / n', \tag{4.17}$$

where Manning's roughness coefficient n' takes typical values in the range 0.01–0.1 $\text{m}^{-1/3}$ s, depending on stream depth, roughness, etc. Manning's law can be derived from an expression for the shear stress of the form (cf. (4.9))

$$\tau = \frac{\rho g n'^2 u^2}{R^{1/3}}. \tag{4.18}$$

For Manning's formula, we have

$$\begin{aligned} Q &\sim A^{4/3} \quad \text{if } R \sim A^{1/2}, \\ Q &\sim A^{5/3} \quad \text{if } l \text{ is width, } R = A/l \sim A. \end{aligned} \tag{4.19}$$

Thus we see that for a variety of stream types and velocity laws, we can pose a relation between discharge and area of the form

$$Q \sim A^{m+1}, \quad m > 0, \tag{4.20}$$

with typical values $m = \frac{1}{4} - \frac{2}{3}$. In practice, for a given stream, one could attempt to fit a law of the form (4.20) by direct measurement.

4.3 The Flood Hydrograph

Suppose in general that

$$Q = \frac{c A^{m+1}}{m+1}. \tag{4.21}$$

We can non-dimensionalise the equation for A so that it becomes

$$\frac{\partial A}{\partial t} + A^m \frac{\partial A}{\partial s} = M, \tag{4.22}$$

Fig. 4.1 Formation of a
shock wave in the solution of
(4.22) (cf. Fig. 1.14)

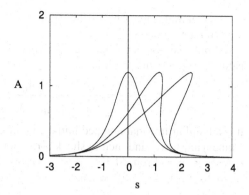

a first-order nonlinear hyperbolic equation, also known as a kinematic wave equation, whose solution can be written down. The source term M is in general a function of s and t, but for simplicity we take it to be constant here. Suppose the initial data are parameterised as

$$A = A_0(\sigma), \qquad s = \sigma > 0, \qquad t = 0. \tag{4.23}$$

Then the characteristic equations are

$$\frac{dA}{dt} = M, \qquad \frac{ds}{dt} = A^m, \tag{4.24}$$

whence

$$A = A_0(\sigma) + Mt, \qquad s = \sigma + \frac{(A_0 + Mt)^{m+1} - A_0^{m+1}}{M(m+1)}, \tag{4.25}$$

thus

$$A = Mt + A_0\left[s - \left\{\frac{A^{m+1} - (A - Mt)^{m+1}}{M(m+1)}\right\}\right] \tag{4.26}$$

determines A implicitly.

We can see from (4.26) that this solution applies for sufficiently small t or large s, since we must have $\sigma > 0$. For larger t, the characteristics are those emanating from $s = 0$, where the boundary data are parameterised by

$$A = 0, \qquad s = 0, \qquad t = \tau, \tag{4.27}$$

and the solution is the steady state

$$\frac{A^{m+1}}{m+1} = Ms. \tag{4.28}$$

This steady state is applicable above the dividing characteristic in the (s, t) plane emanating from the origin, which is

$$s = \frac{M^m t^{m+1}}{m+1}. \tag{4.29}$$

Thus any initial disturbance to the steady state is washed out of the system in a finite time (for any finite s).

From (4.26) we can calculate $\frac{\partial A}{\partial s}$ explicitly in terms of t and the characteristic parameter σ, and the result is

$$\frac{\partial A}{\partial s} = \frac{A_0'}{1 + \frac{A_0'}{M}\{(A_0 + Mt)^m - A_0^m\}}. \tag{4.30}$$

It is a familiar fact that humped initial conditions $A_0(\sigma)$ will lead to propagation of a kinematic wave, and then to shock formation, as shown in Fig. 4.1, when $\partial A/\partial s$ reaches infinity. From (4.30), we see that this occurs on the characteristic through $s = \sigma$ for $t > 0$ if $A_0' < 0$, when

$$t = t_\sigma = \frac{1}{M}\left[\left(-\frac{M}{A_0'} + A_0^m\right)^{1/m} - A_0\right], \tag{4.31}$$

and a shock forms when $t = \min_\sigma t_\sigma > 0$. Thereafter a shock exists at a point $s_d(t)$, and propagates at a rate given, by consideration of the integral conservation law

$$\frac{\partial}{\partial t}\int_{s_1}^{s_2} A\,ds = -[Q]_{s_1}^{s_2} + \int_{s_1}^{s_2} M\,ds, \tag{4.32}$$

by

$$\dot{s}_d = \frac{[Q]_{s_d-}^{s_d+}}{[A]_{s_d-}^{s_d+}}. \tag{4.33}$$

As an application, we consider the flood hydrograph, which measures discharge at a fixed value of s as a function of time. Suppose for simplicity that $M = 0$ (the case $M > 0$ is considered in Question 4.7). As an idealisation of a flood, we consider the initial condition

$$A \approx A^*\delta(s) \quad \text{at } t = 0, \tag{4.34}$$

where $\delta(s)$ is the delta function, representing the input to the river by overland flow after a short period of localised rainfall. Either directly, or by letting $M \to 0$ in (4.26), we have $A = A_0(s - A^m t)$, and it follows that $A \approx 0$ except where $s = A^m t$. The humped initial condition causes a shock to form at $s_d(t)$, with $s_d(0) = 0$, and we have

$$A = 0, \quad s > s_d,$$
$$A = (s/t)^{1/m}, \quad s < s_d, \tag{4.35}$$

as shown in Fig. 4.2.

The shock speed is given by

$$\dot{s}_d = (Q/A)|_{s_d-} = \left.\frac{A^m}{m+1}\right|_{s_d-} = \frac{s_d}{(m+1)t}, \tag{4.36}$$

whence $s_d \propto t^{1/(m+1)}$. To calculate the coefficient of proportionality, we use conservation of mass in the form

$$\int_0^{s_d} A\,ds = A^*, \tag{4.37}$$

Fig. 4.2 Propagation of a shock front

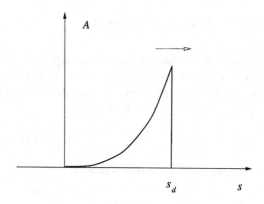

Fig. 4.3 Ideal (*full line*) and observed (*dotted line*) hydrographs

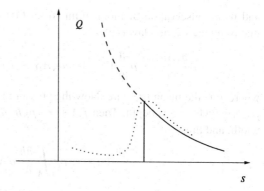

whence, in fact,

$$s_d = \left[\frac{(m+1)A^*}{m} \right]^{m/(m+1)} t^{1/(m+1)}. \qquad (4.38)$$

Denoting $b = [(m+1)A^*/m]^{m/(m+1)}$, the flood hydrograph at a fixed station $s = s^*$ is then as follows. For $t < t^*$, where

$$t^* = (s^*/b)^{m+1}, \qquad (4.39)$$

$Q = 0$. For $t > t^*$, $A = (s^*/t)^{1/m}$, and thus

$$Q = \frac{s^{*(m+1)/m}}{(m+1)} t^{-(m+1)/m}. \qquad (4.40)$$

This result is illustrated in Fig. 4.3, together with a typical observed hydrograph. The smoothed observation can be explained by the fact that a more realistic initial condition would have delivery of the storm flow over an interval of space and time. More importantly, one can expect that a more realistic model will allow for diffusive effects.

4.4 St. Venant Equations

We now re-examine the momentum equation, which we previously assumed to be described by a force balance. Again consider the equations in dimensional form. For the remainder of the chapter we take $M = 0$, largely for simplicity. Conservation of mass can then be written in the form

$$\frac{\partial A}{\partial t} + \frac{\partial}{\partial s}(Au) = 0, \tag{4.41}$$

where the mean velocity u is defined by

$$u = \frac{Q}{A}, \tag{4.42}$$

and then conservation of momentum (from first principles) leads to the equation (adopting the friction law (4.9))

$$\rho \frac{\partial(Au)}{\partial t} + \rho \frac{\partial}{\partial s}(Au^2) = \rho g A S - \rho l f u^2 - \frac{\partial}{\partial s}(A\bar{p}), \tag{4.43}$$

where $\bar{p}$ is the mean pressure. Now the pressure is approximately hydrostatic, thus $p \approx \rho g z$ where z is depth. Then $\bar{p} A \approx \int \frac{1}{2}\rho g h^2\, dx$ where h is total depth and x is width, and thus

$$\frac{\partial}{\partial s}(A\bar{p}) = \rho g \int_A \frac{\partial h}{\partial s}\, dA; \tag{4.44}$$

if we suppose $\partial h/\partial s$ is independent of x, we find[2]

$$\frac{\partial}{\partial s}(A\bar{p}) = \rho g A \frac{\partial \bar{h}}{\partial s}, \tag{4.45}$$

where $\bar{h}$ is the mean depth. Using (4.41), (4.43) reduces to

$$u_t + u u_s = g S - \frac{f l u^2}{A} - g \frac{\partial \bar{h}}{\partial s}. \tag{4.46}$$

Equations (4.41) and (4.46) are known as the St. Venant equations.[3]

---

[2] The assumption that $\partial h/\partial s$ is constant across the stream means that along a transverse section of the river, the surface is horizontal. This is really due to the smallness of the width compared to the length. It is importantly not exactly true for meandering rivers, but is still a very good approximation.

[3] Note that the derivation of (4.46) assumes a constant slope S. If the slope is varying, then the derivation is still valid providing S is the local bed slope. If we then take $\bar{S}$ to be the average downstream slope, and denote the bed by $z = b(s)$ and the surface by $z = \eta(s)$, we have the local slope $S = \bar{S} - b_s$, and thus $S - h_s = \bar{S} - \eta_s$, and thus (4.46) still applies for varying bed slope when S denotes the (constant) mean slope, providing we replace $\bar{h}$ by η. All of this supposes that b does not vary with x, i.e., the channel section is rectangular.

4.4.1 Non-dimensionalisation

We choose scales for $u = Q/A$, t, s, A, R (the hydraulic radius, $= A/l$) and $\bar{h}$ as follows, in keeping with the assumed balances adopted earlier:

$$Au \sim Q, \qquad gS \sim \frac{lfu^2}{A} = \frac{fu^2}{R},$$

$$t \sim \frac{s}{u}, \qquad s \sim \frac{d}{S}, \qquad \bar{h}, R \sim d, \tag{4.47}$$

where we can suppose Q is a typical observed discharge, and d is a typical observed depth. Explicitly, the scales are

$$[\bar{h}], [R] = d, \qquad [s] = \frac{d}{S},$$

$$[u] = \left(\frac{gdS}{f}\right)^{1/2}, \qquad [t] = \left(\frac{fd}{gS^3}\right)^{1/2}, \qquad [A] = Q\left(\frac{f}{gdS}\right)^{1/2}, \tag{4.48}$$

and we put $u = [u]u^*$, etc., and drop asterisks. The resulting equations are

$$A_t + (Au)_s = 0,$$

$$F^2[u_t + uu_s] = 1 - \frac{u^2}{R} - h_s, \tag{4.49}$$

where we would choose $h \sim R \sim A$ for a wide channel, $h \sim R \sim A^{1/2}$ for a rounded channel. In particular, for a wide channel, we have $R = h$, so that the momentum equation can be written

$$(wh)_t + (wuh)_s = 0,$$

$$F^2(u_t + uu_s) = 1 - \frac{u^2}{h} - h_s, \tag{4.50}$$

since $A = wh$, where w is the (dimensionless) width. As before, the Froude number F is given by

$$F = \frac{[u]}{(gd)^{1/2}} = \left(\frac{S}{f}\right)^{1/2}. \tag{4.51}$$

4.4.2 Long Wave and Short Wave Approximation

To estimate some of these scales, we take $d = 2$ m, $u = 1$ m s^{-1} and $S = \sin\alpha = 0.001$, typical lowland valley values. We then have the length scale $[s] = \frac{d}{S} \sim 2$ km, and the time scale $t \sim 33$ minutes, and in some sense these are the natural length and time scales for the dynamic river response. However, it is fairly clear that these scales are not appropriate either for variations over the length of a whole river, or for the shorter length and time scales appropriate to waves generated by passage of a boat, for example. Both of these situations lead to further simplifications, as detailed below.

Long Wave Theory

Suppose we have a river of length $L = 100$ km, and we are concerned with the passage of a flood wave along its length. It is then appropriate to rescale s and t as

$$s \sim \frac{1}{\varepsilon}, \qquad t \sim \frac{1}{\varepsilon}, \qquad \varepsilon = \frac{d}{H}, \qquad H = L \sin \alpha; \qquad (4.52)$$

note that H is the drop in elevation of the river over its length L: in this instance $\varepsilon \sim 0.02 \ll 1$. In this case equations (4.50) become

$$h_t + (uh)_s = 0,$$
$$\varepsilon F^2(u_t + uu_s) = 1 - \frac{u^2}{h} - \varepsilon h_s, \qquad (4.53)$$

and in the limit $\varepsilon \to 0$, we regain the slowly varying flow approximation.

Short Wave Theory

An alternative approximation is appropriate if length scales are much shorter than 2 km. This is often the case, and particularly in dynamically generated waves, as we discuss further below. In this case, it is appropriate to rescale length and time as

$$s \sim \delta, \qquad t \sim \delta, \qquad \delta = \frac{H}{d}, \qquad (4.54)$$

where now $\delta \ll 1$, and then the model equations (4.50) become

$$h_t + (uh)_s = 0,$$
$$F^2(u_t + uu_s) = \delta \left(1 - \frac{u^2}{h} \right) - h_s, \qquad (4.55)$$

and when δ is put to zero, we regain the shallow water equations of fluid dynamics.

4.4.3 The Monoclinal Flood Wave

One of the suggestions made at the end of Sect. 4.3 was that the shocks predicted by the slowly varying flood wave theory would in reality be smoothed out by some higher-order physical effect. This shock structure is called the monoclinal flood wave (because it is a monotonic profile), and it can be understood in the context of the long wave St. Venant theory (4.53). The simplest version is when $F \ll 1$ as well as $\varepsilon \ll 1$, for then we can approximate the momentum equation (4.53)$_2$ by the relation

$$u \approx h^{1/2} \left(1 - \frac{1}{2}\varepsilon h_s \ldots \right), \qquad (4.56)$$

and $(4.53)_1$ becomes

$$\frac{\partial h}{\partial t} + \frac{3}{2}h^{1/2}h_s \approx \frac{1}{2}\varepsilon\frac{\partial}{\partial s}\left(h^{3/2}\frac{\partial h}{\partial s}\right). \tag{4.57}$$

This is a convective diffusion equation much like Burgers' equation, and we expect it to support a monoclinal wave which provides a shock structure joining values h_- upstream to lower values h_+ downstream. We analyse this shock structure by writing

$$s = s_f + \varepsilon X, \tag{4.58}$$

where s_f is the flood wavefront, and X is a local coordinate within the shock structure. To leading order we then obtain the equation

$$-ch_X + \left[h^{3/2}\left(1 - \frac{1}{2}h_X\right)\right]_X = 0, \tag{4.59}$$

where $c = \dot{s}_f$ is the wave speed. Integrating this, we obtain

$$ch = h^{3/2}\left(1 - \frac{1}{2}h_X\right) + K, \tag{4.60}$$

where we require

$$K = ch_- - h_-^{3/2} = ch_+ - h_+^{3/2} \tag{4.61}$$

(which gives the shock speed determined in the usual way by the jump condition $c = [h^{3/2}]_-^+/[h]_-^+$). Hence h is given by the quadrature

$$2X = \int_h^{h_0} \frac{h^{3/2}\,dh}{[ch - h^{3/2}] - K}, \tag{4.62}$$

where the arbitrary choice of $h_0 \in (h_+, h_-)$ simply fixes the origin of X. (4.62) can be simplified to give

$$X = \int_w^{w_0} \frac{w^4\,dw}{(w - w_+)(w_- - w)(w + C)}, \tag{4.63}$$

where $w = h^{1/2}$, and

$$C = \frac{w_+ w_-}{w_+ + w_-}, \tag{4.64}$$

and $X(w)$ can of course be evaluated.

Of particular interest is the small flood limit, in which $\Delta w = w_- - w_+$ is small. In this case $C \approx \frac{1}{2}w_+$, and h can be found explicitly, as the approximation

$$h = \left[\frac{h_+^{1/2} + h_-^{1/2}e^{-X/\Delta X}}{1 + e^{-X/\Delta X}}\right]^2, \tag{4.65}$$

where

$$\Delta X = \frac{2w_+^3}{3\Delta w} = \frac{4h_+^2}{3\Delta h} \tag{4.66}$$

Fig. 4.4 The monoclinal flood wave given by (4.65), with $h_- = 1.5$, $h_+ = 1$, $\Delta X = 1$

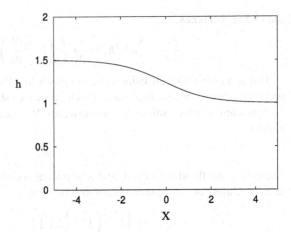

is the shock width. A further simplification (because $\Delta h = h_- - h_+$ is small) is

$$h = h_+ + \frac{\Delta h\, e^{-X/\Delta X}}{1 + e^{-X/\Delta X}}. \tag{4.67}$$

In dimensional terms, the shock width is of order

$$\frac{d^2}{\Delta d \sin\alpha}, \tag{4.68}$$

where d is the depth, and Δd is the change in depth. Following a storm, if a river of depth two metres and bedslope 10^{-3} rises by a foot (thirty centimetres), the shock width is about thirteen kilometres: not very shock-like! Figure 4.4 shows the form of the monoclinal flood wave (as given by (4.65)).

Although (4.57) is useful in indicating the diffusive structure of the long wave theory, the above discussion of the monoclinal flood wave is strictly inaccurate, since the approximation in (4.56) breaks down on short scales. To see that the analysis still holds, we can re-do the analysis on the full system (4.53). Adopting (4.58), we find, approximately,

$$-ch_X + (uh)_X = 0,$$
$$F^2(-cu_X + uu_X) = 1 - \frac{u^2}{h} - h_X, \tag{4.69}$$

with first integral

$$ch = K + uh, \tag{4.70}$$

with K and c determined by (4.61) as before, noting that $u_\pm = \sqrt{h_\pm}$, and thus $u_\pm = w_\pm$, as used in (4.63).

We then find that

$$h_X = \frac{h^3 - (ch - K)^2}{h^3 - K^2 F^2}, \tag{4.71}$$

and (4.63) is replaced by

$$X = \int_{h}^{h_0} \frac{(h^3 - K^2 F^2)\, dh}{(h - h_+)(h_- - h)(h - A)}, \tag{4.72}$$

where

$$A = \frac{K^2}{h_- h_+} = \left(\frac{w_+ w_-}{w_+ + w_-} \right)^2. \tag{4.73}$$

Clearly $A < h_-$, h_+, and thus the flood wave connecting h_- to h_+ as X increases exists (with $h_- > h_+$) if the numerator in (4.72) is positive for all $h > h_+$, which is the case, using the definition of

$$K = \frac{h_+ h_-}{\sqrt{h_+} + \sqrt{h_-}}, \tag{4.74}$$

if

$$F < \frac{h_+}{h_-} + \sqrt{\frac{h_+}{h_-}}. \tag{4.75}$$

Since $h_- > h_+$, the upper limit of the right hand side is two, so that the monoclinal flood wave cannot exist for $F > 2$, consistent with the fact that roll waves then form, as we now show.

4.4.4 Waves and Instability

The monoclinal flood wave is one example of a river wave. More generally, we can expect disturbances to a uniformly flowing stream to cause waves to propagate, and in this section we study such waves. In particular, we will find that if the basic flow is sufficiently rapid, then disturbance waves will grow unstably. Such waves are commonly seen in fast flowing rivulets, for example on steep pavements during rainfall, and even on car windscreens.

To analyse waves on rivers, we take the basic river flow as being (locally) constant, thus in (4.50) (with $R = h$)

$$u = h = 1, \tag{4.76}$$

and we examine its stability by writing

$$u = 1 + v, \qquad h = 1 + H, \tag{4.77}$$

and linearising. We obtain the linear system

$$\begin{aligned} H_t + H_s + v_s &= 0, \\ F^2(v_t + v_s) &= -2v + H - H_s, \end{aligned} \tag{4.78}$$

whence

$$F^2 \left(\frac{\partial}{\partial t} + \frac{\partial}{\partial s} \right)^2 v = -2 \left(\frac{\partial}{\partial t} + \frac{\partial}{\partial s} \right) v - v_s + v_{ss}. \tag{4.79}$$

Fig. 4.5 The function $L(p)$ defined by (4.85), with $\tilde{k} = 1$

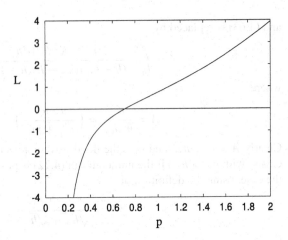

Solutions $v = \exp[iks + \sigma t]$ exist, provided σ satisfies

$$F^2(\sigma + ik)^2 + 2(\sigma + ik) + ik + k^2 = 0, \qquad (4.80)$$

or

$$\tilde{\sigma} = -i\tilde{k} - 1 \pm \left[1 - i\tilde{k} - \tilde{k}^2/F^2\right]^{1/2}, \qquad (4.81)$$

where we write

$$\sigma = \tilde{\sigma}/F^2, \qquad k = \tilde{k}/F^2. \qquad (4.82)$$

There are thus two wave-like disturbances. The possibility of instability exists, if either value of $\tilde{\sigma}$ has positive real part. We define the positive square root in (4.81) to be that with positive real part. Specifically, we define

$$p + ikq = \left\{1 - i\tilde{k} - \frac{\tilde{k}^2}{F^2}\right\}^{1/2}, \qquad (4.83)$$

where we take $p > 0$; thus, the real and imaginary parts of $\tilde{\sigma}$ are given by

$$\tilde{\sigma}_R = \pm p - 1, \qquad -\frac{\tilde{\sigma}_I}{\tilde{k}} = 1 \mp q, \qquad (4.84)$$

and the criterion for instability is that $\tilde{\sigma}_R > 0$, i.e., $p > 1$. In this form, the growth rate of the wave is $\tilde{\sigma}_R/F^2$, while the wave speed is $-\tilde{\sigma}_I/\tilde{k}$. From (4.83), we find

$$q = -\frac{1}{2p}, \qquad L(p) \equiv p^2 - \frac{\tilde{k}^2}{4p^2} = 1 - \frac{\tilde{k}^2}{F^2}. \qquad (4.85)$$

As illustrated in Fig. 4.5, $L(p)$ is a monotonically increasing function of p, and therefore the instability criterion $p > 1$ is equivalent to $L(p) > L(1)$. Since p is determined by $L(p) = 1 - (\tilde{k}^2/F^2)$, while from (4.85), $L(1) = 1 - (\tilde{k}^2/4)$, we see that instability occurs if

$$F > F_c = 2. \qquad (4.86)$$

Thus, for tranquil flow, $F < O(1)$, the flow is stable. For rapid flow, $F > O(1)$, it can be unstable. The wave which goes unstable (when $p = 1$) propagates downstream, because its wave speed is $1 - q = \frac{3}{2}$, and in fact the $p > 0$ wave always propagates downstream. The other wave, always stable, propagates downstream unless $1 + q < 0$, i.e., if and only if $p < 1/2$, or equivalently,

$$F < F_- = \frac{2\tilde{k}}{(3 + 4\tilde{k}^2)^{1/2}}. \tag{4.87}$$

Note that F_- depends on $\tilde{k}$, and that $0 < F_- < 1$. Rewriting this inequality in terms of F and k, it is

$$F^2(1 - F^2) > \frac{3}{4k^2}, \tag{4.88}$$

and upstream propagating waves are possible for short waves with $k > \sqrt{3}$.

We therefore have three distinct ranges for F:

$F > 2$: two waves downstream, one unstable;

$1 < F < 2$: two waves downstream, both stable;

$F < 1$: stable waves can propagate upstream and downstream.

To go further than this requires a study of the nonlinear system (4.49). We see that the transition at $F = 1$ is associated with the ability of waves to propagate upstream. The transition at $F = 2$ is sometimes called a Vedernikov instability and is associated with the formation of downstream propagating *roll waves*.

4.5   Nonlinear Waves

When $F > 2$, linear disturbances will grow, and nonlinear effects become important in limiting their eventual amplitude. Because of the hyperbolic form of the equations, we might then expect shocks to form. To examine this hyperbolic form, we put

$$\gamma = \frac{1}{F}. \tag{4.89}$$

The equations are then

$$h_t + (hu)_s = 0,$$

$$u_t + uu_s + \gamma^2 h_s = \gamma^2 \left[1 - \frac{u^2}{h}\right], \tag{4.90}$$

and they can be written in the form

$$\frac{\partial}{\partial t}\begin{pmatrix} h \\ u \end{pmatrix} + \begin{pmatrix} u & h \\ \gamma^2 & u \end{pmatrix}\frac{\partial}{\partial s}\begin{pmatrix} h \\ u \end{pmatrix} = \begin{pmatrix} 0 \\ \gamma^2[1 - \frac{u^2}{h}] \end{pmatrix}. \tag{4.91}$$

4.5.1 Characteristics

The analysis of characteristics for systems of hyperbolic equations is described in Chap. 1. The eigenvalues of $B = \begin{pmatrix} u & h \\ \gamma^2 & u \end{pmatrix}$ are given by

$$\lambda = u \pm \gamma h^{1/2}, \tag{4.92}$$

and the matrix P of eigenvectors and its inverse P^{-1} are given by

$$P = \begin{pmatrix} \sqrt{h} & \sqrt{h} \\ \gamma & -\gamma \end{pmatrix}, \qquad P^{-1} = \frac{1}{2\gamma\sqrt{h}} \begin{pmatrix} \gamma & \sqrt{h} \\ \gamma & -\sqrt{h} \end{pmatrix}. \tag{4.93}$$

Comparing this with (1.69), we see that the integral

$$\int P^{-1}\, d\mathbf{u} = \int \begin{pmatrix} \frac{dh}{2\sqrt{h}} + \frac{du}{2\gamma} \\ \frac{dh}{2\sqrt{h}} - \frac{du}{2\gamma} \end{pmatrix} = \begin{pmatrix} \sqrt{h} + \frac{u}{2\gamma} \\ \sqrt{h} - \frac{u}{2\gamma} \end{pmatrix} \tag{4.94}$$

is well-defined, and determines the characteristic variables (the *Riemann invariants*, so called because they are constant on the characteristics in the absence of the forcing gravity and friction terms, as in shallow water theory). The equations can thus be compactly written in the characteristic form

$$\left[\frac{\partial}{\partial t} + (u \pm \gamma\sqrt{h}) \frac{\partial}{\partial s} \right] [u \pm 2\gamma\sqrt{h}] = \gamma^2 \left[1 - \frac{u^2}{h} \right]. \tag{4.95}$$

Nonlinear waves propagate downstream if $u/\gamma h^{1/2} > 1$, but one will propagate upstream if $u/\gamma h^{1/2} < 1$. This is consistent with the preceding linear theory (since $u/\gamma h^{1/2}$ is the local Froude number, i.e., the Froude number based on the local values of velocity and depth). Because Eqs. (4.95) are of second order, simple shock wave formation analysis is not generally possible. Equations (4.95) are very similar to those of gas dynamics, or the shallow water equations, and the equations support the existence of propagating shocks in a similar way.

4.5.2 Roll Waves

There is a good deal of evidence that solutions of (4.90) do indeed form shocks, and when these are formed via the instability when $F > 2$, the resultant waves are called *roll waves*. They are seen in steep flows with relatively smooth beds (and thus low friction), but this combination is difficult to find in natural rivers. It is found, however, in artificial spillways, such as that shown in Fig. 4.6, which shows a photograph of roll waves propagating down a spillway in Canada. Roll waves can be found forming on any steep incline. Film flow down steep slopes during heavy rainfall will inevitably form a sequence of periodic waves, and these are also roll waves; see Fig. 4.7. I used to see them frequently at my daughter's school, for example.

Fig. 4.6 Roll waves propagating down a spillway at Lion's Bay, British Columbia. The width of the flow is about 2 m, and the water depth is about 10 cm. Photograph courtesy Neil Balmforth

To describe roll waves, we seek travelling wave solutions to (4.90), in the form $h = h(\xi)$, $u = u(\xi)$, where $\xi = s - ct$ is the travelling wave coordinate, c being the wave speed. Substitution of these into (4.90) yields the two ordinary differential equations

$$-ch' + (uh)' = 0,$$
$$-cu' + uu' = 1 - \frac{u^2}{h} - \gamma^2 h'. \tag{4.96}$$

The first equation has the integral

$$(u - c)h = -K, \tag{4.97}$$

where K is a positive constant. The reason that it must be positive is that the positive characteristics (those with speed $u + \gamma h^{1/2}$) must run into (not away from) the shock, that is,

$$u_+ + \gamma h_+^{1/2} < c < u_- + \gamma h_-^{1/2}, \tag{4.98}$$

where h_+ and h_- are the values of h immediately in front of and immediately behind the shock. Hence

$$\gamma h_+^{3/2} < K < \gamma h_-^{3/2}. \tag{4.99}$$

Fig. 4.7 Laminar roll waves following rainfall at Craggaunowen, Co. Clare, Ireland. The water depth is a few millimetres and the wavelength of the order of twenty centimetres

Substitution of (4.97) into the second equation yields a single first-order equation for u, or h. We choose to write the equation for h, thus

$$h' = \frac{h^3 - (ch - K)^2}{\gamma^2 h^3 - K^2}. \tag{4.100}$$

As indicated in Fig. 4.8, we aim to solve this equation in $(0, L)$, with $h = h_+$ at $\xi = 0$ and $h = h_-$ at $\xi = L$. The quantities involved in this equation and its boundary conditions are L, c, h_-, h_+ and K, and these have to be determined. Solution of the differential equation (4.100) from 0 to L yields one condition,

$$L = \int_{h_+}^{h_-} \frac{\gamma^2 h^3 - K^2}{h^3 - (ch - K)^2} \, dh, \tag{4.101}$$

which determines L in terms of the other quantities. Thus four extra conditions need to be specified to determine these.

There are two jump conditions to apply across the shock. These are conservation of mass, which we omit, as it is automatically satisfied by (4.97), and conservation of momentum, which has the form

$$c = \frac{[hu^2 + \frac{1}{2}\gamma^2 h^2]_-^+}{[hu]_-^+}. \tag{4.102}$$

Fig. 4.8 Schematic form of roll waves

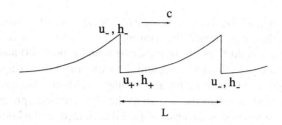

Fig. 4.9 Supercritical and subcritical values of h across a shock: graph of $\frac{1}{2}\gamma^2 h^2 + K^2/h$, $\gamma = K = 1$

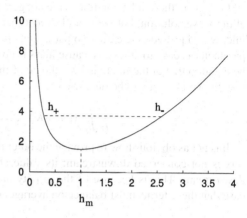

Simplification of this using (4.97) gives

$$\left[\frac{1}{2}\gamma^2 h^2 + \frac{K^2}{h}\right]_-^+ = 0. \tag{4.103}$$

Evidently, consideration of the graph of $\frac{1}{2}\gamma^2 h^2 + \frac{K^2}{h}$ shows that this determines h_+ in terms of h_-, for given K, see Fig. 4.9.

We denote the critical value of h at the minimum in Fig. 4.9 as h_m, thus

$$\gamma^2 h_m^3 = K^2; \tag{4.104}$$

clearly we must have $h_- > h_m$ and $h_+ < h_m$ (this is also implied by (4.99)), that is to say, the flow is subcritical behind the shock and supercritical in front of it. In particular, there is a value of $\xi \in (0, L)$ with $h = h_m$, and in order that the derivative in (4.100) remain finite, it is necessary that the numerator also vanish at this point. Since $K > 0$, this implies

$$c h_m - K = \frac{K}{\gamma}. \tag{4.105}$$

We have added an extra quantity h_m to the other unknowns L, h_-, h_+, K and c. To determine these six quantities, we have the four Eqs. (4.101), (4.103), (4.104) and (4.105). This appears to imply that the roll waves described here form a two parameter family, with (for example) the wavelength and wave speed being arbitrary. This is at odds with our expectation that a sensibly described physical problem will have just the one solution. In order to understand this, we need to reconsider the

hyperbolic form of the describing Eqs. (4.90). A natural domain on which to solve
these equations is the semi-infinite real axis $s > 0$, in which case appropriate boundary conditions are to prescribe h and u on $t = 0$ and $s = 0$. The initial conditions
are prescribed to represent the experimental start-up, and the boundary conditions
at $s = 0$ must represent the inlet conditions. The effect of the initial conditions is
washed out of the system as the characteristics progress down stream, and the roll
waves which are observed are determined by the boundary conditions at $s = 0$.

Of course, these inlet conditions are not generally consistent with a periodic travelling wave solution, but we would expect that prescribed values of u and h at the
inlet would provide the extra two parameters to fix the solution precisely. One such
parameter is easy to assess. Because mass is conserved, the mean volume flux must
be equal to that at the inlet, and by choice of the velocity and depth scales, we can
take the volume flux to be one, whence

$$\frac{1}{L} \int_0^L (ch - K)\,d\xi = 1. \tag{4.106}$$

It is not as obvious how to provide the other recipe, because the mean momentum
flux is not conserved downstream; its value at the inlet does not tell us its value
downstream. This is because of the gravity and friction terms. However, it *is* the
case that these terms must balance on average, that is to say,

$$\int_0^L \left(h - u^2\right) d\xi = 0; \tag{4.107}$$

this actually follows by integrating the momentum equation (written in conservation
form) over a wavelength. The momentum advection and pressure gradient terms
vanish because of (4.103), leaving (4.107). This appears to give a final condition to
close the system: but it does not, as (4.107) actually reduces to (4.103) when the
integration is carried out. An appropriate final condition is not easy to determine;
we provide some further discussion below. Before that, we reduce the conditions
above to a simpler form.

We rewrite the relations (4.101), (4.103), (4.104), (4.105) and (4.106) using h_m
as the defining parameter, and putting

$$h_+ = h_m \phi_+, \qquad h_- = h_m \phi_-; \tag{4.108}$$

then we have K and c given by

$$K = \gamma h_m^{3/2}, \qquad c = h_m^{1/2}(1 + \gamma), \tag{4.109}$$

and L, ϕ_+ and ϕ_- are determined, after some algebra, by

$$\begin{aligned}
L &= \gamma^2 h_m \int_{\phi_+}^{\phi_-} \frac{(\phi^2 + \phi + 1)\,d\phi}{(\phi - \gamma)^2 - \gamma^2 \phi}, \\
1 &= \frac{\gamma^2 h_m^{5/2}}{L} \int_{\phi_+}^{\phi_-} \frac{(\phi^2 + \phi + 1)\{\phi + \gamma(\phi - 1)\}\,d\phi}{(\phi - \gamma)^2 - \gamma^2 \phi}, \\
\left[\frac{1}{2}\phi^2 + \frac{1}{\phi}\right]_-^+ &= 0,
\end{aligned} \tag{4.110}$$

where we have taken $Q = 1$ in (4.106). The second of these can be written independently of L as

$$q = \frac{\int_{\phi_+}^{\phi_-} \frac{(\phi^2+\phi+1)\{\phi+\gamma(\phi-1)\}\,d\phi}{(\phi-\gamma)^2-\gamma^2\phi}}{\int_{\phi_+}^{\phi_-} \frac{(\phi^2+\phi+1)\,d\phi}{(\phi-\gamma)^2-\gamma^2\phi}}, \tag{4.111}$$

where

$$q = \frac{1}{h_m^{3/2}}. \tag{4.112}$$

The profile of ϕ is given by the scaled version of (4.100), which is

$$\phi' = \frac{(\phi-\gamma)^2-\gamma^2\phi}{\gamma^2 h_m(\phi^2+\phi+1)}. \tag{4.113}$$

The numerator must be positive, and since $\phi = 1$ for some ξ, a necessary condition for this to be true is that $\gamma < 1/2$. In terms of the Froude number, this is $F > 2$, which is the condition under which the roll wave instability occurs in the first place. This nicely suggests that the roll waves bifurcate as a non-uniform solution from the steady state at $F = 2$.

It is apparent from the above discussion that the crux of the determination of the roll wave parameters is the solution of (4.110)$_3$ and (4.111) for given positive q. If ϕ_+ and ϕ_- can be found for any such q, then they can be found for any h_m, after which L, K and c follow directly from (4.109) and (4.110)$_1$.

To find the solutions of (4.110)$_3$ and (4.111), we note that ϕ_+ and ϕ_- are uniquely defined in terms of the ordinate of the graph in Fig. 4.9; in fact, for any $\phi_+ \in (0, 1)$, (4.110)$_3$ gives the explicit solution

$$\phi_- = \frac{1}{2}\left[-\phi_+ + \left\{\phi_+^2 + \frac{8}{\phi_+}\right\}^{1/2}\right]; \tag{4.114}$$

then (4.111) gives $q = q(\phi_+; \gamma)$. The other constants are then given explicitly by (4.109), (4.110)$_1$ and (4.111), and in particular, if we define

$$N(\phi_+) = \int_{\phi_+}^{\phi_-} \frac{(\phi^2+\phi+1)\{\phi+\gamma(\phi-1)\}\,d\phi}{(\phi-\gamma)^2-\gamma^2\phi},$$

$$D(\phi_+) = \int_{\phi_+}^{\phi_-} \frac{(\phi^2+\phi+1)\,d\phi}{(\phi-\gamma)^2-\gamma^2\phi} \tag{4.115}$$

(thus $q = N/D$), then using

$$h_m = \left(\frac{D}{N}\right)^{2/3}, \tag{4.116}$$

we have

$$L = \frac{\gamma^2 D^{5/3}}{N^{2/3}}, \qquad c = \frac{(1+\gamma)D^{1/3}}{N^{1/3}}, \qquad K = \frac{\gamma D}{N}. \tag{4.117}$$

Fig. 4.10 Graphs of
$\Delta h = h_- - h_+$ as a function
of ϕ_+ for $\gamma = 0.1$ ($F = 10$),
$\gamma = 0.2$ ($F = 5$) and $\gamma = 0.4$
($F = 2.5$). The *asterisks* mark
the ends of the curves at
$\phi_+ = \alpha_+$

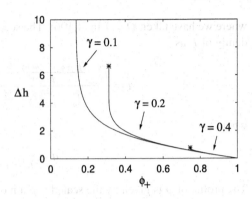

Fig. 4.11 Dimensionless
wavelength L in terms of ϕ_+
for $\gamma = 0.1$ ($F = 10$),
$\gamma = 0.2$ ($F = 5$) and $\gamma = 0.4$
($F = 2.5$). The curves do not
terminate, since
$L \sim -\ln[\phi_+ - \alpha_+]$ as
$\phi_+ \to \alpha_+$

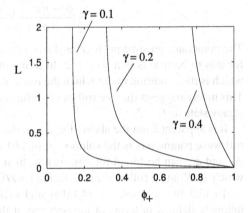

Equations (4.114), (4.116) and (4.117) determine ϕ_-, h_m, L, c and K in terms
of ϕ_+. From these we can find h_- and h_+. Thus it is convenient in computing the
one parameter family of wave solutions to use ϕ_+ as the parameter.

In Figs. 4.10, 4.11, 4.12 we plot the wave height $\Delta h = h_m(\phi_- - \phi_+)$, wavelength
L and speed c (all dimensionless) as a function of the parameter ϕ_+, for various
values of the Froude number F.

Fig. 4.12 Wave speed c in
terms of ϕ_+. The *asterisks*
mark the ends of the curves at
$\phi_+ = \alpha_+$, $c = c_+ = \frac{1+\gamma}{q_+^{1/3}}$

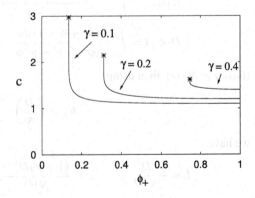

Fig. 4.13 Wave speed c as a function of L, for $\gamma = 0.1$, $\gamma = 0.2$ and $\gamma = 0.4$. The *short dashed lines* at the right ordinate indicate the corresponding asymptotes c_+ for $\gamma = 0.1$ and $\gamma = 0.2$ and $\gamma = 0.4$ at the respective values $c_+ = 2.9717, 2.1495, 1.6216$

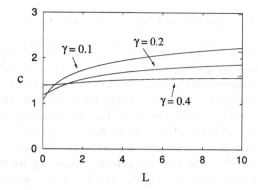

A feature of Fig. 4.10 is the termination of the curves at a finite value. The integrals which define N and D in (4.115) can be explicitly evaluated. If we define the two (positive) roots of $(\phi - \gamma)^2 - \gamma^2 \phi = 0$ to be

$$\alpha_\pm = \frac{\gamma}{2}[2 + \gamma \pm \{\gamma^2 + 4\gamma\}^{1/2}], \tag{4.118}$$

thus $\alpha_+ > \alpha_- > 0$, then we restrict $\phi_+ > \alpha_+$ so that $\phi' > 0$ in (4.113). Consideration of N and D then shows that

$$D = -A \ln(\phi_+ - \alpha_+) + O(1), \qquad N = -C \ln(\phi_+ - \alpha_+) + O(1) \tag{4.119}$$

as $\phi \to \alpha_+$. From this it follows that $q \to q_+$ as $\phi \to \alpha_+$, where $q_+ = C/A$, and is given explicitly by

$$q_+ = (1 + \gamma)\alpha_+ - \gamma. \tag{4.120}$$

These termination points are marked by asterisks at the end of the curves in Fig. 4.10. Because $q = q_+ + O(\frac{1}{-\ln(\phi_+ - \alpha_+)})$, the slope of the curves is infinite at these points. (This also makes it hard to draw the figures. To get within 0.02 of q_+, for example, we can expect to have to take $\phi_+ - \alpha_+ \approx \exp(-50) \approx 10^{-22}$!)

As $\phi_+ \to 1$, then also $\phi_- \to 1$, and hence both N and D are $O(1)$. Direct consideration of (4.115) shows that $q \to 1$ as $\phi_+ \to 1$. As a consequence of these limiting behaviours, $L \to 0$ and c is finite as $\phi_+ \to 1$, while $L \to \infty$ as $\phi_+ \to \alpha_+$, but c tends to a finite limit just as q does. As shown in Figs. 4.10–4.12, all three quantities vary monotonically between $\phi_+ = \alpha_+$ and $\phi_+ = 1$, and consequently c is a monotonically increasing function of L, which tends to a limit c_+ as $L \to \infty$, where

$$c_+ = \frac{(1 + \gamma)}{q_+^{1/3}}. \tag{4.121}$$

This is shown in Fig. 4.13. Analysis of the limit $\phi_+ \to \alpha_+$ shows that $c = c_+ + O(1/L)$ as $L \to \infty$ (Question 4.15), and evidently the approach to the limit is slow, particularly at low γ (high Froude number).

Wavelength Selection and Boundary Conditions

Although it is convenient to compute the properties of the roll waves using the parameter ϕ_+, it is more natural to use the wavelength L as the single parameter. The issue remains how this is selected. This seems to be an open problem, on which we offer some comments, though little further insight.

The first thing to note is that the hyperbolic St. Venant equations (4.90) require two initial conditions at the inlet $s = 0$ if the Froude number $F > 1$. If we imagine flow from a vent below a dam, for example, it is easy to see that prescription of both h and hu (and thus u) can be effected, by having a vent opening of a prescribed height, and adjusting the dam height to control mass flow. From a mathematical point of view, precisely steady inlet conditions $h = u = 1$ lead to uniform downstream flow, provided the St. Venant equations apply precisely. Thus we can see that it is only through the prescription of a time varying inlet velocity, for example, that roll waves can develop downstream. For example, we might prescribe inlet conditions

$$h = 1, \qquad u = 1 + \lambda \cos \omega t \quad \text{at } s = 0, \tag{4.122}$$

where $\lambda \ll 1$. We would then infer that the resulting periodic solution would have frequency ω, and this would prescribe the ratio

$$\frac{L}{c} = \omega, \tag{4.123}$$

which would provide the final prescription of the solution. Consulting Fig. 4.13, we can see that (4.123) would indeed determine a unique value of L.

More generally, we might suppose $u(0, t)$ to be a polychromatic, perhaps stochastic function. We might then expect the wavelength selected to be that of the most rapidly growing mode. Consultation of (4.85), however, indicates that for $F > 2$, p and thus $\mathrm{Re}\,\sigma$ is an increasing function of wave number k, with $p \to F$ as $k \to \infty$. This unbounded growth at large wave number is suggestive of ill-posedness, and in any case is certainly not consistent with the apparent observation that long wavelength roll waves are in practice selected.

A final consideration, and perhaps the most practical one, is that wavelength selection may take place at large times through the interaction of neighbouring wave crests. Larger waves move more rapidly (c is an increasing function of Δh if we plot one in terms of the other), and therefore larger waves will catch smaller ones. This provides a coarsening effect, whereby smaller waves can be removed by larger ones. Since Δh is also an increasing function of L, this coarsening does indeed lead to longer waves. The process should be limited by the fact that very long (and thus flat) waves will be subject to the same Vedernikov instability as is the uniform state.[4] If we supposed that wavelength varied slowly from wave to wave, we can see the beginnings of a kind of nonlinear multiple scales method to describe the evolution of wavelength as a function of space and time. It is less easy to see how

---

[4]This observation is due to Neil Balmforth.

to incorporate the generation of new waves in such a framework, however, and this problem remains open for investigation.

The spectre of ill-posedness described above raises the related issue of how to prescribe the correct boundary conditions for the St. Venant equations. The reason there is an issue is that the equations require two upstream boundary conditions if $F > 1$, but one upstream and one downstream condition if $F < 1$. This makes no sense, insofar as the boundary conditions should be prescribed independently of the solution. A resolution of this conundrum lies in the realisation that the formation of shocks in the hyperbolic system suggests the presence of a missing diffusive term, and this takes the form of a turbulent eddy viscous term.

In our discussion of the basal friction term (4.5), we assumed only the transverse Reynolds stress $-\rho\overline{u'w'} \approx \mu_T \frac{\partial \bar{u}}{\partial z}$ was significant. The longitudinal Reynolds stress $-\rho\overline{u'^2} \approx \mu_T \frac{\partial \bar{u}}{\partial x}$ is small, but provides a crucial diffusive term

$$\frac{\partial}{\partial x}\left(\mu_T A \frac{\partial u}{\partial x}\right) \tag{4.124}$$

to be added to the right hand side of (4.43). Following (B.9) in Appendix B, we suppose

$$\mu_T = \rho \varepsilon_T [u] d, \tag{4.125}$$

and this leads to the corrective term

$$\varepsilon_T F^2 S \frac{1}{A} \frac{\partial}{\partial x}\left(A \frac{\partial u}{\partial x}\right) \tag{4.126}$$

to be added to (4.49)$_2$. Correspondingly, Eqs. (4.90) are modified to

$$h_t + (hu)_s = 0,$$
$$u_t + uu_s + \gamma^2 h_s = \gamma^2 \left[1 - \frac{u^2}{h}\right] + \frac{\kappa}{h} \frac{\partial}{\partial s}\left(h \frac{\partial u}{\partial s}\right), \tag{4.127}$$

where

$$\kappa = \varepsilon_T S \ll 1. \tag{4.128}$$

A typical value of κ is $\sim 10^{-5}$.

Because κ is small, it can be expected to provide a shock structure for the shocks we have described. In addition, the extra derivative suggests that an extra boundary condition for the system (4.127) needs to be prescribed. Most obviously, this is at the outlet, where the river meets the sea. The most obvious such condition might be to prescribe h, or perhaps h_x, but it is more likely that one should prescribe

$$u = 0 \quad \text{at } s = 1, \tag{4.129}$$

indicating the flow of the river into a large reservoir. In any event, the extra condition at the outlet, together with the diffusive term (4.124), can explain the difference in the solutions when $F \lessgtr 1$. The characteristics of (4.90) are the sub-characteristics of (4.127), and the appropriate pair of conditions to apply for (4.90) is determined by the correct way of determining the singular approximation when $\kappa \to 0$.

Fig. 4.14 The Severn bore. This is a famous photograph from 1921, when there were no by-standers, and certainly no surfers. Reproduced from Pugh (1987). The photograph first appears in the book by Rowbotham (1970), where Mr C.W.F. Chubb is acknowledged as the photographer

However, this really sheds no further light on the issue of roll wave length selection. When $F > 2$, clearly two conditions are appropriate at $s = 0$, but how these conspire to select the wavelength is unclear.

4.5.3 Tidal Bores

A bore on a river is a shock-like wave which travels upstream, and it occurs because of forcing at the mouth of the river due to tidal variation in sea level. In England the best known example is the Severn bore, which occurs because of the very high tidal range in the Severn estuary. Large crowds come to view the bore, which manifests itself as a wall of water about a metre high advancing up river at a speed of some four to five metres a second. Figures 4.14 and 4.15 show photographs of the Severn bore. Bores occur on certain rivers due to a confluence of factors. The tidal range has to be very large, and this can be caused by tidal resonance in an estuary; in addition, the river must narrow dramatically upstream, so that the estuary acts like a funnel. The wave then forms because the rapidly rising water level in the estuary causes a large upstream water flux, and with a sufficiently large funnelling effect, a shock wave will be formed. Bores occur all over the world, for example in the Amazon, the Seine, the Petitcodiac river which flows into the Bay of Fundy, and the Tsien Tang river in China. Where they occur, they are spectacular, but relatively few rivers have them, because of the severity of the necessary conditions for their formation.

Fig. 4.15 The Severn bore, viewed from the air in a microlight aircraft by Mark Humpage. The image is copyright Mark Humpage, and is reproduced with his permission. For other photographs, see http://www.markhumpage.com. The undular nature of the bore is very clearly visible (as are the relentless surfers)

Figure 4.16 shows the geometry of the Severn river and estuary. The bore forms near Sharpness, and is best viewed at various places further upstream, notably Minsterworth and Stonebench, where public access is available. Figure 4.17 shows a profile of the river during passage of a bore. There are certain features evident in this figure which are relevant when we formulate a model. The river depth at low stage is about a metre, whereas the tidal range is much greater than this. In the Severn estuary, it can be 14.5 metres, and at Sharpness, it is 9 metres in the figure. The other feature of importance is the apparent alteration in the bedslope as the estuary is approached. As an idealisation of this, Fig. 4.18 shows the basic geometry of a river–estuary system, which we can use to explain bore formation.

The river in Fig. 4.18 flows into a tidal basin, where the water level fluctuates tidally with a period of slightly more than twelve hours. Such fluctuations cause the river/estuary boundary point to migrate back and forth. In particular, approaching high tide this point moves upstream. The idea behind bore formation is that if the upstream velocity of this boundary is faster than the upstream characteristic

Fig. 4.16 A sketch map of
the river Severn

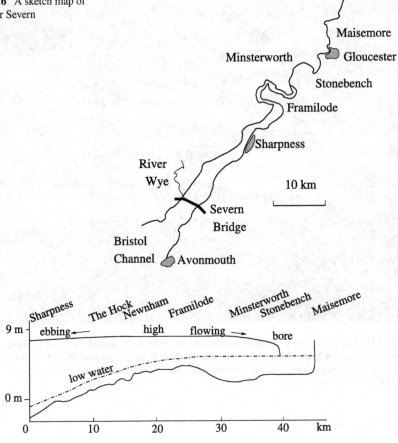

Fig. 4.17 Profile of the Severn during passage of a bore. Note that high water occurs someway below the bore (the tide continues to come in after the passage of the bore), but that the tide near Sharpness already starts to ebb before the bore reaches Maisemore

wave speed,[5] a smooth wave cannot occur, and a shock must form, as indicated in Fig. 4.18.

We want to study this phenomenon in the context of the St. Venant equations (4.49), where for a wide channel, we choose the hydraulic radius and cross sectional area to be

$$R = h, \qquad A = wh, \qquad (4.130)$$

where w is the width, and is taken to be a prescribed function of s. The phenomenon of concern occurs over the length of the river, so that long wave theory is appropriate. From Fig. 4.17, a suitable length scale is of the order of 45 km, where the

---

[5] We assume the Froude number F is less than one at low stage, which is the realistic condition; in that case, one wave travels upstream. If $F > 1$, a standing wave would form at the boundary.

Fig. 4.18 Idealised (and highly exaggerated) river basin geometry

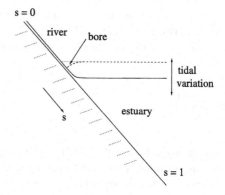

length scale used in writing (4.49) is $d/\sin\alpha$, and is 2 km if we take $d = 2$ m and $S = \sin\alpha = 10^{-3}$. If we take a typical velocity upstream as 2 m s^{-1}, then the corresponding time scale is 10^3 s, or 15 minutes, and the Froude number is about 0.3. The scale up in distance is thus of order 22, while that in time to the half-period of tidal oscillations is similar. This suggests that we rescale both time and space as

$$t \sim \frac{1}{\varepsilon}, \qquad s \sim \frac{1}{\varepsilon}, \qquad (4.131)$$

where a plausible value of ε may be of order 0.05. In this case (4.49) can be written in the form (where now, because u will be negative during inflow, we take the friction term in the corrected form $\propto |u|u$)

$$wh_t + (wuh)_s = 0,$$

$$\varepsilon F^2(u_t + uu_s) = 1 - \frac{|u|u}{h} - \varepsilon h_s, \qquad (4.132)$$

or equivalently in the form

$$\varepsilon\left[\pm F\frac{\partial}{\partial t} + (\sqrt{h} \pm Fu)\frac{\partial}{\partial s}\right][2\sqrt{h} \pm Fu] = 1 - \frac{|u|u}{h} \mp \frac{\varepsilon F w' \sqrt{h}u}{w}, \qquad (4.133)$$

which shows explicitly that the characteristic wave speeds are

$$\pm\frac{\sqrt{h}}{F} + u, \qquad (4.134)$$

as we found before. Finally, we wish to study the situation shown in Fig. 4.18, where the tidal range is significantly larger than the river depth. The simplest choice is to suppose the tidal amplitude is also $O(1/\varepsilon)$, so that appropriate boundary conditions for (4.132) are

$$wuh = 1 \quad \text{at } s = 0,$$

$$h = \frac{H_1(t)}{\varepsilon} \quad \text{at } s = 1, \qquad (4.135)$$

representing a constant upstream volume flux, and a prescribed tidal range.

The assumption that $\varepsilon \ll 1$ allows us to solve (4.132) asymptotically. The solution has two parts, river and estuary, joined at a front which we denote by $s = s_f$. Upstream, for $s < s_f$, the flow is quasi-stationary, and we have, to leading order,

$$wuh \approx 1, \qquad 1 - \frac{u|u|}{h} \approx 1, \tag{4.136}$$

whence

$$u \approx w^{-1/3}, \qquad h = w^{-2/3}. \tag{4.137}$$

The steady solution of (4.132) is appropriate, because the sub-characteristic wave propagates downstream, and after any initial transient, the upstream boundary condition leads to a steady flow.

Downstream, for $s > s_f$, we write

$$h = \frac{H}{\varepsilon}, \tag{4.138}$$

so that

$$w H_t + (wuH)_s \approx 0,$$
$$1 - H_s \approx 0 \tag{4.139}$$

(the surface is flat); from this we have

$$H \approx s - 1 + H_1(t), \tag{4.140}$$

and from this there follows

$$u \approx \frac{-\dot{H}_1 \int_{s_f}^{s} w\, ds}{wH}, \tag{4.141}$$

where we choose the integration constant for matching purposes at s_f. Also to match the solution to that in $s < s_f$, we need to take

$$s_f = 1 - H_1. \tag{4.142}$$

Transition Region

At the front, we define

$$s = s_f + \varepsilon X, \qquad \dot{s}_f = c, \qquad w_f = w[s_f(t)]; \tag{4.143}$$

then to leading order we have

$$-cw_f h_X + (w_f hu)_X = 0,$$
$$F^2(u - c)u_X = 1 - \frac{u|u|}{h} - h_X, \tag{4.144}$$

with boundary conditions

$$h \to h_- = w_f^{-2/3}, \qquad u \to u_- = w_f^{-1/3} \quad \text{as } X \to -\infty,$$
$$h \sim X, \qquad u \sim c \quad \text{as } X \to \infty, \tag{4.145}$$

in order to match to the upstream and downstream solutions. Note that this transition region, like that for the monoclinal flood wave, is mediated by the full St. Venant equations, but without a diffusive term. Only the conditions on h in (4.145) are necessary, those on u following automatically. A first integral of the mass conservation equation (4.144)$_1$ gives

$$(u - c)h = K = \left[w_f^{-1/3} - c\right]w_f^{-2/3},$$ (4.146)

and from this we find

$$h_X = \frac{h^3 - |K + ch|(K + ch)}{h^3 - K^2 F^2}.$$ (4.147)

This can be compared with (4.100). The difference in the present case is that c and K in (4.143) and (4.146) are given, and the question is only whether a solution exists joining $h = h_- = w_f^{-2/3}$ upstream to the downstream solution $h \sim X$. Note that as $X \to -\infty$, $K + ch \to w_f^{-1}$, so that $h \to w_f^{-2/3}$ can consistently be satisfied.

Let us suppose that the tide is coming in, thus $c < 0$. We suspect that a smooth solution in the transition region may not be possible if $-c$ is greater than the upstream wave speed. Using (4.134) and (4.145), this condition can be written in the form

$$-c > w_f^{-1/3}\left(\frac{1}{F} - 1\right)$$ (4.148)

(assuming $F < 1$). If we suppose that the opposite inequality holds, i.e., $-c < w_f^{-1/3}(\frac{1}{F} - 1)$, then a little algebra shows that this is precisely the criterion that

$$h_- = w_f^{-2/3} > (KF)^{2/3},$$ (4.149)

i.e., the denominator of (4.147) is positive. To see that there is a solution of this problem in this case, we need to show that the numerator of the right hand side (4.147) is also positive, for then h will increase indefinitely as required.

The numerator, N, is given by

$$N = \left\{h^3 - w_f^{-2}\right\} - \left\{|w_f^{-1} + c(h - w_f^{-2/3})|[w_f^{-1} + c(h - w_f^{-2/3})] - w_f^{-2}\right\}.$$ (4.150)

Both expressions in curly brackets are zero when $h = h_-$ at $X = -\infty$; for h slightly greater than h_-, the left curly bracketed expression is positive, while the right curly bracketed expression decreases, since $c < 0$. The numerator is thus positive for $h - h_-$ small and positive, and remains so. From this it follows that a solution of the transition problem exists if $-c < w_f^{-1/3}(\frac{1}{F} - 1)$, and thus a bore will not form.

It remains to be shown that no solution exists if the opposite inequality, (4.148), holds. In this case the denominator of the right hand side of (4.147) is initially negative. As before, the numerator is positive if $h > h_-$, and equivalently negative if $h < h_-$, thus implying $h_X < 0$ if $h > h_-$, and $h_X > 0$ if $h < h_-$. This means solutions of (4.147) can only approach h_- as $X \to \infty$, and no transition solution

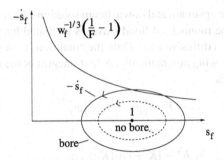

Fig. 4.19 Bore formation occurs for large tides and rapidly widening rivers with reasonably sized Froude numbers. If the tide oscillates sinusoidally and the river slope is constant, then the front position s_f will trace an ellipse as shown in the $(s_f, \dot{s}_f)$ plane. For a funnel-shaped river, the width w decreases as s_f decreases, so that $(\frac{1}{F} - 1)w^{-1/3}$ is a decreasing function of s_f, as shown. Bore formation therefore occurs according to (4.148) for the solid tidal curve, but not for the smaller amplitude dotted one

exists. This suggests another form of solution, one in which a discontinuity forms at the critical condition

$$-\dot{s}_f = w(s_f)^{-1/3}\left(\frac{1}{F} - 1\right), \tag{4.151}$$

and thereafter propagates upstream as a shock front. This is the bore. Figure 4.19 shows a schematic illustration of the criterion (4.151) for bore formation.

Propagation of the Bore

The outer river and estuary solutions (4.137), (4.140) and (4.141) remain valid after the formation of a shock, but the transition region is replaced by a shock at s_f, where the values of h_- and u_- (given by (4.137) with $w = w_f$) jump (up) to values h_+ and u_+, which have to be determined along with s_f. Initially h_+ and u_+ are $O(1)$, and we anticipate that this remains true; in this case s_f is still given by

$$s_f = 1 - H_1 + O(\varepsilon); \tag{4.152}$$

the location of the bore is essentially determined by the tidal range. Jump conditions of mass and momentum across the developing bore then imply that the bore speed $\dot{s}_f = c$ satisfies

$$c = \frac{[hu]_-^+}{[h]_-^+} = \frac{\frac{1}{2}[h^2]_-^+ + F^2[hu^2]_-^+}{[hu]_-^+}, \tag{4.153}$$

and these two relations serve to determine h_+ and u_+, since $c = -\dot{H}_1$.

Shock Structure

We can use the transition equations (4.144), modified by the addition of the diffusive term in (4.127), to study the shock structure of the bore. The equations then take the form

$$- cw_f h_X + (w_f h u)_X = 0,$$

$$F^2 (u - c) u_X = 1 - \frac{u|u|}{h} - h_X + \frac{\kappa F^2}{h} \frac{\partial}{\partial X} \left(h \frac{\partial u}{\partial X} \right), \qquad (4.154)$$

and the boundary conditions are still (4.145). The difference with the preceding analysis is that when a bore forms, we expect the diffusive term to act as a singular perturbation which allows the matching of two distinct outer solutions through an interior shock (the bore). Writing

$$u = c + \frac{K}{h}, \qquad (4.155)$$

we find that h satisfies

$$\frac{\partial h}{\partial X} = \frac{h^3 - |K + ch|(K + ch)}{h^3 - K^2 F^2} - \frac{\kappa F^2 K h^2}{h^3 - K^2 F^2} \frac{\partial}{\partial X} \left[\frac{1}{h} \frac{\partial h}{\partial X} \right]. \qquad (4.156)$$

As discussed before (4.151), the only way h can approach h_- as $X \to -\infty$ in bore-forming conditions is if the outer solution (where $\kappa = 0$) in $X < 0$ is

$$h \equiv h_-, \qquad X < 0. \qquad (4.157)$$

We suppose that h jumps through the shock to a value $h_+ > h_-$. According to the argument following (4.150), the numerator of (4.147) for the outer solution in $X > 0$ is then positive, and so, providing $h_+^3 > K^2 F^2$, the outer solution for h will increase monotonely from h_+, and $h \sim X$ as $X \to \infty$. It only remains to show that a shock structure exists connecting h_- to $h_+ > (KF)^{2/3}$.

Supposing without loss of generality the shock to be at $X = 0$, we define

$$X = \kappa K F^2 \xi \qquad (4.158)$$

(noting that $K > 0$), so that to leading order (4.156) becomes

$$\frac{\partial h}{\partial \xi} = - \frac{h^2}{h^3 - K^2 F^2} \frac{\partial}{\partial \xi} \left[\frac{1}{h} \frac{\partial h}{\partial \xi} \right]. \qquad (4.159)$$

Integrating this, we find

$$\frac{\partial h}{\partial \xi} = - h \left[\frac{1}{2} (h^2 - h_-^2) + K^2 F^2 \left(\frac{1}{h} - \frac{1}{h_-} \right) \right]. \qquad (4.160)$$

Consideration of the right hand side of this equation shows that if $h_-^3 < K^2 F^2$, then $- \frac{h'}{h}$ is zero at $h = h_-$, negative for $h > h_-$ until it becomes positive for large h. Thus there is one further zero of h' at $h_+ > h_-$, and $h' > 0$ between these two values, always assuming that $h_-^3 < K^2 F^2$, which is guaranteed by (4.149). Thus the shock layer structure takes h monotonically from h_- to h_+, given by

$$\frac{1}{2} (h_+^2 - h_-^2) = K^2 F^2 \left(\frac{1}{h_-} - \frac{1}{h_+} \right), \qquad (4.161)$$

and it only remains to check that $h_+ > (KF)^{2/3}$, so that the outer solution to (4.147) in $X > 0$ does indeed increase as $X \to \infty$. This is clear from the definition of $-\frac{h'}{h}$ given by (4.160), which shows that $-\frac{h'}{h}$ is a convex upwards function $G(h)$, and in particular shows that $G'(h_+) > 0$. Since from (4.159),

$$G'(h) = \frac{h^3 - K^2 F^2}{h^2}, \qquad (4.162)$$

we can deduce that indeed $h_+ > (KF)^{2/3}$.

This analysis shows that in bore-forming conditions, the diffusive term in (4.154) does indeed allow a shock structure to exist, and this describes what is known as a turbulent bore, appropriate at reasonably large Froude numbers. The Severn bore shown in Fig. 4.15 is an example of an undular bore, appropriate at lower Froude numbers, and consisting of an oscillatory wave train. The St. Venant equations do not appear to be able to describe this kind of bore, where the oscillations have a wavelength comparable to the depth, and the vertical velocity structure may need to be considered in attempting to model it. This is discussed further below.

4.6 Notes and References

A preliminary version of the material in this chapter is in my own book on modelling (Fowler 1997), although with much less detail than presented here. The general subject of river flow is treated in its contextual, geographical aspect by books on hydrology, such as those of Chorley (1969) or Ward and Robinson (2000). Ward and Robinson's book, for example, deals with precipitation, evaporation, groundwater and other topics as well as the dynamics of drainage basins, but is less concerned with detailed flow processes in rivers. For these, we turn to books on hydraulics, such as those by French (1994) or Chow (1959). A nice book, which bridges the gap, and also includes a discussion of sediment transport and channel morphology and pattern, is that by Richards (1982).

Roll Waves   Flood waves and roll waves have been discussed from the present perspective by Whitham (1974). The linear instability at Froude number greater than two was analysed by Jeffreys (1925), and the finite amplitude form of roll waves was described by Dressler (1949), whose presentation we follow here. The book by Stoker (1957) gives a nice discussion, as well as a useful photograph of roll waves on a spillway in Switzerland. The eddy viscous diffusive term in (4.127) was added by Needham and Merkin (1984). Balmforth and Mandre (2004) provide a thorough review, and also provide a discussion of the mechanics of wavelength selection. They also, following Yu and Kevorkian (1992), provide a weakly nonlinear model for roll wave evolution when $F - 2 \ll 1$; a strongly nonlinear model would be more relevant at higher F. Their experiments are consistent with the idea that the form of the inlet condition is instrumental in determining the roll wavelength.

Tidal Bores The effect of tidal variations on river flow is discussed by Pugh (1987); in particular, he describes the phenomenon of the river bore. Another useful little book is that by Tricker (1965). The literature on bores seems to be rather sparse, although the phenomenon itself has been well known for a (very) long time. Chanson (2005) refers to the fact that the *mascaret* of the Seine river in France was documented in the ninth century. Lord Rayleigh, while president of the Royal Society, wrote down the jump conditions for the bore velocity over a hundred years ago (Rayleigh 1908). There is a very informative article by Lynch (1982), prior to which the principal analysis is that of Abbott and Lighthill (1956), who analyse the St. Venant equations, and apply their results to the Severn bore. The presentation is extremely opaque, however. The little book by Rowbotham (1970) is a gem, and has many other striking photographs besides that shown in Fig. 4.14.

More recently, there has been an upsurge of interest in modelling bores. Su et al. (2001) construct a numerical model of the turbulent bore of the Hangzhou Gulf and Qiantangjiang river in China using the St. Venant equations. In a number of papers, Chanson and co-workers have studied the dynamics of undular bores (Wolanski et al. 2004; Chanson 2005), both observationally and experimentally. Chanson (2009) reviews the observational and experimental literature, with numerous illustrations.

In order to obtain an oscillatory wave train (such as one also finds in capillary waves), it seems that a higher derivative term in (4.160) might be necessary, either as $h_{\xi\xi\xi}$ or from a term u_{XXX} in (4.154). Such terms are commonly found in higher-order approximations to water wave equations, as for example in the Korteweg–de Vries equation. To get a flavour of such an analysis, we consult the derivation of the Korteweg–de Vries equation by Ockendon and Ockendon (2004, pp. 106 ff.). Reverting to dimensional coordinates, their derivation of the Korteweg–de Vries equation takes the form, assuming a backwards travelling wave,

$$u_t + \cdots = \frac{\sqrt{gd}\, d^2}{6} u_{sss}. \tag{4.163}$$

If we simply suppose that such a term can be added to the St. Venant equation, then, using the scales in (4.48), the St. Venant equations (4.50) or (4.127) become

$$wh_t + (wuh)_s = 0,$$

$$F^2(u_t + uu_s) + h_s = 1 - \frac{|u|u}{h} + \frac{\kappa F^2}{h}\frac{\partial}{\partial s}\left(h\frac{\partial u}{\partial s}\right) + \frac{1}{6}FS^2 u_{sss}. \tag{4.164}$$

Repeating the shock structure analysis, (4.156) is replaced by

$$-\frac{1}{6}FKS^2\left(\frac{1}{h}\right)_{XXX} + \kappa F^2 Kh^2\left(\frac{h_X}{h}\right) + P(h)h_X - N(h) = 0, \tag{4.165}$$

where

$$P(h) = h^3 - K^2 F^2, \qquad N(h) = h^3 - |K + ch|(K + ch) \tag{4.166}$$

($N(h)$ is the numerator in (4.147) discussed following (4.150)).

We write

$$h = h_-\phi, \qquad N = h_-^3 n(\phi), \qquad P = h_-^3 p(\phi), \qquad c = -\sqrt{h_-}V, \tag{4.167}$$

Fig. 4.20 Model of a
turbulent bore. Solution of
(4.165) in the form (4.171),
using values $F = 1.5$, $V = 0$,
$\beta = 0.1$, $\delta = 0.01$. The time
step used is 10^{-5}, and the
plot takes $h_- = 1$ in its scales
for X and h

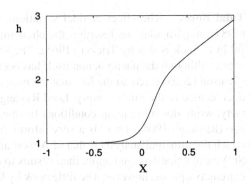

whence (4.145) and (4.146) imply

$$K = h_-^{3/2}(1 + V),$$ (4.168)

and hence

$$n(\phi) = \phi^3 - |1 + V - V\phi|(1 + V - V\phi), \qquad p(\phi) = \phi^3 - (1 + V)^2 F^2. \quad (4.169)$$

Lastly we put

$$X = h_- Z.$$ (4.170)

Then (4.165) becomes

$$-\delta\left(\frac{1}{\phi}\right)_{ZZZ} + \beta\phi^2\left(\frac{\phi_Z}{\phi}\right)_Z + p(\phi)\phi_Z - n(\phi) = 0,$$ (4.171)

where

$$\delta = \frac{F(1 + V)S^2}{6h_-^{11/2}}, \qquad \beta = \frac{\kappa F^2(1 + V)}{h_-^{3/2}},$$ (4.172)

and both are small.

The boundary conditions for ϕ are that

$$\phi \to 1 \quad \text{as } Z \to -\infty, \qquad \phi \sim Z \quad \text{as } Z \to \infty.$$ (4.173)

Figures 4.20 and 4.21 show numerical solutions of the transition equation (4.171) for two different values of β. The first corresponds to a relatively high value of β, when δ is sufficiently small to be ignored, and the preceding shock structure analysis (following (4.154)) is valid. Formally this requires $\delta \ll \beta^2$.

At lower values of β, however, it is inadmissible to neglect the third derivative term. To analyse what happens in this case, write

$$Z = \sqrt{\delta}\zeta,$$ (4.174)

and define

$$\mu = \frac{\beta}{\sqrt{\delta}}.$$ (4.175)

Fig. 4.21 Model of an undular bore. Solution as for Fig. 4.20, except that $\beta = 0.001$

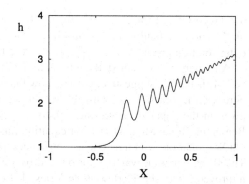

Assuming $\delta \ll 1$, we can neglect the term in n within the transition zone, so that

$$-\left(\frac{1}{\phi}\right)_{\zeta\zeta\zeta} + \mu\phi^2\left(\frac{\phi_\zeta}{\phi}\right)_\zeta + p(\phi)\phi_\zeta \approx 0. \tag{4.176}$$

The turbulent bore is regained if $\mu \gg 1$. For the case $\mu \lesssim 1$, define

$$\psi = 1 - \frac{1}{\phi}, \tag{4.177}$$

whence

$$\psi''' + \frac{\mu}{(1-\psi)^2}\left\{\frac{\psi'}{(1-\psi)}\right\}' + p\left(\frac{1}{1-\psi}\right)\frac{\psi'}{(1-\psi)^2} = 0. \tag{4.178}$$

Suppose first that μ is small; then a first integral of (4.178) with $\mu = 0$ is

$$\psi'' + W'(\psi) = 0, \tag{4.179}$$

where

$$W'(\psi) = \int_1^{\frac{1}{1-\psi}} p(\phi)\,d\phi, \qquad W(0) = 0. \tag{4.180}$$

Integrating and changing the order of integration, we can write

$$W(\psi) = \int_1^{\frac{1}{1-\psi}}\left[\psi - \left(1 - \frac{1}{\phi}\right)\right]p(\phi)\,d\phi. \tag{4.181}$$

As a function of ψ, $W(0) = W'(0) = 0$, and (since $p(1) < 0$, equivalent to the bore-forming condition (4.148)) $W''(0) < 0$; thus W is negative for small $\psi > 0$. Since $W''(\psi) = \frac{p(\phi)}{(1-\psi)^2}$, and p is an increasing function of ϕ, we see that W reaches a negative minimum, and thereafter increases, tending towards ∞ as $\psi \to 1$ and $\phi \to \infty$.

(4.179) is the equation of a nonlinear oscillator, and shows that ϕ increases from zero at $Z = -\infty$, and then oscillates about the minimum of W. In fact with $\mu = 0$, there would be precisely one oscillation, with ϕ returning to zero at $Z = +\infty$. This does not happen for two reasons. The term in μ is a damping term (this is clear in (4.176) if the coefficient ϕ^2 is ignored; alternatively one can view (4.176) as a damped oscillator for ψ), so that the oscillations are damped towards the minimum of W; and the small term in n in (4.171) causes a drift upwards in ϕ towards the outer solution given by $\phi_Z \approx \frac{n(\phi)}{p(\phi)}$. Both these features can be seen in Fig. 4.21.

Although in this context, the introduction of the long wave dispersive term u_{sss} in (4.164) is merely suggestive, it does show that such a term can produce the undular bore seen in practice at relatively low Froude number. The classical approach is given in the paper by Peregrine (1966), who simply writes down as a model the Benjamin–Bona–Mahony (BBM) equation, also called the regularised long wave (RLW) equation, which in essence introduces a term u_{sst} in (4.164) in place of u_{sss}. The BBM equation was (re-)introduced by Benjamin et al. (1972) as a suggested improvement to the Korteweg–de Vries (KdV) equation, on the basis that it has better regularity properties. Specifically, the dispersion relation for modes $e^{ik(s-ct)}$ is $c = 1 + k^2$ for the linearised KdV equation $u_t + u_s = u_{sss}$, while it is $c = \frac{1}{1+k^2}$ for the linearised BBM equation $u_t + u_s = u_{sst}$. The growth of the wave speed at large wave number is associated with ill-posedness. See also Question 9.9.

4.7 Exercises

4.1 Find a relationship between the hydraulic radius R and the area A for triangular (notch shaped) or rectangular (canal shaped) cross sections. Hence show that Chézy's and Manning's laws both lead to a general relationship of the form

$$Q = \frac{cA^{m+1}}{m+1},$$

with $0 < m < 1$, giving explicit prescriptions for c and m. For a canal of depth h, show that the flow is turbulent if

$$h \gtrsim 10^2 \nu^{2/3} \left(\frac{f}{Sg} \right)^{1/3},$$

where ν is the kinematic viscosity, f is the friction factor, S is the slope and g is gravity. Taking $\nu = 10^{-6}$ m^2 s^{-1}, $f = 0.01$, $S = 10^{-3}$, $g = 10$ m s^{-2}, find a critical depth for turbulence. Is the Thames turbulent?

4.2 For flow in a pipe, the friction factor f in the formula $\tau = f\rho u^2$ is often taken to depend on the Reynolds number; for example, Blasius's law of friction has $f \propto Re^{-1/7}$. By taking $Re = UR/\nu$, where R is the hydraulic radius, find modifications to Chézy's law if $f \propto Re^{-\beta}$. Comment on whether you can obtain Manning's flow law this way.

4.3 The cross-sectional area of a river A is assumed to satisfy the wave equation

$$\frac{\partial A}{\partial t} + cA^m \frac{\partial A}{\partial s} = 0,$$

where s is distance downstream. Explain how this equation can be derived from the principle of conservation of mass. What assumptions does your derivation use?

A river admits a steady discharge $Q = Q_+$. At $t = 0$, a tributary at $s = 0$ is blocked, causing a sudden drop in discharge to $Q_- < Q_+$. Solve the equation for A using a characteristic diagram and show that an *expansion fan* branches from $s = 0$, $t = 0$. What is the hydrograph record at a downstream station $s = s_0 > 0$?

Later, the tributary is re-opened, causing a sudden rise from Q_- to Q_+. Draw the characteristic diagram, and show that a shock wave propagates forwards. What is its speed?

4.4 Use the method of characteristics to find the general solution of the equation describing slowly varying flow of a river. Show also that in general shocks will form, and describe in what situations they will not. What happens in the latter case?

Either by consideration of an integral form of the conservation of mass equation, or by consideration from first principles, derive a jump condition which describes the shock speed. In terms of the local water speed, what is the speed of a shock (a) when it first forms; (b) when it advances over a dry river bed?

4.5 A river of rectangular cross section with width w carries a steady discharge Q_0 (m^3 s^{-1}). At time $t = 0$, a rainstorm causes a volume V of water to enter the river at the upstream station $s = 0$. Assuming Chézy's law, find the solution for the resulting flood profile (sketch the corresponding characteristic diagram), and derive a (cubic) equation for the position of the advancing front of the flood. Without solving this equation, find an expression for the discharge Q_l at the downstream station $s = l$.

4.6 Derive the St. Venant equations from first principles, indicating what assumptions you make concerning the channel cross section. Derive a non-dimensional form of these equations assuming Manning's roughness law and a triangular cross section. [*Assume that there is no source term in the equation of mass conservation.*]

A sluice gate is opened at $s = 0$ so that the discharge there increases from Q_- to Q_+. The hydrograph is measured at $s = l$. Using l as a length scale, and with a corresponding time scale $\sim l/u$, derive an approximate expression for the dimensionless discharge in terms of A, if the Froude number is small, and also $\varepsilon = [\bar{h}]/Sl \ll 1$, where $[\bar{h}]$ is the scale for the mean depth and S is the slope.

Hence show that A satisfies the approximate equation

$$\frac{\partial A}{\partial t} + \frac{4}{3} A^{1/3} \frac{\partial A}{\partial s} = \frac{1}{4} \varepsilon \frac{\partial}{\partial s} \left[A^{5/6} \frac{\partial A}{\partial s} \right].$$

What do you think the difference between the hydrographs for $\varepsilon = 0$ and $0 < \varepsilon \ll 1$ might be?

4.7 Why should the equation

$$A_t + c A^m A_s = M$$

represent a better model of slowly varying river flow than that with $M = 0$? Find the general solution of the equation, given that $A = 0$ at $s = 0$, and

$A = A_0(s)$ at $t = 0$, $s > 0$, assuming $M = M(s)$. Find also the steady state solution $A_{eq}(s)$. How would you expect solutions representing disturbances to this steady profile to behave?

Suppose now that M is constant, and $A_0 = A_{eq} + \overline{A}\delta(s)$, representing an initial flood concentrated at $s = 0$. Show that the resulting flood occurs in $s_- < s < s_+$, and show that the profile of A between s_- and s_+ is given implicitly by

$$A^{m+1} - (A - Mt)^{m+1} = \frac{(m+1)Ms}{c},$$

and deduce that

$$s_- = \frac{cM^m t^{m+1}}{(m+1)}.$$

What happens as $M \to 0$?

4.8 A dimensionless long wave model for slowly varying flow of a river of depth h and mean velocity u is given in the form

$$h_t + (uh)_s = M(s),$$

$$0 = 1 - \frac{u^2}{h} - \varepsilon h_s,$$

where $\varepsilon \ll 1$.

How would you physically interpret the positive source term $M(s)$?

Show that for small ε, the model can be reduced to the approximate form

$$h_t + \left(h^{3/2}\right)_s = M(s) + \frac{1}{2}\varepsilon\left[h^{3/2}h_s\right]_s.$$

Show that if $h = 0$ at $s = 0$, then an approximate steady state solution is given by

$$h = \left\{\int_0^s M(s)\,ds\right\}^{2/3}. \qquad (*)$$

Find this approximate solution if $M \equiv 1$. Can you find a function M for which $(*)$ is the exact solution?

Explain why the condition of a horizontal water surface might be an appropriate boundary condition to apply at $s = 1$, and show that in terms of the scaled variables, this implies $h_s = 1/\varepsilon$ at $s = 1$. Show that with this added boundary condition, the approximate solution (when $M \equiv 1$) is still appropriate, except in a boundary layer near the outlet.

Next, suppose that $M = 0$ for large enough s, and that $\int_0^\infty M(s)\,ds = 1$. Write down the linear equation satisfied by small perturbations H to the steady state $h = 1$ when s is large.

By seeking solutions of the form $\exp[\sigma t + iks]$, show that small wave-like disturbances travel at speed $\frac{3}{2}$ and decay on a time scale $t \sim O(1/\varepsilon)$.

Fig. 4.22 $H(s, t)$ plotted at fixed $s = 1$ as a function of t, using values $\varepsilon = 0.03$, $l = 0.005$, $\delta = 1$

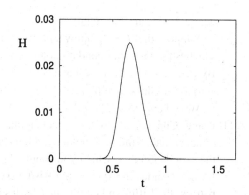

Show that if $\zeta = s - \frac{3}{2}t$, $\tau = \frac{1}{2}\varepsilon t$, then $H_\tau = H_{\zeta\zeta}$, and deduce that if $H = \delta \exp[-s^2/l^2]$ at $t = 0$, then

$$H = \delta \left(\frac{t_0}{t_0 + t}\right)^{1/2} \exp\left[\frac{-\left(s - \frac{3}{2}t\right)^2}{2\varepsilon(t_0 + t)}\right]$$

for $t > 0$, where $t_0 = \frac{l^2}{2\varepsilon}$. (A typical hydrograph described by this function is shown in Fig. 4.22. It is asymmetric, but the steep shock-like rise is limited by the linearity of the model.)

4.9 A dimensionless model for the steady, tranquil flow of a river of depth h, width w and mean velocity u is given in the form

$$(wuh)_s = M,$$

$$F^2 u u_s = 1 - \frac{u^2}{h} - h_s.$$

If $F = 0$, deduce that h satisfies the first-order ordinary differential equation

$$\frac{dh}{ds} = 1 - \frac{Q^2}{w^2 h^3},$$

where

$$Q = \int_0^s M(s)\,ds.$$

Show that if $w = 1$ and $M = 1$, there is no solution of this equation satisfying $h(0) = 0$.

Consider variously and in combinations the cases that $w = s^{1/2}$, $M = (1 + w'^2)^{1/2}$, $M = w$ (motivating these choices physically), and show that a solution with $h(0) = 0$ still cannot be found. Show that this remains true if $F > 0$. What do you conclude?

4.10 A dimensionless model for the steady, tranquil flow of a river of depth h and mean velocity u is given in the form

$$uh = s,$$

$$F^2 s u_s = h - u^2 - hh_s + \delta(hu_s)_s,$$

where $\delta \ll 1$, and we require $h \sim s^{2/3}$ as $s \to \infty$, and $h(0) = 0$.

Suppose that $F = 0$. Show that the leading-order outer solution (with $\delta = 0$) satisfies the far field boundary condition for a unique choice of $\lim_{s \to 0} h = h_0$. By writing $s = e^{\delta X}$, show that a boundary layer exists in which h changes from zero to h_0. Show also that $h \sim s^{h_0^2/2\delta}$ as $s \to 0$.

What happens if $F \neq 0$?

4.11 Using Chézy's law with a rectangular cross section, show how to non-dimensionalise the St. Venant equations, and show how the model depends on the Froude number, which you should define. Choose or guess suitable values for the Thames in London, the Isis/Cherwell in Oxford, the Quoile in Down-patrick, the Liffey in Dublin, the Charles in Boston, the Shannon in Limerick, the Lagan in Belfast (or your own favourite stretch of river), an Alpine (or other) mountain stream, and determine the corresponding natural length and time scales, and the Froude number, for these flows. Show also that in the case of long wave and short wave motions, the equations effectively become those of slowly varying flow and the shallow water equations, respectively.

4.12 The St. Venant equations, assuming Manning's roughness law, zero mass input, and a triangular river cross section, can be written in the dimensionless form

$$A_t + (Au)_s = 0,$$

$$F^2(u_t + uu_s) = 1 - \frac{u^2}{A^{2/3}} - \frac{A_s}{2A^{1/2}}.$$

Show in detail that small disturbances to the steady state $A = u = 1$ can propagate up and down stream if $F < F_1$, but can only propagate downstream if $F > F_1$, and that they are unstable if $F > F_2$. What are the values of F_1 and F_2?

4.13 A river flows through a lowland valley. The river level may fluctuate, so that it lies above or below the local groundwater level. Give a *simple* motivation for the model

$$\frac{\partial A}{\partial t} + cA^m \frac{\partial A}{\partial s} = -r(A - B),$$

$$\frac{\partial B}{\partial t} = r(A - B)$$

to describe the variations of river water (A) and groundwater (B), where B is a measure of the amount of groundwater.

Show that small disturbances to the uniform state $A = B = 1$ exist proportional to $\exp[\sigma t + iks]$ and find the dispersion relation relating σ to k. What do these solutions represent?

4.14 *The hydraulic jump*

Using the dimensionless form of the mass and momentum equations (for a canal), show that discontinuities (shocks) in the channel depth travel at a (dimensionless) speed V given by

$$V = \frac{[Au]_-^+}{[A]_-^+} = \frac{[F^2Au^2 + \frac{1}{2}A^2]_-^+}{[F^2Au]_-^+},$$

where $\pm$ refer to the values on either side of the jump, and F is the Froude number. Show that a stationary jump at $s = 0$ is possible (this can be seen when a tap is run into a flat basin) if $Au = Q$ in $s > 0$ and $s < 0$, and

$$\left[\frac{F^2Q^2}{A} + \frac{A^2}{2}\right]_-^+ = 0.$$

Deduce that for prescribed Q and A_-, a unique choice of $A_+ \neq A_-$ is possible. Show also that the locally defined Froude number is

$$Fr = \frac{FQ}{A^{3/2}},$$

and deduce that the hydraulic jump connects a region of *supercritical* $(Fr > 1)$ flow to a *subcritical* $(Fr < 1)$ one. (In practice, $A_- < A_+$ if $Q > 0$; if $A_- > A_+$, the discontinuity cannot be maintained.)

4.15 The functions $N(\phi_+, \phi_-)$ and $D(\phi_+, \phi_-)$ are defined by

$$N(\phi_+) = \int_{\phi_+}^{\phi_-} \frac{(\phi^2 + \phi + 1)\{\phi + \gamma(\phi - 1)\}\, d\phi}{(\phi - \gamma)^2 - \gamma^2\phi},$$

$$D(\phi_+) = \int_{\phi_+}^{\phi_-} \frac{(\phi^2 + \phi + 1)\, d\phi}{(\phi - \gamma)^2 - \gamma^2\phi},$$

where $\phi_- > \phi_+$, and the quantities L and c are defined by

$$L = \frac{\gamma^2 D^{5/3}}{N^{2/3}}, \qquad c = \frac{(1+\gamma)D^{1/3}}{N^{1/3}},$$

where γ is constant.

Evaluate the integrals to find explicit expressions for N and D, and show that as $\phi_+ \to \alpha_+$,

$$D = -A\ln(\phi_+ - \alpha_+) + D_0 + o(1), \qquad N = -C\ln(\phi_+ - \alpha_+) + N_0 + o(1),$$

and find explicit expressions for A, C, D_0 and N_0. Hence show that as $\phi_+ \to \alpha_+$,

$$\ln\left(\frac{1}{\phi_+ - \alpha_+}\right) \approx b(L + L^*) + O\left(\frac{1}{(L + L^*)}\right),$$

where the constant b should be determined, and deduce that

$$c \approx c_+ - \frac{k}{L + L^*} + O\left(\frac{1}{(L + L^*)^2}\right),$$

where k and L^* should be found. By evaluating k and L^* for different values of γ, show that both quantities increase rapidly as γ is reduced, and hence explain why the convergence of c to c_+ in Fig. 4.13 is so slow. Compare this asymptotic result with a direct numerical evaluation of $c(L)$. How good is the asymptotic result?

Chapter 5
Dunes

The muddy colour of many rivers and the milky colour of glacial melt streams are due to the presence in the water of suspended sediments such as clay and silt. The ability of rivers to transport sediments in this way, and also (for larger particles) by rolling or saltation as *bedload* transport, forms an important constituent of the processes by which the Earth's topography is formed and evolved: the science of geomorphology.

Sediment transport occurs in a variety of different (and violent) natural scenarios. Powder flow avalanches, sandstorms, lahars and pyroclastic flows are all examples of violent sediment laden flows, and the kilometres long black sandur beaches of Iceland, laid down by deposition of ash-bearing floods issuing from the front of glaciers, are testimony to the ability of fluid flows to transport colossal quantities of sediment. In this chapter we will consider some of the landforms which are built through the interaction of a fluid flow with an erodible substrate; in particular we will focus on the formation of *dunes* and *anti-dunes* in rivers, and *aeolian dunes* in deserts.

5.1 Patterns in Rivers

There are two principal types of patterns which are seen in rivers. The first is a pattern of channel form, i.e., the shape taken by the channel as it winds through the landscape. This pattern is known as a meander, and an example is shown in Fig. 5.1.

The second type of pattern consists of variations in channel profile, and there are a number of variants which are observed. A distinction arises between profile variations transverse to the stream flow and those which are in the direction of flow. In the former category are bars; in the latter, dunes and anti-dunes. The formation of lateral bars results in a number of different types of river, in particular the braided and anastomosing river systems (described below).

All of these patterns are formed through an erosional instability of the uniform state when water of uniform depth and width flows down a straight channel. The instability mechanism is simply that the erosive power of the flowing water increases

A. Fowler, *Mathematical Geoscience*, Interdisciplinary Applied Mathematics 36,
DOI 10.1007/978-0-85729-721-1_5, © Springer-Verlag London Limited 2011

Fig. 5.1 A meandering river, the Okavango in Botswana. Photograph supplied by courtesy of Gary Parker, and reprinted with permission of Terence McCarthy

with water speed, which itself increases with water depth. Thus a locally deeper flow will scour its bed more rapidly, forming a positive feedback which generates the instability. The different patterns referred to above are associated with different geometric ways in which this instability is manifested.

River meandering occurs when the instability acts on the banks. A small oscillatory perturbation to the straightness of a river causes a small secondary flow to occur transverse to the stream flow, purely for geometric reasons. This secondary flow is directed outwards (away from the centre of curvature) at the surface and inwards at the bed. As a consequence of this, and also because the stream flow is faster on the outside of a bend, there is increased erosion there, and this causes the bank to migrate away from the centre of curvature, thus causing a meander.

Braided rivers form because of a lateral instability which forms perturbations called bars. This is indicated schematically in Fig. 5.2. A deeper flow at one side

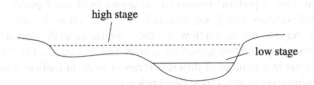

Fig. 5.2 Cross section of a braided river with one lateral bar, which is exposed when the river is at low stage (i.e., the river level is low). The instability which causes the bar is operative in stormflow conditions, when the bar is submerged

Fig. 5.3 A braided river. Image from http://www.braidedriver.net

of a river will cause excess erosion of the bed there, and promote the development of a lateral bar in stormflow conditions. The counteracting (and thus stabilising) tendency is for sediments to migrate down the lateral slope thus generated. Bars commonly form in gravel bed rivers, and usually interact with the meandering tendency to form alternate bars, which form on alternate sides of the channel as the flow progresses downstream. In wider channels, more than one bar may form across the channel, and the resulting patterns are called multiple row bars. In this case the stream at low stage is split up into many winding and connected braids, and the river is referred to as a braided river, as shown in Fig. 5.3.

It is fairly evident that the scouring conditions which produce lateral bars and braiding only occur during bank full discharge, when the whole channel is submerged. Such erosive events are associated with major floods, and are by their nature occasional events. In between such floods, vegetation may begin to colonise the raised bars, and if there is sufficient time, the vegetative root system can stabilise the sediment against further erosion. A further stabilising effect of vegetation is that the plants themselves increase the roughness of the bed, thus diminishing the stress transmitted to the underlying sediment. If the bars become stably colonised by vegetation, then the braided channels themselves become stabilised in position, and the resulting set of channels is known as an anastomosing river system.

The final type of bedform is associated with waveforms in the direction of flow. Depending on the speed of the flow, these are called dunes or anti-dunes. At high values of the Froude number ($F > 1$), anti-dunes occur, and at low values ($F < 1$) dunes occur. A related feature is the ripple, which also occurs at low Froude number. Ripples are distinguished from dunes by their much smaller scale. Indeed, ripples and dunes often co-exist, with ripples forming on the larger dunes. The rest of this chapter focusses on models to describe the formation and evolution of dunes.

5.2 Dunes

Dunes are perhaps best known as the sand dunes of wind-blown deserts. They occur in a variety of shapes, which reflect differences in prevailing wind directions. Where wind is largely unidirectional, transverse dunes form. These are ridges which form at right angles to the prevailing wind. They have a relatively shallow upslope, a sharp crest, and a steep downslope which is at the limiting angle of friction for slip. The air flow over the dune separates at the crest, forming a separation bubble behind the dune. Transverse dunes move at speeds of metres per year in the wind direction.

Linear dunes, or seifs, form parallel to the mean prevailing wind, but are due to two different prevailing wind directions, which alternatively blow from one or other side of the dune. Such dunes propagate forward, often in a snakelike manner.

Other types of dunes are the very large star dunes (which resemble starfish), which form when winds can blow from any direction, and the crescentic barchan dunes, which occur when there is a limited supply of erodible fine sand. They take the shape of a crab-like crescent, with the arms pointing in the wind direction. Barchan dunes have been observed on Mars. (Indeed, it is easier to find images of dunes on Mars than on Earth.) Figure 5.4 shows images of the four principal types of dune described above.

As already mentioned, dunes also occur extensively in river flow. At very low flow rates, ripples form on the bed, and as the flow rate increases, these are replaced by the longer wavelength and larger amplitude dunes. These are regular scarped features, whose steep face points downstream, and which migrate slowly downstream. They form when the Froude number $F < 1$ (the lower regime), and are associated with river surface perturbations which are out of phase, and of smaller amplitude.

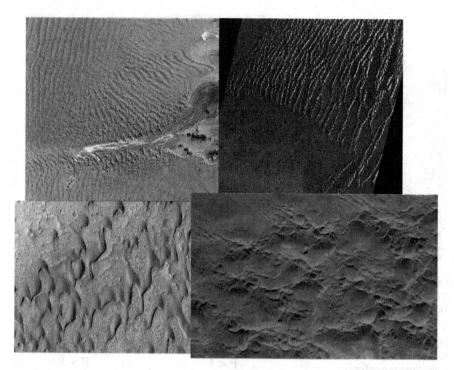

Fig. 5.4 Illustrations of four of the most common types of aeolian dunes: transverse (*top left*), seif (*top right*), barchan (*bottom left*), star (*bottom right*). The satellite view of transverse dunes is in the Namibian desert; source: http://earthasart.gsfc.nasa.gov/images/namib_hires.jpg. The seif dunes are from the Grand Erg Oriental, in the Sahara Desert in Algeria. Image from http://www.eosnap.com/public/media/2009/06/algeria/20090614-algeria-full.jpg, courtesy of Chelys. The barchanoid dunes are on Mars, so-called dark dunes in Herschel Crater. Image courtesy of NASA/JPL/University of Arizona, available at http://hirise.lpl.arizona.edu/PSP_002860_1650. Finally the image of star dunes is an international space station view of the Issaouane Erg, Algeria. Image courtesy of NASA, available at http://www.nasaimages.org (image ISS010-E-13539)

The wavelength of dunes is typically comparable to the river depth, the amplitude is somewhat smaller than the depth.

When the Froude number increases further, the plane bed re-forms at $F \approx 1$, and then for $F > 1$, we obtain the upper regime, wherein anti-dunes occur. Whereas dunes are analogous to shock waves, anti-dunes are typically sinusoidal, and are in phase with the surface perturbations, which can be quite large. They may travel either upstream or (more rarely) downstream. Indeed, for the more rapid flows, backward breaking shocks occur at the surface, and chute and pool sequences form. Anti-dunes can be found on rapid outlet streams on beaches; for example I have seen them on beach streams in Normandy and Ireland, where the velocity is on the order of a metre per second, and the flow depth may be several centimetres. A common observed feature of such flows is their time dependence: anti-dunes form, then migrate upstream as they steepen, leading to hydraulic jumps and collapse of the

Fig. 5.5 Antidunes on a beach stream at Spanish Point, Co. Clare, Ireland. The waves form, migrate slowly upstream (on a time scale determined by slow sediment transport), break and collapse. The process then repeats. Image courtesy of Rosie Fowler

Fig. 5.6 The succession of bedforms which are observed as the Froude number is increased. In the lower regime, where $F < 1$, we see first ripples and then the larger dune features. Surface perturbations are small. In the upper regime, $F > 1$; dunes disappear, giving a flat bed, and then anti-dunes are formed, in phase with surface waves. These are often transient features, occurring in flood conditions, and they are likely to be time dependent also

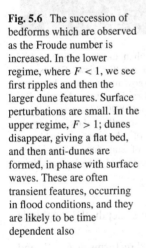

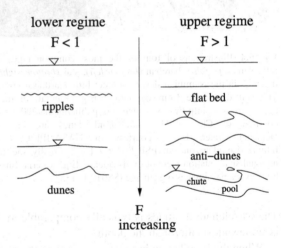

pattern, only for it to re-form elsewhere. An example of such anti-dunes is shown in Fig. 5.5. The succession of bedforms as the Froude number increases is illustrated in Fig. 5.6. Anti-dunes do not form in deserts simply because the Froude number is never high enough.[1]

Dunes and anti-dunes clearly form through the erosion of the underlying bed, and thus mathematical models to explain them must couple the river flow mechanics

---

[1]The Froude number corresponding to a wind of $20 \, \mathrm{m \, s^{-1}} = 45$ miles per hour over a boundary layer depth of 1 km is 0.2.

with those of sediment transport. Sediment transport models are described below. There are two main classes of bedform models. The most simple and appealing is to combine the St. Venant equations with an equation for bedform erosion. There are two ways in which sediment transport occurs, as bedload or as suspended load. Each transport mechanism gives a different model, and we shall find that a suspended load transport model can predict the instability which forms anti-dunes, but not dunes, which indeed may occur in the absence of suspended sediment transport.[2] On the other hand, the St. Venant equations coupled with a simple model of bedload transport cannot predict instability, although such a model can explain the shape and speed of dunes.

The other class of model which has been used describes the variation of stream velocity with depth explicitly. One version employs potential theory, as is customarily done in linearised surface wave theory. At first sight, this appears implausible insofar as the flow is turbulent, and indeed the model can then only explain dunes when the bed stress is artificially phase shifted. In order to deal with this properly, it is necessary to include a more sophisticated description of turbulent flow, and this can be done using an eddy viscosity model, which is then able to explain dune formation. The issue of analysing the model beyond the linear instability regime is more difficult, and some progress in this direction is described in this chapter. In Appendix B, we discuss the use of an eddy viscosity in simple models of turbulent shear flows.

5.2.1 Sediment Transport

Transport of grains of a cohesionless bed occurs as *bedload* or *in suspension*. At a given flow rate, the larger particles will roll along the bed, while the smaller ones are lifted by turbulent eddies into the flow. Clearly there is a transition between the two modes of transport: saltating grains essentially bounce along the bed.

Relations to describe sediment transport are ultimately empirical, though theory suggests the use of appropriate dimensionless groups. The basic quantity is the Shields stress, defined as the dimensionless quantity

$$\tau^* = \frac{\tau}{\Delta \rho g D_s}. \tag{5.1}$$

Here τ is the basal shear stress, $\Delta \rho = \rho_s - \rho_w$ is the excess density of solid grains over water (ρ_s is the density of the solid grains, ρ_w is the density of water), g is gravity, and D_s is the grain size. In general, grain sizes are distributed, and the Shields stress depends on the particle size. The shear stress τ at the bed is usually related to the mean flow velocity u by the semi-empirical relation (4.9), i.e.,

$$\tau = f \rho_w u^2, \tag{5.2}$$

where f is a dimensionless friction factor, of typical value 0.01–0.1. (Larger values correspond to rougher channels.)

---

[2]This also seems to be true of anti-dunes.

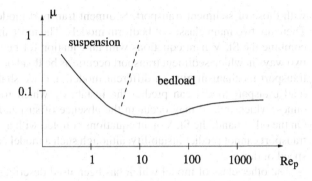

Fig. 5.7 The critical Shields stress for the onset of sediment transport, weakly dependent on the particle Reynolds number $Re_p = u_* D_s / \nu$

Shields found that sediment transport occurred if τ^* was greater than a critical value τ_c^*, which itself depends on flow rate via the particle Reynolds number

$$Re_p = \frac{u_* D_s}{\nu}. \tag{5.3}$$

(The friction velocity is defined to be

$$u_* = (\tau / \rho_w)^{1/2}.) \tag{5.4}$$

Figure 5.7 shows the variation of τ_c^* with $u_* D_s / \nu$; except at low flow rates, $\tau_c^* \approx 0.06$.

5.2.2 Bedload

Various recipes have been given for bedload transport, that due to Meyer-Peter and Müller being popular:

$$q^* = K [\tau^* - \tau_c^*]_+^{3/2}, \tag{5.5}$$

where $[x]_+ = \max(x, 0)$. Here $K = 8$, $\tau_c^* = 0.047$, and q^* is the dimensionless bedload transport rate, defined by

$$q^* = \frac{q_b}{(\Delta \rho g D_s^3 / \rho_w)^{1/2}}, \tag{5.6}$$

q_b being the bedload measured as volume per unit stream width per unit time.

5.2.3 Suspended Sediment

Suspended sediment transport is effected through a balance between an erosion flux v_E and a deposition flux v_D, each having units of velocity. The meaning of these is that $\rho_s v_E$ is the mass of sediment eroded from the bed per unit area per unit time, while $\rho_s v_D$ is the mass deposited per unit area per unit time.

Erosion

It is convenient to define a dimensionless erosion rate E via

$$v_E = v_s E, \tag{5.7}$$

where v_s is the particle settling velocity, given by Stokes's formula

$$v_s = \frac{\Delta \rho g D_s^2}{18 \eta}, \tag{5.8}$$

η being the dynamic viscosity of water. Various expressions for E have been suggested. They share the feature that E is a concave increasing function of basal stress. Typical is Van Rijn's relationship

$$E \propto (\tau^* - \tau_c^*)^{3/2} Re_p^{1/5}; \tag{5.9}$$

typical measured values of E are in the range 10^{-3}–10^{-1}.

Deposition

The calculation of deposition flux v_D is more complicated, as it is analogous to the calculation of basal shear stress in terms of mean velocity via an eddy viscosity model, as indicated in Appendix B. We can define the dimensionless deposition flux D by writing

$$\rho_s v_D = v_s \bar{c} D, \tag{5.10}$$

where $\bar{c}$ is the mean column concentration of suspended sediment, measured as mass per unit volume of liquid, and D depends on a modified Rouse number $R = v_s / \varepsilon_T \bar{u}$. (Here ε_T is related to the eddy viscosity; specifically ε_T^{-1} is the Reynolds number based on the eddy viscosity (see (B.9)), so the Rouse number is a Reynolds number based on particle fall velocity and eddy viscosity.) D increases with R, with $D(0) = 1$, and a typical form for D is

$$D = \frac{R}{1 - e^{-R}} \tag{5.11}$$

(see Appendix B for more details).

5.3 The Potential Model

The first model to explain dune formation dates from 1963, and invoked a potential flow for the fluid, which was assumed inviscid and irrotational. This is somewhat at odds with the fact that it is the basal stress of the fluid which drives sediment transport, but one can rationalise this by supposing that the stress is manifested through a basal turbulent boundary layer. We restrict our attention to two-dimensional motion

Fig. 5.8 Geometry of the
problem

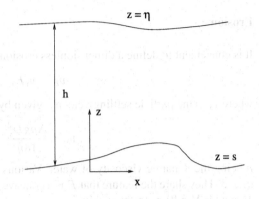

in the (x, z) plane: x is distance downstream, z is vertically upwards. The bed is at
$z = s(x, t)$, the free water surface is at $z = \eta(x, t)$, so that the depth h is given by

$$h = \eta - s; \tag{5.12}$$

the geometry is shown in Fig. 5.8. In the potential flow model, the usual equations
for the fluid flow potential ϕ apply:

$$
\begin{aligned}
\nabla^2 \phi &= 0 \quad \text{in } s < z < \eta, \\
\phi_z &= \eta_t + \phi_x \eta_x \quad \text{on } z = \eta, \\
\phi_t + g\eta + \frac{1}{2}|\nabla \phi|^2 &= \text{constant} \quad \text{on } z = \eta, \\
\phi_z &= s_t + \phi_x s_x \quad \text{on } z = s.
\end{aligned}
\tag{5.13}
$$

The extra equation required to describe the evolution of s is the *Exner equation*:

$$(1 - n)\frac{\partial s}{\partial t} + \frac{\partial q_b}{\partial x} = 0, \tag{5.14}$$

where n is the porosity of the bed; this assumes bedload transport only, and we may
take (see Eqs. (5.5) and (5.2)) $q_b = q_b(u)$, where $q_b'(u) > 0$. Implicitly, we suppose
a (turbulent) boundary layer at the bed, wherein the basal stress develops through a
shear layer; the basal shear stress will then depend on the outer flow velocity. We
define

$$q = \frac{q_b}{1 - n}, \tag{5.15}$$

so that

$$\frac{\partial s}{\partial t} + \frac{\partial q}{\partial x} = 0. \tag{5.16}$$

In the absence of any dynamic effect of the bed shape on the flow, we would expect u, and thus also q, to increase as s increases, due to the constriction of the
flow. If indeed q is an increasing function of the local bed elevation s, then it is
easy to see from (5.16) that perturbations to the uniform state $s = 0$ will persist as
forward travelling waves, and if q is convex ($q''(s) > 0$) then the waves will break

forwards. We interpret slip faces as the consequent shocks, so that this is consistent with observations. However, such a simple model does not allow for instability.

A simple way in which instability can be induced in the model is by allowing the maximum stress to occur upstream of the bed elevation maximum, as is indeed indicated by numerical simulations of the flow. One way to do this is to take

$$q = q(u|_{x-\delta}), \tag{5.17}$$

that is to say, the horizontal velocity $u = \phi_x$ is evaluated at $x - \delta$ and $z = s$, where the phase lag δ is included to model the notion that in shear flow over a boundary, such a lag is indeed present. Of course (5.17) is a crude and possibly dangerous way to model this effect.

To examine the linear stability of a uniform steady state we write $s = 0$, $\eta = h$,

$$\phi = Ux + \Phi, \qquad q = q(U) + Q, \qquad \eta = h + \zeta, \tag{5.18}$$

and then linearise the equations and boundary conditions (which are applied at the unperturbed boundaries $z = 0$ and $z = h$) to obtain

$$\nabla^2 \Phi = 0 \quad \text{in } 0 < z < h;$$
$$\Phi_z = \zeta_t + U\zeta_x, \qquad \Phi_t + g\zeta + U\Phi_x = 0 \quad \text{on } z = h; \tag{5.19}$$
$$\Phi_z = s_t + Us_x, \qquad s_t + Q_x = 0 \quad \text{on } z = 0,$$

where

$$Q = q'(U)\Phi_x|_{x-\delta, z=0}. \tag{5.20}$$

For a mode of wave number k, we put

$$(\zeta, s, Q) = (\bar{\zeta}, \bar{s}, \bar{Q}) \times e^{ikx + \sigma t}, \tag{5.21}$$

and write

$$\Phi = e^{ikx + \sigma t}[A \cosh kz + B \sinh kz], \tag{5.22}$$

so that the boundary conditions together with (5.20) become

$$k[A \sinh kh + B \cosh kh] = (\sigma + ikU)\bar{\zeta},$$
$$(\sigma + ikU)[A \cosh kh + B \sinh kh] + g\bar{\zeta} = 0,$$
$$kB = (\sigma + ikU)\bar{s}, \tag{5.23}$$
$$\sigma\bar{s} + ik\bar{Q} = 0,$$
$$Q = q'ike^{-ik\delta}A.$$

Some straightforward algebra leads to

$$\sigma[(\sigma + ikU)^2 + gk \tanh kh]$$
$$+ (\sigma + ikU)kq'e^{-ik\delta}[(\sigma + ikU)^2 \tanh kh + gk] = 0, \tag{5.24}$$

a cubic for $\sigma(k)$.

Solution of this is facilitated by the observation that we can expect two modes to correspond to upstream and downstream water wave propagation, while the third

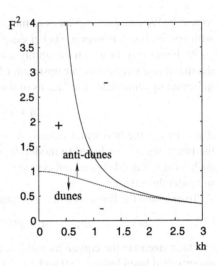

Fig. 5.9 Instability diagram for the potential flow model. The regions marked with a minus sign, above the upper curve and below the lower curve, are regions of instability if $\delta < 0$, more specifically if $\sin k\delta < 0$. The marked distinction between dunes and anti-dunes is based on the surface/bed phase relation (see (5.30)). Wave motion is downstream if $\cos k\delta > 0$, upstream if $\cos k\delta < 0$

corresponding to erosion of the bed may be much smaller, basically if q_b is sufficiently small. Specifically, let us assume (realistically) that $q \ll hu$. Then we may assume $q' \ll h$, and for small q', the roots of (5.24) are approximately the (stable) wave modes

$$\frac{\sigma}{-ik} \approx U \pm \left(\frac{g}{k} \tanh kh \right)^{1/2}, \tag{5.25}$$

and the erosive mode

$$\sigma \approx -k^2 U q' [\sin k\delta + i \cos k\delta] \tanh kh \frac{\left[F^2 - \frac{\coth kh}{kh} \right]}{\left[F^2 - \frac{\tanh kh}{kh} \right]}, \tag{5.26}$$

where we define the Froude number by

$$F = \frac{U}{\sqrt{gh}}. \tag{5.27}$$

For the erosive mode, the growth rate is

$$\mathrm{Re}\,\sigma = -k^2 U q' \sin k\delta \tanh kh \frac{\left[F^2 - \frac{\coth kh}{kh} \right]}{\left[F^2 - \frac{\tanh kh}{kh} \right]}, \tag{5.28}$$

and the wave speed is

$$-\frac{\mathrm{Im}\,\sigma}{k} = k U q' \cos k\delta \tanh kh \left[\frac{F^2 - \frac{\coth kh}{kh}}{F^2 - \frac{\tanh kh}{kh}} \right]. \tag{5.29}$$

This gives us the typical instability diagram shown in Fig. 5.9. For $\delta < 0$ (more specifically, $\sin k\delta < 0$) the regions above and below the two curves are unstable, corresponding to dunes and anti-dunes. The curves are given by $F^2 = \frac{\coth kh}{kh}$ and $F^2 = \frac{\tanh kh}{kh}$, respectively.

The phase relation between surface and bed for the erosive bed is given by

$$\frac{\bar{\zeta}}{\bar{s}} \approx \frac{F^2 \operatorname{sech} kh}{\left[F^2 - \frac{\tanh kh}{kh}\right]},\tag{5.30}$$

and this defines wave forms below the lower curve in Fig. 5.9 as dunes, and those above as anti-dunes.

Figure 5.9 is promising, at least if $\sin k\delta < 0$, as it will predict both dunes and anti-dunes. To get the wave speed positive, we need in fact to have $\cos k\delta > 0$, thus $0 > k\delta > -\pi/2$ (we can take $-\pi < k\delta < \pi$ without loss of generality), whereas we would generally want $k\delta < -\pi/2$ for anti-dunes to migrate backwards.

There is a serious problem with this model, beyond the fact that the phase shift δ is arbitrarily included. The spatial delay is unlikely to provide a feasible model for nonlinear studies; indeed, we see that $\operatorname{Re}\sigma \sim k^2$ at large k, and in the unstable regime this is one of the hallmarks of ill-posedness.

Having said that, it will indeed turn out to be the case that a phase lead ($\delta < 0$) really is the cause of instability. A phase lead means that the stress, and thus the bedload transport, takes its maximum value on the upstream face of a bump in the bed. A phase lead will occur because of the effect of the bump on the turbulent velocity structure above, as we discuss further below. It can also occur through an effect of bedload inertia (see also Question 5.7).

The choice of wave speed in this theory is unclear, since $\cos k\delta$ can be positive or negative. The possibly more likely choice of a positive value implies positive wave speed.

5.4 St. Venant Type Models

Since river flow is typically modelled by the St. Venant equations, it is natural to try using such a model together with a bed erosion equation to examine the possibility of instability. This has the added advantage of being more naturally designed for fully nonlinear studies. A St. Venant/Exner model can be written in the form (cf. the footnote following (4.46))

$$\begin{aligned}
s_t + q_x &= 0,\\
h_t + (uh)_x &= 0,\\
u_t + uu_x &= gS - \frac{fu^2}{h} - g\eta_x,
\end{aligned}\tag{5.31}$$

where S is the downstream slope, $q = q(\tau)$, $\tau = f\rho_w u^2$, and $\eta - s = h$. It is convenient to take advantage of the limit $q \ll hu$, just as we did before, and we do so by first non-dimensionalising the equations. We choose scales as follows:

$$s, x, h, \eta \sim h_0, \qquad u \sim u_0, \qquad q \sim q_0, \qquad t \sim \frac{h_0^2}{q_0},\tag{5.32}$$

Fig. 5.10 $s(u)$ as given by (5.36) for two typical cases of rapid and tranquil flow

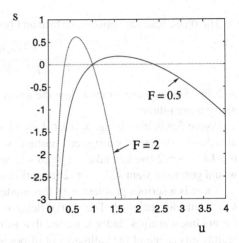

and we choose h_0, u_0 by balancing terms as follows: $uh \sim Q_0$, $gS \sim fu^2/h$; here Q_0 is the (prescribed) volume flow per unit width. We choose q_0 as the size of the bedload transport equation in (5.5).

With these scales, the dimensionless equations corresponding to (5.31) are

$$s_t + q_x = 0,$$
$$\varepsilon h_t + (uh)_x = 0,$$
$$F^2(\varepsilon u_t + uu_x) = -\eta_x + \delta\left(1 - \frac{u^2}{h}\right), \qquad (5.33)$$
$$h = \eta - s,$$

where the parameters are

$$F = \frac{u_0}{\sqrt{gh_0}}, \qquad \varepsilon = \frac{q_0}{Q_0}, \qquad \delta = S. \qquad (5.34)$$

If we now suppose $\varepsilon \ll 1$ and $\delta \ll 1$, both of them realistic assumptions, then we have approximately

$$uh = 1,$$
$$\frac{1}{2}F^2u^2 + \eta = \frac{1}{2}F^2 + 1, \qquad (5.35)$$

supposing that $u, h \to 1$ at large distances. Eliminating h and η, we have

$$s = 1 - \frac{1}{u} + \frac{1}{2}F^2(1 - u^2), \qquad (5.36)$$

whose form is shown in Fig. 5.10. In particular, $s'(1) = (1 - F^2)$, so the basic state $u = 1$ corresponds to the left hand or right hand root of $s(u)$ depending on whether the Froude number $F < 1$ or $F > 1$.

We also have

$$\frac{ds}{d\eta} = \frac{F^2 - h^3}{F^2}, \qquad (5.37)$$

Fig. 5.11 The wave speed $v(q) = 3q^{4/3}/(1 - F^2 q)$ for the tranquil and rapid cases $F = 0.5$ and $F = 1.5$

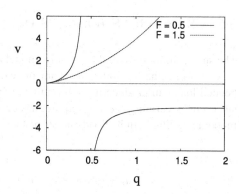

so that small perturbations to $h = 1$ are out of phase (dunes) if $F < 1$ and in phase (anti-dunes) if $F > 1$. If we take the dimensionless bedload transport as $q \approx \tau^{3/2} = u^3$ (the dimensionless basal stress having been scaled with $f\rho_w u_0^2$), so that $u = q^{1/3}$, then we see from (5.36) that $s = s(q)$, and $s(q)$ has the same shape as $s(u)$, as shown in Fig. 5.10.

The whole model reduces to the single first order equation

$$s'(q)q_t + q_x = 0. \tag{5.38}$$

Disturbances to the uniform state $q = 1$ will propagate at speed $v(q) = 1/s'(q)$, where v is shown in Fig. 5.11. For $F < 1$, $v(1) > 0$ and $v'(1) > 0$, thus waves in q (and thus s) propagate downstream and form forward-facing shocks; this is nicely consistent with dunes. For $F > 1$, $v < 0$ and $v'(1)$ is positive if $F < 2$, negative if $F > 2$ (see Question 5.4). Backward-facing shocks form, these are elevations in s if $v' > 0$.

Unfortunately, the hyperbolic equation does not admit instability. It is straightforward to insert a lag as before, by writing $q(x, t) = q[s(x - \delta, t)]$, or equivalently $s(x, t) = s[q(x + \delta, t)]$. Perturbation of

$$\begin{aligned} s_t + q_x &= 0, \\ q &= q\big[s(x - \delta, t)\big], \end{aligned} \tag{5.39}$$

via

$$s = \bar{s}e^{ikx + \sigma t}, \qquad q = 1 + \bar{q}e^{ikx + \sigma t}, \tag{5.40}$$

leads to

$$\begin{aligned} \sigma\bar{s} + ik\bar{q} &= 0, \\ \bar{q} &= q'e^{-ik\delta}\bar{s}, \end{aligned} \tag{5.41}$$

and thus

$$\sigma = kq'[-\sin k\delta - i\cos k\delta]. \tag{5.42}$$

This requires $\sin k\delta < 0$ for instability if $q'(s) > 0$ ($F < 1$) and $\sin k\delta > 0$ if $q'(s) < 0$ ($F > 1$). The long wavelength limit of (5.26) in which $kh \to 0$ is precisely (5.42), bearing in mind that (5.26) is dimensional and that $q' = dq/du$ there, whereas $q' = dq/ds$ in (5.42).

5.5 A Suspended Sediment Model

The shortcoming of both the potential model and the St. Venant/Exner model is the lack of a genuine instability mechanism. We now show that the inclusion of suspended load can produce instability. Ideally, we would hope to predict anti-dunes, since dunes certainly do not require suspended sediment transport. A St. Venant model including both bedload and suspended sediment transport is

$$h_t + (uh)_x = 0,$$

$$u_t + uu_x = g(S - \eta_x) - \frac{fu^2}{h},$$

$$\frac{\partial}{\partial t}(hc) + \frac{\partial}{\partial x}(hcu) = \rho_s(v_E - v_D),$$

$$(1 - n)\frac{\partial s}{\partial t} + \frac{\partial q_b}{\partial x} = -(v_E - v_D),$$

(5.43)

where c is the column average concentration (mass per unit volume) of suspended sediment (written as $\bar{c}$ earlier). The distinction between suspended sediment transport and bedload lies in the source terms due to erosion and deposition, v_E and v_D, and it is these which may enable instability to occur. We have $\eta - s = h$, and we suppose $q_b = q_b(\tau)$, $\tau = f\rho_w u^2$, whence $q = q(u)$. Additionally (see (5.7) and (5.10)), we write

$$v_E = v_s E, \qquad \rho_s v_D = v_s cD,$$

(5.44)

and expect that $E = E(u)$ and $D = D(u)$, with $E' > 0$, $D' < 0$; typically $E < 1$, $D > 1$.

We scale (5.43) as before in (5.32), except that we choose the time scale t_0, downstream length scale x_0, and concentration scale c_0 via

$$c_0 = \rho_s \frac{E_0}{D_0}, \qquad t_0 = \frac{(1 - n)h_0}{v_s E_0}, \qquad x_0 = \frac{Q_0}{v_s D_0},$$

(5.45)

where we write

$$E = E_0 E^*(u/u_0), \qquad D = D_0 D^*(u/u_0),$$

(5.46)

and choose E_0 and D_0 so that E^* and D^* are $O(1)$, and so that these are consistent with typical observed suspended loads of $10 \text{ g} \text{l}^{-1}$. With this choice of scales, we obtain the dimensionless set of equations

$$\eta - s = h,$$

$$\varepsilon h_t + (uh)_x = 0,$$

$$F^2(\varepsilon u_t + uu_x) = \delta\left(1 - \frac{u^2}{h}\right) - \eta_x,$$

(5.47)

$$h(\varepsilon c_t + uc_x) = E^* - cD^*,$$

$$s_t + \beta q_x = -(E^* - cD^*),$$

where the parameters ε, F, δ and β are now given by

$$\varepsilon = \frac{E_0}{(1-n)D_0} = \frac{c_0}{\rho_s(1-n)}, \qquad \delta = \frac{u_0 S}{v_s D_0},$$

$$F = \frac{u_0}{(gh_0)^{1/2}}, \qquad \beta = \frac{q_{b0} D_0}{Q_0 E_0} = \frac{\rho_s q_{b0}}{c_0 Q_0}. \tag{5.48}$$

Here q_{b0} is the scale for q_b rather than $q = q_b/(1-n)$. The Froude number is the same as before, but the parameters ε and δ are different: ε is a measure of the suspended sediment density relative to the bed density, and is always small; δ is the ratio of the (small) bed slope to the ratio of settling velocity to stream velocity. For more rapidly flowing streams, we might expect $\delta \sim 1$. However, if we suppose that wavelengths of anti-dunes are comparable to the depth (so $x_0 \sim h_0$), then (5.45) implies $\delta \sim S \ll 1$. Thus $\delta \sim 1$ implies $x_0 \sim h_0/S \gg h_0$. The parameter β is a direct measure of the ratio of bedload ($\rho_s q_{b0}$) to suspended load ($c_0 Q_0$). For $\beta \gg 1$, we would revert to our preceding bedload model and its scaling, and neglect the suspended load. If we adopt the Meyer-Peter/Müller relation in (5.5) and (5.6), then (noting that $f u_0^2 = g S h_0$)

$$q_{b0} = \frac{K \rho_l}{\Delta \rho}(gh S)^{3/2}, \tag{5.49}$$

and we can write

$$\beta = \left\{ \frac{K \rho_l}{(1-n)\Delta \rho} \right\} \frac{S^{3/2}}{\varepsilon F}; \tag{5.50}$$

both small or large values are possible.

To analyse (5.47), we ignore bedload (put $\beta = 0$) and take $\varepsilon \to 0$. Then

$$\eta = h+s, \qquad uh = 1, \tag{5.51}$$

so that

$$c_x = E^*(u) - cD^*(u) = -s_t,$$

$$\frac{\partial}{\partial x}\left[\frac{1}{2}F^2 u^2 + \frac{1}{u} + s \right] = \delta(1-u^3). \tag{5.52}$$

If, in addition, $\delta \ll 1$, then, taking $s = 0$ when $h = 1$,

$$s = s(u) = \frac{1}{2}F^2(1-u^2) + 1 - \frac{1}{u}, \tag{5.53}$$

and the entire suspended load model is

$$s'(u)\frac{\partial u}{\partial t} = cD^*(u) - E^*(u) = -\frac{\partial c}{\partial x}. \tag{5.54}$$

The function $s(u)$ is the same as we derived before in (5.36) and shown in Fig. 5.10. We can in fact write (5.54) as a single equation for u, by eliminating c; this gives

$$c = \frac{E^*(u)}{D^*(u)} + \frac{s'(u)}{D^*(u)}\frac{\partial u}{\partial t},$$

$$s'(u)\frac{\partial u}{\partial t} + \frac{\partial}{\partial x}\left[\frac{E^*(u)}{D^*(u)} + \frac{s'(u)}{D^*(u)}\frac{\partial u}{\partial t} \right] = 0, \tag{5.55}$$

and the equation for u (or the pair for u, c) is of hyperbolic type. Note that natural initial boundary conditions for (5.54) are to prescribe u at $t = 0$, $x > 0$, and c at $x = 0$, $t > 0$.

Let us examine the stability of the steady state $u = 1$, $c = 1$. We put

$$u = 1 + \text{Re}(U e^{ikx+\sigma t}), \qquad c = 1 + \text{Re}(C e^{ikx+\sigma t}), \tag{5.56}$$

and linearise, to obtain (noting $E^*(1) = D^*(1) = 1$)

$$ikC = [E^{*\prime}(1) - D^{*\prime}(1)]U - C = -\sigma s'(1), \tag{5.57}$$

and thus

$$\sigma = \left[\frac{E^{*\prime}(1) - D^{*\prime}(1)}{s'(1)} \right] \left(\frac{-k^2 - ik}{1 + k^2} \right). \tag{5.58}$$

If we suppose $E^{*\prime} > 0$, $D^{*\prime} < 0$ as previously suggested, then this model implies instability ($\text{Re}\,\sigma > 0$) for $s'(1) < 0$, i.e. $F > 1$, and that the wave speed is $-\text{Im}(\sigma)/k < 0$; thus this theory predicts upstream-migrating anti-dunes.

Two features suggest that the model is not well-posed if $F > 1$. The first is the instability of arbitrarily small wavelength perturbations; the second is that the unstable waves propagate upstream, although the natural boundary condition for c is prescribed at $x = 0$.

Numerical solutions of (5.54) are consistent with these observations. In solving the nonlinear model (5.54) in $0 < x < \infty$, we note that

$$\frac{d}{dt} \int_0^\infty s(u) \, dx = -[c]_0^\infty, \tag{5.59}$$

which simply represents the net erosion of the bed downwards if the sediment flux at infinity is greater than at zero. It thus makes sense to fix the initial boundary conditions so that

$$\begin{aligned} c &= 1 \quad \text{on } x = 0, \\ u &\to 1 \quad \text{as } x \to \infty, \ t = 0. \end{aligned} \tag{5.60}$$

For $F < 1$, numerical solutions are smooth and approach the stable solution $u = c = 1$. However, the solutions are numerically unstable for $F > 1$, and u rapidly blows up, causing breakdown of the solution.

Some further insight into this is gained by consideration of the solution at $x = 0$. If $c = c_0(t)$ on $x = 0$ and $u = u_0(x)$ on $t = 0$, then we can obtain u on $x = 0$ from (5.55), by solving the ordinary differential equation

$$\frac{\partial u}{\partial t} + \frac{E^*(u)}{s'(u)} = \frac{D^*(u)}{s'(u)} c_0(t) \tag{5.61}$$

with $u = u_0(0)$ at $t = 0$. If we suppose that $c = 1$ at $x = 0$, then it is easy to show that if $F < 1$ and $u(0, 0) < 1/F^{2/3}$, then $u(0, t) \to 1$ as $t \to \infty$. If on the other hand, $F > 1$ and $u(0, 0) < 1$, then $u(0, t) \to 1/F^{2/3}$ in finite time, and the solution breaks down as $\partial u/\partial t \to \infty$; if $u(0, 0) > 1$, then $u(0, t) \to \infty$, again in finite time if, for example, $E^* \propto u^3$. More generally, breakdown of the solution when $F > 1$ occurs in one of these ways at some positive value of x. Thus this suspended sediment model shares the same weakness of the phase shift model in not appearing to provide a well-posed nonlinear model.

5.6 Eddy Viscosity Model

The relative failure of the models above to explain dune and anti-dune formation led to the consideration of a full fluid flow model, in which, rather than supposing that the flow is shear free and that viscous effects were confined to a turbulent boundary layer, rotational effects were considered, and a model of turbulent shear flow incorporating an eddy viscosity, together with the Exner equation for bedload transport, was adopted. This allows for a linear stability analysis of the uniform flow over a flat bed via the solution of a suitable Orr–Sommerfeld equation. We shall in fact proceed in somewhat more generality. As an observation, fully-formed dunes have relatively small height to length ratios, and thus the fluid flow over them can be approximately linearised. Although we use a linear approximation to derive the stress at the bed, we may retain the nonlinear Exner equation for example. In this way we may derive a nonlinear evolution equation for bed elevation.

5.6.1 Orr–Sommerfeld Equation

Suppose, therefore, that we have two-dimensional turbulent flow down a slope of gradient S, governed by the Reynolds equations

$$
\begin{aligned}
u_t + u u_x + w u_z &= -\frac{1}{\rho}\frac{\partial p}{\partial x} + v_T \nabla^2 u + g S, \\
w_t + u w_x + w w_z &= -\frac{1}{\rho}\frac{\partial p}{\partial z} + v_T \nabla^2 w - g\left(1 - S^2\right)^{1/2}, \\
u_x + w_z &= 0,
\end{aligned}
\tag{5.62}
$$

where (u, w) are the velocity components and v_T is an eddy viscosity associated with the Reynolds stress terms, such as prescribed in (B.9). In the second equation, we can take $g(1 - S^2)^{1/2} \approx g$ since S is small.

We consider perturbations to a basic shear flow $u(z)$ in $s < z < \eta$ which satisfies (5.62) with v_T taken as constant. (Later, we will study a more realistic eddy viscosity model.) It is convenient first of all to non-dimensionalise the Eqs. (5.62). In the basic uniform state, with $s = 0$ and $\eta = \bar{h}$, the shear flow satisfies

$$
v_T \frac{\partial u}{\partial z} = g S(\bar{h} - z),
\tag{5.63}
$$

whence

$$
u = \frac{g S}{v_T}\left(\bar{h} z - \frac{1}{2} z^2\right),
\tag{5.64}
$$

and the column mean flow is

$$
\bar{u} = \frac{1}{\bar{h}} \int_0^{\bar{h}} u\,dz = \frac{g S}{3 v_T}\bar{h}^2.
\tag{5.65}
$$

Taking $v_T = \varepsilon_T \bar{u} \bar{h}$, we find that the basal shear stress is

$$\tau = \rho_w v_T \frac{\partial u}{\partial z}\bigg|_0 = f \rho_w \bar{u}^2, \tag{5.66}$$

where $f = 3\varepsilon_T$. This gives the relationship between the empirical f and the semi-analytic ε_T. If the bed and hence the flow is perturbed, we would only retain constant v_T if the volume flux per unit width is the same; this we therefore assume.

We now non-dimensionalise the variables by writing

$$(u, w) \sim \bar{u}, \qquad (x, z) \sim \bar{h}, \qquad t \sim \bar{h}/\bar{u}, \qquad p - \rho g(\bar{h} - z) \sim \rho_w \bar{u}^2. \tag{5.67}$$

The dimensionless equations are

$$\begin{aligned}
u_t + u u_x + w u_z &= -p_x + \frac{1}{R}\nabla^2 u + \frac{S}{F^2}, \\
w_t + u w_x + w w_z &= -p_z + \frac{1}{R}\nabla^2 w, \\
u_x + w_z &= 0,
\end{aligned} \tag{5.68}$$

and the parameters are a turbulent Reynolds number and the Froude number:

$$R = \frac{\bar{u}\bar{h}}{v_T}, \qquad F = \frac{\bar{u}}{\sqrt{g\bar{h}}}. \tag{5.69}$$

The dimensionless basic velocity profile is then

$$u = \frac{g S \bar{h}^2}{v_T \bar{u}}\left(z - \frac{1}{2}z^2\right), \tag{5.70}$$

and the dimensionless mean velocity is, by definition of $\bar{u}$,

$$1 = \frac{g S \bar{h}^2}{3 v_T \bar{u}}. \tag{5.71}$$

Since

$$v_T = \varepsilon_T \bar{u}\bar{h} = \frac{1}{3} f \bar{u}\bar{h}, \tag{5.72}$$

this requires

$$\bar{u} = \left(\frac{g S \bar{h}}{f}\right)^{1/2}. \tag{5.73}$$

In particular, the dimensionless basic velocity profile is

$$u = U(z) = 3\left(z - \frac{1}{2}z^2\right). \tag{5.74}$$

We now suppose that s and η are perturbed by small amounts; we may thus linearise (5.68). We put

$$(u, w) = \left(U(z) + \psi_z, -\psi_x\right), \tag{5.75}$$

whence it follows for small ψ that ψ satisfies the steady state Orr–Sommerfeld equation

$$U\nabla^2\psi_x - U''\psi_x = R^{-1}\nabla^4\psi, \tag{5.76}$$

where we assume stationary solutions in view of the anticipated fact that s evolves on a slower time scale.

The condition of zero pressure at $z = \eta$ is linearised to be

$$\eta = 1 + F^2 p\big|_{z=1}. \tag{5.77}$$

If F^2 is small, then we may take η to be constant, and we do so as we are primarily interested in dunes. However, the dimensionless pressure p is only determined up to addition of an arbitrary constant, which implies that the value of the constant η is unconstrained. This represents the vertical translation invariance of the system. If a uniform perturbation to s is made, then the response of the (uniform) stream is to raise the surface by the same amount. We can remove the ambiguity by prescribing $\eta = 1$, with the implication that the mean value of s is required to be zero.

The other boundary conditions on $z = s$ and $z = 1$ are no slip at the base, no shear stress at the top, and the perturbed volume flux is zero. These imply

$$\psi = 0, \qquad \psi_{zz} = 0 \quad \text{on } z = 1,$$

$$\int_0^s U(z)\,dz + \psi = 0, \qquad U + \psi_z = 0 \quad \text{on } z = s. \tag{5.78}$$

Linearisation of this second pair about $z = 0$ gives

$$\psi = 0, \qquad \psi_z = -U_0's \quad \text{on } z = 0, \tag{5.79}$$

where $U_0' = U'(0)$. Our aim is now to solve (5.76) with (5.78) and (5.79) to calculate the perturbed shear stress. The dimensional basal shear stress is then

$$\tau = \rho_w \varepsilon_T \bar{u}^2 U_0'\left[1 + s\frac{U_0''}{U_0'} + \frac{1}{U_0'}\psi_{zz}|0\right], \tag{5.80}$$

and since $f = 3\varepsilon_T = \varepsilon_T U_0'$, we may write this as

$$\tau = f\rho_w \bar{u}^2\left[1 + \frac{sU_0''}{U_0'} + \frac{1}{U_0'}\psi_{zz}|0\right]. \tag{5.81}$$

The problem to solve for ψ is linear and inhomogeneous, and so we suppose that

$$s = \int_{-\infty}^{\infty} \hat{s}(k)e^{ikx}\,dk, \qquad \psi = \int_{-\infty}^{\infty} \hat{\psi}(k)e^{ikx}\,dk. \tag{5.82}$$

(Note that $\hat{s}$ will evolve slowly in time.) For each wave number k, we obtain

$$ik[U(\hat{\psi}'' - k^2\hat{\psi}) - U''\hat{\psi}] = \frac{1}{R}[\hat{\psi}^{iv} - 2k^2\hat{\psi}'' + k^4\hat{\psi}], \tag{5.83}$$

with boundary conditions

$$\hat{\psi} = \hat{\psi}'' = 0 \quad \text{on } z = 1,$$

$$\hat{\psi} = 0, \qquad \hat{\psi}' = -U_0'\hat{s} \quad \text{on } z = 0, \tag{5.84}$$

and thus we finally define

$$\hat{\psi} = -U_0' \hat{s} \Psi(z,k), \tag{5.85}$$

where Ψ satisfies the canonical problem

$$ik\left[U\left(\Psi'' - k^2\Psi\right) - U''\Psi\right] = \frac{1}{R}\left[\Psi^{iv} - 2k^2\Psi'' + k^4\Psi\right],$$

$$\Psi = \Psi'' = 0 \quad \text{on } z = 1, \tag{5.86}$$

$$\Psi = 0, \qquad \Psi' = 1 \quad \text{on } z = 0.$$

In terms of Ψ, the basal (dimensional) shear stress is

$$\tau = f\rho_w \bar{u}^2 \left[1 - s - \int_{-\infty}^{\infty} e^{ikx} \hat{s}(k)\Psi''(0,k)\,dk\right]. \tag{5.87}$$

Using the convolution theorem, this is

$$\tau = f\rho_w \bar{u}^2 \left[1 - s + \int_{-\infty}^{\infty} K(x - \xi)s'(\xi)\,d\xi\right], \tag{5.88}$$

where $s' = \partial s/\partial x$, and

$$K(x) = -\frac{1}{2\pi} \int_{-\infty}^{\infty} \frac{\Psi''(0,k)}{ik} e^{ikx}\,dk. \tag{5.89}$$

Depending on K, we can see how τ may depend on displaced values of s. The form of (5.88) illustrates our previous discussion of the vertical translation invariance of the system. For a possible uniform perturbation $s = \text{constant}$, we would obtain a modification to the basic friction law, $\tau = f\rho_w \bar{u}^2$. This is excluded by enforcing the condition that s has zero mean in x,

$$\lim_{L \to \infty} \frac{1}{2L} \int_{-L}^{L} s(x)\,dx = 0, \tag{5.90}$$

which corresponds (for a periodic bed) to prescribing

$$\hat{s}(0) = 0. \tag{5.91}$$

To determine K, we need to know the solution of (5.86) for all k. In general, the problem requires numerical solution. However, note that $R = 1/\varepsilon_T$, and is reasonably large (for a value $f = 0.005$, $R = 3/f = 600$). This suggests that a useful means of solving (5.86) may be asymptotically, in the limit of large R. The fact that we can obtain analytic expressions for $\Psi''(0,k)$ means this is useful even when R is not dramatically large, as here.

The solution of the Orr–Sommerfeld equation at large R has a long pedigree, and it is a complicated but mathematically interesting problem. We devote Appendix C to finding the solution. We find there that, for $k > 0$,

$$\Psi''(0,k) \approx -3(ikRU_0')^{1/3}\,\text{Ai}(0) + O(1), \tag{5.92}$$

where Ai is the Airy function. For $k < 0$, $\Psi''(0,k) = \overline{\Psi''(0,-k)}$, and this leads to

$$\frac{\Psi''(0,k)}{ik} \approx \begin{cases} -ce^{-i\pi/3}|k|^{-2/3}, & k > 0, \\ -ce^{i\pi/3}|k|^{-2/3}, & k < 0, \end{cases} \tag{5.93}$$

where

$$c = 3(RU_0')^{1/3}\,\mathrm{Ai}(0), \tag{5.94}$$

and $c \approx 1.54 R^{1/3}$ for $U_0' = 3$, as $\mathrm{Ai}(0) = \frac{1}{3^{2/3}\Gamma(\frac{2}{3})} \approx 0.355$. From (5.89), we find

$$K(x) = \frac{c}{\pi} \int_0^\infty \frac{\cos\left[kx - \frac{\pi}{3}\right] dk}{k^{2/3}}. \tag{5.95}$$

Evaluating the integral,[3] we obtain the simple formula

$$K(x) = \frac{\mu}{x^{1/3}}, \quad x > 0,$$
$$K(x) = 0, \quad x < 0, \tag{5.96}$$

where

$$\mu = \frac{3^{2/3} R^{1/3}}{\{\Gamma(\frac{2}{3})\}^2} \approx 1.13 R^{1/3}. \tag{5.97}$$

For stability purposes, note that

$$K = \int_{-\infty}^\infty \hat{K}(k)\, e^{ikx}\, dk, \tag{5.98}$$

where

$$\hat{K} = -\frac{\Psi''(0,k)}{2\pi i k} = \frac{c \exp\left[-\frac{i\pi}{3}\,\mathrm{sgn}\,k\right]}{2\pi |k|^{2/3}}. \tag{5.99}$$

5.6.2 Orr–Sommerfeld–Exner Model

We now reconsider (5.33), which we can write in the form

$$s_t + q_x = 0,$$
$$\varepsilon h_t + (uh)_x = 0,$$
$$F^2(\varepsilon u_t + u u_x) = -\eta_x + \delta\left(1 - \frac{\tau}{h}\right), \tag{5.100}$$
$$h = \eta - s.$$

Here τ is the local basal stress, scaled with $f\rho_w u_0^2$. We suppose $q = q(\tau)$, so that the Exner equation is

$$\frac{\partial s}{\partial t} + q'(\tau)\frac{\partial \tau}{\partial x} = 0. \tag{5.101}$$

---

[3]How do we do that? The blunt approach is to consult Gradshteyn and Ryzhik (1980), where the relevant formulae are on page 420 and 421 (items 4 and 9 of Sect. 3.761). The quicker way, using complex analysis, is to evaluate $\int_0^\infty \theta^{\nu-1} e^{i\theta}\, d\theta$ (after a simple rescaling of k, $k|x| = \theta$) by rotating the contour by $\pi/2$ and using Jordan's lemma. Thus $\int_0^\infty \theta^{\nu-1} e^{i\theta}\, d\theta = \Gamma(\nu) e^{i\pi\nu/2}$.

It is tempting to suppose that, writing $\bar{u} = u_0 u$,

$$\tau = u^2 \left[1 - s + \int_{-\infty}^{\infty} K(x - \xi) \frac{\partial s}{\partial \xi}(\xi, t) d\xi \right]. \tag{5.102}$$

We would then have, with $\varepsilon \ll 1$ and $\eta = 1$, $u \approx \frac{1}{1-s}$ in (5.102). There is a subtle point here concerning the modified stress. Insofar as we may wish to describe different atmospheric or fluvial conditions (e.g., the difference between strong and weak winds at different times of day, or rivers in normal or flood stage), we do want to allow different choices of $\bar{u}$. However, such conditions also imply different values of $\bar{h}$, and the basis of the solution for the perturbed stress is that the mean depth (and thus the mean velocity) do not vary. The value of $u = \frac{1}{1-s}$ is a local column average, whereas the u in (5.102) is in addition a horizontal average. Thus, given $u_0 = \bar{u}$ and $h_0 = \bar{h}$, we define

$$\tau \approx 1 - s + \int_{-\infty}^{\infty} K(x - \xi) \frac{\partial s}{\partial \xi}(\xi, t) d\xi, \tag{5.103}$$

and the model consists of the Exner equation (5.101) and the Orr–Sommerfeld stress formula (5.103). Variable $\bar{u}$ is simply manifested in differing time scales for the Exner equation.

We linearise by writing $\tau = 1 + T$, and then

$$s = \int_{-\infty}^{\infty} \hat{s}(k, t) e^{ikx} dk, \qquad T = \int_{-\infty}^{\infty} \hat{T}(k, t) e^{ikx} dk, \tag{5.104}$$

so that

$$\hat{s}_t + ikq'(1)\hat{T} = 0,$$
$$\hat{T} = -\hat{s} + 2\pi \hat{K} ik\hat{s}, \tag{5.105}$$

and thus, using (5.99), solutions are proportional to $e^{\sigma t}$, where

$$\sigma = q'(1) \left[2\pi k^2 \hat{K} + ik \right]. \tag{5.106}$$

When $\mathrm{Re}\,\hat{K} > 0$, as for (5.99), the steady state is unstable, with $\mathrm{Re}\,\sigma \sim k^{4/3}$ as $k \to \infty$. Specifically, the growth rate is

$$\mathrm{Re}\,\sigma = \frac{1}{2} q'(1) c |k|^{4/3}, \tag{5.107}$$

while the wave speed is

$$-\frac{\mathrm{Im}\,\sigma}{k} = q'(1) \left(\frac{1}{2} \sqrt{3} c |k|^{1/3} - 1 \right); \tag{5.108}$$

thus waves move downstream (except for very long waves).

5.6.3 Well-posedness

The effect of (5.103) is to cause increased τ where s_x is positive, on the upstream slopes of bumps. Since u is in phase with s, this implies τ leads u (i.e., τ is a maximum before s is); it is this phase lead which causes instability. However, the un-bounded growth rate at large wave numbers is a sign of ill-posedness. Without some stabilising mechanism, arbitrarily small disturbances can grow arbitrarily rapidly. In reality, another effect of bed slope is important, and that is the fact that sediment wants to roll downslope: in describing the Meyer-Peter/Müller result, no attention was paid to the variations of bed slope itself.

For a particle of diameter D_s at the bed, the streamflow exerts a force of approx-imately τD_s^2 on it, and it is this force which causes motion. On a slope, there is an additional force due to gravity, approximately $-\Delta \rho g D_s^3 s_x$. Thus the net stress causing motion is actually

$$\tau - \Delta \rho g D_s s_x. \tag{5.109}$$

In dimensionless terms, we therefore modify the bedload transport formula by writing

$$q = q(\tau_e), \quad \tau_e = \tau - \beta s_x, \tag{5.110}$$

where

$$\beta = \frac{\Delta \rho D_s}{\rho_w h_0 S}. \tag{5.111}$$

Typical values in water are $\Delta \rho / \rho_w \approx 2$, $D_s \sim 10^{-3}$ m, $h_0 \sim 2$ m, $S \sim 10^{-3}$, whence $\beta \approx 4$; generally we will suppose that $\beta \sim O(1)$.

The effect of this is to replace the definition of τ in (5.103) by

$$\tau_e = 1 - s + \int_{-\infty}^{\infty} K(x - \xi) \frac{\partial s}{\partial \xi}(\xi, t) \, d\xi - \beta s_x \tag{5.112}$$

(together with (5.101)), and in the stability analysis, $\hat{T} = \hat{s}[-1 + 2\pi i k \hat{K} - ik\beta]$, whence

$$\sigma = q'(1) \left[-ik \left\{ \frac{1}{2} \sqrt{3} c |k|^{1/3} - 1 \right\} + \frac{1}{2} c |k|^{4/3} - \beta k^2 \right]. \tag{5.113}$$

This exhibits the classical behaviour of a well-posed model. The system is stable at high wave number, and the maximum growth rate is at $k = (\frac{c}{3\beta})^{3/2}$. This would be the expected preferred wave number of the instability.

Figure 5.12 shows a numerical solution of the nonlinear Exner equation, showing the growth of dunes from an initially localised disturbance. Because the expression in (5.112) is only valid for small s, we can equivalently write

$$q(\tau_e) = q(\tau - \beta s_x) \approx q(\tau) - D s_x, \tag{5.114}$$

where

$$D = \beta q'(\tau) \approx \beta q'(1), \tag{5.115}$$

and the equation has been solved in this form, with the diffusion coefficient D taken as constant, i.e., s satisfies the equation

$$\frac{\partial s}{\partial t} + \frac{\partial}{\partial x} q[1 - s + K * s_x] = D \frac{\partial^2 s}{\partial x^2}. \tag{5.116}$$

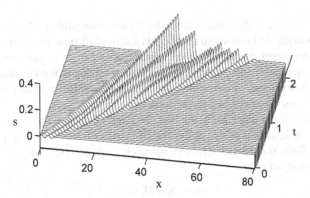

Fig. 5.12 Development of the dune instability from an initial perturbation near $x = 0$ obtained by solving (5.116) using $q(\tau) = \tau^{3/2}$, $K = \frac{\mu}{x^{1/3}}$ when $x > 0$, $K = 0$ otherwise, with parameters $\mu = 9.57$ and $D = 4.3$. Separation occurs in this figure when $t = 0.8$, after which the computation is continued as described in the notes at the end of the chapter. Figure kindly provided by Mark McGuinness

As the dunes grow, the model becomes invalid when $\tau - 1 \sim O(1)$, and this happens when $s \sim \frac{1}{\mu}$. This is a representative value for the elevation of both fluvial and aeolian dunes, and is suggestive of the idea that it is the approach of τ towards zero which controls eventual dune height. Additionally, when τ reaches zero, separation occurs, and the model becomes invalid. Possible ways for dealing with this are outlined in the notes at the end of the chapter. A further issue is that the derivation of (5.112) becomes invalid when $s \sim \frac{1}{\mu}$, because then the thickness of the viscous boundary layer in the Orr–Sommerfeld equation becomes comparable to the elevation of the dunes. This implies that the Orr–Sommerfeld equation should now be solved in a domain where the lower boundary cannot be linearised about $z = 0$, and the Fourier method of solution can no longer be implemented. It is not clear whether this will fundamentally change the nature of the resultant formula for the stress.

The numerical method used to solve (5.116) is a spectral method. Spectral methods for evolution equations of this sort are convenient, particularly when the integral term is of convolution type, but they confuse the issue of what appropriate boundary conditions for such equations should be. In the present case, it is not clear. For aeolian dunes, it is natural to pose conditions at a boundary representing a shore-line, but it is then less clear how to deal with the integral term. The derivation of this term already presumes an infinite sand domain, and it seems this is one of those questions akin to the issue of posing boundary conditions for averaged equations, for example for two-phase flow, where a hidden interchange of limits is occurring.

5.7 Mixing-Length Model for Aeolian Dunes

Measurements of turbulent fluid flow in pipes, as well as air flow in the atmosphere (and also in wind tunnels), show that the assumption of constant eddy viscosity is not a good one, and the basic shear velocity profile is not as simple as assumed in the preceding section. In actual fact, the concept of eddy viscosity introduced by Prandtl was based on the idea of momentum transport by eddies of different sizes, with the transport rate (eddy viscosity) being proportional to eddy size. Evidently, this must go to zero at a solid boundary, and the simplest description of this is Prandtl's mixing-length theory, described in Appendix B. In this section, we generalise the previous approach a little to allow for such a spatially varying eddy viscosity, and we specifically consider the case of aeolian dunes, in which a kilometre deep turbulent boundary layer flow is driven by an atmospheric shear flow.

5.7.1 Mixing-Length Theory

The various forms of sand dunes in deserts were discussed earlier; the variety of shapes can be ascribed to varying wind directions, a feature generally absent in rivers. Another difference from the modelling point of view is that the fluid atmosphere is about ten kilometres in depth, and the flow in this is essentially unaffected by the underlying surface, except in the atmospheric boundary layer, of depth about a kilometre, wherein most of the turbulent mixing takes place. Within this boundary layer, there is a region adjoining the surface in which the velocity profile is approximately logarithmic, and this region spans a range of height from about forty metres above the surface to the 'roughness height' of just a few centimetres or millimetres above the surface.

Consider the case of a uni-directional mean shear flow $u(z)$ past a rough surface $z = 0$, where z measures distance away from the surface. If the shear stress is constant, equal to τ, then we define the friction velocity u_* by

$$u_* = (\tau/\rho)^{1/2}, \tag{5.117}$$

where ρ is density. Observations support the existence near the surface of a logarithmic velocity profile of the form

$$u = \frac{u_*}{\kappa} \ln\left(\frac{z}{z_0}\right), \tag{5.118}$$

where the Von Kármán constant $\kappa \approx 0.4$, and z_0 is known as the roughness length: it represents the effect of surface roughness in bringing the average velocity to zero at some small height above the actual surface.[4] Since z_0 is a measure of actual roughness, a typical value for a sandy surface might be $z_0 = 10^{-3}$ m.

Prandtl's mixing-length theory provides a motivation for (5.118). If we suppose the motion can be represented by an eddy viscosity η, so that

$$\tau = \eta \frac{\partial u}{\partial z}, \tag{5.119}$$

---

[4]A better recipe would be $u = \frac{u_*}{\kappa} \ln(\frac{z+z_0}{z_0})$, which allows no slip at $z = 0$. See also Question 5.11.

then Prandtl proposed

$$\eta = \rho l^2 \left| \frac{\partial u}{\partial z} \right|, \quad l = \kappa z, \tag{5.120}$$

from which, indeed, (5.118) follows. The quantity $l = \kappa z$ is called the mixing length. Prandtl's theory works well in explaining the logarithmic layer, and in extension it explains pipe flow characteristics very well; but it has certain drawbacks. The two obvious ones are that it is not frame-invariant; however, this would be easily rectified by replacing $|\partial u/\partial z|$ by the second invariant $2\dot\varepsilon$, where $2\dot\varepsilon^2 = \dot\varepsilon_{ij}\dot\varepsilon_{ij}$, and $\dot\varepsilon_{ij}$ is the strain rate tensor. Also not satisfactory is the rather loosely defined mixing length, which becomes less appropriate far from the boundary, or in a closed container. Despite such misgivings, we will use a version of the mixing-length theory to see how it deviates from the constant eddy viscosity assumption.

We want to see how to solve a shear flow problem in dimensionless form. To this end, suppose for the moment that we fix $u = U_\infty$ on $z = d$. Then $U_\infty = (u^*/\kappa)\ln(d/z_0)$ determines u_* (and thus τ), and we can define a parameter[5] ε by

$$\varepsilon = \frac{u_*}{U_\infty} = \frac{\kappa}{\ln(d/z_0)}. \tag{5.121}$$

For $d = 10^3$ m, $z_0 = 10^{-3}$ m, $\kappa = 0.4$, $\varepsilon \approx 0.03$. Writing u in terms of U_∞ rather than u_* yields

$$u = U_\infty \left[1 + \frac{\varepsilon}{\kappa} \ln\left(\frac{z}{d}\right) \right]. \tag{5.122}$$

Note also that the basic eddy viscosity is then

$$\eta = \varepsilon \rho U_\infty d \left(\frac{\kappa z}{d}\right), \tag{5.123}$$

and the shear stress is

$$\tau = \varepsilon^2 \rho U_\infty^2. \tag{5.124}$$

We shall use these observations in scaling the equations. For the atmospheric boundary layer, it seems appropriate to assume that U_∞ is prescribed from the large scale model of atmospheric flow (cf. Chap. 3), and that d is the depth of the planetary boundary layer. Of course, this may be an oversimplified description.

---

[5]Note that this definition of ε is unrelated to its previous definition and use, as for example in (5.100).

5.7.2 Turbulent Flow Model

Again we assume a mean two-dimensional flow $(u, 0, w)$ with horizontal coordinate x and vertical coordinate z over a surface topography given by $z = s$. The basic equations are

$$u_x + w_z = 0,$$
$$\rho(uu_x + wu_z) = -p_x + \tau_{1x} + \tau_{3z}, \tag{5.125}$$
$$\rho(uw_x + ww_z) = -p_z + \tau_{3x} - \tau_{1z},$$

where $\tau_1 = \tau_{11}$ and $\tau_3 = \tau_{13}$ are the deviatoric Reynolds stresses, and are defined, we suppose, by

$$\tau_1 = 2\eta u_x, \tag{5.126}$$
$$\tau_3 = \eta(u_z + w_x).$$

We ignore gravity here, so that the pressure is really the deviation from the hydrostatic pressure. Our choice of the eddy viscosity η will be motivated by the Prandtl mixing-length theory (5.120), but we postpone a precise specification for the moment.

The basic flow then dictates how we should non-dimensionalise the variables. We do so by writing

$$u = U_\infty(1 + \varepsilon u^*), \qquad w \sim \varepsilon U_\infty, \qquad x, z \sim d,$$
$$\tau_1, \tau_3 \sim \varepsilon^2 \rho U_\infty^2, \qquad \eta \sim \varepsilon \rho d U_\infty, \qquad p \sim \varepsilon \rho U_\infty^2, \tag{5.127}$$

and then the dimensionless equations are (dropping the asterisk on u^*)

$$u_x + w_z = 0,$$
$$u_x + p_x = \varepsilon[\tau_{1x} + \tau_{3z} - \{uu_x + wu_z\}],$$
$$w_x + p_z = \varepsilon[\tau_{3x} - \tau_{1z} - \{uw_x + ww_z\}], \tag{5.128}$$
$$\tau_3 = \eta(u_z + w_z),$$
$$\tau_1 = 2\eta u_x.$$

5.7.3 Boundary Conditions

The depth scale of the flow d is, we suppose, the depth of the atmospheric boundary layer, of the order of hundreds of metres to a kilometre. Above the boundary layer, there is an atmospheric shear flow, and we suppose that $u \sim u_0(z)$, $w \to 0$, $p \to 0$ as $z \to \infty$.[6] The choice of u_0 is determined for us by the choice of η, as is most easily seen from the case of a uniform flow where $\partial u/\partial z = \tau/\eta$. The correct boundary condition to pose at large z is to prescribe the shear stress delivered by the main

---

[6]The modelling alternative is to specify velocity conditions on a lid at $z = 1$.

atmospheric flow, and this can be taken to be $\tau_3 = 1$ by our choice of stress scale. Thus we prescribe

$$\tau_3 \to 1, \qquad w \to 0, \qquad p \to 0 \quad \text{as } z \to \infty. \tag{5.129}$$

Next we need to prescribe conditions at the surface. This involves two further length scales, the length L and amplitude H of the surface topography. Since we observe dunes often to have lengths in the range 100–1000 m, and heights in the range 2–100 m, we can see that there are two obvious distinguished limits, $L = d$, $H = \varepsilon d$, and it is most natural to use these in scaling the surface s. In fact since dunes are self-evolving it seems most likely that they will select length scales already present in the system. Thus, we suppose that in dimensionless terms the surface is $z = \varepsilon s(x)$, and longer, shorter, taller or smaller dunes can always be introduced as necessary later, by rescaling s. The surface boundary conditions are then taken to be (recalling the definition of the roughness length)

$$u = -\frac{1}{\varepsilon}, \qquad w = 0 \quad \text{on } z = \varepsilon s + z_0^*, \tag{5.130}$$

where

$$z_0^* = \frac{z_0}{d} = e^{-\kappa/\varepsilon}. \tag{5.131}$$

For completeness, we need to specify horizontal boundary conditions, for example at $x = \pm\infty$. We keep these fairly vague, beyond requiring that the variables remain bounded. In particular, we do not allow unbounded growth of velocity or pressure.

5.7.4 Eddy Viscosity

Prandtl's mixing-length theory in scaled units would imply

$$\eta = \kappa^2 (z - \varepsilon s)^2 \left| \frac{\partial u}{\partial z} \right|, \tag{5.132}$$

and we assume this, although other choices are possible. In particular, (5.132) is not frame indifferent, but this is hardly of significance since the eddy viscosity itself is unreliable away from the surface. (We comment further on this in the notes at the end of the chapter.)

To convert to the constant eddy viscosity model of the preceding section, Eq. (5.68), we would rescale u, w, $p \sim 1/\varepsilon$, and choose $\eta = \varepsilon$: thus $\varepsilon^2 = 1/R$.

5.7.5 Surface Roughness Layer

The basic shear flow near a flat surface $z = 0$ is given by (5.122), and in dimensionless terms is

$$u = \frac{1}{\kappa} \ln z; \tag{5.133}$$

we will require similar behaviour when the flow is perturbed. Suppose, more generally, that as $z \to \varepsilon s$,

$$u \sim a + b \ln(z - \varepsilon s) + O(z - \varepsilon s), \tag{5.134}$$

which we shall find describes the solution away from the boundary. We put

$$z = \varepsilon s + \nu Z, \tag{5.135}$$

where

$$\nu = e^{-\kappa/\varepsilon}. \tag{5.136}$$

Additionally, we write

$$u = -\frac{1}{\varepsilon} + U, \qquad w = \varepsilon s_x U + \nu W, \qquad \tau_1 = \varepsilon T_1, \qquad \eta = \nu N. \tag{5.137}$$

Then we find that

$$\begin{aligned}
&U_x + W_Z = 0, \\
&\frac{\partial \tau_3}{\partial Z} - \varepsilon^2 s_x \frac{\partial T_1}{\partial Z} + s_x \frac{\partial p}{\partial Z} \approx 0, \\
&\frac{\partial p}{\partial Z} \approx -\varepsilon^2 \left[s_x \frac{\partial \tau_3}{\partial Z} + \frac{\partial T_1}{\partial Z} \right], \\
&N \approx \kappa^2 Z^2 \frac{\partial U}{\partial Z}, \\
&\tau_3 \approx N \left(1 - \varepsilon^2 s_x^2 \right) \frac{\partial U}{\partial Z}, \\
&T_1 = -2\kappa^2 s_x Z^2 \left(\frac{\partial U}{\partial Z} \right)^2,
\end{aligned} \tag{5.138}$$

where we have neglected transcendentally small terms proportional to ν.

Correct to $O(\varepsilon^2)$, τ_3 is constant through the roughness layer, and equal to its surface value τ, and

$$\tau \approx \kappa^2 Z^2 \left(\frac{\partial U}{\partial Z} \right)^2, \tag{5.139}$$

again correct to $O(\varepsilon^2)$. The boundary conditions on $Z = 1$ (i.e., $z - \varepsilon s = z_0^* = \nu$) are $U = W = 0$, thus

$$U = \frac{\sqrt{\tau}}{\kappa} \ln Z, \tag{5.140}$$

and this must be matched to the outer solution (5.134). Rewriting (5.140) in terms of u and z, we have

$$u \sim \frac{\sqrt{\tau} - 1}{\varepsilon} + \frac{\sqrt{\tau}}{\kappa} \ln(z - \varepsilon s), \tag{5.141}$$

and this is in fact the matching condition that we require from the outer solution. We see immediately that variations of $O(1)$ in u yield small corrections of $O(\varepsilon)$ in τ.

Solving for W, we have

$$W = -\frac{(\sqrt{\tau})'}{\kappa}[Z \ln Z - Z], \qquad (5.142)$$

where $(\sqrt{\tau})' = \partial\sqrt{\tau}/\partial x$, and in terms of w and z, this is written

$$w = s_x + \varepsilon s_x u - \frac{(\sqrt{\tau})'}{\kappa}\left[\ln(z - \varepsilon s) - 1 + \frac{\kappa}{\varepsilon}\right](z - \varepsilon s). \qquad (5.143)$$

Hence the outer solution must satisfy (correct to $O(\varepsilon^2)$)

$$w \approx s_x + \varepsilon s_x u \quad \text{as } z \to \varepsilon s. \qquad (5.144)$$

5.7.6 Outer Solution

We turn now to the solution away from the roughness layer, in the presence of surface topography of amplitude $O(\varepsilon)$ and length scale $O(1)$. The topography has two effects. The $O(1)$ variation in length scale causes a perturbation on a height scale of $O(1)$, but the vertical displacement of the logarithmic layer by $O(\varepsilon)$ causes a shear layer of this thickness to occur. Thus the flow away from the surface consists of an outer layer of thickness $O(1)$, and an inner shear layer of thickness $O(\varepsilon)$. We begin with the outer layer.

We expand the variables as

$$u = u^{(0)} + \varepsilon u^{(1)} + \cdots, \qquad (5.145)$$

etc., so that to leading order, from (5.128),

$$\begin{aligned} u_x^{(0)} + w_z^{(0)} &= 0, \\ u_x^{(0)} + p_x^{(0)} &= 0, \\ w_x^{(0)} + p_z^{(0)} &= 0. \end{aligned} \qquad (5.146)$$

Notice that, at this leading order, the precise form of η in (5.132) is irrelevant, as this outer problem is inviscid. We have

$$u^{(0)} + p^{(0)} = u_0(z), \qquad (5.147)$$

and

$$p_x^{(0)} = w_z^{(0)}, \qquad p_z^{(0)} = -w_x^{(0)}, \qquad (5.148)$$

which are the Cauchy–Riemann equations for $p^{(0)} + iw^{(0)}$, which is therefore an analytic function, and $p^{(0)}$ and $w^{(0)}$ both satisfy Laplace's equation. The matching conditions as $z \to \varepsilon s$ can be linearised about $z = 0$, and if $w^{(0)} = w_0$ and $p^{(0)} = p_0$ on $z = \varepsilon s$, then from (5.144) we have

$$w^{(0)} = s_x \quad \text{on } z = 0. \qquad (5.149)$$

Assuming also that $w^{(0)}, p^{(0)} \to 0$ as $z \to \infty$, we can write the solutions in the form

$$w^{(0)} = \frac{1}{\pi} \int_{-\infty}^{\infty} \frac{z s_\xi \, d\xi}{[(x-\xi)^2 + z^2]}, \qquad p^{(0)} = -\frac{1}{\pi} \int_{-\infty}^{\infty} \frac{(x-\xi) s_\xi \, d\xi}{[(x-\xi)^2 + z^2]}, \qquad (5.150)$$

and in particular, $p^{(0)}$ on $z = \varepsilon s$ is given to leading order by p_0, where

$$p_0 = \frac{1}{\pi} \int_{-\infty}^{\infty} \frac{s_\xi \, d\xi}{\xi - x} = H(s_x); \qquad (5.151)$$

the integral takes the principal value, and H denotes the Hilbert transform.

The shear velocity profile $u_0(z)$ is undetermined at this stage, although we would like it to be the basic shear flow profile; but to justify this, we need to go to the $O(\varepsilon)$ terms. At $O(\varepsilon)$, we have

$$u_x^{(1)} + w_z^{(1)} = 0,$$
$$u_x^{(1)} + p_x^{(1)} = \tau_{1x}^{(0)} + \tau_{3z}^{(0)} - \left\{ u^{(0)} u_x^{(0)} + w^{(0)} u_z^{(0)} \right\},$$
$$w_x^{(1)} + p_z^{(1)} = \tau_{3x}^{(0)} - \tau_{1z}^{(0)} - \left\{ u^{(0)} w_x^{(0)} + w^{(0)} w_z^{(0)} \right\},$$
$$\tau_3^{(0)} = \eta^{(0)} \left[u_z^{(0)} + w_x^{(0)} \right], \qquad (5.152)$$
$$\tau_1^{(0)} = 2 \eta^{(0)} u_x^{(0)},$$
$$\eta^{(0)} = \kappa^2 z^2 \left| \frac{\partial u^{(0)}}{\partial z} \right|.$$

We can use the zeroth order solution to write $(5.152)_2$ in the form

$$u_x^{(1)} + p_x^{(1)} = \frac{\partial \tau_3^{(0)}}{\partial z} + \frac{\partial}{\partial x} \left[\tau_1^{(0)} - \frac{1}{2} \left(u^{(0)2} + w^{(0)2} \right) + u_0'(z) \psi^{(0)} \right], \qquad (5.153)$$

where $\psi^{(0)}$ is the stream function such that $w^{(0)} = -\psi_x^{(0)}$, and specifically, we have

$$\psi^{(0)} = -\frac{1}{2\pi} \int_{-\infty}^{\infty} \ln[(x-\xi)^2 + z^2] p_0(\xi) \, d\xi, \qquad (5.154)$$

which can be found (as can the formulae in (5.150)) by using a suitable Green's function; (this is explained further below when we find $p^{(1)}$).

On integrating (5.153), we have to avoid secular terms which grow linearly in x, and we therefore require the integral of the right hand side of (5.153) with respect to x, from $-\infty$ to ∞, to be bounded. The integral of the derivative term is certainly bounded; thus the secularity condition requires $\int_{-\infty}^{\infty} \tau_{3z}^{(0)} \, dx$ to be bounded, and it is this condition that determines the function of integration $u_0(z)$ in (5.147).

For the particular choice of $\eta^{(0)}$ in (5.152), we have

$$\eta^{(0)} = \kappa^2 z^2 \left(u_0' + w_x^{(0)} \right) \qquad (5.155)$$

(assumed positive), so that

$$\tau_3^{(0)} = \kappa^2 z^2 \left(u_0' + w_x^{(0)} \right) \left(u_0' + 2 w_x^{(0)} \right). \qquad (5.156)$$

The condition that $\partial \tau_3^{(0)} / \partial z$ have zero mean is then

$$\int_{-\infty}^{\infty} \frac{\partial}{\partial z} \left[\kappa^2 z^2 \left(u_0'^2 + 2 w_x^{(0)2} \right) \right] dx = 0, \qquad (5.157)$$

and thus $\overline{\partial \tau_3^{(0)}}/\partial z = 0$, where the overbar denotes the horizontal mean. Thus (with $\overline{\tau_3^{(0)}} = 1$ from the condition at $z = \infty$), u_0 is determined via

$$u_0'^2 + 2\overline{w_x^{(0)2}} = \frac{1}{\kappa^2 z^2}. \tag{5.158}$$

The non-zero quantity $2\overline{w_x^{(0)2}}$ represents the form drag due to the surface topography. Note that the logarithmic behaviour of u_0 near $z = 0$ is unaffected by this extra term, and we can take

$$u_0 = \frac{1}{\kappa} \ln z + O(z^2) \quad \text{as } z \to 0. \tag{5.159}$$

In particular, since $p^{(0)} \approx p_0 + p_z^{(0)}|_{\varepsilon s}(z - \varepsilon s)$ as $z \to \varepsilon s$, and $p_z^{(0)}|_{\varepsilon s} = -w_x^{(0)}|_{\varepsilon s} = -s_{xx}$, we have

$$u^{(0)} \sim -p_0 + \frac{1}{\kappa} \ln z + s_{xx}(z - \varepsilon s) + O(z^2) \quad \text{as } z \to \varepsilon s. \tag{5.160}$$

From (5.156), we now have

$$\tau_3^{(0)} = 1 + 3\kappa^2 z^2 u_0' w_x^{(0)} + \frac{\partial \Phi}{\partial x}, \tag{5.161}$$

where we define

$$\Phi = \int_{-\infty}^{x} 2\kappa^2 z^2 \{ w_x^{(0)2} - \overline{w_x^{(0)2}} \} \, dx. \tag{5.162}$$

Hence from (5.153),

$$u^{(1)} + p^{(1)} = \frac{\partial}{\partial z} [3\kappa^2 z^2 u_0' w^{(0)} + \Phi] + \tau_1^{(0)} - \frac{1}{2}(u^{(0)2} + w^{(0)2})$$
$$+ u_0'(z)\psi^{(0)} + u_1(z), \tag{5.163}$$

where u_1 must be determined at $O(\varepsilon^2)$.

Now $u_0 \sim \frac{1}{\kappa} \ln z + O(z^2)$, $\Phi = O(z^2)$, $\tau_1^{(0)} = O(z)$, $w^{(0)} = s_x + O(z)$, $\psi^{(0)} = -s - zp_0 + O(z^2)$ (this last follows from manipulation of (5.154)). Therefore, as $z \to 0$,

$$u^{(1)} = -p_{10} + 3\kappa s_x - \frac{1}{2}\left[s_x^2 + \left\{-p_0 + \frac{1}{\kappa}\ln z\right\}^2\right]$$
$$+ \frac{1}{\kappa z}(-s - zp_0) + u_1 + O(z), \tag{5.164}$$

where $p_{10} = p^{(1)}|_{z=0}$.

5.7.7 Determination of p_{10}

Define the Green's function

$$K(x, z; \xi, \zeta) = -\frac{1}{4\pi}\left[\ln\{(x-\xi)^2 + (z-\zeta)^2\}\right.$$
$$\left. + \ln\{(x-\xi)^2 + (z+\zeta)^2\}\right]. \tag{5.165}$$

We then have, for example,

$$p^{(0)} = \iint_{\zeta>0}\left[K\nabla^2 p^{(0)} - p^{(0)}\nabla^2 K\right]d\xi\,d\zeta$$

$$= \oint\left(K\frac{\partial p^{(0)}}{\partial n} - p^{(0)}\frac{\partial K}{\partial n}\right)ds$$

$$= -\int_{-\infty}^{\infty}K\frac{\partial p^{(0)}}{\partial\zeta}\,d\xi = \int_{-\infty}^{\infty}K\frac{\partial w^{(0)}}{\partial\xi}\,d\xi = -\int_{-\infty}^{\infty}w^{(0)}\frac{\partial K}{\partial\xi}\,d\xi, \tag{5.166}$$

whence we derive (5.150) for example; the integrals with respect to ξ are taken along $\zeta = 0$.

Next, expanding (5.144) about $z = 0$, we find

$$w^{(1)} \sim \left(su^{(0)}\right)_x \quad \text{as } z \to 0. \tag{5.167}$$

Putting

$$w^{(1)} = s_x u_0 + W, \tag{5.168}$$

we deduce the condition

$$W = -(sp_0)_x \quad \text{on } z = 0, \tag{5.169}$$

and from (5.152)

$$p_x^{(1)} - W_z = R,$$
$$p_z^{(1)} + W_x = S,$$

where

$$R = \tau_{1x}^{(0)} + \tau_{3z}^{(0)} - \{u^{(0)}u_x^{(0)} + w^{(0)}u_z^{(0)} - s_x u_0'\},$$
$$S = \tau_{3x}^{(0)} - \tau_{1z}^{(0)} - \{u^{(0)}w_x^{(0)} + w^{(0)}w_z^{(0)} + s_{xx}u_0\}.$$

Also

$$\nabla^2 p^{(1)} = R_x + S_z, \tag{5.170}$$

and it follows from using the Green's function as in (5.166) that, after some manipulation involving Green's theorem,

$$p^{(1)} = \iint_{\zeta>0}(RK_\xi + SK_\zeta)\,d\xi\,d\zeta - \int_{-\infty}^{\infty}K_\xi W\,d\xi, \tag{5.171}$$

and therefore

$$p_{10} = \frac{1}{\pi}\iint_{\zeta>0}\frac{[(x-\xi)R(\xi,\zeta) - \zeta S(\xi,\zeta)]}{(x-\xi)^2+\zeta^2}\,d\xi\,d\zeta - \frac{1}{\pi}\int_{-\infty}^{\infty}\frac{(sp_0)_\xi\,d\xi}{\xi-x}. \tag{5.172}$$

5.7.8 Matching

Overall, then, the outer solution can be written, as $z \to 0$, in the form (using (5.160))

$$u \sim -p_0 + \frac{1}{\kappa} \ln z + s_{xx}(z - \varepsilon s) + \varepsilon \left[-p_{10} + 3\kappa s_x - \frac{1}{2} s_x^2 - \frac{1}{2} p_0^2 + \frac{p_0}{\kappa} \ln z \right.$$
$$\left. - \frac{1}{2\kappa^2} \ln^2 z - \frac{s}{\kappa z} - \frac{p_0}{\kappa} + u_1 \right]. \tag{5.173}$$

If we define

$$\sqrt{\tau} = 1 + \varepsilon A_1 + \varepsilon^2 A_2 + \cdots, \tag{5.174}$$

then (5.141) takes the form

$$u \sim A_1 + \frac{1}{\kappa} \ln z + \varepsilon \left[-\frac{s}{\kappa z} + \frac{A_1}{\kappa} \ln z + A_2 \right] + \cdots, \tag{5.175}$$

and the leading order term can be matched directly to that of (5.173) by choosing

$$A_1 = -p_0. \tag{5.176}$$

Using (5.176), (5.174) and (5.151), we have

$$\tau \approx 1 + \frac{2\varepsilon}{\pi} \int_{-\infty}^{\infty} \frac{s_\xi \, d\xi}{x - \xi}, \tag{5.177}$$

and this can be compared with (5.103). Whereas in (5.103) $K(x) = 0$ for $x < 0$, the kernel $K(x)$ in (5.177) is proportional to $1/x$ for all x, and thus non-zero for $x < 0$. Consequently, there is no instability, and to find an instability we need to progress to the next order term.

Unfortunately, the $O(\varepsilon)$ terms do not match because the terms $\pm \frac{p_0}{\kappa} \ln z$ in the two expansions are not equal, and also because of the linear term in (5.173). In order to match the expansions to $O(\varepsilon)$, we have to consider a further, intermediate layer: this is the shear layer we alluded to earlier.

5.7.9 Shear Layer

A distinguished limit exists when $z = O(\varepsilon)$, and thus we put

$$z = \varepsilon s + \varepsilon \zeta, \qquad w = s_x + \varepsilon [u s_x + W],$$
$$\eta = \varepsilon N, \qquad \tau_1 = \varepsilon T_1,$$
$$u = -p_0 + \frac{1}{\kappa} \ln(z - \varepsilon s) + \varepsilon v, \tag{5.178}$$

and from (5.173) and (5.141) (using (5.174) and (5.176)), we require

$$v \sim s_{xx}\zeta - p_{10} + 3\kappa s_x - \frac{1}{2}s_x^2 - \frac{1}{2}p_0^2 - \frac{p_0}{\kappa} + \frac{p_0}{\kappa}\ln\varepsilon\zeta$$

$$+ \left(u_1 - \frac{1}{2\kappa^2}\ln^2\varepsilon\zeta\right) \quad \text{as } \zeta \to \infty, \tag{5.179}$$

$$v \sim A_2 - \frac{p_0}{\kappa}\ln\varepsilon\zeta \quad \text{as } \zeta \to 0.$$

It follows from (5.178) that

$$N = \kappa^2\zeta^2\frac{\partial u}{\partial\zeta},$$

$$T_1 = 2N[u_x - s_xu_\zeta],$$

$$\tau_3 = N[u_\zeta + \varepsilon s_{xx} + O(\varepsilon^2)], \tag{5.180}$$

$$u_x + W_\zeta = 0,$$

$$(u + p)_x - s_xp_\zeta = \tau_{3\zeta} - \varepsilon[uu_x + Wu_\zeta] + O(\varepsilon^2),$$

$$p_\zeta = -\varepsilon s_{xx} + O(\varepsilon^2).$$

Since we have $p = p_0 + \varepsilon p_{10}$ and $W = 0$ on $\zeta = 0$, then

$$p = p_0 + \varepsilon(p_{10} - s_{xx}\zeta) + O(\varepsilon^2),$$

$$W = p_0'\zeta + O(\varepsilon), \tag{5.181}$$

and thus v satisfies

$$v_x + p_{10}' - s_{xxx}\zeta + s_xs_{xx} = \frac{\partial}{\partial\zeta}[2\kappa\zeta v_\zeta + \kappa\zeta s_{xx}]$$

$$- \left[-p_0'\left(-p_0 + \frac{1}{\kappa}\ln\varepsilon\zeta\right) + \frac{p_0'}{\kappa}\right] + O(\varepsilon), \tag{5.182}$$

together with (5.179).

The solution of (5.182) is

$$v \approx -p_{10} - \frac{1}{2}s_x^2 - \frac{p_0}{\kappa} - \frac{1}{2}p_0^2 + \frac{p_0}{\kappa}\ln\varepsilon\zeta + s_{xx}\zeta + 3\kappa s_x + V, \tag{5.183}$$

where

$$\frac{\partial V}{\partial x} = \frac{\partial}{\partial\zeta}\left[2\kappa\zeta\frac{\partial V}{\partial\zeta}\right], \tag{5.184}$$

and (5.179) implies

$$V \to 0 \quad \text{as } \zeta \to \infty,$$

$$V \sim A_2^* - \frac{2p_0}{\kappa}\ln\varepsilon\zeta \quad \text{as } \zeta \to 0, \tag{5.185}$$

where

$$A_2 = A_2^* - p_{10} - \frac{1}{2}s_x^2 - \frac{p_0}{\kappa} - \frac{1}{2}p_0^2 + 3\kappa s_x. \tag{5.186}$$

The solution of (5.184) which tends to zero as $\zeta \to \infty$ is

$$V = \int_{-\infty}^{\infty} \hat{V}(\zeta, k) e^{ikx} \, dk, \tag{5.187}$$

where the Fourier transform $\hat{V}$ (as thus defined) is given by

$$\hat{V} = B K_0 \left[\left(\frac{2ik\zeta}{\kappa} \right)^{1/2} \right], \tag{5.188}$$

the square root is chosen so that $\mathrm{Re}(ik)^{1/2} > 0$,[7] and K_0 is a modified Bessel function of order zero. Evidently we require

$$\hat{V} \sim \hat{A}_2^* - \frac{2\hat{p}_0}{\kappa} \ln(\varepsilon \zeta) \quad \text{as } \zeta \to 0, \tag{5.189}$$

where the overhat defines the Fourier transform, in analogy to (5.187). Now $K_0(\xi) \sim -\ln \frac{1}{2}\xi - \gamma$ as $\xi \to 0$, where $\gamma \approx 0.5772$ is the Euler–Mascheroni constant. Also

$$\left(\frac{2ik\zeta}{\kappa} \right)^{1/2} = \left(\frac{2|k|\zeta}{\kappa} \right)^{1/2} \exp\left[\frac{i\pi}{4} \, \mathrm{sgn}\, k \right]; \tag{5.190}$$

therefore (5.188) implies

$$\hat{V} \sim -B \left[\gamma + \frac{1}{2} \ln |k| - \frac{1}{2} \ln 2\kappa + \frac{1}{2} \ln \zeta + \frac{i\pi}{4} \frac{k}{|k|} \right], \tag{5.191}$$

and matching this to (5.189) implies

$$B = \frac{4\hat{p}_0}{\kappa}, \tag{5.192}$$

whence

$$\hat{A}_2^* = \frac{2\hat{p}_0}{\kappa} \ln \varepsilon - \frac{4\hat{p}_0}{\kappa} \left[\gamma + \frac{1}{2} \ln |k| - \frac{1}{2} \ln 2\kappa + \frac{i\pi k}{4|k|} \right]. \tag{5.193}$$

We have $\widehat{s_x} = ik\hat{s}$, $\widehat{H(s_x)} = -|k|\hat{s}$, and $\widehat{J * s_x} = |k|\hat{s} \ln |k|$, where $J * s_x$ is the convolution of J with s_x, and $\hat{J} = -(i/2\pi) \ln |k| \, \mathrm{sgn}\, k$. (The convolution theorem here takes the form $\widehat{f * g} = 2\pi \hat{f} \hat{g}$.) It follows from this that

$$J(x) = -\frac{1}{\pi x} [\gamma + \ln |x|]. \tag{5.194}$$

Thus

$$A_2^* = \frac{2}{\kappa} (\ln 2\varepsilon \kappa - 2\gamma) p_0 + \frac{\pi}{\kappa} s_x + \frac{1}{\pi \kappa} J * s_x, \tag{5.195}$$

and, from (5.186),

---

[7] Assuming the principal branch of the square root, this implies we take $k = |k| e^{-i\pi}$ when k is negative.

$$A_2 = \frac{2}{\kappa}\left(\ln 2\varepsilon\kappa - 2\gamma - \frac{1}{2}\right)p_0 + \left(\frac{\pi}{\kappa} + 3\kappa\right)s_x$$
$$+ \frac{1}{\pi\kappa}J * s_x - p_{10} - \frac{1}{2}p_0^2 - \frac{1}{2}s_x^2, \tag{5.196}$$

where J is given by (5.194), $p_0 = H(s_x)$ ((5.151)), and p_{10} is given by (5.172).

We can summarise our calculation of the basal shear stress as follows. From (5.174), (5.176) and (5.151) we have

$$\tau = 1 + \varepsilon B_1 + \varepsilon^2 B_2 + \cdots, \tag{5.197}$$

where

$$B_1 = 2A_1 = -2H(s_x), \qquad B_2 = 2A_2 + A_1^2. \tag{5.198}$$

Using (5.186) and (5.193), we find after a little algebra that the transform of B_2 is

$$\hat{B}_2 = \frac{\hat{B}_1}{\kappa}\left[-2\ln 2\kappa\varepsilon + 2\ln|k| + i\pi\,\text{sgn}\,k + 4\gamma + 1\right] + \hat{C}, \tag{5.199}$$

where $\hat{C}$ is the transform of

$$C = -2p_{10} - s_x^2 + 6\kappa s_x. \tag{5.200}$$

5.7.10 Linear Stability

The Exner equation is, in appropriate dimensionless form,[8]

$$\varepsilon s_t + q_x = 0, \tag{5.201}$$

and since $q = q(\tau)$,

$$q = q_1 - 2\varepsilon q_1' p_0 + \varepsilon^2\left[(2A_2 + p_0^2)q_1' + 2p_0^2 q_1''\right] + \cdots, \tag{5.202}$$

where $q_1 = q(1)$, $q_1' = q'(1)$, $q_1'' = q''(1)$.

Thus s satisfies the nonlinear evolution equation

$$\frac{\partial s}{\partial t} - 2q_1'\frac{\partial p_0}{\partial x} + \varepsilon\frac{\partial}{\partial x}\left[(2A_2 + p_0^2)q_1' + 2p_0^2 q_1''\right] \approx 0. \tag{5.203}$$

This is

$$\frac{\partial s}{\partial t} - \alpha\frac{\partial p_0}{\partial x} + \varepsilon\frac{\partial}{\partial x}\left[q_1'\left(2\omega s_x + 2\lambda J * s_x - 2p_{10} - s_x^2\right) + 2q_1'' p_0^2\right] = 0, \tag{5.204}$$
$$p_0 = H(s_x),$$

---

[8]Note that the definition of ε here is that pertaining to the mixing-length theory, i.e., (5.121) and not (5.48).

where

$$\alpha = 2q_1'\left[1 - \frac{2\varepsilon}{\kappa}\left(\ln 2\varepsilon\kappa - 2\gamma - \frac{1}{2}\right)\right],$$

$$\omega = \frac{\pi}{\kappa} + 3\kappa, \tag{5.205}$$

$$\lambda = \frac{1}{\pi\kappa}.$$

We linearise (5.204) for small s by neglecting the terms in s_x^2 and p_0^2. Taking the Fourier transform (as defined here in (5.187)), we have

$$\hat{s}_t = ik\alpha\widehat{p_0} - ik\varepsilon q_1'(2\omega ik\hat{s} + 4\pi\lambda ik\hat{J}\hat{s} - 2\widehat{p_{10}}). \tag{5.206}$$

From (5.172),

$$p_{10} = \int_0^\infty (a * R + b * S)\,d\zeta - H\{(sp_0)_x\}, \tag{5.207}$$

where

$$a(x,\zeta) = \frac{x}{\pi(x^2 + \zeta^2)}, \qquad b(x,\zeta) = -\frac{\zeta}{\pi(x^2 + \zeta^2)}. \tag{5.208}$$

Hence, neglecting the quadratic Hilbert transform term,

$$\hat{p}_{10} = 2\pi\int_0^\infty (\hat{a}\hat{R} + \hat{b}\hat{S})\,d\zeta. \tag{5.209}$$

Calculation of $\hat{a}$ and $\hat{b}$ gives

$$\hat{a} = -\frac{i}{2\pi}e^{-|k|\zeta}\,\mathrm{sgn}\,k, \qquad \hat{b} = -\frac{1}{2\pi}e^{-|k|\zeta}, \tag{5.210}$$

so that

$$\hat{p}_{10} = -\int_0^\infty [i\hat{R}\,\mathrm{sgn}\,k + \hat{S}]e^{-|k|\zeta}\,d\zeta. \tag{5.211}$$

Now

$$\begin{aligned}
\tau_3^{(0)} &= 1 + 3\kappa z w_x^{(0)} + 2\kappa^2 z^2 w_x^{(0)2}, \\
\tau_1^{(0)} &= -2\kappa z w_z^{(0)} - 2\kappa^2 z^2 w_x^{(0)} w_z^{(0)}, \\
u^{(0)} &= u_0(z) - p^{(0)}, \\
u_x^{(0)} &= -w_z^{(0)}, \\
u_z^{(0)} &= u_0' + w_x^{(0)},
\end{aligned} \tag{5.212}$$

thus, retaining only the perturbed linear (in s) terms, we have from (5.170)

$$\begin{aligned}
\hat{R} &\approx ik\hat{t}_1 + \hat{t}_{3z} + u_0\hat{w}_z - u_0'[\hat{w} - ik\hat{s}], \\
\hat{S} &\approx ik\hat{t}_3 - \hat{t}_{1z} - iku_0[\hat{w} + ik\hat{s}],
\end{aligned} \tag{5.213}$$

where $\hat{w} = \widehat{w^{(0)}}$, and

$$\hat{t}_1 = -2ik\kappa z\hat{p}, \qquad \hat{t}_3 = 3ik\kappa z\hat{w}, \tag{5.214}$$

where $\hat{p} = \widehat{p^{(0)}}$.

Finally, from (5.150),

$$w^{(0)} = -b(x, z) * s_x, \qquad p^{(0)} = -a(x, z) * s_x, \tag{5.215}$$

whence using (5.210),

$$\hat{w} = ik\hat{s}e^{-|k|z}, \tag{5.216}$$
$$\hat{p} = -|k|\hat{s}e^{-|k|z},$$

and we eventually obtain

$$\hat{p}_{10} = -\hat{s} \int_0^\infty \left[k^2 u_0 \left(1 + 2e^{-|k|\zeta} \right) - |k| u_0' \left(1 - e^{-|k|\zeta} \right) \right.$$
$$\left. - 5i\kappa k |k| e^{-|k|\zeta} \right] e^{-|k|\zeta} \, d\zeta. \tag{5.217}$$

Simplification of this, using the fact that $\int_0^\infty e^{-t} \ln t \, dt = -\gamma$, where $\gamma \approx 0.5772$ is the Euler–Mascheroni constant, yields

$$\hat{p}_{10} = \hat{s} \left[\frac{2|k|}{\kappa} \left(\ln 2|k| + \gamma \right) + \frac{5}{2} i\kappa k \right]. \tag{5.218}$$

Solutions of (5.206) are $\hat{s} = e^{\sigma t}$, where $\sigma = r - ikc$, and after some simplification, we find that the growth rate r is

$$r = 2k^2 \varepsilon q_1' \left(\frac{\pi}{\kappa} + \frac{1}{2}\kappa \right), \tag{5.219}$$

and the wave speed c is

$$c = 2q_1' |k| \left[1 + \frac{2\varepsilon}{\kappa} \left\{ -\left(1 - \frac{1}{2\pi} \right) \ln |k| - \ln 4\varepsilon\kappa + \gamma + \frac{1}{2} \right\} \right]. \tag{5.220}$$

Thus dunes grow, as $r > 0$, on a time scale of $O(1/\varepsilon)$, while the waveforms move downstream at a speed $c \approx 2q_1'|k| = O(1)$.

This apparently more realistic theory for dune-forming instability is less satisfactory than the constant eddy viscosity theory, because the growth rate $r \propto k^2$, and the basic model is again ill-posed. As before, we can stabilise the model by including the downslope force, thus replacing the stress by the effective stress defined using (5.109). The effect of this is to add a term to the stress definition in (5.174), which can then be written as

$$\tau_e = 1 - 2\varepsilon p_0 + 2\varepsilon^2 (A_2 + \cdots) - \hat{\beta} s_x, \tag{5.221}$$

where the definition of $\hat{\beta}$ differs from that in (5.111) because of the different scaling used in the aeolian model. Using (5.124), $x \sim d$ and $s \sim \varepsilon d$, we find

$$\hat{\beta} = \frac{\Delta \rho}{\rho} \frac{D_s}{d} \frac{1}{\varepsilon F^2}, \tag{5.222}$$

where the Froude number is

$$F = \frac{U_\infty}{\sqrt{gd}}. \tag{5.223}$$

Using values $\Delta\rho/\rho = 2.6 \times 10^3$, $D_s/d = 10^{-6}$, $\varepsilon \sim 0.03$, $F^2 \sim 0.04$ (based on $d = 1000$ m and $U_\infty = 20$ m s^{-1}), we find $\hat{\beta} \sim 2.2$.

If we consult (5.196), we see that the destabilising term arises from that proportional to s_x in A_2. Effectively we can write

$$\tau_e = 1 + \cdots + \left[\varepsilon^2 \left(\frac{2\pi}{\kappa} + \kappa \right) - \hat{\beta} \right] \frac{\partial s}{\partial x} + \cdots, \tag{5.224}$$

where the modification of the coefficient ω reflects the effect of the terms in J and p_{10}, as indicated by (5.219). We see that the downslope term stabilises the system if $\hat{\beta} > O(\varepsilon^2)$, and thus practically if $F^2 < 1$. On the Earth, a typical value is $F^2 = 0.04$, so that the instability is removed, at all wave numbers. This is distinct from the constant eddy viscosity case, because the stabilising term has the same wave number dependence as the destabilising one.

If we ignore the stabilising term in $\hat{\beta}$, then the situation is somewhat similar to the rill-forming instability which we will study in Chap. 6. There the instability is regularised at long wavelength by inclusion of singularly perturbed terms. The most obvious modification to make here in a similar direction is to allow for a finite thickness of the moving sand layer. It seems likely that this will make a substantial difference, because the detail of the mixing-length model relies ultimately on the existence of an exponentially small roughness layer through which the wind speed drops to zero. It is noteworthy that the constant eddy viscosity model does not share this facet of the problem.

5.8 Separation at the Wave Crest

The constant eddy viscosity model can produce a genuine instability, with decay at large wave numbers. If pushed to a nonlinear regime, it allows shock formation, although it also allows unlimited wavelength growth. The presumably more accurate mixing-length theory actually fares somewhat worse. It can produce a very slow instability via an effective negative diffusivity, but this is easily stabilised by downslope drift. It is possibly the case that specific consideration of the mobile sand layer will alleviate this result.

A complication arises at this point. Aeolian sand dunes inevitably form slip faces. There is a jump in slope at the top of the slip face, and the wind flow separates, forming a wake (or cavity, or bubble). One authority is of the opinion that no model can be realistic unless it includes a consideration of separation. In this section we will consider a model which is able to do this. Before doing so, it is instructive to consider how such separation arises.

If the constant eddy viscosity model has any validity, it suggests that the uniform flat bed is unstable, and that travelling waves grow to form shocks. If the slope within the shocks is steep enough to exceed the angle of repose of sand grains (some 34°), then a slip face will occur, with the sand resting on the slip face at this angle. The turbulent flow over the dune inevitably separates at the cusp of the dune,

Fig. 5.13 Separation behind
a dune

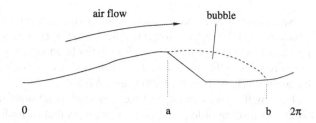

forming a separation bubble, as indicated in Fig. 5.13. The formation of a separation bubble makes the model fundamentally nonlinear, and it provides a possible
mechanism for length scale selection. It is thus an attractive possible way out of the
conundrums concerning instability alluded to above.

It is simplest to treat the separation bubble in the context of the mixing-length
theory, and this we now do, despite our misgivings about its applicability for small
amplitude perturbations. We suppose that there is a periodic sequence of dunes, with
period chosen to be 2π. We suppose that there is a slip face, as shown in Fig. 5.13,
and we suppose the corresponding separation bubble occupies the interval (a, b).
We denote the bubble interval as B, and the corresponding attached flow region
as B'.

Because our method will use complex variables, it is convenient to rechristen the
space coordinates as x and y, and the corresponding velocity components as u and
v. At leading order, the inviscid flow is described by the outer equations (5.146):

$$u_x + v_y = 0,$$
$$u_x + p_y = 0, \qquad (5.225)$$
$$v_x + p_y = 0,$$

and these are valid in $y > \varepsilon s$. From these it follows that p and v satisfy the Cauchy–
Riemann equations, and thus

$$p + iv = f(z) \qquad (5.226)$$

is an analytic function, where $z = x + iy$.

The boundary conditions for p and v are that both tend to zero as $y \to \infty$, and
v satisfies the no flow through condition (5.144), $v = s_x + \varepsilon u s_x$ on $y = \varepsilon s$. These
completely specify the problem in the absence of a separation bubble.

If we suppose that a separation bubble occurs, as shown in Fig. 5.13, then its
upper boundary is unknown, and must be determined by an extra boundary condition. We let $y = \varepsilon s(x)$ denote this unknown upper boundary, and define the ground
surface to be $y = \varepsilon s_0(x)$; thus $s(x) = s_0(x)$ for $x \in B'$.

There are various ways to provide the extra condition. Two such are that the
pressure, or alternatively the vorticity, are constant in the bubble. We shall suppose
the former, and therefore we prescribe

$$p = p_B \quad \text{for } y = \varepsilon s, \ x \in B. \qquad (5.227)$$

The bubble pressure p_B is an unknown constant, and must be determined as part of
the solution.

Separation occurs because the viscous boundary layer (here, the roughness layer) detaches from the surface, forming a free shear layer at the top of the bubble, which rapidly thickens to form a more diffuse upper boundary. The assumption of constant pressure in the bubble is essentially a consequence of this shear layer, implying that mean fluid velocities in the bubble are small.

For small ε, we can expand the boundary conditions at $y = \varepsilon s$ about $y = 0$, so that to leading order, the problem becomes that of finding an analytic function $f(z) = p + iv$ in the upper half plane $\text{Im} \, z > 0$, satisfying

$$
\begin{aligned}
f &\to 0 \quad \text{as } z \to \infty, \\
v &= s_x \quad \text{on } y = 0, \\
p &= p_B \quad \text{on } y = 0, \; x \in B.
\end{aligned}
\tag{5.228}
$$

The extra pressure condition should help determine s in B, but the endpoint locations are not necessarily known. Specification of the behaviour of the solution at the endpoints is necessary to determine these. Firstly, we expect s to be continuous at the end points:

$$
s(a) = s_0(a), \qquad s(b) = s_0(b). \tag{5.229}
$$

A difference now arises depending on whether a slip face occurs or not. If not, then the bed slope is continuous, and at the upstream end point $x = a$, we might surmise that boundary layer separation is associated with the skin friction dropping to zero. Now from (5.174) and (5.176), we have the surface stress defined by

$$
\sqrt{\tau} = 1 - \varepsilon p_0, \tag{5.230}
$$

where p_0 is the surface pressure. The only apparent interpretation of this which we can make in our simplified model is to require that

$$
p \to +\infty \quad \text{on } y = 0 \text{ as } x \to a- \in B'; \tag{5.231}
$$

more detailed consideration of the boundary layer structure near the separation point would be necessary to be more precise than this. We do not pursue this possibility here, mainly because the more relevant situation is when a slip face is present.

If we suppose a slip face is present, then we can presume that separation occurs at its top, and this determines the point $x = a$. In addition, it is natural to suppose that boundary layer detachment occurs smoothly, in the sense that we suppose the slope of s is continuous at a:

$$
s'(a+) = s_0'(a-); \tag{5.232}
$$

this implies that v is continuous at $x = a$. If possible, we would like to have smooth reattachment at b, and in addition (and in fact, because of this) continuity of pressure also:

$$
[p]_{b-}^{b+} = [p]_{a-}^{a+} = 0, \qquad s'(b-) = s_0'(b+). \tag{5.233}
$$

We shall in fact find that all these conditions can be satisfied. This is not always the case in such problems, and sometimes (worse) singularities have to be tolerated. The choice of the behaviour of the solution at the end points actually constitutes the most subtle part of solving Hilbert problems.

5.8.1 Formulation of Hilbert Problem

The first thing we do is to analytically continue $f(z)$ into the lower half plane. Specifically, we define

$$G(z) = \begin{cases} \frac{1}{2}[f(z) - p_B], & \mathrm{Im}\, z > 0, \\ -\frac{1}{2}[\overline{f(\bar{z})} - p_B], & \mathrm{Im}\, z < 0. \end{cases} \tag{5.234}$$

G is analytic in both the upper and lower half planes, and if G_+ and G_- denote the limiting values of G as $z \to x$ from above and below, then

$$\begin{aligned} G_+ + G_- &= is', \\ G_+ - G_- &= p - p_B, \end{aligned} \tag{5.235}$$

everywhere on the real axis.

Because of the assumed periodicity in x, we make the following transformations:

$$\zeta = e^{iz}, \qquad \xi = e^{ix}, \qquad G(z) = H(\zeta). \tag{5.236}$$

The geometry of the problem is then illustrated in Fig. 5.14. The problem to solve is identical to (5.235), replacing G by H, and thus we have the standard Hilbert problem

$$\begin{aligned} H_+ - H_- &= 0 \quad \text{on } B, \\ H_+ + H_- &= i\sigma_0 \quad \text{on } B', \end{aligned} \tag{5.237}$$

where $\sigma_0(\xi) = s_0'(x)$. We have to solve this subject to the supplementary conditions

$$H(0) = -\frac{1}{2}p_B, \qquad H(\infty) = \frac{1}{2}p_B; \tag{5.238}$$

the first of these in fact implies the second automatically. We seek to apply the conditions that both $\frac{1}{2}(p - p_B) = \mathrm{Re}\, H$ and $\frac{1}{2}v = \mathrm{Im}\, H$ are continuous (thus H is continuous) at both endpoints $\xi = \xi_a = e^{ia}$ and $\xi = \xi_b = e^{ib}$. Given H satisfying (5.237), then the separation bubble boundary is given by the solution of

$$s' = -2i H \quad \text{on } B, \qquad s(a) = s_0(a), \tag{5.239}$$

and the pressure on B' is given by

$$p = p_B + H_+ - H_- \quad \text{on } B'. \tag{5.240}$$

Solution

The solution to (5.237), given the location of a and b, is as follows. Define a function $\chi(\zeta)$ such that

$$\chi_+ + \chi_- = 0 \quad \text{on } B' \tag{5.241}$$

Fig. 5.14 B and B' on the unit circle in the complex ζ plane. B' is a branch cut for the solution of the Hilbert problem (5.237)

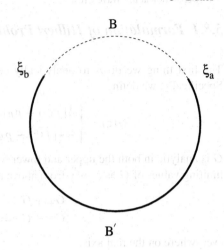

(and χ is analytic away from B'); then

$$\left(\frac{H}{\chi}\right)_+ - \left(\frac{H}{\chi}\right)_- = \frac{i\sigma_0}{\chi_+}, \tag{5.242}$$

and by the discontinuity theorem, we have

$$H = \frac{\chi(\zeta)}{2\pi i}\int_{B'}\frac{i\sigma_0(t)\,dt}{\chi_+(t)(t-\zeta)} + \chi P, \tag{5.243}$$

where P is an as yet undetermined polynomial. To find P, we must specify χ, and this in turn depends on the required singularity structure of the solution.

The smoothness of H is essentially that of χ, and so we will choose the function

$$\chi = \left[(\zeta - \xi_a)(\zeta - \xi_b)\right]^{1/2}. \tag{5.244}$$

The most general choice is $\chi = (\zeta - \xi_a)^{m_a + \frac{1}{2}}(\zeta - \xi_b)^{m_b + \frac{1}{2}}$, where m_a and m_b are integers, but most of these possibilities can in general be eliminated by requirements either of continuity or at least integrability of the solution.

We consider the behaviour of the Cauchy integral

$$\Phi(\zeta) = \frac{1}{2\pi i}\int_{B'}\frac{\phi(t)\,dt}{t-\zeta} \tag{5.245}$$

near the end points of integration. Note that in the present case,

$$\phi(t) = \frac{i\sigma_0(t)}{\chi_+(t)}. \tag{5.246}$$

First suppose that $\phi(t)$ is continuous at an end point.[9] Then we have

$$\Phi(\zeta) = \pm\frac{\phi(c)}{2\pi i}\ln(\zeta - c) + O(1), \tag{5.247}$$

---

[9]More precisely, ϕ should be *Hölder continuous*, that is, $|\phi(t_1) - \phi(t_2)| < K|t_1 - t_2|^\gamma$, for some positive γ.

where c denotes either end point of B', and the upper and lower signs apply at the right (ξ_a) and left (ξ_b) hand ends of B', respectively. (5.247) applies as $\zeta \to c$, with $\zeta \notin B'$.

Similarly, for $\xi \in B'$,

$$\Phi(\xi) = \pm \frac{\phi(c)}{2\pi i} \ln(\xi - c) + O(1), \tag{5.248}$$

where $\Phi(\xi)$ denotes the principal value of the integral (and $\Phi(\xi) = \frac{1}{2}[\Phi_+(\xi) + \Phi_-(\xi)]$). Bearing in mind (5.246), we see that if χ is unbounded at c, and specifically goes algebraically to infinity, then the corresponding Cauchy integral is bounded, and thus H will be unbounded (unless the choice of P can be chosen to remove the singularity). Using the definition

$$H = \chi(\zeta)\big[\Phi(\zeta) + P\big], \tag{5.249}$$

we have from (5.239) and (5.240) that

$$\begin{aligned} s' &= -2i\,H(\xi) = -2i\,\chi(\xi)\big[\Phi(\xi) + P\big] \quad \text{on } B, \\ p - p_B &= 2\chi_+(\xi)\big[\Phi(\xi) + P\big] \quad \text{on } B'. \end{aligned} \tag{5.250}$$

The implication of this is that if χ is unbounded at an end point, then in general both p and s' will also be unbounded, unless the choice of P removes the singularity. The worst singularity we can tolerate is an integrable one, thus $\chi \sim (\zeta - c)^{-1/2}$.

Now suppose that χ is bounded at an end point, and specifically $\chi \sim (\zeta - c)^{1/2}$. (Any higher power causes the Cauchy integral to be undefined, because then ϕ is not integrable.) If we define $\tilde{\phi}$ via

$$\phi(t) \sim \frac{\tilde{\phi}(t)}{(t - c)^{1/2}} \quad \text{as } t \to c, \tag{5.251}$$

then

$$\begin{aligned} \Phi(\zeta) &= \frac{\tilde{\phi}(c)}{2(\zeta - c)^{1/2}} + o\left(\frac{1}{(\zeta - c)^{1/2}}\right), \quad \zeta \in B, \\ \Phi(\xi) &= o\left(\frac{1}{(\xi - c)^{1/2}}\right), \quad \xi \in B'. \end{aligned} \tag{5.252}$$

It then follows from (5.250) that s' is bounded (and in fact continuous) and p is continuous at c. It is because of this that we choose χ as defined in (5.244), in order to satisfy the smoothness conditions (5.232) and (5.233).

In this case, the polynomial P must be zero in order to satisfy the condition at $\zeta = \infty$, and we have

$$H = \frac{\chi(\zeta)}{2\pi i} \int_{B'} \frac{i\sigma_0(t)\,dt}{\chi_+(t)(t - \zeta)}. \tag{5.253}$$

We define the integrals

$$I_0 = \frac{\chi_0}{2\pi i} \int_{B'} \frac{i\sigma_0(t)\,dt}{t\chi_+(t)}, \qquad I_\infty = \frac{1}{2\pi i} \int_{B'} \frac{i\sigma_0(t)\,dt}{\chi_+(t)}; \tag{5.254}$$

we thus have $H(0) = I_0$, $H(\infty) = -I_\infty$, and the conditions in (5.238) correspond to prescribing

$$I_0 = I_\infty = -\frac{1}{2}p_B. \tag{5.255}$$

It is a straightforward exercise in contour integration to show that $\bar{I}_0 = I_\infty$, where the overbar denotes the complex conjugate, therefore (5.255) is tantamount to the single condition $I_0 = -\frac{1}{2}p_B$. Because this is a complex-valued integral, (5.255) actually comprises two conditions for the two unknown quantities p_B and b.

It remains to be seen whether s is continuous at b. Since (5.255) determines b, and s is fully determined by (5.239), it is not obvious that this will be the case. (If it were not, we would have to allow for a singularity in the solution at one of the end points.)

In fact, it is easy to show that (5.255) automatically implies that s is continuous at b. To show this, it is sufficient to show that s is continuous over the periodic domain $[0, 2\pi]$. Equivalently, we need to show that

$$I = \int_0^{2\pi} s' \, dx = \int_{B \cup B'} -i(H_+ + H_-) \frac{d\xi}{i\xi} = 0, \tag{5.256}$$

using (5.239) and (5.237). Denoting contours just inside and outside the unit circle as C_+ and C_- (see Fig. 5.15), we see that

$$I = -\left[\int_{C_+} \frac{H \, d\xi}{\xi} + \int_{C_-} \frac{H \, d\xi}{\xi} \right]. \tag{5.257}$$

H is analytic inside and outside the unit circle. The integral over C_+ is thus just $2\pi i H(0)$ using the residue theorem, while the integral over C_- can be extended by deforming the contour out to infinity, whence we obtain the integral $2\pi i H(\infty)$. Thus

$$I = -2\pi i [I_0 - I_\infty] = 0, \tag{5.258}$$

and continuity of s at b is ensured. We have thus obtained a solution in which the separated streamline leaves and rejoins the surface smoothly, and the pressure is continuous at the end points.

5.8.2 Calculation of the Free Boundary

In order to solve (5.239) for s, we need to evaluate H on B'. There are various ways to do this. One simple one, which may be convenient for subsequent evolution of the bed using spectral methods, is to use a Fourier series representation. Let us suppose that

$$s(x) = \sum_{k=-\infty}^{\infty} a_k e^{ikx}, \tag{5.259}$$

Fig. 5.15 The contours C_+ and C_- lie just inside and outside the unit circle, respectively

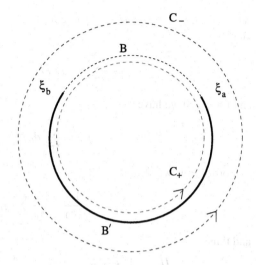

so that

$$i\sigma_0(\xi) = \sum_{k=-\infty}^{\infty} d_k \xi^k,$$ (5.260)

where

$$d_k = -k a_k.$$ (5.261)

We suppose that the Laurent expansion for $i\sigma_0$ extends to the complex plane as an analytic function with singularities only at 0 and ∞. (This is automatically true for any finite such series.) Then we can write the solution for H as

$$H = \frac{1}{2}\left[i\sigma_0(\zeta) - q(\zeta)\chi(\zeta)\right],$$ (5.262)

where q has a Laurent expansion

$$q = \sum_{k=-\infty}^{\infty} l_k \zeta^k.$$ (5.263)

Then we obtain s by solving

$$s' = s_0' + iq\chi$$ (5.264)

on $[a, b]$, with $s(a) = s_0(a)$. In practice, we would obtain b by shooting.

Suppose that

$$\frac{1}{\chi(\zeta)} = \sum_{r=0}^{\infty} \frac{f_r}{\zeta^{r+1}}, \quad |\zeta| > 1$$ (5.265)

(see Question 5.10 for one way to calculate the coefficients); then we can write

$$\frac{H}{\frac{1}{2}\chi} = \sum_{m=-\infty}^{\infty} d_m \zeta^m \sum_{r=0}^{\infty} \frac{f_r}{\zeta^{r+1}} - \sum_{j=-\infty}^{\infty} l_j \zeta^j.$$ (5.266)

As $\zeta \to \infty$, $\chi \sim \zeta$ and $H \to \frac{1}{2}p_B$; equating coefficients of ζ^j in (5.266) for $j \geq 0$ yields

$$l_j = \sum_{r=0}^{\infty} d_{j+r+1} f_r, \quad j \geq 0, \tag{5.267}$$

and for $j = 0$ we have

$$p_B = \sum_{r=0}^{\infty} d_r f_r - l_{-1}. \tag{5.268}$$

For $|\zeta| < 1$, we find

$$\frac{1}{\chi(\zeta)} = \frac{1}{\chi_0} \sum_{r=0}^{\infty} \bar{f}_r \zeta^r, \tag{5.269}$$

and thus

$$\frac{H}{\frac{1}{2}\chi} = \frac{1}{\chi_0} \sum_{r=0}^{\infty} \bar{f}_r \zeta^r \sum_{m=-\infty}^{\infty} d_m \zeta^m - \sum_{j=-\infty}^{\infty} l_j \zeta^j. \tag{5.270}$$

As $\zeta \to 0$, $H \to -\frac{1}{2}p_B$; equating powers of ζ^j for $j \leq -1$, we find

$$l_j = \frac{1}{\chi_0} \sum_{r=0}^{\infty} \bar{f}_r d_{j-r}, \quad j \leq -1, \tag{5.271}$$

and for $j = 0$ we have

$$\frac{p_B}{\chi_0} = \frac{1}{\chi_0} \sum_{r=0}^{\infty} \bar{f}_r d_{-r} - l_0. \tag{5.272}$$

Putting these results together, we find that (5.268) and (5.272) together give (bearing in mind that $d_{-k} = -\bar{d}_k$)

$$p_B = \sum_{r=0}^{\infty} d_r f_r + \bar{\chi}_0 \sum_{r=0}^{\infty} \bar{f}_r \bar{d}_{r+1}, \tag{5.273}$$

with the added constraint that p_B is real.

We can now use the definitions of l_j in (5.267) and (5.271) to evaluate $iq\chi$ in (5.264). Being careful with the arguments, we find that on B,

$$\chi = 2\xi^{1/2} \chi_0^{1/2} R, \tag{5.274}$$

where

$$\chi_0 = \exp\left[\frac{1}{2}(a+b)\right], \quad R = \left[\sin\left(\frac{x-a}{2}\right)\sin\left(\frac{b-x}{2}\right)\right]^{1/2}, \tag{5.275}$$

and after some algebra, we have the differential equation for s on B:

$$s' = s_0' - 4R \, \mathrm{Im}\left[\chi_0^{1/2} \sum_{j=0}^{\infty}\sum_{r=0}^{\infty} f_r d_{j+r+1} \exp\left\{i\left(j+\frac{1}{2}\right)x\right\}\right], \tag{5.276}$$

with initial condition $s(a) = s_0(a)$. To solve this, guess b; we can then calculate the right hand side. Solving for s, we adjust b by decreasing it if s reaches s_0 for $x < b$, and increase it if s remains $> s_0$ for all $x \le b$.

Computational Approaches

Complex analysis is all very elegant, but is probably not an efficient way to compute a time-evolving interface. A direct computational approach would be preferable, but the free boundary nature renders this problematic. Two ways of dealing with this issue have been suggested, and are discussed further in the notes.

5.9 Notes and References

Books describing sediment transport and its effects on river morphology include those by Allen (1985), Ahnert (1996), Knighton (1998) and Goudie (1993). Mention must also be made of Gary Parker's e-book (Parker 2004), which describes in the form of powerpoint lectures a wealth of phenomena and theory concerning river bedforms. The classical book on aeolian dunes is that of Bagnold (1941), and a more recent classic is that of Pye and Tsoar (1990). Both books have recently been reprinted, Bagnold's by Dover in 2005, and Pye and Tsoar's by Springer in 2009.

Linear Stability   The first theory for dune and anti-dune formation which embodied the principle of upstream stress migration was due to Kennedy (1963), as described in Sect. 5.3. Kennedy was motivated by Benjamin's earlier (1959) result on laminar fluid flow over small bumps, but the prescription of a fixed spatial lag is flawed. Parker (1975) suggested that the inertial effect of bedload (i.e., sediment flux relaxes to its equilibrium value over a finite length) could be a causative mechanism for the formation of anti-dunes.

St. Venant-type models were introduced by Reynolds (1965), and the failure of averaged models to locate instability led Engelund (1970) and Smith (1970) to study eddy viscosity type models in which the two-dimensional nature of the flow was paramount. Subsequent developments of the instability theory were made by Fredsøe (1974), Richards (1980) (who extended the theory to the formation of ripples), Engelund and Fredsøe (1982), Sumer and Bakioglu (1984), Colombini (2004) and Charru and Hinch (2006).

Sediment Transport   The Shields stress, and the experimental data in Fig. 5.7, were given in his thesis by Shields (1936). There are a number of empirical estimates for fluvial bedload transport, of which that described by Meyer-Peter and Müller (1948) (see also Einstein 1950) is a popular one, though possibly not the best. Similar relations are found for aeolian sand transport (see, e.g., Bagnold 1936; Pye and Tsoar 1990). Formulae describing the rate of entrainment or erosion of sediments into suspension are given by García and Parker (1991), Van Rijn (1984), and Smith and McLean (1977), for example.

Turbulent Flow and Eddy Viscosity   The use of an eddy viscosity gives the simplest description for a turbulent flow, but as mentioned in Sect. 5.7.4, the choice of eddy viscosity is problematic. Prandtl's mixing-length theory for a shear flow

$$\tau = \kappa^2 z^2 \left| \frac{\partial u}{\partial z} \right| \frac{\partial u}{\partial z} \qquad (5.277)$$

correctly yields the logarithmic velocity profile, but is frame dependent, as well as having an infinite velocity at the wall. Usually (e.g., Schlichting 1979[10]) one retains the no slip condition by specifying a wall roughness, which has the effect of applying the no slip condition at a finite elevation $z = z_0$. An alternative (and preferable) method is to include the small laminar viscosity, thus replacing (5.277) by

$$\tau = \left(\varepsilon + \kappa^2 z^2 \left| \frac{\partial u}{\partial z} \right| \right) \frac{\partial u}{\partial z}, \qquad (5.278)$$

in suitably scaled variables; essentially $\varepsilon = 1/Re$. Solution of a constant shear stress shear flow satisfying $u = 0$ at $z = 0$ shows that the effective roughness concept can be applied, where (for (5.278)), we find (see Question 5.11) $z_0 = \frac{\varepsilon}{2\kappa}$.

The frame indifference issue could be resolved by using Von Kármán's version of the mixing length, which replaces (5.278) with

$$\tau = \left(\varepsilon + \frac{\kappa^2 |u_z|^3}{|u_{zz}|^2} \right) \frac{\partial u}{\partial z}; \qquad (5.279)$$

this is of course also not frame indifferent, but can easily be made so by generalising to, for example,

$$\boldsymbol{\tau} = 2\eta \dot{\boldsymbol{\varepsilon}}, \qquad \dot{\boldsymbol{\varepsilon}} = \frac{1}{2} \left(\boldsymbol{\nabla} \mathbf{u} + \boldsymbol{\nabla} \mathbf{u}^T \right), \qquad (5.280)$$

where the effective viscosity is

$$\eta = \varepsilon + \frac{2\kappa^2 |\dot{\boldsymbol{\varepsilon}}|^3}{|\boldsymbol{\nabla} . \dot{\boldsymbol{\varepsilon}}|^2}. \qquad (5.281)$$

However, (5.279) is also problematical, because it allows η to depend on the second derivative of u, thus artificially raising the order of the equations. Taking the mixing length $l = \frac{\kappa}{|u_z|}$ does not work. The only other possibility along these lines might be to assume a dependence on fractional derivatives of u_z, although there seems little physical justification for this.

Jackson–Hunt Theory   The classic paper describing turbulent flow over a small hill is that by Jackson and Hunt (1975). Further developments of the theory are given by Sykes (1980), Hunt et al. (1988) and (less easy to find) Weng et al. (1991). It is generally acknowledged that the Jackson–Hunt paper is very difficult to read.

---

[10]The Schlichting book went through many reprints, and currently exists in print in a revised edition by Schlichting and Gersten published by Springer in 2000; this new book is quite different from the earlier version, and a good deal of material in the original book has been removed.

The theory is complicated for one thing, but the manner of presentation is not clear. Rather than present a clearly stated boundary value problem, Jackson and Hunt present solutions, model, approximations, scales and limits all mixed together. Sykes (1980) provided a more rational asymptotic treatment of the problem, and pointed out various difficulties in the Jackson–Hunt theory, but like them, Sykes avoided providing a description for the Reynolds stress until late on. Thus, the Jackson–Hunt theory divorces the assumed basic logarithmic velocity from the assumed form for the Reynolds stresses. The version of the theory presented here, in Sect. 5.7.2, adopts a different philosophy: that the logarithmic profile must itself be a consequence of the boundary value problem to be solved. While this may seem a sensible approach, it raises the issue of how best to prescribe the Reynolds stresses. Sykes provides a fairly sophisticated closure scheme, without an indication that the basic solution has the required logarithmic profile.

Hunt et al. (1988) provide an improved version of the theory, which we summarise here. The paper is again difficult to read. The basal shear stress τ is given by Eq. $(3.1)_H$ (all equation numbers with subscripts H refer to Hunt et al. (1988))

$$\tau = \varepsilon^2 \rho U_0^2 (1 + \tau_d).$$ (5.282)

At the top of page 1,439, we find

$$\varepsilon = \frac{u_*}{U_\infty},$$ (5.283)

where u_* is the friction velocity, and U_∞ is the far-field velocity, essentially the same as our definition in (5.121), whereas U_0 is the velocity of the basic profile at a height h_m; however, in Eq. $(2.4c)_H$ we have $\varepsilon = \frac{u_*}{U_0}$, so we will suppose that $U_0 \approx U_\infty$, which is also consistent with the discussion at the very bottom of page 1,438 and the top of page 1,439.

The perturbation shear stress τ_d is defined in $(3.7d)_H$, and using $(3.12a,b)_H$ its Fourier transform[11] is given by

$$\hat{\tau}_d = -\frac{2p_0}{U^2(l)} \sigma(k) \big[1 + \delta(2\ln k + 4\gamma + 1 + i\pi) \big].$$ (5.284)

p_0 is defined in $(2.15)_H$, as is σ, as minus the Hilbert transform of the bed slope. $U(l)$ is the (scaled, with U_0 in $(2.1)_H$) velocity of the undisturbed flow at a height l above the bed, and is defined (at the bottom of page 1,438) by

$$U(l) = \frac{\varepsilon}{\kappa} \ln\left(\frac{l}{z_0}\right),$$ (5.285)

where z_0 is the roughness length; (5.285) assumes $l \ll h_m$ (as is the case), while l is defined in $(3.6)_H$ by

$$l \ln\left(\frac{l}{z_0}\right) = 2\kappa^2 d.$$ (5.286)

---

[11] Defined, as I have also done here in (5.187), via $\int \ldots e^{-ikx}\, dx$, and denoted by an overhat.

(the depth scale of the flow d is denoted L in the Hunt paper). The small parameter δ in (5.284) is defined on page 1,449, two lines above $(3.7a)_H$:

$$\delta = \frac{1}{\ln\left(\frac{L}{z_0}\right)}.$$ (5.287)

It is convenient to define

$$l = 2\kappa\varepsilon\Lambda d,$$ (5.288)

and then we have

$$\Lambda = \frac{1}{1 + \frac{\varepsilon}{\kappa}\ln 2\kappa\varepsilon\Lambda} \approx 1 - \frac{\varepsilon}{\kappa}\ln 2\kappa\varepsilon,$$ (5.289)

and

$$\delta = \frac{\varepsilon\Lambda}{\kappa}.$$ (5.290)

Using these results, we find that the dimensionless Hunt formula for the basal shear stress can be written as

$$\tau = 1 + \varepsilon B_1 + \varepsilon^2 H_2 + \cdots,$$ (5.291)

where

$$B_1 = -2H(s_x),$$ (5.292)

as in (5.198), and the transform of H_2 is

$$\hat{H}_2 = \frac{\hat{B}_1}{\kappa}\{-2\ln 2\kappa\varepsilon + 2\ln k + 4\gamma + 1 + i\pi\}.$$ (5.293)

We can now compare the results with the formula derived in 5.7. The formulae (5.197) and (5.291) differ in the $O(\varepsilon^2)$ coefficient, and these are related, assuming $-k = |k|e^{-i\pi}$ (as is required: see the comment following (5.188) and its accompanying footnote), by

$$B_2 = H_2 + C,$$ (5.294)

where the transform of C is defined in (5.200). The difference between the two versions of the theory lies in the way in which the Reynolds stress terms are treated when they occur at second order. Since it is the second order terms which provide the instability, we see that the matter of their computation is of some importance. The difference presumably arises because Jackson and Hunt do not make explicit their assumption on the Reynolds stress away from the boundary.

The Herrmann Model   The principal exponent of dune modelling is Hans Herrmann, and there is also a thriving French school under the aegis of Bruno Andreotti. The basis of the Herrmann approach is in the papers by Sauermann et al. (2001) and Kroy et al. (2002a, 2002b), which last is simply a more complete exposition of their earlier paper. The Herrmann model is essentially an Exner–Hunt model, that is to say that the Exner model $s_t + q_x = 0$ is combined with a Bagnold-type transport law $q = q_0(\tau)$, in which a lag is included to represent the finite acceleration of the transport, thus, essentially,

$$l_s \frac{\partial q}{\partial x} = q_0 - q, \tag{5.295}$$

and finally the stress is computed using Jackson–Hunt theory. From (5.291), (5.292) and (5.293), we can write the transform of the stress perturbation, $\tau_1 = \tau - 1$, in the form (assuming $k = |k|e^{-i\pi}$ when $k < 0$)

$$\hat{\tau}_1 = \varepsilon (A|k| + i Bk)\hat{s}, \tag{5.296}$$

where

$$A = 2\left[1 + \frac{\varepsilon}{\kappa}\left\{2\ln\left(\frac{|k|}{2\kappa\varepsilon}\right) + 4\gamma + 1\right\}\right],$$
$$B = \frac{2\pi\varepsilon}{\kappa}. \tag{5.297}$$

Kroy et al. (2002b) give the same formula (5.296) (their Eq. (12), the extra ε arising when their formula is made dimensionless), but their definitions of A and B are not quite the same, although also based on the Hunt formula. The values are similar though; based on values $|k| = 1$, $\varepsilon = 0.03$, $\kappa = 0.4$ corresponding to $\frac{d}{z_0} = 0.6 \times 10^6$, we calculate $A = 3.6$, $B = 0.47$, compared to the typical Kroy values $A \approx 4$, $B \approx 0.25$.

The linearised Herrmann model for the transforms of the perturbed variables takes the form (cf. (5.201), (5.295) and (5.296))

$$\mu i k\hat{q} = q_0'\hat{\tau} - \hat{q},$$
$$\varepsilon\hat{s}_t + ik\hat{q} = 0, \tag{5.298}$$
$$\hat{\tau} = \varepsilon(A|k| + i Bk)\hat{s},$$

where the relaxation length parameter μ is

$$\mu = \frac{l_s}{d}, \tag{5.299}$$

and is small. With $\hat{s} \propto e^{\sigma t}$, we obtain

$$\sigma = r - ikc = \frac{-ikq_0'(A|k| + i Bk)}{1 + \mu ik}, \tag{5.300}$$

and thus the growth rate is

$$r = \frac{q_0'k^2(B - \mu A|k|)}{1 + \mu^2 k^2}, \tag{5.301}$$

and the wave speed[12] is

$$c = \frac{q_0'(A|k| + \mu Bk^2)}{1 + \mu^2 k^2}. \tag{5.302}$$

---

[12]The wave speed is $-\operatorname{Im}\sigma/k$ here because the Fourier transform is defined with e^{-ikx}.

Fluvial Versus Aeolian?   The Herrmann version of the theory is very attractive because the relaxation length causes the growth rate to become negative at large wave number. This is likely relevant for aeolian dunes, but less relevant for fluvial dunes, where one might expect μ to be tiny. However, the instability relies on the parameter $B > 0$, and if the downslope term in (5.224) is included, then the definition of B in $(5.297)_2$ is modified to

$$B = \frac{2\pi\varepsilon}{\kappa} - \frac{\hat{\beta}}{\varepsilon}, \tag{5.303}$$

indicating $B < 0$ and stability. The constant eddy viscosity (Benjamin) model does not suffer this defect because then the growth rate is proportional to $k^{4/3}$. On the other hand, we expect the Hunt theory to be more accurate.

There is thus a conundrum in how the models are designed. In aeolian bed transport, the sand grains are transported by saltation in a layer of tens of centimetres depth. It is likely to be the case that this finite thickness has a quantitative effect on the application of the Hunt theory. In addition, the rôle of the downslope term may become essentially irrelevant, if the transport is largely by saltation. Equally, the relaxation length is likely to be important. Kroy et al.'s estimate is $l_s \sim 1$–2 m, and thus $\mu \sim 0.002$. With B being relatively small, the maximal growth rate from (5.301) occurs at $k \sim \frac{2B}{3\mu A}$, corresponding to a wavelength of 300 m, if we take $d = 1,000$ m, $A = 4$, $B = 0.5$, $\mu = 0.002$.

It is not so obvious that the same will be true in fluvial transport. The thickness of the bedload layer is only a few grain diameters, and the relaxation length is likely to be very small. The downslope component of the effective shear stress may be important, and as we have seen, this also provides a stabilising (diffusive) effect. In this case, it is difficult to see how the Hunt model can produce instability.

Separation   The principal difficulty in applying the Jackson–Hunt theory (or indeed any theory) to dune formation lies in the tacit assumption that the flow is attached, and this is almost never the case in practice. Measurements of separated flow have been made by Vosper et al. (2002); numerical computations indicating separation have been made by Parsons et al. (2004), and attempts to model similar flows have been made by O'Malley et al. (1991), and also Cocks (2005), whose work on a complex variable method is described in Sect. 5.8.1. However, the complex variable approach is unwieldy, and in any case not suitable for three-dimensional calculations.

The approach used by Herrmann and his co-workers is to get around this in a plausible but heuristic way. When the lee side slope exceeds $14°$, then separation occurs, and they carry on the calculation by fitting a cubic function for the separation bubble roof. Since a cubic is defined by four parameters, but also the point of reattachment is unknown, this allows Kroy et al. (2002b) to specify five conditions; these are continuity of interface and its slope at the end points, together with a specification that the maximum (negative) slope of the bubble roof is $14°$ (their Eq. (27)). Towards the beginning of the same paragraph, they also say that they require the curvature of the bed to be continuous; indeed, this ensures that the basal stress is continuous at the detachment point, and thus that separation occurs when $\tau = 0$, since the shear stress is zero in the bubble, but it is not clear whether their prescription satisfies this condition.

Insofar as one wants to solve a separation problem in which the shear stress is zero at the bubble roof, there are two apparent problems with the Herrmann approach. The first is that the calculation of the shear stress via (5.291), for example, involves the assumption of a no slip condition, as opposed to a no stress condition. Kroy et al. recognise this (after their Eq. (26)), but think that 'the corresponding errors are expected to be small', although why this should be so is not clear. One might in fact expect the errors to be large. The second problem is that if the bubble roof s is chosen in a prescribed way, there is no particular reason to suppose that the shear stress thus calculated will actually equal zero.

Despite these misgivings, the utilisation of this model gives strikingly interesting results. Schwämmle and Herrmann (2004) studied transverse dunes, Parteli et al. (2007) studied barchan dunes, and Parteli et al. (2009) studied seif dunes. Durán and Herrmann (2006) studied the transition from barchans to parabolic dunes under the effect of vegetation. The computational results which they show are impressive, perhaps suggesting that the details of the model are not that important.

More recently, Fowler et al. (2011) have adopted a different strategy. They use a constant eddy viscosity approach, which leads to the Exner equation

$$\frac{\partial s}{\partial t} + \frac{\partial q}{\partial x} = 0, \qquad (5.304)$$

with $q = q(\tau_e)$, τ_e being an effective basal stress defined by

$$\tau_e = \tau - \beta s_x, \qquad \tau = 1 - s + K * s_x \qquad (5.305)$$

(cf. (5.103)), and they allow a stabilising down slope coefficient β. Numerical solutions of this equation show that τ reaches zero, signalling the onset of separation. Thereafter, the Exner equation becomes redundant in the separation bubble, and is replaced by $\tau = 0$. *Providing* we assume the same formula for the stress applies when there is separation, a convenient mathematical way of formulating the problem in this case is to separately compute the sand bed $z = b(x, t)$ together with the air flow base $z = s(x, t)$ (i.e., s is the sand surface except in the separation bubble, where it is the roof of the bubble). We then solve the pair of equations

$$\begin{aligned} s_t + q_x &= M, \\ b_t + q_x &= 0, \end{aligned} \qquad (5.306)$$

with q given as a function of τ_e, and M to be chosen. For small s, we can approximate q by

$$q \approx q(\tau) - D s_x, \qquad (5.307)$$

where the diffusion coefficient D is

$$D = \beta q'(\tau), \qquad (5.308)$$

and this is more convenient for numerical purposes.

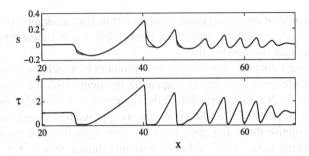

Fig. 5.16 Snapshot of the travelling dune system of Fig. 5.12 at time $t = 2$, found by solving (5.306) and (5.307), using $q = [\tau]_+^{3/2}$ and where $D = 4.3$ is constant, and M is given by (5.309), with $\Lambda_a = 400$ and $\Lambda_s = 20$. The *upper curve* is s, and, where distinct, the *lower* is the sand surface b. Figure courtesy of Mark McGuinness

The choice of M is motivated by the fact that we should have $M = 0$ when $s = b$ and $\tau > 0$, but M is indeterminate when $s > b$ and $\tau = 0$. A suitable computational choice is to define

$$M = \begin{cases} -\Lambda_a(s - b) & \text{if } \tau > 0, \\ -\Lambda_s \tau & \text{if } \tau < 0, \end{cases} \qquad (5.309)$$

where the values of Λ_i are chosen to be large. Since $(s - b)_t = -\Lambda_a(s - b)$ when $\tau > 0$, this forces the air flow to remain attached to the sand surface, while if τ starts to become negative, 'fake' sand is artificially produced to inflate s so as to keep $\tau \approx 0$. (We will find a similar strategy to this bears fruit when modelling drumlin formation in Chap. 10.)

Figure 5.16 shows the result of a computation with this model, corresponding to the evolution from an initial disturbance, as shown in Fig. 5.12. In this figure, we have taken the diffusion coefficient to be a constant, which aids numerical computation. However, this choice allows the stationary sand inside the air bubble to diffuse. More realistically, since $D \to 0$ when $\tau \to 0$, the bubble sand will steepen to form a shock at the lee of the dune, but this itself does not occur in practice because a gradient steeper than about $34°$ cannot be obtained. We can model the resulting slip face by allowing the diffusion coefficient to increase without bound as $-s_x$ approaches the critical slope $S_c = \tan 34° \approx 0.67$, for example by allowing $\beta \to \infty$ as $-s_x \to S_c$. This is awkward to arrange, and largely cosmetic, so long as the diffusing sand in the air bubble does not reach the downstream end of the bubble.

The shapes of the bubbles are also not very realistic, but this may be due to the incorrect calculation of the shear stress in the presence of separation. This is the second difficulty, which no model has yet addressed: the issue of prescribing the shear stress when there is separation. The simplest situation is the constant viscosity model, in which there will now be ordinary Blasius boundary layers which join the attached flow to an outer flow in which there is slip past the boundary. The degree of slip must be calculated as a consistency condition with the boundary layer solution.

This has yet to be done, but the structure of the resulting dune theory is likely to be very different.

5.10 Exercises

5.1 Just as the straightforward St. Venant model is unable to predict the occurrence of transverse dunes, it is also apparently unable to produce lateral bars; at least, this is suggested by the following example.

Show that a two-dimensional form of the St. Venant equations describing flow in a stream of constant width, which allows for downslope sediment transport, can be written in the dimensionless form

$$s_t + \nabla . \mathbf{q} = 0,$$

$$\mathbf{q} = \frac{q(\tau_e)}{\tau_e} \boldsymbol{\tau}_e,$$

$$\boldsymbol{\tau}_e = |\mathbf{u}|\mathbf{u} - \beta \nabla s,$$

$$\varepsilon h_t + \nabla . (h\mathbf{u}) = 0,$$

$$F^2 \left[\varepsilon \mathbf{u}_t + (\mathbf{u}.\nabla)\mathbf{u} \right] = -\nabla \eta + \delta \left(\mathbf{i} - \frac{|\mathbf{u}|\mathbf{u}}{h} \right),$$

$$h = \eta - s.$$

Assume that $\beta \sim O(1)$, $F \sim O(1)$, $\delta \ll 1$, and $\varepsilon \ll 1$. Suppose also that the cross stream width $y \sim v \ll 1$. Show that it is appropriate to rescale the transverse velocity v (i.e., $\mathbf{u} = (u, v)$) as $v \sim v \ll 1$, and then also $s \sim v^2$ and $t \sim v^2$. Assuming that $\varepsilon \ll v^2$ and that $q = \tau_e^{3/2}$, show that a consistent approximate rescaled model is

$$\frac{\partial s}{\partial t} + \frac{\partial (u^3)}{\partial x} + \frac{\partial (u^2 v)}{\partial y} = \beta \frac{\partial}{\partial y} \left(u \frac{\partial s}{\partial y} \right),$$

$$\frac{\partial (hu)}{\partial x} + \frac{\partial (hv)}{\partial y} = 0,$$

$$F^2 (uu_x + vu_y) + \eta_x = 0,$$

and that $\eta \approx \eta(x, t)$, $h \approx \eta$. Deduce that s satisfies the equation

$$\frac{\partial s}{\partial t} = \left(\frac{2u}{F^2} + \frac{u^3}{h} \right) h_x + \beta \frac{\partial}{\partial y} \left(u \frac{\partial s}{\partial y} \right).$$

For small perturbations to the uniform state $h = u = 1$, $s = 0$, show that $F^2 uu_x + h_x \approx 0$ in a linearised approximation, and deduce that $u \approx u(x, t)$. Show that then s relaxes to a steady state, and by considering suitable boundary conditions at the stream margins, show that in fact $h_x = 0$, and hence the uniform state is stable.

Now suppose that the stream is not supposed narrow, so that the rescaling with v is not done. Show that for sufficiently small spanwise perturbations such that we can still take $|\mathbf{u}| \approx u$, $\tau_e \approx u^2$ and $q \approx u^3$, the model may be reduced to

$$\frac{\partial s}{\partial t} + \frac{\partial (u^3)}{\partial x} + \frac{\partial (u^2 v)}{\partial y} = \beta \left[\frac{\partial}{\partial x} \left(u \frac{\partial s}{\partial x} \right) + \frac{\partial}{\partial y} \left(u \frac{\partial s}{\partial y} \right) \right],$$

$$\frac{\partial (hu)}{\partial x} + \frac{\partial (hv)}{\partial y} = 0,$$

$$F^2 (uu_x + vu_y) + h_x + s_x = 0,$$

$$F^2 (uv_x + vv_y) + h_y + s_y = 0.$$

By linearising about the uniform state, show that perturbations proportional to (the real part of) $\exp[\sigma t + ik_1 x + ik_2 y]$ have a growth rate determined by

$$\sigma = -\beta k^2 - \frac{ik_1 (3k_1^2 + k_2^2)}{k_2^2 + (1 - F^2)k_1^2},$$

where $k^2 = k_1^2 + k_2^2$. Deduce that perturbations take the form of decaying travelling waves, and comment on the direction of propagation for purely longitudinal and purely transverse waveforms.

5.2 Write down the Exner equation for bedload transport, and show how it can be used to study the onset of bedform instability, assuming a suitable bedload transport law. Show that in conditions of slow flow, the resultant equation for the bed profile $s(x, t)$ is a first order hyperbolic equation, and deduce that the profile is neutrally stable. Show also that bed waves will form shocks which propagate downstream.

Now suppose that the bedload transport $q_b(x, t)$ is a function of the basal stress τ evaluated at $x - \delta$. Show that instability can occur if $\delta < 0$, i.e., the stress *leads* the bed profile.

Can you think of a physical reason why such a lead should occur?

Do you think such a model would be a good *nonlinear* model?

5.3 The Kennedy model for dune growth leads to the dispersion relation

$$\sigma \left[(\sigma + ikU)^2 + gk \tanh kh \right]$$
$$+ (\sigma + ikU)kq' e^{-ik\delta} \left[(\sigma + ikU)^2 \tanh kh + gk \right] = 0,$$

where σ is the growth rate and k is the wave number.

Show that if q' is small, then $\sigma \approx -ikc_\pm$, where

$$c_\pm = U \pm \sqrt{\frac{g}{k} \tanh kh}.$$

Use this result to show, by considering a correction to this approximate value, that

$$\mathrm{Re}\,\sigma \approx -\frac{gkq' \sin k\delta \, \mathrm{sech}^2 kh}{2c_\pm},$$

and deduce that forwards travelling waves are (weakly) unstable if $\sin k\delta < 0$.

5.4 In Reynolds's model of dune formation, the bed elevation is s, the surface
elevation is η, the water speed is u, and the sediment flux is q, and these are
related by the equations

$$\eta = 1 + \frac{1}{2}(1 - u^2),$$

$$s = \eta - \frac{1}{u},$$

$$u = q^{1/3},$$

and q satisfies the Exner equation in the form

$$q_t + v(q)q_x = 0,$$

where the wave speed $v(q) = \frac{dq}{ds}$.
   Show that

$$v(q) = \frac{3q^{4/3}}{1 - F^2 q},$$

and deduce that for perturbations to the steady state $q = 1$, waves propagate
forwards if $F < 1$ and backwards if $F > 1$. By consideration of $v'(q)$, show
also that for $F < 1$, waves will form forward-facing shocks in q and thus also
s, while if $1 < F < 2$, waves form backward-facing shocks as elevations in s
(and η).
   What happens in this model if $F > 2$? What do you think would happen in
practice?

5.5 In a model of dune formation, the sediment concentration c and bed height s
are modelled by the equations

$$\frac{\partial}{\partial t}(hc) + \frac{\partial}{\partial x}(hcu) = \rho_s(v_E - v_D),$$

$$(1 - n)\frac{\partial s}{\partial t} = -(v_E - v_D),$$

where h is fluid depth, u is mean fluid velocity, ρ_s is sediment density, n is
bed porosity, and v_E and v_D are erosion and deposition rates. Parker (1978)
suggests the following expressions for the erosion and deposition rates in a
stream:

$$v_E = \frac{\beta u_*^3}{v_s^2}, \qquad v_D = \frac{\gamma v_s^2 c}{\rho_s u_*},$$

where c is the sediment concentration (mass per unit volume), v_s is the set-
tling velocity, u_* is the friction velocity $(\tau/\rho_w)^{1/2}$, and β and γ are constants
(≈ 0.007 and 13, respectively).
   Consider the two cases where (i) the surface $\eta = h + s$ is flat, and $\eta = h_0$
is constant; and (ii) where the surface is determined by a local force balance,
thus

$$\tau = \rho_w g h(S - \eta_x),$$

where ρ_w is water density, g is gravity, and S is bed slope.

Assuming $\tau = f\rho_w u^2$ and $uh = q$ is constant, find appropriate scales for x, t and c in cases (i) and (ii) if $h, \eta, s \sim h_0$ and q is the fluid flux per unit width. Hence derive the dimensionless model for slow flow

$$\varepsilon \frac{\partial}{\partial t}(hc) + \frac{\partial c}{\partial x} = \frac{1}{h^3} - ch = -\frac{\partial s}{\partial t},$$

where

$$\varepsilon = \frac{c_0}{\rho_s(1-n)}.$$

Show that in case (i), $h = 1 - s$, while in case (ii),

$$\frac{1}{h^3} = 1 - \Lambda \eta_x,$$

and define the parameter Λ in the second case. By analysing the stability of the basic state $h = c = \eta = 1$, show that, for ε small, the steady state is stable. Show that in case (i), waves propagate downstream, but in case (ii), they can propagate upstream if Λ is small enough.

More generally, derive a stability criterion in case (i) (when ε is small) if $v_E = E(h)$, $v_D = cV(h)$. How is the result affected if ε is *not* small?

5.6 A simple model of bed erosion based on the St. Venant equations can be written in dimensionless form as

$$\varepsilon h_t + (uh)_x = 0,$$

$$F^2(\varepsilon u_t + uu_x) = -\eta_x + \delta\left(1 - \frac{u^2}{h}\right),$$

$$h(\varepsilon c_t + uc_x) = E(u) - c = -s_t,$$

where $h = \eta - s$. Explain a plausible basis for the derivation of this model. By considering the stability of the steady state $u = h = c = 1$ on a time scale t of $O(1)$, and assuming that $\delta \ll 1$, $\varepsilon \ll 1$, show that instability can occur depending on the size of $E'(1)$. Show also that η and s are out of phase if $F < 1$, and in phase if $F > 1$; interpret this in terms of dune and anti-dune formation.

5.7 The Exner equation for bed evolution is written in the form

$$(1 - n)s_t + q_x = 0,$$

and the bedload transport is given by

$$q_x = K[q_0(\tau) - q],$$

where τ is the bed shear stress.

Explain in physical terms why such an equation should be appropriate to describe bedload transport.

Suppose it is assumed that the depth of the flow h is constant. Show that if the bed stress is $\tau = f\rho u^2$, then the momentum equation of St. Venant can be written in the approximate form

$$\frac{h}{2f}\tau_x + \tau = \rho g h(S - s_x).$$

Show how to non-dimensionalise these equations to obtain the set

$$s_t + q_x = 0,$$
$$\delta q_x = q_0(\tau) - q,$$
$$\tau_x + \tau = 1 - s_x,$$

and identify the parameter δ.

Write down a suitable steady state solution, and show that if $q_0(\tau)$ is a monotonically increasing function of τ, then the steady state is linearly unstable if $K > 0$. Show also that the corresponding waves move upstream. Show that the growth rate remains positive as the wave number $k \to \infty$. [*This is an indication of ill-posedness.*]

For what values of the Froude number might the assumption of constant depth be valid?

5.8 Suppose that

$$s = s(u) = \frac{1}{2}F^2(1 - u^2) + 1 - \frac{1}{u},$$

and that

$$s'(u)\frac{\partial u}{\partial t} = cD^*(u) - E^*(u) = -\frac{\partial c}{\partial x}.$$

Assume $D^* = 1$ and $E^* = u^3$. Simplify the equations to the form

$$\frac{\partial u}{\partial t} = f(u, c), \qquad \frac{\partial c}{\partial x} = g(u, c),$$

giving expressions for f and g.

Suppose that $c = 1$ at $x = 0$ and $u = u_0(x)$ at $t = 0$, and that $u_0(\infty) = 1$. Derive an ordinary differential equation for $U(t) = u(0, t)$ in the form $\frac{dU}{dt} = h(U)$, and by consideration of the graphical form of $h(U)$ in the two cases $F < 1$ and $F > 1$, determine the behaviour of U for $t > 0$, explaining in particular how it depends on $U(0)$.

Why is it inadvisable to prescribe $c \to 1$ as $x \to \infty$ instead of the boundary condition at $x = 0$?

5.9 Show that, if $\nu > 0$,

$$\int_0^\infty \theta^{\nu-1} e^{i\theta} \, d\theta = \Gamma(\nu) \exp\left(\frac{i\pi\nu}{2}\right).$$

Hence, if

$$K(\eta) = \begin{cases} \eta^{\nu-1}, & \eta > 0, \\ 0, & \eta < 0, \end{cases}$$

where $0 < \nu < 1$, show that the Fourier transform, defined here as

$$\hat{K}(k) = \int_{-\infty}^\infty K(\eta) e^{-ik\eta} \, d\eta,$$

is given by

$$\hat{K}(k) = \frac{\Gamma(\nu)\exp\left(-\frac{i\pi\nu\operatorname{sgn}k}{2}\right)}{2\pi|k|^\nu}$$

for real values of k.

Now suppose that ϕ satisfies the evolution equation

$$\phi_\tau + \frac{\partial}{\partial\xi}\left[\frac{1}{2}\phi^2 + \alpha K * \phi_\xi - \phi_\xi\right] = 0,$$

where $f * g$ denotes the Fourier convolution of f and g and α is small. Show that the steady state $\phi = 0$ is linearly unstable, and find the wave number of the maximum growth rate.

When this equation is solved numerically, coarsening occurs, with the wavelength of the bedforms increasing with time. Show that if $\xi \sim L \gg 1$, the equation can be approximated by Burgers' equation with small diffusivity. Hence explain the way in which coarsening occurs.

5.10 Suppose that

$$\psi(t) = \sum_{r=0}^{\infty} f_r t^r = (1 - \xi_a t)^{-1/2}(1 - \xi_b t)^{-1/2},$$

and it is desired to calculate the coefficients f_r numerically. By consideration of the power series for ψ^2 (or otherwise!) show that an iterative recipe for f_n is

$$f_0 = 1,$$

$$2f_n = \left(\frac{\xi_a^{n+1} - \xi_b^{n+1}}{\xi_a - \xi_b}\right) - \sum_{s=1}^{n-1} f_s f_{n-s}.$$

5.11 A shear flow is described by the dimensional equation

$$\tau = \rho\left(\nu + \kappa^2 z^2\left|\frac{\partial u}{\partial z}\right|\right)\frac{\partial u}{\partial z},$$

where ν is the kinematic viscosity. Show that a suitable dimensionless form is

$$\tau = \left(\varepsilon + \kappa^2 z^2\left|\frac{\partial u}{\partial z}\right|\right)\frac{\partial u}{\partial z},$$

where $\varepsilon = \frac{1}{Re}$, and Re is the Reynolds number. Use the method of strained coordinates[13] (i.e., write $z = s + \varepsilon z_1(s) + \cdots$, $u \sim u_0(s) + \varepsilon u_1(s) + \cdots$ to show that

$$u \approx \frac{1}{\kappa}\ln\left(\frac{z + s_0}{s_0}\right),$$

where $s_0 = \frac{\varepsilon}{2\kappa}$ in order to suppress higher singularities in u_1.

---

[13] See, for example, Van Dyke (1975).

Chapter 6
Landscape Evolution

Landscape is one of the most obvious features of the environment we live in. Rolling countryside, rugged cliffs, wind-swept mountains; the scenery which surrounds us is formed by topography, and the shapes which the Earth's surface is moulded into forms the subject of geomorphology.

Topography is created by tectonic processes, and removed by erosion. The most striking example on the planet today is the continuing formation of the Himalayas, as India continues to crash into Asia after its separation from Antarctica during the break up of Gondwanaland over a 100 million years ago. This continental collision buckles the Earth's crust and forms mountain belts; in the case of the Himalayas, the mountains rise to a height above sea level of nearly nine kilometres, and they are still rising.[1]

Creation of mountains creates *hillslope*, and precipitation and its subsequent runoff provides an erosive mechanism. At low elevations, this is through a variety of actions of running water: slow processes such as rainsplash and sheetwash, which carry surficial sediments towards a developed stream system, in which more rapid evacuation of sediments occurs. At higher elevations, precipitation as snow may form glaciers, which similarly carve the landscape and erode it. Typical rates of tectonic uplift and erosion are comparable, and very slow in stable, vegetated landscapes: a hundred metres per million years (100 m Ma^{-1}) is a typical rate for each. In unvegetated landscapes such as badlands, erosion and gully formation can be much more rapid.

6.1 Weathering

A necessary pre-condition for sediment removal is the break up of the basement rock into pieces (boulders, cobbles, gravel, silt, clay: sediments of varying grain

---

[1]The formation of the Himalayan mountains is actually more complicated than that which would arise from a simple buckling mechanism. Geological investigations indicate that the Himalayas are formed by a backflow of partially molten rock driven by its own buoyancy; the partial melting occurs in the subducting Indian crust due to extremely high levels of uranium concentration.

A. Fowler, *Mathematical Geoscience*, Interdisciplinary Applied Mathematics 36,
DOI 10.1007/978-0-85729-721-1_6, © Springer-Verlag London Limited 2011

sizes), and this is achieved through a variety of mechanisms. Mechanical weathering refers to the physical disruption of rock. For example, joints and fractures are natural consequences of rock formation; igneous rocks may have shrinkage cracks as they cool, sedimentary rocks have bedding planes and faults formed through failure in tectonic compression. When exposed at the surface, these are subjected to frost action and other thermally induced stresses, which break up the rock, and form an unconsolidated *regolith*.

Chemical weathering refers to a variety of chemical effects which have similar effects. Over long time scales, rain water will react with rocks. Feldspars break down into clays (very fine grained particles) in this way, and carbonate rocks are dissolved by weakly acidic rainwater. This is particularly noticeable in limestone, where dissolution of the rocks can lead to the formation of spectacular underground cave systems as found in *karst* regions. Chemical weathering can also form sediment by attacking intergranular cements.

The eventual result of weathering is the rotting of rock to a decayed state called *saprolite*. Further evolution of the regolith, for example, by the effects of vegetation or through transport, leads to the formation of soils. It is the sediments of the regolith which are subject to erosion through fluvial or glacial transport.

6.2 The Erosional Cycle

Our purpose in this chapter is to describe how the processes of tectonic uplift and subsequent erosion can lead to the formation of the topographic patterns such as that shown in Fig. 6.1, in which we see a typically dendritic pattern of stream channels draining a catchment area.

The framework in which we do this is through a model which describes the evolution of hillslope through the balance of the two processes, uplift and erosion. More or less, this model will be

$$\frac{\partial s}{\partial t} = U - v_A, \tag{6.1}$$

where s is the vertical height of the topography, U is the rate of uplift, and v_A is the rate of abrasion, or erosion, of the hillslope. Our particular approach will be to study first the steady state of this equation, in which uplift balances erosion, and then to seek conditions under which a uniform slope is unstable, and under which the typical pattern of a drainage system can develop.

However, it should be pointed out that this assumption is rather an idealisation, possible but not necessary. Another framework is the cycle of erosion described in the latter part of the nineteenth century by W. M. Davis, thus called the Davisian cycle of erosion.

Davis's idea was that uplift was a relatively sudden thing, so that topography evolved from an initial elevated stage. In a young landscape, slopes are steep and stream erosion is strong, but as the landscape ages, slopes become gentler and erosion is less dramatic. The next orogenic event initiates the next cycle, and so on.

Fig. 6.1 Hillslope topography. Photograph courtesy of Gary Parker

From the modelling point of view, this framework differs from the notion of continual uplift and erosion only in detail, and there is no serious problem in using it as a starting point. The mathematical difficulty it raises is that instability and stream formation then occur on a slowly developing background state, so that an appropriate frozen time hypothesis is required in order to determine stability criteria. Because of the disparity of time scales between stream formation and hillslope erosion, this is in fact hardly an issue.

6.3 River Networks

River networks are (approximate) fractals.[2] That this is so may be inferred from a number of power-law relationships satisfied by river systems. One such is Hack's law, which relates the length l of a stream to its drainage basin area A, as $l \sim A^{0.6}$.

Other power-law relationships are exhibited by quantities associated with the Horton–Strahler ordering system, for example between the number of streams and their order, which is an algorithmic measure of the stream size and importance within the network.

---

[2]A fractal set is one whose dimension is non-integral. Curves have dimension one, areas have dimension two. A way of characterising the dimension of a set is to count the number of boxes N of size ε required to cover the set. If $N \sim \varepsilon^{-D}$ as $\varepsilon \to 0$, then the set has fractal dimension D. This is consistent with our intuitive sense of dimension, but also allows the calculation of non-integer dimension for such exotica as the Koch snowflake, the Sierpinski gasket, and so on. Fractal sets typically exhibit power-law relationships in their description.

Fig. 6.2 Geometry of
overland flow

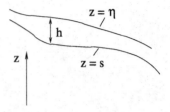

A question of fundamental philosophical significance is to explain how fractal
sets can arise as the response of a continuous, deterministic system to constant forcing. In the study of chaotic (time-dependent) systems, the answer to this is known.
Strange attractors have fractal dimension, and this is because the systems can be
characterised by the action of (Poincaré return) maps, which generate the Cantor
sets of the attractor by a continued process of splitting and mixing. It is less obvious
how such a process might work in generating spatially fractal sets, and this appears
to remain as a challenge for nonlinear dynamicists.

6.4 Denudation Models

The basic structure of erosional theories for channel formation is typically that of a
shallow water (St. Venant) model of overland flow, coupled to an Exner type equation describing hillslope erosion. If we denote the vertical coordinate as z, and the
position of the free water surface and hillslope as η and s, respectively (see Fig. 6.2),
then a St. Venant model can be written as

$$h_t + \nabla.(h\mathbf{u}) = r,$$
$$\mathbf{u}_t + (\mathbf{u}.\nabla)\mathbf{u} = -g\nabla\eta - \frac{f|\mathbf{u}|\mathbf{u}}{h}. \tag{6.2}$$

These equations represent conservation of mass and momentum, and $\mathbf{u}$ is the mean
horizontal (or along slope) velocity. The source term r represents rainfall. The water
depth h is given by

$$h = \eta - s. \tag{6.3}$$

The Exner equation to model the evolution of the hillslope s is usually taken in
the form

$$\rho_s(1 - \phi)s_t + \nabla.\mathbf{q}_b = [-\rho_s v_E + \rho_s v_D] \quad [+\rho_s(1 - \phi)U], \tag{6.4}$$

wherein ϕ is the bed porosity, ρ_s is the sediment density, $\mathbf{q}_b$ represents bedload
transport, and v_E and v_D, if present, represent erosion and deposition of sediments
(measured as velocities: see (5.44)); U would represent tectonic uplift, also measured as a velocity. The erosion and deposition terms arise when sediment can be
carried as suspended load, which is assisted by rapid flow and small particles, and
thus may be relevant in stormflow conditions. In this case the mean concentration c
of suspended sediment is given by

$$(hc)_t + \nabla.(hc\mathbf{u}) = \rho_s v_E - \rho_s v_D. \tag{6.5}$$

v_E and v_D have to be prescribed, as discussed in Chap. 5.

A possible point of confusion which arises when we study models of this type is that some of them include down-cutting or abrasion of the landscape, while others do not, allowing sediment to move from place to place without actually being excavated: do we use (6.2) or (6.4) to evolve the hillslope? The source of this confusion lies in the way in which the bedload transport is described. Conventionally, there is transport but no bedload density, and this masks a conceptual awkwardness in the models. At the outset, we address this awkwardness by modifying (6.4) in the following way.

We suppose that s describes the interface between stationary bed and moving bedload, and we define a bedload density ρ^b and bedload transport $\mathbf{q}_b$, so that

$$\rho_t^b + \nabla . \mathbf{q}_b = -\rho_s v_E + \rho_s v_D + \rho_s (1 - \phi) v_A, \tag{6.6}$$

where v_A is the abrasion or entrainment rate of the bed, measured as a velocity. The bedload density has units of mass per unit area, and is conveniently written in the form

$$\rho^b = \rho_s (1 - \phi) a, \tag{6.7}$$

where a is a length, and can be interpreted as the depth of the active deforming bedload layer. Finally, the Exner equation is

$$\rho_s (1 - \phi) s_t = -\rho_s (1 - \phi) v_A \quad \left[+ \rho_s (1 - \phi) U \right]. \tag{6.8}$$

The earlier model is regained if $\rho^b \to 0$ (or $a \to 0$), which is consistent with the idea of bedload being a surface density. Otherwise, v_A needs to be prescribed, in which case it is natural to suppose it is a function of active depth a; it is this quantity v_A which is really the down-cutting or excavation rate. It remains to be seen how small a actually has to be in order that it be negligible; we will come back to this later. Note that if $a > 0$, then (6.3) must be modified to read

$$\eta = s + a + h. \tag{6.9}$$

6.4.1 Sediment Transport

These equations, or generalisations of them, can be used to study some forms of morphological instability (dunes, braiding, meandering, etc.), and this formed the subject of Chap. 5. They also allow a description of the erosional instability mechanism which is responsible for the formation of channels. The efficacy of this depends on the chosen forms for bedload transport and bed erosion terms. The basic instability mechanism is less subtle than the stress shift analysed in Chap. 5, and is due simply to the fact that a locally increased depth h causes increased flow $\mathbf{u}$, and thus increased erosion, which in turn allows a further increase in h.

The constituents of this mechanism involve prescription of the sediment transport terms, as described in Chap. 5. Both erosion rate v_E and bedload transport $\mathbf{q}_b$ depend on stress delivered by the water flow. This stress is taken to be

$$\boldsymbol{\tau} = f \rho_w |\mathbf{u}| \mathbf{u}. \tag{6.10}$$

Erosion and bedload are increasing functions of τ, except that allowance needs to be made for the local bed slope (it is easier to move particles downhill than uphill). As discussed in Chap. 5 (see the similar discussion at (5.109)), for a particle of diameter D_s at the bed, the streamflow exerts a force of approximately τD_s^2 on it, and it is this force which causes motion. On a slope, there is an additional force due to gravity, approximately $-\Delta\rho_{sw} g D_s \nabla s$, where $\Delta\rho_{sw}$ is the density difference between particle and fluid. Thus the net stress causing motion is actually

$$\tau_e = \tau - \Delta\rho_{sw} g D_s \nabla s. \tag{6.11}$$

This slope effect will turn out to have a (crucial) stabilising effect in the equation for s.

If the stress is below a critical yield stress, then no motion occurs, and both v_E and q_b will be zero. The relevant dimensionless quantity is the Shields stress,

$$\mu = \frac{\tau_e}{\Delta\rho_{sw} g D_s}, \tag{6.12}$$

and sediment transport or erosion occurs if $\mu > \mu_c \approx 0.05$. The critical Shields stress depends somewhat on particle size (through the particle Reynolds number), but can reasonably be approximated as a constant. Denoting the dimensional yield stress as

$$\tau_c = \mu_c \Delta\rho_{sw} g D_s, \tag{6.13}$$

then typical assumptions for bedload transport are the Meyer-Peter and Müller relationship

$$q_b = \left(\frac{\rho_s K}{\rho_w^{1/2} \Delta\rho_{sw} g}\right)(\tau_e - \tau_c)_+^{3/2}, \tag{6.14}$$

where Meyer-Peter and Müller chose values of $K = 8$ and $\mu_c = 0.047$. For bed erosion, the Van Rijn relationship is similar:

$$v_E \propto (\tau_e - \tau_c)_+^{3/2}. \tag{6.15}$$

6.4.2 Non-dimensionalisation

The variables of the model are water depth h, water velocity $\mathbf{u}$, water surface η, suspended load c, active bedload depth a, bed surface s, effective shear stress τ_e, and bedload transport q_b, which is assumed to be a function of τ_e, such as that given by (6.14). The describing equations of the model are (6.2), (6.9), (6.5), (6.6), (6.7), (6.8) and (6.11).

First we scale the constitutive transport terms. These are the erosional, depositional, and abrasional velocities v_E, v_D, and v_A, and the bedload transport q_b. The erosional and depositional velocities are scaled as in Chap. 5, thus

$$v_E = v_e E, \qquad \rho_s v_D = v_s c D, \tag{6.16}$$

where v_e is a typical erosion rate and v_s is the particle settling velocity, and E and D are dimensionless erosion and deposition rates of $O(1)$, which are functions of mean flow velocity. In addition, we write

$$v_A = U_D A, \qquad \mathbf{q}_b = q_b^D \mathbf{Q}_b, \tag{6.17}$$

where U_D is a typical value of the uplift rate U, and q_b^D is determined by a relation such as that of Meyer-Peter and Müller. Since we suppose that v_A is a function of active depth a, the choice $v_A \sim U$ actually provides a scale for a.

We choose scales for the other eight variables h, $\mathbf{u}$, η, s, c, τ_e, as well as $\mathbf{x}$ and t, by balancing water flux with rainfall, gravitational acceleration with friction, erosion rate with deposition rate, hillslope elevation with water surface elevation, bedload transport with abrasion rate, and hillslope rate of change with both abrasion rate and tectonic uplift (if this is present). If we suppose that d is a suitable hillslope height scale and l is a suitable horizontal length scale, then we specifically choose

$$\eta, s \sim d, \qquad \mathbf{x} \sim l, \qquad t \sim [t] = \frac{d}{U_D}, \qquad \tau_e \sim f\rho_w[u]^2,$$

$$\mathbf{u} \sim [u] = \left(\frac{gr_D d}{f}\right)^{1/3}, \qquad c \sim \frac{\rho_s v_e}{v_s}, \qquad h \sim [h] = l\left(\frac{fr_D^2}{gd}\right)^{1/3}, \tag{6.18}$$

where we take r_D as a typical precipitation rate.

There appears as yet to be nothing to constrain the length and depth scales l and d. The first of these might be determined by the implied tectonic setting. The simplest conceptual idea is the continuing uplift of an island (or a mountain belt), with sea level fixed at prescribed boundaries, and this determines a natural length scale l, the scale of the island. The other scale is fixed by the balance of uplift rate with hillslope denudation, which requires

$$l = \frac{q_b^D}{\rho_s(1 - \phi)U_D}. \tag{6.19}$$

This appears to determine l again, but in fact it determines d through the dependence of q_b^D on τ_e and thus $[u]$. For example, if we take $q_b^D \propto \tau_e^{3/2} \propto [u]^3$, then from (6.18), we have $q_b^D \propto d$, and thus $d \propto U_D l.^3$

Using the scaled variables in the model equations (6.2), (6.3), (6.5), (6.6) and (6.8), we obtain the dimensionless set

$$\delta\varepsilon h_t + \nabla.(h\mathbf{u}) = r,$$

$$\delta F^2\big[\delta\varepsilon\mathbf{u}_t + (\mathbf{u}.\nabla)\mathbf{u}\big] = -\nabla\eta - \frac{|\mathbf{u}|\mathbf{u}}{h},$$

$$\eta = s + \delta h + \delta va,$$

$$\delta\varepsilon(hc)_t + \nabla.(hc\mathbf{u}) = \gamma(E - cD), \tag{6.20}$$

$$\delta va_t + \nabla.\mathbf{Q}_b = -\alpha(E - cD) + A,$$

$$s_t = -A + U,$$

$$\tau_e = |\mathbf{u}|\mathbf{u} - \beta\nabla s,$$

---

[3]More precisely, $\frac{d}{l} \sim (\frac{\Delta\rho_{sw}(1-\phi)}{Kf^{1/2}\rho_w})\frac{U_D}{r_D} \sim \frac{U_D}{r_D}$, so high mountains are (in this theory) a consequence of high uplift rate and low rainfall, which makes intuitive sense.

and we suppose for the moment that

$$\mathbf{Q}_b = f(\tau_e)\mathbf{N}, \tag{6.21}$$

where

$$\mathbf{N} = \frac{\tau_e}{\tau_e}, \tag{6.22}$$

and f is no longer used for the friction factor.[4] The parameters are defined by

$$\varepsilon = \frac{U_D}{r_D}, \qquad \delta = \frac{[h]}{d}, \qquad F^2 = \frac{[u]^2}{g[h]}, \qquad \beta = \frac{\Delta \rho_{sw} D_s}{\rho_w [h]}, \tag{6.23}$$

$$\gamma = \frac{v_s}{r_D}, \qquad v = \frac{[a]}{[h]}, \qquad \alpha = \frac{v_e}{(1-\phi)U_D},$$

$[a]$ being the scale for the active depth a. The definitions of ε and Froude number F here correspond to those used in Chap. 5.

It is easy to get a sense of the size of the various parameters in the model, simply from observation. We have seen in Chap. 5 that $\varepsilon \ll 1$, and this is confirmed in the present case; if, for example, we take $U_D \sim 10^{-3}$ m y^{-1} (1 km per million years) and $r_D \sim 1$ m y^{-1}, then $\varepsilon \sim 10^{-3}$.

The important parameter δ is the ratio of stream depth to hillslope height, and is small. An estimate for $[h]$ based on values $r_D \sim 1$ m y^{-1}, $L \sim 10^5$ m, $f \sim 0.1$, $g \sim 10$ m s^{-2}, $d \sim 10^3$ m, is $[h] \sim 2$ cm, thus $\delta \sim 10^{-5}$, and even in stormflow conditions, a transient depth of as much as a metre yields $\delta \sim 10^{-3}$. Note that for this choice of d and U_D, the (erosional) time scale is a million years.

The velocity scale corresponding to the above depth is 0.2 m s^{-1}, and from this we calculate that the Froude number is about 0.4. This is quite vigorous, indicative of the implicit assumption of relatively bare (and thus smooth) ground. Generally, overland flow will have quite small Froude number, except in stormflow conditions; even then, the Froude number does not greatly exceed unity.

The erosional parameters are γ, α and v. Since bedload transport typically occurs over an active depth of at most a few grains, we can safely assume that $v \lesssim 1$.

The particle settling velocity is given by (5.8), which dimensionally gives $v_s \sim [D_s]^2$ m s^{-1} if $D_s = [D_s]$ mm. Thus v_s is typically quite large, and consequently γ is large; a typical estimate is (for particle grain size 1 mm) $\gamma \sim 10^7$. Estimates for v_e are about $10^{-2}v_s$, since typical suspended loads $\rho_s v_e/v_s$ are of order 10 g l^{-1}, and for $U_D \sim 10^{-3}$ m y^{-1}, we have $\alpha \sim 10^8 \gg 1$.

Approximately, $\beta \approx D_s/[h] \lesssim 1$; we will thus generally take $\beta = O(1)$.

In summary then, we estimate

$$\varepsilon, \ \delta \ll 1,$$
$$F^2, \ v, \ \beta \lesssim 1, \tag{6.24}$$
$$\gamma, \ \alpha \gg 1.$$

---

[4]It is an unfortunate feature of applied mathematics that there are not enough letters in the Roman and Greek alphabets, even allowing for capitals, overhats, tildes, asterisks, subscripts and superscripts. Duplication is inevitable, and here, apologetically, we use f for the dimensionless sediment transport function, having just made use of it as the friction factor.

6.4.3 The Issue of Time Scale

The sceptical reader will at this point enjoy some doubts about these 'typical' values. A first obvious point is that for a supposed typical overland flow depth of 2 cm, the turbulent friction law (6.10) may not be appropriate. This is, in fact, easily modified by allowing a laminar component of stress at low Reynolds number, but this is a cosmetic component which does not alter the structure of the model (see also Question 6.2).

A more subtle concern is illustrated by the size of the Froude number. The fact of the matter is that it does not rain all the time, but in occasional showers and storms, and furthermore, we might expect that erosion and bedload transport would hardly occur, except in the most severe storms. Since our erosional time scale is $[t] = d/U \sim 10^6$ y, it seems likely that the principal erosion forming events are such extremely rare storm-induced floods, and that during these, the Froude number may be $O(1)$ or higher. This raises the concern that the estimation of the parameters on the basis of typical 'average' values may be inappropriate.

For example, consider a landscape where is there is an extreme storm for a day every 100 years, and drought at other times (and ignore the fact that this would actually be a desert). In this scenario, water flow and erosion only occur during the storm, and the intervening century is irrelevant for the purpose of calculating water flux and erosion. It then becomes appropriate to use a different time scale (and thus also water depth and velocity scales) to describe the erosive effect of the storm. The evolution of the topography over longer tectonic time scales then occurs through the aggregation of these short time scale discrete erosional events. It is not obvious that averaging in this way (i.e., solve the short time erosional problem and then average over time) yields the same result as the approach we have taken (i.e., average over time and then solve the erosional problem), but we shall proceed on this basis, partly because it may well be appropriate in certain (e.g., temperate) environments, and partly because this is the approach which has implicitly been taken in the literature of the subject. Further consideration of this point is consigned to the exercises.

6.5 Channel-Forming Instability

The basic problem in understanding the development of river networks is the formation of a channel from a uniform overland flow. To see how this can happen, we will begin by considering a simplified version of the model equations (6.20). We assume $\delta \ll 1$, $\varepsilon \ll 1$, $F^2 \lesssim 1$, $v \lesssim 1$, and ignore the corresponding terms in the equations. The last of these (neglect of the term δv in $(6.20)_5$) corresponds to the neglect of the bedload density ρ^b, even in the stormflow case where a significant active layer may be mobilised.

We also make the important assumption of supposing that sediments are coarse, and are transported only as bedload, so that $v_E = 0$ and thus also $c = 0$. This avoids the issue of dealing with the large parameters γ and α (see also Question 6.4). With these approximations, $\eta \approx s$, we apparently do not need to specify A, and the model reduces to

$$\nabla.(h\mathbf{u}) = r,$$
$$|\mathbf{u}|\mathbf{u} = -h\nabla\eta,$$
$$\eta_t = -\nabla.[f\mathbf{N}] + U,$$ (6.25)

where $\mathbf{N}$ is given by (6.22). Putting $\delta = 0$, we get

$$\tau_e \approx -(h + \beta)\nabla\eta.$$ (6.26)

Solving for $\mathbf{u}$ and substituting into the water mass equation, we can write the system (6.25) in the form introduced by Smith and Bretherton:

$$\nabla.[q\mathbf{n}] = r,$$
$$\eta_t = -\nabla.[f\mathbf{n}] + U,$$
$$\mathbf{n} = -\frac{\nabla\eta}{|\nabla\eta|},$$ (6.27)

where the water flux is

$$q = h|\mathbf{u}| = h^{3/2}S^{1/2},$$ (6.28)

the effective stress is

$$\tau_e = (qS)^{2/3} + \beta S,$$ (6.29)

and the magnitude of the slope is

$$S = |\nabla\eta|.$$ (6.30)

The dimensionless Meyer-Peter/Müller relationship, for example, can be written in the form

$$f = \left[(qS)^{2/3} + \beta S - \tau_c^*\right]_+^{3/2},$$ (6.31)

where the dimensionless critical stress is given, using (6.13) and (6.18), by

$$\tau_c^* = \frac{\mu_c \Delta\rho_{sw}}{\rho_w} \frac{D_s}{[h]} \frac{l}{d}.$$ (6.32)

Using estimates $\mu_c = 0.05$, $\Delta\rho_{sw}/\rho_w = 2$, $D_s = 1$ mm, $[h] = 2$ cm, $l = 10^5$ m, $d = 10^3$ m, we have $\tau_c^* \approx 0.5$. We see that (6.27) forms a pair of equations for η and q; the equation for η is essentially parabolic, while that for q is hyperbolic, although the coupling between the equations confuses the situation.

6.5.1 Boundary Conditions

The approximation $\delta \to 0$ which takes us from (6.20) to (6.27) is in fact a singular one. This can be seen through the fact that the approximation of $\mathbf{N}$ by $\mathbf{n}$ involves the loss of a term proportional to $\delta\nabla h$, so that a Laplacian term in h (or equivalently, q)

is missing from (6.27)$_2$. This suggests that we ought to prescribe two conditions on the boundary of the domain. If we consider the uplift of an island continent D with prescribed boundary ∂D, then the natural conditions to apply are

$$\eta = 0 \quad \text{and} \quad \frac{\partial \eta}{\partial n} = 0 \quad \text{on } \partial D. \tag{6.33}$$

These represent the idea that the position of the coastline is known, and that the water surface gradient becomes equal to the ocean gradient (zero) at the coastline.

The loss of the parameter δ precludes us from applying the second boundary condition in (6.33), and in order to do so, we would require a boundary layer of thickness δ at the boundary, over which h will change rapidly in order to satisfy (6.33)$_2$. This is a passive boundary layer and of little consequence. The reduced equation (6.27)$_1$ with $\delta = 0$ is then hyperbolic, and the natural boundary condition on the resulting sub-characteristics is to prescribe q at their upstream end. These occur at summits where $|\nabla s| = 0$, and we expect $q = 0$ there, but since the q equation appears to be degenerate at the summit ($\nabla s / |\nabla s|$ is undefined), it is not entirely obvious whether it is necessary to prescribe q at all. A similar situation applies for a two-dimensional ridge, if the slope is smooth.

A further complication is that it can be expected that the bedload transport function $f(S, q) \to 0$ as both S and $q \to 0$. In particular, since we expect $q \to 0$ at a ridge, this suggests that the diffusion equation (6.27)$_2$ for s is degenerate, so that smoothness may be lost at a ridge, and ∇s may be discontinuous there. In particular, it is not obvious that $S \to 0$ at a ridge. The best that can definitely be said is that the downslope component of both water flux and sediment flux should tend to zero at a ridge or a summit.

6.5.2 Steady State Solution

Our purpose is to study conditions under which uniform overland flow is unstable, and the simplest situation in which we can do this is when the basic steady state flow is one-dimensional. Therefore we consider a basic state consisting of a symmetric hillslope inclining down towards a fixed boundary, as indicated in Fig. 6.3. We take the divide to be at $x = 0$ and the margin at $x = 1$, and we solve the one-dimensional version of (6.27) in $x > 0$, assuming that $\eta_x < 0$, thus $-\nabla \eta / |\nabla \eta| = \mathbf{i}$, where $\mathbf{i}$ is the unit vector in the x-direction:

$$\frac{\partial q}{\partial x} = r, \qquad \frac{\partial f}{\partial x} = U, \tag{6.34}$$

and the solution with zero fluxes at the ridge is

$$q = rx, \qquad f = Ux. \tag{6.35}$$

More generally,

$$q = R = \int_0^x r \, dx, \qquad f = W = \int_0^x U \, dx, \tag{6.36}$$

Fig. 6.3 One-dimensional
hillslope geometry

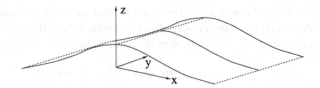

if we suppose rainfall r and uplift U to be functions of the distance x from the
divide. If we take the Meyer-Peter/Müller transport law (6.31), for example (but
written in terms of h and $S = -\eta'(x)$), then we find that h and η are given by

$$\frac{h^3}{h+\beta} = \frac{R^2}{\tau_c^* + W^{2/3}}, \qquad \eta = \int_x^1 \frac{R^2\,dx}{h^3}. \tag{6.37}$$

If τ_c^* and β are small and can be ignored, these are simply

$$h \approx \frac{R}{W^{1/3}}, \qquad \eta \approx \int_x^1 \frac{W\,dx}{R}, \tag{6.38}$$

and the slope is

$$S = \frac{W}{R}. \tag{6.39}$$

The particular choices of constant uplift and rainfall then give

$$h \approx \frac{rx^{2/3}}{U^{2/3}}, \qquad S = \frac{U}{r}, \tag{6.40}$$

and the slope is constant. Note that in this case the hillslope is not zero at the divide,
and this is generally true if the yield stress is non-zero, since the sediment transport
is positive for all $x > 0$.

A crucial feature of the hillslope for stability purposes is the sign of the slope curvature, or the sign of S'. The shape of the hillslope depends on the bedload transport
function $f(S, q)$, and in general must be found numerically. For constant rainfall
and uplift, the sign of the curvature can be ascertained analytically, however. Differentiating f with respect to x, we find (using (6.35)),

$$x\frac{\partial f}{\partial S}S' = f - q\frac{\partial f}{\partial q}. \tag{6.41}$$

Supposing always that $\partial f/\partial S > 0$, we have $S' > 0$ if and only if $f - q\partial f/\partial q > 0$.
We define concavity and convexity of hillslopes as shown in Fig. 6.4, thus convex
slopes have $S' < 0$, concave ones have $S' > 0$. Thus convexity or concavity of a
hillslope is determined only by the dependence of the bedload transport on water
flux, in conditions of constant rainfall and uplift.

In particular, and absolutely confusingly, functions which are *mathematically
convex*, so that, in particular, $\partial f/\partial q > f/q$ everywhere, lead to (geomorphologically) concave hillslopes, and *vice versa*. Worse, a mathematician would call
the 'concave' portion of the graph in Fig. 6.4 convex. Worse still, the criterion
$\partial f/\partial q > f/q$ is satisfied by mathematically convex functions, but not all such functions are mathematically convex everywhere. It is best to forget the mathematical

Fig. 6.4 Convexity and concavity

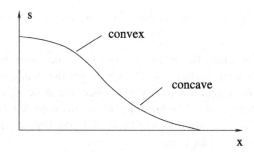

definitions; geomorphologically concave transport laws lead to similarly shaped hillslopes.

It is fairly evident that convex or concave hillslopes can be formed independently of f, if rainfall or uplift are spatially varying. For example, the Meyer-Peter/Müller law with $\beta = \tau_c^* = 0$ is simply $f = qS$, so that $f_q = f/q$, and the slope is constant (as we found) for constant r and U. When they vary, $S = W/R$, and by appropriate choice of r and U, we may have S' being positive or negative in different regions.

6.5.3 Uplift and Denudation

Before we begin our stability analysis of the purely x-dependent state, we should put our model into a physical context. We pose a constant (or perhaps spatially dependent) rate of uplift and purely fluvial erosion mainly in order to establish a clean mathematical problem. In reality, orogenic episodes are more likely to occur in transient events. For example India runs into Asia some 50 million years ago, and starts building the Himalayas, a process which is still continuing. So while our model might apply in the early phases of the orogeny, there is certainly never a steady state, and the fluvial drainage systems which no doubt grew in the early phases of the orogeny did so as instabilities formed on a slowly evolving uplifting topography. Thus, while we emphasise the instability of a steady state as a mathematically precise way of analysing the problem, this is not really the 'right' problem. Equally and more obviously, fluvial erosion becomes irrelevant at greater topographic heights, when glacial erosion and landslides become the erosive mechanisms of choice.

As regards instability, this issue is not a real barrier to understanding, because instability, when it occurs, does so rapidly, so that to all intents and purposes we can still do stability analysis even when the background hillslope topography is changing slowly. In that case, the form of the topography is determined essentially by tectonic processes, and not by the model proposed here. For example we might suppose that tectonic compression could form a series of sinusoidal folds transverse to the direction of compression; such folds are unstable to channel formation in their valleys, as indeed we would expect.

6.5.4 *Geomorphically Concave Slopes are Unstable*

We study the linear stability of the one-dimensional steady state (6.35) of the governing equations (6.27). We denote the steady state with a subscript zero, and will focus on the region $x > 0$, where the slope $\eta_0' < 0$; to extend the discussion to the other side of the hillslope, we simply note that the basic solutions for q and f there are even extensions of those in (6.35). Unless otherwise stated, we will assume that r and U are constant.

Denoting perturbations by an overtilde, thus

$$\eta = \eta_0 + \tilde{\eta}, \tag{6.42}$$

and so on, we find

$$\nabla \eta = \eta_0' \mathbf{i} + \nabla \tilde{\eta},$$
$$S = |\nabla \eta| = S_0 - \tilde{\eta}_x \ldots, \tag{6.43}$$

where $S_0 = -\eta_0'$. From these, it follows that the downslope unit vector is

$$\mathbf{n} = \mathbf{i} - \frac{\tilde{\eta}_y}{S_0}\mathbf{j} + \cdots, \tag{6.44}$$

where $\mathbf{j}$ is the unit vector in the transverse y direction.

This gives the linear approximation for $\mathbf{n}$. If we perturb f and q by quantities $\tilde{f}$ and $\tilde{q}$, then we obtain, on linearising the equations,

$$\frac{\partial \tilde{q}}{\partial x} = \frac{q}{S}\frac{\partial^2 \tilde{\eta}}{\partial y^2},$$
$$\frac{\partial \tilde{\eta}}{\partial t} = \frac{\partial}{\partial x}\left[fs\frac{\partial \tilde{\eta}}{\partial x}\right] - f_q'\tilde{q} + \frac{q}{S}\left(\frac{f}{q} - f_q\right)\frac{\partial^2 \tilde{\eta}}{\partial y^2}, \tag{6.45}$$

where $f_s = \partial f/\partial S$, $f_q = \partial f/\partial q$, and f_q' denotes the x derivative of f_q, all these quantities being evaluated in the basic steady state (we omit the subscripts zero for convenience).

It is worth pausing with these linearised equations. We see in the second that it is of diffusive type, but that the apparent diffusion coefficient in the y direction can be negative, and is so if $f_q > f/q$; returning to (6.41), we see that this condition is met precisely if the steady hillslope is geomorphologically concave (when r and U are constant).

Such an assessment might seem premature, since the term in $\tilde{q}$ is coupled (via $(6.45)_1$) to another $\tilde{\eta}_{yy}$ term. But as we shall see, it is essentially accurate; as an example, if the bedload transport function is taken as $f \propto q^n$ for some positive exponent n, then $f_q' = 0$, and the conclusion above is precise.

We use normal mode analysis, and write $\tilde{\eta}$ and $\tilde{q}$ in the form

$$\tilde{q} = \phi(x)e^{iky+\sigma t}, \qquad \tilde{\eta} = \psi(x)e^{iky+\sigma t}; \tag{6.46}$$

from these definitions it follows that

$$\phi' = -\frac{k^2 q \psi}{S},$$

$$\sigma \psi = [f_S \psi']' - (f_q \phi)' - \frac{k^2 f \psi}{S}.$$

(6.47)

The boundary conditions we would like to apply to these equations are

$$\phi = f_S \psi' = 0 \quad \text{at } x = 0,$$

$$\psi = 0 \quad \text{at } x = 1.$$

(6.48)

The condition on ϕ at the ridge $x = 0$ follows from the prescription of zero water flux there; the condition on ψ at $x = 1$ follows from fixing the (sea-level) topography at the margin.

The condition on ψ at $x = 0$ represents the condition of zero sediment flux at the ridge, and merits some discussion. The actual condition we require is that $f(S, q) = 0$ (or more properly $f(S, q)n_1 = 0$) at $x = 0$. Since we require $q = 0$ there, this is equivalently $f(S, 0) = 0$, which defines a constant, say $S = S_r$. If $S_r \neq 0$ (as for the Meyer-Peter/Müller law with non-zero τ_c^* and β), then the linearisation in (6.43) is valid at $x = 0$, and the appropriate condition is indeed that given in (6.48) (with the f_S included). If $f_S \neq 0$, then $\psi' = 0$ at $x = 0$. If, on the other hand, $f_S = 0$ at $x = 0$, then the second equation in (6.47) has a degenerate second derivative in ψ (i.e., the coefficient of the highest derivative is zero at $x = 0$), and it is well known in such circumstances that no specific boundary condition for ψ following from (6.47)$_2$ can be prescribed, beyond requiring that ψ' be bounded. This needs to be borne in mind when specific transport laws are used.

The situation is more complicated when $S_r = 0$. In this case, linearisation as in (6.43) fails at $x = 0$, and we have $\mathbf{n} = -\nabla \tilde{\eta}/|\nabla \tilde{\eta}|$. The condition $|\nabla \tilde{\eta}| = 0$ would imply both $\tilde{\eta}_x = \tilde{\eta}_y = 0$, and thus $\psi = \psi' = 0$ at $x = 0$, two conditions rather than one. However, we also have $n_1 = -\tilde{\eta}_x/|\nabla \tilde{\eta}|$, so that the condition $f n_1 = 0$ is satisfied by the single requirement that $\tilde{\eta}_x = 0$, and thus $\psi' = 0$ in the linear approximation. Again, (6.47)$_2$ applies since the (linearised) perturbation of f is zero independently of (bounded) ψ' if $f_S = 0$, despite the fact that the slope linearisation breaks down at $x = 0$ in this case.

We eliminate ψ from (6.47) to obtain the third order system

$$\frac{\sigma S}{q} \phi' = \left[f_S \left(\frac{S}{q} \phi' \right)' \right]' + k^2 \left[(f_q \phi)' - \frac{f}{q} \phi' \right],$$

(6.49)

with boundary conditions

$$\phi = f_S \left(\frac{S}{q} \phi' \right)' = 0 \quad \text{at } x = 0,$$

$$\phi' = 0 \quad \text{at } x = 1.$$

(6.50)

Instability occurs if, for any wave number k, the real part of σ is positive. It is straightforward to show (see Exercise 6.8) that for one-dimensional perturbations downslope (i.e., $k = 0$), $\sigma < 0$. Therefore instability requires $k > 0$, i.e., lateral perturbations.

To get some idea of the behaviour of solutions, suppose that the coefficient functions of x in (6.49) (i.e., q, S, f, f_q and f_S) are taken as (positive) constants.[5] The three independent solutions are then $\phi = $ constant, and $\phi = \exp[\pm(\Lambda/f_S)^{1/2}kx]$, where we define

$$\Lambda = \frac{\sigma}{k^2} - \frac{q}{S}\left(f_q - \frac{f}{q}\right). \tag{6.51}$$

The linear combination of the three solutions which satisfies the two boundary conditions at $x = 0$ is

$$\phi = \sin\left[\left(\frac{-\Lambda}{fs}\right)^{1/2}kx\right], \tag{6.52}$$

and the condition on σ to satisfy the final boundary condition at $x = 1$ requires that $\Lambda < 0$, and $k(-\Lambda/f_S)^{1/2} = (m + \frac{1}{2})\pi$ for integral m, and thus

$$\sigma = \frac{q}{S}\left(f_q - \frac{f}{q}\right)k^2 - fs\left(m + \frac{1}{2}\right)^2\pi^2. \tag{6.53}$$

If we take $k = 0$, then $\sigma < 0$, consistent with our earlier statement that one-dimensional downslope perturbations are stable. The most unstable mode when $k \neq 0$ is that for $m = 0$, and (6.52) thus suggests instability if

$$f_q - \frac{f}{q} > \frac{\pi^2 Sfs}{4qk^2}, \tag{6.54}$$

which is identical to our earlier statement following (6.45), as $k \to \infty$.

Ill-posedness

When the criterion for instability is satisfied, the growth rate $\sigma \sim k^2$, and this is the hallmark of a process with negative (lateral) diffusion, and the resultant unbounded increase of σ at high wave number implies ill-posedness. It suggests that there is something fundamentally wrong with this approximate model, and that direct numerical simulations of (6.27) are ill-advised. As we hope that the model is not physically unsound, it is natural to expect that one of the simplifying assumptions we made has removed a stabilising term which can dampen high wave number modes. The obvious candidate is the parameter δ, since as we have seen, its neglect is singular, and that leads to a change of type in the equations.

There is a beneficial effect of this ill-posedness, however. The solution obtained above assumes all the coefficient functions are constant, but we would like to extend the result to x-dependent coefficients. This is not generally possible, but the fact that the most unstable modes have short wavelength implies that, at least for these modes, a WKB analysis is possible, since at high k, the coefficient functions are relatively slowly varying. We now show how to do this.

---

[5]Note that in this case we avoid all the complications of degeneracy at $x = 0$.

6.5.5 WKB Approximation at High Wave Number

We use the definition of Λ in (6.51) so that (6.49) can be written in the form

$$\Lambda\phi' = \frac{q}{S}\left[\frac{1}{k^2}\left\{fs\left(\frac{S}{q}\phi'\right)'\right\}' + f_q'\phi\right], \tag{6.55}$$

and we seek solutions in the limit $k \gg 1$ which satisfy the boundary conditions in (6.50). In the (first order) geometric WKB approximation, the solutions are given approximately by

$$\phi_0 \sim \exp\left[\int_0^x \frac{qf_q'}{S\Lambda}\,dx\right],$$
$$\phi_\pm \sim \exp\left[\pm k\int_0^x \left(\frac{\Lambda}{fs}\right)^{1/2}\,dx\right], \tag{6.56}$$

and we shall suppose that Λ is either positive or negative throughout. The more interesting case where its sign changes is treated in the following section.

An additional complication arises because of the fact that $q \to 0$ as $x \to 0$ in (6.55). It also independently occurs if $fs \to 0$ as $x \to 0$, which is the case for the Meyer-Peter/Müller relation. Because of these degeneracies, the WKB solutions in (6.56) are not uniformly valid near $x = 0$. Despite this, we shall for ease of exposition suppose that they are. As we might expect, the details at the summit do not appear to significantly affect the instability criterion; further consideration of the matter is deferred to Question 6.9.

A suitable choice of transport function allows the assumption of uniform validity of the WKB solutions to be made explicit. If we suppose

$$f = q^\alpha S^{1-\alpha} = \tau^{3\alpha/2} S^{(3-5\alpha)/2}, \tag{6.57}$$

then for any $\alpha \in (0, \frac{3}{5})$, we have a physically meaningful transport law with $fs > 0$ and S/q finite at $x = 0$ (and in fact, constant) so that the second boundary condition in (6.50) reduces to

$$\phi'' = 0 \quad \text{at } x = 0. \tag{6.58}$$

For purposes of illustration, we shall make this assumption, as it provides a direct comparison with the previous, constant coefficient analysis.

The function ϕ_0 is slowly varying, and follows from a regular approximation to (6.55); it generalises the constant solution for the constant coefficient version of (6.49); the functions $\phi_\pm$ are rapidly varying, and constitute the generalisation of the sinusoidal solutions of the constant coefficient model.

The solution of the equation is thus approximately

$$\phi \sim U_0\phi_0 + U_+\phi_+ + U_-\phi_-. \tag{6.59}$$

Satisfaction of the boundary conditions $\phi = \phi'' = 0$ at $x = 0$ and $\phi' = 0$ at $x = 1$ requires

$$U_0 + U_+ + U_- = 0,$$

$$U_0 \phi_0''(0) + k^2 \left(\frac{\Lambda}{fs}\right)\bigg|_0 (U_+ + U_-) = 0,$$

$$U_0 \phi'(1) + k \left(\frac{\Lambda}{fs}\right)^{1/2}\bigg|_1 \left[U_+ \exp\left\{ k \int_0^1 \left(\frac{\Lambda}{fs}\right)^{1/2} dx \right\} \right. \tag{6.60}$$

$$\left. - U_- \exp\left\{ -k \int_0^1 \left(\frac{\Lambda}{fs}\right)^{1/2} dx \right\} \right] = 0.$$

Since k is large, the first two of these give $U_0 \approx 0$, $U_+ \approx -U_-$. Evidently we will require $\Lambda < 0$, and the solution can be written (taking $U_+ = 1/2i$ without loss of generality) as

$$\phi = \sin\left[k \int_0^x \left(\frac{-\Lambda}{fs}\right)^{1/2} dx \right], \tag{6.61}$$

and the condition $\phi' = 0$ at $x = 1$ implies

$$\int_0^1 \left(\frac{-\Lambda}{fs}\right)^{1/2} dx = \frac{(2m + 1)\pi}{2k}, \tag{6.62}$$

where m is an integer. (6.61) is a simple generalisation of (6.52), and the eigenvalue condition (6.62), written in the form

$$\int_0^1 \left[-\frac{\sigma}{k^2 fs} + \frac{q}{Sfs}\left(f_q - \frac{f}{q} \right) \right]^{1/2} dx = \frac{(m + \frac{1}{2})\pi}{k}, \tag{6.63}$$

is a simple generalisation of (6.53).

If $f_q \leq f/q$ everywhere, then (6.63) implies $\sigma < 0$, and the (convex) hillslope is stable. However, if $f_q > f/q$ everywhere, then it is clear that there will be positive values of σ, since for large k, the right hand side of (6.63) can approximately take any positive value. Thus the maximal growth rate for $f_q > f/q$ will be

$$\sigma \approx k^2 \min_{[0,1]} \frac{q}{S}\left(f_q - \frac{f}{q} \right). \tag{6.64}$$

Again the growth rate is unbounded at small wavelength.

This is not the end of the story, because the nature of the instability via a negative lateral diffusion coefficient suggests that if $f_q > f/q$ anywhere, then the hillslope will still be unstable. If this is the case, then Λ must change sign, and the WKB analysis must be modified to allow for turning points, where $\Lambda = 0$. This we now do.

6.5.6 Turning Point Analysis

We will consider the simplest case in which there is a single turning point, at x_0, say, and we suppose that $\Lambda > 0$ upslope in $x < x_0$, and $\Lambda < 0$ downslope in $x > x_0$. At the onset of instability ($\sigma = 0$), this corresponds to a locally stable (convex) slope at the ridge, and a locally concave (unstable) slope at the margin, as illustrated by Fig. 6.4. The solution is much as before, but now we need to include the second order correcting term of physical optics in the approximations for the rapidly varying solutions.

In $x < x_0$, the slowly varying solution is

$$\phi_0 \sim \exp\left[\int_0^x \frac{qf_q'}{S\Lambda} dx\right] + O\left(\frac{1}{k^2}\right), \quad x < x_0, \tag{6.65}$$

and the physical optics (two term) approximations for the rapidly varying solutions can be taken as

$$\phi_\pm \sim \frac{qf_S^{1/4}}{S\Lambda^{3/4}\phi_0^{1/2}} \exp\left[\pm k \int_{x_0}^x \left(\frac{\Lambda}{fs}\right)^{1/2} dx\right], \quad x < x_0; \tag{6.66}$$

these can be compared with (6.56).

In $x > x_0$, we take the corresponding solutions as

$$\phi_0 \sim \exp\left[\int_1^x \frac{qf_q'}{S\Lambda} dx\right] + O\left(\frac{1}{k^2}\right), \quad x > x_0, \tag{6.67}$$

and

$$\phi_\pm \sim \frac{qf_S^{1/4}}{S|\Lambda|^{3/4}\phi_0^{1/2}} \exp\left[\pm ik \int_{x_0}^x \left(\frac{|\Lambda|}{fs}\right)^{1/2} dx\right], \quad x > x_0. \tag{6.68}$$

The solution upslope is

$$\phi = U_0\phi_0 + U_+\phi_+ + U_-\phi_-, \quad x < x_0, \tag{6.69}$$

and that downslope is

$$\phi = D_0\phi_0 + D_+\phi_+ + D_-\phi_-, \quad x > x_0, \tag{6.70}$$

but both approximations break down in the vicinity of $x = x_0$, where $\Lambda = 0$. The object is to solve the problem in this transition region, in order to provide connection formulae relating U_j to D_j; the eigenvalue relation for Λ, and thus σ, can then be established.

The relevant coordinate in the transition region is ζ, defined by

$$x = x_0 - \left(\frac{fs}{|\Lambda_0'|}\right)^{1/3} \frac{\zeta}{k^{2/3}}, \tag{6.71}$$

where Λ_0' is the (negative) gradient of Λ at x_0, i.e.,

$$\Lambda \approx -|\Lambda_0'|(x - x_0). \tag{6.72}$$

Note that ζ points upslope. Evaluating the outer solutions as $x \to x_0$, we find, for $x < x_0$ ($\zeta > 0$),

$$\phi_0 \sim a_0 \zeta^{P-1}\left[1 + O\left(k^{-2/3}\right)\right], \tag{6.73}$$

where

$$p = 1 - \frac{qf_q'}{|\Lambda_0'|S}\Big|_{x_0}, \tag{6.74}$$

and

$$a_0 = \left[\left(\frac{fs}{|\Lambda_0'|}\right)^{1/3} \frac{1}{k^{2/3}x_0}\right]^{p-1} \exp\left[\int_0^{x_0}\left\{\frac{qf_q'}{S\Lambda} - \frac{p-1}{x-x_0}\right\}dx\right]. \tag{6.75}$$

Then also as $x \to x_0-$,

$$\phi_\pm \sim a\zeta^{-\frac{1}{2}p-\frac{1}{4}} \exp\left[\mp\frac{2}{3}\zeta^{3/2}\right]\left[1 + O\left(k^{-2/3}\right)\right], \tag{6.76}$$

where

$$a = \frac{q}{S}\left(\frac{k}{|\Lambda_0'|a_0}\right)^{1/2}, \tag{6.77}$$

evaluated at $x = x_0$.

In a similar way, expansion of the solutions in $x > x_0$ as $x \to x_0+$ yields

$$\phi_0 \sim b_0(-\zeta)^{p-1}\left[1 + O\left(k^{-2/3}\right)\right], \tag{6.78}$$

where

$$b_0 = \left[\left(\frac{fs}{|\Lambda_0'|}\right)^{1/3} \frac{1}{k^{2/3}(1-x_0)}\right]^{p-1} \exp\left[-\int_{x_0}^1\left\{\frac{qf_q'}{S\Lambda} - \frac{p-1}{x-x_0}\right\}dx\right], \tag{6.79}$$

and

$$\phi_\pm \sim b(-\zeta)^{-\frac{1}{2}p-\frac{1}{4}} \exp\left[\pm\frac{2}{3}i(-\zeta)^{3/2}\right]\left[1 + O\left(k^{-2/3}\right)\right], \tag{6.80}$$

where

$$b = \frac{q}{S}\left(\frac{k}{|\Lambda_0'|b_0}\right)^{1/2}; \tag{6.81}$$

all of these coefficients are evaluated at x_0.

Substituting (6.71) into (6.55), and expanding to leading order for large k, we find the transition layer equation for ϕ,

$$\phi''' - \zeta\phi' + (p - 1)\phi = 0. \tag{6.82}$$

Our object is to find asymptotic forms as $\zeta \to \pm\infty$ for the solutions of this equation, and thus connect the two limiting sets of expressions given above for $x \to x+$ and $x \to x-$. In fact, we have seen this equation before, in the context of the preceding chapter. Its solutions are described in Appendix C. Three independent solutions are given by the Laplace integrals

$$A_k(\zeta, p) = \frac{1}{2\pi i}\int_{L_k} t^{-p}e^{\zeta t - \frac{1}{3}t^3}dt, \tag{6.83}$$

where the contour L_k is one of the three shown in Fig. C.1, and for non-integral p, we take a branch cut from 0 to ∞ in the complex t plane.

The asymptotic behaviour of these three solutions for large $|\zeta|$ is determined by the method of steepest descents, and the results can be summarised as follows. We define three functions, $A_\pm$ and B:

$$A_\pm(\zeta, p) = \frac{1}{2\sqrt{\pi}\zeta^{\frac{1}{2}p+\frac{1}{4}}} \exp\left[-\frac{3}{2}p\pi i \pm \left(\xi - \frac{1}{2}p\pi i\right)\right] \sum_{s=0}^{\infty} \frac{(\pm 1)^s a_s(p)}{\xi^s}, \quad (6.84)$$

where $\xi = \frac{2}{3}\zeta^{3/2}$ and, in particular, $a_0(p) = 1$; also,

$$B(\zeta, p) = \sum_{s=0}^{\infty} \frac{(\pm 1)^s \zeta^{p-3s-1}}{3^s s! \Gamma(p - 3s)}. \quad (6.85)$$

We suppose $-\frac{4\pi}{3} < \arg\zeta < \frac{2\pi}{3}$, and define the three sectors

$$T_1: \quad -\frac{4\pi}{3} < \arg\zeta < -\frac{2\pi}{3},$$

$$T_2: \quad 0 < \arg\zeta < \frac{2\pi}{3}, \quad (6.86)$$

$$T_3: \quad -\frac{2\pi}{3} < \arg\zeta < 0,$$

as shown in Fig. C.2. Different asymptotic expansions apply for the functions A_k in the different sectors, whose boundaries are thus Stokes lines for (6.83).

Specifically, we have

$$A_1 \sim A_-, \quad \zeta \in T_2 \cup T_3,$$
$$A_2 \sim iA_+, \quad \zeta \in T_3 \cup T_1,$$
$$A_3 \sim \begin{cases} -A_-, & \zeta \in T_1, \\ -i\exp(2\pi i p)A_+, & \zeta \in T_2, \end{cases} \quad (6.87)$$
$$\begin{pmatrix} A_1 \\ A_2 \\ A_3 \end{pmatrix} \sim \begin{pmatrix} -i & -1 & 1 \\ i & -\exp(-2\pi i p) & -1 \\ -i\exp(2\pi i p) & -1 & -1 \end{pmatrix} \begin{pmatrix} A_+ \\ B \\ A_- \end{pmatrix} \quad \text{for } A_k, \ \zeta \in T_k.$$

From (6.84) and (6.85), we have

$$A_+ \sim \frac{1}{2\sqrt{\pi}\zeta^{\frac{1}{2}p+\frac{1}{4}}} \exp\left(-2p\pi i + \frac{2}{3}\zeta^{3/2}\right),$$

$$A_- \sim \frac{1}{2\sqrt{\pi}\zeta^{\frac{1}{2}p+\frac{1}{4}}} \exp\left(-p\pi i - \frac{2}{3}\zeta^{3/2}\right), \quad (6.88)$$

$$B \sim \frac{\zeta^{p-1}}{\Gamma(p)}$$

as $\zeta \to \infty$ (in any direction), and in particular $A_+ \to \infty$ and $A_- \to 0$ for $\arg\zeta = 0$. Thus crossing the Stokes line at $\arg\zeta = 0$ only introduces an exponentially small term relative to the dominant term for A_2 and A_3, and we have

$$A_1 \sim A_-, \quad A_2 \sim iA_+, \quad A_3 \sim -i\exp(2\pi i p)A_+, \quad \zeta \to +\infty,$$
$$A_1 \sim -iA_+ - B + A_-, \quad A_2 \sim iA_+, \quad A_3 \sim -A_-, \quad \zeta \to -\infty. \quad (6.89)$$

Consulting (6.65) and (6.66), we see that ϕ_0 is order one, ϕ_- is exponentially large, and ϕ_+ is exponentially small at $x = 0$. It follows as previously that in order to satisfy the two boundary conditions at $x = 0$, we must have both U_0 and U_- exponentially small. In particular, consultation of (6.76) then tells us that (bearing in mind (6.89)) the solution we require in the transition region is

$$\phi = \alpha A_1(\zeta, p), \tag{6.90}$$

where α is constant. Matching this to ϕ in $x < x_0$ using (6.76) and (6.88) then gives

$$a = \frac{\alpha e^{-2p\pi i}}{2\sqrt{\pi}}. \tag{6.91}$$

In $\zeta < 0$, we have $\arg \zeta = -\pi$; then from (6.90), using (6.88), (6.89) and (6.91), we have

$$\phi \sim a\left[\frac{2\sqrt{\pi}}{\Gamma(p)}(-\zeta)^{p-1} + \frac{2}{(-\zeta)^{\frac{1}{2}p+\frac{1}{4}}} \sin\left\{\frac{2}{3}(-\zeta)^{3/2} - \left(\frac{1}{2}p - \frac{1}{4}\right)\pi\right\}\right]$$
$$\text{as } \zeta \to -\infty. \tag{6.92}$$

Finally, we match this to the downslope solution (6.70) using the limiting expressions for ϕ_0 and $\phi_\pm$ from (6.78) and (6.80). These give

$$D_0 = \frac{2a\sqrt{\pi}}{b_0 \Gamma(p)}, \qquad \frac{2ae^{-(\frac{1}{2}p - \frac{1}{4})i\pi}}{2i} = D_+ b, \qquad D_- = \overline{D}_+, \tag{6.93}$$

and thus

$$\phi \sim D_0 \exp\left[\int_1^x \frac{qf_q'}{S\Lambda} dx\right] + C \sin\left[k \int_{x_0}^x \left(\frac{|\Lambda|}{fs}\right)^{1/2} dx - \left(\frac{1}{2}p - \frac{1}{4}\right)\pi\right] \tag{6.94}$$

in $x > x_0$, where

$$C = \frac{2aqf_s^{1/4}}{bS|\Lambda|^{3/4}\phi_0^{1/2}}, \tag{6.95}$$

and ϕ_0 is given by (6.67).

We need to choose Λ to satisfy the last boundary condition, (6.50), $\phi' = 0$ at $x = 1$. For large k, we have from (6.75), (6.77), (6.79) and (6.81) that

$$a_0 \sim b_0 \sim k^{-\frac{2}{3}(p-1)}, \qquad a \sim \left(\frac{k}{a_0}\right)^{1/2}, \qquad b \sim \left(\frac{k}{b_0}\right)^{1/2}, \tag{6.96}$$

and thus $C \sim O(1)$, $D_0 \sim k^{p-\frac{1}{2}}$. Consequently, the derivative of the sinusoidal term in (6.94) is $O(k)$, while that of the slowly varying term is $O(k^{p-\frac{1}{2}})$. If $p > \frac{3}{2}$, ϕ' is approximately given by the slowly varying exponential, and the boundary condition at the margin cannot be satisfied. We thus require $p < \frac{3}{2}$, and in this case, $\phi' = 0$ at $x = 1$ when, approximately,

$$\int_{x_0}^1 \left[-\frac{\sigma}{k^2 fs} + \frac{q}{Sfs}\left(f_q - \frac{f}{q}\right)\right]^{1/2} dx = \frac{(m + \frac{1}{2}p + \frac{1}{4})\pi}{k}, \tag{6.97}$$

where m is an integer. This should be compared with (6.63). Note that x_0 also depends on σ, since it is determined by where the integrand is zero. The condition that $p < \frac{3}{2}$ is satisfied if, from (6.74),

$$\frac{qf'_q}{|\Lambda'_0|S}\bigg|_{x_0} > -\frac{1}{2}, \tag{6.98}$$

which is easily satisfied if, for example, $f'_q \geq 0$.

To interpret (6.97), notice first that if $f_q < f/q$ everywhere then necessarily $\sigma < 0$ and the hillslope is stable. Second, we assume $\Lambda > 0$ in $x < x_0$ and $\Lambda > 0$ in $x > x_0$. To be specific, let us suppose that Λ is in fact monotonically decreasing, thus $\frac{q}{S}(f_q - \frac{f}{q})$ is monotonically increasing, at least while it is positive. (We can suppose it is negative near the ridge, for otherwise there is an unstable eigenfunction with $\sigma > 0$ without a turning point.) Let x^* be the (unique) point where $f_q = f/q$. Then we see that $x_0(\sigma)$ is a monotonically increasing function, with $x_0(0) = x^*$ and $x_0(\sigma_{\max}) = 1$, where $\sigma_{\max}$ is given by

$$\sigma_{\max} = k^2 \max\left\{\frac{q}{S}\left(f_q - \frac{f}{q}\right)\right\}, \tag{6.99}$$

here obtained at $x = 1$. Therefore the integral in (6.97), $I(\sigma)$, say, is a monotonically decreasing function of σ with

$$I(0) = \int_{x^*}^{1}\left[\frac{q}{Sfs}\left(f_q - \frac{f}{q}\right)\right]^{1/2} dx \tag{6.100}$$

and $I(\sigma_{\max}) = 0$. For large k, the right hand side of (6.97) is essentially a continuous function of m/k, and thus the equation determines $O(k)$ values of positive σ, the maximum of which corresponds to $m = 0$ (if $p > -1/2$), where $\sigma \approx \sigma_{\max}$; thus the maximum growth rate is given by (6.99) in this case, and instability occurs when $f_q > f/q$.

The only remaining issue is to describe what happens if Λ is not monotonic when it is negative. The case is encapsulated by assuming that $\frac{q}{S}(f_q - \frac{f}{q})$ is negative in $x < x^*$, is zero at x^*, and positive for $x^* < x < 1$ with a maximum at x_m, say, equal to σ_m/k^2. Now for values $0 < \sigma < \sigma_m$, $\Lambda < 0$ for $x_0(\sigma) < x_1(\sigma)$, with a minimum at x_m, and for $0 < \sigma < \sigma_m$, x_0 is monotonically increasing from x^* to x_m, and x_1 is monotonically decreasing from 1 to x_m.

The complication in the solution for $0 < \sigma < \sigma_m$ is that there are two turning points (at x_0 and x_1), and connection formulae for each must be determined. Having done this once, however, it is simple to extend the preceding results. In order to satisfy the boundary conditions, we need the outer solutions where $\Lambda > 0$ to be exponentially small, and thus the relevant solution for both transition regions will be A_1. Thus the outer solution which matches to the upslope turning point will have the form of (6.94), that is,

$$\phi \sim D_0 \exp\left[\int_{x_m}^{x}\frac{qf'_q}{S\Lambda}dx\right] + C_0 \sin\left[k\int_{x_0}^{x}\left(\frac{|\Lambda|}{fs}\right)^{1/2}dx - \left(\frac{1}{2}p_0 - \frac{1}{4}\right)\pi\right] \tag{6.101}$$

(note that the lower limit of the integral in the slow exponential is now x_m, and that the value of p at each turning point will be different, thus p_0 and p_1). Similarly, that which matches to the lower turning point will be of the same form:

$$\phi \sim D_1 \exp\left[\int_{x_m}^x \frac{qf_q'}{S\Lambda} dx\right] + C_1 \sin\left[k \int_{x_1}^x \left(\frac{|\Lambda|}{fs}\right)^{1/2} dx - \left(\frac{1}{2}p_1 - \frac{1}{4}\right)\pi\right]; \quad (6.102)$$

and these two expressions must be the same.

We ensure this by choosing

$$D_0 = D_1, \qquad C_0 = \pm C_1, \qquad (6.103)$$

and

$$\int_{x_0}^{x_1} \left[-\frac{\sigma}{k^2 fs} + \frac{q}{Sfs}\left(f_q - \frac{f}{q}\right)\right]^{1/2} dx = \frac{\left(m + \frac{1}{2}(p_0 - p_1)\right)\pi}{k}, \quad (6.104)$$

which can be compared to (6.97). The same discussion which followed there applies, with the conclusion that the maximum growth rate is still given by (6.99), and instability occurs if $f_q > f/q$.

6.5.7 Rivulet Theory: $\delta \ll 1$

The message from all the above theory is the relatively simple one we gleaned from assuming all the coefficients in (6.45) could be taken as constant, that the instability criterion is that $f_q > f/q$ at some point, and that when this happens, the growth rate $\sigma \propto k^2$, signalling an ill-posedness in the reduced model (6.25). We assume that this is because of the neglect of some of the terms in (6.20), and our first task is to show that by bringing back some of these terms, we can stabilise the system for high wave number perturbations.

The only term we need bring back is the δh which distinguishes η from s; this is suggested both because it is the largest of the neglected terms, and because, as we have already suggested, its neglect represents a singular perturbation to the model. We take $\delta \varepsilon = 0$, $\delta F^2 = 0$, $E = 0$ (thus $c = 0$), and $\delta \nu = 0$ in (6.20); the model thus reduces to

$$\nabla.[q\mathbf{n}] = r,$$
$$\eta_t - \delta h_t = -\nabla.[f\mathbf{N}] + U,$$
$$\mathbf{n} = -\frac{\nabla \eta}{|\nabla \eta|}, \qquad \mathbf{N} = \frac{\boldsymbol{\tau}_e}{|\boldsymbol{\tau}_e|}, \qquad (6.105)$$
$$\boldsymbol{\tau}_e = -(h + \beta)\nabla \eta + \delta\beta \nabla h,$$
$$q = h^{3/2}|\nabla \eta|^{1/2},$$

where f is the sediment transport.

Geometric Linearity

The one-dimensional steady state is exactly as before, since then $\mathbf{N} = \mathbf{n} = \mathbf{i}$. Denoting the steady water surface as $\eta_0(x)$, we put

$$\eta = \eta_0 + \tilde{\eta}, \tag{6.106}$$

and suppose that $\tilde{\eta}$ is small. As in (6.43), we then have

$$\nabla\eta = \eta_0' \mathbf{i} + \nabla\tilde{\eta},$$
$$|\nabla\eta| = S = S_0 - \tilde{\eta}_x + \cdots, \tag{6.107}$$

where again we suppose $x > 0$, and

$$S_0 = |\eta_0'|. \tag{6.108}$$

Thus

$$\mathbf{n} = \mathbf{i} - \frac{\tilde{\eta}_y}{S_0}\mathbf{j} + \cdots,$$
$$q = h^{3/2}S^{1/2}, \tag{6.109}$$

and similarly (if also δh is small)

$$\tau_e = (h + \beta)S + \delta\beta h_x + \cdots,$$
$$\mathbf{N} = \mathbf{i} - \frac{1}{S_0}\left\{\tilde{\eta}_y - \frac{\delta\beta}{h+\beta}h_y\right\}\mathbf{j} + \cdots. \tag{6.110}$$

Adopting for the moment only these approximations (that is, we linearise the geometry only), we derive from (6.105) the following approximate model:

$$\frac{\partial q}{\partial x} - \frac{\partial}{\partial y}\left[\frac{q}{S_0}\frac{\partial\tilde{\eta}}{\partial y}\right] = r,$$
$$\frac{\partial\tilde{\eta}}{\partial t} - \delta\frac{\partial h}{\partial t} = U - \frac{\partial f}{\partial x} + \frac{\partial}{\partial y}\left[\frac{f}{S_0}\left\{\frac{\partial\tilde{\eta}}{\partial y} - \frac{\beta\delta}{h+\beta}\frac{\partial h}{\partial y}\right\}\right], \tag{6.111}$$

with q and τ_e defined in (6.109) and (6.110). Notice that this model is still nonlinear.

We need to retain the terms in δ involving t and y derivatives, because of the growth of short wave perturbations, but it is safe to neglect the term $\delta\beta h_x$ in (6.110). Note that this is the same singular approximation as before, insofar as the highest derivatives of h are lost, and thus we are not able to satisfy the slope condition (6.33)$_2$. We therefore have

$$\tau_e \approx (h + \beta)S, \qquad q = h^{3/2}S^{1/2}, \tag{6.112}$$

and the transport law $f(\tau_e)$ can equivalently be written, as before, as $f(q, S)$.

We linearise (6.111) in the same way as before, by putting

$$q = q_0 + \tilde{q}, \qquad h = h_0 + \tilde{h}, \tag{6.113}$$

and the modified version of (6.45) with the terms in δ included is

$$\frac{\partial \tilde{q}}{\partial x} = \frac{q}{S} \frac{\partial^2 \tilde{\eta}}{\partial y^2},$$

(6.114)

$$\frac{\partial \tilde{\eta}}{\partial t} - \delta \frac{\partial \tilde{h}}{\partial t} = -f'_q \tilde{q} + \frac{\partial}{\partial x}\left[fs \frac{\partial \tilde{\eta}}{\partial x} \right] + \frac{q}{S}\left(\frac{f}{q} - f_q \right) \frac{\partial^2 \tilde{\eta}}{\partial y^2} - \left(\frac{\delta \beta}{h + \beta} \right) \frac{f}{S} \frac{\partial^2 \tilde{h}}{\partial y^2},$$

and we have dropped the suffixes of zero for convenience.

We use the normal mode definitions of ϕ and ψ in (6.46) as before, and also

$$\tilde{h} = \chi \exp[iky + \sigma t].$$

(6.115)

Then (6.112) and (6.107) imply

$$\phi = \frac{3q\chi}{2h} - \frac{q\psi'}{2S},$$

(6.116)

and ϕ and ψ satisfy modified versions of (6.47):

$$\phi' = -\frac{k^2 q \psi}{S},$$

$$\sigma\psi = [fs\psi']' - f'_q\phi + \frac{k^2 q}{S}\left(f_q - \frac{f}{q} \right)\psi + \delta\left[\sigma + \frac{k^2 \beta}{(h+\beta)} \frac{f}{S} \right]\chi.$$

(6.117)

Eliminating ψ and χ, we obtain the third order modification of (6.49):

$$\left[\frac{\sigma}{k^2} - \frac{q}{S}\left(f_q - \frac{f}{q} \right) \right]\phi' = \frac{q}{S}\left[\frac{1}{k^2}\left\{ fs\left(\frac{S\phi'}{q} \right)' \right\}' + f'_q\phi \right]$$

$$- \delta\left[\frac{\sigma}{k^2} + \frac{\beta}{(h+\beta)} \frac{f}{S} \right]\left[\frac{2k^2 h\phi}{3S} - \frac{qh}{3S^2}\left(\frac{S\phi'}{q} \right)' \right],$$

(6.118)

with the same boundary conditions as in (6.50),

$$\phi = fs\left(\frac{S\phi'}{q} \right)' = 0 \quad \text{at } x = 0,$$

$$\phi' = 0 \quad \text{at } x = 1.$$

(6.119)

Our procedure with (6.118) would now be as before: first take the coefficient functions as constant, then use these as the basis for a WKB solution. Here we will simply take the first step, since we are only really interested in whether the terms in δ will provide stabilisation at large wave number. We thus take $f'_q = 0$, and define

$$\lambda = \frac{\sigma}{k^2}, \qquad A = \frac{q}{S}\left(f_q - \frac{f}{q} \right), \qquad B = \frac{\beta}{(h+\beta)} \frac{f}{S},$$

$$\alpha = fs, \qquad \varepsilon = \frac{\delta h}{3S}, \qquad \gamma = \varepsilon k^2.$$

(6.120)

Then solutions to (6.118) of the form $\phi = e^{mx}$ exist, providing m satisfies the polynomial

$$\frac{\varepsilon\alpha}{\gamma}m^3 + \varepsilon(\lambda + B)m^2 - (\lambda - A)m - 2\gamma(\lambda + B) = 0. \tag{6.121}$$

The assumption that $\delta \ll 1$ implies that $\varepsilon \ll 1$. The terms in δ in (6.118) are precisely those proportional to $(\lambda + B)$ in (6.121), and their neglect leads to the same result as before, instability if $A > 0$. In particular, we can suppose $\lambda \sim O(1)$ ($\sigma \sim k^2$).

The effect of the terms in δ at large k is expressed through the parameter γ, and there is clearly a distinguished limit where this parameter becomes important, which is when $\gamma = O(1)$. In this case, the roots of (6.121) are approximately

$$m_0 \approx \frac{2\gamma(\lambda + B)}{A - \lambda}, \qquad m_{\pm} \approx i\theta_{\pm} = \pm i\left(\frac{\gamma(A - \lambda)}{\varepsilon\alpha}\right)^{1/2}, \tag{6.122}$$

and the solution for ϕ (with $\phi = 0$ on $x = 0$) is

$$\phi = U_0\left[e^{m_0 x} - \cos(\theta_+ x)\right] + U_s \sin(\theta_+ x). \tag{6.123}$$

Satisfaction of $\phi''(0) = 0$ requires $U_0(m_0^2 + \theta_+^2) = 0$, and thus $U_0 = 0$. Therefore $\phi = U_s \sin(\theta_+ x)$, and the boundary condition at $x = 1$ requires $\cos\theta_+ = 0$, whence $\theta_+ = (r + \frac{1}{2})\pi$ for some integer r. The case $r = 0$ gives the most unstable mode, and for this

$$\lambda = A - \frac{\alpha\varepsilon\pi^2}{4\gamma}, \tag{6.124}$$

and instability occurs approximately if $A > 0$, as before. The presence of terms in δ does not affect the instability result. (6.124) is in fact independent of δ.

We are interested in what happens to λ at higher k, or higher γ. There is a further distinguished limit when the above discussion breaks down, and this is when $\gamma \sim 1/\varepsilon$. We put

$$\gamma = \frac{g}{\varepsilon}, \qquad m = \frac{M}{\varepsilon}, \tag{6.125}$$

so that

$$\frac{\alpha}{g}M^3 + (\lambda + B)M^2 + (A - \lambda)M - 2g(\lambda + B) = 0. \tag{6.126}$$

The roots are now all comparable, and no easy recipe to find λ is possible. For even larger wave number, g becomes large, and in this case the roots are

$$M_{\pm} \approx \pm\sqrt{2g}, \qquad M_0 \approx -\frac{(\lambda + B)g}{\alpha}, \tag{6.127}$$

and

$$\phi = U_0\left[e^{M_0 x} - \cosh(M_+ x)\right] + U_s \sinh(M_+ x). \tag{6.128}$$

The second boundary condition at $x = 0$ requires $U_0(M_0^2 - M_+^2) = 0$, but $U_0 = 0$ is no longer an option, since the boundary condition at $x = 1$ cannot be satisfied by $\sinh(M_+x)$; therefore we require $M_0^2 = M_+^2$, and this gives us the two possible (leading order) values for λ,

$$\lambda \approx -B \pm \alpha\sqrt{\frac{2}{g}};\qquad(6.129)$$

thus the positive term B arising from the terms in δ, and particularly, from the stabilising coefficient β (see (6.120)), ensures stability at high wave number. In fact (6.129) suggests that stabilisation occurs for $g = O(1)$, that is, $k \sim 1/\varepsilon$.

A more detailed analysis (see Question 6.11) suggests that the maximum growth rate is at $\gamma \sim O(\varepsilon^{-1/2})$, and that (if A is positive), λ is positive for γ in the range $\frac{\alpha\pi^2}{4A} < \gamma < \frac{\pi^2 A}{8\varepsilon B}$, i.e., $O(\frac{1}{\delta^{1/2}}) < k < O(\frac{1}{\delta})$.

6.6 Channel Formation

The discussion above of linear stability when $\delta \ll 1$ suggests that a distinguished lateral length scale of order $\lesssim \delta^{1/2}$ may serve to delineate the unstable growth of rills. Let us now focus on this growth by defining

$$y = \delta^{1/2}Y,\qquad \tilde{\eta} = \delta Z,\qquad t = \delta\tilde{t};\qquad(6.130)$$

the rescaling of $\tilde{\eta}$ and t is motivated by the linear stability result, which suggests that when $y \sim 1/k \ll 1$, then (from (6.47)) $\tilde{\eta} \sim \tilde{q}/k^2$, or more generally $\tilde{\eta} \sim h^{3/2}/k^2$, and $t \sim 1/k^2$. For $k \sim 1/\delta^{1/2}$ and $h \sim O(1)$, we obtain (6.130). Note that if the original time scale $\sim d/U_D$ was 10^6 years, then this new time scale is $[h]/U_D$ (film thickness divided by uplift or erosion rate), of order 10 years.

Equations (6.111) retain their validity based on geometric linearity, and take the form (in $x > 0$)

$$\begin{aligned}\frac{\partial q}{\partial x} - \frac{\partial}{\partial Y}\left[\frac{q}{S}\frac{\partial Z}{\partial Y}\right] &= r,\\[2mm]\frac{\partial Z}{\partial\tilde{t}} - \frac{\partial h}{\partial\tilde{t}} &= U - \frac{\partial f}{\partial x} + \frac{\partial}{\partial Y}\left[\frac{f}{S}\left\{\frac{\partial Z}{\partial Y} - \frac{\beta}{h+\beta}\frac{\partial h}{\partial Y}\right\}\right],\end{aligned}\qquad(6.131)$$

in which $S(x)$ is the steady slope (i.e., such that $Z = 0$ is a solution of (6.131)), and the water flux q and effective driving stress for sediment transport τ_e are given by (6.112):

$$\tau_e \approx (h+\beta)S,\qquad q = h^{3/2}S^{1/2}.\qquad(6.132)$$

To be specific, we pose these equations on a rectangular domain $-L < y < L$ (thus $-L/\delta^{1/2} < Y < L/\delta^{1/2}$) and $0 < x < 1$. In terms of x and y, the no flux and shoreline boundary conditions require

$$\begin{aligned}\frac{\partial h}{\partial y} = \frac{\partial Z}{\partial y} &= \quad \text{on } y = \pm L,\\[2mm]q = f = 0 &\quad \text{on } x = 0,\\[2mm]Z = \frac{\partial Z}{\partial x} = 0 &\quad \text{on } x = 1.\end{aligned}\qquad(6.133)$$

These equations enclose the linear instability of the steady state (on a lateral space scale $Y = O(1)$, and time scale $\tilde{t} = O(1)$); but they are fully nonlinear equations, and may provide a vehicle to understand the nonlinear development of the linear rill instability we have found before.

One possibility is that stable finite amplitude solutions (rills) exist for this model, with $h \sim O(1)$. Such rills have depths of order millimetres or centimetres, and do not correspond to larger river channels, which presumably evolve over longer geological time scales.

To study such channels, we seek solutions in which $h \gg 1$, and is a function of the short length scale $Y \sim O(1)$. Note that a consequence of $(6.131)_1$ is that

$$\int_{-L/\delta^{1/2}}^{L/\delta^{1/2}} q \, dY = 2Lrx/\delta^{1/2}, \tag{6.134}$$

which serves as a constraint on the channel depth. In particular, (6.132) suggests a distinguished limit $h \sim 1/\delta^{1/3}$ when most of the rainfall finds its way into the channel. Thus we rescale the variables as

$$h = \frac{H}{\delta^{1/3}}, \qquad q = \frac{Q}{\delta^{1/2}}, \qquad f = \frac{F}{\delta^{1/2}}, \qquad \tau_e = \frac{T_e}{\delta^{1/3}}, \qquad \tilde{t} = \delta^{1/6}T. \tag{6.135}$$

(This assumes that $f \sim \tau_e^{3/2}$ for large τ_e.) With $\delta \approx 10^{-5}$, then $1/\delta^{1/3} \approx 46$, and the new depth scale is of the order of a metre, sensible for a developed stream. The choice of time scale (corresponding dimensionally to a year) is so that the time derivative of h in $(6.131)_2$ is balanced. On the other hand, we expect the water surface to remain flat, so that we do not seek to rescale Z: as we will see, this is consistent with the model equations.

Introducing (6.135) into (6.131) and (6.132), we obtain

$$\frac{\partial Q}{\partial x} - \frac{\partial}{\partial Y}\left[\frac{Q}{S}\frac{\partial Z}{\partial Y}\right] = \delta^{1/2}r,$$

$$\delta^{1/2}\frac{\partial Z}{\partial T} - \frac{\partial H}{\partial T} = \delta^{1/2}U - \frac{\partial F}{\partial x} + \frac{\partial}{\partial Y}\left[\frac{F}{S}\left\{\frac{\partial Z}{\partial Y} - \frac{\beta}{H + \delta^{1/3}\beta}\frac{\partial H}{\partial Y}\right\}\right], \tag{6.136}$$

$$T_e \approx (H + \delta^{1/3}\beta)S, \qquad Q = H^{3/2}S^{1/2}. \tag{6.137}$$

The rescaled sediment transport function F is only $O(1)$ with this rescaling if $F \sim \tau_e^{3/2}$, which is of course precisely true for the Meyer-Peter/Müller law:

$$F = \left[T_e - \delta^{1/3}\tau_c^*\right]_+^{3/2}. \tag{6.138}$$

Any other choice of transport law would require a more contorted rescaling.

We can use (6.137) to write (6.138) in the form

$$F = QS + \frac{3}{2}(\delta QS)^{1/3}(\beta S - \tau_c^*) + \cdots, \tag{6.139}$$

from which it follows that the instability parameter A defined in (6.120) is given by

$$A \approx -\delta^{1/3}(\beta S - \tau_c^*)\left(\frac{H}{S}\right)^{1/2}. \tag{6.140}$$

It is a peculiarity of the Meyer-Peter/Müller law that $A = 0$ to leading order, so that the steady state is approximately neutrally linearly stable. Simplification of $(6.136)_2$ now yields

$$-\delta^{1/2}\frac{\partial Z}{\partial T} + \frac{\partial H}{\partial T} = S'S^{1/2}H^{3/2} + S^{1/2}\frac{\partial}{\partial Y}\left[\beta H^{1/2}\frac{\partial H}{\partial Y}\right] + A\frac{\partial^2 Z}{\partial Y^2}, \qquad (6.141)$$

with inessential error terms of $O(\delta^{1/3})$.

(6.141) reveals the essence of linear instability and its nonlinear development. Linear instability is associated with the negative diffusion coefficient of Z if $A > 0$, i.e.,

$$S < S_c = \frac{\mu_c l}{d}, \qquad (6.142)$$

using $(6.20)_7$ and (6.32). In dimensional terms, this suggests instability if the slope is less than μ_c, which occurs precisely at the shoreline. If the resulting rills are able to grow to significant depth, then the nonlinear evolution of H is described approximately by

$$\frac{\partial H}{\partial T} = S'S^{1/2}H^{3/2} + S^{1/2}\frac{\partial}{\partial Y}\left[\beta H^{1/2}\frac{\partial H}{\partial Y}\right], \qquad (6.143)$$

and Z then follows from (6.136) by quadrature.

(6.143) is a degenerate nonlinear diffusion equation, about which a good deal is known. The source term is suggestive (if $S' > 0$, i.e., on the (upper) convex portion of the hillslope) of blow-up, and the possibility that H could reach ∞ at a finite time. The degenerate diffusion coefficient is suggestive of solutions of compact support.[6] If such solutions develop, then the integral constraint (6.134) can be written in the form

$$\int_{-\infty}^{\infty} H^{3/2}\, dY = \frac{2Lrx}{S^{1/2}}. \qquad (6.144)$$

Note that this constraint is independent of Eq. (6.143), which is derived from sediment conservation, whereas (6.144) is a condition of water mass flow. More precisely, the right hand side is an upper bound for the left hand side, since there may be a non-zero flux arising from the outer solution.

Suitable boundary conditions for (6.143) follow from matching to an outer film flow, where $Y \sim 1/\delta^{1/2}$ and $H \sim \delta^{1/3}$. Consequently, we require

$$H \to 0 \quad \text{as } Y \to \pm\infty. \qquad (6.145)$$

A suitable initial condition is less easy to provide, other than stating that H is initially small (since we suppose it arises from an instability of the steady state $H \sim \delta^{1/3}$). The reason for this is that we have omitted an intermediate discussion of the nonlinear stability of the steady state. The long time evolution of an arbitrary (small) perturbation to the steady state can be described by consideration of a

---

[6]Meaning that they are non-zero only on finite interval(s).

Fourier integral over normal modes of wave number k. The upshot of this is that the emerging linear solution is a monochromatic oscillation whose wave number is that with maximum growth rate, and this would serve as a suitable initial condition for the resulting nonlinear equations in (6.131). However, to obtain an appropriate initial condition for (6.143), we really need to know how solutions to (6.131) behave. In seeking solutions at larger amplitude, we are motivated by the fact that developed river channels do attain depths on the order of a metre, and thus we implicitly assume that the nonlinear equations (6.131) do not have bounded stable solutions for H.

6.6.1 Channel Solutions

Suppose the channel depth satisfies (6.143), with coefficients of $O(1)$.[7]

Suppose first that $S' < 0$ (a concave slope, linearly unstable if it is a steady state). The nonlinear algebraic term in (6.143) is thus negative, and we can expect solutions to decay towards zero.[8] This suggests that large channels are not viable, although paradoxically the uniform film state is unstable. The implication is that in this case, finite amplitude rill solutions of (6.131) exist and are stable.

If $S' > 0$, then channels can grow. If we define

$$Y = \left(\frac{2\beta}{3S'}\right)^{1/2}\xi, \qquad T = \frac{\tau}{S'S^{1/2}}, \tag{6.146}$$

then H satisfies

$$H_\tau = H^{3/2} + \left[H^{3/2}\right]_{\xi\xi}, \tag{6.147}$$

with the constraint

$$\int_{-\infty}^{\infty} H^{3/2}\,d\xi = \bar{Q} = \left(\frac{6S'}{\beta S}\right)^{1/2} Lrx. \tag{6.148}$$

With this constraint, there is a unique steady state corresponding to a single isolated channel,

$$H^{3/2} = \frac{\bar{Q}}{2}\cos\xi. \tag{6.149}$$

6.6.2 Bank Migration, Stability and Blow-up

The hallmark of nonlinear diffusion equations such as (6.147) is that we expect to have solutions with compact support, and we expect the margins (where $H = 0$) to

---

[7]For constant uplift U and rainfall r and the Meyer-Peter/Müller transport law (6.139), this is not true for the steady state slope, since then $S = \frac{U}{r} + O(\delta^{1/3})$. Nevertheless it is reasonable to suppose $S' = O(1)$, either because the base state is not in equilibrium, or because uplift and/or rainfall are not uniform.

[8]This is easy to show, by consideration of the time derivative of $\int_{-\infty}^{\infty} H^2\,dY$.

move at finite rates. If $H \approx a(\xi_m - \xi)^\nu$ near a margin $\xi = \xi_m$, then a local balance in (6.147) shows that either $\nu = 2$ (H is smooth) and $\dot{\xi}_m \approx 3a^{1/2}$ (the margin advances) or $\nu = 2/3$ and the margin is stationary (see also Question 6.12). Actually this latter result is inconclusive, since if $H \approx a(\xi_m - \xi)^{2/3} + b(\xi_m - \xi)^{4/3} + \cdots$, then $\dot{\xi}_m \approx \frac{15b}{2a^{1/2}}$, and can be zero, positive or negative. This appears to be the more general result, and it also shows that great care needs to be taken in solving (6.147) numerically.

Let us define

$$u = H^{3/2}, \tag{6.150}$$

so that

$$\frac{2}{3u^{1/3}} u_\tau = u + u_{\xi\xi}, \tag{6.151}$$

with steady state solution

$$u = \frac{\bar{Q}}{2} \cos \xi. \tag{6.152}$$

We write

$$u = \frac{\bar{Q}}{2} \cos \xi + V, \tag{6.153}$$

and linearise the equation on the basis that V is small, so that

$$\frac{2}{3 \cos^{1/3} \xi} V_\tau = V + V_{\xi\xi}. \tag{6.154}$$

If V is zero and analytic at the margins, then the margins are stationary at $\pm \pi/2$, in view of the above discussion on margin migration. Multiplying by V and integrating,

$$\frac{1}{3} \frac{d}{d\tau} \int_{-\pi/2}^{\pi/2} \frac{V^2}{\cos^{1/3} \xi} d\xi = \int_{-\pi/2}^{\pi/2} \left(V^2 - V_\xi^2 \right) d\xi \tag{6.155}$$

(since $V = 0$ at $\pm \pi/2$). Now the variational principle for the right hand side integral as a functional of V tells us that it is maximised when $V \propto \cos \xi$, in which case the right hand side is zero. $V = \cos \xi$ is not an admissible solution to (6.155), since the integral constraint (6.148) also implies that

$$\int_{-\pi/2}^{\pi/2} V \, d\xi = 0, \tag{6.156}$$

and therefore it follows that

$$\frac{d}{d\tau} \int_{-\pi/2}^{\pi/2} \frac{V^2}{\cos^{1/3} \xi} d\xi < 0. \tag{6.157}$$

Equally one can show that all normal modes for V decay. This indicates that the steady state is linearly stable. Numerical computations confirm that the steady state is indeed globally stable. Further discussion of the solutions of (6.147), and in particular of the issue of blow-up, is given in the notes (see also Question 6.13).

Fig. 6.5 Expected solution structure for (6.158)

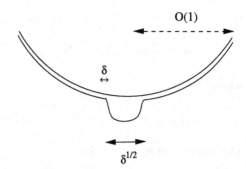

6.7 Channels and Hillslope Evolution

As the channels described by (6.143) evolve, a transverse flow will develop on the hillslope, causing erosion and thus subsidence. The effect of this on the hillslope is that the assumption of a basic one-dimensional downhill slope becomes unrealistic, and a channel will be flanked on either side by different hillslopes. We now show how this situation can be described.

We revert to the model (6.105), using the Meyer-Peter/Müller transport law:

$$\mathbf{V}.[q\mathbf{n}] = r,$$
$$q = h^{3/2}|\mathbf{V}\eta|^{1/2},$$
$$\eta_t - \delta h_t = U - \mathbf{V}.[f\mathbf{N}],$$
$$\mathbf{n} = -\frac{\mathbf{V}\eta}{|\mathbf{V}\eta|},$$
$$\boldsymbol{\tau}_e = -(h+\beta)\mathbf{V}\eta + \delta\beta\mathbf{V}h, \qquad (6.158)$$
$$\mathbf{N} = -\frac{\left[\mathbf{V}\eta - \frac{\delta\beta}{(h+\beta)}\mathbf{V}h\right]}{\left|\mathbf{V}\eta - \frac{\delta\beta}{(h+\beta)}\mathbf{V}h\right|},$$
$$f = [\tau_e - \tau_c^*]_+^{3/2}.$$

Our key to the solution behaviour stems from how we expect channels and hillslope to behave; this is shown in Fig. 6.5. We have already described channels, of width $O(\delta^{1/2})$. On the hillslope, we now seek solutions in which $\eta \sim O(1)$, but η depends on both x and y (i.e., the gradient is not simply in the x direction).

On the hillslope, appropriate scales are $\eta \sim 1$, $h \sim 1$, $q \sim 1$, and the approximate model for $\eta \approx s$ and h is just (6.27), as before. In the channel, which we take to be at $y = 0$, we have $h = H/\delta^{1/3}$, $q = Q/\delta^{1/3}$, $y = \delta^{1/2}Y$, $\eta = \eta_0(x) + \delta Z$ (where $\eta_0(x) = \eta(x, 0)$), $t = \delta T$, and the channel depth satisfies (6.143), and we suppose $H \to 0$ at $Y = \pm Y_0$.

It is clear that if the outer hillslope depends on y (and thus feeds water to the channel), then $\mathbf{V}\eta$ must have a sharp jump between hillslope and channel, and in order for this to occur, there must be a singular transition region near the channel margin (i.e., the river bank). In this bank layer, we put

$$y = \delta^{1/2}Y_0 + \delta\zeta, \qquad \eta = \eta_0(x) + \delta Z, \qquad h \sim 1, \qquad q \sim 1. \qquad (6.159)$$

Again, we may linearise the geometry, and we obtain

$$\mathbf{n} = -\frac{1}{S}(\eta_0', Z_\zeta), \qquad \mathbf{N} \approx \frac{1}{E}\left[-\eta_0', \frac{\beta}{(h+\beta)}h_\zeta - Z_\zeta\right], \tag{6.160}$$

where[9]

$$S = (\eta_0'^2 + Z_\zeta^2)^{1/2}, \qquad E = \left[\eta_0'^2 + \left(\frac{\beta}{(h+\beta)}h_\zeta - Z_\zeta\right)^2\right]^{1/2}. \tag{6.161}$$

The effective shear stress is then

$$\tau_e \approx (h + \beta)E. \tag{6.162}$$

At leading order, the water and sediment conservation equations are simply

$$\frac{\partial}{\partial \zeta}\left[\frac{qZ_\zeta}{S}\right] \approx 0,$$
$$\frac{\partial}{\partial \zeta}\left[\frac{F}{E}\left(\frac{\beta}{(h+\beta)}h_\zeta - Z_\zeta\right)\right] \approx 0, \tag{6.163}$$

which state that the water and sediment fluxes are continuous across the bank. Thus

$$\frac{h^{3/2}Z_\zeta}{S^{1/2}} = K, \qquad \frac{F}{E}\left(\frac{\beta}{(h+\beta)}h_\zeta - Z_\zeta\right) = -C, \tag{6.164}$$

where K and C are constants (the water and sediment fluxes to the channel). The existence of a satisfactory bank transition layer relies on the solutions of (6.164) being able to match both to the hillslope film flow and the deep channel flow.

If the outer hillslope limits of film thickness and normal slope are $h_0 = h(x, 0)$ and $\mu = \frac{\partial \eta(x,0)}{\partial y} > 0$ (for $y > 0$), then we require

$$Z_\zeta \to \mu, \qquad h \to h_0 \quad \text{as } \zeta \to \infty. \tag{6.165}$$

The steady channel solution (6.149) is still applicable, and from this we find that the appropriate matching conditions to the channel are

$$h^{3/2} \sim -A\zeta, \qquad Z_\zeta \to 0 \quad \text{as } \zeta \to -\infty. \tag{6.166}$$

(Since $h \to \infty$ as $\zeta \to -\infty$, (6.163) implies $Z_\zeta \to 0$.) If $\bar{Q}$ is the volume flux in the channel, then

$$A = \frac{3S_0'\bar{Q}}{4\beta S_0^{1/2}}, \tag{6.167}$$

where $S_0 = |\eta_0'|$.

Let S_+ be the slope at the channel of the hillslope, i.e.,

$$S_+ = (S_0^2 + \mu^2)^{1/2}; \tag{6.168}$$

---

[9] E here is unrelated to the dimensionless erosion rate used a long time ago in (6.16).

then as $\zeta \to \infty$, $E \to S_+$, $\tau_e \to (h_0 + \beta)S_+$, and comparison of equations (6.164) with the matching condition (6.165) then shows that

$$K = \frac{h_0^{3/2}\mu}{S_+^{1/2}}, \qquad C = \frac{f\{(h_0 + \beta)S_+\}\mu}{S_+}. \tag{6.169}$$

Equally, as $\zeta \to -\infty$, $S \to S_0$, $E \to S_0$, thus $\tau_e \sim S_0 h$, $f \sim S_0^{3/2} h^{3/2}$, $h^{3/2} \sim -A\zeta$, $h_\zeta/h \sim 2/3\zeta$ and $Z_\zeta \sim K S_0^{1/2}/h^{3/2}$; it follows from (6.166) that

$$K S_0 + \frac{2\beta S_0^{1/2} A}{3} = C. \tag{6.170}$$

It now remains to see whether a connecting solution of (6.164) exists. Using the definition of S in (6.161), we can solve (6.164)$_1$ to find

$$Z_\zeta = U(h) \equiv \frac{K}{\sqrt{2}h^3}\left[K^2 + \{K^4 + 4S_0^2 h^6\}^{1/2}\right]^{1/2}. \tag{6.171}$$

$U(h)$ is a monotonically decreasing function of h, with

$$\begin{aligned} U &\sim \frac{K^2}{h^3} \quad \text{as } h \to 0, \\ U &\sim \frac{K S_0^{1/2}}{h^{3/2}} \quad \text{as } h \to \infty. \end{aligned} \tag{6.172}$$

Next we define

$$L = Z_\zeta - \frac{\beta}{(h+\beta)}h_\zeta. \tag{6.173}$$

We have the Meyer-Peter/Müller law $f = [(h+\beta)E - \tau_c^*]^{3/2}$, where $E = [S_0^2 + L^2]^{1/2}$, and (6.164)$_2$ is simply $fL = EC$. This implies $L > 0$, and simplification then shows that

$$h + \beta = \frac{\tau_c^* + C^{2/3}\left(1 + \frac{S_0^2}{L^2}\right)^{1/3}}{(S_0^2 + L^2)^{1/2}}. \tag{6.174}$$

This defines h in terms of L as a monotonically decreasing function, with $h \to -\beta$ as $L \to \infty$, and $h \sim (\frac{C^2}{S_0 L^2})^{1/3}$ as $L \to 0$. Thus $L(h)$ is a monotonically decreasing positive function, with $L(0)$ being finite and

$$L \sim \frac{C}{S_0^{1/2} h^{3/2}} \quad \text{as } h \to \infty. \tag{6.175}$$

It now follows from (6.173) that h is given by the solution of the ordinary differential equation

$$\frac{\beta}{(h+\beta)}h_\zeta = U(h) - L(h). \tag{6.176}$$

The function $U - L$ is positive and decreasing for small h, and as $h \to \infty$, (6.170), (6.172) and (6.175) imply that $U - L \sim -\frac{2\beta A}{3h^{3/2}}$. Since this is negative, it implies

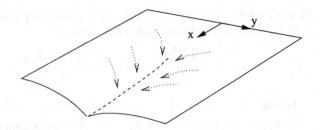

that there is a zero[10] h_0 of $U - L$ and that $U - L < 0$ for $h > h_0$. Consequently, the solution of (6.176) takes h from ∞ (and specifically $h^{3/2} \sim -A\zeta$) as $\zeta \to -\infty$ to h_0 as $\zeta \to \infty$, as required. Thus the bank singular layer engineers the transition we seek in adjusting the hillslope to the channel.

6.7.1 Hillslope Evolution

The analysis of the preceding section provides a series of recipes which relate the hillslope to the channel. Now we show how this combination provides a way to compute the long term evolution of the hillslope without having to resolve (computationally) the details of the bank and channel.

The situation we have in mind is shown in Fig. 6.6. Suppose that we know the channel location, i.e., $\eta_0(x)$; then we also know S_0. We can therefore solve (6.27) for $s \approx \eta$ and h, given zero flux conditions on the adjoining ridges, and in this way μ and h_0 are determined.

Equations (6.167) and (6.169) provide three prescriptions for A, K and C, and then (6.170) determines an ordinary differential equation for S_0. If we take $f = [\tau_e - \tau_c^*]^{3/2}$, $\tau_e \approx (h + \beta)S_+$, where (6.168) gives S_+ in terms of S_0, then we have

$$\bar{Q}S_0' = \frac{2\mu}{S_+}\left[(h_0 + \beta)^{3/2}S_+^{3/2} - \tau_c^* - h_0^{3/2}S_+^{1/2}S_0\right]. \tag{6.177}$$

In addition the water flux to the channel is just $2K$ (if we suppose that a symmetric slope lies in $y < 0$). Therefore the water flux $\bar{Q}$ in the channel satisfies

$$\bar{Q}' = \frac{2\mu h_0^{3/2}}{S_+^{1/2}}. \tag{6.178}$$

(The prime denotes differentiation with respect to x.) A similar equation for the sediment flux in the channel is satisfied, $\frac{d}{dx}\int_{-Y_0}^{Y_0} F\, dY = 2C$. Since F (rescaled) in the channel is approximately QS_0, it follows from this that $(S_0\bar{Q})' = 2C$, but in fact this is already implied by (6.177) and (6.178).

---

[10]Probably only one, but if there were more, then h_0 is taken as the largest zero.

Equations (6.177) and (6.178) provide two ordinary differential equations for channel water flux $\bar{Q}$ and channel slope S_0. Since $S_0 = -\eta_0'$, we therefore have a second order differential equation for channel elevation η_0, with boundary conditions specifying its source (contiguous with the hillslope) and terminus. In general, these equations must be solved numerically (see also Question 6.14). Note that $S_0' > 0$, consistent with the basic assumption in the channel description.

6.7.2 Detachment Limited Erosion

One of the worrying features of *Smith–Bretherton* model and our discussion of it is that channels form when $S' > 0$, i.e., the hillslope is convex, and indeed the resulting channel long profile is also convex. This is inconsistent with what we expect from the linear stability results, and also in practice. Consulting (6.136), we can see that the source term in the nonlinear diffusion equation for channel depth h, (6.143), arises from the term $\frac{\partial F}{\partial x}$, and specifically from its S component, i.e., $\frac{\partial F}{\partial S} S'$. Since $\frac{\partial F}{\partial S} > 0$, positivity of the source requires $S' > 0$.

This is odd, because we would expect sediment flux to increase with distance downstream irrespective of whether the slope is convex or concave. In fact, there is something not quite right with the use of the sediment transport law in the form $f \approx q\,S$, because it implies that there is a sediment flux even if there is no sediment! The resolution of this paradox lies in the formulation of the original sediment transport model in the form (6.20). There we allowed for the existence of a non-zero thickness a of the bedload layer. In order to pose a more physically realistic transport law, we need to retain the dependence of sediment flux on a, and therefore we modify our definition of the sediment flux from that in (6.21) to be

$$\mathbf{Q}_b = f\mathbf{N}, \tag{6.179}$$

where

$$f = a\hat{v}. \tag{6.180}$$

Because of the specific inclusion of a, we see that $\hat{v}$ is the mean bedload speed, and evidently this will be in the direction of the mean effective stress, $\mathbf{N}$. We expect that $\hat{v}$ will depend on flow speed and slope, and indeed that it will be consistent with measured transport rates.

Equally, we suppose the abrasion rate A in (6.20) must depend on a. Erosion or mobilisation of an underlying rock or compacted sediment due to the rubbing of a mobile overlying layer of thickness can be expected to decrease as a increases, and to be specific, we will suppose that

$$A = \hat{A}[1 - a]_+, \tag{6.181}$$

where $\hat{A}$ will depend on the stress, and the (dimensionless) thickness at which bed abrasion ceases can be taken to be $a = 1$ by the choice of scale for a.[11] In order that f give the Meyer-Peter/Müller result when $a = 1$, we take

$$\hat{v} = [\tau_e - \tau_c^*]_+^{3/2}, \tag{6.182}$$

and to be specific, we will suppose that

$$\hat{A} = \alpha\hat{v}, \tag{6.183}$$

since the driving process for bedload transport and abrasion is the same. The dimensionless parameter α is thus the ratio of abrasion rate to uplift rate.

Consulting (6.20), we see that in a channel, with $s = \eta - \delta h - \delta va$, $\eta = \eta_0 + \delta Z$, $h = H/\delta^{1/3}$, $t = \delta^{7/6}T$, $f = F/\delta^{1/2}$, $\tau_e = T_e/\delta^{1/3}$, and if we put $A = \tilde{A}/\delta^{1/2}$, $\hat{v} = V/\delta^{1/2}$, then

$$\delta^{1/3}[Z_T - va_T] - H_T = \delta^{1/2}U - \tilde{A},$$
$$\delta^{1/3}va_T + \mathbf{\nabla}.[F\mathbf{N}] = \tilde{A}, \tag{6.184}$$

where

$$F = aV, \qquad \tilde{A} = \alpha V(1-a), \qquad V = [T_e - \delta^{2/3}\tau_c^*]_+^{3/2}. \tag{6.185}$$

At leading order in a channel, we thus have

$$H_T = \tilde{A} = \mathbf{\nabla}.[F\mathbf{N}]. \tag{6.186}$$

The first thing to do is to see whether this prescription for sediment transport can be consistent with the observation that bedload transport depends only on bed stress. Suppose that in a one-dimensional flume, the flow and depth are constant, so that also V is constant. Then we have

$$Va_x = \alpha V(1-a), \tag{6.187}$$

whence

$$a = 1 - \exp(-\alpha x). \tag{6.188}$$

This immediately identifies two limiting régimes, *transport limited* and *detachment limited*. The transport limited régime occurs when $\alpha \gg 1$. Then $a \to 1$ rapidly, and the sediment transport $F \approx V$, which gives the Meyer-Peter/Müller result. If, however, $\alpha \ll 1$, we have detachment limited transport, in which $F \approx \alpha Vx$, and transport is limited by the rate that sediment can be eroded from the underlying regolith. Arguably, this is the more likely situation in mature mountainous terrain.

Let us now revisit the derivation of the deep channel equation, allowing for the variable bedload density. In the channel, $V \approx QS$, and thus $\tilde{A} \approx \alpha QS(1-a)$, and

---

[11]This choice is distinct from our earlier assumption in (6.20), which was that $U \sim A$. By selecting the scale for a to be the presumed value where abrasion ceases, we cannot guarantee $A \sim U$, and therefore A is not necessarily $O(1)$.

$F \approx aQS$. Using the linearised geometric description of $\mathbf{N}$ in (6.110), we find that (6.186) becomes

$$H_T = \alpha(1-a)QS = \frac{\partial F}{\partial x} - \frac{\partial}{\partial Y}\left(\frac{F}{S}\frac{\partial Z}{\partial Y}\right) + \frac{\partial}{\partial Y}\left[\frac{\beta F}{SH}\frac{\partial H}{\partial Y}\right], \tag{6.189}$$

while the water flow equation is approximately

$$SQ_x = \frac{\partial}{\partial Y}\left(Q\frac{\partial Z}{\partial Y}\right), \tag{6.190}$$

and $Q \approx H^{3/2}S^{1/2}$.

We can use (6.190) and the definition of Q to write (6.189) in the form

$$\alpha S - a\left(\alpha S + S' + \frac{D}{Q}\right) = Sa_x - \left(Z_Y - \frac{\beta H_Y}{H}\right)a_Y, \tag{6.191}$$

where D is the diffusion term we had before for H,

$$D = \frac{\partial}{\partial Y}\left[\beta H^{1/2}S^{1/2}\frac{\partial H}{\partial Y}\right]. \tag{6.192}$$

Obviously it is no longer simple to solve this equation for a. If we suppose that S and Q are known, then (6.191) is a hyperbolic equation for a. The characteristics come into the channel from the bank and turn downstream. A reasonable estimate for the solution follows from neglecting the term proportional to a_Y. In this case, the solution satisfying $a = 0$ at $x = x_0$ is

$$a = \frac{\alpha}{S}\int_{x_0}^{x} S(X)\exp\left[-\int_{X}^{x}\left(\alpha + \frac{D}{QS}\right)dX'\right]dX. \tag{6.193}$$

There is a fundamental distinction between the solutions depending on whether $\alpha QS + D \lessgtr 0$. If $\alpha QS + D > 0$, then a increases towards a saturation value. Since we expect $D < 0$ (because of the steady solutions of (6.147)), this (transport limited) case occurs if α is large enough. If $\alpha \gg 1$, then direct asymptotic solution of (6.191) shows that $a = 1 + O(1/\alpha)$, thus $S(1-a) \approx \frac{1}{\alpha}(S' + \frac{D}{Q})$, and therefore

$$\frac{\partial H}{\partial T} \approx QS' + D, \tag{6.194}$$

which is the same result as (6.143), and this is therefore the transport limiting channel equation.

If, on the other hand, $\alpha \ll 1$, then we can expect $\alpha QS + D < 0$, and the active bedload continues to increase, but is small: $a \sim \alpha$. No major simplification of (6.193) is possible, but we do have

$$\frac{1}{\alpha}\frac{\partial H}{\partial T} = QS + O(\alpha). \tag{6.195}$$

Evolution of channel depth is on the longer time scale $T \sim O(1/\alpha)$, and the source term is $QS \approx H^{3/2}S^{3/2}$ and is independent of curvature. We might suppose that the diffusive term is included in the $O(\alpha)$ term, and this is borne out by (6.193) if we make an additional assumption that S, H and therefore also Q vary slowly with x. In that case (bearing in mind that α is small),

$$a \approx -\frac{\alpha Q S}{\mathcal{D}}\left[\exp\left\{-\frac{\mathcal{D}(x - x_0)}{Q S}\right\} - 1\right]. \tag{6.196}$$

If $\mathcal{D}$ is small, we can take a Taylor expansion of (6.196), and then

$$a \approx \alpha(x - x_0) - \frac{\alpha \mathcal{D}}{Q S}\frac{(x - x_0)^2}{2}\cdots, \tag{6.197}$$

and thus

$$\frac{1}{\alpha}\frac{\partial H}{\partial T} = Q S + \alpha\left\{\frac{(x - x_0)^2}{2}\right\}\mathcal{D}, \tag{6.198}$$

with relative error of $O(\alpha)$. The small diffusion term is retained because it represents a singular approximation.

Equation (6.198) perhaps represents the simplest version of the channel evolution equation in detachment limited conditions. It is similar to (6.194), but importantly the source term does not rely on the convexity (or concavity) of the hillslope. The derivation of a pair of equations for channel profile and channel flux analogous to (6.177) and (6.178) is left as an exercise.

6.7.3 Headward Erosion

More or less, the Smith–Bretherton model, or others of similar type, gives us a recipe to evolve a landscape. It is then a computational matter to compare the results of such evolutions with actual landscapes. In our discussion, we have jumped abruptly from the initial instability which will form rills to the evolution of deep channels in a mature landscape.

In a computational experiment (but these have yet to be done), an initially up-lifting island continent will first develop rills at the margin, where the slope bound-ary condition forces concavity. In conditions of plentiful sediment supply, channels cannot apparently form on lower slopes, and it is not until detachment limitation becomes effective that channels will grow, attaining a locally stable equilibrium shape described by steady solutions of, for example, (6.198). The outer hillslope then evolves according to (6.27), with the channel flux and long profile being deter-mined by two equations similar to (6.177) and (6.178). Conditions on the water flux $\bar{Q}$ and the hillslope S at the channel head are that they are determined by continuity with the outer flow. Specifically, the channel head is a singularity for the hillslope, in the sense that upstream of the head, the flow vectors $\mathbf{n}$ must all converge at the channel head, as indicated in Fig. 6.7. In this case, there is a non-zero water flux at the channel head. The alternative possibility is that only a single flow line enters the channel, $\nabla \eta$ is smooth at the head, and the water flux is zero there.

At least for (6.177), the condition $\bar{Q} = 0$ at $x = x_H$ (the channel head position) provides both an initial condition for $\bar{Q}$, and a condition on the channel slope S_0.

Fig. 6.7 Schematic
indication of flow line vectors
n on the hillslope if the head
channel flux is non-zero. A
fan of flow lines reaches the
channel head

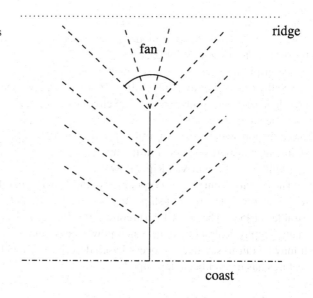

The other condition on the channel profile is that η_0 be continuous with the hillslope
at the head. Thus for zero head flux, the channel head position (and thus also its rate
of migration) is determined by the extra condition on η_0.

Whether or when the situation depicted in Fig. 6.7 can occur mathematically
is less clear. The position of the channel head is then apparently determined by
the condition of intersection of a fan of characteristics of the hillslope water flow
equation. It seems difficult for the hillslope to maintain this ability, so that the zero
flux may be the preferred option.

6.7.4 Side-Branching

The other ingredient for a computational model is the formation of side branches.
This is equivalent to the problem of channel initiation. As a channel develops, the
hillslope on either side will steepen, resulting in instability, rill formation, and eventual channel development. As for a primary channel, the model suggests that tributary channels will form when sediment is depleted sufficiently for detachment limited transport to be appropriate.

6.8 Notes and References

Hillslope Processes   The Davisian cycle of erosion was developed by Davis towards the end of the nineteenth century (e.g., Davis 1899), and the rise and fall of the
theory and its historical context has been voluminously described by Orme (2007),

who incidentally gives a much wider description of the early history of the geo-sciences in general. A more model-based discussion of uplift and erosion is given by Willett and Brandon (2002).

Description of erosion and sedimentation is given in the books by Julien (1995) and Selby (1993), for example. Julien's book is largely concerned with stream flows and their sediment transport, while Selby's is more concerned with hillslope erosion. Between them, they give a good survey of the physical processes of erosion and transport which occur in different types of landscape. Various relationships for sediment transport discussed in the text are given by Shields (1936), Meyer-Peter and Müller (1948) and Van Rijn (1984).

The development of geomorphometry, and in particular the various ordering systems and power-law relationships, are described by Hack (1957), Horton (1945) and Strahler (1952). The book by Turcotte (1992) gives a wide-ranging description of many geophysical contexts in which power laws, and thus fractal processes, arise. It may be fair to say that a process-based descriptive model which can predict such features has not yet been developed.

Channel-Forming Instability   The seminal paper in the theory of channel development is that by Smith and Bretherton (1972). As discussed above, the St. Venant–Exner model is studied (suspended load is neglected, $E = D = 0$), and instability is found to occur for concave surfaces (centre of curvature above ground).

Smith and Bretherton's result was predicated on the basis of the negative diffusion coefficient manifested in (6.45), for example. Loewenherz (1991) carried out a formal linear stability analysis using normal modes, and extended this to convex/concave slopes using WKB theory at high wave number. She also considered the problem of regularisation as $k \to \infty$, by introduction of a (fairly arbitrary) modification to the transport law.

Later (Loewenherz-Lawrence 1994), she treated the whole problem again, but now starting from the hydrodynamic theory, much as we have done here. The rôle of the parameter δ (representing film thickness) becomes critical. She also takes a more general form of sediment transport law,

$$\mathbf{Q}_b = f_D \mathbf{m} + f_A \mathbf{n}, \tag{6.199}$$

where f_A and f_D signify advective and diffusive fluxes, and $\mathbf{m} = -\nabla s/|\nabla s|$ is the downslope unit vector. The result of this is that perturbations are damped at large wave number, essentially for the same reason as described here.

Izumi and Parker (1995, see also 2000) were also concerned with the lack of wavelength selection in the Smith–Bretherton theory, and sought to rectify this by explicitly including the yield stress in the transport law. They find a selected wavelength of 33 m in their (hydrodynamic) theory, which they compare favourably (in order of magnitude) with observed mature spacings in the range 60–130 m. In our discussion of the stabilising effect of the slope parameter β, we similarly find a preferred wavelength (see (6.120) and Question 6.11) of order $l/k \sim \delta^{3/4} l \sim 18$ m, for $l = 100$ km and $\delta = 10^{-5}$. Izumi and Parker (1995) also share the idea in their discussion that hillslope development will be accompanied by a coarsening process, whereby drainage basins will gradually enlarge, with larger streams capturing smaller ones.[12]

Relatively few authors have attempted to describe the evolution of channels within the confines of a general hillslope evolution model. One example of this type of model is that of Kramer and Marder (1992), also discussed by Fowler (1997, pp. 266 ff.), and see Question 6.1. This model is essentially that of Eqs. (6.2)–(6.4), with acceleration neglected in the momentum equation (small Froude number), bed abrasion included but uplift omitted. Sediment transport is modelled by a term $\mathbf{q}_b \propto -\nabla s$, thus sediment moves down the bed slope. This model exhibits channel-forming instability which does not depend on hillslope curvature. Kramer and Marder discuss possible nonlinear solutions for these channels. With certain assumptions (see Question 6.1), one can derive a nonlinear evolution equation for (dimensionless) channel depth H in the form

$$H_t = H^{3/2} - 1 + \left[H^{3/2} H_Y\right]_Y, \tag{6.200}$$

where Y is the cross stream spatial variable. Depending on the boundary conditions, this model allows finite amplitude steady states. A similar nonlinear development for the Smith–Bretherton model as discussed in this chapter was carried out by Winstanley (2001).

Smith et al. (1997a, 1997b) use what we have called the 'outer' approximation (essentially (6.27) with no uplift) to represent mature hillslope evolution, and find certain classes of separable solutions, whose stability they then examine. They also compute solutions numerically, but a representation of channels is not included. A recent paper by Smith (2010) gives a thorough review of the development of the continuum-based theories, and discusses the instability theory in great detail.

WKB Theory WKB theory was applied by Loewenherz (1991) and Loewenherz-Lawrence (1994) to the study of rill-forming instability of flow over erodible slopes. Succinct treatments of the theory are in the books by Carrier et al. (1966) and Bender and Orszag (1978). The basic idea is very simple. The equation

$$w'' + k^2 q(x)w = 0, \tag{6.201}$$

where $w' = dw/dx$ and k is large, has solutions $\exp[\pm ikq^{1/2}x]$ if q is constant. These oscillate rapidly if $q > 0$. If q varies, we can still expect rapidly oscillating solutions, but the period will vary. The *ansatz*

$$w \sim \exp[kw_0 + w_1 + \cdots] \tag{6.202}$$

and subsequent expansion in powers of k lead to the solutions

$$w_0 = \pm i \int^x q^{1/2}\, dx, \qquad w_1 = -\frac{1}{4}\ln|q|, \tag{6.203}$$

---

[12]Such coarsening is familiar in the equally dendritic environment of a solidifying alloy, see for example Marsh and Glicksman (1996).

and thus

$$w \sim \frac{1}{|q|^{1/4}} \exp\left[\pm i \int^x q^{1/2}\,dx\right]. \qquad (6.204)$$

The one term approximation is called geometric optics, the two term approximation (i.e., as in (6.204)) is called physical optics. Evidently, something unusual happens if q goes through zero. Near such transition points, w approximately satisfies the Airy equation, and the resulting Airy function solutions enable a connection to be made between oscillatory solutions (where $q > 0$) and exponential solutions (where $q < 0$) on either side of the transition point.

The Equation $\phi''' - \zeta\phi' + (p-1)\phi = 0$ The properties of the solutions to this equation have been extensively investigated by Bill Reid in connection with asymptotic solutions of the Orr–Sommerfeld equation. See for example Reid (1972) and Lakin et al. (1978), as well as the book by Drazin and Reid (1981). The asymptotic properties used here are given succinctly by Hershenov (1976) and Baldwin (1985).

Nonlinear Diffusion and Blow-up for $H_T = H^{3/2} + (H^{1/2}H_Y)_Y$ Properties of the solutions of this equation (and generalisations) are extensively discussed in the book by Samarskii et al. (1995). If we solve the equation of the heading in an interval $[-L, L]$, with $H = 0$ on $\pm L$ and $H \geq 0$ at $T = 0$, then the solution blows up in finite time if $L > \pi/\sqrt{6}$ (see Question 6.13). In general, solutions of compact support remain so, thus the margins (where H reaches zero) move at finite speed, as expected for such degenerate diffusion equations. One can even construct certain exact solutions which blow up.

So why do we claim that blow-up does not occur? The proof of blow-up involves the essential assumption that $H^{1/2}H_Y = 0$ when $H = 0$, thus there is no efflux of H from the channel, and because of this and the superlinear source, it is almost obvious that blow-up will occur. In our case, $H^{1/2}H_Y \neq 0$ at the margins, and the production of H can be balanced by its loss at the margins. Indeed, the extra integral constraint $\int_{-Y_0}^{Y_0} H^{3/2}\,dY = \bar{Q}$ which we pose to complete the formulation of the problem for H almost implies boundedness of the solution for all time. Fowler et al. (2007) provide detailed numerical analysis of the equation, which shows that the steady state is globally stable, and the solution attains compact support instantaneously, even if the initial condition has infinite support.

Lattice Models   As Willgoose et al. (1991) remark concerning hillslope models, "the difficulty of the problem is such that the number of researchers that have attempted to unify the geomorphology and the hydrology is small". And in fact, the subject took a right angled turn at the beginning of the 1990s. Faced with a relatively simple pair of equations for hillslope and water flow, and a successful characterisation of the hillslope instability criterion, efforts to compute the solutions directly simply foundered. Kramer and Marder (1992) state the consequence as follows:

> The computer time to study [the model] on a large scale is formidable. In order to learn about larger-scale features of river networks, we have abstracted from our previous work a simple lattice model ...

Essentially, this has been the standard approach in the last 20 or so years. Kramer and Marder used a lattice with rules based on sediment and water conservation to prescribe transport between adjacent nodes in each time step. In the lattice model, channels are interpreted to be where flow is maximum. In Willgoose et al.'s (1991) approach, a channel indicator function is used as an artificial switch to decide when and where a channel appears.

The direction in which these lattice models have gone is exemplified by the book of Rodríguez-Iturbe and Rinaldo (1997), which relentlessly emphasises the fractal character of river networks, and the lattice-type models which simulate them. A related 'principle' which is sometimes invoked is that of optimality, the idea that river networks organise themselves in order to minimise energy expenditure. As with the long-standing idea that turbulent convection maximises heat transport, there appears to be no deductive basis to this idea.

Howard (1994) also uses a cellular model to compute simulations, but he emphasises the importance of detachment limited transport: indeed, he argues that fluvial erosion of natural slopes and headwater channels is dominantly detachment limited. This idea nicely dovetails with our apparently similar deduction from the (generalised) Smith–Bretherton transport model. However, his model does not explicitly model weathering (i.e., what we have called abrasion) in the way described here.

Tucker and Slingerland (1994) use a cellular model to simulate landscape evolution, with a view to explaining the formation and retreat of escarpments in high elevation rifted continental margins, such as in southern Africa or eastern Australia. Their model, in its differential equation form, is similar to that presented here, insofar as the weathering rate (A) is specifically included, so that conservation of rock and conservation of sediment provide two equations equivalent to (6.8) and (6.6), and a variable equivalent to sediment thickness a is included. As here, the weathering rate is a monotonically decreasing function of a. In their simulations, they note that there is a profound difference between transport limited and supply (detachment) limited erosion, with the latter providing a mechanism for scarp formation and retreat. A recent survey of lattice-type models is given by Willgoose (2005).

6.9 Exercises

6.1 Kramer and Marder (1992) propose a model for hillslope evolution as follows. With the same notation as in this chapter, their equations for water flow and hillslope evolution are

$$\frac{\partial h}{\partial t} + \nabla.(h\mathbf{v}) = r,$$

$$\frac{\partial s}{\partial t} = -\frac{\rho_w g}{P_0} \frac{h|\mathbf{v}|[1 + |\nabla s|^2]^{1/2}}{[1 + |\nabla \eta|^2]} + \nu \nabla.[h|\mathbf{v}|\nabla s],$$

$$\eta = s + h,$$

where ρ_w is water density, P_0 is an erodibility constant, and the water velocity is

$$\mathbf{v} = -\frac{(2g)^{1/2}h^{1/2}\nabla\eta}{[1+|\nabla\eta|^2]^{1/2}}.$$

Compare this model with that derived in the text. Is it possible to derive this model by making suitable constitutive assumptions?

Non-dimensionalise the model using scales

$$h \sim [h], \qquad \eta, \, s \sim [H], \qquad \mathbf{x} \sim l, \qquad \mathbf{v} \sim [v], \qquad r \sim [r], \qquad t \sim [t],$$

where $[H]$, l and $[r]$ are prescribed. Show that by suitable choice of $[t]$, $[h]$ and $[v]$, the model can be written in the non-dimensional form

$$\frac{\partial h}{\partial t} = \nabla \cdot \left\{ \frac{h^{3/2}\nabla\eta}{[1+\varepsilon^2|\nabla\eta|^2]^{1/2}} \right\} + r,$$

$$\frac{\partial s}{\partial t} = -\frac{\beta h^{3/2}|\nabla\eta|[1+\varepsilon^2|\nabla s|^2]^{1/2}}{[1+\varepsilon^2|\nabla\eta|^2]^{3/2}} + \gamma \nabla \cdot \left\{ \frac{h^{3/2}|\nabla\eta|\nabla s}{[1+\varepsilon^2|\nabla\eta|^2]^{1/2}} \right\},$$

$$\eta = s + \delta h,$$

and show that the parameters are given by

$$\beta = \frac{\rho_w g [h] l}{P_0 [H]}, \qquad \gamma = \frac{v[h]}{l}, \qquad \varepsilon = \frac{[H]}{l}, \qquad \delta = \frac{[h]}{[H]},$$

where

$$[h] = \left\{ \frac{l^2[r]}{(2g)^{1/2}[H]} \right\}^{2/3}.$$

Use values $v = 0.1$ and $P_0 = 3 \times 10^{10}$ kg m^{-1} s^{-2} along with other typical values for $[H]$, l and $[r]$ to find estimates for ε, δ, β and γ. Deduce that the model may reasonably be approximated as

$$\frac{\partial h}{\partial t} = \nabla \cdot \left\{ h^{3/2}\nabla\eta \right\} + r,$$

$$\frac{\partial s}{\partial t} = -\beta h^{3/2}|\nabla\eta| + \gamma \nabla \cdot \left\{ h^{3/2}|\nabla\eta|\nabla s \right\},$$

explaining why the terms in β and γ are important.

Now suppose that $r = 0$ and a uniform flow descends a uniform slope in the x direction. Show that an exact solution representing this exists in the form

$$\eta = -\beta t - x, \qquad h = 1.$$

Show further that if $\eta = -\beta t - x$, then the sediment transport equation admits solutions in which $h = h(y, t)$, where y is the cross stream coordinate. By choosing a suitable rescaling of y and t, show that h satisfies the equation

$$\frac{\partial h}{\partial T} = h^{3/2} - 1 + \frac{\partial}{\partial Y} \left[h^{3/2} \frac{\partial h}{\partial Y} \right].$$

Show that the uniform solution $h = 1$ is unstable, and show that steady channels $h = h(Y)$ exist satisfying $h(\pm\infty) = 0$, where

$$h_Y^2 = \frac{K}{h^3} + \frac{4}{5h^{1/2}} - \frac{1}{2h},$$

and K is a non-negative constant. Show that solutions of this equation reach zero at a finite value of $Y = Y_0$, say, and show that the maximum depth h_m (at $Y = 0$) is defined (uniquely) in terms of the non-negative K by

$$K = \frac{1}{2}h_m^4 - \frac{4}{5}h_m^{5/2}.$$

Show that a unique channel shape is identified (by its maximum depth) if the channel flux $Q = \int_{-Y_0}^{Y_0} h^{3/2}\,dY$ is prescribed, and show in this case that h_m is determined by

$$Q = 2h_m^{11/4}\sqrt{10}\int_0^1 \frac{u^3\,du}{[5h_m^{3/2}(1 - u^4) - 8(1 - u^{5/2})]^{1/2}}.$$

Does this determine h_m uniquely?

6.2 In two-dimensional laminar flow of a film of depth h down an incline, the horizontal velocity u satisfies

$$\mu\frac{\partial^2 u}{\partial z^2} = \rho_w g\eta'(x),$$

where μ is the viscosity, x is the downstream coordinate and $z = \eta$ is the top surface. Suitable boundary conditions for the flow are of no stress, $\partial u/\partial z = 0$, at $z = \eta$, and no slip, $u = 0$, at $z = s = \eta - h$. Assuming h is constant, show that the mean velocity is

$$\bar{u} = -\frac{\rho_w g h^2\eta'}{3\mu},$$

and explain why the horizontal momentum equation for averaged laminar film flow can be taken to be of the form

$$\mathbf{u}_t + (\mathbf{u}.\nabla)\mathbf{u} = -g\nabla\eta - \frac{k\mathbf{u}}{h^2},$$

where $k = \frac{3\mu}{\rho_w}$. Show also that the boundary stress is

$$\tau = \frac{\rho_w k\mathbf{u}}{h},$$

and that mass conservation for water flow takes the form

$$h_t + \nabla.(h\mathbf{u}) = r,$$

where r is rainfall.

Non-dimensionalise the momentum and mass flow equations for water appropriately using the laminar definition of the stress, and show that the corresponding velocity and depth scales are

$$[h] = \left(\frac{r_D l^2 k}{gH}\right)^{1/3}, \qquad [u] = \left(\frac{r_D^2 lgH}{k}\right)^{1/3},$$

where l is the topographic horizontal length scale and H is the topographic elevation scale. Use values $r_D \sim 1$ m y^{-1}, $l \sim 10^5$ m, $k \sim 3 \times 10^{-6}$ m^2 s^{-1}, $g \sim 10$ m s^{-2}, $H \sim 10^3$ m, to find typical values for $[u]$ and $[h]$. Hence show that the Reynolds number for the flow is $Re = [u][h]/k \sim 10^3$.

Show that the effective dimensionless driving stress for sediment transport is

$$\boldsymbol{\tau}_e = \mathbf{u} - \gamma \nabla s,$$

where

$$\gamma = \frac{\Delta \rho_{sw}\, g\, D_s\, H[h]}{\rho_w l k [u]}.$$

Using the additional values $\Delta \rho_{sw}/\rho_w \sim 1.6$, $D_s \sim 1$ mm, show that $\gamma \approx 0.34$.

If the yield stress is defined by

$$\tau_c = \mu_c \Delta \rho_{sw} g D_s,$$

show that the dimensionless yield stress is

$$\tau_c^* = \frac{\mu_c \gamma l}{H} \approx 1.7,$$

if $\mu = 0.05$.

Hence show that an approximate model of hill slope erosion is provided by the dimensionless equations

$$-\nabla.\left[h^3 \nabla s\right] = r,$$

$$s_t = U + \nabla.\left[F(\tau_e)\nabla s\right],$$

$$\boldsymbol{\tau}_e = \left(h^2 + \gamma\right)S,$$

where $S = |\nabla s|$, and F is the dimensionless sediment transport rate.

6.3   Suppose that in the dimensionless model for hillslope erosion

$$\delta \varepsilon h_t + \nabla.(h\mathbf{u}) = r,$$

$$\delta F^2\left[\delta \varepsilon \mathbf{u}_t + (\mathbf{u}.\nabla)\mathbf{u}\right] = -\nabla \eta - \frac{|\mathbf{u}|\mathbf{u}}{h},$$

$$\eta = s + \delta h + \delta v a,$$

$$\delta v a_t + \nabla.\mathbf{Q}_b = A,$$

$$s_t = -A + U,$$

$$\boldsymbol{\tau}_e = |\mathbf{u}|\mathbf{u} - \beta \nabla s,$$

rainfall occurs in brief, severe storms. Suppose that the rainfall intensity during these storms is r_S, so that in dimensionless terms we have (why?) $r \sim 1/\omega$ for periods of $O(\omega)$ at intervals of $O(1)$, where $\omega = r_D/r_S \ll 1$, r_D being the scale for the rainfall.

Show that between storms, $s_t \approx U$, while during the storms, the approximations $\delta, \varepsilon \to 0$ can be made if $\delta \varepsilon \ll \omega^{2/3}$ and $\delta F^2 \ll \omega^{2/3}$. Use the definitions

$$\varepsilon = \frac{U_D}{r_D}, \quad \delta = \frac{[h]}{d}, \quad F^2 = \frac{[u]^2}{g[h]}, \quad [u] = \left(\frac{g r_D d}{f}\right)^{1/3}, \quad [h] = l\left(\frac{f r_D^2}{g d}\right)^{1/3},$$

to show that

$$\frac{\delta F^2}{\omega^{2/3}} = \left(\frac{r_S}{f\sqrt{gd}}\right)^{2/3},$$

and

$$\frac{\delta \varepsilon}{\omega^{2/3}} = f\left\{\frac{U_D}{r_D}\frac{l}{d}\right\}\left(\frac{r_S}{f\sqrt{gd}}\right)^{2/3}.$$

On the assumption that $H/l \sim U_D/r_D$, deduce that the same model can be applied for time varying storm flow provided the maximum rainfall intensity satisfies $r_S \ll f\sqrt{gH}$, and show that this is always the case.

6.4 Suppose that there is a suspended sediment mean concentration $c > 0$, and that in the model

$$\delta \varepsilon h_t + \mathbf{\nabla}.(h\mathbf{u}) = r,$$

$$\delta F^2\left[\delta \varepsilon \mathbf{u}_t + (\mathbf{u}.\mathbf{\nabla})\mathbf{u}\right] = -\mathbf{\nabla}\eta - \frac{|\mathbf{u}|\mathbf{u}}{h},$$

$$\eta = s + \delta h + \delta v a,$$

$$\delta \varepsilon (hc)_t + \mathbf{\nabla}.(hc\mathbf{u}) = \gamma(E - cD),$$

$$\delta v a_t + \mathbf{\nabla}.\mathbf{Q} = -\alpha(E - cD) + A,$$

$$s_t = -A + U,$$

$$\tau_e = |\mathbf{u}|\mathbf{u} - \beta \mathbf{\nabla}s,$$

we have $\alpha \sim \gamma \gg 1$, $\varepsilon, \delta \ll 1$, E is an increasing function of u, and D is a decreasing function of u. Show that, approximately, $c = c(u)$ is an increasing function of stream speed, and that the hillslope erosion equation is approximately

$$\frac{\partial s}{\partial t} = U - \mathbf{\nabla}.\mathbf{Q} - \frac{\alpha}{\gamma}\mathbf{\nabla}.[hc\mathbf{u}].$$

Deduce that the suspended load term acts as a stabilising, diffusive influence on the hillslope evolution.

How do solutions behave if $\alpha \gg \gamma \gg 1$?

6.5 In one space dimension (with x pointing downhill), a model for overland flow and hillslope erosion takes the form

$$q_x = r, \qquad s_t + F_x = U. \qquad (*)$$

Assuming that $F = F(q, S)$ and $S = -s_x$, write the system in the form

$$A\begin{pmatrix} s \\ q \end{pmatrix}_t + B\begin{pmatrix} s \\ q \end{pmatrix}_x = \mathbf{c},$$

specifying the matrices A and B, and the vector $\mathbf{c}$.

Hence show that the characteristic speeds satisfying $\det(\lambda A - B) = 0$ are $\lambda = \frac{F_S}{F_q}$ and $\lambda = \infty$, and thus that the system is a mixed hyperbolic/parabolic system.

[To say that one root of the equation is infinity means that the normal quadratic one would expect has degenerated to a linear equation, because the rank of A is only one. This is because of the presence of only one time derivative in the pair of equations.]

Show that if a term εq_t is added to the first equation in $(*)$ above, where ε is small, then a second finite root of $O(1/\varepsilon)$ occurs. [*This root tends to infinity as $\varepsilon \to 0$.*]

6.6 Suppose that in a model of hillslope erosion, sediment transport is given by $F = q^\alpha S^\beta$, where q is water flux and S is surface slope, and α and β are constant, and that uplift U and rainfall r are constant. Show that if x measures distance from a ridge of a one-dimensional hillslope (i.e., whose elevation depends only on x), and the equations for q and F are

$$q_x = r, \qquad F_x = U,$$

then the slope $S = cx^v$ for values of c and v which you should determine. Hence show that (geomorphically) concave slopes occur if $\alpha > 1$, and concave slopes occur if $\alpha < 1$.

6.7 A tectonic province is uplifted rapidly, and thereafter erodes according to the equation

$$\frac{\partial s}{\partial t} = \nabla. \left[\frac{F \nabla s}{S} \right],$$

where $S = |\nabla s|$ is the terrain slope. Suppose that $F = qS$, and that

$$\nabla. \left[\frac{q \nabla s}{S} \right] = -p,$$

where p represents precipitation.

If the topography varies only in the x direction, show that a similarity solution can be found in the form $s = \frac{1}{t} f(\frac{x}{t})$, and hence show that $f(\sigma) = B \exp(-|\sigma|)$. What determines the constant B?

Find a comparable (cylindrically symmetric) similarity solution in two dimensions, and show in this case that

$$s = \frac{B}{t^2} \exp \left[-\frac{pr}{2t} \right].$$

Determine B in terms of the initial uplifted volume V.

6.8 Suppose that in the stability equations

$$\phi' = -\frac{k^2 q \psi}{S},$$

$$\sigma \psi = [f_S \psi']' - (f_q \phi)' - \frac{k^2 f \psi}{S},$$

subject to

$$\phi = f_S \psi' = 0 \quad \text{at } x = 0,$$

$$\psi = 0 \quad \text{at } x = 1,$$

we take $k = 0$, i.e., perturbations are in the x direction only. Show that $\phi = 0$ and that ψ satisfies the equation

$$\sigma \psi = [f_S \psi']',$$

with

$$f_S \psi' = 0 \quad \text{at } x = 0,$$

$$\psi = 0 \quad \text{at } x = 1.$$

Allowing for the fact that ψ and σ may be complex, show that

$$\sigma \int_0^1 |\psi|^2 \, dx = -\int_0^1 f_S |\psi'|^2 \, dx,$$

and deduce that all eigenvalues σ are negative, i.e., one-dimensional perturbations are stable. What does the assumption $f_S > 0$ mean physically?

6.9 Suppose that ϕ satisfies the differential equation

$$\Lambda \phi' = \frac{q}{S} \left[\frac{1}{k^2} \left\{ F_S \left(\frac{S}{q} \phi' \right)' \right\}' + F_q' \phi \right],$$

with boundary conditions

$$\phi = F_S \left(\frac{S}{q} \phi' \right)' = 0 \quad \text{at } x = 0,$$

$$\phi' = 0 \quad \text{at } x = 1.$$

Show that if $F = qS$ and S is constant, then $\Psi = S\phi'/q$ satisfies the equation

$$\Lambda \Psi = \frac{1}{k^2} [q \Psi']',$$

with

$$q \Psi' = 0 \quad \text{on } x = 0,$$

$$\Psi = 0 \quad \text{on } x = 1,$$

and then

$$\phi = \int_0^x \frac{q\Psi}{S} \, dx.$$

Suppose that $q = x$. Show that the solution for Ψ which satisfies both boundary conditions is

$$\Psi = J_0 \left[2k(-\Lambda)^{1/2} x \right],$$

providing

$$\Lambda = -\left(\frac{j_{0,n}}{2k}\right)^2,$$

where $j_{0,n}$ is the nth (positive) zero of the Bessel function J_0.

Compare this result with that obtained from a WKB analysis assuming $q \neq 0$.

[*The zeros of the zeroth order Bessel function satisfy* $j_{0,n} \sim (n - \frac{1}{4})\pi$ *for large n.*]

6.10 The function $A(\zeta, p)$ is defined by the integral

$$A(\zeta, p) = \frac{1}{2\pi i} \int_L t^{-p} e^{\zeta t - \frac{1}{3}t^3} \, dt,$$

where the contour L goes from $\infty \, e^{4\pi i/3}$ to $\infty \, e^{2\pi i/3}$ in the complex t plane (it is the contour L_1 in Fig. C.1). Use the method of steepest descents to find the asymptotic behaviour of $A(\zeta, p)$ as $\zeta \to \pm\infty$.

6.11 Show that for $\gamma, \alpha, \lambda, A$ and B of $O(1)$, and $\varepsilon \ll 1$, the roots of the polynomial

$$\frac{\varepsilon\alpha}{\gamma}m^3 + \varepsilon(\lambda + B)m^2 - (\lambda - A)m - 2\gamma(\lambda + B) = 0$$

are given by

$$m = m_0 \approx \frac{2\gamma(\lambda + B)}{A - \lambda}, \qquad m = m_\pm \approx \pm\frac{i\theta}{\sqrt{\varepsilon}} - \nu + \cdots,$$

where

$$\theta = \left[\frac{\gamma(A - \lambda)}{\alpha}\right]^{1/2}, \qquad \nu = \gamma(\lambda + B)\left(\frac{1}{2\alpha} + \frac{1}{A - \lambda}\right).$$

Hence show that if

$$\phi(x) = U_0 e^{m_0 x} + U_+ e^{m_+ x} + U_- e^{m_- x}$$

(where one of the coefficients U_k can be chosen arbitrarily), then satisfaction of the boundary conditions $\phi(0) = \phi''(0) = \phi'(1) = 0$ requires

$$U_0 = \frac{2\sqrt{\varepsilon}\nu}{\theta_0} + \cdots, \qquad U_\pm = \pm\frac{1}{2i} - \frac{\sqrt{\varepsilon}\nu}{\theta_0} + \cdots$$

(if U_- is chosen in this way), and

$$\theta = \theta_0 + \frac{\varepsilon\nu}{\theta_0} + \cdots,$$

where $\theta_0 = (r + \frac{1}{2})\pi$, for integral values of r.

Hence show that

$$\lambda \approx \frac{A - \frac{\alpha\theta_0^2}{\gamma} - \varepsilon B\left(1 + \frac{2\gamma}{\theta_0^2}\right) + \cdots}{1 + \varepsilon\left(1 + \frac{2\gamma}{\theta_0^2}\right) + \cdots}.$$

Deduce that

$$\lambda'(\gamma) \approx \frac{\alpha\theta_0^2(1-\varepsilon)}{\gamma} - \frac{2\varepsilon(B+A)}{\theta_0^2},$$

and therefore that λ has a maximum as γ varies at

$$\gamma \approx \left[\frac{\alpha}{2\varepsilon(B+A)}\right]^{1/2}\theta_0^2.$$

6.12 Suppose that u is non-negative and of compact support, and that

$$\frac{2}{3u^{1/3}}u_\tau = u + u_{\xi\xi}.$$

Suppose that $u \approx \alpha[\xi_m(\tau) - \xi]^\nu$ near a margin $\xi = \xi_m(\tau)$ (where $u = 0$), and α and ν are positive. Show by balancing terms that possible choices for ν and $\dot{\xi}_m$ are

$$\nu = 3, \qquad \dot{\xi}_m = 3\alpha^{1/3},$$

and

$$\nu = 1, \qquad \dot{\xi}_m = 0.$$

Show further in the latter case that if

$$u = \alpha(\xi_m - \xi) + \beta(\xi_m - \xi)^{5/3} + \cdots,$$

then

$$\dot{\xi}_m \approx \frac{5\beta}{3\alpha^{2/3}},$$

and deduce that both marginal advance or retreat are possible.

6.13 Suppose that u satisfies the equation

$$u_t = u^{3/2} + \left(u^{3/2}\right)_{\xi\xi}$$

on an interval Ω, with $u = 0$ on $\partial\Omega$, and $u \geq 0$ at $t = 0$. Explain why the solution u will be non-negative.

Let ψ and λ be the first eigenfunction and eigenvalue of the equation

$$\psi'' + \lambda\psi = 0, \qquad \psi = 0 \quad \text{on } \partial\Omega,$$

(thus ψ is of one sign, let us say positive, in Ω: without loss of generality, choose $\int_\Omega \psi \, d\xi = 1$). Show that the positive quantity $E(t) = \int_\Omega u\psi \, d\xi$ satisfies the equation

$$\dot{E} = (1-\lambda)\int_\Omega u^{3/2}\psi \, d\xi.$$

Use the measure $d\omega = \psi \, d\xi$ and Jensen's inequality[13] to deduce that if $\lambda < 1$, $\dot{E} \geq (1-\lambda)E^{3/2}$, and therefore that the solution blows up in finite time if the interval length $|\Omega| > \pi$.

---

[13]See (1.163).

Equally, show that E decreases if $\lambda > 1$ ($|\Omega| < \pi$), so that the solution is apparently bounded.[14]

Show that an exact blow-up solution of the form

$$u^{1/2} = \frac{B \cos^2 \alpha \xi}{t_0 - t}, \quad |\xi| < \frac{\pi}{2\alpha},$$

$u = 0$ otherwise, exists in $\Omega = (-L, L)$ providing $L > \pi/2\alpha$, and if $B = \frac{12}{5}$ and $\alpha = \frac{1}{6}$.

6.14 The equations

$$QS' = \frac{2\mu}{S_+}[(h_0 + \beta)^{3/2} S_+^{3/2} - \tau_c^* - h_0^{3/2} S_+^{1/2} S]$$

and

$$Q' = \frac{2\mu h_0^{3/2}}{S_+^{1/2}},$$

where $S_+ = (S^2 + \mu^2)^{1/2}$, describe the long profiles of channel slope and water flux in the Smith–Bretherton model. Suppose that $\tau_c^* = 0$ and define

$$\Gamma = \left(1 + \frac{\beta}{h_0}\right)^{3/2}.$$

Show that

$$\ln Q = \int^S \frac{dS}{\Gamma(S^2 + \mu^2)^{1/2} - S}.$$

By integrating this, show that

$$Q \propto (S_+ + S)^{\Gamma/(\Gamma^2 - 1)} (\Gamma S_+ - S)^{1/(\Gamma^2 - 1)}.$$

[*The substitution $S = \mu \sec \theta$ may help.*]

6.15 Let $Q = H^{3/2} S^{1/2}$ and Z satisfy the equation

$$S \frac{\partial Q}{\partial x} = \frac{\partial}{\partial Y}\left(Q \frac{\partial Z}{\partial Y}\right)$$

in $0 < Y < Y_0$, where $S = S(x)$ and Y_0 is constant, and where $\frac{\partial Z}{\partial Y} = 0$ at $Y = 0$. Suppose that $H^{3/2} = \frac{\bar{Q}(x)}{2} \cos \xi$, where $\xi = \frac{\pi Y}{2 Y_0}$.

Show that

$$\frac{\partial Z}{\partial Y} - \frac{\beta}{H} \frac{\partial H}{\partial Y} = \frac{2 Y_0}{\pi} g(x),$$

---

[14]Using the Cauchy–Schwarz inequality, the boundedness of E implies the boundedness of the $L^2(\Omega)$ norm. This is not the same as proving the boundedness of u (or equivalently, that of the $L^\infty(\Omega)$ norm).

where $g(x)$ should be determined, and thus show that the characteristics of the equation

$$\alpha S - a\left(\alpha S + S' + \frac{D}{Q}\right) = Sa_x - \left(Z_Y - \frac{\beta H_Y}{H}\right)a_Y,$$

when written in terms of x and ξ, are

$$\sin \xi = \exp\left[-\int_\sigma^x \frac{g(x)}{S(x)}\,dx\right],$$

where $x = \sigma$ when $Y = Y_0$. Draw the shape of the characteristics in the (x, Y) plane, assuming $g > 0$.

Now suppose that

$$D = \frac{\partial}{\partial Y}\left[\beta H^{1/2}S^{1/2}\frac{\partial H}{\partial Y}\right];$$

show that

$$\frac{D}{QS} = -\frac{\pi\beta^2}{6Y_0^2 S},$$

and deduce that on a characteristic,

$$\frac{da}{dx} = \alpha - b(x)a,$$

and determine $b(x)$.

Show that if S is constant, then $b(x)$ is constant, and deduce that

$$a = \frac{\alpha}{b}[1 - \exp\{-b(x - \sigma)\}].$$

Illustrate the behaviour of the solution graphically in the two cases $\alpha \lessgtr \alpha_c$, where

$$\alpha_c = \frac{\pi^2\beta}{6Y_0^2 S}.$$

Show that if also $\bar{Q}$ is constant, then

$$\sigma = x - \frac{1}{\alpha_c}\ln\operatorname{cosec}\xi,$$

and deduce that

$$a = \frac{\alpha}{b}[1 - \sin^{b/\alpha_c}\xi].$$

Chapter 7
Groundwater Flow

Groundwater is water which is stored in the soil and rock beneath the surface of the Earth. It forms a fundamental constituent reservoir of the hydrological system, and it is important because of its massive and long lived storage capacity. It is the resource which provides drinking and irrigation water for crops, and increasingly in recent decades it has become an unwilling recipient of toxic industrial and agricultural waste. For all these reasons, the movement of groundwater is an important subject of study.

Soil consists of very small grains of organic and inorganic matter, ranging in size from millimetres to microns. Differently sized particles have different names. Particularly, we distinguish clay particles (size <2 microns) from silt particles (2–60 microns) and sand (60 microns to 1 mm). Coarser particles still are termed gravel.

Viewed at the large scale, soil thus forms a continuum which is granular at the small scale, and which contains a certain fraction of pore space, as shown in Fig. 7.1. The volume fraction of the soil (or sediment, or rock) which is occupied by the pore space (or void space, or voidage) is called the *porosity*, and is commonly denoted by the symbol ϕ; sometimes other symbols are used, for example n, as in Chap. 5.

As we described in Chap. 6, soils are formed by the weathering of rocks, and are specifically referred to as soils when they contain organic matter formed by the rotting of plants and animals. There are two main types of rock: igneous, formed by the crystallisation of molten lava, and sedimentary, formed by the cementation of sediments under conditions of great temperature and pressure as they are buried at depth.[1] Sedimentary rocks, such as sandstone, chalk, shale, thus have their porosity built in, because of the pre-existing granular structure. With increasing pressure, the grains are compacted, thus reducing their porosity, and eventually intergranular cements bond the grains into a rock. Sediment compaction is described in Sect. 7.11.

Igneous rock tends to be porous also, for a different reason. It is typically the case for any rock that it is fractured. Most simply, rock at the surface of the Earth

---

[1] There are also *metamorphic* rocks, which form from pre-existing rocks through chemical changes induced by burial at high temperatures and pressures; for example, marble is a metamorphic form of limestone.

A. Fowler, *Mathematical Geoscience*, Interdisciplinary Applied Mathematics 36,   387
DOI 10.1007/978-0-85729-721-1_7, © Springer-Verlag London Limited 2011

Fig. 7.1 A granular porous
medium

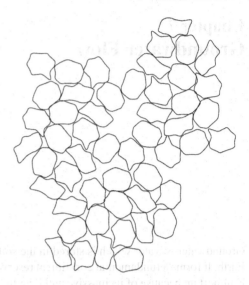

is subjected to enormous tectonic stresses, which cause folding and fracturing of rock. Thus, even if the rock *matrix* itself is not porous, there are commonly faults and fractures within the rock which act as channels through which fluids may flow, and which act on the large scale as an effective porosity. If the matrix is porous at the grain scale also, then one refers to the rock as having a dual porosity, and the corresponding flow models are called double porosity models.

In the subsurface, whether it be soil, underlying regolith, a sedimentary basin, or oceanic lithosphere, the pore space contains liquid. At sufficient depth, the pore space will be saturated with fluid, normally water. At greater depths, other fluids may be present. For example, oil may be found in the pore space of the rocks of sedimentary basins. In the near surface, both air and water will be present in the pore space, and this (unsaturated) region is called the unsaturated zone, or the vadose zone. The surface separating the two is called the piezometric surface, the phreatic surface, or more simply the water table. Commonly it lies tens of metres below the ground surface.

7.1 Darcy's Law

Groundwater is fed by surface rainfall, and as with surface water it moves under a pressure gradient driven by the slope of the piezometric surface. In order to characterise the flow of a liquid in a porous medium, we must therefore relate the flow rate to the pressure gradient. An idealised case is to consider that the pores consist of uniform cylindrical tubes of radius a; initially we will suppose that these are all aligned in one direction. If a is small enough that the flow in the tubes is laminar (this will be the case if the associated Reynolds number is $\lesssim 1000$), then Poiseuille flow in each tube leads to a volume flux in each tube of $q = \frac{\pi a^4}{8\mu}|\nabla p|$, where μ is the liquid viscosity, and ∇p is the pressure gradient along the tube. A more realistic porous medium is *isotropic*, which is to say that if the pores have this tubular shape, the tubules will be arranged randomly, and form an interconnected network. However, between nodes of this network, Poiseuille flow will still be appropriate, and an appropriate generalisation is to suppose that the volume flux vector is given by

$$\mathbf{q} \approx -\frac{a^4}{\mu X}\nabla p, \tag{7.1}$$

where the approximation takes account of small interactions at the nodes; the numerical tortuosity factor $X \gtrsim 1$ takes some account of the arrangement of the pipes.

To relate this to macroscopic variables, and in particular the porosity ϕ, we observe that $\phi \sim a^2/d_p^2$, where d_p is a representative particle or grain size so that $\mathbf{q}/d_p^2 \sim -(\frac{\phi^2 d_p^2}{\mu X})\nabla p$. We define the volume flux per unit area (having units of velocity) as the discharge $\mathbf{u}$. Darcy's law then relates this to an applied pressure gradient by the relation

$$\mathbf{u} = -\frac{k}{\mu}\nabla p, \tag{7.2}$$

where k is an empirically determined parameter called the *permeability*, having units of length squared. The discussion above suggests that we can write

$$k = \frac{d_p^2 \phi^2}{X}; \tag{7.3}$$

the numerical factor X may typically be of the order of 10^3.

To check whether the pore flow is indeed laminar, we calculate the (particle) Reynolds number for the porous flow. If $\mathbf{v}$ is the (average) fluid velocity in the pore space, then

$$\mathbf{v} = \frac{\mathbf{u}}{\phi}. \tag{7.4}$$

If a is the pore radius, then we define a particle Reynolds number based on grain size as

$$Re_p = \frac{2\rho v a}{\mu} \sim \frac{\rho|\mathbf{u}|d_p}{\mu\sqrt{\phi}}, \tag{7.5}$$

since $\phi \sim a/d_p$. Suppose (7.3) gives the permeability, and we use the gravitational pressure gradient ρg to define (via Darcy's law) a velocity scale[2]; then

$$Re_p \sim \frac{\phi^{3/2}}{X}\left(\frac{\rho\sqrt{gd_p}\,d_p}{\mu}\right)^2 \sim 10[d_p]^3, \tag{7.6}$$

---

[2]This scale is thus the hydraulic conductivity, defined below in (7.9).

where $d_p = [d_p]$ mm, and using $\phi^{3/2}/X = 10^{-3}$, $g = 10 \, \text{m s}^{-2}$, $\mu/\rho = 10^{-6} \, \text{m}^2 \, \text{s}^{-2}$. Thus the flow is laminar for $d < 5$ mm, corresponding to a gravel. Only for free flow through very coarse gravel could the flow become turbulent, but for water percolation in rocks and soils, we invariably have slow, laminar flow.

In other situations, and notably for forced gas stream flow in fluidised beds or in packed catalyst reactor beds, the flow can be rapid and turbulent. In this case, the Poiseuille flow balance $-\nabla p = \mu \mathbf{u}/k$ can be replaced by the *Ergun equation*

$$-\nabla p = \frac{\rho |\mathbf{u}| \mathbf{u}}{k'}; \qquad (7.7)$$

more generally, the right hand side will be a sum of the two (laminar and turbulent) interfacial resistances. The Ergun equation reflects the fact that turbulent flow in a pipe is resisted by *Reynolds stresses*, which are generated by the fluctuation of the inertial terms in the momentum equation. Just as for the laminar case, the parameter k', having units of length, depends both on the grain size d_p and on ϕ. Evidently, we will have

$$k' = d_p E(\phi), \qquad (7.8)$$

with the numerical factor $E \to 0$ as $\phi \to 0$.

7.1.1 Hydraulic Conductivity

Another measure of flow rate in porous soil or rock relates specifically to the passage of water through a porous medium under gravity. For free flow, the pressure gradient downwards due to gravity is just ρg, where ρ is the density of water and g is the gravitational acceleration; thus the water flux per unit area in this case is just

$$K = \frac{k \rho g}{\mu}, \qquad (7.9)$$

and this quantity is called the *hydraulic conductivity*. It has units of velocity. A hydraulic conductivity of $K = 10^{-5} \, \text{m s}^{-1}$ (about 300 m y^{-1}) corresponds to a permeability of $k = 10^{-12} \, \text{m}^2$, this latter unit also being called the *darcy*.

7.1.2 Homogenisation

The 'derivation' of Darcy's law can be carried out in a more formal way using the method of homogenisation. This is essentially an application of the method of multiple (space) scales to problems with microstructure. Usually (for analytic reasons) one assumes that the microstructure is periodic, although this is probably not strictly necessary (so long as local averages can be defined).

Consider the Stokes flow equations for a viscous fluid in a medium of macroscopic length l, subject to a pressure gradient of order $\Delta p / l$. If the microscopic (e.g., grain size) length scale is d_p, and $\varepsilon = d_p / l$, then if we scale velocity with $d_p^2 \Delta p / l \mu$ (appropriate for local Poiseuille-type flow), length with l, and pressure with Δp, the Navier–Stokes equations can be written in the dimensionless form

$$\mathbf{\nabla}.\mathbf{u} = 0,$$
$$0 = -\mathbf{\nabla}p + \varepsilon^2 \nabla^2 \mathbf{u}, \tag{7.10}$$

together with the no-slip boundary condition,

$$\mathbf{u} = 0 \quad \text{on } S : f(\mathbf{x}/\varepsilon) = 0, \tag{7.11}$$

where S is the interfacial surface. We put $\mathbf{x} = \varepsilon \boldsymbol{\xi}$ and seek solutions in the form

$$\mathbf{u} = \mathbf{u}^{(0)}(\mathbf{x}, \boldsymbol{\xi}) + \varepsilon \mathbf{u}^{(1)}(\mathbf{x}, \boldsymbol{\xi})\ldots,$$
$$p = p^{(0)}(\mathbf{x}, \boldsymbol{\xi}) + \varepsilon p^{(1)}(\mathbf{x}, \boldsymbol{\xi})\ldots. \tag{7.12}$$

Expanding the equations in powers of ε and equating terms leads to $p^{(0)} = p^{(0)}(\mathbf{x})$, and $\mathbf{u}^{(0)}$ satisfies

$$\mathbf{\nabla}_{\boldsymbol{\xi}}.\mathbf{u}^{(0)} = 0,$$
$$0 = -\mathbf{\nabla}_{\boldsymbol{\xi}} p^{(1)} + \nabla_{\boldsymbol{\xi}}^2 \mathbf{u}^{(0)} - \mathbf{\nabla}_x p^{(0)}, \tag{7.13}$$

equivalent to Stokes' equations for $\mathbf{u}^{(0)}$ with a forcing term $-\mathbf{\nabla}_x p^{(0)}$. If $\mathbf{w}^j$ is the velocity field which (uniquely) solves

$$\mathbf{\nabla}_{\boldsymbol{\xi}}.\mathbf{w}^j = 0,$$
$$0 = -\mathbf{\nabla}_{\boldsymbol{\xi}} P + \nabla_{\boldsymbol{\xi}}^2 \mathbf{w}^j + \mathbf{e}_j, \tag{7.14}$$

with periodic (in $\boldsymbol{\xi}$) boundary conditions and $\mathbf{u} = 0$ on $f(\boldsymbol{\xi}) = 0$, where $\mathbf{e}_j$ is the unit vector in the ξ_j direction, then (since the equation is linear) we have (summing over j)[3]

$$\mathbf{u}^{(0)} = -\frac{\partial p^{(0)}}{\partial x_j} \mathbf{w}^j. \tag{7.15}$$

We define the average flux

$$\langle \mathbf{u} \rangle = \frac{1}{V} \int_V \mathbf{u}^{(0)} \, dV, \tag{7.16}$$

---

[3] In other words, we employ the summation convention which states that summation is implied over repeated suffixes, see for example Jeffreys and Jeffreys (1953).

where V is the volume over which S is periodic.[4] Averaging (7.15) then gives

$$\langle \mathbf{u} \rangle = -\mathbf{k}^*.\nabla p, \tag{7.17}$$

where the (dimensionless) permeability tensor is defined by

$$k_{ij}^* = \langle w_i^j \rangle. \tag{7.18}$$

Recollecting the scales for velocity, length and pressure, we find that the dimensional version of (7.17) is

$$\langle \mathbf{u} \rangle = -\frac{\mathbf{k}}{\mu}.\nabla p, \tag{7.19}$$

where

$$\mathbf{k} = \mathbf{k}^* d_p^2, \tag{7.20}$$

so that $\mathbf{k}^*$ is the equivalent in homogenisation theory of the quantity ϕ^2 / X in (7.3).

7.1.3 Empirical Measures

While the validity of Darcy's law can be motivated theoretically, it ultimately relies on experimental measurements for its accuracy. The permeability k has dimensions of (length)^2, which as we have seen is related to the mean 'grain size'. If we write $k = d_p^2 C$, then the number C depends on the pore configuration. For a tubular network (in three dimensions), one finds $C \approx \phi^2 / 72\pi$ (as long as ϕ is relatively small). A different and often used relation is that of Carman and Kozeny, which applies to pseudo-spherical grains (for example sand grains); this is

$$C \approx \frac{\phi^3}{180(1 - \phi)^2}. \tag{7.21}$$

The factor $(1 - \phi)^2$ takes some account of the fact that as ϕ increases towards one, the resistance to motion becomes negligible. In fact, for media consisting of uncemented (i.e., separate) grains, there is a critical value of ϕ beyond which the medium as a whole will deform like a fluid. Depending on the grain size distribution, this value is about 0.5 to 0.6. When the medium deforms in this way, the description of the intergranular fluid flow can still be taken to be given by Darcy's law, but this now constitutes a particular choice of the interactive drag term in a two-phase flow model. At lower porosities, deformation can still occur, but it is elastic not viscous (on short time scales), and given by the theory of consolidation or compaction, which we discuss later.

---

[4]Specifically, we take V to be the soil volume, but the integral is only over the pore space volume, where $\mathbf{u}$ is defined. In that case, the average $\langle \mathbf{u} \rangle$ is in fact the Darcy flux (i.e., volume fluid flux per unit area).

Table 7.1 Different grain size materials and their typical permeabilities

k (m^2)	Material
10^{-8}	gravel
10^{-10}	sand
10^{-12}	fractured igneous rock
10^{-13}	sandstone
10^{-14}	silt
10^{-18}	clay
10^{-20}	granite

In the case of soils or sediments, empirical power laws of the form

$$C \sim \phi^m \qquad (7.22)$$

are often used, with much higher values of the exponent (e.g. $m = 8$). Such behaviour reflects the (chemically derived) ability of clay-rich soils to retain a high fraction of water, thus making flow difficult. Table 7.1 gives typical values of the permeability of several common rock and soil types, ranging from coarse gravel and sand to finer silt and clay.

An explicit formula of Carman–Kozeny type for the turbulent Ergun equation expresses the 'turbulent' permeability k', defined in (7.7), as

$$k' = \frac{\phi^3 d_p}{175(1 - \phi)}. \qquad (7.23)$$

7.2 Basic Groundwater Flow

Darcy's equation is supplemented by an equation for the conservation of the fluid phase (or phases, for example in oil recovery, where these may be oil and water). For a single phase, this equation is of the simple conservation form

$$\frac{\partial}{\partial t}(\rho\phi) + \nabla.(\rho\mathbf{u}) = 0, \qquad (7.24)$$

supposing there are no sources or sinks within the medium. In this equation, ρ is the material density, that is, mass per unit volume *of the fluid*. A term ϕ is not present in the divergence term, since $\mathbf{u}$ has already been written as a volume flux (i.e., the ϕ has already been included in it: cf. (7.4)).

Eliminating $\mathbf{u}$, we have the parabolic equation

$$\frac{\partial}{\partial t}(\rho\phi) = \nabla.\left[\frac{k}{\mu}\rho\nabla p\right], \qquad (7.25)$$

and we need a further equation of state (or two) to complete the model. The simplest assumption corresponds to incompressible groundwater flowing through a rigid porous medium. In this case, ρ and ϕ are constant, and the governing equation reduces (if also k is constant) to Laplace's equation

$$\nabla^2 p = 0. \tag{7.26}$$

This simple equation forms the basis for the following development. Before pursuing this, we briefly mention one variant, and that is when there is a compressible pore fluid (e.g., a gas) in a non-deformable medium. Then ϕ is constant (so k is constant), but ρ is determined by pressure and temperature. If we can ignore the effects of temperature, then we can assume $p = p(\rho)$ with $p'(\rho) > 0$, and

$$\rho_t = \frac{k}{\mu\phi}\nabla.\left[\rho p'(\rho)\nabla\rho\right], \tag{7.27}$$

which is a nonlinear diffusion equation for ρ, sometimes called the *porous medium equation*. If $p \propto \rho^\gamma$, $\gamma > 0$, this is degenerate when $\rho = 0$, and the solutions display the typical feature of finite spreading rate of compactly supported initial data.

7.2.1 Boundary Conditions

The Laplace equation (7.26) in a domain D requires boundary data to be prescribed on the boundary ∂D of the spatial domain. Typical conditions which apply are a no flow through condition at an impermeable boundary, $\mathbf{u}.\mathbf{n} = 0$, whence

$$\frac{\partial p}{\partial n} = 0 \quad \text{on } \partial D, \tag{7.28}$$

or a permeable surface condition

$$p = p_a \quad \text{on } \partial D, \tag{7.29}$$

where for example p_a would be atmospheric pressure at the ground surface. Another example of such a condition would be the prescription of oceanic pressure at the interface with the oceanic crust.

A more common application of the condition (7.29) is in the consideration of flow in the saturated zone below the water table (which demarcates the upper limit of the saturated zone). At the water table, the pressure is in equilibrium with the air in the unsaturated zone, and (7.29) applies. The water table is a free surface, and an extra kinematic condition is prescribed to locate it. This condition says that the phreatic surface is also a material surface for the underlying groundwater flow, so that its velocity is equal to the average fluid velocity (*not* the flux): bearing in mind (7.4), we have

$$\frac{\partial F}{\partial t} + \frac{\mathbf{u}}{\phi}.\nabla F = 0 \quad \text{on } \partial D, \tag{7.30}$$

if the free surface ∂D is defined by $F(\mathbf{x}, t) = 0$.

7.2.2 Dupuit Approximation

One of the principally obvious features of mature topography is that it is relatively flat. A slope of 0.1 is very steep, for example. As a consequence of this, it is typically also the case that gradients of the free groundwater (phreatic) surface are also small, and a consequence of this is that we can make an approximation to the equations of groundwater flow which is analogous to that used in shallow water theory or the lubrication approximation, i.e., we can take advantage of the large aspect ratio of the flow. This approximation is called the Dupuit, or Dupuit–Forchheimer, approximation.

To be specific, suppose that we have to solve

$$\nabla^2 p = 0 \quad \text{in } 0 < z < h(x, y, t), \tag{7.31}$$

where z is the vertical coordinate, $z = h$ is the phreatic surface, and $z = 0$ is an impermeable basement. We let $\mathbf{u}$ denote the horizontal (vector) component of the Darcy flux, and w the vertical component. In addition, we now denote by $\nabla = (\frac{\partial}{\partial x}, \frac{\partial}{\partial y})$ the horizontal component of the gradient vector. The boundary conditions are then

$$p = 0, \quad \phi h_t + \mathbf{u}.\nabla h = w \quad \text{on } z = h,$$
$$\frac{\partial p}{\partial z} + \rho g = 0 \quad \text{on } z = 0; \tag{7.32}$$

here we take (gauge) pressure measured relative to atmospheric pressure. The condition at $z = 0$ is that of no normal flux, allowing for gravity.

Let us suppose that a horizontal length scale of relevance is l, and that the corresponding variation in h is of order d, thus

$$\varepsilon = \frac{d}{l} \tag{7.33}$$

is the size of the phreatic gradient, and is small. We non-dimensionalise the variables by scaling as follows:

$$x, y \sim l, \quad z \sim d, \quad p \sim \rho g d,$$
$$\mathbf{u} \sim \frac{k\rho g d}{\mu l}, \quad w \sim \frac{k\rho g d^2}{\mu l^2}, \quad t \sim \frac{\phi \mu l^2}{k\rho g d}. \tag{7.34}$$

The choice of scales is motivated by the same ideas as lubrication theory. The pressure is nearly hydrostatic, and the flow is nearly horizontal.

The dimensionless equations are

$$\mathbf{u} = -\nabla p, \quad \varepsilon^2 w = -(p_z + 1),$$
$$\nabla.\mathbf{u} + w_z = 0, \tag{7.35}$$

with

$$p_z = -1 \quad \text{on } z = 0,$$
$$p = 0, \qquad h_t = w + \nabla p.\nabla h \quad \text{on } z = h. \tag{7.36}$$

At leading order as $\varepsilon \to 0$, the pressure is hydrostatic:

$$p = h - z + O(\varepsilon^2). \tag{7.37}$$

More precisely, if we put

$$p = h - z + \varepsilon^2 p_1 + \cdots, \tag{7.38}$$

then (7.35) implies

$$p_{1zz} = -\nabla^2 h, \tag{7.39}$$

with boundary conditions, from (7.36),

$$p_{1z} = 0 \quad \text{on } z = 0,$$
$$p_{1z} = -h_t + |\nabla h|^2 \quad \text{on } z = h. \tag{7.40}$$

Integrating (7.39) from $z = 0$ to $z = h$ thus yields the evolution equation for h in the form

$$h_t = \nabla.[h\nabla h], \tag{7.41}$$

which is a nonlinear diffusion equation of degenerate type when $h = 0$.

This is easily solved numerically, and there are various exact solutions which are indicated in the exercises. In particular, steady solutions are found by solving Laplace's equation for $\frac{1}{2}h^2$, and there are various kinds of similarity solution. (7.41) is a second order equation requiring two boundary conditions. A typical situation in a river catchment is where there is drainage from a watershed to a river. A suitable problem in two dimensions is

$$h_t = (hh_x)_x + r, \tag{7.42}$$

where the source term r represents recharge due to rainfall. It is given by

$$r = \frac{r_D}{\varepsilon^2 K}, \tag{7.43}$$

where r_D is the rainfall rate and $K = k\rho g/\mu$ is the hydraulic conductivity. At the divide (say, $x = 0$), we have $h_x = 0$, whereas at the river (say, $x = 1$), the elevation is prescribed, $h = 1$ for example. The steady solution is

$$h = \left[1 + r - rx^2\right]^{1/2}, \tag{7.44}$$

and perturbations to this decay exponentially. If this value of the elevation of the water table exceeds that of the land surface, then a seepage face occurs, where water seeps from below and flows over the surface. This can sometimes be seen in steep mountainous terrain, or on beaches, when the tide is going out.

The Dupuit approximation is not uniformly valid at $x = 1$, where conditions of symmetry at the base of a valley would imply that $u = 0$, and thus $p_x = 0$. There is therefore a boundary layer near $x = 1$, where we rescale the variables by writing

$$x = 1 - \varepsilon X, \qquad w = \frac{W}{\varepsilon}, \qquad h = 1 + \varepsilon H, \qquad p = 1 - z + \varepsilon P. \tag{7.45}$$

Substituting these into the two-dimensional version of (7.35) and (7.36), we find

$$u = P_X, \qquad W = -P_z, \qquad \nabla^2 P = 0 \quad \text{in } 0 < z < 1 + \varepsilon H, \ 0 < X < \infty, \tag{7.46}$$

with boundary conditions

$$P = H, \quad \varepsilon H_t + P_X H_X = \frac{W}{\varepsilon} + r \quad \text{on } z = 1 + \varepsilon H,$$

$$P_X = 0 \quad \text{on } X = 0, \tag{7.47}$$

$$P_z = 0 \quad \text{on } z = 0,$$

$$P \sim H \sim rX \quad \text{as } X \to \infty.$$

At leading order in ε, this is simply

$$\nabla^2 P = 0 \quad \text{in } 0 < z < 1, \ 0 < X < \infty,$$

$$P_z = 0 \quad \text{on } z = 0, 1,$$

$$P_X = 0 \quad \text{on } X = 0, \tag{7.48}$$

$$P \sim rX \quad \text{as } X \to \infty.$$

Evidently, this has no solution unless we allow the incoming groundwater flux r from infinity to drain to the river at $X = 0$, $z = 1$. We do this by having a singularity in the form of a sink at the river,

$$P \sim \frac{r}{\pi} \ln\{X^2 + (1-z)^2\} \quad \text{near } X = 0, \ z = 1. \tag{7.49}$$

The solution to (7.48) can be obtained by using complex variables and the method of images, by placing sinks at $z = \pm(2n + 1)$, for integral values of n. Making use of the infinite product formula (Jeffrey 2004, p. 72)

$$\prod_1^\infty \left(1 + \frac{\zeta^2}{(2n+1)^2}\right) = \cosh\frac{\pi\zeta}{2}, \tag{7.50}$$

where $\zeta = X + iz$, we find the solution to be

$$P = \frac{r}{\pi} \ln\left[\cosh^2\frac{\pi X}{2}\cos^2\frac{\pi z}{2} + \sinh^2\frac{\pi X}{2}\sin^2\frac{\pi z}{2}\right]. \tag{7.51}$$

Fig. 7.2 Groundwater flow
lines towards a river at $X = 0$,
$z = 1$

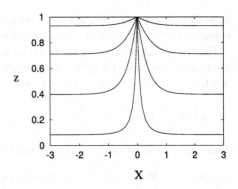

The complex variable form of the solution is

$$\phi = P + i\psi = \frac{2r}{\pi} \ln \cosh \frac{\pi \zeta}{2}, \qquad (7.52)$$

which is convenient for plotting. The streamlines of the flow are the lines $\psi =$ constant, and these are shown in Fig. 7.2.

This figure illustrates an important point, which is that although the flow towards a drainage point may be more or less horizontal, near the river the groundwater seeps upwards from depth. Drainage is not simply a matter of near surface recharge and drainage. This means that contaminants which enter the deep groundwater may reside there for a very long time.

A related point concerns the recharge parameter r defined in (7.43). According to Table 7.1, a typical permeability for sand is 10^{-10} m^2, corresponding to a hydraulic conductivity of $K = 10^{-3}$ m s^{-1}, or 3×10^4 m y^{-1}. Even for phreatic slopes as low as $\varepsilon = 10^{-2}$, the recharge parameter $r \lesssim O(1)$, and shallow aquifer drainage is feasible.

However, finer-grained sediments are less permeable, and the calculation of r for a silt with permeability of 10^{-14} m^2 ($K = 10^{-7}$ m s$^{-1} = 3$ m y^{-1} suggests that $r \sim 1/\varepsilon^2 \gg 1$, so that if the Dupuit approximation applied, the groundwater surface would lie above the Earth's surface everywhere. This simply points out the obvious fact that if the groundmass is insufficiently permeable, drainage cannot occur through it but water will accumulate at the surface and drain by overland flow. The fact that usually the water table is below but quite near the surface suggests that the long term response of landscape to recharge is to form topographic gradients and sufficiently deep sedimentary basins so that this *status quo* can be maintained.

7.3 Unsaturated Soils

Let us now consider flow in the unsaturated zone. Above the water table, water and air occupy the pore space. If the porosity is ϕ and the water volume fraction per unit volume of soil is W, then the ratio $S = W/\phi$ is called the *relative saturation*. If

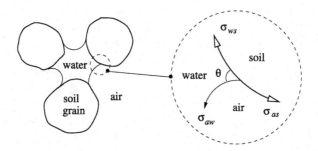

Fig. 7.3 Configuration of air and water in pore space. The contact angle θ measured through the water is acute, so that water is the *wetting* phase. σ_{ws}, σ_{as} and σ_{aw} are the surface energies of the three interfaces

$S = 1$, the soil is saturated, and if $S < 1$ it is unsaturated. The pore space of an unsaturated soil is configured as shown in Fig. 7.3. In particular, the air/water interface is curved, and in an equilibrium configuration the curvature of this interface will be constant throughout the pore space. The value of the curvature depends on the amount of liquid present. The less liquid there is (i.e., the smaller the value of S), then the smaller the pores where the liquid is found, and thus the higher the curvature. Associated with the curvature is a suction effect due to surface tension across the air/water interface. The upshot of all this is that the air and water pressures are related by a *capillary suction characteristic* function which expresses the difference between the pressures as a function of mean curvature, and hence, directly, S:

$$p_a - p_w = f(S). \tag{7.53}$$

The suction characteristic $f(S)$ is equal to $2\sigma\kappa$, where κ is the mean interfacial curvature: σ is the surface tension. For air and water in soil, f is positive as water is the *wetting phase*, that is, the *contact angle* at the contact line between air, water and soil grain is acute, measured through the water (see Fig. 7.3). The resulting form of $f(S)$ displays hysteresis as indicated in Fig. 7.4, with different curves depending on whether drying or wetting is taking place.

7.3.1 The Richards Equation

To model unsaturated flow, we have the conservation of mass equation in the form

$$\frac{\partial(\phi S)}{\partial t} + \nabla.\mathbf{u} = 0, \tag{7.54}$$

where we take ϕ as constant. Darcy's law for an unsaturated flow has the form, now with gravitational acceleration included,

$$\mathbf{u} = -\frac{k(S)}{\mu}[\nabla p + \rho g\hat{\mathbf{k}}], \tag{7.55}$$

Fig. 7.4 Capillary suction
characteristic. It displays
hysteresis in wetting and
drying

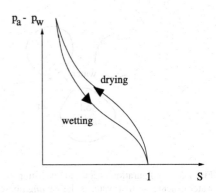

where $\hat{\mathbf{k}}$ is a unit vector upwards, and the permeability k depends on S. If $k(1) = k_0$ (the saturated permeability), then one commonly writes $k = k_0 k_{\mathrm{rw}}(S)$, where k_{rw} is the *relative permeability*. The most obvious assumption would be $k_{\mathrm{rw}} = S$, but this is rarely appropriate, and a better representation is a convex function, such as $k_{\mathrm{rw}} = S^3$. An even better representation is a function such as $k_{\mathrm{rw}} = (\frac{S-S_0}{1-S_0})^3_+$, where S_0 is known as the residual saturation. It represents the fact that in fine-grained soils, there is usually some minimal water fraction which cannot be removed. It is naturally associated with a capillary suction characteristic function $p_a - p = f(S)$ which tends to infinity as $S \to S_0+$, also appropriate for fine-grained soils.

In one dimension, and if we take the vertical coordinate z to point *downwards*, we obtain the *Richards equation*

$$\phi \frac{\partial S}{\partial t} = -\frac{\partial}{\partial z}\left[\frac{k_0}{\mu} k_{\mathrm{rw}}(S) \left\{ \frac{\partial f}{\partial z} + \rho g \right\} \right]. \tag{7.56}$$

We are assuming $p_a = $ constant (and also that the soil matrix is incompressible).

7.3.2 Non-dimensionalisation

We choose scales for the variables as follows:

$$f = \frac{\sigma}{d_p}\psi, \qquad z \sim \frac{\sigma}{\rho g d_p}, \qquad t \sim \frac{\phi \mu z}{\rho g k_0}, \tag{7.57}$$

where d_p is grain size and σ is the surface tension, assumed constant. The Richards equation then becomes, in dimensionless variables,

$$\frac{\partial S}{\partial t} = -\frac{\partial}{\partial z}\left[k_{\mathrm{rw}}\left(\frac{\partial \psi}{\partial z} + 1 \right) \right]. \tag{7.58}$$

To be specific, we consider the case of soil wetting due to surface infiltration: of rainfall, for example. Suitable boundary conditions for infiltration are

$$S = 1 \quad \text{at } z = 0 \tag{7.59}$$

if surface water is ponded, or

$$k_{rw}\left(\frac{\partial \psi}{\partial z} + 1\right) = u^* = \frac{\mu u_0}{k_0 \rho_w g} = \frac{u_0}{K_0}, \tag{7.60}$$

if there is a prescribed downward flux u_0; K_0 is the saturated hydraulic conductivity. In a dry soil we would have $S \to 0$ as $z \to \infty$, or if there is a water table at $z = z_p$, $S = 1$ there.[5] For silt with $k_0 = 10^{-14}$ m^2, the hydraulic conductivity $K_0 \sim 10^{-7}$ m s^{-1} or 3 m y^{-1}, while average rainfall in England, for example, is ≤ 1 m y^{-1}. Thus *on average* $u^* \leq 1$, but during storms we can expect $u^* \gg 1$. For large values of u^*, the desired solution may have $S > 1$ at $z = 0$; in this case ponding occurs (as one observes), and (7.60) is replaced by (7.59), with the pond depth being determined by the balance between accumulation, infiltration, and surface run-off.

7.3.3 Snow Melting

An application of the unsaturated flow model occurs in the study of melting snow. In particular, it is found that pollutants which may be uniformly distributed in snow (e.g. SO_2 from sulphur emissions via acid rain) can be concentrated in melt water run-off, with a consequent enhanced detrimental effect on stream pollution. The question then arises, why this should be so. We shall find that uniform surface melting of a dry snowpack can lead to a meltwater spike at depth.

Suppose we have a snow pack of depth d. Snow is a porous aggregate of ice crystals, and meltwater formed at the surface can percolate through the snow pack to the base, where run-off occurs. (We ignore effects of re-freezing of meltwater.) The model (7.58) is appropriate, but the relevant length scale is d. Therefore we define a parameter

$$\kappa = \frac{\sigma}{\rho g d d_p}, \tag{7.61}$$

and we rescale the variables as $z \sim 1/\kappa$, $t \sim 1/\kappa$. To be specific, we will also take

$$k_{rw} = S^3, \tag{7.62}$$

and

$$\psi(S) = \frac{1}{S} - S, \tag{7.63}$$

based on typical experimental results.

Suitable boundary conditions in a melting event might be to prescribe the melt flux u_0 at the surface, thus

$$k_{rw}\left(\frac{\partial \psi}{\partial z} + 1\right) = u^* = \frac{u_0}{K_0} \quad \text{at } z = 0. \tag{7.64}$$

---

[5] With constant air pressure, continuity of S follows from continuity of pore water pressure.

If the base is impermeable, then

$$k_{rw}\left(\frac{\partial \psi}{\partial z} + 1\right) = 0 \quad \text{at } z = h. \tag{7.65}$$

This is certainly not realistic if S reaches 1 at the base, since then ponding must occur and presumably melt drainage will occur via a channelised flow, but we examine the initial stages of the flow using (7.65). Finally, we suppose $S = 0$ at $t = 0$. Again, this is not realistic in the model (it implies infinite capillary suction) but it is a feasible approximation to make.

Simplification of this model now leads to the dimensionless Darcy–Richards equation in the form

$$\frac{\partial S}{\partial t} + 3S^2 \frac{\partial S}{\partial z} = \kappa \frac{\partial}{\partial z}\left[S(1 + S^2)\frac{\partial S}{\partial z}\right]. \tag{7.66}$$

If we choose $\sigma = 70$ mN m^{-1}, $d_p = 0.1$ mm, $\rho = 10^3$ kg m^{-3}, $g = 10$ m s^{-2}, $d = 1$ m, then $\kappa = 0.07$. It follows that (7.66) has a propensity to form shocks, these being diffused by the term in κ over a distance $O(\kappa)$ (by analogy with the shock structure for the Burgers equation, see Chap. 1).

We want to solve (7.66) with the initial condition

$$S = 0 \quad \text{at } t = 0, \tag{7.67}$$

and the boundary conditions

$$S^3 - \kappa S(1 + S^2)\frac{\partial S}{\partial z} = u^* \quad \text{on } z = 0, \tag{7.68}$$

and

$$S^3 - \kappa S(1 + S^2)\frac{\partial S}{\partial z} = 0 \quad \text{at } z = 1. \tag{7.69}$$

Roughly, for $\kappa \ll 1$, these are

$$\begin{aligned} S &= S_0 \quad \text{at } z = 0, \\ S &= 0 \quad \text{at } z = 1, \end{aligned} \tag{7.70}$$

where $S_0 = u^{*1/3}$, which we initially take to be $O(1)$ (and <1, so that surface ponding does not occur).

Neglecting κ, the solution is the step function

$$\begin{aligned} S &= S_0, \quad z < z_f, \\ S &= 0, \quad z > z_f, \end{aligned} \tag{7.71}$$

and the shock front at z_f advances at a rate $\dot{z}_f$ given by the jump condition

$$\dot{z}_f = \frac{[S^3]_-^+}{[S]_-^+} = S_0^2. \tag{7.72}$$

Fig. 7.5 $S(Z)$ given by (7.78); the shock front terminates at the origin

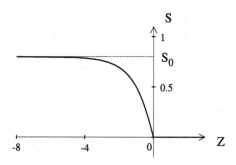

In dimensional terms, the shock front moves at speed $u_0/\phi S_0$, which is in fact obvious (given that it has constant S behind it).

The shock structure is similar to that of Burgers' equation. We put

$$z = z_f + \kappa Z, \tag{7.73}$$

and S rapidly approaches the quasi-steady solution $S(Z)$ of

$$-VS' + 3S^2S' = \left[S(1 + S^2)S'\right]', \tag{7.74}$$

where $V = \dot{z}_f$; hence

$$S(1 + S^2)S' = -S(S_0^2 - S^2), \tag{7.75}$$

in order that $S \to S_0$ as $Z \to -\infty$, and where we have chosen

$$V = S_0^2, \tag{7.76}$$

(as $S_+ = 0$), thus reproducing (7.72). The solution is a quadrature,

$$\int^S \frac{(1 + S^2)\,dS}{(S_0^2 - S^2)} = -Z, \tag{7.77}$$

with an arbitrary added constant (amounting to an origin shift for Z). Hence

$$S - \frac{(1 + S_0^2)}{2S_0} \ln\left[\frac{S_0 + S}{S_0 - S}\right] = Z. \tag{7.78}$$

The shock structure is shown in Fig. 7.5; the profile terminates where $S = 0$ at $Z = 0$. In fact, (7.75) implies that $S = 0$ or (7.78) applies. Thus when S given by (7.78) reaches zero, the solution switches to $S = 0$. The fact that $\partial S/\partial Z$ is discontinuous is not a problem because the diffusivity $S(1 + S^2)$ goes to zero when $S = 0$. This degeneracy of the equation is a signpost for fronts with discontinuous derivatives: essentially, the profile can maintain discontinuous gradients at $S = 0$ because the diffusivity is zero there, and there is no mechanism to smooth the jump away.

Suppose now that $k_0 = 10^{-10}$ m^2 and $\mu/\rho = 10^{-6}$ m^2 s^{-1}; then the saturated hydraulic conductivity $K_0 = k_0\rho g/\mu = 10^{-3}$ m s^{-1}. On the other hand, if a metre thick snow pack melts in ten days, this implies $u_0 \sim 10^{-6}$ m s^{-1}. Thus $S_0^3 = u_0/K_0 \sim 10^{-3}$, and the approximation $S \approx S_0$ looks less realistic. With

$$S^3 - \kappa S(1 + S^2)\frac{\partial S}{\partial z} = S_0^3, \qquad (7.79)$$

and $S_0 \sim 10^{-1}$ and $\kappa \sim 10^{-1}$, it seems that one should assume $S \ll 1$. We define

$$S = \left(\frac{S_0^3}{\kappa}\right)^{1/2} s; \qquad (7.80)$$

(7.79) becomes

$$\beta s^3 - s\left[1 + \frac{S_0^3}{\kappa}s^2\right]\frac{\partial s}{\partial z} = 1 \quad \text{on } z = 0, \qquad (7.81)$$

and we have $S_0^3/\kappa \sim 10^{-2}$, $\beta = (S_0/\kappa)^{3/2} \sim 1$.
   We neglect the term in S_0^3/κ, so that

$$\beta s^3 - s\frac{\partial s}{\partial z} \approx 1 \quad \text{on } z = 0, \qquad (7.82)$$

and substituting (7.80) into (7.66) leads to

$$\frac{\partial s}{\partial \tau} + 3\beta s^2\frac{\partial s}{\partial z} \approx \frac{\partial}{\partial z}\left[s\frac{\partial s}{\partial z}\right], \qquad (7.83)$$

if we define $t = \tau/(\kappa S_0^3)^{1/2}$. A simple analytic solution is no longer possible, but the development of the solution will be similar. The flux condition (7.82) at $z = 0$ allows the surface saturation to build up gradually, and a shock will only form if $\beta \gg 1$ (when the preceding solution becomes valid).

7.3.4 Similarity Solutions

If, on the other hand, $\beta \ll 1$, then the saturation profile approximately satisfies

$$\frac{\partial s}{\partial \tau} = \frac{\partial}{\partial z}\left[s\frac{\partial s}{\partial z}\right],$$

$$-s\frac{\partial s}{\partial z} = \begin{cases} 1 & \text{on } z = 0, \\ 0 & \text{on } z = 1. \end{cases} \qquad (7.84)$$

At least for small times, the model admits a similarity solution of the form

$$s = \tau^a f(\eta), \quad \eta = z/\tau^b, \qquad (7.85)$$

Fig. 7.6 Schematic representation of the evolution of S for both large and small β

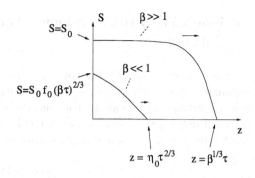

where satisfaction of the equations and boundary conditions requires $2a = b$ and $2b = 1 = a$, whence $a = \frac{1}{3}$, $b = \frac{2}{3}$, and f satisfies

$$(ff')' - \frac{1}{3}(f - 2\eta f') = 0, \tag{7.86}$$

with the condition at $z = 0$ becoming

$$-ff' = 1 \quad \text{at } \eta = 0. \tag{7.87}$$

The condition at $z = 1$ can be satisfied for small enough τ, as we shall see, because Eq. (7.86) is degenerate, and f reaches zero in a finite distance, η_0, say, and $f = 0$ for $\eta > \eta_0$. As $\eta = 1/\tau^{2/3}$ at $z = 1$, then this solution will satisfy the no flux condition at $z = 1$ as long as $\tau < \eta_0^{-3/2}$, when the advancing front will reach $z = 1$.

To see why f behaves in this way, integrate once to find

$$f\left(f' + \frac{2}{3}\eta\right) = -1 + \int_0^\eta f\, d\eta. \tag{7.88}$$

For small η, the right hand side is negative, and f is positive (to make physical sense), so f decreases (and in fact $f' < -\frac{2}{3}\eta$). For sufficiently small $f(0) = f_0$, f will reach zero at a finite distance $\eta = \eta_0$, and the solution must terminate. On the other hand, for sufficiently large f_0, $\int_0^\eta f\, d\eta$ reaches 1 at $\eta = \eta_1$ while f is still positive (and $f' = -\frac{2}{3}\eta_1$ there). For $\eta > \eta_1$, then f remains positive and $f' > -\frac{2}{3}\eta$ (f cannot reach zero for $\eta > \eta_1$ since $\int_0^\eta f\, d\eta > 1$ for $\eta > \eta_1$). Eventually f must have a minimum and thereafter increase with η. This is also unphysical, so we require f to reach zero at $\eta = \eta_0$. This will occur for a range of f_0, and we have to select f_0 in order that

$$\int_0^{\eta_0} f\, d\eta = 1, \tag{7.89}$$

which in fact represents global conservation of mass. Figure 7.6 shows the schematic form of solution both for $\beta \gg 1$ and $\beta \ll 1$. Evidently the solution for $\beta \sim 1$ will have a profile with a travelling front between these two end cases.

7.4 Immiscible Two-Phase Flows: The Buckley–Leverett Equation

In some circumstances, the flow of more than one phase in a porous medium is important. The type example is the flow of oil and gas, or oil and water (or all three!) in a sedimentary basin, such as that beneath the North Sea. Suppose there are two phases; denote the phases by subscripts 1 and 2, with fluid 2 being the *wetting fluid*, and S is its saturation. Then the capillary suction characteristic is

$$p_1 - p_2 = p_c(S), \tag{7.90}$$

with the capillary suction p_c being a positive, monotonically decreasing function of saturation S; mass conservation takes the form

$$-\phi \frac{\partial S}{\partial t} + \mathbf{\nabla}.\mathbf{u}_1 = 0,$$
$$\phi \frac{\partial S}{\partial t} + \mathbf{\nabla}.\mathbf{u}_2 = 0, \tag{7.91}$$

where ϕ is (constant) porosity, and Darcy's law for each phase is

$$\mathbf{u}_1 = -\frac{k_0}{\mu_1} k_{r1} [\mathbf{\nabla} p_1 + \rho_1 g \hat{\mathbf{k}}],$$
$$\mathbf{u}_2 = -\frac{k_0}{\mu_2} k_{r2} [\mathbf{\nabla} p_2 + \rho_2 g \hat{\mathbf{k}}], \tag{7.92}$$

with k_{ri} being the relative permeability of fluid i.

For example, if we consider a one-dimensional flow, with z pointing upwards, then we can integrate (7.91) to yield the total flux

$$u_1 + u_2 = q(t). \tag{7.93}$$

If we define the mobilities of each fluid as

$$M_i = \frac{k_0}{\mu_i} k_{ri}, \tag{7.94}$$

then it is straightforward to derive the equation for S,

$$\phi \frac{\partial S}{\partial t} = -\frac{\partial}{\partial z} \left[M_{\text{eff}} \left\{ \frac{q}{M_1} + \frac{\partial p_c}{\partial z} + (\rho_1 - \rho_2) g \right\} \right], \tag{7.95}$$

where the effective mobility is determined by

$$M_{\text{eff}} = \left(\frac{1}{M_1} + \frac{1}{M_2} \right)^{-1}. \tag{7.96}$$

Fig. 7.7 Graph of
dimensionless wave speed
$V(S)$ as a function of wetting
fluid saturation, indicating the
speed and direction of wave
motion ($V > 0$ means waves
move upwards) if the wetting
fluid is more dense. The
viscosity ratio μ_r (see
(7.100)) is taken to be 30

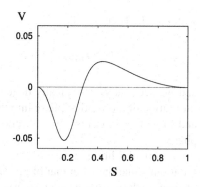

This is a convective-diffusion equation for S. If suction is very small, we obtain
the *Buckley–Leverett equation*

$$\phi \frac{\partial S}{\partial t} + \frac{\partial}{\partial z}\left[M_{\text{eff}}\left\{\frac{q}{M_1} + (\rho_1 - \rho_2)g\right\}\right] = 0, \tag{7.97}$$

which is a nonlinear hyperbolic wave equation. As a typical situation, suppose
$q = 0$, and $k_{r2} = S^3$, $k_{r1} = (1 - S)^3$. Then

$$M_{\text{eff}} = \frac{k_0 S^3 (1 - S)^3}{\mu_1 S^3 + \mu_2 (1 - S)^3}, \tag{7.98}$$

and the wave speed $v(S)$ is given by

$$v = -(\rho_2 - \rho_1)g M'_{\text{eff}}(S) = v_0 V(S), \tag{7.99}$$

where

$$v_0 = \frac{(\rho_2 - \rho_1)g k_0}{\mu_2}, \qquad V(S) = \frac{\chi'(S)}{\chi(S)^2},$$

$$\chi(S) = \frac{\mu_r}{(1 - S)^3} + \frac{1}{S^3}, \qquad \mu_r = \frac{\mu_1}{\mu_2}. \tag{7.100}$$

The variation of V with S is shown in Fig. 7.7. For $\rho_2 > \rho_1$ (as for oil and water,
where water is the wetting phase), waves move upwards at low water saturation and
downwards at high saturation.

Shocks will form, but these are smoothed by the diffusion term $-\frac{\partial}{\partial z}[M_{\text{eff}}p'_c \frac{\partial S}{\partial z}]$,
in which the diffusion coefficient is

$$D = -M_{\text{eff}}p'_c. \tag{7.101}$$

As a typical example, take

$$p_c = \frac{p_0 (1 - S)^{\lambda_1}}{S^{\lambda_2}} \tag{7.102}$$

with $\lambda_i > 0$. Then we find

$$D = k_0 p_0 S^{2-\lambda_2}(1-S)^{2+\lambda_1}\left[\frac{\lambda_1 S + \lambda_2(1-S)}{\mu_1 S^3 + \mu_2(1-S)^3}\right], \qquad (7.103)$$

and we see that D is typically degenerate at $S = 0$. In particular, if $\lambda_2 < 2$, then infiltration of the wetting phase into the non-wetting phase proceeds at a finite rate, and this always occurs for infiltration of the non-wetting phase into the wetting phase.

A particular limiting case is when one phase is much less dense than the other, the usual situation being that of gas and liquid. This is exemplified by the problem of snow-melt run-off considered earlier. In that case, water is the wetting phase, thus $\rho_2 - \rho_1 = \rho_w - \rho_a$ is positive, and also $\mu_w \approx 10^{-3}$ Pa s, $\mu_a \approx 10^{-5}$ Pa s, whence $\mu_a \ll \mu_w$ ($\mu_r \ll 1$), so that, from (7.98),

$$M_{\text{eff}} \approx \frac{k_0 S^3}{\mu_w}, \qquad (7.104)$$

at least for saturations not close to unity. Shocks form and propagate downwards (since $\rho_2 > \rho_1$). The presence of non-zero flux $q < 0$ does not affect this statement. Interestingly, the approximation (7.104) will always break down at sufficiently high saturation. Inspection of $V(S)$ for $\mu_r = 0.01$ (as for air and water) indicates that (7.104) is an excellent approximation for $S \lesssim 0.5$, but not for $S \gtrsim 0.6$; for $S \gtrsim 0.76$, V is positive and waves move upwards. As $\mu_r \to 0$, the right hand hump in Fig. 7.7 moves towards $S = 1$, but does not disappear; indeed the value of the maximum increases, and is $V \sim \mu_r^{-1/3}$. Thus the single phase approximation for unsaturated flow is a singular approximation when $\mu_r \ll 1$ and $1 - S \ll 1$.

7.5 Heterogeneous Porous Media

Perhaps the major concern in groundwater studies concerns the permeability. Whereas we tend to think of the permeability as a well-defined quantity which reflects the local soil or rock properties, in reality it varies over many orders of magnitude on very small length scales. The consequence of this is that the value of the permeability itself needs to be averaged in some way.

Permeability is so variable because of soil and rock heterogeneity. Because it scales with the square of the constituent grain size, clay and sand permeabilities are vastly different. And because sediments are lain down so slowly, over millions of years, sand and clay layers often lie in close proximity. The same is true for sedimentary rocks, which are simply the same sand and clay layers cemented together after burial and consequent subjection to high pressure and temperature.

In seeking to quantify porous medium flow at the large scale, we need to average the permeability in some way at the mesoscale: larger than the pore scale, but less than the macroscale. The simplest approach is to suppose that the permeability in a mesoscale block has a random distribution, often assumed to be a lognormal distribution. An averaged permeability can be derived by supposing, for example, that fluctuations have small amplitude. One finds that the consequent averaged permeability is a tensor, whose components depend on the direction of flow. In the following section, we consider a more specific model of the mesoscale structure, where the heterogeneity is related to the occurrence of fractures in the medium. This leads to the idea of a secondary porosity associated with the fractures.

7.5.1 Dual Porosity Models

Take a walk on exposed basement rock: at the seaside, in the mountains. Rocks are not uniform, but are inevitably fractured, or jointed. There are numerous reasons for this. Sedimentary rocks are lain down over millions of years via the deposition of outwash clays, sands or calcareous microfossils in marine environments. Over this time the deposition rate may average a millimetre or less per year. A metre of rock may take a million, or ten million years, to accumulate. In this time, sea level may rise or fall by tens or more of metres, and the land itself rises or falls because of tectonic processes: the crashing of continents, the uplift of mountains, the burial of sedimentary basins.

It is no surprise that in an exposed sedimentary sequence, such as one sees in coastal cliffs, rocks form stratigraphic layers separated by unconformities marking different sedimentary epochs. These unconformities are layers of weakness, and when the rocks are later subjected to tectonic compression and folding, fractures will form.

It is not only sedimentary rocks which tear as they are stressed. Igneous rocks fracture as they solidify because of solidification shrinkage. They also form intrusions such as dikes and sills, whose different erosional properties can cause subsequent voidage.

The occurrence of faulting or jointing in rocks leads to a particular problem in the description of groundwater flow through them. The rock itself is porous, and admits a Darcy flow through its pore space; but the fractures act as a second porosity, admitting a secondary flow which would occur even if the rock itself was completely impermeable. The situation is illustrated in Fig. 7.8. It is because of this configuration that the system is called a double, or dual, porosity system, and the resulting model to describe the flow is called a dual porosity model.

In order to characterise porous flow through such a medium, we distinguish between the blocks of the matrix and the cross-cutting fractures. We suppose the fractures are tabular, or planar, of width h, and the blocks are of dimension d_B, and that $h \ll d_B$. We denote the blocks by the domain M, and the fractures as ∂M. Because the fractures are narrow, ∂M essentially represents the external surfaces of the blocks. We also suppose that $d_B \ll l$, where l is a relevant macroscopic length

Fig. 7.8 A doubly porous
system. Porous matrix blocks
are transected by (here) two
sets of transverse fractures

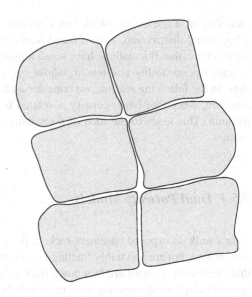

scale. For these tabular cracks, we can define a fracture porosity

$$\phi_f = \frac{h}{d_B + h}. \tag{7.105}$$

We define a matrix pore pressure p_m, which is the locally averaged pore pressure
in the matrix blocks, and a fracture pressure p_f. There is then a matrix volume flux
per unit area $\mathbf{u}_m$, and in the usual way we have Darcy's law in the form

$$\mathbf{u}_m = -\frac{k_m}{\mu} \nabla p_m, \tag{7.106}$$

where k_m is the permeability of the fine-grained matrix. We have

$$k_m = \frac{d_p^2}{\tau_m}, \tag{7.107}$$

where d_p is grain size and τ_m is a tortuosity factor.

We suppose that flow in the fractures is essentially Poiseuille flow, and this leads
to a prescription of fracture volume flux per unit transverse width of crack, $\mathbf{q}_f$,
(through a single fracture) as

$$\mathbf{q}_f = -\frac{h^3}{\tau_f \mu} \nabla p_f, \tag{7.108}$$

where for a plane walled crack of width h, the fracture tortuosity $\tau_f = 12$; for rough
cracks, one can expect a higher value to be appropriate. The mean fracture velocity

(which is also the fracture volume flux per unit area of fracture) is thus

$$\mathbf{u}_f = \frac{\mathbf{q}_f}{h} = -\frac{k_f}{\mu}\nabla p_f, \qquad (7.109)$$

where we define the fracture permeability parameter k_f as

$$k_f = \frac{h^2}{\tau_f}. \qquad (7.110)$$

Now if we consider the total (averaged) Darcy flux $\mathbf{u}$ through such a doubly porous medium, it is straightforward to show that

$$\mathbf{u} = (1 - \phi_f)\langle \mathbf{u}_m \rangle_M + \phi_f \langle \mathbf{u}_f \rangle_{\partial M}, \qquad (7.111)$$

where the angle brackets denote averages: for $\mathbf{u}_m$, a volume average over the matrix blocks; for $\mathbf{u}_f$, an average over the fractures. Since $h \ll d_B$, we can effectively consider the average of $\mathbf{u}_f$ to be a surface average over the fracture surface denoted by ∂M, the external boundary of the matrix blocks M. Note that each fracture has two walls, and thus provides two external surfaces to M.

Our object is to characterise these averages in terms of macroscopic variables, if possible. Notice that we have already carried out a primary averaging in defining the fluxes $\mathbf{u}_m$ and $\mathbf{u}_f$ in the first place: $\mathbf{u}_m$ is averaged over the grain scale of the matrix, and $\mathbf{u}_f$ is averaged over the width of the fractures. However, these fluxes still represent values at a point within the larger block/fracture system. In particular, note that by its definition the fracture flux is parallel to the fracture, and this carries the implication that

$$\frac{\partial p_f}{\partial n} = 0, \qquad (7.112)$$

where $\mathbf{n}$ is the normal to ∂M (and we take it to point from the matrix into the fracture).

We now want to average over the larger block scale d_B. The 'point' fluxes $\mathbf{u}_m$ and $\mathbf{u}_f$ satisfy the conservation of mass equations

$$\nabla.\mathbf{u}_m = 0 \qquad (7.113)$$

and

$$\nabla.(h\mathbf{u}_f) = \mathbf{u}_m.\mathbf{n}|_{\partial M}, \qquad (7.114)$$

where in (7.114) there is a flux $\mathbf{u}_m.\mathbf{n}$ at the fracture surfaces from the matrix to the fractures. Some comment on this equation is necessary. It takes this form because of the fact that the fracture flux as defined in (7.108) is already averaged over the cross section of the fracture. Continuity of the fluxes at the block fracture interface produces the source term in (7.114) through the integration of the fracture *point* transverse velocity across the fracture.

To use the ideas of homogenisation, we first define dimensionless variables. We scale the variables as

$$\mathbf{u}_k \sim U, \qquad p_k \sim P, \qquad \mathbf{x} \sim l, \tag{7.115}$$

where l is the macroscopic length scale, and we define a second (now dimensionless) spatial variable $\mathbf{X}$ by putting

$$\mathbf{x} = \varepsilon \mathbf{X}, \tag{7.116}$$

where

$$\varepsilon = \frac{d_B}{l}. \tag{7.117}$$

The blocks thus have size $\mathbf{X} \sim O(1)$. We write, with an obvious notation, the dimensionless gradient operator in the form

$$\nabla = \nabla_{\mathbf{x}} + \frac{1}{\varepsilon}\nabla_{\mathbf{X}}, \tag{7.118}$$

where we are now using $\mathbf{x}$ and $\mathbf{X}$ as multiple spatial scales. We suppose that the block structure is periodic in $\mathbf{X}$, although this is inessential for the methodology.

Now the requirement of (7.112) that $\frac{\partial p_f}{\partial n} = 0$ implies

$$\mathbf{n}.\left[\nabla_{\mathbf{x}} p_f + \frac{1}{\varepsilon}\nabla_{\mathbf{X}} p_f\right] = 0, \tag{7.119}$$

whence it follows that we can write, approximately,

$$p_f = p(\mathbf{x}) + \varepsilon \tilde{p}_f(\mathbf{X}), \tag{7.120}$$

and then

$$\frac{\partial p}{\partial n} \equiv \mathbf{n}.\nabla_{\mathbf{x}} p = -\mathbf{n}.\nabla_{\mathbf{X}} \tilde{p}_f \equiv -\frac{\partial \tilde{p}_f}{\partial N}. \tag{7.121}$$

p is the macroscopic average pressure variable, and we may impose periodicity in $\mathbf{X}$ of $\tilde{p}_f$ with zero mean.

We have continuity of matrix and fracture pressure at ∂M, and therefore we can write

$$p_m = p(\mathbf{x}) + \varepsilon \tilde{p}_m(\mathbf{X}), \tag{7.122}$$

and the matrix pressure satisfies

$$\nabla_{\mathbf{X}}^2 \tilde{p}_m = 0 \quad \text{in } M, \tag{7.123}$$
$$\tilde{p}_m = \tilde{p}_f \quad \text{on } \partial M.$$

To find the solution $\tilde{p}_m$ of (7.123), define a Green's function $G(\mathbf{X}, \mathbf{Y})$ which satisfies

$$\nabla_{\mathbf{Y}}^2 G = \delta(\mathbf{X} - \mathbf{Y}) \quad \text{in } M, \qquad G = 0 \quad \text{for } \mathbf{Y} \in \partial M. \tag{7.124}$$

Then $\tilde{p}_m$ is given by

$$\tilde{p}_m = \int_{\partial M} \frac{\partial G(\mathbf{X}, \mathbf{Y})}{\partial N_Y} \tilde{p}_f(\mathbf{Y}) \, dS(\mathbf{Y}), \tag{7.125}$$

where $\frac{\partial}{\partial N_Y} = \mathbf{n}(\mathbf{Y}).\nabla_{\mathbf{Y}}$, and it follows from this that (for $\mathbf{X} \in \partial M$)

$$\frac{\partial \tilde{p}_m}{\partial N} = \int_{\partial M} K(\mathbf{X}, \mathbf{Y}) \, \tilde{p}_f(\mathbf{Y}) \, dS(\mathbf{Y}), \tag{7.126}$$

where

$$K(\mathbf{X}, \mathbf{Y}) = \frac{\partial^2 G(\mathbf{X}, \mathbf{Y})}{\partial N_X \partial N_Y}. \tag{7.127}$$

It remains to determine the fracture pressure perturbation $\tilde{p}_f$. This involves solving (7.109) and (7.114). Supposing h and k_f are constant, these reduce, at leading order in ε, to

$$\nabla_{\mathbf{X}}^2 \tilde{p}_f = \left(\frac{d_B k_m}{h k_f} \right) \left[\frac{\partial \tilde{p}_m}{\partial N} + \frac{\partial p}{\partial n} \right], \tag{7.128}'$$

subject to conditions of periodicity in $\mathbf{X}$ and zero mean. Note that the Laplacian in (7.128) is defined on the surface ∂M. Using (7.126), we can write (7.128) in the form

$$\nabla_{\mathbf{X}}^2 \tilde{p}_f = \alpha \left[\int_{\partial M} K(\mathbf{X}, \mathbf{Y}) \, \tilde{p}_f(\mathbf{Y}) \, dS(\mathbf{Y}) + \frac{\partial p}{\partial n} \right], \tag{7.129}$$

where

$$\alpha = \frac{d_B k_m}{h k_f} = \frac{\tau_f d_B d_p^2}{\tau_m h^3}. \tag{7.130}$$

The canonical microscale fracture problem to be solved on ∂M is thus

$$\nabla_{\partial M}^2 \mathbf{q} = \alpha \left[\int_{\partial M} K(\mathbf{X}, \mathbf{Y}) \mathbf{q}(\mathbf{Y}) \, dS(\mathbf{Y}) + \mathbf{n} \right], \tag{7.131}$$

with periodic boundary conditions and zero mean, and then

$$\tilde{p}_f = \mathbf{q}(\mathbf{X}).\nabla_{\mathbf{x}} p. \tag{7.132}$$

We now use these results to find the effective permeability of the medium. Averaging (7.106) over the matrix blocks yields (dimensionlessly)

$$U \langle \mathbf{u}_m \rangle_M = -\frac{k_m P}{\mu l} \nabla_{\mathbf{x}} p, \tag{7.133}$$

since $\tilde{p}_m$ is continuous across fractures and periodic in $\mathbf{X}$. In a similar way,

$$U \langle \mathbf{u}_f \rangle_{\partial M} = -\frac{k_f P}{\mu l} [\nabla_{\mathbf{x}} p + \langle \nabla_{\mathbf{X}} \tilde{p}_f \rangle_{\partial M}], \tag{7.134}$$

but the surface average term in this expression does not obviously vanish. We have

$$\nabla_X \tilde{p}_f = [\nabla_X \tilde{p}_f - \mathbf{n}(\mathbf{n}.\nabla_X \tilde{p}_f)] + \mathbf{n}(\mathbf{n}.\nabla_X \tilde{p}_f); \tag{7.135}$$

the term in square brackets is a tangential derivative of $\tilde{p}_f$ along ∂M, and we separate the terms in this way because $\tilde{p}_f$ is only defined on ∂M. In addition, because $\tilde{p}_f = \tilde{p}_m$ on ∂M, we could replace the subscript f by m in the square-bracketed expression. Because of (7.121), we have

$$\mathbf{n}(\mathbf{n}.\nabla_X \tilde{p}_f) = -\mathbf{n}(\mathbf{n}.\nabla_X p), \tag{7.136}$$

and thus

$$\langle \mathbf{n}(\mathbf{n}.\nabla_X \tilde{p}_f) \rangle_{\partial M} = -\mathbf{e}_i \langle n_i n_j \rangle_{\partial M} \frac{\partial p}{\partial x_j}, \tag{7.137}$$

where $\mathbf{e}_i$ is the unit vector in the x_i direction. We therefore have

$$U \langle \mathbf{u}_f \rangle_{\partial M} = -\frac{k_f P}{\mu l} [(\mathbf{I} - \langle \mathbf{nn} \rangle_{\partial M}).\nabla_X p + \langle \nabla_X \tilde{p}_f - \mathbf{n}(\mathbf{n}.\nabla_X \tilde{p}_f) \rangle_{\partial M}], \tag{7.138}$$

where $\mathbf{I}$ is the unit tensor, and $\mathbf{nn}$ is the tensor with elements $n_i n_j$.

We now substitute the expression for $\tilde{p}_f$ in (7.131) into (7.138); averaging over ∂M, we finally derive the expression for the mean fracture velocity in the dimensionless form

$$U \langle \mathbf{u}_f \rangle_{\partial M} = -\frac{k_f P}{\mu l} [(\mathbf{I} - \langle \mathbf{nn} \rangle_{\partial M}) + \langle \{(\mathbf{I} - \mathbf{nn}).\nabla_X \} \mathbf{q} \rangle].\nabla_X p. \tag{7.139}$$

Rewriting this in dimensional form, we have

$$\langle \mathbf{u}_f \rangle_{\partial M} = -\frac{k_f}{\mu} \mathbf{k}^*.\nabla p, \tag{7.140}$$

where the fracture relative permeability tensor $\mathbf{k}^*$ is defined by

$$k_{ij}^* = \left\langle (\delta_{ik} - n_i n_k) \left(\delta_{jk} + \frac{\partial q_j}{\partial X_k} \right) \right\rangle_{\partial M}. \tag{7.141}$$

Equally, the dimensional matrix flux is, from (7.133),

$$\langle \mathbf{u}_m \rangle_M = -\frac{k_m}{\mu} \nabla p. \tag{7.142}$$

(7.140), (7.141) and (7.142) give the recipes for the averaged matrix and fracture fluxes in terms of the macroscopic pressure gradient and the solution of the block scale fracture pressure problem (7.131).

If we take a representative volume consisting of many blocks, and integrate (7.113) over the matrix volume, and (7.114) over the fracture volume, we obtain the averaged (dimensional) equations for the averaged fluxes in the form

$$\nabla.\left[\phi_f\langle\mathbf{u}_f\rangle_{\partial M}\right] = s_f\langle\mathbf{u}_m.\mathbf{n}\rangle_{\partial M},$$
$$\nabla.\left[(1-\phi_f)\langle\mathbf{u}_m\rangle_M\right] = -s_f\langle\mathbf{u}_m.\mathbf{n}\rangle_{\partial M},$$

(7.143)

where s_f is the specific fracture surface area (i.e., surface area per unit volume: here $s_f \sim d_B^{-1}$). Note that the source term in (7.143) is just

$$s_f\langle\mathbf{u}_m.\mathbf{n}\rangle_{\partial M} = -\frac{k_m s_f P}{\mu l}\left\langle\frac{\partial\tilde{p}_m}{\partial N}\right\rangle_{\partial M},$$

(7.144)

because $\langle\frac{\partial p}{\partial n}\rangle_{\partial M} = \langle\mathbf{n}\rangle_{\partial M}.\nabla_\mathbf{x}p = 0,$[6] and in the present case this is just zero because of (7.123).

The usefulness of all this methodology is that it carries across to other, more complicated averaging problems (as we see below), but in the present case of incompressible double porosity flow, it may be somewhat unnecessary. The reason for this is that fracture relative permeability depends on the solution of (7.131), and thus on the fracture geometry and the single dimensionless parameter α given by (7.130). Assuming small fracture porosity, the ratio of matrix flow to fracture flow is, from (7.111), (7.109), (7.105) and (7.106), of the order of

$$\frac{u_m}{\phi u_f} \sim \frac{k_m d_B}{k_f h} = \alpha.$$

(7.145)

If α is large, then very little flow occurs through the fractures anyway, and the secondary porosity is of little concern. If α is small, the blocks are essentially impermeable, and the fracture network is crucial; but then the solution of (7.131) is just $\mathbf{q} = O(\alpha)$, and the relative permeability is simply

$$\mathbf{k}^* = \mathbf{I} - \langle\mathbf{nn}\rangle_{\partial M},$$

(7.146)

which only differs from the unit tensor if the medium is anisotropic. It is only in the case $\alpha = O(1)$ that the competition between the two systems becomes important. If we use values $\tau_f = 10^2$, $d_B = 1$ m, $h = 10^{-3}$ m and $k_m = 10^{-12}$ m^2 (cf. Table 7.1), we get $\alpha \approx 0.01$. This might be appropriate for a fractured sandstone on a regional scale. Generally, the primary and secondary (fracture) permeability will only be comparable if the host rock is itself quite permeable.

7.6 Contaminant Transport

Much of the interest in modelling groundwater flow lies in the prediction of solute transport, in particular in understanding how pollutants will disperse: for example,

---

[6]The average of $\mathbf{n}$ over ∂M is zero because the normals on opposite sides of a fracture are in opposite directions. More specifically, $\int_{\partial M}\phi\mathbf{n}\,dS = \int_M\nabla\phi\,dV$, thus $\int_{\partial M}\mathbf{n}\,dS = 0$.

how do nitrates used for agricultural purposes disperse via the local groundwater system? Mostly simply, one would simply add a diffusion term to the advection of the solute concentration c:

$$\phi c_t + \mathbf{u}.\nabla c = \nabla.[\phi D \nabla c]. \tag{7.147}$$

The diffusive width Δl of a sharp front travelling at speed u after it has travelled a distance l is of the order of $\Delta l \sim (Dl/u)^{1/2}$; if we take $D \sim 10^{-9} \ \mathrm{m^2\,s^{-1}}$, $u \sim 10^{-6} \ \mathrm{m\,s^{-1}}$ ($30 \ \mathrm{m\,y^{-1}}$), $l = 10^3$ m, then $\Delta l \sim 1$ m, and the diffusion zone is relatively narrow. For a more porous sand, the diffusion width is even smaller.

In fact, as velocity increases, the effect of diffusion increases. That this is so is due to a remarkable phenomenon called *Taylor dispersion*, described by G.I. Taylor in 1953. Consider the diffusion of a solute in a tube of circular cross section through which a Poiseuille flow passes. If the mean velocity is U and the tube is of radius a, then the velocity is $2U(1 - r^2/a^2)$, and the concentration satisfies the equation

$$c_t + 2U\left(1 - r^2/a^2\right)c_x = D\left(c_{rr} + \frac{1}{r}c_r + c_{xx}\right), \tag{7.148}$$

where x is measured along the tube, and r is the radial coordinate. Taylor showed, rather ingenuously, that when the Péclet number $Pe = aU/D$ is large, then the effect of the diffusion term in (7.148) is to *disperse* the mean solute concentration diffusively about the position of its centre of mass, $x = Ut$, with a dispersion coefficient of $a^2 U^2/48D$. Aris later improved this to

$$D^T = \frac{a^2 U^2}{48D} + D, \tag{7.149}$$

which is asymptotically valid for $x \gg a$. The dispersive mechanism is due to the radial variation of the velocity profile, which can disperse the solute even if the diffusion coefficient is very small.

Typically, this is generalised for porous media (where we think of the pores as being like Taylor's tube) by writing the dispersion coefficient as

$$D^T = D^* + D_\parallel, \tag{7.150}$$

where D^* represents molecular diffusion and $D_\parallel$ dispersion in the direction of flow. The tortuosity of the flow paths and the possibility of adsorption on to the solid causes D^* to be less than D, and ratios D^*/D between 0.01 and 0.5 are commonly observed. In porous media, remixing at pore junctions causes the dependence of $D_\parallel$ on the flow velocity to be less than quadratic, and a relation of the form

$$D_\parallel = \alpha u^m, \tag{7.151}$$

where $\mathbf{u}$ is the Darcy flux, fits experimental data reasonably well for values $1 < m < 1.2$. A common assumption is to take $m = 1$. Mixing at junctions also causes transverse dispersion to occur, with a coefficient $D_\perp$ which is measured to be less than $D_\parallel$ by a factor of order 10^2 when $Pe \gg 1$. Dispersion is thus a tensor property.

If we write

$$D_{\|} = \alpha_{\|}|\mathbf{u}| \tag{7.152}$$

for the longitudinal dispersion coefficient, and

$$D_{\perp} = \alpha_{\perp}|\mathbf{u}| \tag{7.153}$$

for the lateral dispersion coefficient, then a suitable tensor generalisation is

$$D_{ij}^T = \alpha_{\perp}|\mathbf{u}|\delta_{ij} + (\alpha_{\|} - \alpha_{\perp})\frac{u_i u_j}{|\mathbf{u}|}, \tag{7.154}$$

where δ_{ij} is the Kronecker delta.

The conservation of solute equation is then

$$\phi\frac{\partial c}{\partial t} + \mathbf{u}.\nabla c = \nabla.(\phi\mathbf{D}^T.\nabla c] = \frac{\partial}{\partial x_i}\left(\phi D_{ij}^T \frac{\partial c}{\partial x_j}\right). \tag{7.155}$$

For a one-dimensional flow in the x direction, c satisfies

$$\frac{\partial c}{\partial t} + v\frac{\partial c}{\partial x} = \frac{\partial}{\partial x}\left[D_{\|}\frac{\partial c}{\partial x}\right] + \frac{\partial}{\partial y}\left[D_{\perp}\frac{\partial c}{\partial y}\right] + \frac{\partial}{\partial z}\left[D_{\perp}\frac{\partial c}{\partial z}\right] \tag{7.156}$$

($v = u/\phi$ is the linear velocity) and if the dispersivities are constant, then the solution for release of a mass M at the origin at $t = 0$ is

$$c = \frac{M}{8\phi(\pi D_{\|})^{1/2}D_{\perp}t^{3/2}}\exp\left[-\frac{(x - vt)^2}{4D_{\|}t} - \frac{r^2}{4D_{\perp}t}\right], \tag{7.157}$$

where $r^2 = y^2 + z^2$.

7.6.1 Reactive Dual Porosity Models

Let us now consider the reactive transport of a contaminant of concentration c within a fractured soil or rock which has dual porosity. We follow the ideas of averaging and homogenisation in Sect. 7.5.1. The point forms of the equations are taken in the form

$$\frac{\partial c}{\partial t} + \nabla.(c\mathbf{u}) = \nabla.[D\nabla c] + S_f,$$

$$\frac{\partial c_m}{\partial t} + \nabla.(c_m\mathbf{u}) = \nabla.[\mathbf{D}_m.\nabla c_m] + S_m, \tag{7.158}$$

where S_f, S_m represent source or sink terms due to chemical reaction, D is the molecular diffusion coefficient of the contaminant within the fractures, and $\mathbf{D}_m$ is the dispersivity within the matrix blocks. The concentrations within the fractures and matrix are denoted by c and c_m, respectively.

The first thing to do is to average the fracture concentration equation across the width of the fracture. When we do this, we effectively regain the Taylor dispersion equation, with the addition of the reaction terms, and also a solute flux delivered from the matrix:

$$\frac{\partial c_f}{\partial t} + \nabla.(c_f \mathbf{u}_f) = \nabla.[\mathbf{D}_f.\nabla c_f] + S_f - \frac{1}{h}\left[(\mathbf{n}.\mathbf{D}_m.\nabla c_m) - c_m \mathbf{u}_m.\mathbf{n}\right]\big|_{\partial M}. \quad (7.159)$$

This equation is analogous to (7.114). It differs from (7.158) in that c_f is the cross-sectional average concentration (actually, in the derivation of the Taylor dispersion equation, one finds the concentration is cross-sectionally uniform, so that $c_f = c$); $\mathbf{u}_f$ is the cross-sectionally averaged fracture velocity, just as before; and $\mathbf{D}_f$ is the *local* dispersion coefficient: it will be modified again at the larger macroscale. The reaction term S_f depends on cross-sectionally averaged concentrations, which equal their point forms, so that S_f is unchanged.

Now we write down the equivalents of (7.143). We define the block averaged concentrations

$$\bar{c}_m = \langle c_m \rangle|_M, \qquad \bar{c}_f = \langle c_f \rangle|_{\partial M}, \qquad (7.160)$$

where as before $\langle c \rangle|_M$ denotes an average of c over the matrix blocks M, and $\langle c \rangle|_{\partial M}$ denotes the average of c over the fracture surface ∂M. Then we have the equation for $\bar{c}_f$:

$$\frac{\partial}{\partial t}(\phi_f \bar{c}_f) + \nabla.[\phi_f \bar{c}_f \mathbf{u}_f] = \nabla.[\phi_f \mathbf{D}_f.\nabla \bar{c}_f] + \phi_f \bar{S}_f$$
$$+ s_f \langle \mathbf{n}.\{c_m \mathbf{u}_m - \mathbf{D}_m.\nabla c_m\}\rangle\big|_{\partial M}. \quad (7.161)$$

We can make use of (7.143)$_1$, and the fact that $c_f = c_m$ on ∂M, to simplify this to

$$\phi_f\left[\frac{\partial \bar{c}_f}{\partial t} + \mathbf{u}_f.\nabla \bar{c}_f\right] = \nabla.[\phi_f \mathbf{D}_F.\nabla \bar{c}_f] + \phi_f \bar{S}_f - s_f \langle \mathbf{n}.\mathbf{D}_m.\nabla c_m \rangle|_{\partial M}. \quad (7.162)$$

The specific fracture surface area is defined as s_f, as before. The macroscopic fracture dispersivity $\mathbf{D}_F$ here is distinct from $\mathbf{D}_f$, in the same way that Taylor dispersion in a tube is distinct from that in a porous medium; in this case it is because of remixing of fracture fluid at the junctions between fractures at the block boundaries. The formal averaging assumption which is made is

$$\langle \mathbf{D}_f.\nabla c_f - c_f \mathbf{u}_f \rangle|_{\partial M} = \mathbf{D}_F.\nabla \bar{c}_f - \bar{c}_f \langle \mathbf{u}_f \rangle|_{\partial M}, \quad (7.163)$$

and this has some justification insofar as it is just this result which emerges in the study of Taylor dispersion.

In a similar way to the derivation of the average fracture concentration equation, the matrix averaged concentration satisfies

$$(1 - \phi_f)\left[\frac{\partial \bar{c}_m}{\partial t} + \mathbf{u}_m . \nabla \bar{c}_m\right]$$

$$= \nabla . \left[(1 - \phi_f)\mathbf{D}_m . \nabla \bar{c}_m\right] + (1 - \phi_f)\bar{S}_m + s_f \langle \mathbf{n}.\mathbf{D}_m . \nabla c_m \rangle|_{\partial M}. \quad (7.164)$$

The result of averaging is the two Eqs. (7.162) and (7.164) for the average fracture and matrix concentrations. In principle, the block average fracture dispersivity $\mathbf{D}_F$ should be calculable by solving the local block problem, although in practice one would assume a value by analogy with assumptions about porous medium dispersion coefficients. However, unlike the incompressible dual porosity mass flow equations (7.143), the source term $s_f \langle \mathbf{n}.\mathbf{D}_m . \nabla c_m \rangle|_{\partial M}$ is non-zero, and this must be constituted, ideally by solving the block scale problem, which is given by Eqs. (7.158)$_2$ and (7.159); these can be slightly simplified to the forms

$$\frac{\partial c_m}{\partial t} + \mathbf{u}_m . \nabla c_m = \nabla . [\mathbf{D}_m . \nabla c_m] + S_m,$$

$$\frac{\partial c_f}{\partial t} + \mathbf{u}_f . \nabla c_f = \nabla . [\mathbf{D}_f . \nabla c_f] + S_f - \frac{1}{h}(\mathbf{n}.\mathbf{D}_m . \nabla c_m)|_{\partial M}. \quad (7.165)$$

The boundary condition for c_m is that

$$c_m = c_f \quad \text{on } \partial M, \quad (7.166)$$

and c_f is periodic over ∂M.

The basis for the method of homogenisation is the expansion of the local block problem in terms of the parameter $\varepsilon = d_B/l$, where d_B is the block scale and l is the macroscopic length scale. However, because of the complexity of the equations to be solved, the application of this method must be done judiciously.

To illustrate this point, suppose that the dispersion coefficient tensors are all isotropic, and equal to $D_T \mathbf{I}$, where D_T is constant, and that suitable (macroscopic) scales for the variables are $c \sim c_0$, $\mathbf{x} \sim l$, $\mathbf{u} \sim U$, $t \sim l/U$, and suppose to be precise that the reaction kinetics are first order, i.e., $S = -rc$, and that the specific fracture surface area is constant, $s_f = 1/d_B$. The dimensionless equation for the matrix average $\bar{c}_m$ is then

$$(1 - \phi_f)Pe\left[\frac{\partial \bar{c}_m}{\partial t} + \mathbf{u}_m . \nabla_\mathbf{x} \bar{c}_m\right]$$

$$= (1 - \phi_f)\nabla_\mathbf{x}^2 \bar{c}_m - Pe\Lambda(1 - \phi_f)\bar{c}_m + \frac{1}{\varepsilon^2}\left\langle \frac{\partial c_m}{\partial N}\right\rangle\bigg|_{\partial M}, \quad (7.167)$$

where $\frac{\partial c_m}{\partial N}$ denotes $\mathbf{n}.\nabla_\mathbf{x} c_m$, and

$$Pe = \frac{Ul}{D_T}, \qquad \Lambda = \frac{rl}{U}. \quad (7.168)$$

The local block equation $(7.165)_1$ rewritten in the block variables

$$\mathbf{X} = \frac{\mathbf{x}}{\varepsilon}, \qquad T = \frac{t}{\varepsilon}, \tag{7.169}$$

is then

$$\varepsilon Pe \left[\frac{\partial c_m}{\partial T} + \mathbf{u}_m . \nabla_{\mathbf{X}} c_m \right] = \nabla_{\mathbf{X}}^2 c_m - \varepsilon^2 Pe \Lambda c_m. \tag{7.170}$$

What is obvious is that all the terms cannot balance in both local and global problems, and this leads to simplifications. The simplest case is where the macroscopic Péclet number and reaction number Λ are both taken to be $O(1)$; then (7.167) suggests that in (7.170) we put

$$c_m = \bar{c}(\mathbf{x}) + \varepsilon^2 \tilde{c}_m(\mathbf{X}) \tag{7.171}$$

(and thus also $c_f = \bar{c}(\mathbf{x}) + \varepsilon^2 \tilde{c}_f(\mathbf{X})$); both inner and outer problems are well scaled, and the advective term can be neglected in solving the inner problem, which in fact reduces approximately to Poisson's equation. Specifically, using multiple scales in both $\mathbf{x}$ and t, we have at leading order in ε,

$$\nabla_{\mathbf{X}}^2 \tilde{c}_m = Pe \left[\frac{\partial \bar{c}}{\partial t} + \mathbf{u}_m . \nabla_{\mathbf{x}} \bar{c} \right] + Pe \Lambda \bar{c} - \nabla_{\mathbf{x}}^2 \bar{c} \tag{7.172}$$

subject to $\tilde{c}_m = \tilde{c}_f$ on ∂M and conditions of periodicity on M. (If we integrate (7.172) over M, we regain the averaged equation (7.164) as an integrability condition for (7.172), as expected via homogenisation.)

In a similar way, the leading order local equation for the fracture perturbed concentration $\tilde{c}_f$ is

$$\nabla_{\mathbf{X}}^2 \tilde{c}_f - \frac{(1 - \phi_f)}{\phi_f} \frac{\partial \tilde{c}_m}{\partial N} \Big|_{\partial M} = Pe \left[\frac{\partial \bar{c}}{\partial t} + \mathbf{u}_f . \nabla_{\mathbf{x}} \bar{c} \right] + Pe \Lambda \bar{c} - \nabla_{\mathbf{x}}^2 \bar{c} \tag{7.173}$$

(note that a term $\frac{(1-\phi_f)}{\varepsilon \phi_f} \frac{\partial \bar{c}}{\partial n}|_{\partial M} \propto \nabla_{\mathbf{x}} \bar{c} . \mathbf{n}|_{\partial M}$ is identically zero, because the normals $\mathbf{n}$ on the two faces of a fracture cancel each other). $\tilde{c}_f$ satisfying (7.173) is subject to periodicity in ∂M and zero mean. As for the fluid flow, we can solve (7.172) using a Green's function for the Laplacian in M, so that the boundary derivative term in (7.173) becomes an integral convolution in terms of $\tilde{c}_f$. We then solve (7.173) using a Green's function for the Laplacian on ∂M. This allows us to determine the homogenised equation for $\bar{c}$ (see Question 7.13).

In fact, it is rarely the case that Pe and Λ are $O(1)$: more commonly they are both large. Suppose for example that $\varepsilon = 10^{-4}$, $Pe = 10^4$, $\Lambda = 10^4$: not unreasonable values (as we shall see below). Putting

$$p = \varepsilon Pe, \qquad \lambda = \varepsilon \Lambda, \tag{7.174}$$

it seems natural to take $\lambda \sim p \sim O(1)$. The block problem becomes (in terms of T and $\mathbf{X}$ only)

$$p\left[\frac{\partial c_m}{\partial T} + \mathbf{u}_m \cdot \nabla_{\mathbf{X}} c_m\right] = \nabla_{\mathbf{X}}^2 c_m - \lambda p c_m, \qquad (7.175)$$

which implies that reaction occurs on the block scale. This is a linear equation for c_m which can in principle be solved to give the boundary flux term in the c_f equation as a convolution integral in terms of c_f. In a strict sense, this distinguished limit describes the structure of a reaction front in which both reaction and dispersion are important. Outside this front, reaction and dispersion are negligible, and the reactant simply advects with the flow.[7] Importantly, it implies that the speed of the front is determined by the local diffusion and reaction within the block.

If Λ is even larger than this, so that $\lambda \gg 1$, then the reaction is fast at the block scale, and occurs in a thin rind within the block; for a single species, as here, this rind must be on the boundary, as the interior reactant concentration reaches zero rapidly. Bearing in mind that the normal coordinate n at ∂M points into the fracture, the boundary layer solution for c_m is just

$$c_m \approx c_f \exp\left[(\lambda p)^{1/2} n\right], \qquad (7.176)$$

and thus the flux term in the local fracture equation derived from $(7.165)_2$ is

$$-\frac{1}{\phi_f} \frac{\partial c_m}{\partial n}\bigg|_{\partial M} = -\frac{1}{\phi_f}(\lambda p)^{1/2} c_f. \qquad (7.177)$$

Hence c_f satisfies the local equation

$$\nabla_{\mathbf{X}}^2 c_f - \frac{1}{\phi_f}(\lambda p)^{1/2} c_f = p\left[\frac{\partial c_f}{\partial T} + \mathbf{u}_f \cdot \nabla_{\mathbf{X}} c_f\right] + \lambda p c_f. \qquad (7.178)$$

As is perhaps obvious, the reaction is fast in the fractures also and c_f rapidly approaches zero.

7.7 Environmental Remediation

Environmental pollution is now an endemic phenomenon. All over every industrialised country, spills of hydrocarbons, effluents, and industrial waste have caused pollution of underlying groundwater. The resulting 'plumes' move slowly with the prevailing background groundwater flow, and in course of time will enter streams, rivers and lakes, with consequent health risk via the contamination of drinking water. Short of more drastic measures such as pumping out polluted groundwater and

---

[7] Actually, this only makes sense for multi-species reactions of the form $A + B \to P$, for then the reaction rate $\propto AB$, and can be zero by virtue of $A = 0$ on one side of the front, and $B = 0$ on the other.

treating it, natural bioremediation anticipates that microbial action will eventually break down most pollutants, rendering them harmless. The issue for the environmental scientist is to predict the future movement of the plume, and the likelihood of microbial breakdown before it reaches drinking water sources.

In so doing, groundwater flow modelling is essential, because of its ability to predict into the future, and also because accurate monitoring of subsurface contamination is expensive and not straightforward. Against this, subsurface soil and rock is usually an extremely heterogeneous medium, both physically and chemically, and the validation of computational results is difficult.

7.7.1 Reactive Groundwater Flow

The general context of many subsurface pollution problems of concern is similar, and we therefore begin with some generalities. Contaminants may be aqueous, in which case they mix with the groundwater, or non-aqueous, in which case they do not. Hydrocarbons, for example, are non-aqueous. Amongst the non-aqueous phase liquids (NAPLs), one distinguishes dense liquids (DNAPLs) from light ones (LNAPLs). DNAPLs will sink into the saturated zone, whereas LNAPLs, such as hydrocarbons, will sink to the base of the unsaturated zone, and there sit on the water table, from where their constituents may diffuse downwards.

The contaminant plume will typically consist of a cocktail of different chemicals, which flow with the local groundwater flow, disperse within it, and react with oxygen and other substances in the soil via the agency of microbial action. The typical sort of model of concern is thus the reaction-advection-dispersion equation

$$R\frac{\partial c}{\partial t} + \nabla.(c\mathbf{u}) = \nabla.(\mathbf{D}.\nabla c) + \mathcal{S}, \qquad (7.179)$$

where c is one of a sequence of reactants, $\mathbf{u}$ is the local groundwater flux (given by Darcy's law), $\mathbf{D}$ represents dispersion, and is typically anisotropic, in the sense that dispersion in the longitudinal direction is larger than lateral dispersion. Dispersion itself is partly due to molecular diffusion, but more importantly (at high grain scale Péclet number) is due to grain scale shear-induced distortion of the fluid associated with Taylor dispersion, together with remixing at pore junctions. Typically, the longitudinal dispersion coefficient $D_\parallel \sim d_p|\mathbf{u}|$, where d_p is grain size, while lateral dispersion $D_\perp$ is a factor of 10–100 smaller.

The coefficient R is the retardation factor, and it is a slowing rate due to the adsorption of aqueous phase concentration on solid particles. Specifically, we actually have two separate conservation laws for solid and aqueous concentrations c_s and c_l, respectively, thus

$$\frac{\partial c_l}{\partial t} + \nabla.(c_l\mathbf{u}) = \nabla.(\mathbf{D}.\nabla c_l) + \mathcal{S} + k_d c_s - k_a c_l,$$

$$\frac{\partial c_s}{\partial t} = -k_d c_s + k_a c_l, \qquad (7.180)$$

where k_d and k_a are desorption and adsorption rates, respectively. (7.180) specifically assumes that the reaction occurs only in the aqueous phase, by way of example.

Now the point is that if the sorption rates are very large (and constant), then the second equation in (7.180) tells us that

$$c_s \approx \frac{k_a c_l}{k_d}, \qquad (7.181)$$

and the sum of the two equations then gives us (7.179), with the retardation factor

$$R = 1 + \frac{k_a}{k_d}. \qquad (7.182)$$

The source term S represents reaction driven sources and sinks. It is typically the case that there are many, many reactions and reactants. Equally typically, many of the reaction rates are not well known, and the rates may be widely disparate. In general this will imply that many reactions can be taken to be in equilibrium, with only the slowest (rate-controlling) reaction being of dynamical importance.

7.7.2 Biomass Modelling

The reaction terms S in (7.180) are mediated by bacteria within soil, which consume the various nutrients provided through metabolic reactions. Like all living things, microbes survive by generating energy from nutrients through a variety of such reactions. This process involves a network of redox (oxidation–reduction) reactions, and involves the overall exchange of electrons between two distinct chemical fuels which are consumed in the reactions; the metabolic process is in this case called respiration. While there may be a number of such fuels, there is a hierarchy in their use. Dissolved oxygen is commonly the terminal electron acceptor (as the externally sourced oxidant is typically referred to), while an organic carbon compound is the electron donor. When these preferred substrates are absent or depleted, other compounds can be used instead. When the same organic compound is used as both donor and acceptor, the metabolic process is called fermentation.

Many bacteria are able to use several reaction pathways independently, giving them a degree of flexibility to differing conditions. This capability is very species-dependent, and competition ensures that the species which are best adapted to local conditions become dominant. Microbial growth depends heavily on energy metabolism but also requires the uptake of other substrates needed to generate new biomass. Growth rate is generally limited by the supply of one or more substrates, but saturates to a maximum growth rate in conditions of ample supply. The dependence of bacterial growth rate is commonly taken, by analogy with simple enzyme kinetic uptake rates, as proportional to $\frac{c}{K+c}$, where c is the relevant nutrient concentration, and K is a constant; such kinetics are called Monod kinetics. When two nutrients control growth, as in respiration, it is usual to take the growth rate as proportional to the product of two Monod factors, thus (for example)

$$S = -r_0 X \frac{c_1}{K_1 + c_1} \frac{c_2}{K_2 + c_2}, \tag{7.183}$$

in which r_0 is a reaction rate constant, and X is the biomass density.

More complex models also consider the growth of the microbial population X in terms of the nutrient supply, in particular in the form of a microbial mat which becomes attached to the soil grains, and usually called a biofilm. In particular, the concept of biodegradation, for example of oil spills in sea water, or of factory effluent in groundwater, commonly involves the development of bacterial colonies which are able to efficiently use the offending contaminant to promote their own growth.

7.7.3 Non-dimensionalisation

It is commonly the case that the water table is at a depth of 10–20 metres, whereas a plume may have spread (or we are concerned with whether it will spread) a distance of order kilometres. Therefore these plumes generally have high aspect ratio, a cause both of computational stress and analytical simplification. The latter may be offset by the decreased lateral dispersion coefficient, as we now show.

Let us consider the scalar contaminant equation (7.179) in two dimensions, with horizontal and vertical coordinates x and z, corresponding Darcy fluxes u and w, and longitudinal (horizontal) and transverse (vertical) dispersivities $D_\parallel$ and $D_\perp$, which we take to be constant. We suppose l and d are suitable horizontal and vertical length scales, U is a horizontal Darcy flux scale (and therefore mass conservation implies that hU/l is a suitable vertical flux scale), Rl/U is then the convective time scale, and we take $\mathcal{S}_0$ to be a measure of the reaction rate term, and c_0 to be a typical contaminant concentration. The units of concentration are $\mathrm{mol\,l}^{-1}$ (moles per litre), and the units of $\mathcal{S}_0$ are $\mathrm{mol\,l}^{-1}\,\mathrm{s}^{-1}$.

We define dimensionless variables by writing

$$u = U u^*, \qquad w = \frac{hU}{d} w^*, \qquad c = c_0 c^*, \qquad t = \frac{Rl}{U} t^*,$$

$$x = l x^*, \qquad z = d z^*, \qquad S = \mathcal{S}_0 S; \tag{7.184}$$

substituting these into (7.179) (and forthwith dropping the asterisks), we obtain the dimensionless equation

$$\frac{\partial c}{\partial t} + u \frac{\partial c}{\partial x} + w \frac{\partial c}{\partial z} = \frac{1}{Pe_\parallel} \frac{\partial^2 c}{\partial x^2} + \frac{1}{Pe_\perp} \frac{\partial^2 c}{\partial z^2} + \Lambda S, \tag{7.185}$$

where

$$Pe_\parallel = \frac{Ul}{D_\parallel}, \qquad Pe_\perp = \frac{Ud^2}{D_\perp l}, \qquad \Lambda = \frac{\mathcal{S}_0 l}{U c_0}. \tag{7.186}$$

The aspect ratio d/l and transverse dispersivity ratio $D_\perp/D_\parallel$ compete against each other, and generally we might suppose $Pe_\perp \sim Pe_\parallel$; but also if $D_\parallel \sim d_p U$, then we can expect that usually $Pe_\parallel \gg 1$, heralding the existence of thin boundary layers in which dispersion is effective. The parameter Λ is the ratio of advection time to reaction time, and will often be very large for microbially mediated reactions of interest. These two observations cause numerical difficulties in solving (7.186), but can aid analytical insight. In the following section, we discuss three specific groundwater contamination problems of recent concern.

7.8 Three Specific Remediation Problems

Four Ashes   Four Ashes is a site in the Midlands of England where a fifty year old plume of phenol (C_6H_6O) and other contaminants, about 500 metres long, lies within the saturated zone. The plume is thought to lie between depths of 10 m and 30 m at 130 m from the source, but is sinking as it moves west (perhaps because of surface recharge), reaching depths 21–44 m at 350 m distance (this information comes from two boreholes drilled at the site). Degradation is very slow: perhaps only 5% of the total contaminant load has so far been degraded. This is thought to be due to toxicity of the phenol within the plume, and to supply limitation of electron acceptors (oxygen and nitrate) at the plume fringe.

The overall reaction scheme which is considered to apply is that the phenol is oxidised at the plume boundary by oxygenated groundwater, producing TIC (total inorganic carbon). Within the plume, phenol is fermented to produce hydrogen amongst other things, which then reduces various oxides within the soil via microbial agency. A suitable set of reactions to describe the situation consists of the following:

$$C_6H_6O + 7O_2 + 3H_2O \xrightarrow{r_1} 6HCO_3^- + 6H^+,$$

$$C_6H_6O + 5.6NO_3^- + 0.2H_2O \xrightarrow{r_2} 6HCO_3^- + 0.4H^+ + 2.8N_2,$$

$$C_6H_6O + 5H_2O \xrightarrow{r_3} 3CH_3COOH + 2H_2,$$

$$C_6H_6O + 17H_2O \xrightarrow{r_4} 6HCO_3^- + 6H^+ + 14H_2,$$

$$H_2 + MnO_{2(s)} + 2H^+ \xrightarrow{r_5} 2H_2O + Mn^{2+},$$

$$H_2 + 2FeOOH_{(s)} + 4H^+ \xrightarrow{r_6} 4H_2O + 2Fe^{2+},$$

$$H_2 + SO_4^{2-} + 0.25H^+ \xrightarrow{r_7} H_2O + 0.25HS^-,$$

$$H_2 + CO_3^{2-} + 0.5H^+ \xrightarrow{r_8} 0.75H_2O + 0.25CH_4,$$

(7.187)

and r_1–r_8 are the reaction rates, defined using Monod kinetics. In terms of the reaction rates, the source term in each reactant equation is then

$$S_j = \sum_l s_{jl} r_l,$$

(7.188)

where s_{jl} is the (sign dependent) stoichiometric coefficient for reactant j in reaction l.

The maximum (saturated) rates k_j of the Monod rates r_j are thought to range from about 10^{-13} mol l^{-1} s^{-1} to about 10^{-10} mol l^{-1} s^{-1}. The background phenol concentration p_0 is of the order of 10^{-10} mol l^{-1}, and the aqueous dissolved oxygen level c_0 is of order 10^{-4} mol l^{-1}. The horizontal Darcy flux is estimated to be $U = 10$ m y^{-1}. The dispersivities are taken in the form $D = \alpha U$, and values of $\alpha_\parallel = 1$ m and $\alpha_\perp = 4 \times 10^{-4}$ m are assumed.[8] Also, we take $l = 500$ m, $d = 20$ m, and $R = 1$. With all of these values, we have

$$Pe_\parallel = \frac{l}{\alpha_\parallel} = 500, \qquad Pe_\perp = \frac{d^2}{\alpha_\perp l} = 2000, \qquad \Lambda \sim 10^6 - 10^9, \qquad (7.189)$$

where the definition of Λ is based on p_0, i.e.,

$$\Lambda = \frac{S_0 l}{U p_0}. \qquad (7.190)$$

As suggested, $Pe_\parallel \sim Pe_\perp \gg 1$, and the reaction rate parameter Λ is *extremely* large. To get an idea of the solution behaviour when $Pe \gg 1$ and $\Lambda \gg 1$, we consider the simpler reaction

$$C_6H_6O + \sigma O_2 \xrightarrow{r} \text{products}, \qquad (7.191)$$

where σ is a suitable stoichiometric coefficient. The corresponding version of (7.185) is simply[9]

$$\frac{\partial p}{\partial t} + \nabla.(p\mathbf{u}) = \frac{1}{Pe}\nabla^2 p + \Lambda S,$$

$$\frac{\partial c}{\partial t} + \nabla.(c\mathbf{u}) = \frac{1}{Pe}\nabla^2 c + \nu \Lambda S, \qquad (7.192)$$

where p and c are the concentrations of the two species (phenol and oxygen), and the reaction term is just

$$S = -\left(\frac{p}{k_p + p}\right)\left(\frac{c}{k_c + c}\right); \qquad (7.193)$$

the parameter ν is given by

$$\nu = \frac{\sigma p_0}{c_0} \sim 10^{-6}. \qquad (7.194)$$

---

[8] For both theoretical and experimental reasons, we expect $\alpha_\parallel \sim d_p$, the pore or grain size (Bear 1972, p. 609), with $\alpha_\perp / \alpha_\parallel \sim 0.01$–$0.05$ (Sahimi 1995, p. 225). The large value of $\alpha_\parallel$ here suggests flow in sub-parallel fractures with spacing on the order of a metre.

[9] We can take $Pe_\parallel = Pe_\perp$ by choosing d appropriately.

The asymptotic structure of the solution of (7.192) is easily given. For $\Lambda \gg 1$, the reaction region is very thin, and will occur only at the fringe of the plume. Outside the plume, $p = 0$, while inside the plume $c = 0$, so that $S = 0$ everywhere apart from the reaction front, which is located at the plume boundary. Denoting the plume by P and its boundary by ∂P, we thus have to solve

$$
\begin{aligned}
\frac{\partial p}{\partial t} + \nabla.(p\mathbf{u}) &= \frac{1}{Pe}\nabla^2 p, \quad \mathbf{x} \in P, \\
\frac{\partial c}{\partial t} + \nabla.(c\mathbf{u}) &= \frac{1}{Pe}\nabla^2 c, \quad \mathbf{x} \notin P,
\end{aligned}
\tag{7.195}
$$

and p and c both satisfy $p = c = 0$ on ∂P. The location of P has to be determined (it is a free boundary), but this is simply done because of the precise conservation law for $c - \nu p$, which follows from (7.192):

$$
\frac{\partial(c - \nu p)}{\partial t} + \nabla.\left[(c - \nu p)\mathbf{u}\right] = \frac{1}{Pe}\nabla^2(c - \nu p).
\tag{7.196}
$$

Integrating this across the reaction front yields the extra condition[10] to determine the front ∂P:

$$
\frac{\partial c}{\partial n} = \nu \frac{\partial p}{\partial n}.
\tag{7.197}
$$

Note that the normal derivative here has to be treated with some attention when non-dimensionalised. If the plume fringe is at dimensionless position $z = \zeta(x, t)$, then the *dimensional* normal component of the dispersive flux of oxygen (in terms of dimensionless variables) is just

$$
(\mathbf{D}.\nabla c).\mathbf{n} = \frac{c_0 U d}{l(1 + \delta^2 \zeta_x^2)^{1/2}}\left[\frac{1}{Pe_\perp}\frac{\partial c}{\partial z} - \frac{1}{Pe_\parallel}\frac{\partial c}{\partial x}\frac{\partial \zeta}{\partial x}\right],
\tag{7.198}
$$

where δ is the aspect ratio. There is a similar expression for $\partial p/\partial n$. It follows from this that more generally, the conservation boundary condition (7.197) can be expressed as

$$
\frac{1}{Pe_\perp}\frac{\partial c}{\partial z} - \frac{1}{Pe_\parallel}\frac{\partial c}{\partial x}\frac{\partial \zeta}{\partial x} = \nu\left[\frac{1}{Pe_\perp}\frac{\partial p}{\partial z} - \frac{1}{Pe_\parallel}\frac{\partial p}{\partial x}\frac{\partial \zeta}{\partial x}\right],
\tag{7.199}
$$

and to interpret this as (7.197) requires the unit normal $\mathbf{n}$ to be suitably defined.

The problem is even easier when (as here) $Pe \gg 1$. Then diffusive boundary layers of (dimensionless) thickness $O(1/Pe^{1/2})$ adjoin the plume boundary, and away from these, the reactants are simply advected with the flow.

In the present case, the parameter $\nu \ll 1$. This modifies the discussion as follows. Reaction must still occur in thin regions, outside which the reactants obey (7.195),

---

[10]In this particular case, it is even simpler, since we just integrate the convective-diffusion equation (7.196) with appropriate external boundary conditions, and ∂P is located by the curve where $c - \nu p = 0$.

but there cannot be a front (in the sense of a moving boundary) at which $c = \frac{\partial c}{\partial n} = 0$. If, for example, we have the one-dimensional problem

$$c_t + u c_x = \frac{1}{Pe} c_{xx},$$ (7.200)

with $c = 1$ at $t = 0$ and as $x \to \infty$, and $c = 0$ at $x = 0$, then if $Pe \sim O(1)$, reaction occurs near $x = 0$, oxygen supply being sufficient to remove the phenol. If $Pe \gg 1$, then a front moves away from $x = 0$ at speed u. Behind this front, $c \approx 1$, ahead of it $c \approx 0$, and the front itself is purely diffusional, of width $O\{(t/Pe)^{1/2}\}$, having an error function profile for solution.

Rexco   The site was a former coal carbonisation plant (between 1935 and 1970) where ammonium liquor had been allowed to drain on site. The liquid drained through the unsaturated zone of some 18 metres depth to the saturated zone, where it is spreading as a plume of some 25 metres depth in a sandstone aquifer. At the plume fringe, ammonium ions (NH_4^+) ions are subject to oxidation, releasing nitrogen. Phenol is also present in front of the ammonium, which is highly retarded (with a retardation factor of $R = 5$). Well extraction by a new factory on site is causing the groundwater flow field to be time dependent, and this is an issue in modelling studies.

The modelling scheme is similar to the Four Ashes site, but less is known of the reaction rates. The reactions which are considered to be important for the ammonium are

$$NH_4^+ + 2O_2 \xrightarrow{r_1} NO_3^- + H_2O + 2H^+,$$

$$NO_3^- + 1.25CH_2O + H^+ \xrightarrow{r_2} 0.5N_2 + 1.25CO_2 + 1.75H_2O,$$ (7.201)

$$NH_4^+ + 0.6NO_3^- \xrightarrow{r_3} 0.8N_2 + 1.8H_2O + 0.4H^+.$$

The kinetics (only) of these reactions is described by the five ordinary differential equations for the five reactants, ammonium NH_4^+, oxygen O_2, nitrate NO_3^-, hydrogen ion H^+ and organic carbon CH_2O:

$$\frac{dc_{NH_4^+}}{dt} = -r_1 - r_3,$$

$$\frac{dc_{O_2}}{dt} = -2r_1,$$

$$\frac{dc_{NO_3^-}}{dt} = r_1 - r_2 - 0.6r_3,$$ (7.202)

$$\frac{dc_{H^+}}{dt} = 2r_1 - r_2 + 0.4r_3,$$

$$\frac{dc_{CH_2O}}{dt} = -1.25r_2.$$

To estimate Péclet numbers, we take $d = 25$ m, $l = 1000$ m, $R = 5$, $U = 100$ m y^{-1}. Values of dispersion coefficients are uncertain, so we choose for illustration $D_\parallel = 100$ m^2 y^{-1}, $D_\perp = 10$ m^2 y^{-1}, i.e., $\alpha_\parallel = 1$ m (like Four Ashes) and $\alpha_\perp = 0.1$ m. The (unretarded) time scale l/U is then 10 years, and $Pe_\parallel \sim 10^3$, $Pe_\perp \sim 6$. These values are not too reliable, and perhaps are consistent with a highly fractured aquifer. For a homogeneous medium, we would expect much smaller α, and thus much larger Pe.

It is possible to extend the Four Ashes discussion of the simple phenol/oxygen reaction to the more complicated scheme above. Reaction rates are not well known. Most of the reactants are present at the level of mmol l^{-1} (millimole per litre). Specific estimates are

$$c_{NH_4^+} \sim 12 \text{ mmol l}^{-1}, \qquad c_{O_2} \sim 0.3 \text{ mmol l}^{-1}, \qquad c_{NO_3^-} \sim 0.15 \text{ mmol l}^{-1},$$
$$c_{H^+} \sim 10^{-4} \text{ mmol l}^{-1}, \qquad c_{CH_2O} \sim 1 \text{ mmol l}^{-1}. \tag{7.203}$$

If we use $\hat{c}_{NH_4^+} = 12$ mmol l^{-1} as a concentration scale in the definition of Λ in (7.186) and arbitrarily pick a reaction rate scale of $S_0 = 10^{-10}$ mol l^{-1} s^{-1} (comparable to phenol degradation rate at Four Ashes), then we have $\Lambda \sim O(1)$, and the simple description for Four Ashes is inappropriate.

Despite this, it is thought that reactions are fast, and we will suppose that in fact $\Lambda \gg 1$. In particular, there are data from a borehole measurement which appear to be consistent with the idea that there is a thin reaction zone at the upper plume boundary. Oxygen (presumably) diffuses to this front from above, and ammonium from below; in the reaction zone there is a huge spike of nitrate (see Fig. 7.9). We wish to see whether the existence of this nitrate spike is consistent with the model, assuming the measurement was realistic. (Figure 7.9 actually suggests the existence of two fronts, with nitrate produced at the upper front diffusing down to the second front, but we will focus only on the upper front.)

The dimensionless equations equivalent to (7.192) for the reaction scheme (7.201) are

$$R\frac{\partial c_{NH_4^+}}{\partial t} + \nabla.(c_{NH_4^+}\mathbf{u}) = \frac{1}{Pe_\perp}\frac{\partial^2 c_{NH_4^+}}{\partial z^2} + \frac{1}{Pe_\parallel}\frac{\partial^2 c_{NH_4^+}}{\partial x^2} - \Lambda(r_1 + r_3),$$

$$\frac{\partial c_{O_2}}{\partial t} + \nabla.(c_{O_2}\mathbf{u}) = \frac{1}{Pe_\perp}\frac{\partial^2 c_{O_2}}{\partial z^2} + \frac{1}{Pe_\parallel}\frac{\partial^2 c_{O_2}}{\partial x^2} - 2\Lambda\nu_{O_2}r_1,$$

$$\frac{\partial c_{NO_3^-}}{\partial t} + \nabla.(c_{NO_3^-}\mathbf{u}) = \frac{1}{Pe_\perp}\frac{\partial^2 c_{NO_3^-}}{\partial z^2} + \frac{1}{Pe_\parallel}\frac{\partial^2 c_{NO_3^-}}{\partial x^2}$$
$$+ \Lambda\nu_{NO_3^-}(r_1 - r_2 - 0.6r_3), \tag{7.204}$$

$$\frac{\partial c_{H^+}}{\partial t} + \nabla.(c_{H^+}\mathbf{u}) = \frac{1}{Pe_\perp}\frac{\partial^2 c_{H^+}}{\partial z^2} + \frac{1}{Pe_\parallel}\frac{\partial^2 c_{H^+}}{\partial x^2} + \Lambda\nu_{H^+}(2r_1 - r_2 + 0.4r_3),$$

$$\frac{\partial c_{CH_2O}}{\partial t} + \nabla.(c_{CH_2O}\mathbf{u}) = \frac{1}{Pe_\perp}\frac{\partial^2 c_{CH_2O}}{\partial z^2} + \frac{1}{Pe_\parallel}\frac{\partial^2 c_{CH_2O}}{\partial x^2} - 1.25\Lambda\nu_{CH_2O}r_2,$$

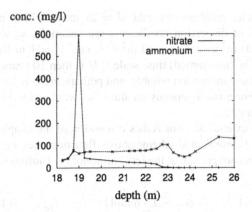

Fig. 7.9 Data from borehole 102 at Rexco, May 2003, courtesy of David Lerner and Arnë Huttmann, GPRG, University of Sheffield. Units are metres for depth, mg l^{-1} for concentrations. Since the molecular weights of ammonium and nitrate are 18 and 62 (g mole^{-1}), respectively, then 1 mg NH_4^+ l^{-1} = $\frac{1}{18}$ mmol NH_4^+ l^{-1}, 1 mg NO_3^- l^{-1} = $\frac{1}{62}$ mmol NO_3^- l^{-1}. Nitrate shows a sharp spike (of about 10 mmol l^{-1}) at a depth of 19 m. There appears to be a second front at 23 m: the nitrate produced at 19 m diffuses there and takes out (according to (7.201)) either the inorganic carbon CH_2O or the acid H^+

where the parameters ν_i are defined analogously to (7.194), i.e.,

$$\nu_i = \frac{\hat{c}_i}{\hat{c}_{NH_4^+}},\qquad(7.205)$$

$\hat{c}_i$ being the scale for c_i. From the values given above, we can suppose that all the $\nu_i \lesssim 1$.

We now mirror the Four Ashes discussion, supposing that $\Lambda \gg 1$. The reaction front at the plume fringe should then be thin as before, with oxygen diffusing to the front from outside the plume, and ammonium diffusing to it from inside, and both concentrations being zero at the front. The fringe position is unknown, and we seek a conservation law to provide an extra condition for it analogous to (7.197). Denote the right hand sides of (7.202) by e_1-e_5. There are five reactants but only three reactions, therefore there are two (linear) relationships between the e_i: these provide suitable jump conditions across the fringe. The equations $\sum_i \beta_i e_i = 0$ are satisfied for any β_1 and β_2 provided $\beta_3 = -\beta_1 + \frac{1}{2}\beta_2$, $\beta_4 = \beta_1 + \frac{3}{4}\beta_2$, and $\beta_5 = -\beta_2$. Selecting $(\beta_1, \beta_2) = (1, 0)$ and $(0, 1)$, we thus have

$$e_1 - e_3 + e_4 = 0,$$

$$e_2 + \frac{1}{2}e_3 + \frac{3}{4}e_4 - e_5 = 0,\qquad(7.206)$$

and it follows from this that in consideration of (7.204) there are two conserved quantities across the fringe (i.e., flux in equals flux out); explicitly (with the same *caveat* concerning the normal $\mathbf{n}$ as expressed in (7.199)), we have

$$\left[\frac{\partial c_{NH_4^+}}{\partial n} - \nu_{NO_3^-}\frac{\partial c_{NO_3^-}}{\partial n} + \nu_{H^+}\frac{\partial c_{H^+}}{\partial n}\right]_-^+ = 0,$$

$$\left[\nu_{O_2}\frac{\partial c_{O_2}}{\partial n} + \frac{1}{2}\nu_{NO_3^-}\frac{\partial c_{NO_3^-}}{\partial n} + \frac{3}{4}\nu_{H^+}\frac{\partial c_{H^+}}{\partial n} - \nu_{CH_2O}\frac{\partial c_{CH_2O}}{\partial n}\right]_-^+ = 0,$$

(7.207)

where $[j]_-^+$ denotes the jump in j across the fringe.

Consulting (7.201), we can write the dimensionless reaction rates r_i in terms of Monod rates

$$M_i = \frac{c_i}{K_i + c_i}$$

(7.208)

in the following way:

$$r_1 = M_{NH_4^+}M_{O_2}, \qquad r_2 = k_2 M_{NO_3^-}M_{H^+}M_{CH_2O}, \qquad r_3 = k_3 M_{NH_4^+}M_{NO_3^-},$$

(7.209)

where k_2 and k_3 are dimensionless constants (the ratios of the saturated maxima k_{max} of reactions 2 and 3 in (7.201) to that of reaction 1). The requirement that reaction rates vanish on either side of the fringe is satisfied if $c_{NH_4^+} = 0$ outside the plume (then $r_1 = r_3 = 0$), $c_{O_2} = 0$ within the plume (so $r_1 = 0$) and if $c_{NO_3^-} = 0$ on both sides (then $r_2 = r_3 = 0$). This simplifies the situation to one where pre-existing soil nitrate concentrations are low, and assumes all the nitrate produced in the reaction front by reaction 1 is consumed there by reaction 2. An alternative would be that the produced nitrate diffuse away on either side, but this is not consistent with a fast reaction 2 unless there is a natural source for nitrate production (e.g., from surface composting).[11] Since from (7.203), $\nu_{H^+} \sim 10^{-5}$, this suggests that $\frac{\partial c_{NH_4^+}}{\partial n}$ is approximately continuous across the fringe, which is thus diffusive in character (providing $Pe \gg 1$). In that case the fringe is located simply by advection of ammonium.

The second jump condition then provides a flux boundary condition for c_{CH_2O}, which is not constrained to be zero at the front. In addition, existence of a spike of nitrate within the front must be determined by solution of the reaction equations within the front. The reaction front is of thickness $\sim(1/\Lambda Pe)^{1/2}$ (taking $Pe_\parallel = Pe_\perp$ for simplicity), and within this front, the nitrate concentration is given approximately by

$$\frac{\partial^2 c_{NO_3^-}}{\partial N^2} + \nu_{NO_3^-}(r_1 - r_2 - 0.6r_3) = 0,$$

(7.210)

---

[11] Alternatively, there may be more than one reaction front. This is suggested by Fig. 7.9.

where N is a suitably rescaled normal variable. We can expect this to be solvable for $c_{NO_3^-}$ subject to $c_{NO_3^-} \to 0$ as $N \to \pm\infty$, since for example if we linearise the nitrate Monod coefficient, $M_{NO_3^-} = c_{NO_3^-}$, then the solution can be written using a Green's function which depends also on all the other reactant concentrations.

There remains the issue of determining a boundary condition for H^+ at the front. Counting conditions, the first flux condition in (7.207) determines the location of the fringe. We already have boundary conditions for $c_{NH_4^+}$ and c_{O_2} (both equal zero), and the nitrate spike is determined from (7.210). Thus, in principle we know the jump in flux of ammonium, oxygen and nitrate, and by integrating (7.210) and its equivalents for ammonium and oxygen through the front, we know that

$$\int_{-\infty}^{\infty} (r_1 + r_3)\, dN = \left[\frac{\partial c_{NO_3^-}}{\partial N}\right]_-^+,$$

$$\nu_{O_2} \int_{-\infty}^{\infty} 2 r_1 \, dN = \left[\frac{\partial c_{O_2}}{\partial N}\right]_-^+, \tag{7.211}$$

$$\nu_{NO_3^-} \int_{-\infty}^{\infty} -(r_1 - r_2 - 0.6 r_3)\, dN = 0.$$

Thus we know the values of $\int_{-\infty}^{\infty} r_i\, dN$ for $i = 1, 2, 3$ and this tells us (by integrating through the reaction front) the values of $\int_{-\infty}^{\infty} r_i\, dN$ for $i = 4, 5$, and thus the jump in flux of $\frac{\partial c_{H^+}}{\partial N}$; this provides the extra boundary condition we seek. The jump in flux of c_{CH_2O} is also given by $\int_{-\infty}^{\infty} 1.25 \nu_{CH_2O} r_2\, dN$, but this is equivalent to (7.207)$_2$. In this way, the approximate model provides all the conditions necessary to determine the solution.

St. Alban's   At St. Alban's, there has been a petroleum spillage (at a filling station) into an underlying chalk aquifer. The fluids are LNAPLs: hydrocarbons, BTEX (a cancer-forming aromatic hydrocarbon[12] and MTBE.[13] BTEX is retarded compared to MTBE and thus forms a secondary plume within the MTBE plume. The LNAPLs have seeped through the unsaturated zone and sit on top of the chalk aquifer, acting as a source (via dissolution) of contaminant to the underlying groundwater flow.

The sequence of reactions appears to be similar to those of the other examples, with oxidation by oxygen and nitrate at the plume fringe, and by Mn (manganese), Fe (iron) and SO_4^{2-} (sulphate) in the plume core. The sequence of reactions which

---

[12]More specifically, BTEX refers to a suite of volatile hydrocarbons, the acronym referring to benzene, toluene, ethylbenzene and xylene, with chemical formulae C_6H_6 (benzene), C_7H_8 (toluene), C_8H_{10} (ethylbenzene and xylene); we use toluene in the chemical reaction model.

[13]Methy tert-butyl ether, $C_5H_{12}O$.

are modelled is the following:

$$C_7H_8 + 9O_2 + 3H_2O \xrightarrow{r_1} 7CO_3^{2-},$$

$$C_7H_8 + 7.5H_2O \xrightarrow{r_2} 2.5CO_3^{2-} + 4.5CH_4 + 5H^+,$$

$$C_7H_8 + 36Fe^{3+} \xrightarrow{r_3} 7CO_3^{2-} + 36Fe^{2+} + 50H^+,$$

$$CH_4 + 2O_2 \xrightarrow{r_4} CO_3^{2-} + 2H^+ + H_2O,$$

$$CH_4 + 8Fe^{3+} + 3H_2O \xrightarrow{r_5} CO_3^{2-} + 8Fe^{2+} + 10H^+, \tag{7.212}$$

$$Fe^{2+} + O_2 + H^+ \xrightarrow{r_6} Fe^{3+} + 0.5H_2O,$$

$$C_7H_8 + 36FeOOH_{(s)} + 58H^+ \xrightarrow{r_7} 36Fe^{2+} + 7CO_3^{2-} + 51H_2O,$$

$$CH_4 + 8FeOOH_{(s)} + 7H^+ \xrightarrow{r_8} 8Fe^{2+} + CO_3^{2-} + 5H_2O.$$

Maximum reaction rates vary over about four orders of magnitude, from about 10^{-12} mol l^{-1} s^{-1} (reactions 2, 3, 7) to 10^{-8} mol l^{-1} s^{-1} (reaction 8). The principal oxidising reaction rate $r_1 \sim 10^{-10}$ mol l^{-1} s^{-1}. Contaminant levels are of order 10^{-7} mol l^{-1}, with oxygen level of order 10^{-6} mol l^{-1}. Plume depth and length are $d \sim 20$ m, $l \sim 300$ m, dispersivity parameters are taken as $\alpha_{\parallel} \sim 1$ m, $\alpha_{\perp} \sim 0.2$ m. The hydraulic conductivity of the upper chalk aquifer is 25 m d^{-1} (that of the lower part is only about 0.1 m d^{-1}), and the hydraulic gradient is 1.75×10^{-3}, thus the longitudinal flux scale is $U \sim 16$ m y^{-1}. With these values, the parameters defined by (7.186) have values

$$Pe_{\parallel} \sim 300, \qquad Pe_{\perp} \sim 10, \qquad \Lambda \sim 10^6. \tag{7.213}$$

As now seems monotonously to be the case, reactions are fast, longitudinal dispersion is small, but transverse dispersion may be effective because of the high aspect ratio l/d.

The distinguishing feature of this particular site is that, although the chalk is very porous ($\phi \gtrsim 0.3$), it has a very low effective permeability, presumably due to chemical adsorption by the chalk. On the other hand, the chalk matrix is dissected into blocks by numerous fractures, and thus acts as a dual porosity system. Contaminant can diffuse into the pore space of the matrix, and the issue of concern is whether and how fast this happens, since storage in matrix blocks will act as a residual source of contamination after the fracture system has been flushed.

We have considered this problem before, in Sect. 7.6.1. Let us suppose for simplicity that dispersivities of matrix and fractures are constant and isotropic but not necessarily equal. (Additionally, we suppose the local fracture dispersivity D_f is equal to the macroscopic fracture dispersivity D_F.) The point forms of the equations describing a single reactant are given by (7.165), and the block scale and averaged equations for the matrix concentration are given by (7.167) and (7.170):

$$Pe_m \left[\frac{\partial \bar{c}_m}{\partial t} + \bar{\mathbf{u}}_m . \nabla_{\mathbf{x}} \bar{c}_m \right] = \nabla_{\mathbf{x}}^2 \bar{c}_m + Pe_m \, \Lambda \bar{S}_m + \frac{1}{\varepsilon^2} \langle \mathbf{n} . \nabla_{\mathbf{X}} c_m \rangle |_{\partial M},$$

$$(7.214)$$

$$\varepsilon Pe_m \left[\frac{\partial c_m}{\partial T} + \mathbf{u}_m . \nabla_{\mathbf{X}} c_m \right] = \nabla_{\mathbf{X}}^2 c_m + \varepsilon^2 Pe_m \, \Lambda S_m,$$

where

$$Pe_m = \frac{Ul}{D_m}, \qquad \Lambda = \frac{S_0 l}{U c_0}. \tag{7.215}$$

Analogous dimensionless forms for the fracture average and block scale equations are, from (7.162) and (7.165)$_2$,

$$Pe_f \left[\frac{\partial \bar{c}_f}{\partial t} + \bar{\mathbf{u}}_f . \nabla_{\mathbf{x}} \bar{c}_f \right] = \nabla_{\mathbf{x}}^2 \bar{c}_f + Pe_f \, \Lambda \bar{S}_f - \frac{D_m}{\varepsilon^2 D_f \phi_f} \langle \mathbf{n} . \nabla_{\mathbf{X}} c_m \rangle |_{\partial M},$$

$$(7.216)$$

$$\varepsilon Pe_f \left[\frac{\partial c_f}{\partial T} + \mathbf{u}_f . \nabla_{\mathbf{X}} c_f \right] = \nabla_{\mathbf{X}}^2 c_f + \varepsilon^2 Pe_f \, \Lambda S_f - \frac{D_m}{D_f \phi_f} (\mathbf{n} . \nabla_{\mathbf{X}} c_m) |_{\partial M},$$

and $Pe_f = Ul/D_f$.

The question is how the interfacial transport term should be modelled. In principle we solve the block equation for c_m with $c_m = c_f$ on ∂M, yielding the interfacial term as an integral convolution of c_f. Putting this into the block equation for c_f and solving this then determines c_f and thus gives the interfacial term, which closes the description of the averaged equations.

The upshot of our earlier discussion was that the details of the homogenisation process depend on the relation between the parameters $\varepsilon = d_B/l$, Pe and Λ. Classical homogenisation theory as in Sect. 7.6.1 assumes all the parameters are $O(1)$ apart from ε, but this is unlikely ever to be appropriate. Estimates for St. Alban's may be $Pe_f \sim 10^2$, $\Lambda \sim 10^6$, $\varepsilon \sim 10^{-4}$; in addition, $u_m \ll u_f$ and $D_m \ll D_f$. Thus the reaction terms in the macroscopic equations are always large, and this suggests that, as before, reactions will be restricted to thin fronts. A complication in (7.214) is that the interfacial transport term may be large also, so this now needs to be determined.

To be specific, let us consider a simple two component reaction similar to (7.191), i.e.,

$$C_7H_8 + \sigma O_2 \xrightarrow{r} \text{products}, \tag{7.217}$$

and let p denote dimensionless BTEX concentration and c denote dimensionless oxygen concentration. Just as in (7.192), we can put $S_m = \nu S$ in (7.221), and there is a corresponding reaction term S in the equivalent equation for p: S can be taken as the product of Monod rates in (7.193), and for simplicity we take the linear rates, thus $S = -pc$.

The critical parameter in $(7.214)_2$ is $\varepsilon^2 Pe_m \Lambda$; we write

$$B_m = \varepsilon^2 Pe_m \Lambda, \qquad B_f = \varepsilon^2 Pe_f \Lambda; \qquad (7.218)$$

if $Pe_f \sim 10^2$ then $B_f \sim O(1)$, and $B_m \gg 1$: reaction in the block is fast.

Just as in (7.196), $c_m - vp_m$ satisfies an advective-diffusion conservation equation in the block, diffusion acting on a time scale $t \sim \varepsilon^2 Pe_m$, where t is the macroscopic time variable. The parameter $\varepsilon^2 Pe_m$ is crucial. If we suppose that molecular diffusion applies in the blocks, then $D_m \sim 10^{-9}$ m^2 s^{-1}, $Pe_m \sim 10^5$, and $\varepsilon^2 Pe_m \sim 10^{-3}$ is small. Therefore on the long macroscopic time scale, diffusion smoothes $c_m - vp_m$, and we can take it to be locally constant in the blocks (and thus also the fractures). Since reaction is fast in the blocks, this implies

$$c_m \approx vp_m, \qquad (7.219)$$

and therefore the block reaction rate for c_m can be written as

$$\varepsilon^2 Pe_m \Lambda S_m \approx -B_m c_m^2. \qquad (7.220)$$

When B_m is large there is a reaction boundary layer like a rind at the block surface, within which c_m satisfies

$$\frac{\partial^2 c_m}{\partial N^2} - B_m c_m^2 \approx 0. \qquad (7.221)$$

In the interior of a block inside the plume the oxygen is depleted. and we have $c_m \to 0$ as $N \to -\infty$. The first integral of (7.221) determines the flux to the blocks as

$$\frac{\partial c_m}{\partial N} = \sqrt{\frac{2 B_m}{3}} c_f^{3/2}, \qquad (7.222)$$

using the fact that $c_m = c_f$ on ∂M. The matrix concentration is negligible, and the average fracture concentration satisfies the Eq. $(7.216)_1$

$$Pe_f \left[\frac{\partial \bar{c}_f}{\partial t} + \bar{\mathbf{u}}_f . \nabla_x \bar{c}_f \right] = \nabla_x^2 \bar{c}_f - \frac{B_f}{\varepsilon^2} \bar{c}_f^2 - \frac{1}{\varepsilon^2 \phi_f} \sqrt{\frac{2 B_f Pe_f}{3 Pe_m}} \bar{c}_f^{3/2}, \qquad (7.223)$$

where we take $S_f = -c_f^2$ as in the matrix, and we have supposed that

$$\overline{c_f^2} = \bar{c}_f^2, \qquad \overline{c_f^{3/2}} = \bar{c}_f^{3/2}. \qquad (7.224)$$

With $B_f \sim O(1)$, $Pe_f \sim 10^2$, $Pe_m \sim 10^5$, the flux term to the blocks, $\frac{B_f}{\varepsilon^2} \bar{c}_f^2 \gg 1$ and the fracture reaction term $\frac{1}{\varepsilon^2 \phi_f} \sqrt{\frac{2 B_f Pe_f}{3 Pe_m}} \bar{c}_f^{3/2} \gg 1$; the implication is that the fracture concentration of oxygen within the plume rapidly decreases within both blocks and fractures to very small levels. The dual porosity appears to have little effect on the characteristics of the reactant distributions.

7.9 Precipitation and Dissolution

A sedimentary basin, as illustrated in Fig. 7.10, refers to an accumulation of sediments (of typical depth 10 km) derived from river outwash sands and silts, or marine deposited muds and microfossils. Sedimentary basins are everywhere in continental rocks: we have the Paris basin, the London basin, the North Sea, and so on. As sediments accumulate and are buried, they are subjected to increasing heat and pressure, and these two factors enable the formation of intergranular cements, which thus convert the sediments to rock (sand to sandstone; clay to shale, marine organisms to limestone). As they are buried, the sediments also compact, expelling pore water. Depending on the permeability, this can lead to pore pressures above hydrostatic, a situation which is of concern in oil-drilling operations (and is discussed further in Sect. 7.11).

Diagenesis refers generally to the process of chemical alteration to rock, and in this section we study the effects of diagenesis on the groundwater flow of accumulating sediments within a sedimentary basin. The particular type of diagenesis which we discuss is the conversion of smectite (a form of hydrated clay) to illite (a dehydrated clay) via a dewatering reaction. The resultant release of water is also a potential cause of excess pore pressures, but the main purpose of the present discussion is to show how the use of an approximation which we may call the weak solubility limit allows enormous simplification of quite complicated reaction schemes.

One view of the smectite–illite reaction is to treat it using first order kinetics, thus

$$S^S \rightarrow I^S + nH_2O, \tag{7.225}$$

where each mole conversion yields n moles of water: S denotes smectite, I denotes illite, and the superscript ^{S} denotes the solid phase (likewise, ^{L} will denote an aqueous phase). Such a scheme is not inconsistent with at least some experimental data, and the rate factor involved depends on temperature, with an activation energy in the range 60–80 kJ mol^{-1}.

However, it is likely that the transformation of smectite to illite occurs through a compound sequence of precipitation and dissolution reactions; one possible de-

Fig. 7.10 Schematic of a sedimentary basin. Sediment accumulates from outflow from rivers, and also through the settlement of marine organisms

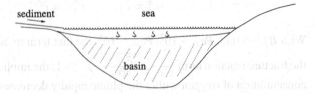

scription is the following:

$$S^S \xrightarrow{R_1} X^L + nH_2O,$$

$$KFs \xrightarrow{R_2} K^{+L} + AlO_2^{-L} + s\,SiO_2^L,$$

$$K^{+L} + AlO_2^{-L} + fX^L \xrightarrow{R_3} fI^S + SiO_2^L, \qquad (7.226)$$

$$SiO_2^L \underset{R_4^-}{\overset{R_4^+}{\rightleftharpoons}} Qz.$$

The smectite S^S dissolves to form a hydrous silica combination X^L, such as $Si_4O_{10}(OH)_2$. Additionally, illite precipitation requires potassium ions, and these may be obtained from the dissolution of potassium feldspar (in the second reaction); the aluminium hydroxyl ions AlO_2^{-L} act in the same way. The hydrous silica combination now combines with the potassium and aluminium to form illite precipitate, together with aqueous silica SiO_2^L, which itself precipitates as quartz Qz. Taking suitable multiples of the reactions and adding to eliminate the aqueous phases, the overall reaction is found to be

$$S + f^{-1}KFs \xrightarrow{R} I + nH_2O + f^{-1}(s+1)Qz. \qquad (7.227)$$

Ideally we would like to be able to write kinetics for (7.227) analogously to (7.225), with a recipe for the effective reaction rate R. Note that all the reactions in (7.226) are precipitation or dissolution reactions, and therefore the reaction rates are proportional to grain surface area.

It turns out, at least for this reaction scheme (but we might suspect more generally), that the weak solubility limit allows such a recipe to be found. To illustrate the method, note that conservation equations for the concentrations S, X, F, K, A, L, Q, I of the substances smectite, hydrous silica, feldspar, potassium ions, aluminium hydroxyl ions, aqueous silica, quartz and illite satisfy equations of the type

$$\frac{\partial}{\partial t}[(1-\phi)S] + \nabla.[(1-\phi)Su^s] = -r_1,$$

$$\frac{\partial}{\partial t}(\phi X) + \nabla.(\phi Xu^l) - \nabla.(\phi D.\nabla X) = r_1 - fr_3, \qquad (7.228)$$

and so on, where the reaction rates r_i are scaled with respect to the surface rates R_i by the specific interfacial surface areas Σ_i (thus $r_i = \Sigma_i R_i$). u^l and u^s are the liquid and solid Darcy fluxes (i.e., the volume fluxes per unit area). We allow a non-zero solid velocity in order to cater for the effects of compaction of the porous matrix. There are six other equations of the type in (7.228), together with a water conservation equation (this is the equation for ϕ). Of the total of nine equations, four are for aqueous concentrations. That for A is identical to that for K, and we ignore it henceforth.

The weak solubility limit is associated with the observations that the aqueous dissolved species X (hydrous silica), K (potassium) and L (silica) are typically present in trace quantities of the order of 10–100 ppm (1 ppm $= 10^{-3}$ kg m^{-3}),[14] and thus the concentrations of the aqueous phases are much less than those of the solid phases. When the model is suitably non-dimensionalised, the result is that the transport terms for the aqueous phases are very small, so that the corresponding reaction terms can be taken to be in equilibrium. The reaction terms in the equations for X, K and L are, respectively, $r_1 - fr_3$ (as above), $r_2 - r_3$ and $sr_2 + r_3 - r_4^+ + r_4^-$. From this, we obtain the three relationships

$$r_1 \approx fr_3,$$

$$r_2 \approx r_3, \qquad\qquad\qquad\qquad (7.229)$$

$$r_4^+ - r_4^- \approx (s+1)r_3.$$

Since the rate of dissolution of S is r_1 and the rate of precipitation of I is fr_3, this immediately shows that first order kinetics of the form (7.225) does apply, with the reaction rate being fr_3. Since r_4^+ is a precipitation rate, and r_4^- the dissolution rate of the same mineral pair SiO$_2^L \leftrightarrow$ Qz, and since *either* $r_4^+ = 0$ or $r_4^- = 0$, it is apparent that (7.229)$_3$ determines r_4^+ and r_4^- together. Thus all the reaction rates can be written in terms of r_3, and this then determines the aqueous phase pseudo-equilibrium concentrations of X, L and K in terms of r_3 and various temperature dependent rate factors, since the kinetic rates r_i are prescribed in terms of these.

If the smectite equation (7.228)$_1$ is written in terms of smectite volume fraction ϕ_S, then it becomes

$$\frac{\partial \phi_S}{\partial t} + \nabla.[\phi_S \mathbf{u}^s] = -\frac{M_S}{\rho_S} fr_3, \qquad\qquad (7.230)$$

where M_S is the molecular weight of smectite, and ρ_S its density. An equivalent equation for illite is

$$\frac{\partial \phi_I}{\partial t} + \nabla.[\phi_I \mathbf{u}^s] = f\frac{M_I}{\rho_I} r_3, \qquad\qquad (7.231)$$

and there are two similar equations for ϕ_F and ϕ_Q (with the right hand sides being proportional to r_3). In addition, the porosity ϕ satisfies

$$\frac{\partial \phi}{\partial t} + \nabla.(\phi \mathbf{u}^l) = n\frac{M_w}{\rho_w} fr_3. \qquad\qquad (7.232)$$

It remains to determine the rate constant r_3 in terms of the reactant concentrations.

---

[14] Actually, 1 ppm $= 1$ mg kg^{-1}, but 1 m^3 of water weighs 10^3 kg, so for aqueous solutions, this is equivalent to 10^{-3} kg m^{-3}.

We assume that the reaction rates take the general form (in which D denotes dissolution, and P precipitation)

$$r_i^D = \Sigma_i R_i \left[1 - \frac{c_i}{c_{is}} \right]_+ ,$$

$$r_i^P = \Sigma_i R_i \left[\frac{c_i}{c_{is}} - 1 \right]_+ ,$$

(7.233)

where c_i is the relevant aqueous concentration of phase i and c_{is} is the associated solubility limit, i.e., the saturation concentration of aqueous phase i in the presence of solid phase s. The rate factor R_i generally depends on temperature. (7.233) states that precipitation occurs when a solution is oversaturated, and dissolution occurs if it is undersaturated.

For the specific case of the precipitation/dissolution scheme in (7.226), we suppose that

$$r_1^D = \phi_S R_{SD} [1 - \psi_X]_+ ,$$

$$r_2^D = \phi_f R_{FD} [1 - \psi_K]_+ [\theta_L - \psi_L]_+ ,$$

$$r_3^P = \phi_I R_{IP} [\psi_K - \theta_K]_+ [\psi_X - \theta_X]_+ ,$$

(7.234)

$$r_4^P = \phi_Q R_{QP} [\psi_L - 1]_+ ,$$

$$r_4^D = \phi_Q R_{QD} [1 - \psi_L]_+ .$$

In these equations we have assumed that specific surface area is proportional to volume fraction, thus $\Sigma_S = \phi_S / d_p$, and we have absorbed the grain size d_p into the reaction rates. The reaction rate subscripts describe what they represent: SD, smectite dissolution; FD, feldspar dissolution; IP, illite precipitation; QP, quartz precipitation; QD, quartz dissolution.

The quantities ψ_D represent dimensionless aqueous concentrations (of phase D), scaled with a suitable solubility limit: c_X is scaled with c_{XS}, the solubility limit for hydrous silica in the presence of smectite; c_L is scaled with c_{LQ}, the solubility limit for silica in the presence of quartz. Potassium is more complicated, because in (7.226), we see that dissolution of feldspar produces two aqueous phases, and precipitation of illite requires two. We can in fact define four further solubility limits: c_{LF}, that of silica in the presence of feldspar; c_{KF}, that of potassium in the presence of feldspar; c_{KI}, that of potassium in the presence of illite; and c_{XI}, that of hydrous silica in the presence of illite. The assumption is then that, for example, feldspar will dissolve only if both potassium and silica are undersaturated, i.e., $c_K < c_{KF}$ and $c_L < c_{LF}$. In writing (7.234), we have scaled c_K with c_{KF}, and the three solubility ratios in (7.234) are therefore defined by

$$\theta_L = \frac{c_{LF}}{c_{LQ}}, \qquad \theta_K = \frac{c_{KI}}{c_{KF}}, \qquad \theta_X = \frac{c_{XI}}{c_{XS}}.$$

(7.235)

Various possibilities can now occur depending on the values of the different solubility limits. We suppose that the solubility of hydrous silica X with respect to illite is much less than that with respect to smectite (thus $\theta_X \ll 1$), and we suppose this is also true for potassium, that is to say, $\theta_K \ll 1$. On the other hand, we suppose that the solubilities of silica with respect to feldspar or quartz are comparable, so that $\theta_L \sim 1$, and in fact we will take $\theta_L > 1$ as appears to be appropriate (indeed otherwise the smectite–illite transition will not occur in this model).

The equilibrium equations (7.229) are

$$\phi_S R_{SD}[1 - \psi_X]_+ = f r_3,$$
$$\phi_F R_{FD}[1 - \psi_K]_+[\theta_L - \psi_L]_+ = r_3,$$
$$\phi_Q R_{QP}[\psi_L - 1]_+ - \phi_Q R_{QD}[1 - \psi_L]_+ = (s + 1)r_3, \tag{7.236}$$
$$r_3 = \phi_I R_{IP}[\psi_K - \theta_K]_+[\psi_X - \theta_X]_+,$$

and these are four equations for the reaction rate r_3 and the three aqueous concentrations ψ_X, ψ_L and ψ_K. Assuming non-zero reaction rates, and taking θ_K, $\theta_X \ll 1$, and $\theta_L > 1$, we deduce from (7.236) that $\psi_L > 1$ (thus quartz precipitates from solution), and

$$r_3 = \phi_I R_{IP} \psi_K \psi_X,$$

$$\psi_X = 1 - \frac{f r_3}{\phi_S R_{SD}},$$

$$\psi_L = 1 + \frac{(s + 1)r_3}{\phi_Q R_{QP}}, \tag{7.237}$$

$$\psi_K = 1 - \frac{r_3}{\phi_F R_{FD}\left\{\theta_L - 1 - \frac{(s+1)r_3}{\phi_Q R_{QP}}\right\}},$$

whence the basic reaction rate r_3 is determined by

$$r_3 = \phi_I R_{IP}\left[1 - \frac{f r_3}{\phi_S R_{SD}}\right]\left[1 - \frac{r_3}{\phi_F R_{FD}\left\{\theta_L - 1 - \frac{(s+1)r_3}{\phi_Q R_{QP}}\right\}}\right]. \tag{7.238}$$

The right hand side of this expression is a decreasing function of r_3 if it is positive, while the left hand side is increasing; therefore this expression defines the rate r_3 uniquely in terms of the solid phase concentrations of feldspar, quartz, smectite and illite.

A general complication in solving for ϕ_S and the other reactant porosities is that r_3 depends on ϕ_S, ϕ_I, ϕ_Q and ϕ_F. Given $\mathbf{u}^s$, (7.230) is a hyperbolic equation for ϕ_S with boundary condition $\phi_S = \phi_S^0$ on the upper surface $z = h$ of a sedimentary basin $b < z < h$. The equations for the solid fractions ϕ_Y, $Y = S, I, Q, F$ are all of the form

$$\frac{\partial \phi_Y}{\partial t} + \nabla.[\phi_Y \mathbf{u}^s] = \alpha_Y r_3. \tag{7.239}$$

for certain constants α_Y, and if $\phi_Y = \phi_Y^0$ on $z = h$, then ϕ_Y can be written as a linear combination of ϕ_S and ϕ_I,

$$\phi_Y = \frac{(\alpha_Y \phi_S^0 + \phi_Y^0)\phi_I + (\alpha_I \phi_Y^0 - \alpha_Y \phi_I^0)\phi_S}{\alpha_I \phi_S^0 + \phi_I^0}. \tag{7.240}$$

Therefore, the reaction rate r_3 can generally be written explicitly as a function of ϕ_S and ϕ_I, and the diagenesis model collapses to equations for ϕ_S, ϕ_I and ϕ, together with Darcy's law.

Diagenesis (at least in this theory) turns out to have a minor quantitative effect on groundwater flow, essentially because the source term in (7.232) is relatively small. What is perhaps of more interest is that a fairly complicated sequence of precipitation/dissolution steps can be reduced, in the limit of weak solubility, to a model with first order kinetics, albeit with a complicated (but explicitly defined) reaction rate. In fact, this observation is likely to be true in general. Suppose we have a sequence of precipitation and dissolution steps for solids S_i and liquids L_j:

$$L_1 + \cdots \xrightarrow{R_1} S_1 + \cdots,$$
$$S_2 + \cdots \xrightarrow{R_2} L_2 + \cdots. \tag{7.241}$$

Each reaction step necessarily involves at least one aqueous phase component, and thus all the reaction rates $R_1, \ldots, R_n$ occur in the conservation equations for the aqueous phase components. Since these can all be taken to be in equilibrium, then if there are k different aqueous phase components, we obtain k relations for the n reactions. If $k = n - 1$, then all the reaction rates can be written in terms of the overall production rate, and first order kinetics will apply.

In the present example (7.226), there are five reaction steps, and three aqueous components (lumping K^{+L} and $Al(OH)_4^{-L}$ together), but the precipitation/dissolution of quartz is effectively one reaction (either but not both at once can occur), and so the condition $n = k + 1$ is effectively met. More generally, we see that the production of solid precipitate P from solid substrates S through a sequence of intermediate dissolution/precipitation steps may often lead to this situation.

7.10 Consolidation

Consolidation refers to the ability of a granular porous medium such as a soil to compact under its own weight, or by the imposition of an overburden pressure. The grains of the medium rearrange themselves under the pressure, thus reducing the porosity and in the process pore fluid is expelled. Since the porosity is no longer constant, we have to postulate a relation between the porosity ϕ and the pore pressure p. In practice, it is found that soils, when compressed, obey a (non-reversible) relation between ϕ and the *effective pressure*

$$p_e = P - p, \tag{7.242}$$

where P is the overburden pressure.

Fig. 7.11 Form of the
relationship between porosity
and effective pressure. A
hysteretic decompression-re-
consolidation loop is
indicated. In soil mechanics
this relationship is often
written in terms of the *void
ratio* $e = \phi/(1 - \phi)$, and
specifically
$e = e_0 - C_c \log p_e$, where C_c
is the *compression index*

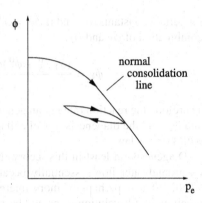

The concept of effective pressure, or more generally effective stress, is an extremely important one. The idea is that the total imposed pressure (e.g., the overburden pressure due to the weight of the rock or soil) is borne by both the pore fluid and the porous medium. The pore fluid is typically at a lower pressure than the overburden, and the extra stress (the effective stress) is that which is applied through grain to grain contacts. Thus the effective pressure is that which is transmitted through the porous medium, and it is in consequence of this that the medium responds to the effective stress; in particular, the characteristic relation between ϕ and p_e represents the nonlinear pseudo-elastic effect of compression.

As p_e increases, so ϕ decreases, thus we can write (ignoring irreversibility)

$$p_e = p_e(\phi), \qquad p_e'(\phi) < 0. \tag{7.243}$$

Taking the fluid density ρ to be constant, we obtain from the conservation of mass equation the nonlinear diffusion equation

$$\phi_t = \nabla \cdot \left[\frac{k(\phi)}{\mu} |p_e'(\phi)| \nabla \phi \right], \tag{7.244}$$

assuming Darcy's law with a permeability k, ignoring gravity, and taking P as constant. This is essentially the same as the Richards equation for unsaturated soils.

The dependence of the effective pressure on porosity is non-trivial and involves hysteresis, as indicated in Fig. 7.11. Specifically, a soil follows the *normal consolidation line* providing consolidation is occurring, i.e. $\dot{p}_e > 0$. However, if at some point the effective pressure is reduced, only a partial recovery of ϕ takes place. When p_e is increased again, ϕ more or less retraces its (overconsolidated) path to the normal consolidation line, and then resumes its normal consolidation path. Here we will ignore effects of hysteresis, as in (7.243).

When modelling groundwater flow in a consolidating medium, we must take account also of deformation of the medium itself. In turn, this requires prescription of a constitutive rheology for the deformable matrix. This is often a complex matter, but luckily in one dimension, the issue does not arise, and a one-dimensional model is often what is of practical interest. We take z to point vertically upwards, and let v^l

and v^s be the linear (or *phase-averaged*) velocities of liquid and solid, respectively. Then $u^l = \phi v^l$ and $u^s = (1 - \phi)v^s$ are the respective fluxes, and conservation of mass of each phase requires

$$\frac{\partial \phi}{\partial t} + \frac{\partial(\phi v^l)}{\partial z} = 0,$$
$$-\frac{\partial \phi}{\partial t} + \frac{\partial\{(1 - \phi)v^s\}}{\partial z} = 0; \tag{7.245}$$

Darcy's law is then

$$\phi(v^l - v^s) = -\frac{k}{\mu}\left[\frac{\partial p}{\partial z} + \rho_l g\right], \tag{7.246}$$

while the overburden pressure is

$$P = P_0 + [\rho_s(1 - \phi) + \rho_l \phi]g(h - z); \tag{7.247}$$

here $z = h$ represents the ground surface and P_0 is the applied load. (7.247) assumes variations of ϕ are small. More generally, we would have $\partial P/\partial z = -[\rho_s(1 - \phi) + \rho_l \phi]g$. The effective pressure is then just $p_e = P - p$.

We suppose these equations apply in a vertical column $0 < z < h$, for which suitable boundary conditions are

$$v^l = v^s = 0 \quad \text{at } z = 0,$$
$$p = 0, \qquad \dot{h} = v^s \quad \text{at } z = h, \tag{7.248}$$

and with an initial condition for p (or ϕ).

The two mass conservation equations imply

$$v^s = -\frac{\phi v^l}{1 - \phi}. \tag{7.249}$$

Substituting this into (7.246), we derive, using (7.245),

$$\frac{\partial \phi}{\partial t} = \frac{\partial}{\partial z}\left[\frac{k}{\mu}(1 - \phi)\left\{\frac{\partial p}{\partial z} + \rho_l g\right\}\right]. \tag{7.250}$$

If we assume the normal consolidation line takes the commonly assumed form (see Fig. 7.11)

$$\frac{\phi}{1 - \phi} = e_0 - C_c \ln(p_e/p_e^0), \tag{7.251}$$

then we derive the consolidation equation

$$\frac{\partial p_e}{\partial t} = \frac{p_e}{C_c(1 - \phi)^2}\frac{\partial}{\partial z}\left[\frac{k}{\mu}(1 - \phi)\left\{\frac{\partial p_e}{\partial z} + \Delta\rho(1 - \phi)g\right\}\right], \tag{7.252}$$

where $\Delta\rho = \rho_s - \rho_l$.

If C_c is small (and typical values are in the range $C_c \leq 0.1$) then ϕ varies little, and the consolidation equation takes the simpler form

$$\frac{\partial p_e}{\partial t} = c_v \frac{\partial^2 p_e}{\partial z^2}, \tag{7.253}$$

where

$$c_v = \frac{k}{\mu} \frac{p_e}{C_c(1 - \phi)} \tag{7.254}$$

is the *coefficient of consolidation*.

Suitable boundary conditions are

$$\frac{\partial p_e}{\partial z} + \Delta\rho(1 - \phi)g = 0 \quad \text{at } z = 0,$$

$$p_e = P_0 \quad \text{at } z = h, \tag{7.255}$$

and if the load is applied at $t = 0$, the initial condition is

$$p_e = \Delta\rho(1 - \phi)g(h - z) \quad \text{at } t = 0. \tag{7.256}$$

The equation is trivially solved. The consolidation time is

$$t_c \sim \frac{h^2}{c_v} = \frac{\mu C_c(1 - \phi)h^2}{k p_e}, \tag{7.257}$$

and depends primarily on the permeability k. If we take $k \sim 10^{-14}$ m^2 (silt), $C_c = 0.1$, $\phi = 0.3$, $\mu = 10^{-3}$ Pa s, $P_0 = 10^5$ Pa (a small house), then $c_v \sim 10^{-5}$ m^2 s^{-1}, and $t_e \sim 1$ year for $h \sim 10$ m.

7.11 Compaction

Compaction is the same process as consolidation, but on a larger scale. Other mechanisms can cause compaction apart from the rearrangement of sediments: pressure solution in sedimentary basins, grain creep in partially molten mantle (see Chap. 9). The compaction of sedimentary basins is a problem which has practical consequences in oil-drilling operations, since the occurrence of abnormal pore pressures can lead to blow-out and collapse of the borehole wall. Such abnormal pore pressures (i.e., above hydrostatic) can occur for a variety of reasons, and part of the purpose of modelling the system is to determine which of these are likely to be realistic causes. A further distinction from smaller scale consolidation is that the variation in porosity (and, particularly, permeability) is large.

The situation we consider was shown in Fig. 7.10. Sediments, both organic and inorganic, are deposited at the ocean bottom and accumulate. As they do so, they compact under their weight, thus expelling pore water. If the compaction is fast (i.e., the rate of sedimentation is greater than the hydraulic conductivity of the sediments) then excess pore pressure will occur.

Sedimentary basins, such as the North Sea or the Gulf of Mexico, are typically hundreds of kilometres in extent and several kilometres deep. It is thus appropriate to model the compacting system as one-dimensional. A typical sedimentation rate is 10^{-11} m s^{-1}, or 300 m My^{-1}, so that a 10 kilometre deep basin may accumulate in 30 My (30 million years). On such long time scales, tectonic processes are important, and in general accumulation is not a monotonic process. If tectonic uplift occurs so that the surface of the basin rises above sea level, then erosion leads to denudation and a negative sedimentation rate. Indeed, one purpose of studying basin porosity and pore pressure profiles is to try and infer what the previous subsidence history was—an inverse problem.

The basic mathematical model is that of slow two-phase flow, where the phases are solid and liquid, and is the same as that of consolidation theory. The effective pressure p_e is related, in an elastic medium, to the porosity by a function $p_e = p_e(\phi)$. In a soil, or for sediments near the surface up to depths of perhaps 500 m, the relation is elastic and hysteretic. At greater depths, more than a kilometre, pressure solution becomes important, and an effective viscous relationship becomes appropriate, as described below. At greater depths still, cementation occurs and a stiffer elastic rheology should apply.[15] In addition, the permeability is a function $k = k(\phi)$ of porosity, with k decreasing to zero fairly rapidly as ϕ decreases to zero.

Let us suppose the basin overlies an impermeable basement at $z = 0$, and that its surface is at $z = h$; then suitable boundary conditions are

$$v^s = v^l = 0 \quad \text{at } z = 0,$$
$$p_e = 0, \quad \dot{h} = \dot{m}_s + v^s \quad \text{at } z = h, \tag{7.258}$$

where v^s and v^l are solid and liquid average velocities, and $\dot{m}_s$ is the prescribed sedimentation rate, which we take for simplicity to be constant.

If we assume a specific elastic compactive rheology of the form

$$p_e = p_0\{\ln(\phi_0/\phi) - (\phi_0 - \phi)\}, \tag{7.259}$$

then non-dimensionalisation (using a depth scale $d = \dfrac{p_0}{(\rho_s - \rho_l)g}$ and a time scale $\dfrac{d}{\dot{m}_s}$) and simplification of the model leads to the nonlinear diffusion equation, analogous to (7.250),

$$\frac{\partial \phi}{\partial t} = \lambda \frac{\partial}{\partial z}\left\{\tilde{k}(1-\phi)^2\left[\frac{1}{\phi}\frac{\partial \phi}{\partial z} - 1\right]\right\}, \tag{7.260}$$

where the permeability is defined to be

$$k = k_0\tilde{k}(\phi), \tag{7.261}$$

k_0 being a suitable scale for k.

---

[15] Except that at elevated temperatures, creep deformation will start to occur.

The dimensionless parameter λ is given by

$$\lambda = \frac{K_0}{\dot{m}_s},$$
(7.262)

where $K_0 = k_0(\rho_s - \rho_l)g/\mu$ is essentially the surface hydraulic conductivity, and we can then distinguish between slow compaction ($\lambda \ll 1$) and fast compaction ($\lambda \gg 1$). Typical values of λ depend primarily on the sediment type. For $\dot{m}_s = 10^{-11}$ m s^{-1}, we have $\lambda \approx 0.1$ for the finest clay, $\lambda \approx 10^9$ for coarse sands. In general, therefore, we can expect large values of λ. The associated boundary conditions for the model become

$$\phi_z - \phi = 0 \quad \text{at } z = 0,$$

$$\phi = \phi_0, \qquad \dot{h} = 1 + \lambda \tilde{k}(1 - \phi)\left[\frac{1}{\phi}\frac{\partial \phi}{\partial z} - 1\right] \quad \text{at } z = h.$$
(7.263)

Slow Compaction, $\lambda \ll 1$ When λ is small, overpressuring occurs. A boundary layer analysis is easy to do, and shows that $\phi \approx \phi_0$ in the bulk of the (uncompacted) sediment, while a compacting boundary layer of thickness $\sqrt{\lambda t}$ exists at the base.

Fast Compaction, $\lambda \gg 1$ The more realistic case of fast compaction is also the more mathematically interesting. Most simply, the solution when $\lambda \gg 1$ is the equilibrium profile

$$\phi = \phi_0 \exp[h - z];$$
(7.264)

the exponential decline of porosity with depth is sometimes called an Athy profile, but it only applies while $\lambda \tilde{k} \gg 1$. If we assume a power law for the dimensionless permeability of the form

$$\tilde{k} = (\phi/\phi_0)^m,$$
(7.265)

then we find that $\lambda \tilde{k}$ reaches one when ϕ decreases to a value

$$\phi^* = \phi_0 \exp\left[-\frac{1}{m}\ln \lambda\right],$$
(7.266)

and this occurs at a dimensionless depth

$$\Pi = \frac{1}{m}\ln \lambda$$
(7.267)

and time

$$t^* = \frac{\Pi - \phi_0(1 - e^{-\Pi})}{1 - \phi_0}.$$
(7.268)

Typical values $m = 8$, $\lambda = 100$, $\phi_0 = 0.5$, give values $\phi^* = 0.28$, $\Pi = 0.58$, $t^* = 0.71$. In particular, for a reasonable depth scale of 1 km (corresponding to $p_0 = 2 \times 10^7$ Pa $= 200$ bars), this would correspond to a depth of 580 m. Below this, the

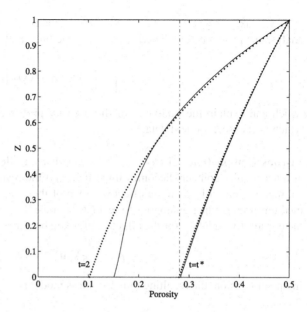

Fig. 7.12 Solution of (7.260) for $\lambda = 100$ at times $t = t^* \approx 0.71$ and at $t = 2$. The porosity (*horizontal axis*) is plotted as a function of the scaled vertical height $z/h(t)$. The *solid lines* are numerical solutions, whereas the *dotted lines* are the large λ equilibrium profiles. There is a clear divergence at depth for $t > t^*$

profile is not equilibrated, and the pore pressure is elevated. Figure 7.12 shows the resulting difference in the porosity profiles at $t = t^*$ and $t > t^*$, and Fig. 7.13 shows the effect on the pore pressure, whose gradient changes abruptly from hydrostatic to lithostatic at the critical depth.

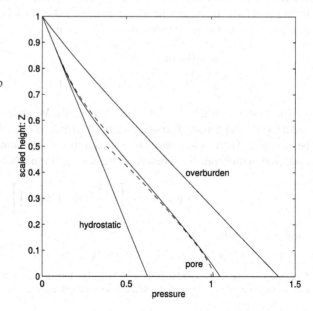

Fig. 7.13 Hydrostatic, overburden (lithostatic) and pore pressures at $t = 5$ and $\lambda = 100$, as functions of the scaled height $z/h(t)$. The transition from equilibrium to non-equilibrium compaction at the critical depth is associated with a transition from normal to abnormal pore pressures. The *dashed lines* represent two distinct approximations to the pore pressure profile, respectively, valid above and below the transition region

If we take $\phi^* = O(1)$ and $\lambda \gg 1$, then formally $m \gg 1$, and it is possible to analyse the profile below the critical depth. One finds that

$$\phi = \phi^* \exp\left[-\frac{1}{m}\{\ln m + O(1)\}\right], \qquad (7.269)$$

which can explain the flattening of the porosity profile evident in Fig. 7.12, and which is also seen in field data.

Viscous Compaction   Below a depth of perhaps a kilometre, pressure solution at intergranular contacts becomes important, and the resulting dissolution and local reprecipitation leads to an effective creep of the grains (and hence of the bulk medium) in a manner analogous to regelation in ice. For such viscous compaction, the constitutive relation for the effective pressure becomes

$$p_e = -\xi \mathbf{\nabla}.\mathbf{u}^s. \qquad (7.270)$$

In one dimension, the resulting dimensionless model is

$$-\frac{\partial\phi}{\partial t} + \frac{\partial}{\partial z}\big[(1-\phi)u\big] = 0,$$

$$u = -\lambda\tilde{k}\left[\frac{\partial p}{\partial z} + 1 - \phi\right], \qquad (7.271)$$

$$p = -\varXi\frac{\partial u}{\partial z},$$

where p is the scaled effective pressure. The compaction parameter is the same as before, and the extra parameter $\varXi$ can be taken to be of $O(1)$ for typical basin depths of kilometres. Boundary conditions for (7.271) are

$$u = 0 \quad \text{on } z = 0,$$
$$p = 0, \qquad \phi = \phi_0, \qquad \dot{h} = 1 + u \quad \text{at } z = h. \qquad (7.272)$$

This system can also be studied asymptotically. When $\lambda \ll 1$, compaction is slow and a basal compaction layer again forms. When $\lambda \gg 1$, explicit solutions can again be obtained. There is an upper layer at equilibrium, but now the porosity decreases concavely with depth.[16] As before, there is a transition when $\phi = \phi^*$, and below this

$$\phi = \phi^* \exp\left[-\frac{2}{m}\{\ln m + O(1)\}\right], \qquad (7.273)$$

similar to (7.269).

---

[16]In view of Chap. 6, we need to be careful here. The function is mathematically concave, i.e., the rate of decrease of porosity with depth increases as depth increases.

Fig. 7.14 Evolution of the porosity as a function of depth $h - z$, with a viscous rheology, at $\lambda = 100$. The upper concave part is in equilibrium, while overpressuring occurs where the profile is flatter below this

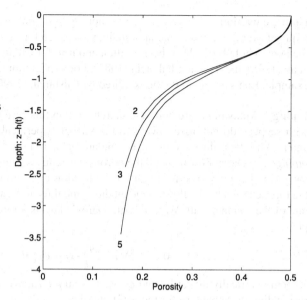

The main distinction between viscous and elastic compaction is thus in the form of the rapidly compacted equilibrium profile near the surface (Fig. 7.14). The concave profile is not consistent with observations, but we need not expect it to be, as the viscous behaviour of pressure solution only becomes appropriate at reasonable depths. A more general relation which allows for this is a viscoelastic compaction law of the form

$$\nabla . \mathbf{u}^s = -\frac{1}{K_e}\frac{dp_e}{dt_s} - \frac{p_e}{\xi}. \tag{7.274}$$

7.12 Notes and References

Flow in porous media is described in the books by Bear (1972) and Dullien (1979). More recent versions, for example by Bear and Bachmat (1990) have developed a taste for more theoretical, deductive treatments based on homogenisation (see below) or averaging, with a concomitant loss of readability. The classic geologists' book on groundwater is by Freeze and Cherry (1979) and the classic engineering text is by Polubarinova-Kochina (1962). A short introduction, of geographical style, is by Price (1985). A more mathematical survey, with a variety of applications, is by Bear and Verruijt (1987). The book edited by Cushman (1990) contains a wealth of articles on topics of varied and current interest, including dispersion, homogenisation, averaging, dual porosity models, multigrid methods and heterogeneous porous media. Further information on the concepts of soil mechanics can be found in Lambe and Whitman (1979).

Homogenisation   The technique of homogenisation is no more than the technique of averaging in the spatial domain, most often formulated as a multiple scales method. Whole books have been written about it, for example those by Bensoussan et al. (1978) and Sanchez-Palencia (1983). For application to porous media, see, for example, Ene's article in the book edited by Cushman (1990).

Piping   Many dams are built of concrete, and in this case the problems associated with seepage do not arise, owing to the virtual impermeability of concrete. Earth and rockfill dams do exist, however, and are liable to failure by a mechanism called *piping*. The Darcy flow through the porous dam causes channels to form by eroding away fine particles. The resultant channelisation concentrates the flow, increasing the force exerted by the flow on the medium and thus increasing the erosion/collapse rate of the channel wall. We can write Darcy's law as a force balance on the liquid phase,

$$0 = -\phi \nabla p - \frac{\phi \mu}{k} \mathbf{v}_l - \phi \rho_l g \mathbf{k} \tag{7.275}$$

(**k** being vertically upwards) and $\phi \mu \mathbf{v}_l / k$ is an *interactive drag term*; then the corresponding force balance for the solid phase is

$$0 = -(1 - \phi) \nabla p_s + \frac{\phi \mu}{k} \mathbf{v}_l - (1 - \phi) \rho_s g \mathbf{k}, \tag{7.276}$$

where p_s is the pressure in the solid. For a granular solid, we can expect grain motion to occur if the interactive force is large enough to overcome friction and cohesion; the typical kind of criterion is that the shear stress τ satisfies

$$\tau \geq c + p_e \tan \phi, \tag{7.277}$$

but in view of the large confining pressure and the necessity of dilatancy for soil deformation, the piping criterion will in practice be satisfied at the toe of the dam (i.e. the front), and piping channels will eat their way back into the dam, in much the same way that river drainage channels eat their way into a hillslope. A simpler criterion at the toe then follows from the necessity that the effective pressure on the grains be positive. A lucid discussion by Bear and Bachmat (1990, p. 153) indicates that the solid pressure is related to the *effective* pressure p_e which controls grain deformation by

$$p_e = (1 - \phi)(p_s - p), \tag{7.278}$$

and in this case the piping criterion at the toe is that $p_e < 0$ in the soil there, or $\partial p_e / \partial z > 0$. From (7.275), (7.276) and (7.278), this implies piping if

$$\frac{\mu v}{k} > (\rho_s - \rho_l)(1 - \phi)g, \tag{7.279}$$

where v is the vertical component of $\mathbf{v}_l$. This criterion is given by Bear (1972). More generally, piping can be expected to occur if p_e reaches 0 in the soil interior (ignoring cohesion). Sellmeijer and Koenders (1991) develop a model for piping.

Taylor Dispersion   Taylor dispersion is named after its investigation by Taylor (1953), who carried out experiments on the dispersal of solute in flow down a tube. The dispersion is enabled by the combination of differential axial advection by the down tube velocity, typically a Poiseuille flow, and the rapid cross stream diffusion which renders the cross-sectional concentration profile radially uniform. The theory of Taylor is somewhat heuristic; it was later elaborated by Aris (1956). For a formal derivation using asymptotic methods, see Fowler (1997, p. 222, Exercise 2).

Its application to porous media stems from the conceptual idea that the pore space consists of a network of narrow tubules connected at pore junctions. If the tubes are of radius a and length d_p, the latter corresponding to grain size, then the Darcy flux $|\mathbf{u}| \sim \pi \phi U$, while the pore radius $a \sim d_p \sqrt{\phi}$. This would suggest a Taylor dispersion coefficient of

$$D_T \approx \frac{a^2 U^2}{48 D} \sim \frac{d_p^2 |\mathbf{u}|^2}{48 \pi^2 D \phi}, \tag{7.280}$$

as opposed to the measured values which more nearly have $D_T \sim |\mathbf{u}|$. Taylor dispersion in porous media has been studied by Saffman (1959), Brenner (1980) and Rubinstein and Mauri (1986), the latter using the method of homogenisation.

Biofilm Growth   Monod kinetics was described by Monod (1949), by way of analogy with enzyme kinetics, where one considers the uptake of nutrients as occurring through a series of fast intermediary reactions; when two nutrients control growth, as in respiration, it is usual to take the growth rate as proportional to the product of two Monod factors (Bader 1978). A variety of enhancements to this simple model have also been proposed to account for nutrient consumption due to maintenance, inactivation of cells in adverse conditions, and other observed effects (Beeftink et al. 1990; Wanner et al. 2006).

Bacteria in soils commonly grow as attached biofilms on soil grains, with a thickness of the order of 100 μ. A variety of models to describe biofilm growth have been presented, with an ultimate view of being able to parameterise the uptake rate of contaminant species in soils and other environments (Rittmann and McCarty 1980; Picioreanu et al. 1998; Eberl et al. 2001; Dockery and Klapper 2001; Cogan and Keener 2004).

Remediation Sites   The three sites described in Sect. 7.8 are under study by the Groundwater Restoration and Protection Group at the University of Sheffield, led by Professor David Lerner. The site at Four Ashes is described by Mayer et al. (2001), that at Rexco by Hüttmann et al. (2003), and that at St. Alban's by Wealthall et al. (2001).

The description of the two species reaction front given by (7.192) is similar to that for a *diffusion flame* (Buckmaster and Ludford 1982) in combustion, and also corrosion in alloys (Hagan et al. 1986). It is not conceptually difficult to extend this approach to an arbitrary number of reactions, although it may become awkward when multiple reaction fronts are present (see, for example, Dewynne et al. 1993).

Diagenesis   The first order reaction kinetics (7.225) for the smectite–illite transition was proposed by Eberl and Hower (1976). Information on solubility limits is given by Aagaard and Helgeson (1983) and Sass et al. (1987). The asymptotic approximation called here the weak solubility limit is called solid density asymptotics by Ortoleva (1994). Details of the use of the weak solubility approximation can be found in Fowler and Yang (2003).

Compaction   Interest in compaction is motivated by its occurrence in sedimentary basins, and also by issues of subsidence due to groundwater or natural gas extraction (see, for example, Baú et al. 2000). The constitutive law used here for effective pressure is that of Smith (1971); it mimics the normal consolidation behaviour of compacting sediments (such as soils), and is further discussed by Audet and Fowler (1992) and Jones (1994).

Athy's law comes from the paper by Athy (1930). Smith (1971) advocates the use of the high exponent $m = 8$ in (7.265). Further details of the asymptotic solution of the compaction profiles are given by Fowler and Yang (1998). Freed and Peacor (1989) show examples of the flattened porosity profiles at depth.

Early work on pressure solution in sedimentary basins was by Angevine and Turcotte (1983) and Birchwood and Turcotte (1994). More recently, Fowler and Yang (1999) derived the viscous compaction law. An extension to viscoelastic compaction has been studied by Yang (2000).

Seals   One process which we have not described is the formation of high pressure seals. In certain circumstances, pore pressures undergo fairly rapid jumps across a 'seal', typically at depths of 3000 m. Such jumps cannot be predicted within the confines of a simple compaction theory, and require a mechanism for pore-blocking. Mineralisation is one such mechanism, as some seals are found to be mineralised with calcite and silica (Hunt 1990). In fact, a generalisation of the clay diagenesis model to allow for calcite precipitation could be used for this purpose. As it stands, (7.232) predicts a source for ϕ, but mineralisation would cause a corresponding sink term. Reduction of ϕ leads to reduction of diffusive transport, and the feedback is self-promoting. Problems of this type have been studied by Ortoleva (1994), for example.

7.13 Exercises

7.1 Show that for a porous medium idealised as a cubical network of tubes, the permeability is given (approximately) by $k = d_p^2 \phi^2 / 72\pi$, where d_p is the grain size. How is the result modified if the pore space is taken to consist of planar sheets between identical cubical blocks? (The volume flux per unit width between two parallel plates a distance h apart is $-h^3 p'/12\mu$, where p' is the pressure gradient.)

7.2 A sedimentary rock sequence consists of two types of rock with permeabilities k_1 and k_2. Show that in a unit with two horizontal layers of thickness d_1 and d_2, the effective horizontal permeability (parallel to the bedding plane) is

$$k_\| = k_1 f_1 + k_2 f_2,$$

where $f_i = d_i / (d_1 + d_2)$, whereas the effective vertical permeability is given by

$$k_\perp^{-1} = f_1 k_1^{-1} + f_2 k_2^{-1}.$$

Show how to generalise this result to a sequence of n layers of thickness $d_1, \ldots, d_n$.

Hence show that the effective permeabilities of a thick stratigraphic sequence containing a distribution of (thin) layers, with the proportion of layers having permeabilities in $(k, k + dk)$ being $f(k)\, dk$, are given by

$$k_\| = \int_0^\infty k f(k)\, dk, \qquad k_\perp^{-1} = \int_0^\infty \frac{f(k)\, dk}{k}.$$

7.3 Groundwater flows between an impermeable basement at $z = h_b(x, y, t)$ and a phreatic surface at $z = z_p(x, y, t)$. Write down the equations governing the flow, and by using the Dupuit approximation, show that the saturated depth h satisfies

$$\phi h_t = \frac{k \rho g}{\mu} \nabla . [h \nabla z_p],$$

where $\nabla = (\partial/\partial x, \partial/\partial y)$. Deduce that a suitable time scale for flows in an aquifer of typical depth h_0 and extent l is $t_{gw} = \phi \mu l^2 / k \rho g h_0$.

I live a kilometer from the river, on top of a layer of sediments 100 m thick (below which is impermeable basement). What sort of sediments would those need to be if the river responds to rainfall at my house within a day; within a year?

7.4 A two-dimensional earth dam with vertical sides at $x = 0$ and $x = l$ has a reservoir on one side ($x < 0$) where the water depth is h_0, and horizontal dry land on the other side, in $x > l$. The dam is underlain by an impermeable basement at $z = 0$.

Write down the equations describing the saturated groundwater flow, and show that they can be written in the dimensionless form

$$u = -p_x, \qquad \varepsilon^2 w = -(p_z + 1),$$
$$p_{zz} + \varepsilon^2 p_{xx} = 0,$$

and define the parameter ε. Write down suitable boundary conditions on the impermeable basement, and on the phreatic surface $z = h(x, t)$.

Assuming $\varepsilon \ll 1$, derive the Dupuit–Forchheimer approximation for h,

$$h_t = (h h_x)_x \quad \text{in } 0 < x < 1.$$

Show that a suitable boundary condition for h at $x = 0$ (the dam end) is

$$h = 1 \quad \text{at } x = 0.$$

Now define the quantity

$$U = \int_0^h p \, dz,$$

and show that the horizontal flux

$$q = \int_0^h u \, dz = -\frac{\partial U}{\partial x}.$$

Hence show that the conditions of hydrostatic pressure at $x = 0$ and constant (atmospheric) pressure at $x = 1$ (the seepage face) imply that

$$\int_0^1 q \, dx = \frac{1}{2}.$$

Deduce that, if the Dupuit approximation for the flux is valid all the way to the toe of the dam at $x = 1$, then $h = 0$ at $x = 1$, and show that in the steady state, the (dimensional) discharge at the seepage face is

$$q_D = \frac{k\rho g h_0^2}{2\mu l}.$$

Supposing the above description of the solution away from the toe to be valid, show that a possible boundary layer structure near $x = 1$ can be described by writing

$$x = 1 - \varepsilon^2 X, \qquad h = \varepsilon H, \qquad z = \varepsilon Z, \qquad p = \varepsilon P,$$

and write down the resulting leading order boundary value problem for P.

7.5 I get my water supply from a well in my garden. The well is of depth h_0 (relative to the height of the water table a large distance away) and radius r_0. Show that the Dupuit approximation for the water table height h is

$$\phi \frac{\partial h}{\partial t} = \frac{k\rho g}{\mu} \frac{1}{r} \frac{\partial}{\partial r}\left(r h \frac{\partial h}{\partial r}\right).$$

If my well is supplied from a reservoir at $r = l$, where $h = h_0$, and I withdraw a constant water flux q_0, find a steady solution for h, and deduce that my well will run dry if

$$q_0 > \frac{\pi k\rho g h_0^2}{\mu \ln[l/r_0]}.$$

Use plausible values to estimate the maximum yield (litres per day) I can use if my well is drilled through sand, silt or clay, respectively.

7.6 A volume V of effluent is released into the ground at a point ($r = 0$) at time t. Use the Dupuit approximation to motivate the model

$$\phi \frac{\partial h}{\partial t} = \frac{k\rho g}{\mu} \frac{1}{r} \frac{\partial}{\partial r} \left(rh \frac{\partial h}{\partial r} \right),$$

$$h = h_0 \quad \text{at } t = 0, r > 0,$$

$$\int_0^\infty r(h - h_0)\, dr = V/2\pi, \quad t > 0,$$

where h_0 is the initial height of the water table above an impermeable basement. Find suitable similarity solutions in the two cases (i) $h_0 = 0$, (ii) $h_0 > 0$, $h - h_0 \ll h_0$, and comment on the differences you find.

7.7 Fluid flows through a porous medium in the x direction at a linear velocity U. At $t = 0$, a contaminant of concentration c_0 is introduced at $x = 0$. If the longitudinal dispersivity of the medium is D, write down the equation which determines the concentration c in $x > 0$, together with suitable initial and boundary conditions. Hence show that c is given by

$$\frac{c}{c_0} = \frac{1}{2} \left[\text{erfc} \left\{ \frac{x - Ut}{2\sqrt{Dt}} \right\} + \exp \left(\frac{Ux}{D} \right) \text{erfc} \left\{ \frac{x + Ut}{2\sqrt{Dt}} \right\} \right],$$

where

$$\text{erfc}\, \xi = \frac{2}{\sqrt{\pi}} \int_\xi^\infty e^{-s^2}\, ds.$$

[*Hint: you might try Laplace transforms, or else simply verify the result.*]

Show that for large ξ, $\text{erfc}\, \xi = e^{-\xi^2}[\frac{1}{\sqrt{\pi}\xi} + \cdots]$, and deduce that if $x = Ut + 2\sqrt{Dt}\, \eta$, with $\eta = O(1)$, then

$$\frac{c}{c_0} \approx \frac{1}{2} \text{erfc}\, \eta + O \left(\frac{1}{\sqrt{t}} \right).$$

Hence show that at a fixed station $x = X$ far downstream, the measured profile is approximately given by

$$c \approx c_0 \left[1 - \frac{1}{2} \text{erfc} \left\{ \frac{1}{2} \left(\frac{U^3}{DX} \right)^{1/2} \left(t - \frac{X}{U} \right) \right\} \right].$$

This is called the breakthrough curve, and indicates that dispersion causes breakthrough to occur over a time interval (at large distance) of order $\Delta t_b = (DX/U^3)^{1/2}$. If $D \approx aU$, show that the ratio of Δt_b to $t_b = X/U$ is $\Delta t_b / t_b \sim (a/X)^{1/2}$.

7.8 Rain falls steadily at a rate q (volume per unit area per unit time) on a soil of saturated hydraulic conductivity K_0 ($= k_0 \rho_w g/\mu$, where k_0 is the saturated

permeability). By plotting the relative permeability k_{rw} and suction characteristic $\sigma \psi/d$ as functions of S (assuming a residual liquid saturation S_0), show
that a reasonable form to choose for $k_{rw}(\psi)$ is $k_{rw} = e^{-c\psi}$. If the water table
is at depth h, show that, in a steady state, ψ is given as a function of the dimensionless depth $z^* = z/z_c$, where $z_c = \sigma/\rho_w g d$ (σ is the surface tension, d
the grain size), by

$$h^* - z^* = \frac{1}{2}\psi - \frac{1}{c}\ln\left[\frac{\sinh\{\frac{1}{2}(\ln\frac{1}{q^*} - c\psi)\}}{\sinh\{\frac{1}{2}\ln\frac{1}{q^*}\}}\right],$$

where $h^* = h/z_c$, providing $q^* = q/K_0 < 1$. Deduce that if $h \gg z_c$, then $\psi \approx$
$\frac{1}{c}\ln\frac{1}{q^*}$ near the surface. What happens if $q > K_0$?

7.9 Derive the Richards equation

$$\rho_w \phi \frac{\partial S}{\partial t} = -\frac{\partial}{\partial z}\left[\frac{k_0}{\mu}k_{rw}(S)\left\{\frac{\partial p_c}{\partial z} + \rho_w g\right\}\right]$$

for one-dimensional infiltration of water into a dry soil, explaining the meaning of the terms, and giving suitable boundary conditions when the surface flux
q is prescribed. Show that if the surface flux is large compared with $k_0 \rho_w g/\mu$,
where k_0 is the saturated permeability, then the Richards equation can be approximated, in suitable non-dimensional form, by a nonlinear diffusion equation of the form

$$\frac{\partial S}{\partial t} = \frac{\partial}{\partial z}\left[D\frac{\partial S}{\partial z}\right].$$

Show that, if $D = S^m$, a similarity solution exists in the form

$$S = t^\alpha F(\eta), \quad \eta = z/t^\beta,$$

where $\alpha = \frac{1}{m+2}$, $\beta = \frac{m+1}{m+2}$, and F satisfies

$$(F^m F')' = \alpha F - \beta \eta F', \qquad F^m F' = -1 \quad \text{at } \eta = 0, \qquad F \to 0 \quad \text{as } \eta \to \infty.$$

Deduce that

$$F^m F' = -(\alpha + \beta)\int_\eta^{\eta_0} F\,d\eta - \beta \eta F,$$

where η_0 (which may be ∞) is where F first reaches zero. Deduce that $F' < 0$,
and hence that η_0 must be finite, and is determined by

$$\int_0^{\eta_0} F\,d\eta = \frac{1}{\alpha + \beta}.$$

What happens for $t > F(0)^{-1/\alpha}$?

7.10 Write down the equations describing one-dimensional consolidation of wet sediments in terms of the variables ϕ, v^s, v^l, p, p_e, these being the porosity, solid and liquid (linear) velocities, and the pore and effective pressures. Neglect the effect of gravity.

Saturated sediments of depth h lie on a rigid but permeable (to water) basement, through which a water flux W is removed. Show that

$$v^s = \frac{k}{\mu}\frac{\partial p}{\partial z} - W,$$

and deduce that ϕ satisfies the equation

$$\frac{\partial \phi}{\partial t} = \frac{\partial}{\partial z}\left[(1-\phi)\left\{\frac{k}{\mu}\frac{\partial p}{\partial z} - W\right\}\right].$$

If the sediments are overlain by water, so that $p = $ constant (take $p = 0$) at $z = h$, and if $\phi = \phi_0 + p/K$, where the compressibility K is large (so $\phi \approx \phi_0$), show that a suitable reduction of the model is

$$\frac{\partial p}{\partial t} - W\frac{\partial p}{\partial z} = c\frac{\partial^2 p}{\partial z^2},$$

where $c = K(1-\phi_0)k/\mu$, and $p = 0$ on $z = h$, $p_z = \mu W/k$. Non-dimensionalise the model using the length scale h, time scale h^2/c, and pressure scale $\mu W h/k$. Hence describe the solution if the parameter $\varepsilon = \mu W h/k$ is small, and find the rate of surface subsidence. What has this to do with Venice?

7.11 Write down a model for vertical flow of two immiscible fluids in a porous medium. Deduce that the saturation S of the wetting phase satisfies the equation

$$\phi\frac{\partial S}{\partial t} + \frac{\partial}{\partial z}\left[M_{\text{eff}}\left\{\frac{q}{M_{nw}} + g\Delta\rho\right\}\right] = -\frac{\partial}{\partial z}\left[M_{\text{eff}}\frac{\partial p_c}{\partial z}\right],$$

where z is a coordinate pointing *downwards*,

$$p_c = p_{nw} - p_w, \qquad \Delta\rho = \rho_w - \rho_{nw}, \qquad M_{\text{eff}}^{-1} = \left(M_w^{-1} + M_{nw}^{-1}\right),$$

q is the total downward flux, and the suffixes w and nw refer to the wetting and non-wetting fluid, respectively. Define the phase mobilities M_i. Give a criterion on the capillary suction p_c which allows the Buckley–Leverett approximation to be made, and show that for $q = 0$ and $\mu_w \gg \mu_{nw}$, waves typically propagate downwards and form shocks. What happens if $q \neq 0$? Is the Buckley–Leverett approximation realistic—e.g. for air and water in soil? (Assume $p_c \sim 2\gamma/r_p$, where $\gamma = 70$ mN m^{-1}, and r_p is the pore radius: for clay, silt and sand, take $r_p = 1$ μ, 10 μ, 100 μ, respectively.)

7.12 A model for snow-melt run-off is given by the following equations:

$$u = \frac{k}{\mu}\left[\frac{\partial p_c}{\partial z} + \rho_l g\right],$$

$$k = k_0 S^3,$$

$$\phi \frac{\partial S}{\partial t} + \frac{\partial u}{\partial z} = 0,$$

$$p_c = p_0 \left(\frac{1}{S} - S \right).$$

Explain the meaning of the terms in these equations, and describe the assumptions of the model.

The intrinsic permeability k_0 is given by

$$k_0 = 0.077 \, d^2 \exp[-7.8 \rho_s / \rho_l],$$

where ρ_s and ρ_l are snow and water densities, and d is grain size. Take $d = 1$ mm, $\rho_s = 300$ kg m^{-3}, $\rho_l = 10^3$ kg m^{-3}, $p_0 = 1$ kPa, $\phi = 0.4$, $\mu = 1.8 \times 10^{-3}$ Pa s, $g = 10$ m s^{-2}, and derive a non-dimensional model for melting of a one metre thick snow pack at a rate (i.e. u at the top surface $z = 0$) of 10^{-6} m s^{-1}. Determine whether capillary effects are small; describe the nature of the model equation, and find an approximate solution for the melting of an initially dry snowpack. What is the (meltwater flux) run-off curve?

7.13 Consider the following model, which represents the release of a unit quantity of groundwater at $t = 0$ in an aquifer $-\infty < x < \infty$, when the Dupuit approximation is used:

$$h_t = (h h_x)_x,$$

$$h = 0 \quad \text{at } t = 0, \, x \neq 0,$$

$$\int_{-\infty}^{\infty} h \, dx = 1$$

(i.e., $h = \delta(x)$ at $t = 0$). Show that a similarity solution to this problem exists in the form

$$h = t^{-1/3} g(\xi), \quad \xi = x / t^{1/3},$$

and find the equation and boundary conditions satisfied by g. Show that the water body spreads at a finite rate, and calculate what this is.

Formulate the equivalent problem in three dimensions, and write down the equation satisfied by the similarity form of the solution, assuming cylindrical symmetry. Does this solution have the same properties as the one-dimensional solution?

7.14 The tensor D_{ij} ($i, j = 1, 2, 3$) has three invariants

$$D_I = D_{ii}, \qquad D_{II} = D_{ij} D_{ij}, \qquad D_{III} = D_{ij} D_{jk} D_{ki}.$$

(Summation over repeated indices is implied.) Show that the invariants of the tensor

$$D_{ij} = \alpha_\perp u \delta_{ij} + (\alpha_\| - \alpha_\perp) \frac{u_i u_j}{u},$$

where $u = |\mathbf{u}|$ and δ_{ij} is the Kronecker delta ($= 1$ if $i = j$, $= 0$ if $i \neq j$), are the same as those of the tensor

$$\mathbf{D} = \begin{pmatrix} \alpha_\| u & 0 & 0 \\ 0 & \alpha_\perp u & 0 \\ 0 & 0 & \alpha_\perp u \end{pmatrix}.$$

7.15 Suppose that a doubly porous medium consists of a periodic sequence of blocks M with boundaries (fractures) ∂M. The concentration of a chemical reactant c is taken to be a function of the fast space variable $\mathbf{X}$ and the slow space variable $\mathbf{x} = \varepsilon \mathbf{X}$, and we assume that $c = \bar{c}(\mathbf{x}) + \varepsilon^2 c_k(\mathbf{X})$, where the suffix k refers to fractures (f) or matrix block (m).

Let $G(\mathbf{X}, \mathbf{Y})$ be a Green's function satisfying

$$\nabla_\mathbf{Y}^2 G = \delta(\mathbf{X} - \mathbf{Y}) \quad \text{in } M, \qquad G = 0 \quad \text{for } \mathbf{Y} \in \partial M,$$

and suppose that

$$\nabla_\mathbf{X}^2 \psi = 0 \quad \text{in } M,$$
$$\psi = \chi \quad \text{on } \partial M.$$

Show that

$$\psi = \int_{\partial M} \frac{\partial G(\mathbf{X}, \mathbf{Y})}{\partial N_Y} \chi(\mathbf{Y}) \, dS(\mathbf{Y}),$$

where $\frac{\partial G}{\partial N_Y} = \mathbf{n} . \nabla_\mathbf{Y} G(\mathbf{X}, \mathbf{Y})$, and hence show that

$$\frac{\partial \psi}{\partial N}\bigg|_{\partial M} = \int_{\partial M} K(\mathbf{X}, \mathbf{Y}) \chi(\mathbf{Y}) \, dS(\mathbf{Y}),$$

where

$$K(\mathbf{X}, \mathbf{Y}) = \frac{\partial^2 G(\mathbf{X}, \mathbf{Y})}{\partial N_X \partial N_Y}.$$

Now suppose that the fluctuating matrix and fracture concentrations of the chemical reactant are given by

$$\nabla_\mathbf{X}^2 c_m = Pe \left[\frac{\partial \bar{c}}{\partial t} + \mathbf{u}_m . \nabla_\mathbf{x} \bar{c} \right] + Pe \Lambda \bar{c} - \nabla_\mathbf{x}^2 \bar{c} \equiv R_m \quad \text{in } M,$$

subject to $c_m = c_f$ on ∂M, and

$$\nabla_\mathbf{X}^2 c_f - \frac{(1 - \phi_f)}{\phi_f} \frac{\partial c_m}{\partial N}\bigg|_{\partial M}$$
$$= Pe \left[\frac{\partial \bar{c}}{\partial t} + \mathbf{u}_f . \nabla_\mathbf{x} \bar{c} \right] + Pe \Lambda \bar{c} - \nabla_\mathbf{x}^2 \bar{c} \equiv R_f \quad \text{on } \partial M,$$

subject to conditions of periodicity with zero mean.

Show that, if we define c^* to be the solution of

$$\nabla_{\mathbf{X}}^2 c^* = 1 \quad \text{in } M,$$

with

$$c^* = 0 \quad \text{on } \partial M,$$

then

$$c_m = \int_{\partial M} \frac{\partial G(\mathbf{X}, \mathbf{Y})}{\partial N_Y} c_f(\mathbf{Y}) \, dS(\mathbf{Y}) + R_m c^*,$$

and deduce that, for $\mathbf{X} \in M$,

$$\phi_f \nabla_{\mathbf{X}}^2 c_f - (1 - \phi_f) \int_{\partial M} K(\mathbf{X}, \mathbf{Y}) c_f(\mathbf{Y}) \, dS(\mathbf{Y})$$

$$= \phi_f R_f + (1 - \phi_f) R_m \left. \frac{\partial c^*}{\partial N} \right|_{\partial M}.$$

By integrating this equation over ∂M, show that the condition of periodicity of c_f implies that the equation to determine $\bar{c}$ is

$$Pe \left[\frac{\partial \bar{c}}{\partial t} + \mathbf{u}.\nabla_{\mathbf{x}} \bar{c} \right] = \nabla_{\mathbf{x}}^2 \bar{c} - Pe \Lambda \bar{c},$$

where $\mathbf{u} = \phi_f \mathbf{u}_f + (1 - \phi_f) \mathbf{u}_m$.

7.16 The reaction rates in the reactions

$$S^S \xrightarrow{r_1} X^L + n H_2 O,$$

$$\text{KFs} \xrightarrow{r_2} K^{+L} + A l O_2^{-L} + s \, SiO_2^L,$$

$$K^{+L} + A l O_2^{-L} + f X^L \xrightarrow{r_3} f I^S + SiO_2^L,$$

$$SiO_2^L \underset{r_4^-}{\overset{r_4^+}{\rightleftharpoons}} Qz,$$

are related by

$$r_1 \approx f r_3,$$

$$r_2 \approx r_3,$$

$$r_4^+ - r_4^- \approx (s + 1) r_3.$$

The reaction rate r_3 is given by

$$r_3 = \phi_I R_3 \left[1 - \frac{f r_3}{\phi_S R_1} \right] \left[1 - \frac{r_3}{\phi_F R_2 \{ \theta_L - 1 - \frac{(s+1) r_3}{\phi_Q R_4^+} \}} \right],$$

where ϕ_i are porosities, R_k are rate factors (such that $r_k \propto R_k$), and the stoichiometric constants f and s, and the constant θ_L, may be taken as $O(1)$ (and $\theta_L > 1$). Show that r_3 can be written explicitly in the form

$$\frac{2}{r_3} = \left\{ \frac{1}{(\theta_L - 1)\gamma_F} + \frac{s+1}{(\theta_L - 1)\gamma_Q} + \frac{f}{\gamma_S} + \frac{1}{\gamma_I} \right\}$$
$$+ \left[\left\{ \frac{1}{(\theta_L - 1)\gamma_F} + \frac{s+1}{(\theta_L - 1)\gamma_Q} + \frac{f}{\gamma_S} + \frac{1}{\gamma_I} \right\}^2 \right.$$
$$\left. + \frac{4(s+1)}{(\theta_L - 1)\gamma_Q} \left(\frac{f}{\gamma_S} + \frac{1}{\gamma_I} \right) \right]^{1/2},$$

where the coefficients γ_Y represent the porosity weighted rate factors, i.e.,

$$\gamma_I = \phi_I R_3, \qquad \gamma_S = \phi_S R_1, \qquad \gamma_Q = \phi_Q R_4^+, \qquad \gamma_F = \phi_F R_2.$$

Deduce that the slowest reaction of the four (as measured by γ_Y) controls the overall rate, and give explicit approximations for r_3 for each of the consequent four possibilities.

Chapter 8
Mantle Convection

It is now virtually common knowledge, extending into the school curriculum, that the Earth's surface is constructed of a number of 'plates', and the grinding of these against each other is the principal cause of much of the Earth's volcanism, its earthquakes, and is also the geometric cause of mid-ocean rises and oceanic trenches. Scientific television programmes abound with the story of how Iceland lies at the boundary of two of these plates, how it is being formed by their creation, or how the Hawaiian Islands are being created by volcanism beneath the overriding Pacific plate.

The thrusting of one plate into another tells us why the Alps or Himalayas have been formed; the sinking of plates in subduction zones is what causes oceanic trenches to occur, and the frictional rubbing of the subducting plates against the overlying mantle generates the back-arc volcanism which builds the Andes and the Rockies. If plates move laterally past each other, then the situation is that of the San Andreas fault; motion is 'stick–slip', just like frictional sliding, and the slip phases generate large earthquakes.

The concepts of plate motion also tell us about intra-plate processes. At mid-ocean rises, where two plates peel apart, volcanism occurs, and at particular places this is manifested at the surface; thus Iceland. In fact, Iceland, like Hawaii, is thought to be the surface expression of a much deeper mantle excrescence, the mantle plume. The interplay between plumes and plates is fundamental to the understanding of how plate tectonics works.

8.1 Plate Tectonics

Plate tectonics describes the division of the Earth's surface into between thirteen and twenty plates, and it describes how these plates move relative to each other. It is essentially a kinematic theory, and the dynamic theory which supports it is the theory of mantle convection; we will come to that later.

The history of the development of plate tectonics is a fascinating illustration of the way in which science lurches forwards via the provision of dogma and conflict,

A. Fowler, *Mathematical Geoscience*, Interdisciplinary Applied Mathematics 36, DOI 10.1007/978-0-85729-721-1_8, © Springer-Verlag London Limited 2011

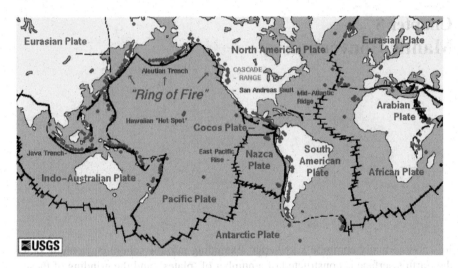

Fig. 8.1 The tectonic plates of the Earth. Image courtesy USGS, see http://vulcan.wr.usgs.gov

and their eventual resolution as the knowledge base increases. Plate tectonics is, in fact, a very old theory. Its origins lie in the nineteenth century, but it was perhaps first properly formulated as a consistent theory by Alfred Wegener, a German meteorologist, whose book on the subject was published in 1915.

The basic idea is very simple. Inspection of a map of the globe (see Fig. 8.1) leads to the observation that the coastlines of the Americas and Africa fit together very well. (The fit is in fact improved if boundaries are drawn at the edge of the continental shelves.) This seemingly fortuitous observation gains credibility when one finds that Europe and Greenland slot into the jigsaw, and in fact one can plug the other continents in as well, forming a global continent which now is known as Pangaea. However, the resulting concept that there used to be a super-continent, which then broke up into a number of continental 'plates', requires more than simply putting a jigsaw together. Wegener's contribution lay in supporting the basic concept with the evidence from a wide variety of separate disciplines, that there were paleontological, paleogeological and paleoclimatic continuities between the now widely separated parts of the proto-continent. For example, the continuation of the great rift fault which defines Loch Ness in Scotland lies in Newfoundland. There are many examples of such continuations, so much so that geologists had invented the concept of 'land bridges' which had previously joined different continents in order to explain these matches. The presumption was that the bridges had subsequently disappeared when sea levels rose. This is a good example of a scientific explanation which is not mathematically coherent, in the following sense: a process is invoked to explain an observation which is no less mysterious than the observation itself. Contradistinctively, Wegener's explanation is an example of Occam's razor; one simple (but mysterious) hypothesis can explain a whole host of seemingly disconnected observations.

Wegener's theory was built around observations, but his discussion of mechanism was less sophisticated. He imagined continental blocks ploughing through a viscous substratum, driven by gravitational 'pole-displacement' forces. The strength of his theory was that it unified a variety of different observations, but his suggestion that tidal forces were responsible was ridiculed by Jeffreys. It is at this point in the story that scientific enquiry lost its sense of rational progress and became polarised, almost religiously.

Apart from his lack of a satisfactory theoretical mechanism, Wegener's main problem appears to have been that he was not a geophysicist, but a meteorologist. For those trying to do interdisciplinary science a century later, the story is familiar; scientists are territorial, at the expense, sometimes, of rationality. Jeffreys' rejection of Wegener's hypothesis on the basis of a lack of realistic mechanism was in itself reasonable (the Earth is made of solid rock for the most part, why should it deform like a fluid?) but it is also hugely unimaginative. Much of the dogma which pervades scientifically accepted thought is based on the premise that things must behave in ways that we expect. Since what we expect is based on what we experience, mostly on relatively short time scales, it is difficult to imagine the possibility of unfamiliar events occurring. Quite simply, it is difficult to imagine the 3000 kilometre deep mantle being stirred by convection just like a bowl of porridge.

However, there is an analogy for the fluid-like behaviour of rocks. A similar controversy had raged over fifty years earlier in the nascent field of glaciology. Louis Agassiz championed the idea that great ice sheets had covered large parts of Europe, transporting sediments and, particularly, large ('erratic') boulders. Scientists were well aware that glaciers flowed like huge rivers, despite consisting of apparently solid ice. The fact that ice could flow when its temperature is raised near to its melting point should have been a clue to a comparable behaviour of crystalline rocks, as indeed might be the common observation of deeply folded rock strata.

Arthur Holmes was one of a number of people who suggested that thermal convection could be the causative mechanism for continental drift, and this is the accepted mechanism today. Holmes's original concept of how convection organised the continents and oceans was flawed—not surprisingly, since at that time very little was known about how convection works. But Jeffreys would not accept the concept, and the dogmatic debate carried on till the early 1960s.

The concept of mantle convection which is now accepted is that the Earth's mantle (i.e., the solid part of the Earth consisting of silicate rocks to a depth of 3000 km) undergoes thermal convection. There are two possible heat sources to drive this convection: firstly chondritic (i.e., primitive) rocks contain trace elements of radioactive material which releases sufficient heat to drive convection. Of course, the precise distribution of the heat source is not known. It *is* known that radiogenic rocks are concentrated in the continents, but the extent of such internal heating at great depth is speculative. In addition, the Earth is cooling. The release of heat from the core is also sufficient in itself to drive vigorous convection in the mantle.

If we heat a pan of frying oil, two forms of convection can be observed. There is a cellular structure consisting of boxes of circulating fluid. Hot fluid rises in the interior of the boxes and cold fluid sinks at their borders. In addition, we can see isolated mushroom shaped 'plumes'. When the lower surface is heated, there is a tendency for hot (thus buoyant) fluid to accumulate and rise through the surrounding cooler fluid, looking much like a sprouting fungus. These plumes are affected by the convective circulation, but can be thought of as largely independent.

The Earth's mantle is a convecting fluid (in that it flows; it is not a liquid). The plates which comprise the outer surface are the surface expression of an underlying convective circulation. The plates are not uniform, and this is consistent with vigorous convection, where we see the cell boundaries migrating irregularly (for example, in the pan of oil). In the mantle, we also have thermal plumes, and where these impinge on the surface, volcanism occurs. Iceland and Hawaii are the two best known examples of this.

The rôle of the continents in this story is slightly confusing. The continents are not themselves the plates, but rather they are carried by the plates. One can think of them as a kind of residue or scum which is formed on melting mantle rocks; to understand this we need to discuss petrogenesis briefly.[1]

All rocks in the mantle are silicate rocks, i.e., they contain silica (SiO_2), in proportions ranging from about 40% to about 70%. At the lower end of this range, the rocks are magnesium or iron rich and are called 'mafic' (or, extremely, 'ultramafic'). The typical such rock is basalt. At the other extreme the rocks are silicic, the type example being granite. Roughly speaking, mafic rocks have higher melting points and are less viscous when molten; they are also heavier. Hence, when ascending mantle rock melts to form magma, it is the lighter, more silicic rocks which remain as residue. Thus where volcanism occurs, so also does chemical segregation of the parent rocks. It is through this segregation that the lighter continental rocks are formed, and remain buoyantly on the surface. It ought to be emphasised that the petrology of the igneous[2] rocks is extremely complicated chemically; for example, it is clearly *wrong* to think of granite as a simple residue when more mafic components have been melted first and removed.

The concept of plates and plumes convection works best for sub-oceanic convection; indeed, there is no clear idea of what sub-continental convection is like. We shall come back to this later (in the notes). The oceanic plates are the surface expressions of a convecting cell; the mantle rock rises beneath mid-ocean ridges, which are long, pseudo-linear features (examples are the mid-Atlantic ridge or the East Pacific rise). At the ridge the flow diverges on either side, and the cell terminates normally at a downwelling, where the surface plate is subducted below an adjoining continental plate, as for example off the west coast of South America. This circulation is essentially self-organising; plumes such as those beneath Hawaii and Iceland act independently. They are normally thought to originate at the core–mantle boundary, and can occur in a plate interior (Hawaii) or at a mid-ocean ridge (Iceland).

Plates imply rigidity, and the astute reader will be wondering why the surface plates should behave more or less rigidly (as they are generally conceived to do) if

---

[1] Further discussion of petrology is given in Chap. 9.

[2] Meaning born of fire.

the mantle behaves viscously; surely there is a contradiction here? It is at this point in the discussion that we begin to look at the problem from the point of view of an applied mathematician. Although early theories of mantle convection provided good quantitative predictions of such things as plate velocity and oceanic heat flux, the story is not as simple as was at first thought; indeed, from the perspective of modelling, fundamental features of convection are still not yet completely understood.

We begin with the resolution of the query above, why should the plates which constitute the *lithosphere* be rigid if they are convecting? The answer to this lies in the rheology of crystalline rocks. The rate at which they creep depends on temperature, and this rate dependence has been measured. Theoretical considerations suggest that the process is thermally activated, so that the temperature dependence is of Arrhenius form (specifically, $\eta \propto \exp(E^*/RT)$), where E^* is the activation energy, R is the gas constant, T is absolute temperature, and η is the viscosity. Measured values of E^* for mantle rocks such as olivine are of typical values around 500 kJ mol^{-1}, while the gas constant is 8.3 J mol^{-1} K^{-1}. It is a simple and illuminating exercise to see that the temperature dependence of the viscosity is very strong. For example, with the values above, and if we take a pre-multiplicative factor of 10^2 Pa s (for illustrative purposes), then we find that the viscosities at temperatures 1500 K, 1000 K, and 500 K, respectively, are approximately 2.8×10^{19} Pa s, 1.5×10^{28} Pa s and 2.1×10^{52} Pa s. The viscosity variation is vast, and it is this extreme increased stickiness which causes the lithospheric plates to act as if they are rigid.

We chose the temperature of 1500 K and the corresponding pre-multiplicative factor in order to give a realistic estimate at the base of the lithosphere, where various lines of evidence point to temperatures and viscosities of this order. The surface temperature is 300 K, but even at 1000 K there is a nine order of magnitude increase in viscosity above that below the lithosphere. It is true that crustal near-surface rocks are weaker, and that elastic or brittle behaviour may become more important than creep there, but it cannot alter the observation that in general the lithosphere must be much stiffer than the underlying *asthenosphere*.

Hence the rigidity of the plates.[3] We now immediately have what may be the central conundrum of mantle convection. The temperature dependence of the viscosity makes the cold lithosphere rigid. How does convection operate for such a fluid? We will find out in the next section that the onset of convection is controlled by a dimensionless number called the *Rayleigh number*, which is a measure of the ratio of the destabilising buoyancy force to the stabilising viscosity and thermal diffusivity. More generally, the size of the Rayleigh number measures the vigour of convection. The effect of the convective circulation in a cell is to stir the fluid. In the absence of convection, a heated lower boundary and cooled upper boundary will cause a conductive (linear, in a Cartesian geometry) temperature gradient in the cell.

---

[3]In effect, of course, the plates are not rigid. The ocean floor is littered with transform faults, which can be seen as the offsets in Fig. 8.1. The lithosphere deforms in an effectively viscous way by means of these fractures, which indicates a plastic behaviour of the surface.

As convection increases, this linear gradient is distorted. The interior fluid becomes more isothermal, steeper thermal gradients develop near the boundaries.

Now suppose the viscosity varies with temperature, so that the cooler fluid is more viscous. The cooler fluid is at the top of the cell, and is therefore less mobile. Because it is less mobile, conduction becomes more important, thus the temperature gradient decreases, and the less mobile lid becomes thicker. The opposite occurs at the base, whence the hot basal layer becomes thinner. In fact the bulk of the temperature drop across the cell becomes focussed in the upper lid. If the viscosity contrast across the cell is large enough, then the upper lid becomes virtually stagnant, and the convection (which can still be vigorous) is confined to a relatively isothermal core flow beneath.

This behaviour is what we would expect, and it is what we see in laboratory experiments involving, say, golden syrup. But: it is *not* what we see in the Earth. The plates may be (relatively) rigid, but they move. This is clearest with the oceanic plates (for continental plates, it is less clear, as mentioned above). That in itself is not a problem, but there is a problem when we come to consider how the plates descend into the mantle at subduction zones. The oceanic lithosphere (meaning the thermal lithosphere, i.e., the cool, 'rigid' upper boundary layer of the mantle) is about 100 km thick when it descends at a subduction zone. The uppermost part of this lithosphere is elastic rather than viscous, and the detailed bathymetry at a subduction zone can indeed be understood via the mechanics of a buckling beam. However, this does not explain how the lithosphere can creep round the corner. A worse problem is the initiation of subduction. If convection can occur below a stagnant lid, what is the mechanism whereby the lid begins to subduct?

This might seem a rather abstruse question, since presumably the Earth has evolved from a much hotter initial state in which the problem of the stagnant lid may not have been important, and the question of interest is how (and if) subduction ceases. It becomes less abstruse when we come to consider tectonics on Venus. Venus is a planet much like the Earth (except for its atmosphere), and we should presumably expect a similar tectonic habit. But plate tectonics is apparently absent from Venus. And so we have the question: why should subduction occur on the Earth but not on Venus?

Actually, things are worse (and thus more interesting) than this. It appears (from meteor impact counts) that the surface of Venus is of relatively uniform age, about 300–500 million years, much younger than the planet. A possible inference is that planetary resurfacing occurred in that previous epoch through a rapid plate tectonic cycle, and then ceased.[4] Has it ceased for ever, or is plate tectonics episodic on Venus? What will happen on Earth? We will provide glimpses of possible answers to these questions in this chapter, but first, we begin at the beginning with a discussion of the mathematical basis for the theory of thermal convection.

---

[4] Another possibility is that volcanic activity caused the resurfacing.

8.2 Rayleigh–Bénard Convection

The classical study of thermal convection considers the motion of a fluid bounded by two horizontal plates, at $z = 0$ and $z = d$, say, and heated from below, for example by prescribing the upper and lower temperatures to be T_s and T_b, where the temperature difference across the cell $T_b - T_s > 0$. The Navier–Stokes equations which describe the motion are

$$\rho_t + \nabla.(\rho \mathbf{u}) = 0,$$

$$\rho[\mathbf{u}_t + (\mathbf{u}.\nabla)\mathbf{u}] = -\nabla p + \nabla.\boldsymbol{\tau} - \rho g \hat{\mathbf{k}}, \tag{8.1}$$

where ρ is the density, $\mathbf{u}$ is the fluid velocity, p is the pressure, $\boldsymbol{\tau}$ is the deviatoric stress tensor, and g is the downwards acceleration due to gravity. In the mantle, one might suppose g would vary, but in fact it is relatively constant. The linear decrease with depth that one would find for a constant density is almost exactly compensated by the fact that the Earth's core is much denser than the mantle. The assumption of a Cartesian geometry is obviously not appropriate for the mantle, but it is not likely that this will make a substantial qualitative difference to the results.

In mantle convection, it is often held that the heat source should be internal (radioactive) heating. The reason for this supposition appears to be largely historical, following Holmes's original suggestion. While it is true that radiogenic heating may be important, it is not necessary, since the Earth is cooling, and the heat released from the core is also well able to provide the necessary heating from below.

In Eqs. (8.1), we require constitutive relations for ρ as a function of T and p (the equation of state) and $\boldsymbol{\tau}$ (the rheology). We suppose that ρ is given by

$$\rho = \rho_0[1 - \alpha T + \beta p], \tag{8.2}$$

where T is temperature, α is the thermal expansion coefficient, and β is the compressibility; ρ_0 is a reference density. In the mantle, the value of α is $3 \times 10^{-5}\,\text{K}^{-1}$, and it decreases somewhat with depth. The assumption of a constant compressibility coefficient β is an oversimplification but will suffice. With a change in pressure of some 1200 kbar, the density changes from about $3.5 \times 10^3\,\text{kg m}^{-3}$ to $5.6\,\text{kg m}^{-3}$, which implies $\beta \rho_0 g d \approx 1.6$ (since pressure is essentially lithostatic).

We suppose that the rheologic equation of state is defined by a viscosity η, so that the components of the stress tensor are given by (we use the summation convention)

$$\tau_{ij} = \eta \left(\frac{\partial u_i}{\partial x_j} + \frac{\partial u_j}{\partial x_i} - \frac{2}{3}\delta_{ij}\frac{\partial u_k}{\partial x_k} \right) = 2\eta \left(\dot{\varepsilon}_{ij} - \frac{1}{3}\delta_{ij}\,\text{div}\,\mathbf{u} \right), \tag{8.3}$$

where $\dot{\varepsilon}_{ij}$ is the strain rate tensor. In general, η is a function of temperature, pressure and stress, but to begin with we will take it as a constant, η_0. (Later we will use η_0 as a viscosity scale.)

The final equation we require is the energy equation. This is

$$\rho c_p \frac{dT}{dt} - \alpha T \frac{dp}{dt} = \nabla.[k\nabla T] + \rho Q + \frac{\tau_{ij}\tau_{ij}}{2\eta}, \tag{8.4}$$

where d/dt denotes the material derivative $\partial/\partial t + \mathbf{u}.\nabla$, c_p is the specific heat at constant pressure, k is the thermal conductivity, Q is the (radiogenic) internal heating (units are W kg^{-1}), and the last term is the viscous heat dissipation. The advection and conduction terms will be familiar, the others being neglected in normal laboratory circumstances. In the Earth, we cannot necessarily assume they are negligible, however.

8.2.1 Boundary Conditions

The Earth's mantle lies above the liquid outer core and below the ocean or asthenosphere. Since the viscosity of the mantle is enormous, the effective boundary conditions which are appropriate are those of no shear stress. More generally, the stresses are continuous, and in particular the jump in the shear stress is zero at top and bottom. The mantle shear stress is effectively zero if the viscosity above or below is very small; but strictly this also relies on the fact that velocity gradients in the ocean or core are not correspondingly large. Without doubt this is the case.

For most of our discussion we will consider only two-dimensional motion in the (x, z) plane; generalisation to the third dimension is easy to effect. In this case we also define the velocity components via $\mathbf{u} = (u, 0, w)$. The boundary conditions we thus apply are the following:

$$\frac{\partial u}{\partial z} = w = 0, \qquad T = T_b \quad \text{at } z = 0;$$
$$\frac{\partial u}{\partial z} = w = 0, \qquad T = T_s \quad \text{at } z = d. \tag{8.5}$$

Lateral conditions (in x) are also strictly necessary. Since our initial concerns are with stability, where we select modes of various horizontal wave numbers, it is of less importance to enunciate these.

8.2.2 Non-dimensionalisation

We choose the obvious space scale d and the temperature scale T_b. One (but not the only) natural velocity scale is the thermal velocity scale $\frac{\kappa_0}{d}$, where the thermal diffusivity scale κ_0 is given by

$$\kappa_0 = \frac{k_0}{\rho_0 c_p}, \tag{8.6}$$

and k_0 is a reference value of the thermal conductivity; the corresponding thermal time scale is d^2/κ_0. Finally, the pressure and stress scales are chosen to balance the pressure gradient with the viscous terms, of order $\eta u/d \sim \eta_0 \kappa_0/d^2$ (where η_0 is a reference value for the viscosity), except that the lithostatic pressure is subtracted off first. This lithostatic pressure is that which occurs in the absence of motion, and is not directly relevant to the convective motion; we will determine it shortly.

Thus we write

$$\mathbf{x} = d\mathbf{x}^*, \qquad T = T_b T^*, \qquad \eta = \eta_0 \eta^*, \qquad \mathbf{u} = \frac{\kappa_0}{d}\mathbf{u}^*, \qquad t = \frac{d^2}{\kappa_0}t^*,$$

$$k = k_0 k^*, \qquad p = \rho_0 g d\, \bar{p}(z^*) + \frac{\eta_0 \kappa_0}{d^2}p^*, \qquad \rho = \rho_0 \rho^*, \qquad \tau = \frac{\eta_0 \kappa_0}{d^2}\tau^*, \qquad (8.7)$$

substitute these variables into the governing equations, and hence derive the following dimensionless model, written in full vector form, and where we omit the asterisks for convenience:

$$\rho = \bar{\rho}(z) - BT + \frac{BC}{R}p,$$

$$\bar{\rho} = 1 + C\bar{p},$$

$$\rho_t + \nabla.(\rho\mathbf{u}) = 0,$$

$$\frac{1}{\sigma}\left[\rho\{\mathbf{u}_t + (\mathbf{u}.\nabla)\mathbf{u}\}\right] = -\nabla p + \nabla.\boldsymbol{\tau} + \hat{\mathbf{k}}[RT - Cp], \qquad (8.8)$$

$$\frac{\partial\bar{p}}{\partial z} = -(1 + C\bar{p}),$$

$$\rho\frac{dT}{dt} - DT\left(w\bar{p}' + \frac{B}{R}\frac{dp}{dt}\right) = \nabla.[k\nabla T] + \rho H + \frac{D}{R}\tau^2,$$

$$\tau_{ij} = \eta\left[\frac{\partial u_i}{\partial x_j} + \frac{\partial u_j}{\partial x_i} - \frac{2}{3}\delta_{ij}\nabla.\mathbf{u}\right],$$

where the second stress invariant is defined as

$$2\tau^2 = \tau_{ij}\tau_{ij}. \qquad (8.9)$$

The parameters are a Boussinesq number

$$B = \alpha T_b, \qquad (8.10)$$

a barometric number

$$C = \beta\rho_0 gd, \qquad (8.11)$$

the Rayleigh number[5]

$$R = \frac{\alpha T_b \rho_0 gd^3}{\eta_0 \kappa_0}, \qquad (8.12)$$

---

[5]The Rayleigh number is normally defined with $\Delta T = T_s - T_b$, on the basis of scaling the temperature as $T = T_s + \Delta T\, T^*$, but we avoid this here because the later introduction of a temperature-dependent viscosity makes the present definition less cumbersome. For the mantle, we have $T_b \sim \Delta T$ in any case.

the Prandtl number

$$\sigma = \frac{\eta_0}{\rho_0 \kappa_0}, \tag{8.13}$$

the dissipation number

$$D = \frac{\alpha g d}{c_p}, \tag{8.14}$$

and an internal heating number

$$H = \frac{\rho_0 Q d^2}{k_0 T_b}. \tag{8.15}$$

The solutions will therefore depend on the six parameters above, together with the temperature ratio

$$\theta_s = \frac{T_s}{T_b}. \tag{8.16}$$

8.2.3 Boussinesq Approximation

The values of α and T_b for the mantle are about 3×10^{-5} K^{-1} and 3500 K. (The core–mantle boundary is thought to be at a temperature of 3500 K, or greater.) For these values $B \approx 0.1$, and as α decreases with depth, B will decline somewhat with depth. The Boussinesq approximation is associated with the limit $B \to 0$, and provides a major simplification of the model. We have already estimated the parameter C as being of order 1.6, while the Rayleigh number for the mantle is of order 10^6 (and generally is always large when convection occurs in environmental flows). Therefore also $BC/R \ll 1$, and so

$$\rho \approx \bar{\rho}(z). \tag{8.17}$$

The Boussinesq approximation actually goes further; it assumes that variations in density can be neglected, except in the buoyancy term (this is the term RT in the momentum equation). In particular, it assumes C is small, which is inevitably true in laboratory experiments, but is not accurate for the mantle. Nevertheless, we proceed by supposing that C is small; then

$$\bar{\rho} = 1, \qquad \bar{p} = 1 - z \tag{8.18}$$

and thus

$$\nabla.\mathbf{u} \approx 0; \tag{8.19}$$

the fluid is approximately incompressible.

The Prandtl number is the ratio of the kinematic viscosity η_0/ρ_0 to the thermal diffusivity. In the Earth's mantle, this is huge (of order 10^{23}) and the corresponding acceleration terms are utterly negligible. For liquids and gases, the Prandtl number is generally of $O(1)$; it is quite large for some oils, and low for metals (e.g., mercury). We suppose that $\sigma \gg 1$. The dissipation number D is negligible in the laboratory. In the mantle, with $\alpha = 3 \times 10^{-5}$ K^{-1}, $g = 10$ m s^{-2}, $d = 3000$ km, $c_p = 10^3$ J kg^{-1} K^{-1}, we have $D \approx 0.9$. The decrease of α with depth causes D to decrease also, but it is clearly not negligible in the Earth. Again, we begin by neglecting it. In any event, the viscous heating term of order D/R can generally be safely neglected.

The heating parameter H depends on the assumed distribution of radioactive elements in the mantle. For a value of $Q = 1$ pW kg^{-1} (1 picowatt $= 10^{-12}$ W), and $\rho_0 = 4 \times 10^3$ kg m^{-3}, $d = 3000$ km, $k_0 = 4$ W m^{-1} K^{-1}, $T_b = 3500$ K, we find $H \sim 10$; essentially of $O(1)$, but as we shall see in due course, relatively small when convection is vigorous.

If we take the scaled viscosity and thermal conductivity to be constant ($\eta = k = 1$), and put the parameters B, C, σ^{-1}, D and H equal to zero, then we obtain the Boussinesq equations of thermal convection:

$$\nabla.\mathbf{u} = 0,$$
$$\nabla p = \nabla^2 \mathbf{u} + RT\hat{\mathbf{k}}, \tag{8.20}$$
$$T_t + \mathbf{u}.\nabla T = \nabla^2 T,$$

with associated boundary conditions for free slip:

$$
\begin{aligned}
T = 1, & \quad \mathbf{u}.\mathbf{n} = \tau_{nt} = 0 \quad \text{on } z = 0, \\
T = 0, & \quad \mathbf{u}.\mathbf{n} = 0 \quad \text{on } z = 1,
\end{aligned}
\tag{8.21}
$$

where τ_{nt} represents the shear stress, and we can take the surface temperature to be zero by a mild rescaling of the temperature.

8.2.4 Linear Stability

In the absence of motion, $\mathbf{u} = \mathbf{0}$, the temperature profile is linear,

$$T = 1 - z, \tag{8.22}$$

and the lithostatic pressure is modified by the addition of

$$p = -\frac{R}{2}(1 - z)^2. \tag{8.23}$$

To study small perturbations to this basic state, we define a stream function ψ for two-dimensional motion via

$$u = -\psi_z, \qquad w = \psi_x. \tag{8.24}$$

(The sign is opposite to the usual convention; for $\psi > 0$ this describes a clockwise circulation.) We also define the temperature perturbation θ by

$$T = 1 - z + \theta. \tag{8.25}$$

Eliminating the pressure gradient by cross differentiation, we find

$$\nabla^4 \psi + R\theta_x = 0,$$
$$\theta_t - \psi_x - \psi_z\theta_x + \psi_x\theta_z = \nabla^2\theta, \tag{8.26}$$

and the boundary conditions are

$$\psi_{zz} = \psi = \theta = 0 \quad \text{on } z = 0, 1. \tag{8.27}$$

The linear stability of the basic state is determined by neglecting the nonlinear advective terms in the heat equation. We then seek normal modes of wave number k in the form

$$\psi = f(z)e^{\sigma t + ikx},$$
$$\theta = g(z)e^{\sigma t + ikx}, \tag{8.28}$$

whence f and g satisfy

$$\left(D^2 - k^2\right)^2 f + ikRg = 0,$$
$$\sigma g - ikf = \left(D^2 - k^2\right)g, \tag{8.29}$$

where $D = d/dz$, and

$$f = g = 0 \quad \text{on } z = 0, 1. \tag{8.30}$$

By inspection, solutions are

$$f = \sin n\pi z, \qquad g = b\sin n\pi z, \tag{8.31}$$

$(n = 1, 2, \ldots)$ providing

$$\sigma = \frac{k^2 R}{(n^2\pi^2 + k^2)^2} - \left(n^2\pi^2 + k^2\right), \tag{8.32}$$

which determines the growth rate for the nth mode of wave number k.

Since σ is real, instability is characterised by a positive value of σ. We can see that σ decreases as n increases; therefore the value $n = 1$ gives the most unstable value of σ. Also, σ is negative for $k \to 0$ or $k \to \infty$, and has a single maximum. Since σ increases with R, we see that $\sigma > 0$ (for $n = 1$) if $R > R_{ck}$, where

$$R_{ck} = \frac{(\pi^2 + k^2)^3}{k^2}. \tag{8.33}$$

In turn, this value of the Rayleigh number depends on the selected wave number k. Since an arbitrary disturbance will excite all wave numbers, it is the minimum value of R_{ck} which determines the absolute threshold for stability. The minimum is obtained when

$$k = \frac{\pi}{\sqrt{2}}, \tag{8.34}$$

and the resulting critical value of the Rayleigh number is

$$R_c = \frac{27\pi^4}{4} \approx 657.5; \tag{8.35}$$

That is, the steady state is linearly unstable if $R > R_c$.

For other boundary conditions, the solutions are still exponentials, but the coefficients, and hence also the growth rate, must be found numerically. The resultant critical value of the Rayleigh number is higher for no-slip boundary conditions, for example, (it is about 1707), and in general, thermal convection is initiated at values of $R \gtrsim O(10^3)$.

8.3 Nonlinear Stability; Planforms

Linear stability analysis does not inform us as to the subsequent development of the unstable convective mode. This is the domain of nonlinear stability theory, or bifurcation theory. There are three questions we can use this theory to address. Firstly, does the convective mode saturate at a small but finite amplitude (supercritical bifurcation) or is there a sudden transition to a large amplitude solution? The basis for the alternatives lies in the solution of the Landau equation

$$\frac{dA}{dt} = k_1 A + k_2 |A|^2 A, \tag{8.36}$$

which we expect to derive via a multiple scales analysis (and we will do so below). In (8.36), A is the (complex) amplitude of the principal convective mode at $R = R_c$. When the steady state is unstable ($\operatorname{Re} k_1 > 0$), then supercritical bifurcation occurs if $\operatorname{Re} k_2 < 0$.

In Bénard's original experiments, he found hexagonal convective cells. More generally, two-dimensional convective rolls (as we have been describing) can be unstable to three-dimensional motions; for example, hexagons are a superposition of three sets of rolls, square cells are a superposition of two sets. It is clear that the two-dimensional linear stability analysis has a degeneracy, as the roll orientation is unconstrained. In particular, we might examine the evolution of two or more sets of rolls. We would expect that their amplitudes $A_1, A_2, \ldots$ would then be described by coupled sets of Landau equations, whose solution would determine the preference for rolls, hexagons, or other planforms.

The final question which multiple scales analysis can address is the evolution of the plan form over long space scales. Vagaries of initial conditions can lead to the onset of (for example) rolls of differing orientations in different parts of a large fluid container. The subsequent evolution of the resultant dislocations can be described by a spatial version of the Landau equation, which in some circumstances produces the famous Ginzburg–Landau equation.

These issues are certainly relevant to convection theory, but they are something of a sideshow here, as convection in the Earth's mantle (and any other large scale geophysical system: the Earth's core, magma chambers, the oceans, the atmosphere) occurs at high Rayleigh number, well beyond the range of applicability of the small amplitude theories. We will begin by deriving the Landau equation in some detail, and then discuss more briefly its generalisations to study planforms and dislocations.

8.3.1 Landau Equation

We begin by returning to the nonlinear Boussinesq model of two-dimensional convection in (8.26) and (8.25):

$$\nabla^4 \psi + R\theta_x = 0,$$
$$\theta_t - \psi_x - \psi_z \theta_x + \psi_x \theta_z = \nabla^2 \theta,$$
(8.37)

together with $\theta = \psi = 0$ on $z = 0$ and $z = 1$. The procedure of nonlinear stability theory is straightforward and we do not dwell on it here. Essentially, when R is close to (and above) R_c, so that $R - R_c = \varepsilon^2$, say, with $\varepsilon \ll 1$, then a convective mode of amplitude $O(\varepsilon)$ will grow on a time scale of $O(\varepsilon^2)$. The critical mode which is excited has wave number $k_c = \pi/\sqrt{2}$, but in fact a side band in which $k - k_c = O(\varepsilon)$ (since R_{ck} has a quadratic minimum at k_c) is excited; it is then natural to expect slow variation also on space scales of $O(1/\varepsilon)$. We return to slow spatial variation later.

The method of multiple scales now proceeds by defining

$$R = R_c + s\varepsilon^2,$$
$$t = \varepsilon^2 \tau$$
(8.38)

(where $s = \pm 1$ to allow for supercritical or subcritical bifurcations), and seeking perturbation expansions for θ and ψ in the form

$$\psi = \varepsilon \psi_1 + \varepsilon^2 \psi_2 + \cdots,$$
$$\theta = \varepsilon \theta_1 + \varepsilon^2 \theta_2 + \cdots.$$
(8.39)

In general, the functions depend on both the fast and slow scales t and τ. Since the basic convective mode is steady, we can here ignore the fast time scale altogether, thus we suppose ψ_i and θ_i are functions of τ, as well as x and z.

Substituting the expansions into (8.37), we obtain a sequence of problems for ψ_i and θ_i when we equate terms of order ε^n, $n = 1, 2, 3, \ldots$. In succession, these are

$$\mathcal{L} \begin{pmatrix} \psi_1 \\ \theta_1 \end{pmatrix} = 0, \tag{8.40}$$

where the linear operator $\mathcal{L}$ is defined by

$$\mathcal{L} = \begin{pmatrix} \nabla^4 & R_c \frac{\partial}{\partial x} \\ \frac{\partial}{\partial x} & \nabla^2 \end{pmatrix}, \tag{8.41}$$

and then

$$\mathcal{L} \begin{pmatrix} \psi_2 \\ \theta_2 \end{pmatrix} = \mathbf{N}_{20} \equiv \begin{pmatrix} 0 \\ \psi_{1x}\theta_{1z} - \psi_{1z}\theta_{1x} \end{pmatrix},$$

$$\mathcal{L} \begin{pmatrix} \psi_3 \\ \theta_3 \end{pmatrix} = \mathbf{N}_{30} \equiv \begin{pmatrix} -s\theta_{1x} \\ \theta_{1\tau} + \{\psi_{1x}\theta_{2z} - \psi_{1z}\theta_{2x} + \psi_{2x}\theta_{1z} - \psi_{2z}\theta_{1x}\} \end{pmatrix}, \tag{8.42}$$

and so on. Suitable boundary conditions are that $(\psi_i, \theta_i)^T = (0, 0)^T$ on $z = 0$ and $z = 1$.

The marginally stable solution of wave number k_c can be written as (suppression of the fast time scale t suppresses all the stable modes)

$$\begin{pmatrix} \psi_1 \\ \theta_1 \end{pmatrix} = \frac{1}{2} A(\tau) e^{ik_c x} \begin{pmatrix} -i \\ \beta \end{pmatrix} \sin \pi z + (cc), \tag{8.43}$$

where (cc) denotes the complex conjugate, and β is given from (8.29) when $\sigma = 0$ by

$$\beta = \frac{k_c}{\pi^2 + k_c^2} = \frac{\sqrt{2}}{3\pi} = \frac{1}{3k_c}. \tag{8.44}$$

Here, $A(\tau)$ is the unknown amplitude of the motion; in general it may be complex, but we anticipate that here it is in fact real. Specifically, then,

$$\psi_1 = A \sin k_c x \sin \pi z,$$
$$\theta_1 = \beta A \cos k_c x \sin \pi z, \tag{8.45}$$

and thus

$$\psi_{1x}\theta_{1z} - \psi_{1z}\theta_{1x} = \frac{1}{3}\pi A^2 \sin \pi z \cos \pi z. \tag{8.46}$$

In order to solve (8.42)$_1$, note that if

$$\mathbf{v} = \begin{pmatrix} a \\ b \end{pmatrix} e^{ipk_c x} \sin q\pi z, \tag{8.47}$$

then

$$\mathcal{L}\mathbf{v} = L_{pq}\mathbf{v},\tag{8.48}$$

where

$$L_{pq} = \begin{pmatrix} (p^2 k_c^2 + q^2 \pi^2)^2 & ipk_c R_c \\ ipk_c & -(p^2 k_c^2 + q^2 \pi^2) \end{pmatrix}.\tag{8.49}$$

Now $(\psi_2, \theta_2)^T$ satisfies

$$\mathcal{L}\begin{pmatrix} \psi_2 \\ \theta_2 \end{pmatrix} = \mathbf{N}_{20} = \begin{pmatrix} 0 \\ \frac{1}{6}\pi A^2 \end{pmatrix} \sin 2\pi z,\tag{8.50}$$

and therefore

$$\begin{pmatrix} \psi_2 \\ \theta_2 \end{pmatrix} = L_{02}^{-1} \begin{pmatrix} 0 \\ \frac{1}{6}\pi A^2 \end{pmatrix} \sin 2\pi z$$

$$= \begin{pmatrix} 0 \\ -\frac{1}{24\pi} A^2 \end{pmatrix} \sin 2\pi z.\tag{8.51}$$

(There is also an arbitrary multiple of $(\psi_1, \theta_1)^T$, but this term is irrelevant in calculating $A(\tau)$.)

What is important in solving for $(\psi_3, \theta_3)^T$ is to determine the secular terms on the right hand side of (8.42), which are those proportional to $e^{ik_c x} \sin \pi z$. Note that $\psi_2 = 0$, and the remnant nonlinear term in (8.42)$_2$ is (since $\theta_{2x} = 0$)

$$\psi_{1x}\theta_{2z} = \frac{1}{24} k_c A^3 (\sin \pi z - \sin 3\pi z) \cos k_c x,\tag{8.52}$$

and thus

$$\mathcal{L}\begin{pmatrix} \psi_3 \\ \theta_3 \end{pmatrix} = \frac{1}{2} \begin{pmatrix} -\frac{1}{3}isA \\ \beta A' + \frac{1}{24} k_c A^3 \end{pmatrix} e^{ik_c x} \sin \pi z + (cc) + \text{harmonics}.\tag{8.53}$$

In general, this has no bounded solution; thus in order to obtain such a solution, the right hand side must be orthogonal to the adjoint of the linear operator $\mathcal{L}$. Equivalently, we can find a bounded solution to (8.53) of the form

$$\begin{pmatrix} \psi_3 \\ \theta_3 \end{pmatrix} = \frac{1}{2} \begin{pmatrix} a \\ b \end{pmatrix} e^{ik_c x} \sin \pi z + (cc) + \text{harmonics}\tag{8.54}$$

providing

$$L_{11} \begin{pmatrix} a \\ b \end{pmatrix} = \begin{pmatrix} -\frac{1}{3}isA \\ \beta A' + \frac{1}{24} k_c A^3 \end{pmatrix},\tag{8.55}$$

and this can be solved only if the right hand side is orthogonal to the null vector of L_{11}^*, the adjoint of L_{11} (since det $L_{11} = 0$).

We have

$$L_{11}^* = \begin{pmatrix} (k_c^2 + \pi^2)^2 & -ik_c \\ -ik_c R_c & -(k_c^2 + \pi^2) \end{pmatrix},$$

(8.56)

with null vector

$$\eta = \begin{pmatrix} ik_c \\ (k_c^2 + \pi^2)^2 \end{pmatrix};$$

(8.57)

hence the orthogonality condition for solvability at $O(\varepsilon^2)$ requires (since the inner product $(\mathbf{v}, \mathbf{w}) = \mathbf{v}^T \bar{\mathbf{w}}$, where $\bar{\mathbf{w}}$ is the complex conjugate of $\mathbf{w}$)

$$\beta \frac{dA}{d\tau} = \left\{ \frac{sk_c}{3(k_c^2 + \pi^2)^2} \right\} A - \frac{1}{24} k_c A^3.$$

(8.58)

When $s = 1$ (thus $R = R_c + \varepsilon^2$), this can be written as the Landau equation

$$\frac{dA}{d\tau} = \frac{2}{9\pi^2} A - \frac{\pi^2}{16} A^3;$$

(8.59)

the bifurcation is supercritical, and the equilibrium amplitude is $A = \frac{4\sqrt{2}}{3\pi^2}$.

8.3.2 Plan Forms

Bénard's original experiments showed cellular convection occurring in hexagons, formed by the superposition of three sets of rolls at 120° to each other. In other situations, one sees square cells. Since the orientation of rolls is arbitrary, an obvious problem is to extend the preceding nonlinear analysis to the consideration of several sets of skewed rolls.

To do this, we need to write the equations of motion in a suitable three-dimensional form. By analogy with the definition of a stream function, we define the vector potential $\mathbf{A}$ via

$$\mathbf{u} = \operatorname{curl} \mathbf{A}, \qquad \operatorname{div} \mathbf{A} = 0;$$

(8.60)

$\mathbf{A}$ is defined up to the addition of the gradient of a harmonic function. Denote the components of $\mathbf{A}$ via

$$\mathbf{A} = (-\phi, \psi, \omega),$$

(8.61)

so that

$$\mathbf{u} = (\omega_y - \psi_z, -\phi_z - \omega_x, \psi_x + \phi_y).$$

(8.62)

Taking the curl of $(8.20)_2$ and writing $T = 1 - z + \theta$, we find

$$\nabla^4 \phi + R\theta_y = 0,$$
$$\nabla^4 \psi + R\theta_x = 0,$$
$$\nabla^4 \omega = 0,$$
$$\phi_x = \psi_y + \omega_z,$$
$$\nabla^2 \theta + \psi_x + \phi_y = \theta_t + (\psi_x \theta_z - \psi_z \theta_x)$$
$$+ (\phi_y \theta_z - \phi_z \theta_y) + (\omega_y \theta_x - \omega_x \theta_y),$$

(8.63)

and the boundary conditions are

$$\theta = 0,$$
$$[\tau_{13} \propto] \omega_{zy} - \psi_{zz} + \psi_{xx} + \phi_{xy} = 0,$$
$$[\tau_{23} \propto] - \omega_{xz} - \phi_{zz} + \phi_{yy} + \psi_{xy} = 0,$$
$$\psi_x + \phi_y = 0,$$

(8.64)

on $z = 0$ and $z = 1$.

We note from $(8.63)_{1-3}$ that $\nabla^4(-\phi_x + \psi_y + \omega_z) = 0$. Therefore the compatibility condition $-\phi_x + \psi_y + \omega_z = 0$ is automatically satisfied if we apply the boundary conditions

$$\phi_x = \psi_y + \omega_z,$$
$$\phi_{xz} = \psi_{yz} + \omega_{zz},$$

(8.65)

on $z = 0, 1$. In particular, we can eliminate ω from the other boundary conditions so that (using also $(8.64)_4$)

$$\nabla^2 \phi = \nabla^2 \psi = \theta = \psi_x + \phi_y = 0 \quad \text{on } z = 0, 1,$$

(8.66)

and the model reduces to the form

$$\mathcal{L} \begin{pmatrix} \psi \\ \phi \\ \theta \end{pmatrix} = \begin{pmatrix} -\varepsilon^2 \theta_x \\ -\varepsilon^2 \theta_y \\ \theta_t + \mathbf{u}.\nabla\theta \end{pmatrix},$$

(8.67)

where

$$\varepsilon^2 = R - R_c,$$

(8.68)

and

$$\mathcal{L} = \begin{pmatrix} \nabla^4 & 0 & R_c \frac{\partial}{\partial x} \\ 0 & \nabla^4 & R_c \frac{\partial}{\partial y} \\ \frac{\partial}{\partial x} & \frac{\partial}{\partial y} & \nabla^2 \end{pmatrix},$$

(8.69)

and $\mathbf{u}.\nabla\theta$ is given from the right hand side of the last equation in (8.63). We can then find ω by quadrature from the compatibility condition.

The linearised system at $R = R_c$ is simply $\mathcal{L}(\psi, \phi, \theta)^T = 0$, and particular solutions are rolls aligned in the y and x directions:

$$
\begin{aligned}
\phi = \omega = 0, && \psi = A \sin k_c x \sin \pi z, && \theta = \beta A \cos k_c x \sin \pi z, \\
\psi = \omega = 0, && \phi = B \sin k_c y \sin \pi z, && \theta = \beta B \cos k_c y \sin \pi z.
\end{aligned}
\tag{8.70}
$$

Of course other oblique rolls exist, having

$$
\begin{aligned}
\omega = 0, && \theta = \beta C \cos\big[k_c(x \cos\alpha + y \sin\alpha)\big] \sin \pi z, \\
\psi = C \cos\alpha \sin\big[k_c(x \cos\alpha + y \sin\alpha)\big] \sin \pi z, \\
\phi = C \sin\alpha \sin\big[k_c(x \cos\alpha + y \sin\alpha)\big] \sin \pi z,
\end{aligned}
\tag{8.71}
$$

but we suppose now that the leading order solution consists of just two sets of orthogonal rolls. Specifically, we write as before

$$
\begin{aligned}
R = R_c + \varepsilon^2, && t = \varepsilon^2 \tau, \\
\psi = \varepsilon \psi_1 + \varepsilon^2 \psi_2 + \cdots, \\
\phi = \varepsilon \phi_1 + \varepsilon^2 \phi_2 + \cdots, \\
\omega = \varepsilon \omega_1 + \varepsilon^2 \omega_2 + \cdots, \\
\theta = \varepsilon \theta_1 + \varepsilon^2 \theta_2 + \cdots,
\end{aligned}
\tag{8.72}
$$

and by equating powers of ε, we obtain the equations for the vectors $\mathbf{w}_i = (\psi_i, \phi_i, \theta_i)^T$,

$$
\begin{aligned}
\mathcal{L}\mathbf{w}_1 &= \mathbf{0}, \\
\mathcal{L}\mathbf{w}_2 &= (0, 0, N_2)^T, \\
\mathcal{L}\mathbf{w}_3 &= (-\theta_{1x}, -\theta_{1y}, \theta_{1\tau} + N_3)^T,
\end{aligned}
\tag{8.73}
$$

where the nonlinear terms are

$$
N_2 = [\psi_{1x}\theta_{1z} - \psi_{1z}\theta_{1x}] + [\phi_{1y}\theta_{1z} - \phi_{1z}\theta_{1y}] + [\omega_{1y}\theta_{1x} - \omega_{1x}\theta_{1y}],
\tag{8.74}
$$

and

$$
\begin{aligned}
N_3 = {}& [\psi_{1x}\theta_{2z} - \psi_{1z}\theta_{2x}] + [\psi_{2x}\theta_{1z} - \psi_{2z}\theta_{1x}] \\
& + [\phi_{1y}\theta_{2z} - \phi_{1z}\theta_{2y}] + [\phi_{2y}\theta_{1z} - \phi_{2z}\theta_{1y}] \\
& + [\omega_{1y}\theta_{2x} - \omega_{1x}\theta_{2y}] + [\omega_{2y}\theta_{1x} - \omega_{2x}\theta_{1y}].
\end{aligned}
\tag{8.75}
$$

The two roll solution we select is

$$\omega_1 = 0, \qquad \psi_1 = A(\tau)\sin k_c x \sin \pi z,$$

$$\theta_1 = \beta[A\cos k_c x + B\cos k_c y]\sin \pi z, \qquad \phi_1 = B(\tau)\sin k_c y \sin \pi z, \tag{8.76}$$

representing two orthogonal sets of rolls.

We now omit the subscript c on k_c for convenience. From (8.74) and (8.76), we find

$$N_2 = \beta k\pi\left[\frac{1}{2}(A^2 + B^2) + AB\cos kx \cos ky\right]\sin 2\pi z. \tag{8.77}$$

Let us denote

$$v_{pqr} = e^{ipkx + iqky}\sin r\pi z; \tag{8.78}$$

then if $\mathbf{c}$ is constant,

$$\mathcal{L}(\mathbf{c}v_{pqr}) = (L_{pqr}\mathbf{c})v_{pqr}, \tag{8.79}$$

where

$$L_{pqr} = \begin{pmatrix} \delta^2 & 0 & ipkR_c \\ 0 & \delta^2 & iqkR_c \\ ipk & iqk & -\delta \end{pmatrix}, \tag{8.80}$$

and

$$\delta = (p^2 + q^2)k^2 + r^2\pi^2. \tag{8.81}$$

From (8.73) and (8.77), $\mathbf{w}_2$ satisfies

$$L\mathbf{w}_2 = \frac{1}{2}\beta k\pi\left(A^2 + B^2\right)\hat{\boldsymbol{\theta}}v_{002} + \frac{1}{4}\beta k\pi AB\hat{\boldsymbol{\theta}}\left[v_{112} + v_{1,-1,2} + (cc)\right], \tag{8.82}$$

where

$$\hat{\boldsymbol{\theta}} = (0, 0, 1)^T, \tag{8.83}$$

and hence we find the particular solution (using the fact that $\beta k = \frac{1}{3}$)

$$\mathbf{w}_2 = \frac{1}{6}\pi\left(A^2 + B^2\right)\mathbf{a}_{002}v_{002} + \frac{1}{12}\pi AB\left[\mathbf{a}_{112}v_{112} + \mathbf{a}_{1,-1,2}v_{1,-1,2} + (cc)\right], \tag{8.84}$$

where $\mathbf{a}_{pqr} = L_{pqr}^{-1}\hat{\boldsymbol{\theta}}$. The inverse of the matrix L_{pqr} is

$$L_{pqr}^{-1} = \frac{1}{|L_{pqr}|}\begin{pmatrix} q^2k^2R_c - \delta^3 & -pqk^2R_c & -ipkR_c\delta^2 \\ -pqk^2R_c & p^2k^2R_c - \delta^3 & -iqkR_c\delta^2 \\ -ipk\delta^2 & -iqk\delta^2 & \delta^4 \end{pmatrix}, \tag{8.85}$$

where

$$|L_{pqr}| = \delta^2\left[(p^2 + q^2)k^2R_c - \delta^3\right]. \tag{8.86}$$

and thus

$$\mathbf{a}_{pqr} = \frac{\delta^2}{|L_{pqr}|} \begin{pmatrix} -ipk R_c \\ -iqk R_c \\ \delta^2 \end{pmatrix}. \tag{8.87}$$

We write $\mathbf{a}_{pqr} = (\alpha_{pqr}, \beta_{pqr}, \gamma_{pqr})^T$, and for the particular values of p, q, r in (8.84), we have

$$\mathbf{a}_{002} = \begin{pmatrix} \alpha_{002} \\ \beta_{002} \\ \gamma_{002} \end{pmatrix} = \begin{pmatrix} 0 \\ 0 \\ -\frac{1}{4\pi^2} \end{pmatrix},$$

$$\mathbf{a}_{112} = \begin{pmatrix} \alpha_{112} \\ \beta_{112} \\ \gamma_{112} \end{pmatrix} = \frac{1}{573\pi^2} \begin{pmatrix} 27ik \\ 27ik \\ -100 \end{pmatrix}, \tag{8.88}$$

$$\mathbf{a}_{1,-1,2} = \begin{pmatrix} \alpha_{112} \\ -\beta_{112} \\ \gamma_{112} \end{pmatrix}.$$

In particular,

$$\phi_{2x} = \frac{1}{12}\pi AB \sin 2\pi z \left[ik\beta_{112}(v_{112} - v_{1,-1,2}) + (cc) \right],$$
$$\tag{8.89}$$
$$\psi_{2y} = \frac{1}{12}\pi AB \sin 2\pi z \left[ik\alpha_{112}(v_{112} - v_{1,-1,2}) + (cc) \right],$$

which are equal since $\alpha_{112} = \beta_{112}$; hence $\omega_{2z} = \phi_{2x} - \psi_{2y} = 0$, and

$$\omega_2 = 0. \tag{8.90}$$

Next, we calculate the secular terms in (8.75). We have

$$\psi_1 = A \sin kx \sin \pi z,$$
$$\phi_1 = B \sin ky \sin \pi z, \tag{8.91}$$
$$\theta_1 = \beta[A \cos kx + B \cos ky] \sin \pi z,$$

and also

$$\begin{pmatrix} \psi_2 \\ \phi_2 \\ \theta_2 \end{pmatrix} = \left[\frac{1}{6}\pi (A^2 + B^2) \begin{pmatrix} 0 \\ 0 \\ \gamma_{002} \end{pmatrix} + \frac{1}{3}\pi AB \begin{pmatrix} -a \cos ky \sin kx \\ -a \cos kx \sin ky \\ \gamma_{112} \cos ky \cos kx \end{pmatrix} \right] \sin 2\pi z,$$
$$\tag{8.92}$$

where we have written

$$\alpha_{112} = \beta_{112} = ia, \tag{8.93}$$

and $a = 9k/191\pi^2$ is real. We are only interested in the secular terms in N_3, and these are terms in $e^{ikx} \sin \pi z$ or $e^{iky} \sin \pi z$. Successively, we find

$$\psi_{1x}\theta_{2z} = -\frac{1}{6}\pi^2 kA \sin \pi z\big[(A^2 + B^2)\gamma_{002}\cos kx + \gamma_{112}AB\cos ky\big] + \cdots ,$$

$$-\psi_{1z}\theta_{2x} = \frac{1}{12}\pi^2 k\gamma_{112}A^2 B \sin \pi z \cos ky + \cdots ,$$

$$\psi_{2x}\theta_{1z} = -\frac{1}{12}\beta k\pi^2 a AB \sin \pi z(A\cos ky + B\cos kx) + \cdots ,$$ (8.94)

$$-\psi_{2z}\theta_{1x} = \frac{1}{6}\beta k\pi^2 a A^2 B \sin \pi z \cos ky + \cdots ,$$

and the corresponding ϕ terms can be found by swapping A with B and x with y.

Summing the terms in (8.94) and adding the corresponding ϕ terms, we find that the secular terms are

$$N_3 = -[N_{101}v_{101} + N_{011}v_{011}] + (cc) + \cdots ,$$ (8.95)

where

$$N_{101} = \frac{1}{12}\pi^2 k\gamma_{002}A(A^2 + B^2) + \frac{1}{24}\pi^2 k\gamma_{112}AB^2,$$

$$N_{011} = \frac{1}{12}\pi^2 k\gamma_{002}B(A^2 + B^2) + \frac{1}{24}\pi^2 k\gamma_{112}A^2 B.$$ (8.96)

Noting that

$$\theta_1 = \frac{1}{2}\beta A v_{101} + \frac{1}{2}\beta B v_{011} + (cc),$$ (8.97)

the equation for $\mathbf{w}_3$ can be written (from (8.73))

$$\mathcal{L}\mathbf{w}_3 = \begin{pmatrix} -\frac{1}{2}ik\beta A \\ 0 \\ \frac{1}{2}\beta A' - N_{101} \end{pmatrix} v_{101} + \begin{pmatrix} 0 \\ -\frac{1}{2}ik\beta B \\ \frac{1}{2}\beta B' - N_{011} \end{pmatrix} v_{011} + (cc) + \cdots ,$$ (8.98)

and the particular solution is

$$\mathbf{w}_3 = \mathbf{w}_{101}v_{101} + \mathbf{w}_{011}v_{011} + (cc) + \cdots ,$$ (8.99)

providing

$$L_{101}\mathbf{w}_{101} = \begin{pmatrix} -\frac{1}{2}ik\beta A \\ 0 \\ \frac{1}{2}\beta A' - N_{101} \end{pmatrix}, \qquad L_{011}\mathbf{w}_{011} = \begin{pmatrix} 0 \\ -\frac{1}{2}ik\beta B \\ \frac{1}{2}\beta B' - N_{011} \end{pmatrix}.$$ (8.100)

Now

$$L_{101} = \begin{pmatrix} \delta^2 & 0 & ikR_c \\ 0 & \delta^2 & 0 \\ ik & 0 & -\delta \end{pmatrix}, \qquad L_{011} = \begin{pmatrix} \delta^2 & 0 & 0 \\ 0 & \delta^2 & ikR_c \\ 0 & ik & -\delta \end{pmatrix}, \qquad (8.101)$$

where

$$\delta = \pi^2 + k^2. \qquad (8.102)$$

The adjoint matrices are

$$L_{101}^* = \begin{pmatrix} \delta^2 & 0 & -ik \\ 0 & \delta^2 & 0 \\ -ikR_c & 0 & -\delta \end{pmatrix}, \qquad L_{011}^* = \begin{pmatrix} \delta^2 & 0 & 0 \\ 0 & \delta^2 & -ik \\ 0 & -ikR_c & -\delta \end{pmatrix}, \qquad (8.103)$$

and the respective null vectors are

$$\eta_{101} = \begin{pmatrix} ik \\ 0 \\ \delta^2 \end{pmatrix}, \qquad \eta_{011} = \begin{pmatrix} 0 \\ ik \\ \delta^2 \end{pmatrix}. \qquad (8.104)$$

Therefore the solvability condition for (8.100), that the right hand sides are orthogonal to the corresponding null vectors of the adjoint, implies the Landau equations (remember that $(\eta, \mathbf{f}) = \eta^T \bar{\mathbf{f}}$, do not forget to take the conjugate of $\mathbf{f}$)

$$-\frac{1}{2}\beta k^2 A + \delta^2 \left(\frac{1}{2}\beta A' - N_{101} \right) = 0,$$

$$\qquad (8.105)$$

$$-\frac{1}{2}\beta k^2 B + \delta^2 \left(\frac{1}{2}\beta B' - N_{011} \right) = 0.$$

Simplification, using

$$k = \frac{\pi}{\sqrt{2}}, \qquad \beta k = \frac{1}{3}, \qquad \gamma_{002} = -\frac{1}{4\pi^2}, \qquad \gamma_{112} = -\frac{100}{573\pi^2}, \qquad (8.106)$$

leads to

$$\frac{dA}{d\tau} = \frac{2}{9\pi^2}A - \frac{\pi^2}{16}A(A^2 + cB^2),$$

$$\qquad (8.107)$$

$$\frac{dB}{d\tau} = \frac{2}{9\pi^2}B - \frac{\pi^2}{16}B(B^2 + cA^2),$$

which generalises (8.59). The value of c is

$$c = \frac{773}{573}. \qquad (8.108)$$

It is straightforward to analyse (8.107) using phase plane analysis. There are two solutions corresponding to rolls

$$(A, B) = (A^*, 0) \text{ or } (0, A^*),$$ (8.109)

where

$$A^* = \frac{4\sqrt{2}}{3\pi^2},$$ (8.110)

and a square cell solution

$$A = B = \frac{A^*}{(1+c)^{1/2}}.$$ (8.111)

The square cell solution is stable if $c < 1$ and unstable if $c > 1$ (see Question 8.9). Hence it is (just) unstable. This is in line with conventional wisdom, which has it that only rolls are stable, and the breakdown to three-dimensional motion occurs at higher Rayleigh number. It is also thought that non-Boussinesq effects (e.g., temperature dependence of fluid properties) can cause square cells or hexagons to be stable at the onset of convection.

8.3.3 Dislocations and Chaos

A different use of nonlinear stability theory is in tracking the evolution of convection patterns over long space scales. In general, we would want to allow the orientation of rolls to vary slowly in space, but here we restrict ourselves to the two-dimensional case where the orientation is fixed, and only the cell size, velocity and temperature vary slowly in space as well as time. To be specific, we consider rolls pointing in the y direction, but we generalise the treatment at the beginning of this section to allow for slow variation in the x direction.

We define the slow space variable

$$X = \varepsilon x,$$ (8.112)

so that x derivatives become

$$\frac{\partial}{\partial x} \rightarrow \frac{\partial}{\partial x} + \varepsilon \frac{\partial}{\partial X},$$ (8.113)

and the Laplacian and biharmonic operators become

$$\nabla^2 \rightarrow \nabla^2 + 2\varepsilon \frac{\partial^2}{\partial X \partial x} + \varepsilon^2 \frac{\partial^2}{\partial X^2},$$

$$\nabla^4 \rightarrow \nabla^4 + 4\varepsilon \nabla^2 \frac{\partial^2}{\partial X \partial x} + \varepsilon^2 \left(2\nabla^2 + 4\frac{\partial^2}{\partial x^2} \right) \frac{\partial^2}{\partial X^2} + \cdots.$$ (8.114)

With the operator $\mathcal{L}$ still defined by (8.41), we have

$$\mathcal{L} \to \mathcal{L}_0 + \varepsilon \mathcal{L}_1 + \varepsilon^2 \mathcal{L}_2 + \cdots, \tag{8.115}$$

where $\mathcal{L}_0$ is just as in (8.41), and

$$\mathcal{L}_1 = \begin{pmatrix} 4\frac{\partial}{\partial x}\nabla^2 & R_c \\ 1 & 2\frac{\partial}{\partial x} \end{pmatrix} \frac{\partial}{\partial X}, \qquad \mathcal{L}_2 = \begin{pmatrix} 2\frac{\partial^2}{\partial z^2} + 6\frac{\partial^2}{\partial x^2} & 0 \\ 0 & 1 \end{pmatrix} \frac{\partial^2}{\partial X^2}. \tag{8.116}$$

Next we expand $\mathbf{w} = (\psi, \theta)^T$ as

$$\mathbf{w} = \varepsilon \mathbf{w}_1 + \varepsilon^2 \mathbf{w}_2 + \cdots, \tag{8.117}$$

and we define $\mathbf{N}_{20}$ and $\mathbf{N}_{30}$ as in (8.42), and we define

$$\mathbf{N}_{21} = \begin{pmatrix} 0 \\ \psi_{1X}\theta_{1z} - \psi_{1z}\theta_{1X} \end{pmatrix}; \tag{8.118}$$

expanding in powers of ε, we derive the sequence of equations

$$\mathcal{L}_0 \mathbf{w}_1 = \mathbf{0},$$
$$\mathcal{L}_0 \mathbf{w}_2 = \mathbf{N}_{20} - \mathcal{L}_1 \mathbf{w}_1, \tag{8.119}$$
$$\mathcal{L}_0 \mathbf{w}_3 = \mathbf{N}_{30} + \mathbf{N}_{21} - \mathcal{L}_1 \mathbf{w}_2 - \mathcal{L}_2 \mathbf{w}_1.$$

The solution for $\mathbf{w}_1$ is given by (8.43), and with $\mathbf{w}_{20}$ defined by (8.51), we have

$$\mathcal{L}_0 \mathbf{w}_{20} = \mathbf{N}_{20}. \tag{8.120}$$

Now we need to find a particular solution to the equation

$$\mathcal{L}_0 \mathbf{w}_{21} = -\mathcal{L}_1 \mathbf{w}_1 = \mathbf{r}_{11} A x e^{ikx} \sin \pi z + (cc), \tag{8.121}$$

where

$$\mathbf{r}_{11} = \begin{pmatrix} \frac{3}{4}\pi^2 k \\ \frac{1}{6}i \end{pmatrix}. \tag{8.122}$$

This is given by

$$\mathbf{w}_{21} = \mathbf{v}_{21} A x e^{ikx} \sin \pi z + (cc), \tag{8.123}$$

where

$$L_{11} \mathbf{v}_{21} = \mathbf{r}_{11}, \tag{8.124}$$

L_{11} being defined through (8.49). This is only possible if $\mathbf{r}_{11} \perp \boldsymbol{\eta}$ given by (8.57), and, on doing the calculation, we find that it is(!). The solution is then given up to addition of a multiple of $\mathbf{w}_1$ which we can ignore; the particular solution we use is

$$\mathbf{w}_{21} = \begin{pmatrix} 0 \\ -\dfrac{i}{9\pi^2} \end{pmatrix} A_X e^{ikx} \sin \pi z + (cc). \tag{8.125}$$

We can thus define $\mathbf{w}_2$ as

$$\mathbf{w}_2 = \mathbf{w}_{20} + \mathbf{w}_{21}. \tag{8.126}$$

It remains to find the secular-producing terms proportional to $e^{ikx} \sin \pi z$ (and its conjugate) on the right hand side of $(8.119)_3$. We easily find that there are no such terms in $\mathcal{L}_1 \mathbf{w}_{20}$, and that $\mathbf{N}_{21} = \mathbf{0}$. As before (cf. (8.53)),

$$\mathbf{N}_{30} = \begin{pmatrix} -\frac{1}{6} i A \\ \frac{1}{2}\beta A_\tau + \frac{1}{48} k A^3 \end{pmatrix} e^{ikx} \sin \pi z + \cdots, \tag{8.127}$$

while

$$\mathcal{L}_1 \mathbf{w}_{21} = \begin{pmatrix} -\frac{3}{4} i \pi^2 \\ \frac{2k}{9\pi^2} \end{pmatrix} A_X e^{ikx} \sin \pi z + \cdots, \tag{8.128}$$

and finally

$$\mathcal{L}_2 \mathbf{w}_1 = \begin{pmatrix} \frac{5}{2} i \pi^2 \\ \frac{1}{2}\beta \end{pmatrix} A_{XX} e^{ikx} \sin \pi z + \cdots. \tag{8.129}$$

The coefficient of $e^{ikx} \sin \pi z$ on the right hand side of (8.119) must be orthogonal to $\boldsymbol{\eta}$ given by (8.57). Computing the inner product, we find that this requires that A satisfy the Ginzburg–Landau equation

$$\frac{\partial A}{\partial \tau} = \frac{2}{9\pi^2} A - \frac{\pi^2}{16} A^3 + 4 \frac{\partial^2 A}{\partial X^2}. \tag{8.130}$$

In deriving (8.130), we made the tacit assumption that A was real. More generally, one derives the same equation, but in the general form

$$\frac{\partial A}{\partial \tau} = k_1 A + k_2 |A|^2 A + D \frac{\partial^2 A}{\partial X^2}, \tag{8.131}$$

and in other systems, the coefficients may be complex.

8.4 High Rayleigh Number Convection

We have seen that convection occurs if the Rayleigh number is larger than $O(10^3)$ in general, depending on the precise boundary conditions which apply. In the Earth's mantle, suitable values of the constituent parameters are $\alpha = 3 \times 10^{-5}\,\mathrm{K}^{-1}$, $T_b = 3500\,\mathrm{K}$, $\rho_0 = 3 \times 10^3\,\mathrm{kg\,m}^{-3}$, $g = 10\,\mathrm{m\,s}^{-2}$, $d = 3000\,\mathrm{km}$, $\eta_0 = 10^{21}\,\mathrm{Pa\,s}$, $\kappa_0 =$

Fig. 8.2 Schematic
representation of boundary
layer convection

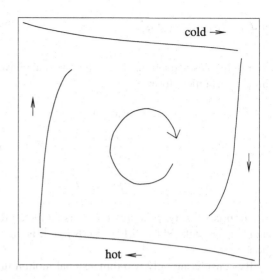

$10^{-6}\ \mathrm{m}^2\,\mathrm{s}^{-1}$, and for these values, the Rayleigh number is about 0.85×10^8. Thus
the Rayleigh number is much larger than the critical value, and as a consequence we
can expect the convection to be vigorous (if velocities of centimetres per year can
be said to be vigorous).

There are various intuitive ways in which we can get a sense of the likely behaviour of the convective solutions of the Boussinesq equations when $R \gg 1$. Since
R multiplies the buoyancy term, any $O(1)$ lateral temperature gradient will cause
enormous velocities (this is best seen in (8.26) rather than (8.12)). One might thus
expect the flow to organise itself so that either horizontal temperature gradients are
small, or they are confined to thin regions, or both. Since $O(1)$ temperature variations are enforced by the boundary conditions, the latter is more plausible, and thus
we have the idea of the *thermal plume*, a localised upwelling of hot fluid which will
be instantly familiar to glider pilots and seabirds.

A mathematically intuitive way of inferring the same behaviour follows from
the expectation that increasing R drives increasing velocities; then large R should
imply large velocity, and the conduction term in the heat equation $\mathbf{u}.\nabla T = \nabla^2 T$ is
correspondingly small. Since the conduction term represents the highest derivative
in the equation, its neglect would imply a reduction of order, and correspondingly we
would expect *thermal boundary layers* to exist at the boundaries of the convecting
cell. This is in fact what we will find: a hot thermal boundary layer adjoins the
lower boundary, and a cold one adjoins the upper boundary, and a rapid circulation
in the interior of the cell detaches these as upwelling and downwelling plumes. The
general structure of the resulting flow is shown in Fig. 8.2. We analyse this structure
in the following sections.

8.4.1 Boundary Layer Theory

The Boussinesq equations describing thermal convection are written in the following dimensionless form:

$$\nabla . \mathbf{u} = 0,$$

$$\frac{1}{\sigma}\frac{d\mathbf{u}}{dt} = -\nabla p + \nabla^2 \mathbf{u} + RT\mathbf{k}, \qquad (8.132)$$

$$\frac{dT}{dt} = \nabla^2 T,$$

where $\mathbf{u}$ is velocity, p is pressure, T is temperature, and the Rayleigh and Prandtl numbers are defined by (8.12) and (8.13); $\mathbf{k}$ is the unit vector in the vertical direction.

By considering only two-dimensional motion in the (x, z) plane, we define the stream function ψ by

$$u = -\psi_z, \qquad v = \psi_x; \qquad (8.133)$$

the vorticity is then $(0, \omega, 0)$, where $\omega = -\nabla^2 \psi$. Taking the curl of the momentum equation, we derive the set

$$\omega = -\nabla^2 \psi,$$

$$\frac{dT}{dt} = T_t + \psi_x T_z - \psi_z T_x = \nabla^2 T, \qquad (8.134)$$

$$\frac{1}{\sigma}\frac{d\omega}{dt} = -RT_x + \nabla^2 \omega,$$

which are supplemented by the boundary conditions

$$\psi, \omega = 0 \quad \text{on } x = 0, a, \quad z = 0, 1,$$

$$T = \frac{1}{2} \quad \text{on } z = 0, \qquad T = -\frac{1}{2} \quad \text{on } z = 1, \qquad (8.135)$$

$$T_x = 0 \quad \text{on } x = 0, a;$$

here a is the aspect ratio, and we have chosen free slip (no stress) conditions at the cell boundaries.

Rescaling   The idea is that when $R \gg 1$, thermal boundary layers of thickness $\delta \ll 1$ will form at the edges of the flow, and both ψ and ω will be $\gg 1$ in the flow. To scale the equations properly, we rescale the variables as

$$\psi, \omega \sim \frac{1}{\delta^2}, \qquad (8.136)$$

and define

$$\delta = R^{-1/3}. \tag{8.137}$$

Rescaled, the equations are thus, in the steady state,

$$\omega = -\nabla^2 \psi,$$

$$\psi_x T_z - \psi_z T_x = \delta^2 \nabla^2 T, \tag{8.138}$$

$$\nabla^2 \omega = \frac{1}{\delta} T_x + \frac{1}{\sigma \delta^2} \frac{d\omega}{dt}.$$

In order that the inertia terms be unimportant, we require $\sigma \delta^2 \gg 1$, i.e. $\sigma \gg R^{2/3}$. This assumption is easily vindicated in the Earth's mantle, but is difficult to achieve in the laboratory. We also see that if internal heating is included, then we should add a term $\delta^2 H$ to the right hand side of (8.138). For typical Earth parameters, this is of order 10^{-4}, and in practice negligible.

As in any singular perturbation procedure, we now examine the flow region by region, introducing special rescalings in regions where boundary conditions cannot be satisfied.

Core Flow   The temperature equation is linear in T, and implies $T = T_0(\psi) + O(\delta^2)$. For a flow with closed streamlines, the Prandtl–Batchelor theorem then implies $T_0 = $ constant (this follows from the exact integral $\oint_C \frac{\partial T}{\partial n} ds = 0$, where the integral is around a streamline, whence $T_0'(\psi) \oint_C \frac{\partial \psi}{\partial n} ds = 0$); it then follows that T is constant to all (algebraic) orders of δ, and is in fact zero by the symmetry of the flow. Thus

$$T = 0,$$

$$\nabla^4 \psi = 0, \tag{8.139}$$

and clearly the core flow cannot have $\psi = \omega = 0$ at the boundaries, for non-zero ψ. In fact, ω jumps at the side-walls where the plume buoyancy generates a non-zero vorticity. We examine the plumes next.

Plumes   Near $x = 0$, for example, we rescale the variables as

$$x \sim \delta, \qquad \psi \sim \delta, \tag{8.140}$$

and denote rescaled variables by capital letters. At leading order, we then have

$$\Psi_{XX} \approx 0, \tag{8.141}$$

whence $\Psi \sim v_p(z)X$, and to match to the core flow, we define $v_p = \psi_x|_{x=0}$ as the core velocity at $x = 0$. Also

$$\Psi_X T_z - \Psi_z T_X \approx T_{XX},$$

$$\omega_{XX} \approx T_X. \tag{8.142}$$

the latter of which integrates to give

$$\omega = \int_0^X T \, dX, \qquad \omega_p = \int_0^\infty T \, dX, \qquad (8.143)$$

where matching requires ω_p to be the core vorticity at $x = 0$. Integration of $(8.142)_1$ gives

$$\int_0^\infty T \, d\Psi = C, \qquad (8.144)$$

where C is constant, and it follows that the core flow must satisfy the boundary condition $\omega \psi_x = C$ on $x = 0$. In summary, the effective boundary conditions for the core flow are

$$\psi = 0 \quad \text{on } x = 0, a, \text{ and } z = 0, 1,$$

$$\psi_{zz} = 0 \quad \text{on } z = 0, 1, \qquad (8.145)$$

$$\psi_x \psi_{xx} = -C \quad \text{on } x = 0, \qquad \psi_x \psi_{xx} = C \quad \text{on } x = a,$$

and the solution can be found as $\psi = C^{1/2} \hat{\psi}$, where $\hat{\psi}$ is determined numerically. It thus remains to determine C. This requires consideration of the thermal boundary layers.

Thermal Boundary Layers   Near the base, for example, we rescale the variables

$$z \sim \delta, \qquad \psi \sim \delta, \qquad \omega \sim \delta, \qquad (8.146)$$

to find the leading order rescaled equations as

$$\Psi_{ZZ} \approx 0, \qquad (8.147)$$

whence $\Psi \sim u_b(x)Z$, and $-u_b$ is the core value of the basal velocity. Then $\Omega_{ZZ} \sim T_x$ determines Ω (with $\Omega = 0$ on $Z = 0$, and $\Omega \sim \omega_b(x)Z$ as $Z \to \infty$, where ω_b is the core value of the basal vorticity), and T satisfies

$$\Psi_x T_Z - \Psi_Z T_x \approx T_{ZZ}. \qquad (8.148)$$

In Von Mises coordinates x, Ψ, the equation is

$$-T_x \approx \frac{\partial}{\partial \Psi} \left[\Psi_Z \frac{\partial T}{\partial \Psi} \right], \qquad (8.149)$$

and putting $\xi = \int_x^a u_b(x) \, dx$ (so ξ marches from right to left in the direction of flow), this is just the diffusion equation

$$T_\xi = T_{\Psi\Psi}, \qquad (8.150)$$

with

$$T = \frac{1}{2} \quad \text{on } \Psi = 0, \qquad T \to 0 \quad \text{as } \Psi \to \infty. \tag{8.151}$$

A quantity of interest is the Nusselt number, defined as

$$Nu = -\int_0^1 \frac{\partial T}{\partial z}(x, 0)\, dx, \tag{8.152}$$

and from the above, this can be written as

$$Nu \approx \left[\int_0^\infty T|_{z=0}\, d\Psi\right]_{x=a}^{x=0} R^{1/3}. \tag{8.153}$$

Notice that the plume temperature equation can also be written as (8.150), where ξ is extended as $\int_0^z v_p(z)\, dz$, etc.

Corner Flow   The core flow has a singularity in each corner, where (if r is distance from the corner), then $\psi \sim r^{3/2}$, $\omega \sim r^{-1/2}$, and (for the corner at $x = 0$, $z = 0$, for example) $x, z \sim r$. There must be a region where this singularity is alleviated by the incorporation of the buoyancy term. This requires $\omega/r^2 \sim 1/\delta r$, whence $r \sim \delta^{2/3}$. Rescaling the variables as indicated ($x, z \sim \delta^{2/3}$, $\psi \sim \delta$, $\omega \sim \delta^{-1/3}$) then gives the temperature equation as

$$\Psi_X T_Z - \Psi_Z T_X \approx \delta \nabla^2 T, \tag{8.154}$$

which shows that (since the ψ scale, δ, is the same as that of the boundary layers adjoining the corner) the boundary layer temperature field is carried through the corner region. The corner flow has $T \sim T(\Psi)$, so that

$$\nabla^4 \Psi + T'(\Psi)\Psi_X = 0, \tag{8.155}$$

with appropriate matching conditions. Jimenez and Zufiria (1987) claim that the equivalent problem for the case of no-slip boundary conditions has no solution, but do not adduce details. Their inference is that the boundary layer approximation fails: this seems a hazardous conclusion.

Solution Strategy   The temperature equation (8.150) must now be solved in the four regions corresponding to the boundary layer at $z = 0$, plume at $x = 0$, boundary layer at $z = 1$, plume at $x = a$, with T being continuous at each corner, and

$$T \to 0 \quad \text{as } \Psi \to \infty,$$

$$T = \frac{1}{2} \quad \text{on } \Psi = 0 \ [z = 0, \text{base}],$$

$$\frac{\partial T}{\partial \Psi} = 0 \quad \text{on } \Psi = 0 \ [x = 0, \text{left}], \tag{8.156}$$

$$T = -\frac{1}{2} \quad \text{on } \Psi = 0 \ [z = 1, \text{top}],$$

$$\frac{\partial T}{\partial \Psi} = 0 \quad \text{on } \Psi = 0 \ [x = a, \text{right}];$$

in addition, T is periodic in ξ. Beginning from $x = a$, $z = 0$, denote the values of ξ at the corners as ξ_A $(x = 0, z = 0)$, ξ_B $(x = 0, z = 1)$, ξ_C $(x = a, z = 1)$. From the definition of ξ, we have $\xi_k = C^{1/2} \hat{\xi}_k$, where $\hat{\xi}_k$ are independent of C. Putting

$$\xi = C^{1/2} \hat{\xi}, \qquad \Psi = C^{1/4} \hat{\Psi}, \tag{8.157}$$

then the problem for $T(\hat{\xi}, \hat{\Psi})$ is independent of C. If we can solve this numerically, then $\int T \, d\Psi = C^{1/4} \int T \, d\hat{\Psi}$, thus

$$Nu \approx C^{1/4} \left[\int_0^\infty T \, d\hat{\Psi} \right]_0^{\hat{\xi}_A} R^{1/3}, \tag{8.158}$$

and lastly, C is determined from

$$C = \left[\int_0^\infty T \, d\hat{\Psi} \right]^{4/3}, \tag{8.159}$$

where the integral is evaluated at $\hat{\xi}_A$. Since also $-C^{3/4} = \int_0^\infty T \, d\hat{\Psi}$ at $\xi = 0$, (8.158) can be written as

$$Nu \approx 2C R^{1/3}. \tag{8.160}$$

The necessary numerical results to compute C are given by Roberts (1979) and Jimenez and Zufiria (1987). The results are slightly different, with the latter paper considering Roberts's numerical results to be wrong. For $a = O(1)$, we have $2C \approx 0.2$. Figure 8.3 shows the typical isotherm profile for high Rayleigh number constant viscosity convection.

No-slip Boundary Conditions   For no-slip boundary conditions, the necessary preliminary rescaling is $\psi \sim 1/\delta^3$, $\omega \sim 1/\delta^3$, where $\delta = Ra^{-1/5}$. Thus the Nusselt number $Nu \sim R^{1/5}$. There is no longer parity between the thermal boundary layers and plumes, as the former are slowed down by the no-slip conditions. The rescaled

Fig. 8.3 Temperature
isotherms of a calculation of
constant viscosity convection
at Rayleigh number
0.9×10^6. The thermal
plumes and boundary layers
are clearly indicated. Figure
courtesy Mike Vynnycky

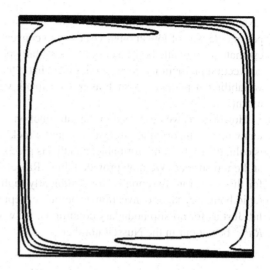

equations are

$$\omega = -\nabla^2\psi,$$

$$\psi_x T_z - \psi_z T_x = \delta^3 \nabla^2 T, \qquad (8.161)$$

$$\nabla^2\omega = \frac{1}{\delta^2}T_x.$$

The core flow is as before; the thermal boundary layers have $\psi \sim \delta^2$, $\omega \sim 1$, $z \sim \delta$, so that vorticity balances buoyancy (an omission in Roberts's 1979 paper, thus precluding his similarity solution), and all three equations are necessary to solve for T; it is still the case that $\int T\, d\psi$ is conserved at corners, but now in the plume $x \sim \delta^{3/2}$, $\psi \sim \delta^{3/2}$, and $T \sim \delta^{1/2}$. The initial plume profile is effectively a delta function, and the plume temperature is just the resultant similarity solution. The remainder of the structure must be computed numerically, something which has not been done.

8.5 Variable Viscosity

If we try and apply the above theory to the convection of the Earth's mantle, we would predict a surface velocity of order

$$u \approx C^{1/2}Ra^{2/3}\frac{\kappa_0}{d}, \qquad (8.162)$$

and with $\kappa_0 = 10^{-6}$ m^2 s^{-1}, $d = 3000$ km, $C^{1/2} = 0.3$, $Ra = 7 \times 10^7$, we find $u \approx 53$ cm y^{-1}. This is in remarkably good agreement with observed plate velocities 1–10 cm y^{-1}. In fact, when the theory was first proposed in 1967 by Turcotte and

Oxburgh, they used a depth d of 700 km, since it was then thought that only the upper mantle was of low enough viscosity to convect. Since $u \propto d$, the corresponding estimate for u would be 12.5 cm y^{-1}: essentially perfect! But even the whole-mantle convection prediction is very good, given that effects of cell size, sphericity, and the variability of parameters such as α with depth will all modify the result to some extent.

However, as we described in the introduction to the chapter, there is a real problem with this theory: the viscosity of mantle rock is highly variable, so that where it is cold, the rock is undeformable; and this is precisely in the thermal boundary layer at the top surface. We now provide a high Rayleigh number boundary layer theory for this situation. Essentially, for sufficiently high Rayleigh number, we will have rapid boundary layer convection as before, except that this occurs *below* a stagnant lid. Just as for no-slip boundary conditions, we will find that the rigid lid causes an $Ra^{1/5}$ behaviour in the Nusselt number.

8.5.1 Rheology of Polycrystalline Rocks

Many experiments on crystalline rocks lead to an expression for the viscosity of the form

$$\eta = \frac{1}{2A\tau^{n-1}} \exp\left[\frac{E^* + pV^*}{RT}\right], \tag{8.163}$$

where T is (absolute) temperature, p is pressure, τ is the second invariant of the deviatoric stress tensor ($2\tau^2 = \tau_{ij}\tau_{ij}$), and the constants are a rate factor A, the gas law constant $R = 8.3$ J mol^{-1} K^{-1}, the activation energy E^*, and the activation volume V^*. Typical values of these constants are

$$A = 10^5 \text{ MPa}^{-n}\text{s}^{-1}, \quad n = 3.5,$$
$$E^* = 533 \text{ kJ mol}^{-1}, \quad V^* = 1.7 \times 10^{-5} \text{ m}^3 \text{mol}^{-1}. \tag{8.164}$$

In writing the equations in dimensionless form, we now have to choose representative values of the absolute temperature, in order to have a meaningful viscosity scale. Because the viscosity is so variable, it is not obvious how to do this. It turns out that the right temperature to choose is the 'rate-controlling' value in the asthenosphere, which is the region just below the (cold, rigid) lithosphere. It is rate-controlling in the sense that the viscosity is minimal there, so that the velocity scale is controlled by the asthenospheric viscosity.

However, we do not know the value of the asthenospheric temperature (although we know what a reasonable value may be, i.e., 1500 K); and even if we did, we do not know the viscosity as we do not know the appropriate scales for τ. So non-dimensionalising the equations has to be done 'blindly', as it were, with the proper choice of scales being determined after the fact.

8.5.2 Governing Equations

We consider two-dimensional convection in a Cartesian box. A Boussinesq form of the governing equations is

$$\frac{\partial u}{\partial x} + \frac{\partial w}{\partial z} = 0,$$

$$\frac{\partial p}{\partial x} = \frac{\partial \tau_1}{\partial x} + \frac{\partial \tau_3}{\partial z},$$

$$\frac{\partial p}{\partial z} = \frac{\partial \tau_3}{\partial x} - \frac{\partial \tau_1}{\partial z} - \rho g,$$

$$\tau_1 = 2\eta \frac{\partial u}{\partial x},$$

$$\tau_3 = \eta \left(\frac{\partial u}{\partial z} + \frac{\partial w}{\partial x} \right),$$

$$\frac{dT}{dt} = \kappa \nabla^2 T. \tag{8.165}$$

Here, τ_1 ($= \tau_{11}$) and τ_3 ($= \tau_{13}$) are the longitudinal and shear components of the deviatoric stress tensor. In addition, the viscosity is defined by (8.163), the second stress invariant is

$$\tau^2 = \tau_1^2 + \tau_3^2, \tag{8.166}$$

and we suppose the density is

$$\rho = \rho_a \left[1 - \alpha (T - T_a) \right]. \tag{8.167}$$

We have ignored inertia terms, and also have put isothermal compressibility and internal heating to zero.

8.5.3 Boundary Conditions

At the base, $z = 0$, we prescribe

$$T = T_b, \qquad w = 0, \qquad \tau_3 = 0; \tag{8.168}$$

at the top surface $z = d$,

$$T = T_s, \qquad w = 0, \qquad \tau_3 = 0; \tag{8.169}$$

and at the sides $x = 0$ and $x = ad$ (say):

$$\frac{\partial T}{\partial x} = 0, \qquad u = 0, \qquad \tau_3 = 0. \tag{8.170}$$

In addition, the normal stress should be continuous. In practice this is used to prescribe the uplift. For convection under a free surface at $z = d\Delta$, say, we prescribe

$$p + \tau_1 = p_s, \tag{8.171}$$

where p_s is the surface loading (zero if atmospheric pressure; or hydrostatic pressure if the mantle is sub-oceanic), and this extra condition will give Δ if p_s is prescribed.

8.5.4 Boundary Layer Analysis

We begin by non-dimensionalising the terms as follows:

$$p - \rho_a g(d - z), \tau_1, \tau_3, \tau \sim \frac{\eta_0 \kappa}{d^2} \equiv \tau_0, \qquad T \sim T_a,$$

$$\eta \sim \eta_0, \qquad t \sim \frac{d^2}{\kappa}, \qquad (x, z) \sim d, \qquad (u, w) \sim \frac{\kappa}{d}. \tag{8.172}$$

At this point we do not know either T_a or η_0: they must be determined later.

We introduce a stream function ψ satisfying

$$u = -\frac{\partial \psi}{\partial z}, \qquad w = \frac{\partial \psi}{\partial x}; \tag{8.173}$$

then the resulting dimensionless equations are these:

$$\frac{\partial p}{\partial x} = \frac{\partial \tau_1}{\partial x} + \frac{\partial \tau_3}{\partial z},$$

$$\frac{\partial p}{\partial z} = \frac{\partial \tau_3}{\partial x} - \frac{\partial \tau_1}{\partial z} - Ra(1 - T),$$

$$\tau_1 = -2\eta \frac{\partial^2 \psi}{\partial x \partial z},$$

$$\tau_3 = \eta \left(\frac{\partial^2 \psi}{\partial x^2} - \frac{\partial^2 \psi}{\partial z^2} \right), \tag{8.174}$$

$$\frac{\partial \psi}{\partial x} \frac{\partial T}{\partial z} - \frac{\partial \psi}{\partial z} \frac{\partial T}{\partial x} = \nabla^2 T,$$

$$\tau^2 = \tau_1^2 + \tau_3^2,$$

$$\eta = \frac{\Lambda}{\tau^{n-1}} \exp \left[\frac{1 - T + \mu\{1 - z + Bp/Ra\}}{\varepsilon T} \right],$$

where the parameters are given by

$$Ra = \frac{\alpha \rho_a g T_a d^3}{\eta_0 \kappa}, \qquad \varepsilon = \frac{R T_a}{E^*},$$

$$\mu = \frac{\rho_a g d V^*}{E^*}, \qquad B = \alpha T_a, \tag{8.175}$$

$$\Lambda = \frac{1}{2\eta_0 A \tau_0^{n-1}} \exp \left(\frac{E^*}{R T_a} \right).$$

We expect that η_0 will be roughly the asthenospheric viscosity, and we proceed on the basis that $Ra \gg 1$, and also that $\varepsilon \ll 1$, since if we take $E^* = 533$ kJ mol^{-1}, $R = 8.3$ J mol^{-1}, $T_a = 1500$ K, then $\varepsilon \approx 0.023$. The Boussinesq number $B \approx 0.05$, so we may neglect the term Bp/Ra in the viscosity. The other parameter μ takes the approximate value 2.8 if $V^* = 1.7 \times 10^{-5}$ m^3 mol^{-1}, $\rho = 3 \times 10^3$ kg m^{-3}, $g = 10$ m s^{-2}, $d = 3 \times 10^6$ m, and is clearly important; however, we will first study the simpler problem in which $\mu = 0$, only adding some comments in the notes (Sect. 8.8) about the possible structure if $\mu = O(1)$. Thus we take the viscosity to be

$$\eta = \frac{\Lambda}{\tau^{n-1}} \exp\left[\frac{1-T}{\varepsilon T}\right].$$ (8.176)

The structure we anticipate is this. There is a cold, rigid lid of thickness $vs(x)$ (say) adjoining the top surface, in which $T < 1$ and η is exponentially large. Hence $\psi \approx 0$ there and $\nabla^2 T = 0$. It is formally convenient to suppose $v \ll 1$ (with $s = O(1)$), so that the temperature profile is approximately linear with depth, and we shall make this assumption.

Below the lid is a well-stirred, rapidly convecting region, in which $T \approx$ constant (and thus $T \approx 1$) and there are thermal boundary layers at the base and beneath the lid, and plumes at the side. In these layers, $T = 1 + O(\varepsilon)$ (otherwise η would be exponentially small, and ψ exponentially large), and in particular this tells us (with $\mu = 0$) that $T_a \approx T_b$.

Suppose that the thermal boundary layer (sometimes called the delamination layer) which joins the rigid lid to the rapidly convecting core is of dimensionless thickness $\delta \ll 1$. We write

$$T = 1 + \varepsilon\theta$$ (8.177)

in this region; then continuity of heat flux into the lid implies $T_z \sim 1/v \sim \varepsilon/\delta$, so we choose

$$v = \delta/\varepsilon.$$ (8.178)

Next suppose that the plumes are of thickness δ_p (this will also be the thickness of the basal boundary layer). As for the isoviscous situation where the top condition is no slip, we anticipate that $\delta_p \ll \delta$. Since the flow below the lid has $T = 1 + O(\varepsilon)$ everywhere, the same scales as for the isoviscous case should apply, and this implies that in the core $p, \tau_1, \tau_3, \psi \sim 1/\delta_p^2$, and therefore we choose

$$\Lambda = \frac{1}{\delta_p^{2(n-1)}},$$ (8.179)

and a balance of shear stress with buoyancy in the plume implies

$$\frac{1}{\delta_p^3} = Ra\varepsilon\theta_p,$$ (8.180)

if $\theta \sim \theta_p$ there (note also that $\psi \sim 1/\delta_p$ in the plume).

In the delamination layer, a balance of advection with conduction implies $\psi \sim 1/\delta$, and because buoyancy ($\int \theta \, d\psi$) is advected round the corner, we have

$$\frac{\theta_p}{\delta_p} = \frac{1}{\delta}. \tag{8.181}$$

Finally, we must balance shear stress with buoyancy in the delamination layer (by analogy with the boundary layer beneath a rigid lid). Thus $p_x \sim \tau_{3z}$, $p_z \sim Ra\varepsilon\theta$, and since $\partial/\partial x \sim 1/\varepsilon$ (since $v = \delta/\varepsilon$) and $\partial/\partial z \sim 1/\delta$, we have in the delamination layer

$$\tau_3 \sim \delta^2 Ra, \qquad p \sim \varepsilon\delta Ra, \tag{8.182}$$

and together with $\psi \sim 1/\delta$, $\partial/\partial z \sim 1/\delta^2$, the definition $\tau_3 \approx \eta\psi_{zz}$ ($\gg \tau_1$) implies $\tau_3 \sim \tau \sim \eta/\delta^3 \sim \Lambda/\tau^{n-1}\delta^3$, hence $\tau_3 \sim \tau \sim (\Lambda/\delta^3)^{1/n}$, and combining this with (8.182) and (8.179), we find

$$\delta^2 Ra = \left(\frac{1}{\delta^3 \delta_p^{2(n-1)}}\right)^{1/n}. \tag{8.183}$$

From (8.178), (8.179), (8.180), (8.181) and (8.183), we finally find

$$\delta = \left(\frac{\varepsilon^{\frac{n-1}{n+1}}}{Ra}\right)^{1/5}, \qquad \Lambda = Ra^{\frac{3(n-1)}{5}} \varepsilon^{\frac{(2n+3)(n-1)}{5(n+1)}}. \tag{8.184}$$

The second of these defines η_0, and ensures that the sub-lithospheric viscosity is $O(\eta_0)$. From (8.175), we have

$$\Lambda = \frac{(d^2/\kappa)^{n-1}}{2A} \cdot \frac{1}{\eta_0^n} \exp(1/\varepsilon). \tag{8.185}$$

Combining this with (8.184), we find

$$\eta_0 = \left[\left\{\frac{(d^2/\kappa)^{n-1}}{2A} \exp\left(\frac{E^*}{RT_a}\right)\right\}^5 \left(\frac{E^*}{RT_a}\right)^{\frac{(2n+3)(n-1)}{(n+1)}} \left(\frac{\kappa}{\alpha\rho_a g T_a d^3}\right)^{3(n-1)}\right]^{\frac{1}{2n+3}}. \tag{8.186}$$

As discussed above, we have $T_a \approx T_b$, and more precisely, we can define

$$T_a = \frac{T_b}{\left(1 + \frac{\varepsilon\delta_p}{\delta}\phi_b\right)}, \tag{8.187}$$

where the unknown $O(1)$ constant ϕ_b is chosen in an analysis of the basal boundary layer (cf. (8.200) below). In effect we can take T_a as known. Taking all the values for the constants which we have used earlier, we find that the expected value of η_0 for the Earth is 1.4×10^{20} Pa s. Extraordinarily, this is exactly the sort of value which is thought to be appropriate in the Earth's asthenosphere. The theory smells right.

Delamination Layer   We would now progress through the separate regions: core, basal boundary layer, plumes, etc. These are much the same as for the constant viscosity case, but the delamination layer (and the slab above) is different, and we begin with it.

In the delamination layer, we rescale the variables as follows, based on the discussion above:

$$z = 1 - vs - \delta\zeta, \qquad T = 1 + \varepsilon\theta, \qquad \tau = \delta^2 Ra T^*, \qquad \eta = \varepsilon^{\frac{n-1}{n+1}} N,$$

$$\psi = \frac{\Psi}{\delta}, \qquad p = \varepsilon\delta Ra P, \qquad \tau_3 = \delta^2 Ra T_3, \qquad \tau_1 = \frac{\delta^3 Ra}{\varepsilon} T_1; \qquad (8.188)$$

thus

$$\frac{\partial}{\partial x} \to \frac{\partial}{\partial x} - \frac{s'}{\varepsilon}\frac{\partial}{\partial \zeta}, \qquad \frac{\partial}{\partial z} \to -\frac{1}{\delta}\frac{\partial}{\partial \zeta}, \qquad (8.189)$$

and at leading order we find (assuming $\delta \ll \varepsilon$)

$$-s'\frac{\partial P}{\partial \zeta} = -\frac{\partial T_3}{\partial \zeta},$$

$$-\frac{\partial P}{\partial \zeta} = \theta,$$

$$T_1 = -2N\frac{\partial^2 \Psi}{\partial \zeta^2},$$

$$T_3 = -N\frac{\partial^2 \Psi}{\partial \zeta^2}, \qquad (8.190)$$

$$\frac{\partial \Psi}{\partial \zeta}\frac{\partial \theta}{\partial x} - \frac{\partial \Psi}{\partial x}\frac{\partial \theta}{\partial \zeta} = \frac{\partial^2 \theta}{\partial \zeta^2},$$

$$T^* = |T_3|,$$

$$N = \frac{1}{T^{*n-1}}e^{-\theta}.$$

We write $T_3 = -S$; then (8.190) can be shrunk to the three coupled equations

$$\frac{\partial S}{\partial \zeta} = s'\theta,$$

$$\frac{\partial^2 \Psi}{\partial \zeta^2} = |S|^{n-1}Se^{\theta}, \qquad (8.191)$$

$$\frac{\partial \Psi}{\partial \zeta}\frac{\partial \theta}{\partial x} - \frac{\partial \Psi}{\partial x}\frac{\partial \theta}{\partial \zeta} = \frac{\partial^2 \theta}{\partial \zeta^2}.$$

We anticipate suitable boundary conditions for these equations as follows. As $\zeta \to -\infty$, the viscosity increases exponentially as we enter the lid, and we anticipate negligible flow and a conductive temperature gradient. Thus

$$\Psi \to 0, \qquad \theta_\zeta \to \Gamma \qquad (8.192)$$

as $\zeta \to -\infty$. Below the lid lies the isothermal core, and hence we can suppose

$$S \to S_\infty, \qquad \theta \to 0 \quad \text{as } \zeta \to +\infty, \tag{8.193}$$

where S_∞ must be chosen. Because $p, \tau_1, \tau_3 \sim 1/\delta_p^2$ in the core, and because (from (8.180) and (8.181))

$$\frac{1}{\delta_p^4} = \frac{Ra\,\varepsilon}{\delta}, \tag{8.194}$$

we find, using the definition of δ in (8.184), that the delamination layer scaled variables P and T_3 take values

$$P \sim \nu\varepsilon^{\frac{1}{n+1}}, \qquad T_3 \sim \varepsilon^{\frac{1}{n+1}} \tag{8.195}$$

in the core. Thus T_3 must reach $O(\varepsilon^{\frac{1}{n+1}}) \ll 1$ in the core, and an obvious choice for S_∞ is thus $S_\infty = 0$. Despite its appeal, we shall find that this is incorrect, because of a further buoyant layer below the delamination layer which arises from the upwelling plume. Before considering the plume structure, we solve for the temperature in the lid.

Stagnant Lid Temperature   Anticipating that $\psi \ll 1$ (fuller discussion follows later), then $\nabla^2 T = 0$ in the lid. We rescale

$$z = 1 - \nu Z, \tag{8.196}$$

so that

$$T_{ZZ} + \nu^2 T_{xx} = 0, \tag{8.197}$$

and at leading order

$$T = T_0 + (1 - T_0)\frac{Z}{s}, \tag{8.198}$$

where $T_0 = T_s/T_a$ is the non-dimensional surface temperature. It follows from the definition of (8.192) that

$$\Gamma = \left|\frac{\partial T}{\partial Z}\right|_{Z=s} = \frac{1 - T_0}{s}. \tag{8.199}$$

Sub-lithospheric Flow   In the plumes and the basal thermal boundary layer, $\theta \sim \delta_p/\delta \ll 1$, so that viscosity is approximately constant, and the flow is directly comparable to the isoviscous case. However, the upper boundary for the core flow is now one of no slip, so that the flow is not symmetric, and the buoyancy in each plume may be different.

Put

$$\theta = \frac{\delta_p}{\delta}\phi, \tag{8.200}$$

Fig. 8.4 The style of
stagnant lid convection for a
strongly variable viscosity
fluid

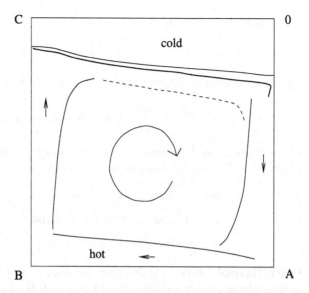

and write

$$\psi = \frac{1}{\delta_p} \Psi \tag{8.201}$$

in plumes and basal boundary layer. Suppose that

$$\int_0^\infty \phi \, d\Psi = C_l, \ -C_r, \tag{8.202}$$

give the values of plume buoyancy in the left and right plumes ($C_l, C_r > 0$).

Note that if we choose T_a such that $\theta \equiv 0$ in the core, then the value of ϕ at the base is undetermined, let us say $\phi = \phi_b$ at $z = 0$. (This just reflects the fact that the internal core temperature T_a is not exactly known.) Thus we have to determine ϕ_b, C_l and C_r.

The core flow problem for $\hat{\psi} = \delta_p^2 \psi$ is the following:

$$\nabla^4 \hat{\psi} = 0, \tag{8.203}$$

with

$$
\begin{aligned}
\hat{\psi} = 0, && \hat{\psi}_x \hat{\psi}_{xx} = C_l && \text{on } x = 0, \\
\hat{\psi} = 0, && \hat{\psi}_x \hat{\psi}_{xx} = -C_r && \text{on } x = a, \\
\hat{\psi} = 0, && \hat{\psi}_{zz} = 0 && \text{on } z = 0, \\
\hat{\psi} = 0, && \hat{\psi}_z = 0 && \text{on } z = 1.
\end{aligned}
\tag{8.204}
$$

Given the solution to this, then the anti-clockwise boundary velocity on the edges OABC (see Fig. 8.4) is $u = \partial \hat{\psi} / \partial n$, where n is the inward normal, and if we define

τ (no longer the stress!) by

$$\tau = \int_0^s u \, ds, \tag{8.205}$$

where s measures distance anti-clockwise along OABC from O, then ϕ satisfies

$$\frac{\partial \phi}{\partial \tau} = \frac{\partial^2 \phi}{\partial \Psi^2} \quad \text{in } 0 < \Psi < \infty, 0 < \tau < \tau_C, \tag{8.206}$$

where $\Psi = \delta_p \psi$ is the boundary layer coordinate, and if $\tau = 0, \tau_A, \tau_B, \tau_C$ at the points O, A, B, C, then suitable boundary conditions are

$$\phi \to 0 \quad \text{as } \Psi \to \infty,$$

$$\phi_\Psi = 0 \quad \text{on } \Psi = 0, 0 < \tau < \tau_A \text{ and } \tau_B < \tau < \tau_C, \tag{8.207}$$

$$\phi = \phi_b \quad \text{on } \Psi = 0, \tau_A < \tau < \tau_B.$$

Outer Thermal Layer   At this point we must enquire what happens to the upwelling plume after it impinges the lid at $x = 0$. We have already described the delamination layer, in which $\psi \sim 1/\delta$, $\theta \sim 1$ and $1 - vs - z \sim \delta$. In the plume, however, $\psi \sim 1/\delta_p$ and $\theta \sim \delta_p/\delta$. The plume turns the corner and is carried across in an outer thermal layer below the delamination layer. In this layer, we write (from the delamination layer variable)

$$\zeta = \left(\frac{\delta}{\delta_p}\right)^{1/2} \eta, \tag{8.208}$$

which is appropriate if $S_\infty \neq 0$ in (8.193).

We then find that at leading order,

$$-s' \frac{\partial P}{\partial \eta} = -\frac{\partial T_3}{\partial \eta},$$

$$-\frac{\partial P}{\partial \eta} = \phi,$$

$$T_3 = -N \frac{\partial^2 \Psi}{\partial \eta^2},$$

$$T_1 = -2Ns' \frac{\partial^2 \Psi}{\partial \eta^2}, \tag{8.209}$$

$$T^* = |T_3|,$$

$$N = \frac{1}{T^{*n-1}},$$

$$\Psi_\eta \phi_x - \Psi_x \phi_\eta = 0,$$

provided $\varepsilon \ll (\delta_p/\delta)^{1/2}$, which we assume.

It firstly follows that $\phi \approx \phi(\Psi)$, and the plume buoyancy structure is advected across the top surface unchanged. Secondly, $S = -T_3$ and Ψ satisfy the equations

$$\frac{\partial^2 \Psi}{\partial \eta^2} = |S|^{n-1} S,$$

$$\frac{\partial S}{\partial \eta} = s' \phi(\Psi),$$

$$(8.210)$$

with matching conditions

$$\Psi = \Psi_\eta = 0, \qquad S = S_\infty \quad \text{on } \eta = 0,$$

$$S \to 0 \quad \text{as } \eta \to \infty.$$

$$(8.211)$$

The extra condition determines the value of S_∞, which is then used to solve the delamination layer equations (8.191).

Plume Circulation   Returning now to the model for ϕ in the plumes, (8.206) and (8.207), we see that if the upwelling plume value at $x = 0$, $\tau = \tau_C$ is $\phi_C(\Psi)$, then

$$\int_0^\infty \phi_C(\Psi) \, d\Psi = -C_r.$$

$$(8.212)$$

The change in the buoyancy $\int_0^\infty \phi \, d\Psi$ across the top surface is $(C_l + C_r)$, but this is manifested in the delamination layer (where $\Psi \sim \delta_p/\delta \ll 1$). Therefore the initial condition for the plume at $x = a$, $\tau = 0$, may be approximately represented as

$$\phi|_{\tau=0} = \phi_C(\Psi) - (C_r + C_l)\delta_+(\Psi),$$

$$(8.213)$$

where the half-range delta function $\delta_+(\Psi)$ is zero for $\Psi > 0$, and $\int_0^\infty \delta_+(\Psi) \, d\Psi = 1$. In addition, we have the heat flux conditions

$$C_l + C_r = - \int_{\tau_A}^{\tau_B} \frac{\partial \phi}{\partial \Psi}\bigg|_{\Psi=0} d\tau$$

$$(8.214)$$

at the base, and similarly at the top

$$C_l + C_r = \int_0^a \Gamma \, dx = (1 - T_0) \int_0^a \frac{dx}{s}.$$

$$(8.215)$$

We can simplify the model as follows. Define

$$\gamma = \frac{C_l}{C_l + C_r}, \quad 2C = C_l + C_r.$$

$$(8.216)$$

Given s, (8.215) defines C. The core stream function can be written as

$$\hat{\psi} = C^{1/2} \psi^\gamma,$$

$$(8.217)$$

where ψ^γ (γ is merely a labelling superscript, not an exponent) depends only on γ, the relevant plume conditions being

$$\psi^\gamma \psi_{xx}^\gamma = 2\gamma \quad \text{on } x = 0,$$
$$\psi^\gamma \psi_{xx}^\gamma = 2(1-\gamma) \quad \text{on } x = a. \tag{8.218}$$

Next we define σ by

$$\phi = 2C\sigma - \frac{2C}{\sqrt{\pi \tau}} \exp\left(-\frac{\psi^2}{4\tau}\right), \tag{8.219}$$

so that σ satisfies

$$\sigma_\tau = \sigma_{\psi\psi} \quad \text{in } 0 < \psi < \infty, 0 < \tau < \tau_C, \tag{8.220}$$

with boundary conditions

$$\sigma \to 0 \quad \text{as } \psi \to \infty,$$
$$\sigma_\psi = 0 \quad \text{on } \psi = 0, 0 < \tau < \tau_A,$$
$$\sigma = \alpha + \frac{1}{\sqrt{\pi \tau}} \quad \text{on } \psi = 0, \tau_A < \tau < \tau_B, \tag{8.221}$$
$$\sigma_\psi = 0 \quad \text{on } \psi = 0, \tau_B < \tau < \tau_C,$$

where

$$\phi_b = 2\alpha C, \tag{8.222}$$

and we require, from (8.212)–(8.214),

$$\sigma|_{\tau=0} = \sigma|_{\tau=\tau_C} - \frac{1}{\sqrt{\pi \tau_C}} \exp\left(-\frac{\psi^2}{4\tau_C}\right), \tag{8.223}$$

$$-\int_{\tau_A}^{\tau_B} \frac{\partial \sigma}{\partial \psi}\bigg|_{\psi=0} d\tau = 1, \tag{8.224}$$

$$\int_0^\infty \sigma|_{\tau=0} d\psi = \gamma. \tag{8.225}$$

Now since $\hat{\psi} = C^{1/2}\psi^\gamma$, it follows that

$$\tau = C^{1/2} \int_0^s \frac{\partial \psi^\gamma}{\partial n} ds, \tag{8.226}$$

so we define

$$\tau = C^{1/2}\hat{\tau}, \qquad \psi = C^{1/4}\chi, \qquad \sigma = \Sigma(\chi, \hat{\tau})/C^{1/4}, \tag{8.227}$$

and $\tau_A = C^{1/2}\hat{\tau}_A$, etc. Σ satisfies

$$\Sigma_{\hat{\tau}} = \Sigma_{\chi\chi}, \tag{8.228}$$

with

$$\Sigma \to 0 \quad \text{as } \chi \to \infty,$$
$$\Sigma_\chi = 0 \quad \text{on } \chi = 0, 0 < \hat{\tau} < \hat{\tau}_A,$$
$$\Sigma = \hat{\alpha} + \frac{1}{\sqrt{\pi\hat{\tau}}} \quad \text{on } \chi = 0, \hat{\tau}_A < \hat{\tau} < \hat{\tau}_B, \tag{8.229}$$
$$\Sigma_\chi = 0 \quad \text{on } \chi = 0, \hat{\tau}_B < \hat{\tau} < \hat{\tau}_C,$$

where $\hat{\alpha} = \alpha C^{1/4}$, and the constraints

$$\Sigma|_{\hat{\tau}=\hat{\tau}_C} = \Sigma|_{\hat{\tau}=0} + \frac{1}{\sqrt{\pi\hat{\tau}_C}} \exp\left(-\frac{\chi^2}{4\hat{\tau}_C}\right), \tag{8.230}$$

$$-\int_{\hat{\tau}_A}^{\hat{\tau}_B} \frac{\partial \Sigma}{\partial \chi}\bigg|_{\chi=0} d\hat{\tau} = 1, \tag{8.231}$$

$$\int_0^\infty \Sigma|_{\tau=0}\, d\chi = \gamma. \tag{8.232}$$

The value of $\hat{\alpha}$ is chosen so that (8.231) is satisfied. Given an initial function $\Sigma_0(\chi) = \Sigma|_{\hat{\tau}=0}$, we solve (8.229) with (8.231) till $\hat{\tau} = \hat{\tau}_c$, which determines $\Sigma|_{\hat{\tau}=\hat{\tau}_c}$ as a linear functional of Σ_0. (8.230) is then a linear inhomogeneous integral equation, with (so we suppose) a unique solution, which depends on γ, since $\hat{\tau}_A$, $\hat{\tau}_B$ and $\hat{\tau}_C$ depend on γ. (8.232) then provides an equation for γ. Note that the plume head value of $\phi_C(\Psi)$ used in (8.210) is given by

$$\phi_C(\Psi) = 2C^{3/4}\Sigma_0\big[\Psi/C^{1/4}\big]. \tag{8.233}$$

It remains to determine s and thus C, via (8.215):

$$2C = (1 - T_0)\int_0^a \frac{dx}{s}. \tag{8.234}$$

This appears to require that s can be fully determined through the solution of (8.191). We now consider whether this can be true.

Delamination Layer: Similarity Solution   The delamination equations (8.191) form a fifth order set of differential equations. One can argue that the boundary conditions in (8.192) and (8.193) actually constitute six conditions, since the implied conditions $\Psi \to 0$ as $\zeta \to -\infty$ and $\theta_\zeta \to 0$ as $\zeta \to \infty$ are understood. If that is the case, then we can expect that s' in (8.191) is a nonlinear eigenvalue for the model, and thus (since also Γ depends on s), that the solution will implicitly determine a differential equation of the form $s' = \Upsilon(s)$ for s.

If that is the case, then the question arises what is the appropriate boundary condition for s? One possibility is that $s(0) = 0$; another would be that $s'(0) = 0$, which would determine $s(0)$ as the root (if it exists) of $\Upsilon = 0$. Without detailed matching of the delamination layer to the corner flow, it is difficult to be more precise.

Some insight into this question can be gained if we suppose that $S_\infty = 0$ in (8.193). This would follow if the function $\phi_C(\Psi)$ used in the solution of (8.210) is identically zero. In this case, Eqs. (8.191) have a similarity solution, and this is given by

$$\xi = \Gamma(x)\zeta, \qquad \theta = g(\xi), \qquad T_3 = \frac{-s'}{\Gamma}h(\xi), \qquad \Psi = \frac{s'}{\Gamma^3}\left|\frac{s'}{\Gamma}\right|^{n-1}f(\xi).$$

$$(8.235)$$

The functions f, g, h then satisfy the equations

$$h' = g,$$

$$f'' = |h|^{n-1}he^g,$$

$$g'' + B_n fg' = 0,$$

$$(8.236)$$

with

$$g(\infty) = h(\infty) = 0, \qquad f(-\infty) = 0, \qquad g'(-\infty) = 1, \qquad (8.237)$$

providing s satisfies the equation

$$\left\{\frac{s'}{\Gamma^3}\left|\frac{s'}{\Gamma}\right|^{n-1}\right\}' = B_n\Gamma. \qquad (8.238)$$

Given Γ, this equation determines the location of the (unknown) lithosphere base.

To solve (8.236), we use a shooting method for the equation with $B_n = 1$:

$$H' = G,$$

$$F'' = |H|^{n-1}He^G,$$

$$G'' + FG' = 0,$$

$$(8.239)$$

with

$$F = F' = 0, \qquad G' = c, \qquad G = -cM, \qquad H = H_0 + \frac{1}{2}cM^2 \quad \text{at } \xi = -M,$$

$$(8.240)$$

and M is chosen to be large ($M = 20$ is adequate). The values of H_0 and c are adjusted via Newton iteration at $\xi = M$ until $G = H = 0$ there. Once a solution is found, then f, g, h are determined by

$$g(\xi) = G(\xi/c), \qquad h(\xi) = cH(\xi/c), \qquad f(\xi) = c^{n+2}F(\xi/c), \qquad (8.241)$$

Table 8.1 Values of B_n

n	B_n
1	0.087
2	4.74×10^{-3}
3	1.39×10^{-4}
3.5	1.98×10^{-5}
4	2.53×10^{-6}
5	3.15×10^{-8}

and

$$B_n = \frac{1}{c^{n+3}}. \tag{8.242}$$

As shown in Table 8.1, B_n is small and decreases rapidly as n increases.

Lithosphere Base   We substitute (8.199) into (8.238), so that

$$\left(s^{n+2} s'^n\right)' = \frac{B_n (1 - T_0)^{n+3}}{s} \tag{8.243}$$

(since $\Gamma > 0$, and thus we can take $s' > 0$). In order to complete the solution for $s(x)$, we require two boundary conditions. It is not entirely obvious where these are to come from. One reasonable choice would appear to be $s'(0) = 0$, which ensures that $\Psi \to 0$ as $x \to 0$ (as would appear necessary, not only because of the vertical boundary at $x = 0$, but also because $\Psi \sim \delta_p / \delta \ll 1$ in the plume). If $s = s_0$ at $x = 0$, then the solution of (8.243) as

$$\int_{s_0}^{s} \frac{s^{\frac{n+2}{n}} \, ds}{\left[s^{\frac{n+2}{n}} - s_0^{\frac{n+2}{n}}\right]^{\frac{1}{n+1}}} = bx, \tag{8.244}$$

where

$$b = \left\{ \left(\frac{n+1}{n+2}\right) B_n (1 - T_0)^{n+3} \right\}^{\frac{1}{n+1}}. \tag{8.245}$$

More likely, however, is that we should choose $s(0) = 0$, on the basis that the similarity variable is $\xi = \frac{(1 - T_0)\zeta}{s}$, and identification of $x \to 0$ with $\zeta \to \infty$, as usual in similarity solutions, requires this condition. In that case, the solution for s is

$$s = \left[\frac{(2n + 3)bx}{n + 1}\right]^{\frac{n+1}{2n+3}}, \tag{8.246}$$

and the determination of the solution is complete.

Slab Stress   As $\zeta \to -\infty$ in the delamination layer, we have, from (8.191),

$$\theta \sim \Gamma\zeta, \qquad T_3 \sim -\frac{1}{2}s'\Gamma\zeta^2,$$

$$\Psi \sim \left(\frac{1}{2}s'\zeta^2\right)^n \Gamma^{n-2}e^{\Gamma\zeta}. \tag{8.247}$$

If we define

$$z = 1 - \nu Z \tag{8.248}$$

in the lid, so that the lid base is at $Z = s$, then (8.247) becomes

$$\theta \sim -\frac{\Gamma(s-Z)}{\varepsilon}, \qquad T_3 \sim -\frac{1}{2}s'\Gamma\left(\frac{s-Z}{\varepsilon}\right)^2,$$

$$\Psi \sim \left(\frac{s'(s-Z)^2}{2\varepsilon^2}\right)^n \Gamma^{n-2} \exp\left[-\frac{\Gamma(s-Z)}{\varepsilon}\right], \tag{8.249}$$

and in addition, from (8.190),

$$P \sim -\frac{1}{2}\Gamma\left(\frac{s-Z}{\varepsilon}\right)^2, \qquad T_1 \sim -s'\Gamma\left(\frac{s-Z}{\varepsilon}\right)^2,$$

$$N \sim \left\{\frac{2\varepsilon^2}{|s'\Gamma|(s-Z)^2}\right\}^{n-2} \exp\left[\frac{\Gamma(s-Z)}{\varepsilon}\right]. \tag{8.250}$$

We thus rescale the variables as follows:

$$\theta = -\frac{(1-T)}{\varepsilon}, \qquad P = \frac{\tilde{P}}{\varepsilon^2}, \qquad T_1 = \frac{\tilde{T}_1}{\varepsilon^2}, \qquad N = \varepsilon^{2(n-1)}\exp\left[\frac{(1-T)}{\varepsilon T}\right]\tilde{N},$$

$$T_3 = \frac{\tilde{T}_3}{\varepsilon^2}, \qquad \Psi = \frac{1}{\varepsilon^{2n}}\exp\left[-\frac{(1-T)}{\varepsilon T}\right]\tilde{\Psi} + \frac{\lambda}{\varepsilon^{2n}}\exp\left[-\frac{(1-T_0)}{\varepsilon T_0}\right]Zu_s(x). \tag{8.251}$$

The re-definition of Ψ merits some discussion. We add the term in λ to allow for the small non-zero surface velocity. We anticipate that λ is of algebraic order in ε, and so it will only be important at the surface. We include the term separately because it is distinct from the other part of the stream function, proportional to $\exp(\theta/\varepsilon)$.

Substitution of (8.251) into (8.174) (via (8.188)) leads to (on omitting the over-tildes)

$$P_x = v^2 T_{1x} - T_{3Z},$$

$$- P_Z = v^2(T_{3x} + T_{1Z}) - (1 - T),$$

$$T_1 = 2\varepsilon^2 N \exp\left[\frac{(1-T)}{\varepsilon T}\right] \frac{\partial^2}{\partial x \partial Z}\left[\exp\left\{-\frac{(1-T)}{\varepsilon T}\right\}\Psi + \cdots\right],$$

$$T_3 = -\varepsilon^2 N \exp\left[\frac{(1-T)}{\varepsilon T}\right]\left(\frac{\partial^2}{\partial Z^2} - v^2 \frac{\partial^2}{\partial x^2}\right)\left[\exp\left\{-\frac{(1-T)}{\varepsilon T}\right\}\Psi + \cdots\right], \quad (8.252)$$

$$T^{*2} = T_3^2 + v^2 T_1^2,$$

$$N = \frac{1}{T^{*n-1}},$$

$$\frac{1}{\varepsilon^{2n+1}}\left[\frac{\partial T}{\partial x}\frac{\partial}{\partial Z} - \frac{\partial T}{\partial Z}\frac{\partial}{\partial x}\right]\left[\Psi \exp\left\{-\frac{(1-T)}{\varepsilon T}\right\} + \cdots\right] = \frac{\partial^2 T}{\partial Z^2} + v^2 \frac{\partial^2 T}{\partial x^2},$$

where the dots denote the extra term $\lambda \exp[-(1 - T_0)/\varepsilon T_0]Zu_s$, which can be neglected except near the top surface $Z = 0$.

Ignoring λ, the heat equation is

$$\frac{1}{\varepsilon^{2n+1}} \exp\left[-\frac{(1-T)}{\varepsilon T}\right][\Psi_Z T_x - \Psi_x T_Z] = \frac{\partial^2 T}{\partial Z^2} + v^2 \frac{\partial^2 T}{\partial x^2}, \quad (8.253)$$

and we see that convection is formally negligible as $\varepsilon \to 0$, as we suppose. In practice, the value of $Pe = \varepsilon^{-(2n+1)} \exp[-(1 - T)/\varepsilon T]$ (a kind of Péclet number) at the surface, where $T = T_0 = 0.2$ for the Earth, is approximately $\varepsilon^{-8} e^{-4/\varepsilon}$ if $n = 3.5$, and for our preferred value of 0.023, this is about 4×10^{-63}. The value of T where $Pe = \Delta$ is

$$T_\Delta = \frac{1}{1 + (2n + 1)\varepsilon \ln\left(\frac{1}{\varepsilon}\right) + \varepsilon \ln\left(\frac{1}{\Delta}\right)}, \quad (8.254)$$

and for moderate $\Delta < 1$ and $n = 3.5$, $T_\Delta \approx 0.59$. Thus the high value of n and low ε combine to yield a viscous structure which is only truly rigid for $T < T_\Delta$. Figure 8.5 illustrates this by plotting the approximating viscosity function

$$N_{\text{app}}(T) = \left(\frac{\varepsilon}{1-T}\right)^{n-1} \exp\left[\frac{1-T}{\varepsilon T}\right] \quad (8.255)$$

(gleaned from (8.250)) for $\varepsilon = 0.023$, $n = 3.5$. Evidently, the formal limit $N_{\text{app}} = \exp[O(1/\varepsilon)]$ only occurs practically for $T \lesssim 0.7$, and advection is only truly negligible for $T \lesssim 0.6$. Some inaccuracy can be expected in the theory.

Neglecting advection, we have the approximate conductive profile (for $v \ll 1$)

$$T \approx T_0 + \frac{(1 - T_0)Z}{s} \quad (8.256)$$

Fig. 8.5 The
pseudo-viscosity function
$N_{app}(T) = (\frac{\varepsilon}{1-T})^{2n-1} \exp\{\frac{1-T}{\varepsilon T}\}$,
for $n = 3.5$ and $\varepsilon = 0.023$

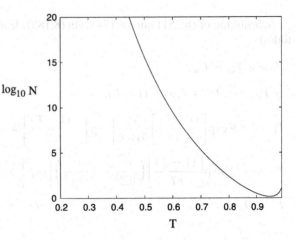

and thus P and T_3, satisfying

$$P_x \approx -T_{3Z},$$
$$-P_Z \approx -(1-T),$$
(8.257)

together with the matching conditions (from (8.249)) $P, T_3 \to 0$ as $Z \to s$, are given
by

$$P = -\frac{(1-T_0)}{2s}(s-Z)^2,$$

$$T_3 = -\frac{(1-T_0)s'}{2s^2}\left[\frac{2}{3}s^3 - s^2Z + \frac{1}{3}Z^3\right].$$
(8.258)

We are unable to satisfy the condition of zero shear stress at the top surface $Z = 0$,
since there is a residual shear stress there given by

$$T_3^0 = -\frac{1}{3}(1-T_0)ss'.$$
(8.259)

In order to enforce $T_3 = 0$ at $Z = 0$, the shear stress must change rapidly in a
boundary layer near $Z = 0$. Why should such a boundary layer exist? Notice that
we also require the stream function to vanish, i.e. $\Psi = 0$ at $Z = 0$. To solve for Ψ,
we expand the expressions for T_1 and T_3 in (8.252). If we put

$$V(T) = \frac{1-T}{T} \approx \frac{1}{T_0 + \frac{(1-T_0)Z}{s}} - 1,$$
(8.260)

then we find

$$
\begin{aligned}
T_1 &= 2N\big[V_Z V_x \Psi - \varepsilon(V_{Zx}\Psi + V_Z\Psi_x + V_x\Psi_Z) + \varepsilon^2\Psi_{xZ} \\
&\quad + \varepsilon^2\lambda \exp[(V - V_0)/\varepsilon]u'_s\big], \\
T_3 &= -N\big[V_Z^2\Psi - \varepsilon(2V_Z\Psi_Z + V_{ZZ}\Psi) + \varepsilon^2\Psi_{ZZ}\big] \\
&\quad + \nu^2 N\big[V_x^2\Psi - \varepsilon(2V_x\Psi_x + V_{xx}\Psi) + \varepsilon^2\Psi_{xx} + \varepsilon^2\lambda \exp\{(V - V_0)/\varepsilon\}Zu''_s\big],
\end{aligned}
\tag{8.261}
$$

where $V = V_0$ at $T = T_0$. Away from $Z = 0$, we have

$$
T_1 \approx 2N V_x V_Z\Psi, \qquad T_3 \approx -N V_Z^2\Psi, \tag{8.262}
$$

thus $\Psi \neq 0$ at $Z = 0$. The presence of a term $\varepsilon^2\Psi_{ZZ}$ in (8.261) suggests the existence of a boundary layer of thickness ε over which Ψ can drop to zero. It is in order to get the shear stress T_3 to zero as well that we introduced the extra stress term in λ.

Define a boundary layer variable ξ by

$$
Z = \varepsilon\xi. \tag{8.263}
$$

Noting that V_x and V_Z are $O(1)$, we have to leading order, from (8.261),

$$
\begin{aligned}
T_1 &= O(1) + 2N\varepsilon^2\lambda \exp[V_Z^0\xi]u'_s, \\
T_3 &= -N\big[V_Z^2\Psi - 2V_Z\Psi_\xi + \Psi_{\xi\xi}\big] + O(\nu^2, \nu^2\varepsilon^3\lambda),
\end{aligned}
\tag{8.264}
$$

where

$$
V_Z^0 = V_Z|_{Z=0} = -\frac{(1 - T_0)}{sT_0^2}. \tag{8.265}
$$

Now we rescale P and T_1 in this stress boundary layer in order that T_3 can change. This necessitates writing

$$
P = \frac{1}{\varepsilon}P^*, \qquad T_1 = \frac{1}{\varepsilon\nu^2}T_1^*, \tag{8.266}
$$

so that

$$
\begin{aligned}
P_x^* &= T_{1x}^* - T_{3\xi}, \\
-P_\xi^* &= \delta^2 T_{3x} + T_{1\xi} - \varepsilon^2(1 - T),
\end{aligned}
\tag{8.267}
$$

and at leading order we have

$$
P^* = -T_1^* \tag{8.268}
$$

(to match to $P^*, T_1 \to 0$ as $\xi \to \infty$), and thus

$$
T_{3\xi} = 2T_{1x}^*. \tag{8.269}
$$

We can enforce the scaling $T_1 \sim 1/\varepsilon v^2$ by choosing $\varepsilon^2 \lambda N \sim 1/\varepsilon v^2$ in $(8.264)_1$, and then, since $vT_1 \sim 1/\varepsilon v = 1/\delta \gg T_3 = O(1)$, $N = 1/T^{*n-1} \approx (1/vT_1)^{n-1} \sim \delta^{n-1}$; thus we choose

$$\lambda = \frac{1}{\varepsilon \delta^{n+1}}. \tag{8.270}$$

This is algebraic in ε as we required (provided one does not examine a limit in which $Ra = \exp[O(1/\varepsilon)]$ or $n \sim 1/\varepsilon$).

It follows from (8.264) that, since $N\varepsilon^2 \lambda \sim 1/\varepsilon v^2 \gg 1$,

$$T_1^* \approx \frac{2}{|T_1^*|^{n-1}} \exp[-|V_Z^0|\xi] u_s', \tag{8.271}$$

whence

$$T_1^* \approx 2^{\frac{1}{n}} \exp[-|V_Z^0|\xi/n] |u_s'|^{\frac{1}{n}-1} u_s'. \tag{8.272}$$

Using (8.270), this gives

$$T_1^* = 2^{\frac{1}{n}} \left\{ |u_s'|^{\frac{1}{n}-1} u_s' \right\} \exp\left[-\frac{(1-T_0)\xi}{nT_0^2 s} \right], \tag{8.273}$$

and the solution for T_3 is

$$T_3 = \frac{2}{B}(sf)'\left[1 - e^{-B\xi/s}\right] - \frac{2fs'}{s}\xi e^{-B\xi/s}, \tag{8.274}$$

where

$$B = \frac{1-T_0}{nT_0^2}, \qquad f = 2^{\frac{1}{n}} \left\{ |u_s'|^{\frac{1}{n}-1} u_s' \right\}. \tag{8.275}$$

The matching condition is that $T_3 \rightarrow T_3^0$ given by (8.259). This implies $2(sf)' = BT_3^0$, hence

$$\left\{ s|u_s'|^{\frac{1}{n}-1} u_s' \right\}' = -\left\{ \frac{(1-T_0)^2}{3nT_0^2 2^{1+\frac{1}{n}}} \right\} ss'. \tag{8.276}$$

We require $u_s = 0$ at $x = 0$ and $x = a$. A first integral is

$$s|u_s'|^{\frac{1}{n}-1} u_s' = A\left(s_c^2 - s^2\right), \tag{8.277}$$

where

$$A = \frac{(1-T_0)^2}{12nT_0^2 2^{\frac{1}{n}}}, \tag{8.278}$$

and $s = s_c$ where $u_s' = 0$. It follows that $u_s > 0$, and

$$u_s = \int_0^x \left(\frac{A}{s}\right)^{\frac{1}{n}} |s_c^2 - s^2|^{\frac{1}{n}-1} \left(s_c^2 - s^2\right) dx, \tag{8.279}$$

and we choose s_c so that $u_s = 0$ at $x = a$, i.e.

$$\int_0^a \left(\frac{A}{s}\right)^{\frac{1}{n}} |s_c^2 - s|^{\frac{1}{n}-1} (s_c^2 - s^2) \, dx = 0. \tag{8.280}$$

Skin Stresses Looking back, we see from (8.279), (8.188), (8.251) and (8.266) that in the stagnant lid, the dimensional shear stress

$$\tau_3^D \sim \frac{\delta^2 Ra}{\varepsilon^2} \frac{\eta_0 \kappa}{d^2}, \tag{8.281}$$

but within the (boundary layer) skin, the pressure and normal stresses are higher,

$$p^D, \tau_1^D \sim \frac{\delta Ra}{\varepsilon^2} \frac{\eta_0 \kappa}{d^2}. \tag{8.282}$$

Since $P^* + T_1^* \approx 0$, the largest stress is thus the horizontal stress $-p^D + \tau_1^D \approx 2\tau_1^D$. Specifically, at the surface, (8.250) implies that the horizontal stress σ_S is

$$\begin{aligned}
\sigma_S = 2\tau_1^D|_{\xi=0} &= \frac{\delta Ra}{\varepsilon^2} \left(\frac{\eta_0 \kappa}{d^2}\right) 2^{1+\frac{1}{n}} \left\{ |u_s'|^{\frac{1}{n}-1} u_s' \right\} \\
&= \frac{\delta Ra}{\varepsilon^2} \left(\frac{\eta_0 \kappa}{d^2}\right) \frac{(1-T_0)^2}{6n T_0^2} \frac{(s_c^2 - s^2)}{s}.
\end{aligned} \tag{8.283}$$

It is extensional ($\sigma_S > 0$) near the upwelling ($x = 0$) and compressive near the downwelling.

We can calculate an estimate for the size of the stress,

$$[\sigma_S] = \frac{\delta (1-T_0)^2}{6n T_0^2 \varepsilon^2} \alpha \rho_a g d T_a \tag{8.284}$$

using our previous estimates. We take $n = 3.5$, $T_0 = 0.2$, $\alpha = 3 \times 10^{-5} \, \text{K}^{-1}$, $T_a = 1500 \, \text{K}$, $\rho_a = 3 \times 10^3 \, \text{kg m}^{-3}$, $g = 10 \, \text{m s}^{-2}$, $d = 3000 \, \text{km}$; if we choose $\kappa = 10^{-6} \, \text{m}^2 \, \text{s}^{-1}$ and $\eta_0 = 10^{20} \, \text{Pa s}$, then $Ra \approx 3.6 \times 10^8$; if $\varepsilon = 0.023$, then $\delta \approx 0.013$, and we compute $[\sigma_S] = 8 \, \text{kbar}$ (1 kilobar $= 10^3 \, \text{bar} = 10^8 \, \text{Pa}$). This is a huge stress, comparable to the breaking strength of rock, and is suggestive of the idea that, within the confines of a purely viscous rheology such as we have here, some realistic adaptation of the model should be made.

8.5.5 Summary

What have we discovered? For a purely viscous fluid with strong Arrhenius dependence on temperature, high Rayleigh number convection occurs as vigorous flow driven by small excess temperatures below a stagnant lid. This is clearly seen in

Fig. 8.6 Stream function
contours of a
temperature-dependent
viscosity convection
calculation at Rayleigh
number 10^6, the viscosity
scaled to that at the basal
temperature. The rheology is
that of (8.176), with $\Lambda = 1$,
$n = 1$, $\varepsilon = 0.2$. Basal and
surface dimensionless
temperatures are 1 and 0.1,
respectively. The absence of
contours towards the top
indicates the stagnant lid.
Figure courtesy Mike
Vynnycky

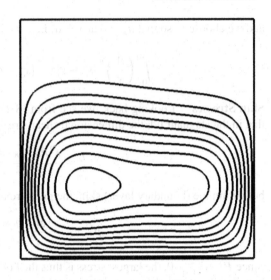

Fig. 8.6. In this lid, the stresses are high, and they increase rapidly in a narrow layer
near the surface. But, as with constant viscosity convection, the analysis is complicated, and in parts unresolved.

At the outset, we do not even know the appropriate value of the rate-controlling
internal viscosity η_0 and temperature T_a in the asthenosphere; a convoluted argument leads to their determination by (8.186) and (8.187), in terms of the unknown
temperature excess ϕ_b at the base. The delamination layer below the lithosphere
is described by (8.191), and this introduces further unknowns: the lithosphere base
$s(x)$, the lithospheric temperature gradient Γ, and the far field stress $S_\infty(x)$. Of
these, Γ is given by the slab temperature field via (8.199), and S_∞ is given (in
principle) by the extra condition for the outer thermal layer in (8.211).

The core flow requires the determination of two further quantities, the plume
buoyancy constants C_l and C_r. Given these, then the plume/thermal boundary layer
equation (8.206) with boundary conditions (8.207) can be solved, subject to the
pseudo-periodicity condition (8.213), which in addition yields the plume head temperature $\phi_C(\Psi)$ used in the outer thermal layer equations. Three extra conditions
are necessary to determine ϕ_b, C_l and C_r, and in addition s must be prescribed.

We may suppose (8.212) determines C_r; given s, (8.215) determines C_l, and
finally (8.214) determines ϕ_b. Of course, all of these relations are coupled, hence
the intent of the untangling discussion following (8.215).

The determination of s apparently forms part of the solution of the delamination equations (8.191). Further study of these equations is undoubtedly warranted.
The existence of the approximating similarity solution suggests that this is the case,
although the appropriate boundary condition for s at $x = 0$ is less clear.

8.6 Subduction and the Yield Stress

We have now arrived at the central conundrum of plate tectonics, alluded to in Sect. 8.1. At high values of Ra and low values of ε, vigorous convection occurs beneath a stagnant lid. Active plate tectonics does not occur in the model, as it does on the Earth. In the lid, the stresses become extremely large, of the order of kilobars, and this provides the clue to resolve the conundrum. We argue that at cold temperatures and at such high stresses, viscous behaviour breaks down, and the flow becomes plastic. In order to describe this, we need to reconsider the rheology of mantle rocks in the vicinity of the Earth's surface.

8.6.1 Near-Surface Mantle Rheology

We have already described, in Sect. 8.5.1, the viscous rheology of polycrystalline rocks. At low temperatures, and for short time behaviour (associated with seismic waves, for example), the mantle is elastic, and a common description of both elastic and creeping behaviours can be represented by assuming a Maxwell viscoelastic fluid, whose constitutive law can be represented schematically by the equation

$$\dot{\varepsilon} = \frac{\tau}{2\eta} + \frac{\dot{\tau}}{E_M}, \qquad (8.285)$$

where η is the viscosity given by (8.163), τ is the stress, and E_M is an elastic modulus. (We dispense with the details of tensor representation.) The ratio

$$t_M = \frac{2\eta}{E_M} \qquad (8.286)$$

thus defines a Maxwell time scale, such that for changes on a time scale $t \ll t_M$, the behaviour is elastic, and on longer time scales $t \gg t_M$, the behaviour is viscous. Note that the Maxwell time depends on temperature and stress.

Into this mixture we now add the concept of failure. Brittle failure is associated with the coalescence of microcracks within the rock, and is classically associated with an internal friction of the material. Thus brittle failure is usually associated with the attainment of a failure stress

$$\tau_f = Kp, \qquad (8.287)$$

where p is the lithostatic pressure, and K is a dimensionless coefficient of friction of $O(1)$.

Geophysicists often combine these two ideas of viscous creep and brittle failure to propose a failure diagram such as that shown in Fig. 8.7. This identifies the yield stress with the minimum of two values, one of which is the brittle failure stress τ_f,

Fig. 8.7 Brittle (*straight full line*) and ductile (*curved full line*) yield stress τ_c as function of depth. The *dashed line* represents a typical corresponding lithospheric shear stress using the boundary layer theory for a purely viscous mantle, with highly temperature-dependent viscosity

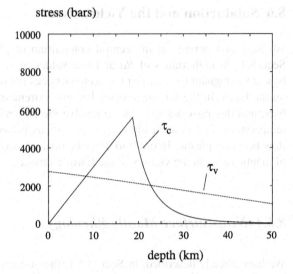

and the other is a so-called ductile failure stress τ_d determined from the viscous rheology (8.163), thus

$$\tau_d = \left[\frac{\dot{\varepsilon}}{A} \exp\left\{ \frac{E^* + pV^*}{RT} \right\} \right], \qquad (8.288)$$

and τ_d depends on temperature and strain rate. Figure 8.7 shows a typical yield curve $\tau_c = \min(\tau_f, \tau_d)$ using a typical mantle strain rate and temperature profile as a function of depth.

The interpretation of Fig. 8.7 is that if the stress τ (often taken to be defined by the second stress invariant $2\tau^2 = \tau_{ij}\tau_{ij}$) is less than τ_c, then deformation is elastic, but if the stress reaches τ_c, then it remains on the yield surface, and the deformation is plastic. The diagram in Fig. 8.7 then looks reassuringly familiar to the yield surface of critical state soil mechanics, with the brittle yield resembling the tension failure or Hvorslev yield surface, and the ductile yield resembling the Roscoe yield surface. It has to be said that the use of this diagram in this way represents a misleading misinterpretation of the classical ideas of yield and plasticity, and it should really be outlawed.

Firstly, the use of (8.287) is based on Byerlee's law, which was developed to describe rock friction and not yield in the classical plastic sense of continuing deformation at a critical stress. It is actually clear that the Earth's mantle exhibits yield at much lower stresses. For example, the motion of deep subducting slabs is by a stick–slip motion facilitated by slip events between the descending slab and the overlying mantle. These slip events indicate yield at a stress τ_c which can be calculated, since it is due to the buoyant excess weight of the subducting slab, and such estimates suggest $\tau_c \lesssim 300$ bars, much less than one would calculate from (8.287).

Secondly, the ductile part of the curve in Fig. 8.7 is not a yield curve at all, since it describes flow behaviour at a pre-assigned strain rate. Since the strain rate is part of the flow problem requiring solution, this part of the curve is truly meaningless.

Fig. 8.8 An illustration of
the region of plastic yield
within the lithosphere,
assuming a viscoplastic
rheology

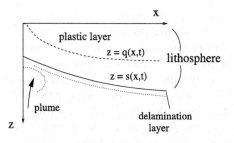

Despite this, the 'yield' curve is often used to divide the upper mantle into a plastic upper part (with brittle yielding), an elastic middle part, and a ductile lower part. Such inferences appear to be groundless.

I want to propose a different kind of rheology which is consistent with observations of fault motion by earthquakes. This is that plastic yield should occur at a yield stress τ_c, which we will take to be typical of stress release in earthquakes, and much less than the brittle Byerlee value. In essence, we associate this kind of failure with subcritical crack propagation, and we do not distinguish necessarily between elastic and viscous behaviour for stresses less than τ_c. We can in fact allow a viscoelastic deformation for $\tau < \tau_c$, but it turns out that the elastic deformation is inessential to the description, and we henceforth omit it.

Our rheology is thus viscoplastic, and takes the form

$$\tau_{ij} = 2\eta \dot{\varepsilon}_{ij}, \tag{8.289}$$

where the viscosity η is given by (8.163) if $\tau < \tau_c$, and is determined in the plastic case by the requirement that $\tau = \tau_c$ on the yield surface.

8.6.2 The Plastic Lid: Failure and Subduction

It is now possible to carry forward the boundary layer analysis of Sect. 8.5 to allow the description of a plastic lid within the lithosphere, but we forgo the dubious pleasure of tormenting the reader further with this, interesting and intricate though the analysis is. We confine ourselves to a description of the results.

The essential novelty is indicated in Fig. 8.8, which indicates that where the stresses in the lithosphere exceed the yield stress, there is a plastic lid of dimensionless (scaled with vd) depth q. In this lid, the material behaves plastically, while the part of the lithosphere below the plastic lid is viscous. Intricacies include the fact that there are boundaries at top and bottom of the lid (of thickness $O(\varepsilon)$ relative to the lid scale) in which the stresses jump.

From the analysis, we obtain relations for s as before, and also for the plastic lid depth q. The analysis assumes a stagnant lid, and is thus self-consistent if $q < s$. The reason for this has to do with the effective viscosity in the plastic lid. Our assumption for the flow rule when the yield stress is reached is that increments of strain are proportional to increments of stress, with the factor of proportionality being such that the yield stress is not exceeded. What this means is that the plastic deformation is also effectively viscous, but with an effective viscosity which must be computed as part of the solution.

At the plastic–viscous transition boundary, the viscosity is denoted by η_q, and is given by the viscous formula, in terms of the stress and temperature there. What we then find is that the effective viscosity throughout the plastic lid is approximately equal to η_q. For example, if we consider the Newtonian rheology with $n = 1$, so that (8.176) gives the dimensionless viscosity

$$\eta = \exp\left(\frac{1-T}{\varepsilon T}\right), \tag{8.290}$$

then, using the linear temperature (8.198), we have

$$\eta_q \approx \exp\left[\frac{(1-T_0)\left(1-\frac{q}{s}\right)}{\varepsilon\{T_0 + (1-T_0)\frac{q}{s}\}}\right]. \tag{8.291}$$

This gives the dimensionless viscosity in the plastic lid, and is also the ratio of the dimensional viscosity to that in the asthenosphere. The effective plastic viscosity of the lid is very large, but drops abruptly to values near one when the ratio q/s approaches one. Thus if q reaches s, the lithospheric column at that point has an effective reduced viscosity equal to that of the underlying asthenosphere. Consequently, the heavy lithosphere will convectively sink into the underlying mantle. This initiates the process of subduction.

It is possible to calculate the variation of s and q with distance x. We calculate s using (8.243). With $n = 1$, and supposing that $s = 0$ at $x = 0$, we find

$$s = k(1 - T_0)^{4/5} x^{2/5}, \tag{8.292}$$

where

$$k = \left(\frac{25 B_1}{6}\right)^{1/5} \approx 0.82. \tag{8.293}$$

Calculation of q requires solution for the plastic stresses in the lid, but a simple approximation which is quite accurate is

$$q \approx \left[\frac{(1-T_0)^{13/5} k^2}{12c}\right] x^{4/5}, \tag{8.294}$$

where

$$c = \frac{\varepsilon^2 d^2 \tau_c}{\eta_0 \kappa Ra^{3/5}}, \tag{8.295}$$

and τ_c is the yield stress.

We can see from this that q/s is an increasing function of x, so that failure will occur at the right hand side of the convecting cell $x = 1$ if $q \geq s$ there, and thus if $c < c^*$, where the approximation (8.294) would suggest $c^* \approx 0.046$. Accurate numerical determination of q in fact shows that $c^* \approx 0.056$. The failure criterion is thus

$$\tau_c \lesssim \tau^* = \frac{c^* \eta_0 \kappa Ra^{3/5}}{\varepsilon^2 d^2}. \tag{8.296}$$

If we use the values $c^* = 0.056$, $\eta_0 = 1.4 \times 10^{20}$ Pa s, $\kappa = 10^{-6}$ m^2 s^{-1}, $Ra = 3 \times 10^8$, $\varepsilon = 0.023$, $d = 3000$ km, then we find $\tau^* \approx 2$ kbar. Although this estimate would change for a more realistic value of $n = 3.5$, it indicates that for apparent yield stresses of the order of 300 bar, lithospheric failure will indeed occur. In principle, this provides a satisfactory dynamical explanation for the occurrence of subduction, and thus active plate tectonics, on the Earth.

8.7 Tectonics on Venus

Venus is a planet which is very similar to the Earth in many respects. Its sulphurous, carbon rich atmosphere is very different of course, causing the hot surface temperature of 750 K, but the planet is of a similar size, and is generally presumed to have a similar structure, with a silicate mantle sheathing an iron core. It is a tectonically active planet, with many different kinds of large scale surface features: tesserae, wrinkle ridges, chasmatae, coronae. There is much evidence of past volcanism. From this we can infer that there is (unsurprisingly) active mantle convection on Venus. But there is no active plate tectonics. There is no system of linear ridges and subduction zones which indicates that the lithosphere takes part in mantle convection.

This might seem perplexing at first, but armed with our new understanding of how convection works in temperature-dependent viscous fluids, the explanation is apparently simple. Mantle convection on Venus operates below a stagnant lid, in just the same way as it presumably does on Mars (another volcanic planet with no plate tectonics), and the stresses generated are simply not large enough to cause lithospheric failure and hence subduction. One contributing factor in the difference between the planets might be the absence of water, which has a weakening effect on the rheology.

There is, however, another twist to this story. The surface of Venus, if it is stagnant, should be as old as the planet, presumably some 4 billion years. However, counts of meteor craters indicate that in fact the planetary surface is of a uniform age of some 500 million years. Old, but significantly younger than might be expected. How can this be?

The most obvious answer is that the planet was resurfaced 500 million years ago in a global resurfacing event, caused by a transient major plate tectonic subduction event on a planetary scale. Such a hypothetical event is not inconsistent with what we know about convection at high Rayleigh number. In fact, as the Rayleigh number increases, convection becomes oscillatory and increasingly intermittent. In a vivid picture developed by Lou Howard, convection in such oscillatory régimes consists alternatively of long tranquil periods, where stagnant conductive boundary layers grow at the base and the surface, and violent overturning events, where these unstable boundary layers rapidly detach and mix the flow.

Could this happen in a planetary mantle? To imagine how, suppose that an overturning event occurs, in which the lithosphere fails, leading to massive subduction and the resurfacing of the entire planetary mantle. The previous cold lithosphere sinks to the base of the mantle, where it forms a cold dense layer. Without any convectively induced stresses to make it plastic, it is stagnant. At the core–mantle boundary, a hot, low viscosity thermal boundary layer grows, but is unable to penetrate the stagnant slab above. This will continue, either until the thermal boundary layer penetrates through the slab, or until the buoyant stresses it generates cause plastic failure of the cold barrier above it. But eventually the thermal boundary layer will break through, causing massive thermal plumes to rise through the mantle to the surface, where they will impinge at the base of the newly forming lithosphere, which in the meantime has been growing conductively downwards from the surface of the planet.

It should be noted that the plastic failure of the lithosphere which we have discussed above relies on an underlying convective flow, which causes the variation in lid thickness, which in turn causes the horizontal temperature variations which are the origin of the stresses in the lid. After an overturn, however, interior convection is weak, and horizontal lid thickness variation will not be induced until the massive plumes arrive at the base of the lid.

The arrival of one of these plumes beneath the newly formed lithosphere will cause uplift and thermal erosion at its centre, and a radial outflow. Thus the situation is similar to that analysed in Sect. 8.6, with the difference that the flow in the delamination layer is radial, and there is essentially no flow in the interior. This latter feature makes little difference, since our analysis of the developing lid and delamination layer is essentially uncoupled from the underlying mantle flow.

The analysis of the model thus proceeds similarly to that already presented, except that the temperature in the lid depends on time. In the same dimensionless variables as before, T in the lid satisfies

$$T_t = T_{zz},$$

$$T = T_0 \quad \text{on } z = 0, \tag{8.297}$$

$$T = 1, \qquad T_z = \Gamma \quad \text{on } z = s.$$

The temperature gradient at the base of the lid Γ is now unknown, but it serves to determine the position of the base s through the equation

$$\omega^3 \left(\frac{rs'}{\Gamma^3} \left| \frac{s'}{\Gamma} \right|^{n-1} \right)' = B_n r \Gamma, \tag{8.298}$$

where $s' = \frac{\partial s}{\partial r}$; this is the equivalent in polar coordinates of (8.238); note that we allow $n \neq 1$ in this analysis. The coefficient ω^3 is introduced here because a slightly different choice of viscosity scale has been used; see Question 8.9, which indicates the connection between the two choices. Essentially, the choice of ν in the steady state is dictated by interior flow driven by the plume stress; in the present case, this is less relevant.

Solution of (8.297) and (8.298) provides T and s, and then it can be shown that the plastic lid base q satisfies

$$2\left(r^{1/2}q\right)' = -\frac{r^{1/2}}{C}\int_0^s zT_r\,dz, \tag{8.299}$$

where $q' = \frac{\partial q}{\partial r}$ and

$$C = \frac{\tau_c}{\alpha\rho_a g T_a vd}. \tag{8.300}$$

As before, failure occurs if q reaches s. In fact, this is a little glib when $n \neq 1$, since the effective viscosity at the lid base depends also on the stress. Taking this into account suggests that failure will occur when

$$\frac{q}{s} \approx \frac{1 - T_0}{T_c - T_0}, \tag{8.301}$$

where

$$T_c \approx \frac{1}{1 + 2(n-1)\varepsilon\ln(1/\varepsilon)}. \tag{8.302}$$

Values appropriate to Venus suggest failure then occurs when $\frac{q}{s} \approx 0.41$.

Numerical solution of the complicated free boundary problem (8.297) and (8.298) indicates that failure first occurs at a time t_f at a radial distance r_f, where these values depend explicitly but complicatedly on the parameters of the problem, which are, however, essentially known. The two main uncertainties are the value of the yield stress τ_c and the asthenospheric temperature T_a. Figure 8.9 shows the result of the calculation yielding the time of failure t_f and radial position of failure r_f as functions of τ_c for two plausible values $T_a = 1500$ K and $T_a = 1700$ K. We see that for a failure time of 500 Ma (million years), the shear stress would be about 160 bars, and the corresponding failure distance is between 200 and 800 km. This value of the yield stress is about half what one might expect on the Earth.

A notable feature of the solutions is the variation of the plastic effective viscosity η_q with radial distance. Computation of this shows that at the point of failure, η_q drops precipitately to the asthenospheric value when the radial distance is about $\frac{1}{2}r_f$, and remains close to this (within a factor of ten) thereafter. This suggests that at the time of failure, there will be a central lithospheric plug of essentially rigid material, surrounded by a uniformly failing concentric exterior.

Coronae on Venus are quasi-circular uplift features having typical radii in the range 100–300 km. They consist of a central domed plateau bounded by an escarpment which descends into a trench. These trenches have the topographic and gravity signatures of oceanic subduction trenches on the Earth, while it is thought that the coronae themselves are the consequence of the impingement of mantle plumes on the Venusian lithosphere. We thus see that the inferred nature of coronae is exactly

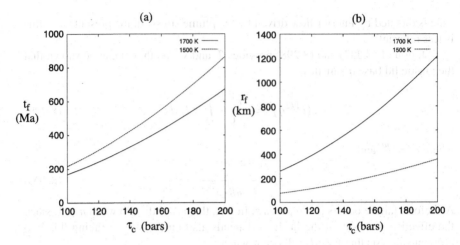

Fig. 8.9 Variation of (**a**) time of yield t_f and (**b**) radial location of failure r_f with yield stress, for values of $T_a = 1500$ K and $T_a = 1700$ K, with other parameters as appropriate for Venus; in particular, the rheological exponent is $n = 3.5$ and the surface temperature is 750 K

that associated with the beginning of subduction via the mechanism described here. Moreover, the time scale and failure radius are what one would expect, provided the long term effective yield stress for Venusian mantle rocks is of the order of 150 bars.

8.8 Notes and References

Mantle convection is described in a number of books, such as those by Davies (1999), Schubert et al. (2001) and Bercovici (2009a). The first of these tells the story, and treads lightly through the mush of equations which besets the presentation here. The second is much more theoretically inclined, while the third (part of the series known as the *Treatise on geophysics*) is a recent and up-to-date comprehensive summary of the current state of the art.

Low Rayleigh Number Convection   Rayleigh–Bénard convection was described experimentally by Bénard in 1900 and 1901, although it later transpired that his experimental results were actually due to Marangoni convection (Pearson 1958), and the phenomenon of convection had been described earlier by Thomson in 1882 and Count Rumford in 1797 (see Chandrasekhar 1981, which, apart from its description of the theory, also contains a useful historical summary; a more thorough historical review is given by Bercovici 2009b). The mechanism of instability was described by Rayleigh (1916). The nonlinear amplitude equation was first described by Malkus and Veronis (1958), thus ushering in one of the major areas of exploration for applied mathematicians in the 1960s and 1970s, the study of nonlinear stability, Hopf bifurcations, and their progeny of phase chaos, weak turbulence, and the like. The Ginzburg–Landau equation was derived in the context of convection by Newell and

Whitehead (1969) and Segel (1969). Later expositions are given by Balmforth et al. (2001) and Ribe (2009); Eq. (8.130) takes the same form (*A* is defined slightly differently), but differing versions of the diffusion coefficient are reported. The value here (4) is the same as that given by Newell and Whitehead (at infinite Prandtl number).

The leading figure in the analysis of finite amplitude convection and its bifurcation in the vicinity of its onset is Fritz Busse; a summary of his results dating back to 1965 is in his review (Busse 1985).

The Theory of Continental Drift   Wegener's book on continental drift, *The origin of continents and oceans*, was published in German in 1915, and went through four editions, the last published in 1929, a year before his premature death during an expedition on the Greenland ice sheet (McCoy 2006). The third edition was translated into French, English, Russian, Swedish and Spanish, and the fourth edition was translated into English and published by Dover in 1966, and for English-speaking audiences this is the most accessible version (Wegener 1966). Wegener was not the only scientist who proposed continental drift, for example the American scientist F.B. Taylor also proposed a version.

Wegener propounded his thesis by weight of observations, but lacked a credible mechanism. The hypothesis that convection could be this mechanism was largely due to Arthur Holmes, who proposed it as an explanation in a series of papers in the 1920s and 1930s. His thesis is summarised in his book, Principles of physical geology, whose first edition appeared in 1944, in the midst of the period of geological unbelief; the second edition appeared in 1965, when the plate tectonic revolution had occurred. The third edition, edited and revised by his widow Doris Reynolds, was published in 1978 (Holmes 1978). As mentioned in the preface, this book surveys almost the whole field of geoscience.

The mystery remains, why did Wegener's hypothesis and Holmes's theory not gain acceptance until the 1960s, and even then (and now), geophysicists still draw a screen over their predecessors' failings, suggesting that proper geophysical evidence did not appear until the sea floor palaeomagnetism studies of the 1960s, as if all the evidence that Wegener had accumulated was not good enough. The study of this denial is an interesting subject in itself for the history and philosophy of science, similar in many ways to the Copernican revolution (Koestler 1964), the transition from scriptural science to geology at the beginning of the nineteenth century (Winchester 2001; Cadbury 2000), and many other past and ongoing controversies, mentioned elsewhere in these pages. Two particular books detailing the acceptance of plate tectonics in a historical context are those by Le Grand (1988) and Oreskes (1999). Sub-controversies within the study of mantle convection include the importance of radioactive heating, the nature of the plate-driving forces, and the plume hypothesis. At least some of the disagreements concerning these latter topics arise implicitly through misunderstanding of the way in which mathematical models of the processes should be interpreted.

High Rayleigh Number Convection   The study of boundary layer convection as $R \to \infty$ was initiated by Pillow (1952), and in the geophysical context in a seminal paper by Turcotte and Oxburgh (1967). Other early analyses were by Robinson (1967) and Wesseling (1969). There was some disagreement between these various results, and it awaited the comprehensive papers of Roberts (1977, 1979) to essentially resolve the differences. Roberts's 1979 paper is analytically correct, except in one point which we mentioned before, but his work has been criticised for numerical inaccuracy by Jimenez and Zufiria (1987). Olson and Corcos (1980) adapted Turcotte and Oxburgh's analysis to the case where the surface (plate) moves at constant velocity.

Much of the confusion in the different analyses may be considered to lie with the fact that the approaches have been more or less heuristic, and have not used explicit asymptotic expansions. The exception is the work of Jimenez and Zufiria (1987).

Various numerical results confirm the trends of these analyses, for example Moore and Weiss (1973) and Jarvis and McKenzie (1982). It is worth emphasizing that the estimated error in $Nu/R^{1/3}$ may not be that small, however, and that it is unlikely that numerical computations have been done at sufficiently high R to deliver adequate quantitative agreement. Nor have such computations ever been carried out in a way that would indicate numerical agreement, for example by plotting $Nu/R^{1/3}$ versus R.

Variable Viscosity   The analytic study of strongly temperature-dependent viscosity at high Rayleigh number was done by Morris and Canright (1984) and Fowler (1985a). The two studies are essentially similar, but differ in detail. Morris and Canright assume the base of the stagnant lid is flat, and thus do not encounter the large stresses which occur in the lid. Fowler considered this case, but thought it less likely than the case where the lid base is sloped, and this latter case seems to be more like the numerically computed results.

The development of numerical methods for strongly temperature-dependent viscous convection owes its inspiration to Christensen and co-workers (see, e.g., Christensen 1984a, 1984b, 1985; Christensen and Harder 1991), but these early results were limited to viscosity variations of 10^6 or so. It is not until the later computations of Solomatov and co-workers (Moresi and Solomatov 1995; Solomatov 1996; Reese et al. 1999; Reese and Solomatov 2002) that larger viscosity variations were obtained, which are more appropriate for inferring mantle behaviour, and for making comparisons with the asymptotic results. Nataf and Richter (1982) conducted laboratory experiments.

If strong pressure dependence is included as well, there is little to indicate what the appropriate limiting behaviour is. Figure 8.10 shows the results of a computation which shows the stagnant lid clearly enough, but there is no clear asymptotic structure visible. Apparently, there have been no clear computational results which suggest an appropriate limiting behaviour, and there has been no asymptotic analysis able to treat the situation where the temperature and pressure dependence are equally strong. The viscosity, when scaled, takes the approximate form (from (8.174))

$$\eta = \frac{\Lambda}{\tau^{n-1}} \exp\left[\frac{1 - T + \mu(1 - z)}{\varepsilon T}\right], \tag{8.303}$$

Fig. 8.10 Stream function contours of a temperature- and pressure-dependent viscosity convection calculation at Rayleigh number 10^6, the viscosity scaled to that at the basal temperature. The rheology is that of (8.303), with $\Lambda = 1$, $n = 1$, $\varepsilon = 0.077$ and $\mu = 0.5$. Basal and surface dimensionless temperatures are 1 and 0.1, respectively. The absence of contours towards the top indicates the stagnant lid. Figure courtesy Mike Vynnycky

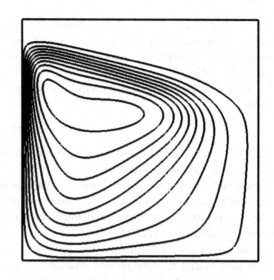

where z is scaled height above the core–mantle boundary, and representative values are $\varepsilon \sim 0.023$, $\mu \sim 2.8$.

Conventional wisdom has it that the vigorous, high Rayleigh number flow in the mantle is adiabatic; by analogy with the atmosphere (see Chap. 3) we balance the advective terms in $(8.8)_6$:

$$\rho \frac{dT}{dt} - DT \frac{dp}{dt} \approx 0, \tag{8.304}$$

which gives below the lithosphere (even allowing that $C > 0$)

$$T = T_{\text{ad}} \approx \exp[DZ], \tag{8.305}$$

where $Z = 1 - z$ is the dimensionless depth; thus the adiabatic temperature increases roughly exponentially with depth. Earlier we found $D \approx 0.9$, but this value probably decreases with increasing depth because of the decreasing value of α. In contrast, we may define an isoviscous profile in which (from (8.303), and ignoring stress dependence),

$$T = T_{\text{iso}} \approx 1 + \mu Z. \tag{8.306}$$

(8.306) tacitly assumes $C = 0$ in $(8.8)_5$. If we take $C > 0$, then

$$T = T_{\text{iso}} \approx 1 + \frac{\mu}{C}\left(e^{cZ} - 1\right)Z. \tag{8.307}$$

The isoviscous and adiabatic profiles are quite different in general, and $T_{\text{iso}} > T_{\text{ad}}$, even if one allows for the decrease of both D and μ with increasing depth. Since, if $T \approx T_{\text{ad}}$,

$$\eta \sim \exp\left[\frac{T_{\text{iso}} - T_{\text{ad}}}{\varepsilon T_{\text{ad}}}\right], \tag{8.308}$$

we see that a sub-asthenospheric adiabatic temperature will cause the mantle viscosity to increase dramatically and exponentially, and this must have the effect of reducing the velocity, thus removing the reason why the temperature was adiabatic in the first place.

What this suggests is that the sub-asthenospheric temperature is close to isoviscous,

$$T \approx T_{\mathrm{iso}} + O(\varepsilon), \tag{8.309}$$

with the small correction providing both the unknown viscosity and the buoyancy term, while the energy equation determines the vertical velocity, and mass conservation then gives the horizontal velocity. Various shear and thermal boundary layers would be necessary to complete the description of the flow, but determining these seems to be quite a hard problem, and has not yet been done. See also Fowler (1993a) and Quareni et al. (1985).

An isoviscous mantle temperature provides a nice explanation of why independent post-glacial rebound studies invariably indicate relatively constant estimates of mantle viscosity below the lithosphere (Cathles 1975), and it explains how the mantle temperature can reach a value of some 4,000 K at the core–mantle boundary (Jaupart et al. 2009), despite a relatively low adiabatic temperature there, and the impossibility of basal thermal boundary layer jumps of more than a few hundred degrees (because of the strongly variable viscosity) (Fowler 1983). Figure 8.11 shows a temperature- and pressure-dependent viscous computation, corresponding to Fig. 8.10, wherein only the viscosity variation in the convecting core is indicated. As is the typical case, the parameters are not extreme enough to indicate what the appropriate asymptotic régime is.

The Issue of Subduction   It was realised fairly early on that variable viscosity convection caused a stagnant lid to occur, and that subduction would only occur if some form of weakening was made. Initially in numerical models, artificial weak zones were introduced, but more sophisticated strain-weakening rheologies were later introduced, and shown to produce self-consistent subduction-like behaviour (Bercovici 1993; Tackley 1998, 2000a, 2000b: see also the review by Bercovici 2003). These authors sometimes use a rheology in which stress τ is a function of strain rate $\dot{\varepsilon}$ which first increases to a maximum and then decreases (a 'pseudo-plastic' rheology), or a 'visco-plastic' rheology with a specific yield stress, similar to that used here. The present discussion follows Fowler (1993b) and Fowler and O'Brien (1996, 2003); illuminating further insight on the transition between stagnant and mobile lids is given by Moresi and Solomatov (1998), who show that the inclusion of a yield stress in a numerical model of strongly temperature-dependent viscous convection can lead either to a fixed lid, a mobile lid, or a cycling between the two, depending on the value of the yield stress. This resembles the putative behaviour apparently observed on Venus.

A hallmark of most of these failure theories is the relatively low value of yield stress which is necessary to initiate subduction. The values here, about 100–200 bars (10–20 MPa) for Venus, and somewhat higher for Earth, are much lower than

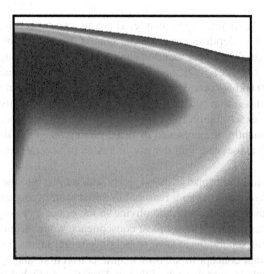

Fig. 8.11 Viscosity contours of a temperature- and pressure-dependent viscosity convection calculation at Rayleigh number 10^6, the viscosity scaled to that at the basal temperature. The rheology is that of (8.303), with $\Lambda = 1, n = 1, \varepsilon = 0.077$ and $\mu = 0.5$. Basal and surface dimensionless temperatures are 1 and 0.1, respectively. The viscosity variation from top to bottom is of order 10^{20}, so that the upper part of this range (in the lid) is excised. The resulting colour indicates a variation from (dimensionless) viscosity 10^{-2} (*blue*) to 10^4 (*red*). Figure courtesy Mike Vynnycky

the brittle strength of near-surface crustal rocks. However, it is not thought that brittle failure is relevant in the lower parts of the lithosphere, but rather that various mechanisms, such as dynamic recrystallisation and void formation (Tackley 1998), may promote the formation of weak shear zones. Once these exist, analogously to faults such as the San Andreas fault, it is easy to suppose that they remain weak and promote slip. It is less easy to imagine the process whereby they form initially, at the onset of subduction.

Sub-continental Convection   The success of mantle convection theory is most obvious for sub-oceanic convection. The plates move away from mid-ocean ridges, causing a square root of age decrease in heat flux, essentially as observed (Parsons and Sclater 1977), because of the similarity solution of the thermal boundary layer equation. There is no such comparable law for continents, which do not fit the concept of active plate convection. Continental lithosphere is often supposed thicker, reflecting the lower value of heat flux despite the concentration of heat-producing elements (Davies and Davies 2010). In fact, the simplest interpretation of sub-continental convection is to suppose that it is of the stagnant lid type. If that is the case, then the Earth's mantle consists of adjoining convective cells of very different types.

Phase Changes and Geochemistry   There is a good deal concerning mantle convection which we have simply passed over. Perhaps the most serious omission is that of the chemical structure of the mantle. There are two aspects to this. The most fundamental is that as depth increases in the mantle, there are a number of phase changes which occur. The upper mantle below the lithosphere is largely thought to consist of olivine, $(Mg,Fe)_2SiO_4$, and this undergoes a transformation to a spinel phase between 400 and 500 km depth. A further transition occurs at 650 km, where the spinel dissociates into a perovskite phase $(Mg,Fe)SiO_3$ and wüstite $(Mg,Fe)O$. Depending on the exact composition, other phase changes may occur at different pressures. In the mantle, the presence of these phase changes is detected seismically, and they are associated with density increases of several percent (Anderson 2007).

(Some) descending lithospheric slabs clearly sink (at least) to the vicinity of the 650 km seismic discontinuity, so a reasonable initial simplifying assumption may be that of a two layer mantle, with the olivine in the upper mantle separated from the perovskite lower mantle. The simplest consequence would seem to be that convection might occur separately in two, or possibly more, layers. In fact, this idea underlies the original concept of shallow mantle convection, and underpinned the choice of depth scale in models such as that of Turcotte and Oxburgh (1967). Later investigations raised doubts that the density jump would be sufficient to prevent whole-mantle convection, and many studies now assume this. The issue revolves round the magnitudes of the relative density jump across a phase change boundary, $\frac{\Delta\rho}{\rho}$, the corresponding buoyancy term $\alpha\Delta T$, and the slope of the Clapeyron curve relating the phase change pressure to the temperature. At least for the descending lithospheric slabs, we may take $\Delta T \sim 10^3$ K, and thus $\alpha\Delta T \sim 3 \times 10^{-2}$. This may be comparable to the density increase across the 650 km boundary. Numerical studies (Christensen and Yuen 1984, 1985) suggest that at least some form of penetration is likely, and this is consistent with ideas of hot spots associated with plumes originating at the core–mantle boundary.

However, layered convection seems the easiest explanation for the inference, from geochemical studies of different magmas at the surface of the Earth, that the mantle consists of at least two, and possibly more, distinct reservoirs, which have been chemically isolated for much of the Earth's history (Tackley 2009). This inference is based on the different trace elements present in the erupted magmas. Mid-ocean ridge basalts (MORB) are depleted in the so-called incompatible trace elements, while the continental crust is enriched in these same elements, which leads us to suppose that the continents form as the residue from continual melting of the MORB source region. Mass balance calculations suggest that this source only occupies around half the mantle (the estimates vary a good deal). The easiest vision is to suppose that the phase change at 650 km causes a form of 'leaky' (see below) layered convection, thus providing the separate mantle reservoirs.

However, this simple picture is increasingly complicated by further geochemical signatures. One such is that ocean island basalts (OIB, such as Hawaii) have anomalous helium isotope ratios, suggesting that they originate from a primordial reservoir. Most simply, this is construed to be the lower mantle, and that is consistent with the idea that they come from mantle plumes, which presumably penetrate the barrier at 650 km. Various other complications arise, and have led to various different conceptualisations of how convection works in practice. It is important to note that these discussions of geochemical signatures ignore the possible importance of transport processes associated with eruptions.

Two further seismic zones warrant mention. The first is a low velocity zone just below the lithosphere. The easiest interpretation for this is that it represents the convective overshoot in temperature which typically arises in high Rayleigh number convection, possibly associated with the presence of partial melt. The other anomalous region is the D'' layer of some 200 km thickness above the core–mantle boundary. There are several ways this layer can be construed. One is that it arises from a phase change from perovskite to a 'post-perovskite' phase. Since the D'' layer is very variable in thickness, this may not be sufficient. A dynamical interpretation is that it represents the remnants of foundered subducted lithospheric slabs; this is attractive to some parts of the geochemical story, as well as to seismological inference that slabs do in some places penetrate to the lower mantle. Such penetration does not argue against an essentially layered style of convection, since the negative buoyancy associated with slabs is much larger than elsewhere in the mantle. Foundering slabs also suggest (à la Howard) a nice explanation for the semi-regular massive flood basaltic eruptions associated with superplumes (Courtillot 1999; Yuen et al. 2007).

Lastly, the interface between the molten iron (or iron oxide) of the core and the mantle is a surface of phase change with an associated phase diagram which must describe the reactions which occur there. If we suppose that the primordial Earth consists of a liquid iron Fe core surrounded by a perovskite $(Mg,Fe)SiO_3$ mantle, then Knittle and Jeanloz (1991) found experimentally that reactions occurred in which FeO and SiO_2 were formed:

$$(Mg_xFe_{1-x})SiO_3^S + a_1Fe^L \rightarrow a_2MgSiO_4^S + a_3FeO^L + a_4FeSi^L + a_5SiO_2^S; \quad (8.310)$$

here a_i are various stoichiometric coefficients. As with all such reactions where the solidus and liquidus temperatures depend on pressure and concentrations, it is possible or likely that a region of mixed phase may exist in the solid. Knittle and Jeanloz's suggestion is that such a putative region corresponds to the D'' layer. This reaction incidentally provides a nice reason why the liquid outer core of the Earth is an alloy of iron with some lighter element. The core–mantle boundary reaction produces the iron oxide which dissolves in the outer core, while the inner core solidifies at its expense, releasing oxygen-rich liquid which will tend to pond at the top of the core (because the motion is so sluggish).

How does all this tie in to our notion of variable viscosity convection? For stagnant lid convection, the lid thickness is quite substantial, both in numerical computations and as predicted by (8.178) and (8.184), and the whole upper mantle might well be in the stagnant lid, so that the phase change would have little effect. And if, when subduction is initiated, the slabs do indeed penetrate the lower mantle in places, then this will induce (locally) rapid convection in the lower mantle, as well as deposition of the slab at the base, and the subsequent eruption of a superplume.

One objection to layered convection which has been made in the past is that 'there would be a strong thermal boundary layer at the base of the upper mantle' (Tackley 2009). This is an unfounded inference based on constant viscosity convection, but inappropriate for variable viscosity convection. Indeed, it seems that variable viscosity convection with subduction can provide a picture which is fairly consistent with most of the basic features which have been observed.

Howard's 1966 Convection Paper   Lou Howard's vision of convection (Howard 1966), discussed in Sect. 8.7, was presented at the International Congress of Applied Mechanics in Münich in 1964. For the most part it is a discursive review of earlier work on turbulent convection by Malkus, Spiegel, and others, now rather dated. It is only in the final two pages that Howard advances his conceit of turbulent thermal convection as consisting of long, quiescent phases where thermal boundary layers grow from the heated and cooled surfaces into the interior, interrupted by rapid convective eruption of plumes as the boundary layers become unstable. As a paradigm, it is essentially identical to the concept of episodic Venusian convection.

Howard's conception, together with the publication in 1963 of Lorenz's famous paper (Lorenz 1963), led to a very productive sequence of research ideas based round the activity at the GFD summer program at the Woods Hole Oceanographic Institution. In particular, Howard and Malkus invented their famous water wheel (e.g., Matson 2007), which provides a bridge between the Lorenz equations and the essentially similar behaviour in the Howard convection model.[6] Malkus's experiment is described by Sparrow (1982), and an attempt to draw an explicit comparison between Howard's rough description and a formal asymptotic solution of the Boussinesq convection equations was made by Fowler (1992b).

8.9 Exercises

8.1 It is required to show that a vector potential $\mathbf{A}$ for an incompressible velocity field $\mathbf{u}$ in a domain D can be defined so that

$$\mathbf{u} = \operatorname{curl} \mathbf{A}, \qquad \operatorname{div} \mathbf{A} = 0.$$

Suppose that $\boldsymbol{\psi}$ is chosen to satisfy

$$\nabla^2 \boldsymbol{\psi} = -\mathbf{u} \quad \text{in } D,$$

with boundary condition

$$\boldsymbol{\nabla}.\boldsymbol{\psi} = 0 \quad \text{on } \partial D.$$

Using the Cartesian identity

$$\nabla^2 \equiv \operatorname{grad} \operatorname{div} - \operatorname{curl} \operatorname{curl},$$

---

[6]Indeed, one can show that the behaviour of the Lorenz equations in the limit of large 'Rayleigh' and Prandtl numbers r and σ is equivalent to Howard's physical description (Fowler and McGuinness 1982).

show that $\nabla^2(\nabla.\psi) = 0$ in D, and deduce that $\nabla.\psi = 0$. Deduce that a suitable vector potential is $\mathbf{A} = \nabla \times \psi$.

Show that if $D = \mathbf{R}^3$, the vorticity is $\omega = \text{curl}\,\mathbf{u}$ and $\mathbf{u} \to 0$ as $|\mathbf{r}| \to \infty$, then

$$\mathbf{u} = \frac{1}{4\pi} \int_{\mathbf{R}^3} \frac{\text{curl}\,\omega(\mathbf{r}')\,dV(\mathbf{r}')}{|\mathbf{r} - \mathbf{r}'|}.$$

Show also that a general representation of the velocity is of the form

$$\mathbf{u} = \text{curl}\,\psi\mathbf{j} + \text{curl}\,\text{curl}\,\chi\mathbf{j},$$

and derive the form of the Boussinesq equations in terms of ψ, χ and T.

8.2   Explain how high Rayleigh number, variable viscosity convection can be used to explain the styles of mantle convection on Venus and Earth, and use it to explain and interpret the following observations on the Earth:

volcanism occurs at plate boundaries;
earthquakes occur at plate boundaries;
the continents drift relative to each other;
earthquakes occur regularly on the San Andreas fault;
oceanic trenches occur at convergent plate boundaries;
the oceans are relatively shallow at mid-ocean ridges;
black smokers occur at mid-ocean rises;
chains of islands occur in the Pacific ocean (e.g., the Hawaiian islands);
ocean island basalts are distinct from mid-ocean ridge basalts.

8.3   Explain what is meant by post-glacial rebound, and how it can be used to infer values for the mantle viscosity.

Explain how the mantle can behave like a fluid even though it is solid.

The viscosity of mantle rock is measured to be of the form

$$\eta = \frac{1}{2A\tau^{n-1}} \exp\left[\frac{E^* + pV^*}{RT}\right],$$

where typical values are (with large error bars) $A = 10^5$ MPa^{-n} s^{-1}, $n = 3.5$, $E^* = 525$ kJ mol^{-1}, $R = 8.3$ J mol^{-1} K^{-1}, and $V^* = 17$ cm^3 mol^{-1}. Use these values to infer likely values of the mantle viscosity at the base of the lithosphere if $T = 1500$ K, $\tau = 10^6$ Pa, and the depth is 100 km (assume pressure is lithostatic, i.e. $p = \rho g z$, where z is depth, $g \approx 10$ m s^{-2}, and $\rho \approx 3.3 \times 10^3$ kg m^{-3}). Is this value consistent with the post-glacial rebound value of 10^{21} Pa s? If not, how big would the error bars on E^* need to be to make it consistent?

8.4   Write down the Boussinesq equations describing convection of a constant viscosity fluid heated from below (assume acceleration terms are negligible). Explain what the Boussinesq approximation means. What are suitable boundary conditions for convection in the Earth's mantle? By choosing suitable scales for the variables, write the model in non-dimensional form, and deduce that the flow depends only on a single dimensionless parameter, the Rayleigh number, and define what this is.

If the acceleration term had been included (thus, dimensionally, $\rho \frac{d\mathbf{u}}{dt} = -\nabla p + \cdots$), show that the size of the inertial acceleration terms is given by $1/Pr$, where $Pr = \eta c_p/k$ is the Prandtl number. Assuming $c_p \approx 10^3 \, \text{J kg}^{-1} \, \text{K}^{-1}$, $k \approx 4 \, \text{W m}^{-1} \, \text{K}^{-1}$, and $\eta \approx 10^{21} \, \text{Pa s}$, estimate the importance of the inertia term.

8.5 Write down a dimensionless set of equations describing Boussinesq convection of a high Prandtl number, constant viscosity fluid in a horizontal layer of fluid, and define the Rayleigh number R.

Using suitable boundary conditions for the flow describing convection in the Earth's mantle, show in detail that convection will occur if $R > R_c$, where you should define R_c.

Use suitable values for the Earth's mantle to show that $R \gg R_c$ (assume $\alpha = 3 \times 10^{-5} \, \text{K}^{-1}$, $\Delta T = 3000 \, \text{K}$, $\rho = 4 \times 10^3 \, \text{kg m}^{-3}$, $g = 10 \, \text{m s}^{-2}$, $\eta = 10^{21} \, \text{Pa s}$, $\kappa = 10^{-6} \, \text{m}^2 \, \text{s}^{-1}$, and that the depth to the core–mantle boundary is 3000 km).

8.6 The amplitudes of orthogonal sets of weakly nonlinear convective rolls are described by the equations

$$\dot{A} = \frac{2}{9\pi^2} A - \frac{\pi^2}{16} A(A^2 + cB^2),$$

$$\dot{B} = \frac{2}{9\pi^2} B - \frac{\pi^2}{16} B(B^2 + cA^2),$$

where an overdot indicates differentiation with respect to the slow time variable τ.

Show by rescaling the variables that the model can be reduced to the form

$$\dot{A} = A - A(A^2 + cB^2),$$

$$\dot{B} = B - B(B^2 + cA^2).$$

Find the steady states, and calculate their stability in the two cases $c < 1$ and $c > 1$, and draw the consequent phase planes in each case.

8.7 The slowly varying amplitude of weakly nonlinear convective rolls satisfies the scaled equation

$$A_T = A - A^3 + A_{XX},$$

where T is a slow time scale and X is a long space scale. Write down the equation satisfied by a travelling wave solution $A = f(\xi)$, where $\xi = X - cT$.

Assuming $c > 0$ and $g = -f'$, write down the equations satisfied by f and g, and, by analysing the phase plane, show that there is always a solution connecting $(1, 0)$ to $(0, 0)$. [It may help to consider the function $E = \frac{1}{2}g^2 + \frac{1}{2}f^2 - \frac{1}{4}f^4$.]

Show that if $c < 2$, the origin in the (f, g) plane is a stable focus, while if $c > 2$, it is a stable node. Deduce that for $c < 2$, travelling wave solutions have oscillatory tails in which $A < 0$.

Now consider the slope $s = \frac{g}{f}$ of the chord joining a point on the travelling wave trajectory in the phase space to the origin. Show that initially s is positive and increasing. As long as $0 < f < 1$, show by consideration of $\frac{ds}{d\xi}$ as a function of s that $s < \frac{c}{2}$, providing $c \geq 2$, and deduce that in this case f (and thus A) remains positive as it approaches zero.[7]

What is the implication for convection in a large pan, if the motion is initiated locally?

8.8 Describe what is meant by *continental drift*, and how the theory of plate tectonics and mantle convection can be used to explain it. How do mid-plate volcanoes, such as Hawaii, fit in with this theory?

The scaled Boussinesq equations for two-dimensional thermal convection at infinite Prandtl number and large Rayleigh number R in $0 < x < a, 0 < z < 1$, can be written in the form

$$\omega = -\nabla^2 \psi,$$

$$\nabla^2 \omega = \frac{1}{\delta} T_x,$$

$$\psi_x T_z - \psi_z T_x = \delta^2 \nabla^2 T,$$

where $\delta = R^{-1/3}$. Explain what is meant by the Boussinesq approximation, and explain what the equations represent. Explain why suitable boundary conditions for these equations which represent convection in a box with stress free boundaries, as appropriate to convection in the Earth's mantle, are given by

$$\psi = 0, \qquad \omega = 0, \quad \text{on } x = 0, a, \quad z = 0, 1,$$

$$T = \frac{1}{2} \text{ on } z = 0, \qquad T = -\frac{1}{2} \text{ on } z = 1, \qquad T_x = 0 \text{ on } x = 0, a.$$

Show that, if $\delta \ll 1$, there is an interior 'core' in which $T \approx 0$, $\nabla^4 \psi = 0$.

By writing $1 - z = \delta Z$, $\psi = \delta \Psi$ and $\omega = \delta \Omega$, show that $\Psi \approx u_s(x) Z$, and deduce that the temperature in the thermal boundary layer at the surface is described by the approximate equation

$$u_s T_x - Z u_s' T_Z \approx T_{ZZ},$$

with

$$T = -\frac{1}{2} \text{ on } Z = 0, \qquad T \to 0 \text{ as } Z \to \infty.$$

---

[7] The same technique can be used for the Fisher equation.

If u_s is constant, find a similarity solution, and show that the scaled surface heat flux $q = \partial T/\partial Z|_{Z=0}$ is given by

$$q = \frac{1}{2}\sqrt{\frac{u_s}{\pi x}}.$$

8.9 Suitable equations to describe mantle convection in a radial geometry are

$$\frac{1}{r}\frac{\partial(ru)}{\partial r} + \frac{\partial w}{\partial z} = 0,$$

$$\frac{\partial p}{\partial r} = \frac{1}{r}\frac{\partial(r\tau_{rr})}{\partial r} + \frac{\partial \tau_{rz}}{\partial z} + \frac{\tau_{rr} + \tau_{zz}}{r},$$

$$\frac{\partial p}{\partial z} = \frac{1}{r}\frac{\partial(r\tau_{rz})}{\partial r} + \frac{\partial \tau_{zz}}{\partial z} + \rho_a\big[1 - \alpha(T - T_a)\big]g,$$

$$\tau_{rr} = 2\eta\frac{\partial u}{\partial r},$$

$$\tau_{zz} = 2\eta\frac{\partial w}{\partial z},$$

$$\tau_{rz} = \eta\left[\frac{\partial u}{\partial z} + \frac{\partial w}{\partial r}\right],$$

$$\eta = \frac{1}{2A\tau^{n-1}}\exp\left[\frac{E^*}{RT}\right],$$

$$2\tau^2 = 2\tau_{rz}^2 + \tau_{rr}^2 + \tau_{zz}^2 + (\tau_{rr} + \tau_{zz})^2,$$

$$\frac{\partial T}{\partial t} + u\frac{\partial T}{\partial r} + w\frac{\partial T}{\partial z} = \kappa\nabla^2 T.$$

Here, r and z are radial and vertical coordinates, u and w the corresponding velocity components, and τ_{rr}, τ_{zz} and τ_{rz} the radial normal deviatoric stress, vertical normal deviatoric stress and shear stress, respectively.

Scale the equations by writing

$$p - \rho_a gz, \tau_{rr}, \tau_{rz}, \tau_{zz} \sim \frac{\eta_a \kappa}{d^2},$$

$$T \sim T_a, \qquad \eta \sim \eta_a, \qquad t \sim \frac{d^2}{\kappa}, \qquad x \sim d, \qquad u \sim \frac{\kappa}{d},$$

and show that the resulting dimensionless equations depend on the parameters

$$Ra^* = \frac{\alpha\rho_a g T_a d^3}{\eta_a\kappa}, \qquad \Lambda = \frac{1}{2A\eta_a^n}\left(\frac{d^2}{\kappa}\right)^{n-1}\exp\left(\frac{E^*}{RT_a}\right), \qquad \varepsilon = \frac{RT_a}{E^*}.$$

$$(\dagger)$$

Now suppose that the dimensionless thickness of the lithosphere is v. Rescale the variables by writing

$$z \sim v, \qquad u \sim Ra^{*\,3/5}v, \qquad w \sim Ra^{*\,3/5}v^2, \qquad t \sim v^2,$$

$$\eta \sim \left(v\varepsilon Ra^{*\,1/5}\right)^2, \qquad \tau \sim \tau_{rz} \sim v^2 Ra^*, \qquad p \sim v Ra^*,$$

$$\tau_{rr} \sim \tau_{zz} \sim v^3 Ra^*,$$

and write down the rescaled equations in terms of the parameters ε, v, and ω, where

$$\omega = v\varepsilon Ra^{*\,1/5}, \qquad (*)$$

and we have defined

$$\Lambda = Ra^{*\,3(n-1)/5}\omega^{2n}. \qquad (\ddagger)$$

Show that this defines an asthenospheric viscosity scale $\eta_A = \eta_a \omega^2$ given by

$$\eta_A = \frac{1}{2A}\left[\frac{E^{*2}}{\alpha \rho_a g d v^2 R^2 T_a^3}\right]^{n-1} \exp\left(\frac{E^*}{R T_a}\right),$$

and that if

$$Ra = \frac{\alpha \rho_a g T_a d^3}{\eta_A \kappa},$$

then

$$\omega^3 = (v\varepsilon)^{1/5} Ra.$$

Now compare this to the scales used in Sect. 8.5, where a viscosity scale η_0 was used, and parameters Ra_0 and Λ_0 defined as above using η_0 instead of η_a. There we defined

$$(v\varepsilon)^5 Ra_0 = \varepsilon^{\frac{n-1}{n+1}}. \qquad (**)$$

With this definition, show that, using $(*)$ above,

$$\omega^3 = \varepsilon^{\frac{n-1}{n+1}} \frac{Ra}{Ra_0}.$$

Use Eqs. $(\dagger)$, $(\ddagger)$, (8.184) and (8.185) to show that

$$\omega^{6(n-1)} = \left(\varepsilon^{\frac{n-1}{n+1}} \frac{Ra}{Ra_0}\right)^{2n+3},$$

and deduce that if v is defined using $(**)$, then $\omega = 1$.

8.10 Two-dimensional convection of a temperature- and pressure-dependent Newtonian viscous fluid at infinite Prandtl number is described by the dimensionless equations

$$p_x = \tau_{1x} + \tau_{3z},$$

$$p_z = \tau_{3x} - \tau_{1z} - R(1 - T),$$

$$\tau_1 = -2\eta\psi_{xz},$$

$$\tau_3 = \eta(\psi_{xx} - \psi_{zz}),$$

$$\psi_x T_z - \psi_z T_x = \nabla^2 T,$$

$$\tau^2 = \tau_1^2 + \tau_3^2,$$

$$\eta = \exp\left[\frac{1 - T + \mu(1 - z)}{\varepsilon T}\right],$$

where

$$R \gg 1, \qquad \varepsilon \ll 1, \qquad \mu \sim 1,$$

and the velocity is given in terms of the stream function by

$$\mathbf{u} = (-\psi_z, \psi_x).$$

Define the isoviscous temperature

$$T_{\text{iso}} = 1 + \mu(1 - z),$$

and put

$$T = T_{\text{iso}} + \varepsilon\phi;$$

show that the model can be reduced to the form

$$\left[\frac{\partial^2}{\partial z^2} - \frac{\partial^2}{\partial x^2}\right]\{\eta(\psi_{zz} - \psi_{xx})\} + 4\frac{\partial^2}{\partial x\partial z}(\eta\psi_{xz}) = R\varepsilon T_{\text{iso}}\frac{\eta_x}{\eta},$$

$$-\mu\psi_x = \varepsilon[\nabla^2\phi + \psi_z\phi_x - \psi_x\phi_z].$$

Deduce that a possible structure for the flow in the interior of the convection cell is a horizontal flow with $\psi = \psi(z)$, in which case the viscosity is
determined by solution of

$$\left[\frac{\partial^2}{\partial z^2} - \frac{\partial^2}{\partial x^2}\right](\psi''\eta) = R\varepsilon T_{\text{iso}}\frac{\eta_x}{\eta}.$$

Chapter 9
Magma Transport

Rocks which are formed at the Earth's surface are of two main types, igneous and sedimentary. A third type of rock, metamorphic, is one which has been subject to post-formation processes of thermal and chemical alteration, often due to the effects of elevated temperature and pressure on burial.

Sedimentary rocks are formed where sediments accumulate, at the bottom of lakes and oceans, and the different rock types are associated with different types of sediments. Shale is formed from clay, sandstone from sand, limestone from carbonate muds, often comprised of skeletal fragments of small marine organisms. The sediment grains form a rock when squeezed together, through the precipitation of cementing phases such as calcite in the intergranular pore space.

Igneous rocks, on the other hand, form when molten magma crystallises. This may occur deep within the Earth's crust in magma chambers, or else at the Earth's surface, when magma is extruded in volcanic eruptions. Such eruptions occur in many different ways, and the study of the dynamics of volcanic eruptions is a problem of interest in its own right. In this chapter, however, we will be more concerned with the processes whereby magma is able to move from deep in the Earth's mantle to regions near or at the Earth's surface.

All rocks in the Earth are silicates, that is to say the principal constituent is silica, SiO_2, and this typically provides more than half of the composition, usually combined with other compounds. Silicate rocks are formed by reaction of silica with various metal oxides, in particular the oxides of aluminium (Al_2O_3), iron (FeO and Fe_2O_3), magnesium (MgO), calcium (CaO), sodium (Na_2O) and potassium (K_2O), as well as water. Because all of these compounds occur in varying proportions, the subject of geochemistry which describes the chemistry of terrestrial rocks is an extremely complex one. The presence of many different components in mantle rocks means, for example, that phase diagrams to describe melting and solidification are extremely complicated, and the stability regions of different phases depends on the geochemical composition as well as temperature and pressure. The subject of igneous petrology has thus been principally concerned with the geochemical evolution of magmas, and it is only relatively recently that the corresponding physical evolution has been considered. In this chapter we are primarily concerned with the

A. Fowler, *Mathematical Geoscience*, Interdisciplinary Applied Mathematics 36,
DOI 10.1007/978-0-85729-721-1_9, © Springer-Verlag London Limited 2011

physical problem of how magma is transported to the Earth's surface, but it is wise to be aware of the underlying geochemical complexity.

Within the igneous rocks, the primary distinction lies in the silica content. Rocks with high (70%) silica content are called acid, and the principal such rock is granite. Rocks with low (50%) silica are called basic, and the principal type is basalt. Below this simple description, there are many, many further layers of sub-division.

Silica content has two major effects on physical properties. First, the viscosity of silicic magma is typically much higher than that of basalt. Typical values for basalt are 10^2 Pa s, while for rhyolite (the term refers to siliceous magma) it may be 10^5 Pa s, or higher. Thus rhyolite is hard to move. Second, the melting temperature of basalt is generally higher. The result of this is that when mantle rocks melt, they *differentiate*, and to a large extent, the different observed compositions of surface rocks reflect their different thermochemical histories during melting.

9.1 The Magmatic Cycle

Just as there is a hydrological cycle and a carbon cycle in the Earth, so there is a magmatic cycle. The magmatic cycle is bound up with the convection of the Earth's mantle, discussed in Chap. 8. The upwelling parts of the mantle circulation are of two types: the oceanic ridges are associated with the upwelling parts of essentially two-dimensional convection cells, while hot spots such as Hawaii are associated with isolated convective mantle plumes, which most probably originate at the core-mantle thermal boundary layer. Both of these features cause partial melting of the mantle, because of the excess heat and depressurisation associated with the upwelling. The depressurisation is the more important of the two, as the precepts of variable viscosity convection tell us that likely excess temperatures in plumes or upwellings are of the order of 100 K. A third origin for magma is behind subduction zones, where the frictional heat associated with the motion of the subducting slab causes partial melting of the mantle.

The magma thus generated flows upwards under its own buoyancy. It percolates through the asthenosphere, trickling as in a porous medium, and then collecting into rivulets and channels which allow the flow of magma through the lithosphere towards the continental crust. The buoyant driving force for the magma flow is the density difference between the liquid and the surrounding country rock. The continental crust is much lighter than the underlying mantle, and so the situation commonly occurs that the liquid reaches a level of neutral buoyancy, beyond which it cannot rise. At this point, which may be kilometres deep in the crust, the magma may spread out in horizontal sills, and in so doing it may uplift or melt the overlying crustal rocks to form lopoliths or laccoliths, large magma chambers tens of kilometres in extent, which then subsequently cool and crystallise over hundreds of thousands of years. Subsequent erosion of the overlying rocks has exposed examples of such igneous intrusions at the Earth's surface. A particularly dramatic example is that of the huge granite batholiths, which can be even larger.

This description is slightly misleading. The propagation upwards of magma in vertical conduits, or dykes, is accommodated by the extension of the dykes as opening fractures. While it is indeed the buoyancy which drives this upwards motion, and the accompanying fluid flow, the continuing propagation of the fracture requires the fracture toughness at the crack tip to be exceeded. One way upwards motion can be halted is by the arrival of the crack tip at a compositional unconformity, when the weakness of the unconformity may allow preferential propagation sideways to form a laccolith.

While it is possible for magma chambers to cool without further ado, it is frequently the case that continued filling of the chamber leads to fracture of the overlying crust, and consequent release of the magma at the surface. This is the mechanism of volcanic eruption. Volcanic eruptions occur in a number of different ways, and are described in Sect. 9.7. They are in essence periodic phenomena, being driven by the continual replenishment of the magma chamber and its consequent pressurisation, and the periodic depressurisation caused by eruptive release. The periodicity is driven by the freezing of the magma in the conduit to the surface between eruptions.

The eruptive style itself is extremely variable, ranging from the relatively mild and relatively continuous Strombolian and Hawaiian eruptions, to the explosive Plinian eruptions. The essential difference between eruptive styles is largely associated with the viscosity of the magma, and also its volatile content. As one might expect, larger volume flows are associated with larger periods between eruptions, and these can lead to very extensive lava flows.

In the geological past, there have been occasional episodes during which truly massive flood basalt eruptions have occurred. One such episode occurred some 66 My (million years) ago, and caused the formation of the kilometre thick Deccan Traps in India. These and other such massive eruptions have been associated with major extinctions of biota, through the effect on climate of the effused ash and gases in the atmosphere. In particular, the formation of the Deccan Traps occurs at much the same time as the final extinction of the dinosaurs. It is thought that such massive eruptive sequences may be associated with the arrival below the lithosphere of mantle 'super-plumes'. Such plumes would be likely to occur following foundering of subducted lithosphere at the core-mantle boundary, in a way similar to that describing the supposed style of episodic re-surfacing on Venus (Sect. 8.7).

The magmatic cycle is closed in disparate ways. Exsolution of gases such as water vapour and carbon dioxide to the atmosphere return to the surface via rainfall, reacting to form carbonate rocks for example. Ultimately, weathering and sediment transport returns these substances to the ocean, and the resulting sediments are subducted back into the Earth's mantle.

Not all magmatic products are returned, however. Magma forms by partial melting of mantle rocks in the asthenosphere. The so-called incompatible elements are those which are melted first, and thus melting and subsequent eruption causes a differentiation of the mantle composition. Additionally, the earlier melting components are lighter than the residue, and thus when erupted and crystallised, they form a lighter crust which remains at the surface. This, in fact, is how the continents were formed, and why they remain perched—floating, really—at the surface of the mantle. Although oceanic products of volcanism are returned directly at subduction zones, the products of continental volcanism are only slowly returned by weathering, and there is a long term continuing chemical differentiation of the mantle taking place.

9.2 Mechanisms of Magma Ascent

The basic physical problem which confronts us in igneous petrology is that of understanding why volcanoes occur at all; how does molten rock find itself at the Earth's surface, where the ambient temperature is about 300 K, far below the melting temperature of any mantle or crustal rock? Indeed, why does the mostly solid mantle melt at all?

The answer to this has two ingredients. The first is the circulation of the Earth's mantle. Mantle convection causes rock to rise beneath mid-ocean ridges, and also in isolated locations (hot spots) due to thermal plumes, such as at Hawaii and Iceland. Recirculation is associated with the subduction of oceanic lithosphere at oceanic trenches. In regions of upwelling, the mantle rocks are relatively hot. In subduction zones, the mantle rocks are wet (through their reaction with sea-water).

The second ingredient is the Clapeyron relation between melting temperature and pressure. The melting temperature of mantle rocks increases steeply as pressure increases. The effect of this is the following: under Hawaii, for example, the hot upwelling mantle cools adiabatically as it rises; but the melting temperature decreases more steeply with pressure, and thus melting can occur through the effect of decompression, as illustrated in Fig. 9.1.

Partial melting occurs as the rock rises, and it is usually thought that the melt fraction is no more than a few per cent. Partial (and not complete) melting occurs because the effective heat is supplied volumetrically—there is nothing to force a jump in heat flux, such as would be necessary to melt rock completely.

Melting is also associated with subduction zones, although the reasons are less clear. Volcanoes occur behind subduction zones (the Andes being a prime example), and the volcanism is presumably associated with the subducting lithospheric slab. Dehydration reactions in the slab as temperature and pressure increase may allow diffusion of water into the surrounding mantle rock, which has the effect of dramatically lowering its melting temperature. Partial melting may occur for this reason, but the nature of progress of the magma to the Earth's surface is enigmatic.

Fig. 9.1 Melting occurs when the pressure melting temperature lies below the mantle geotherm

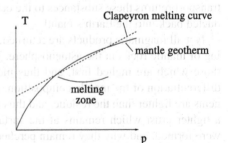

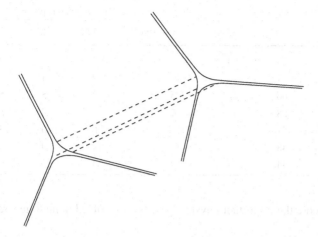

Fig. 9.2 Three-grain junctions form a connected network

Partial melting because of decompression occurs at depths of about 100 km, but it can be seen from Fig. 9.1 that the melt would typically refreeze a good way below the surface. How then does it get to the surface? The answer to this has two stages. For thermodynamic reasons, melt is first formed at four grain intersections, and then spreads along three-grain junctions as indicated in Fig. 9.2. These junctions form a connected tubular network, which allows the lighter melt to drain upwards through the now porous rock matrix.

A model describing how this melt transport occurs in the asthenosphere is described in Sect. 9.4. However, porous melt flow is unable to deliver magma to the Earth's surface (because the melt would simply freeze on to the base of the lithosphere, a process called underplating). In fact, we know that magma is supplied to magma chambers within the Earth's crust through cracks or fissures called dikes; we know this because erosion allows old basaltic dikes to be observed. They typically have widths of the order of a metre. In Sect. 9.5, we provide a model which describes the resulting rapid ascent of magma from an asthenospheric source region.

These parts of the story are relatively coherent. It is less clear how the conversion from porous medium flow to crack hydraulic flow takes place, and also how magma chambers are formed, but we will offer some thoughts on those processes too.

The process of mantle convection with partial melting allows us to explain the formation of the crust and the continents. An initially homogeneous mantle will become differentiated when it melts, because of the characteristics of melting and solidification for multi-component materials. The description of these characteristics is the domain of geochemistry, and we need to provide the rudiments of such a description in order to be able to discuss such issues as the problems of continent formation, or the emplacement of granite batholiths. It is also important to do so, because geochemical observations (for example, of differing chemical characteristics of mid-ocean ridge basalt (MORB) and ocean island basalts (OIB)) have been used to infer geodynamical constraints on the nature of mantle convection; mostly, these

Table 9.1 Approximate compositions (weight per cent) of common igneous rocks

	SiO_2	Al_2O_3	Fe_2O_3/FeO	MgO	CaO	Na_2O/K_2O
Granite	74	14	2	<1	1	9
Granodiorite	67	16	4	2	4	7
Quartz diorite	66	16	5	2	5	5
Andesite	58	17	7	3	7	5
Tholeiitic basalt	51	14	12	6	10	3
Olivine basalt	48	12	12	14	9	2
Peridotite	44	4	12	34	3	1

inferences ignore the important physical processes involved in melt generation and emplacement.

9.3 Phase Diagrams and Geochemistry

The subject of igneous petrology—the geochemical classification of different rocks—is one of fundamental difficulty, particularly for applied mathematicians. Our drive is to consider the simplest models of reality, but the petrologist's ambition is to draw ever more excruciating distinctions between rock types, on the basis of variations in chemistry, pressure, solid-state phase, texture, grain size, and so on. To quote Krauskopf and Bird (1995), "the resulting long catalogue of names ... continues to appal beginners in petrology". Thus the present description will be as simple as possible, although it will appear to be complicated enough.

Rocks are composed of grains of different minerals, which are themselves formed of various oxides. The composition of the Earth or of the crust is sometimes written as percentages of these oxides, or of the mineral assemblages. Table 9.1 shows a condensed version of oxide compositions; Fig. 9.3 shows a breakdown by mineral of the composition of the commoner types of igneous rocks. In keeping with the petrologist's need for complexity, the liquid form (magma) of a rock has a different name from its solid form: rhyolite is liquid granite, for example. Note also that rock types are composed of different minerals: thus granite is composed of feldspar, quartz and plagioclase, and varying minor amounts of other minerals.

The principal observation to take away from Table 9.1 is that the rocks show a more or less linear trend in composition. The acid rocks (granite, granodiorite) have high silica (SiO_2) content, while the basic rocks (olivine, peridotite) are silica poor. So the very simplest description of geochemistry would associate rock type with silica content.

In fact, Table 9.1 shows that this linear trend is more or less reflected in other oxides. As silica content decreases, iron and magnesium content increases (and so Mg/Fe rich magmas are called mafic, while silica-rich magmas are called felsic); aluminium stays the same, calcium increases, sodium and potassium decrease. Only peridotite replaces some aluminium and calcium with magnesium.

Fig. 9.3 Mineralogical composition (per cent by volume) of igneous rocks (after Mason and Moore 1982, p. 96). Minerals are nepheline (Ne), orthoclase (Or), quartz (Qz), biotite (Bt), hornblende (Hb), plagioclase (sodic (Na) to calcic (Ca)), pyroxene (Px), including clinopyroxene (Cpx, denoted C–) and orthopyroxene (Opx, denoted O–), and olivine (Ol)

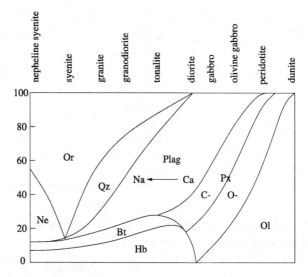

Figure 9.3 shows that this linear trend is more or less reflected in a replacement of minerals as silica content increases: olivine $\to$ pyroxene $\to$ plagioclase $\to$ quartz, K-feldspar, and the important point is that this sequence more or less reflects the sequence in which minerals crystallise from a magma as the temperature is reduced. (Although our concern in this chapter is with melting, it is normal to discuss phase changes from the point of view of crystallisation, and we shall do so, reversing our stance later on.)

A more realistic description of the crystallisation process as temperature is reduced is Bowen's reaction series:

$$\begin{matrix} \text{Ol} \to \text{Px} \to \text{Hb} \to \text{Bt} \\ \text{Ca–Plag} \to \text{Na–Plag} \end{matrix} \quad [\to \text{Qz, Fs}]. \tag{9.1}$$

The abbreviations represent the discontinuous reaction series olivine $\to$ pyroxene $\to$ hornblende $\to$ biotite, and the continuous series calcium plagioclase $\to$ sodium plagioclase. The mafic minerals of the discontinuous series are formed in turn, each crystallised phase reacting with the melt to form the next phase, while those of the continuous series are formed continually from the melt (with less calcium as crystallisation proceeds). Typically both phases (and perhaps others also) may be crystallising at the same time.

It can be seen from Table 9.2 that density also decreases essentially uniformly as silica content increases.

9.3.1 Phase Diagrams

Apart from all the names, the story so far seems relatively simple; but, there is a lot of detail. We begin by describing the use of phase diagrams.

Table 9.2 Density of mineral phases

Mineral	Density (gm cm^{-3})
K-feldspar	2.6
Quartz	2.65
Plagioclase	~ 2.7
Pyroxene	~ 3.5
Olivine	~ 3.8
Nepheline	2.6
Biotite	3.0
Hornblende	3.2

Plagioclase is a solid solution of albite (Ab) with anorthite (An). The chemical formulae for these two 'end-members' are

$$Ab = NaAlSi_3O_8,$$
$$An = CaAl_2Si_2O_8; \tag{9.2}$$

so Ab is 'sodium plagioclase' and An is 'calcium plagioclase'. To say that a solid solution exists means that in the crystal structure of the solid, atomic replacement may occur: NaSi in Ab can simply replace CaAl in An. Thus, in general, we can have any formula intermediate between the two, thus a general formula for the solution is

$$plag = (NaSi, CaAl)AlSi_2O_8 = Na_xCa_{1-x}Al_{2-x}Si_{2+x}O_8, \tag{9.3}$$

which represents the fact (in the first formula) that CaAl can replace NaSi, or in the second that we have a fraction x of Ab, and a fraction $1 - x$ of An. Often one writes An$_{40}$ to represent a composition of 40% An, 60% Ab, as in Table 9.3.

Table 9.3 Approximate compositions (weight per cent) of common minerals

Mineral		SiO$_2$	Al$_2$O$_3$	Fe$_2$O$_3$/FeO	MgO	CaO	Na$_2$O/K$_2$O
Nepheline		44	32	< 1	0	< 1	21
Hornblende		45	11	16	10	12	2
Biotite		37	17	21	9	< 1	9
Pyroxene (e.g., hypersthene)		53	3	18	23	2	< 1
Orthoclase		64	19	< 1	–	< 1	15
Quartz		100	–	–	–	–	–
Plagioclase	An$_{20}$	63	23	–	–	4	10
	An$_{80}$	48	33	–	–	16	2
Olivine	Fo$_{96}$	41	< 1	4	54	–	–
	Fo$_3$	30	< 1	65	1	2	–

Fig. 9.4 Solid solution of the
plagioclase system, between
the end members albite
(Al-sodium plagioclase) and
anorthite (An-calcium
plagioclase). Melting
temperatures shown in
degrees Celsius

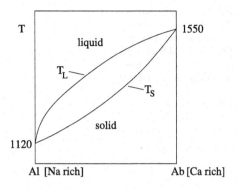

Pure An and pure Ab have different melting points, and for a composition of
the two substances, there is no longer a single melting temperature, but a solidus
and a liquidus. The solidus is the temperature at which melting of a solid begins,
and the liquidus is where crystallisation of a liquid begins. The two temperatures
are not equal, and for a solid solution such as Ab–An, the corresponding phase
diagram is as shown in Fig. 9.4. This is an equilibrium diagram. For temperatures
T greater than the liquidus T_L, only liquid can occur. On the liquidus, liquid can
be in equilibrium with solid, but the compositions of solid and liquid are different.
At a fixed temperature, the liquid composition is that of the liquidus, but the solid
composition is that of the solidus.

The phase diagram can tell us why plagioclase crystallisation becomes more
sodium-rich as it proceeds. Consulting Fig. 9.4, we see that when plagioclase crystallises, the resulting solid is enriched in calcium. As crystallisation proceeds, the
residual melt therefore becomes enriched in sodium, and continued crystallisation
requires decreasing temperatures, so that the crystals become increasingly sodium-
rich. As cooling proceeds, the melt composition travels, in temperature-composition
space, down the liquidus.

Plagioclase is one of the feldspar group of minerals, of which the other member is
orthoclase, or simply K-feldspar, which substitutes potassium for sodium/calcium.
Thus orthoclase is

$$Or = KAlSi_3O_8, \qquad (9.4)$$

formed from Ab by replacing Na with K. Ab and Or also form a solid solution,
but not at high water pressure. Figure 9.5 shows the resulting typical phase diagram
which occurs in this case. The liquidus and solidus curves have the same interpretation, but now there is a eutectic point E and a lower 'solvus' curve comprising the
two portions AA$'$ and KK$'$. Suppose we cool a potassium-rich melt with composition to the right of E. On the liquidus, potassium-rich feldspar (KFs) is precipitated,
thus the residual melt is driven down the liquidus until it reaches E. At E, separate
crystals co-exist stably within the solvus curves AA$'$ and KK$'$. (The solvus curve
represents a solid-state unmixing curve; if solid KFs is cooled through KK$'$ for example, then the crystals become unstable, and further cooling separates the solid
into crystals of separate compositions on KK$'$ and AA$'$, as indicated in Fig. 9.6.)

Fig. 9.5 Water-saturated phase diagram for albite (Ab)-orthoclase (Or), after Krauskopf and Bird (1995, p. 455). The symbol KFs denotes potassium feldspar

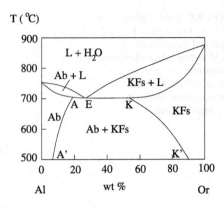

Fig. 9.6 Solid-state unmixing of alkali feldspar

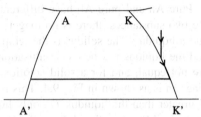

Figure 9.5 illustrates another important effect of water. Water is typically present in rocks at a level of one per cent or less, but the presence of such quantities can depress the liquidus temperatures by hundreds of degrees centigrade to the sorts of temperatures (800–1100°C) which are commonly observed. In fact, the late stage minerals of the discontinuous crystallisation series, hornblende (an amphibole) and biotite (a mica) are specifically hydrous minerals which incorporate water.

9.3.2 Ternary Phase Diagrams

To extend our understanding of phase diagrams, we need now to consider how a three-phase liquid crystallises. To be specific, we will consider how plagioclase interacts with pyroxene (px). Pyroxenes are characterised by the formula $[(Ca,Mg,Fe,Na,Al)SiO_3]_2$, thus we have

$$
\begin{aligned}
&\text{diopside (Di)} && CaMgSi_2O_6, \\
&\text{enstatite (En)} && Mg_2Si_2O_6, \\
&\text{hypersthene} && (Mg,Fe)_2Si_2O_6, \\
&\text{augite} && Ca(Mg,Fe)Si_2O_6, \\
&\text{hedenbergite} && CaFeSi_2O_6, \\
&\text{jadeite} && NaAlSi_2O_6, \\
&\text{acmite} && NaFeSi_2O_6,
\end{aligned}
\tag{9.5}
$$

Fig. 9.7 Ternary phase diagram for the system Di–An–Ab (diopside–anorthite–albite). The isotherms are in degrees centigrade. From the diopside field, a cooling magma evolves towards the cotectic line, and then follows it

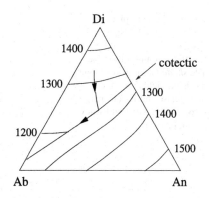

and so on. Augite is a solid solution of diopside and hedenbergite, and hypersthene is a solid solution of enstatite and ferrosilite $Fe_2Si_2O_6$. To be specific, we consider the three-component system Di–An–Ab (diopside–anorthite–albite), the latter two being the calcium-rich and sodium-rich plagioclases. In order to present the phase diagram, we must represent the three compositions in two-space. This is done by plotting compositions on an equilateral triangle, as shown in the ternary phase diagram for Di–An–Ab (Fig. 9.7).

If we take the origin at the centre of the triangle, and denote the vertices by the vectors D_i, A_b and A_n (thus $D_i + A_b + A_n = 0$) then any point in the triangle can be written uniquely in the form $\lambda D_i + \mu A_b + \nu A_n$, where $\lambda + \mu + \nu = 1$; then λ, μ, ν represent the fractional composition of Di, Ab and An. Equivalently, λ is the distance from the point to the line Ab–An, measured as a fraction of the altitude of the triangle. Thus the vertices represent pure components, and the sides represent two component mixtures.

Really, the phase diagram needs to have a temperature axis coming out of the paper, so that liquidus and solidus surfaces are plotted. Usually one does not try to do this. The temperatures of the liquidus surface are plotted as contours (as in Fig. 9.7), and the solidus surface is suppressed entirely—because we are thinking of crystallisation rather than melting. But one should study ternary phase diagrams like Fig. 9.7 with the understanding that diagrams such as those in 9.4 or 9.5 are the suppressed sideways-on view.

Ternary diagrams of three components all with solid solutions are featureless, but also not very interesting. However, pyroxene and plagoclase do not form a solid solution, and this is indicated by the *cotectic* line within the triangle, which is simply the continuation in the three-component diagram of the eutectic points which exist for Di–An and Di–Ab. If we take a slice through temperature-composition space orthogonal to the composition triangle, we will see a phase diagram like Fig. 9.5, and the eutectic point of Fig. 9.5 moves in composition space (becoming the cotectic line) as we move from An to Ab.

We can now read the crystallisation history of a plagioclase-pyroxene melt. Suppose that we start with a melt of 40% Di, 30% Ab, 30% An, which places us on the diopside side of the cotectic. The liquidus temperature is about 1300°C, and on reaching this we begin to crystallise diopside-rich crystals, because the solidus

Fig. 9.8 Ternary phase
diagram in the feldspar
system
anorthite–albite–orthoclase.
The *shaded regions* represent
the stable fields for solid
plagioclase and K-feldspar,
respectively. After McBirney
(1984, p. 107)

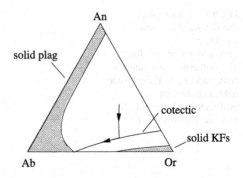

must decrease more rapidly away from the melting point of pure diopside then the
liquidus. (In fact, essentially pure Di is produced.) As cooling proceeds, the residual melt is thus driven towards the cotectic, and subsequent crystallisation occurs
on the cotectic, with the plagioclase Ab–An concentration being determined by the
intersection of the cotectic temperature with the Ab–An solidus.

These general principles apply for more components. For example a four component system would have a compositional tetrahedron to represent compositions.
Phases that do not form solid solutions would have eutectic points on the edges,
extending to cotectic lines on the surfaces, and cotectic surfaces in the interior. As
crystallisation proceeds, the residual melt would drive down temperature gradients
towards cotectic surfaces, then along these (with two phases crystallising) to cotectic lines where two cotectic surfaces intersect, and three phases precipitate, and so
on. Mostly, one sticks with the easier ternary diagrams, though.

Let us now return to the feldspars. We recall that plagioclase is a solid solution of
albite and anorthite, $Ab = NaAlSi_3O_8$, $An = CaAl_2Si_2O_8$, and the other principal
feldspar, orthoclase or potassium feldspar, $Or = KAlSi_3O_8$, forms a eutectic mixture with Ab at elevated water vapour pressures (Fig. 9.5). Just as we considered
the crystallisation history of plagioclase-pyroxene melts, it is natural to wonder how
orthoclase-plagioclase melts crystallise, since we know that K-feldspar is a prominent component of crustal rocks (Fig. 9.3).

We thus consider the three-component feldspar system Or–Ab–An, whose
ternary phase diagram is shown in Fig. 9.8. There is a single cotectic line dividing
plagioclase from K-feldspar. The diagram also shows (hatched) the regions of solid
phase stability (i.e., the projections of the solidus surfaces for Plag and Or). They
are well separated. For a liquid on the plagioclase side of the cotectic, plagioclase
will crystallise first, and then plagioclase and K-feldspar together.

Because the plagioclase is a solid solution, one can add silica to this diagram relatively simply. The corresponding phase diagram for wet rock is shown in Fig. 9.9.
Now each two component sub-system has a eutectic, so there are three cotectic lines,
and these intersect at the cotectic point C. When the residual melt reaches the cotectic point, then all three phases will crystallise. If we more properly allow Ab and An
as separate components, then each section of the four component phase tetrahedron
looks like Fig. 9.9, with the cotectic lines shifting somewhat as Ab/An varies. The
solid which crystallises at the cotectic point has a composition resembling granite

Fig. 9.9 'Ternary' phase diagram for the system quartz–plagioclase–orthoclase. Because plagioclase is a solid solution of albite and anorthite, the base of the triangle is really a side view of the base of a tetrahedron, the base being that shown in Fig. 9.8. *C* denotes the cotectic point, where all three minerals crystallise

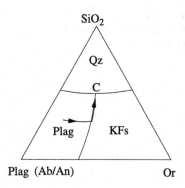

(but without the late stage hornblende and biotite of the discontinuous crystallisation series).

9.3.3 Olivine

We have fought our way through from pyroxene to feldspar; we need to add olivine to the story. Olivine $(Mg,Fe)_2SiO_4$ is a solid solution of forsterite $(Fo = Mg_2SiO_4)$ and fayalite $(Fa = Fe_2SiO_4)$, with a phase diagram like that of plagioclase, forsterite having a higher melting temperature than fayalite.

We can now explain what is meant by the discontinuous crystallisation series. We illustrate this first by consideration of the $Fo–SiO_2$ system, whose phase diagram is shown in Fig. 9.10. If an Fo-rich liquid is cooled, it begins by crystallising Fo. On cooling further, however, Fo can react with the liquid to form the pyroxene enstatite $En = Mg_2Si_2O_6$ via

$$\underset{Fo}{Mg_2Si_2O_4} + SiO_2 \rightleftharpoons \underset{En}{Mg_2Si_2O_6} . \tag{9.6}$$

If the Fo liquidus dips underneath the En liquidus (as in Fig. 9.10), then the enstatite is said to melt *incongruently* (since when it melts, it forms a liquid of a different

Fig. 9.10 Phase diagram for Fo–SiO$_2$

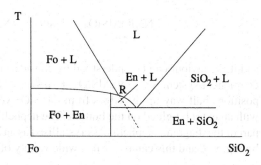

Fig. 9.11
Forsterite–silica–plagioclase
system

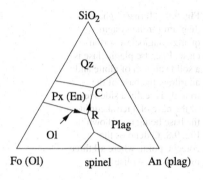

composition), and the consequence on solidification is that when the melt composition reaches the reaction point R in Fig. 9.10, the Fo dissolves in the melt, and En is precipitated. It is in this sense that the crystallisation is discontinuous.

If we now add anorthite to the system, then the phase diagram is as in Fig. 9.11. The incongruent melting region extends to a pyroxene stability field, and in particular olivine and quartz cannot precipitate together; at least if the olivine is Mg-rich: the pyroxene stability field disappears at high Fe in the Fo–Fa–SiO$_2$ system.

9.3.4 Summary

We now have a reasonable concept of how crystallisation of a magma may occur. Olivine crystallises first, which enriches the melt in silica. As the melt cools, the olivine reacts with it to form pyroxene. This drives the melt towards the feldspar field, until also plagioclase (first calcium-rich, then sodium-rich) begins to crystallise, followed by K-feldspar, and finally quartz. Or at least, this is one possible sequence. It should be very clear from the degree of variability present in the discussion that there are many different possible crystallisation sequences. A part of the story which we have not mentioned concerns the three minerals in Fig. 9.3 not so far mentioned: nepheline, biotite and hornblende.

Nepheline (Ne = (Na,K)AlSiO$_4$) is a silica-poor alkaline rock called a feldspathoid. One can think of it as a silica-poor feldspar, indeed one can think of it as a reactant in forming albite or orthoclase:

$$(Na,K)AlSiO_4 + 2SiO_2 \rightleftharpoons (Na,K)AlSi_3O_8,$$
$$Ne \qquad\qquad\qquad Ab, Or \qquad\qquad\qquad (9.7)$$

and its behaviour can be understood by considering the phase diagram of the three-component system SiO$_2$–NaAlSiO$_4$–KAlSiO$_4$, which (since Ab and Or have compositions half way up the edges from the SiO$_2$ vertex) is a bit like that of Fig. 9.9 with an extra bit glued on the bottom. The nepheline stability field lies in this lower part of the diagram. Nepheline is crystallised as an alternative to quartz, as suggested by Fig. 9.3, and this can occur in a wide variety of conditions.

Hornblende and biotite are hydrous minerals; hornblende (Hb) is an amphibole, and biotite (Bt) is a mica. They both have bewildering chemical formulae, but their main characteristic is that they incorporate water. They are silica poor, and they are increasingly stable at higher water vapour pressure (unlike the anhydrous minerals we have been discussing, whose solidus and liquidus temperatures decrease dramatically with increasing water vapour pressure, amphiboles and micas have the opposite behaviour). Thus as the anhydrous minerals crystallise, water is concentrated in the residual melt, which lowers the liquidus of, say, pyroxene, while raising that of amphibole. Eventually they become equal and amphibole will precipitate. Essentially this sequence can be understood in terms of a suitable ternary phase diagram of the system En–Hb–H_2O, for example, via the migration of a cooling path towards a cotectic dividing the enstatite and hornblende fields. In this way, one can add amphibole and mica to the story of late stage crystallisation through the action of water.

9.3.5 Melting

Mostly we have discussed crystallisation. What happens on melting a mantle rock? On the face of it, it is just the opposite of crystallisation. If we warm a solid solution to the solidus, it will form a melt with an enriched composition on the liquidus (Fig. 9.12a). Subsequent heating drives the solid composition up the solidus and the melt composition up the liquidus, becoming less enriched. For a eutectic mixture (Fig. 9.12b) such as diopside–anorthite, if we heat a diopside-rich rock to the eutectic temperature, then the melt will be of eutectic composition, and the solid will be driven towards diopside. Only when all the anorthite is melted will the temperature be able to rise along the diopside liquidus.

It is easy to generalise this to more complicated rocks. The first melt will be of cotectic composition, containing all the phases present. As melting proceeds, the melt will remain of this composition until one of the phases has completely melted; subsequently the melt composition will migrate up cotectic lines. This simple story is slightly soiled by the possibility of reactions between minerals, for example in the incongruent melting of enstatite through the reaction of forsterite with silica ((9.6) and Fig. 9.10). An olivine-pyroxene rock will begin to melt as shown in Fig. 9.12c; the pyroxene will melt to form melt of peritectic composition at the point R.

There are two complications. One is geochemical: the phase diagrams are complicated! However, the basic picture that we have is that silicate rocks are compounds formed by reactions of metal oxides with silica, and to some extent, the different oxide-silica phase diagrams can be glued together. Figure 9.13 shows the 'basalt tetrahedron' which glues together three tetrahedra (corresponding to alkali basalt, olivine tholeiite and quartz basalt) comprising nepheline, diopside, silica and olivine, with the intermediate reactive products albite and hypersthene. The only obvious ingredient missing is potassium, which can be glued on via the residual system nepheline–kalsilite–silica ($NaAlSiO_4$–$KAlSiO_4$–SiO_2), which also contains the intermediate reaction products Ab = $NaAlSi_3O_9$ and Or = $KAlSi_3O_8$, and is

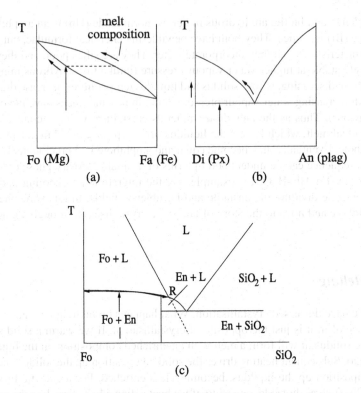

Fig. 9.12 Melting of (**a**) a solid solution; (**b**) a eutectic solid; (**c**) an incongruent system

closely associated with felsic rocks such as granite. Oceanic basalts contain very little potassium, and Fig. 9.13 is a reasonable synopsis.

The second complication is this: physics. Our entire discussion assumes melting and crystallisation occurs at equilibrium. This is unlikely to be the case! As melting proceeds, equilibrium requires solid-state diffusion to be effective in allowing solid phase concentrations to remain on the solidus, at least for solid solutions; whether this can be so remains to be seen.

But the more obvious point is that as soon as partial melting occurs, the buoyant melt wants to move. As soon as partially molten rock is permeable, melt will rise and be replaced by melt from below. In general, changing pressure will alter the location of cotectic points, so that the replacement melt will have a different composition (and temperature) than its predecessor. We can suppose local melting or refreezing will occur in order to keep the liquid at the liquidus, but one can no longer assume that the melt rate is purely controlled by the rise velocity of adiabatic mantle, nor that the melt composition is that of the local cotectic point. In order to be able to consider the effects of melt transport on melt composition, we need to be able to describe the physics of magma transport, and this will commence in the following section. Before doing so, we make some geochemically informed comments on various types of differentiation in the Earth.

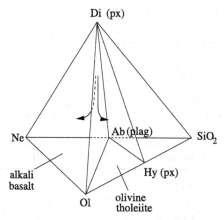

Fig. 9.13 Olivine–pyroxene–quartz–nepheline system. Because albite (a plagioclase) has a composition intermediate to nepheline and silica, and similarly hypersthene (a pyroxene) is intermediate to olivine and silica, the system acts like three separate subsystems glued together. In particular, similar compositions close to the olivine–pyroxene–plagioclase system can evolve into olivine tholeiites, enriched in silica, or the silica-poor alkali basalts, as indicated by the arrowed curves

9.3.6 Continental Crust

The average composition of continental crust is andesitic, while the upper mantle is essentially peridotite (see Table 9.1). The differentiation of the light crust from the heavier mantle occurred near the beginning of Earth's history, and can be understood in various ways. If the Earth was covered by a magma ocean, as has been suggested, then as it crystallised, one would expect the heavier crystals to sink towards the base. With olivine and then pyroxene crystallising first, we would obtain an olivine rich mantle and a more silicic 'crust'.

Instead, or as well, subsequent partial melting of a silica-poor (relative to cotectic) convecting solid mantle would cause silica-enriched melt to be erupted. This provides a means to establish a chemical discontinuity (the Mohorovičić discontinuity, or 'Moho') above which lies the crust which is (presumably) enriched in minerals lying in the cotectic direction from a primitive mantle composition. So it is not difficult to see how to form the crust. Crust is still forming, in the sense that one has continental volcanism, but this is offset by the return of continental crust to the mantle via erosion, sedimentation and subsequent subduction of ocean floor sediments. It may be the case that crust production and removal has produced a steady state, but this is not really known. The break-up of the crust to form the continents is presumably entirely associated with active plate tectonics.

9.3.7 MORB, OIB, CFB

Tables 9.4 and 9.5 show the major oxide and silicate chemistry associated with some mid-ocean ridge basalts (MORB), ocean island basalts (OIB), continental

Table 9.4 Oxide assemblages of different basalts

Basalt	SiO_2	Al_2O_3	Fe_2O_3/FeO	MgO	CaO	Na_2O	K_2O	Ba (ppm)
MORB	50	16	11	8	11	3	< 1	10
OIB	49	14	11	8	10	2	< 1	125
CFB	48	15	11	7	10	3	< 1	280
AlkB	46	16	11	10	9	4	< 1	48

Table 9.5 Mineral assemblages of different basalts

Basalt	Plag	Px	Ol	Ne
MORB	55	29	11	–
OIB	48	43	–	–
CFB	55	30	8	–
AlkB	50	14	24	5

flood basalts (CFB), and alkali basalts. It can be seen that the overall oxide and silica content is similar, suggesting similar source rocks, but the crystallisation history has produced different mineral assemblages. In principle, it is easy to understand how this can be so, since different temperature-pressure histories will cause such variations due to the dependence of the phase diagrams on both variables.

Table 9.4 also shows the concentration (in ppm, parts per million by weight, or μg/g) of one of the trace elements, Ba (barium). Trace elements are those present in minute quantities in mantle rocks, and there are many of them (rubidium Rb, thorium Th, uranium U, etc.). These are all incompatible elements, in the sense that the atoms do not fit well into the crystalline silicate structure, and melt preferentially in the mantle; they are mostly found in the continental crust for this reason. It is noticeable that MORB is depleted (relative to OIB) of Ba, and this is also true of other incompatible elements. Together with the analysis of radioactive isotope ratios, this has been taken to infer that the sources of OIB and MORB have been chemically separated for a period of the order of 2 Ga (2×10^9 y), and this in turn has been interpreted as evidence for some form of layered convection. Since OIB presumably comes from mantle plumes from the lower mantle, while MORB comes from subducting plate driven recirculation of the upper mantle, this is not too shocking, and the phase change at 670 km, perhaps with a density increase of several per cent, may help to maintain a separate circulation in the lower mantle.

9.3.8 Granite

As indicated by Table 9.1 and Fig. 9.3, granite is a highly silicic rock with very little magnesium or iron. It primarily consists of feldspar (orthoclase), quartz, plagioclase and mica, and most simply can be viewed as the crystallised end product of a magma which is itself the differentiated melt of more basic magma. That is to say, complete melting of a mean mantle composition followed by crystallisation (and settling out) of olivines and pyroxenes would leave a residual melt of potentially granitic type.

However, there are two (related) problems with this idea. The first is that as the silica content of a magma increases, the viscosity increases enormously. For example, plagioclase magma viscosity changes from about 2 Pa s to 10^5 Pa s as the composition changes from anorthite (43% SiO_2) to albite (69% SiO_2). This raises the physical question of whether granitic magma can separate from its earlier forming crystals at all before freezing. For example, one does not find granite dikes, presumably for this reason (they would freeze before they could form).

The second problem is that granites are often emplaced as enormous batholiths with linear dimensions of order 50 km, which are abundantly in evidence at the Earth's surface. Dartmoor in Devon, Southwest England, is a granite of some 40 km diameter and about 290 Ma age, and batholiths of this size litter the Earth. Such batholiths are far too large to be explained by differentiation of more basic magma, and must have formed by anatexis—the melting of crustal rock by (for example) an upwelling mantle plume. Such batholiths may then have solidified in situ, if they were too viscous to rise as diapirs or via crack formation through the crust. Their exposure at the Earth's surface then owes its occurrence to the effects of crustal erosion, together with the fact that granite is hard rock and difficult to erode. The erosion of 40 km of continental crust over 300 Ma indicates a surface erosion rate of 0.13 mm a^{-1}, which is comparable to the average inferred rate measured from fluvial sedimentary discharge, of 0.03 mm a^{-1}.

In the following sections we will examine some of the physical processes associated with generation and transport of magma.

9.4   Melt Transport in the Asthenosphere

As mantle rock rises at mid-ocean ridges or in hot spots such as that underneath Hawaii, it begins to melt. The melt should first form at the cotectic temperature, yielding a cotectic melt which is lighter than the surrounding rock. For thermodynamic reasons, it is thought that this melt forms at three-grain junctions, forming a connected tubular network (see Fig. 9.2) which is therefore permeable, and the melt will tend to rise.

The cotectic temperature will depend on pressure, and the cotectic fluid composition will also, but we begin by supposing that the liquid composition is constant. We then assume that the cotectic temperature is given by

$$T = T_0 + \Gamma p_l, \qquad (9.8)$$

where p_l is the pressure in the connected liquid phase, and T is the absolute temperature. (This allows for the possibility of differential stress in the solid grains, since the Clapeyron relation (9.8) must actually relate melting temperature to the normal interfacial stress, which in the liquid can be taken to be simply the pressure.)

Let us suppose the solid and liquid densities are ρ_s and ρ_l, and are constant (at least in the sense of the Boussinesq approximation). If the solid and liquid have velocities **u** and **v**, and the volumetric melt fraction is ϕ, then conservation of mass of solid and liquid take the form

$$\rho_l[\phi_t + \nabla.(\phi\mathbf{v})] = S,$$
$$\rho_s[-\phi_t + \nabla.\{(1-\phi)\mathbf{u}\}] = -S, \tag{9.9}$$

where S is the rate of melting (which is not known *a priori*).

It is customary to suppose that the fluid motion is described by Darcy's law, thus

$$\phi(\mathbf{v} - \mathbf{u}) = -\frac{k}{\eta_l}[\nabla p_l + \rho_l g\hat{\mathbf{k}}], \tag{9.10}$$

where k is permeability, η_l is liquid viscosity, g is gravity, and $\hat{\mathbf{k}}$ is a unit vector directed vertically upwards. Because of the tubular nature of the liquid phase network, it is appropriate to take

$$k = \frac{d_g^2\phi^2}{b}, \tag{9.11}$$

where d_g is grain size, and b is a constant taking a value in the range $b \approx 100$–1000. We define a permeability coefficient

$$\Pi_0 = \frac{d_g^2}{\eta_l b}. \tag{9.12}$$

Before proceeding, we step back for a moment. We can write (9.10) in the form

$$0 = -\phi\nabla p_l - \rho_l\phi g\hat{\mathbf{k}} - \frac{\eta_l b}{d_g^2}(\mathbf{v} - \mathbf{u}). \tag{9.13}$$

More generally, the averaged momentum equations for two-phase flow (with no inertia terms) are

$$0 = \nabla.(\phi\bar{\sigma}_l) - \rho_l\phi g\hat{\mathbf{k}} + \mathbf{M},$$
$$0 = \nabla.[(1-\phi)\bar{\sigma}_s] - \rho_s(1-\phi)g\hat{\mathbf{k}} - \mathbf{M}, \tag{9.14}$$

where $\bar{\sigma}_l$ and $\bar{\sigma}_s$ are the locally phase-averaged liquid and solid stress tensors, and **M** is the interfacial momentum source, defined as

$$\mathbf{M} = -\overline{\sigma_l.\nabla X} = -\overline{\sigma_s.\nabla X}, \tag{9.15}$$

where X is the phase function, $X(\mathbf{x}) = 1$ if **x** is in the liquid phase, $X = 0$ if **x** is in the solid phase. The two expressions in (9.15) are equal, because

$$\overline{\mathbf{f}.\nabla X} = \langle f_n \rangle, \tag{9.16}$$

where $\langle f_n \rangle$ is the interfacial average of the normal component of **f**, where **n** points from solid to liquid, and because $(\sigma_s - \sigma_l).\mathbf{n} = 0$ at the interface (due to equal and opposite reactions). It is reasonable to suppose that the liquid pressure does not vary on the microscale, so that, if $\sigma = -p_l\delta + \tau_l$, then, since also $\overline{\nabla X} = \nabla\phi$,

$$\mathbf{M} = p_l\nabla\phi + \mathbf{M}', \qquad (9.17)$$

where

$$\mathbf{M}' = -\overline{\tau_l.\nabla X} \qquad (9.18)$$

is the interactive drag. Then (9.14) becomes

$$0 = -\phi\nabla p_l + \nabla[\phi\bar{\tau}_l] - \rho_l\phi g\hat{\mathbf{k}} + \mathbf{M}',$$
$$0 = \nabla.\big[(1-\phi)\bar{\sigma}_s\big] - \rho_s(1-\phi)g\hat{\mathbf{k}} - p_l\nabla\phi - \mathbf{M}'. \qquad (9.19)$$

Darcy's law follows from the assumption that $\bar{\tau}_l = 0$ (which is reasonable given the long time scale and low melt viscosity), and

$$\mathbf{M}' = -\frac{\eta_l b}{d_g^2}(\mathbf{v} - \mathbf{u}), \qquad (9.20)$$

which is the drag due to viscous intergranular pore flow.

The second momentum equation describes the bulk motion of the solid matrix. Note that alternatively we may consider the equation of total momentum conservation,

$$0 = -\nabla(\phi p_l) + \nabla.\big[(1-\phi)\bar{\sigma}_s\big] - \big[\rho_l\phi + \rho_s(1-\phi)\big]g\mathbf{k}. \qquad (9.21)$$

In order to solve this, we need to prescribe a constitutive relation for the solid stress.

First, note that in general the solid pressure will not be equal to the liquid pressure. Indeed, we can expect solid pressure to approach lithostatic values, but liquid pressure to approach (lesser) hydrostatic values, and the resultant tendency must be to drive melt upwards (and thus solid downwards). However, the resultant solid compaction is unlike the settling of a suspension since the grains are in contact. Differential pressure between solid and liquid will cause viscous deformation on the long time scales of relevance here. Just as we did for the liquid, it seems reasonable to propose for the solid phase that $\sigma_s = -p_s\delta + \tau_s$, where p_s is the locally-averaged pressure, so that only the deviatoric stress τ_s varies at the granular microscale.

Similarly, just as Darcy's law (i.e., (9.20)) can be posited on the basis of a local Poiseuille flow in the cylindrical pores, so a constitutive equation to relate p_s to p_l can be posited on the basis of a model of closure of a cylindrical void in a viscous medium.

To be specific, let us consider a circular void of radius $r = a$ in an infinite viscous medium; the void pressure is p_l and the far field medium pressure is p_s. Suppose the grain has (constant) viscosity η_s. Then mass conservation of the grain implies that the radial velocity in the grain is

$$u = \frac{C}{r}, \qquad (9.22)$$

while the momentum equation is

$$p_r = \eta_s \left[u_{rr} + \frac{1}{r} u_r - \frac{u}{r^2} \right], \tag{9.23}$$

whence $p = p_s$ is constant, and at $r = a$, a force balance implies $-p_s + 2\eta_s u_r = -p_l$, thus

$$C = -\frac{(p_s - p_l)a^2}{2\eta_s}. \tag{9.24}$$

The constitutive law for $p_s - p_l$ now follows from consideration of the kinematic boundary condition for the solid flow which applies at the pore boundary. To relate our cylindrical void to macroscopic quantities, we suppose it is encased in a grain of dimension l (specifically, a cylinder of length l and radius l) so that $\phi = a^2/l^2$. The rate of melting is S (mass per unit volume per unit time). This corresponds to a rate of removal (or ablation) at the pore boundary of

$$v = \frac{l^2 S}{2\rho_s a} \tag{9.25}$$

(because S/ρ_s is the volume removed per unit volume per unit time, and this is equal to $2\pi a l v/\pi l^3$). Finally, the kinematic boundary condition for a is that

$$\dot{a} = u|_a + v, \tag{9.26}$$

and in terms of macroscopic variables, this becomes (since $\dot{\phi} = 2a\dot{a}/l^2$)

$$\dot{\phi} = \frac{S}{\rho_s} - \frac{(p_s - p_l)\phi}{\eta_s}. \tag{9.27}$$

The time derivative $\dot{\phi}$ represents the derivative following the grain, i.e. $\dot{\phi} = \phi_t + \mathbf{u}.\nabla\phi$.

(9.31) applies if the grain viscosity is constant, i.e., diffusion creep prevails. For the more likely dislocation creep with a flow law given by

$$\dot{\varepsilon}_{ij} = A\tau^{n-1}\tau_{ij}, \tag{9.28}$$

and $2\tau^2 = \tau_{ij}\tau_{ij}$, we find that (9.24) is replaced by

$$u|_a = \frac{C}{a} = -\frac{Aa}{n^n}|p_s - p_l|^{n-1}(p_s - p_l). \tag{9.29}$$

Note that this reduces to (9.24) when $n = 1$, since then $A = 1/2\eta_s$. The corresponding form of (9.27) is

$$\dot{\phi} = \frac{S}{\rho_s} - \frac{2A\phi}{n^n}|p_s - p_l|^{n-1}(p_s - p_l). \tag{9.30}$$

A viscous prescription for the solid matrix would be

$$\bar{\sigma}_s = -p_s \delta + 2\eta_s \dot{e}_s, \tag{9.31}$$

where $\dot{e}_s$ is the solid strain rate. In (9.21), we effectively regain the mantle momentum equation

$$\nabla.\bar{\sigma}_s = [\rho_s(1-\phi) + \rho_l \phi]g\hat{k}, \tag{9.32}$$

because (as we shall find) $\phi \ll 1$, and we can suppose $p_s - p_l \lesssim \bar{\sigma}_s$. The main effect of partial melting on the circulation is in the buoyancy term in (9.32), where 1% partial melting ($\phi = 0.01$) is equivalent to a buoyancy temperature difference of 300 K (if the thermal expansion coefficient is $\alpha = 3 \times 10^{-5}$ K^{-1}).

Lastly, there is an energy equation for each phase. These equations are analogous to the two momentum equations, with a huge interfacial heat transport term, which consequently implies the two phases have locally equal temperatures. The other equation then follows from the equation of total energy conservation. The simplest form of this is

$$LS + \rho_s c_s (1-\phi)\frac{dT}{dt_s} + \rho_l c_l \phi \frac{dT}{dt_l} + \alpha T\left[\phi\frac{dp_l}{dt_l} + (1-\phi)\frac{dp_s}{dt_s}\right] = K\nabla^2 T, \tag{9.33}$$

where L is the latent heat, and we take the specific heats c_s and c_l, and averaged thermal conductivity $K = K_s(1-\phi) + K_l \phi$ to be constant; the time derivatives d/dt_l and d/dt_s are equal to $\partial/\partial t + \mathbf{v}.\nabla$ and $\partial/\partial t + \mathbf{u}.\nabla$, respectively. Various small dissipative terms have been neglected in (9.33).

9.4.1 Summary

The equations we propose to describe the partial melt region are thus

$$T = T_0 + \Gamma p_l,$$

$$\rho_l[\phi_t + \nabla.(\phi\mathbf{v})] = S,$$

$$\rho_s[-\phi_t + \nabla.\{(1-\phi)\mathbf{u}\}] = -S,$$

$$\mathbf{v} - \mathbf{u} = -\Pi_0 \phi[\nabla p_l + \rho_l g\hat{k}],$$

$$0 = -\nabla(\phi p_l) + \nabla.[(1-\phi)\bar{\sigma}_s] - [\rho_l \phi + \rho_s(1-\phi)]g\hat{k}, \tag{9.34}$$

$$\phi_t + \mathbf{u}.\nabla\phi = \frac{S}{\rho_s} - \frac{\phi(p_s - p_l)}{\eta_s},$$

$$\bar{\sigma}_s = -p_s \delta + 2\eta_s \dot{e}_s,$$

$$LS + (1-\phi)\left[\rho_s c_s \frac{dT}{dt_s} - \alpha T\frac{dp_s}{dt_s}\right] + \phi\left[\rho_l c_l \frac{dT}{dt_l} - \alpha T\frac{dp_l}{dt_l}\right] = K\nabla^2 T.$$

Boundary conditions for these equations will be described in due course.

9.4.2 Simplification

We anticipate that $\phi \ll 1$. Then we have, from $(9.34)_3$,

$$\mathbf{V}.\mathbf{u} \approx 0, \qquad (9.35)$$

and

$$\mathbf{V}.\bar{\sigma}_s \approx \left[\rho_l\phi + \rho_s(1-\phi)\right]g\hat{\mathbf{k}}. \qquad (9.36)$$

Apart from the added buoyancy term, the solid matrix flow equations are contiguous with those of the mantle outside the partial melt zone, and it is convenient to suppose that $\mathbf{u}$ and p_s can thus be determined independently of ϕ. In particular, p_s is in any case approximately lithostatic, and thus

$$\mathbf{V}p_s \approx -\rho_s g\hat{\mathbf{k}}. \qquad (9.37)$$

In order to proceed, we want to non-dimensionalise these equations. We guess, or anticipate, that the adiabatic terms are less important than the advective terms in the energy equation, and that heat conduction is small. We also anticipate that $\mathbf{v} \gg \mathbf{u}$ (melt velocity is rapid) but $\phi\mathbf{v} \ll \mathbf{u}$ (melt flux is small). Then in the energy equation, $LS \approx -\rho_s c_s \Gamma \, dp_l/dt_s$. The constitutive relation for the effective pressure $p_s - p_l$ suggests that it is balanced by the melt rate, thus $p_s - p_l \sim \eta_s S/\rho_s\phi$. We anticipate that this stress is less than lithostatic stress, and this in turn suggests $p_l \approx p_s$. With these assumptions we scale the variables as

$$\mathbf{x} \sim d, \qquad \mathbf{u} \sim u_m, \qquad \mathbf{v} \sim [v], \qquad t \sim d/[v], \qquad \phi \sim [\phi], \qquad S \sim [S], \qquad (9.38)$$

where d is a suitable length scale for the partial melt zone, and we write

$$p_l = p_s - [p]\psi, \qquad (9.39)$$

and put

$$p_s = p_0 - \rho_s gz, \qquad (9.40)$$

where z is the height above a convenient reference level.

Our choice of balances in the equations then suggests

$$\frac{\rho_l[\phi][v]}{d} = [S],$$

$$[v] = \Pi_0[\phi](\rho_s - \rho_l)g,$$

$$[p] = \frac{\eta_s[S]}{\rho_s[\phi]}, \qquad (9.41)$$

$$[S] = \frac{\rho_s^2 c_s g u_m \Gamma}{L},$$

Table 9.6 Assumed values of constants

Symbol	Meaning	Typical value
g	gravity	$10 \, \mathrm{m\,s^{-2}}$
ρ_s	matrix density	$3 \times 10^3 \, \mathrm{kg\,m^{-3}}$
ρ_l	melt density	$2.5 \times 10^3 \, \mathrm{kg\,m^{-3}}$
L	latent heat	$3 \times 10^5 \, \mathrm{J\,kg^{-1}}$
Γ	Clapeyron slope	$10^{-7} \, \mathrm{K\,Pa^{-1}}$
α	thermal expansion coefficient	$3 \times 10^{-5} \, \mathrm{K^{-1}}$
T_m	ambient melting temperature	$1500 \, \mathrm{K}$
K	thermal conductivity	$4 \, \mathrm{W\,m^{-1}\,K^{-1}}$
η_s	matrix viscosity	$10^{19} \, \mathrm{Pa\,s}$
b	tortuosity	10^3
d_g	grain size	$2 \times 10^{-3} \, \mathrm{m}$
η_l	melt viscosity	$1 \, \mathrm{Pa\,s}$
c_s, c_l	specific heats	$10^3 \, \mathrm{J\,kg^{-1}\,K^{-1}}$
u_m	mantle ascent velocity	$10^{-9} \, \mathrm{m\,s^{-1}} \; (3 \, \mathrm{cm\,y^{-1}})$
d	melt zone depth	$10^4 \, \mathrm{m} \; (10 \, \mathrm{km})$
$\Delta\rho$	$\rho_s - \rho_l$	$0.5 \times 10^3 \, \mathrm{kg\,m^{-3}}$

and these determine the unknowns $[\phi]$, $[v]$, $[S]$ and $[p]$; explicitly

$$[\phi] \equiv \mu = \left[\frac{\rho_s^2}{\rho_l \Delta\rho} \left\{ \frac{c_s \Gamma u_m d}{\Pi_0 L} \right\} \right]^{1/2}, \qquad [v] = \rho_s g \left[\frac{\Delta\rho c_s \Gamma u_m d \Pi_0}{\rho_l} \right]^{1/2}, \quad (9.42)$$

where $\Delta\rho = \rho_s - \rho_l$. We can now examine whether the assumption that $[\phi][v] \ll u_m \ll [v]$ is valid.

We use the values of the constants in Table 9.6. From these we find

$$\mu = [\phi] \sim 0.24 \times 10^{-2}, \qquad [v] \sim 4.8 \times 10^{-8} \, \mathrm{m\,s^{-1}} \; (1.5 \, \mathrm{m\,y^{-1}}),$$
$$[S] \sim 3 \times 10^{-11} \, \mathrm{kg\,m^{-3}\,s^{-1}}, \tag{9.43}$$

and thus $[\phi][v] \sim 1.2 \times 10^{-10} \, \mathrm{m\,s^{-1}}$, $u_m \sim 10^{-9} \, \mathrm{m\,s^{-1}}$, $[v] \sim 4.8 \times 10^{-8} \, \mathrm{m\,s^{-1}}$, vindicating our assumptions.

Dimensionless equations for the partial melt variables are then

$$\phi_t + \nabla \cdot (\phi \mathbf{v}) = S,$$

$$\mathbf{v} - \delta \mathbf{u} = \phi[\hat{\mathbf{k}} + \nu \nabla \psi],$$

$$(1+r)[\phi_t + \delta \mathbf{u}.\nabla \phi] = S - \phi\psi, \tag{9.44}$$

$$S + (1 - \mu\phi)\{(\delta^{-1}\partial_t + \mathbf{u}.\nabla)(-z - \tilde{\nu}\psi) + \lambda_s \theta w\}$$

$$+ \frac{1}{St}(1 - \lambda_l \theta)(\partial_t + \mathbf{v}.\nabla)(-z - \tilde{\nu}\psi) = -\varepsilon \nabla^2 \psi,$$

where $w = \mathbf{u}.\hat{\mathbf{k}}$, and the parameters are defined by

$$St = \frac{L}{c_l \Delta T}, \quad \Delta T = \rho_s g d \Gamma,$$

$$\delta = \frac{u_m}{[v]}, \quad \varepsilon = \frac{K[p]}{\rho_s^2 c_s g u_m d^2},$$

$$v = \frac{[p]}{\Delta \rho g d}, \quad \hat{v} = \frac{[p]}{\rho_s g d},$$

$$\lambda_s = \frac{\alpha T_m}{\rho_s c_s \Gamma}, \quad \lambda_l = \frac{\alpha T_m}{\rho_l c_l \Gamma}, \quad r = \frac{\Delta \rho}{\rho_l},$$

(9.45)

and also $\theta = T/T_m$. Using Table 9.6, we have $\Delta T \sim 30$ K, $[p] \sim 4 \times 10^7$ Pa, and thus

$$St \sim 10, \quad \delta \sim 0.02, \quad \varepsilon \sim 1.2 \times 10^{-3}, \quad v \sim 0.8,$$

$$\tilde{v} \sim 0.12, \quad \lambda_{s,l} \sim 0.15, \quad r \sim 0.2.$$

(9.46)

On this basis, we neglect relatively small terms proportional to μ, λ_s, St^{-1} in the energy equation (the last two mainly for simplicity):

$$\phi_t + \delta \nabla.(\phi \mathbf{u}) + \nabla.[\phi^2 \{\hat{\mathbf{k}} + v \nabla \psi\}] = S,$$

$$(1 + r)[\phi_t + \delta \mathbf{u}.\nabla \phi] + \phi \psi = S,$$

$$\left\{ \delta^{-1} \frac{\partial}{\partial t} + \mathbf{u}.\nabla \right\} (z + \tilde{v} \psi) - \varepsilon \nabla^2 \psi = S.$$

(9.47)

Further simplification results from the equation for conservation of mass of solid (9.34)₃, which can be written in the dimensionless form

$$\nabla.\mathbf{u} = \mu \nabla.(\phi \mathbf{u}) + \frac{\mu}{\delta} \left[(1 + r)\phi_t - \frac{S}{1 + r} \right].$$

(9.48)

Neglecting terms of $O(\mu)$, (9.47)₁ can be written as

$$\phi_t + \delta \mathbf{u}.\nabla \phi + \nabla.[\phi^2 \{\hat{\mathbf{k}} + v \nabla \psi\}] = S,$$

(9.49)

and combining this with (9.47)₂ gives

$$2\phi \phi_z + v \nabla.\{\phi^2 \nabla \psi\} = \phi \psi + r \phi_t,$$

(9.50)

where we also neglect the advective derivative of ϕ, of $O(\delta r)$.

Finally, combination of (9.49) with (9.47)₃ yields

$$\left(\delta^{-1} \frac{\partial}{\partial t} + \mathbf{u}.\nabla \right) (z + \tilde{v} \psi - \delta \phi) - \nabla.[\phi^2 \{\hat{\mathbf{k}} + v \nabla \psi\}] = \varepsilon \nabla^2 \psi.$$

(9.51)

Fig. 9.14 Schematic
illustration of the domain D.
The lower part of the
boundary ∂D_- is indicated;
on this we would expect to
prescribe $\phi = 0$

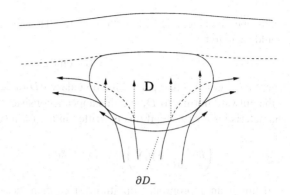

A comment on the parameters ε, ν and $\tilde{\nu}$ is in order. From (9.45), these are
proportional to $[p]$, which in turn is proportional (via (9.41)) to η_s. We took
$\eta_s = 10^{19}$ Pa s in Table 9.6, reflecting the typical inferred asthenospheric viscosity (e.g., from postglacial rebound studies). The consequent inferred grain stress is
then large (400 bars), which for typical dislocation creep laws would suggest effective viscosities about three to four orders of magnitude lower. Such lower viscosities
are not inconsistent with global asthenospheric values, since melting occurs locally.
In turn they would suggest lower values of $[p]$, which in turn would reduce the lowering of viscosity via nonlinear creep. A more consistent way to guess grain stress
values is to use the nonlinear closure law (9.30), which implies, for $n \neq 1$,

$$[p] = n \left(\frac{[S]}{2A\rho_s[\phi]} \right)^{1/n}. \tag{9.52}$$

If we identify $\eta_s = 1/2A\tau^{n-1}$, where τ is a typical asthenospheric shear stress, say
10 bars, then (9.52) can be written as

$$\frac{[p]_n}{\tau} = n \left(\frac{[p]_1}{\tau} \right)^{1/n}, \tag{9.53}$$

where $[p]_n$ is the pressure scale for flow exponent n. For $n = 3.5$, for example,
$[p]_{3.5} \approx 100$ bars, and ν, $\tilde{\nu}$ and ε would be a factor of four lower. It is therefore not
unreasonable to think of ν as being small also, and we will adopt this strategy in our
analysis below.

9.4.3 Boundary Conditions

Equations (9.47)$_{3,4}$ are to be solved for ϕ and ψ in a domain which we can expect
to resemble that shown in Fig. 9.14.

Both (9.50) and (9.51) are hyperbolic equations for ϕ if ψ is known. It is therefore natural to suppose that we should prescribe ϕ on those parts of the boundary

where the characteristics enter the domain. The physically appropriate condition
would seem to be

$$\phi = 0 \quad \text{on } \partial D_-,$$ (9.54)

where ∂D_- denotes that part of the boundary ∂D of D on which $\mathbf{u.n} < 0$, where $\mathbf{n}$
is the outward normal to D, and this appears consistent with the choice of apparent
characteristics for (9.51), if this is written in the form (using (9.50))

$$\left(\delta^{-1}\frac{\partial}{\partial t} + \mathbf{u.\nabla}\right)(z + \tilde{v}\psi - \delta\phi) = \phi\psi + r\phi_t + \varepsilon\nabla^2\psi.$$ (9.55)

If this is an appropriate condition for ϕ, then the elliptic (9.47)$_3$ suggests that
ψ should be prescribed on ∂D, except where $\phi = 0$ and the elliptic term for ψ is
degenerate. Since the melting temperature is $T_0 + \Gamma p_l$ in the partially molten region,
while it is $T_0 + \Gamma p_s$ in the cold mantle, it seems we should ensure continuity of
temperature by having $p_l = p_s$ on ∂D, i.e. $\psi = 0$ on ∂D ($\phi \neq 0$). The two boundary
conditions can be combined in the form

$$\phi\psi = 0 \quad \text{on } \partial D.$$ (9.56)

While this is plausible, it is by no means certain, and one might expect the non-
degenerate elliptic ψ term in (9.55) to require ψ to be prescribed everywhere. This
is an example of a model where some functional analysis would actually be useful.

A further difficulty is that to say the melting temperature in the cold mantle is
$T_0 + \Gamma p_s$ carries in itself no meaning, since equilibrium melting temperature is
defined through thermodynamic equilibrium at the interface. In addition, we should
add that (9.56) has not been systematically derived from any physical principle. We
now attempt to resolve this latter issue.

9.4.4 Thermodynamic Equilibrium

At the microscopic interface between partially molten and cold mantle ∂D, we require that the jump in temperature, pressure and Gibbs free energy be zero:

$$[T]_c^m = [p]_c^m = [G]_c^m = 0.$$ (9.57)

The first of these is standard, the second is a force balance (it should properly be a
balance of normal stress, but we omit deviatoric stresses for simplicity), while the
third is the condition of thermodynamic equilibrium, and it is *this* condition which
we need to prescribe, instead (perhaps) of (9.56).

Consider first a reference state in which $p_s = p_l$ in the partial melt region. The
Gibbs free energy of each separate phase depends on temperature and pressure, and
we can write for the partial melt side of ∂D, $G = G_m$, where

$$G_m = \phi(h_l - TS_l) + (1 - \phi)(h_s - TS_s),$$ (9.58)

in which h_k and S_k are the specific enthalpies and entropies of each phase. In this reference state, $p = p_s$ on the cold side, and therefore on this side $G = G_c$, where

$$G_c = h_s - T S_s. \tag{9.59}$$

The enthalpies and entropies are evaluated at the interfacial temperature T and interfacial pressure $p = p_s = p_l$. The jump in G across the boundary is

$$\Delta G = [G]_c^m = \phi[\Delta h - T \Delta S], \tag{9.60}$$

where $\Delta h = h_l - h_s$, $\Delta S = S_l - S_s$, and this is zero since the latent heat $L = \Delta h = T \Delta S$; $[\ ]_c^m$ represents the jump across ∂D from cold to partially molten.

Now suppose we change p_s, p_l and ϕ away from the reference state, by amounts δp_s, δp_l and $\delta \phi$. On the cold side, the force balance condition implies

$$\delta p_c = \delta p_s - \delta[\phi(p_s - p_l)], \tag{9.61}$$

and thus

$$\begin{aligned} \delta G_c &= v_s \delta p_c - S_s \delta T \\ &= v_s \delta p_s - S_s \delta T - v_s \delta[\phi(p_s - p_l)], \end{aligned} \tag{9.62}$$

whereas on the molten side

$$\begin{aligned} \delta G_m &= \delta\phi[\Delta h - T \Delta S] + \phi(v_l \delta p_l - S_l \delta T) \\ &\quad + (1 - \phi)(v_s \delta p_s - S_s \delta T). \end{aligned} \tag{9.63}$$

Using the fact that $\delta G_c = \delta G_m$ and $\Delta h = T \Delta S$, we get

$$-v_s \delta[\phi(p_s - p_l)] = \phi[(v_l \delta p_l - S_l \delta T) - (v_s \delta p_s - S_s \delta T)]. \tag{9.64}$$

Now the right hand side is $\phi[\delta G_l - \delta G_s]$, where G_l and G_s are the local specific Gibbs free energies of liquid and solid within the mush. We are assuming that thermodynamic equilibrium prevails locally at the pore-grain interface so that $\delta G_l = \delta G_s$ (which is in fact equivalent to the Clapeyron relation $T = T_0 + \Gamma p_l$); therefore the assumption of macroscopic thermodynamic equilibrium $[G]_c^m = 0$ implies $\delta[\phi(p_s - p_l)] = 0$, and since $\phi(p_s - p_l) = 0$ in the reference state, the condition we prescribe is

$$\phi(p_s - p_l) = 0 \quad \text{on } \partial D. \tag{9.65}$$

This condition is identical to (9.56). It is deduced from the principle of macroscopic thermodynamic equilibrium, rather than being an *ad hoc* proposition. It remains to be seen whether it is sufficient for the determination of ϕ and ψ if the domain is known.

9.4.5 Stefan Condition

The location of ∂D is not known *a priori*. Its determination is effected from a second thermal boundary condition, which represents the energy balance across ∂D. This is known as the Stefan condition, and takes the form

$$\rho_l L \phi (\mathbf{v} - \mathbf{V}).\mathbf{n} = \left[K \frac{\partial T}{\partial n} \right]_c^m, \tag{9.66}$$

where $\mathbf{V}$ is the velocity of ∂D. Scaling $T - T_0$ with $\Delta T = \rho_s g d \Gamma$ and $\mathbf{V}$ with $[\mathbf{v}]$ leads to the dimensionless version of (9.66),

$$\tilde{v} \phi (v_n - V_n) = -\varepsilon \left[\frac{\partial}{\partial n}(z + \tilde{v}\psi) - \frac{\partial T}{\partial n} \right] \quad \text{on } \partial D. \tag{9.67}$$

In particular the part of the boundary where $\phi = 0$ is determined by the condition

$$\frac{\partial T}{\partial n} = \frac{\partial}{\partial n}(z + \tilde{v}\psi), \tag{9.68}$$

and to leading order (neglecting $\tilde{v}$), this is independent of the partial melt dynamics.

9.4.6 Steady State Solution, One Dimension

Our confidence in the prescription of the boundary conditions will be enhanced by showing that solutions can be found which satisfy them. One way of doing this is to find explicit approximate solutions. To this end, we seek asymptotic solutions for (9.50) and (9.55) based on the assumptions that ε, δ and $\tilde{v}$ are small, but that $v = O(1)$.

Specifically, we will consider steady state, one-dimensional solutions, which will be appropriate if $\mathbf{u} = w\mathbf{k}$. In this case, the equations reduce to

$$2\phi\phi_z + v\left[\phi^2\psi_z\right]_z = \phi\psi,$$
$$w[1 + \tilde{v}\psi_z - \delta\phi_z] = \phi\psi + \varepsilon\psi_{zz}, \tag{9.69}$$

with anticipated boundary conditions

$$\phi = 0 \quad \text{on } z = 0, \qquad \psi = 0 \quad \text{on } z = 1. \tag{9.70}$$

In what follows we will suppose that w is constant.

9.4.7 Outer Solution

We put $\varepsilon, \tilde{v}$ and δ to zero. Then $\phi\psi \approx w$, and a first integral of (9.69)$_1$ is

$$\phi^2 + v\phi^2\psi_z = w(z + C), \tag{9.71}$$

where C is constant. With $\psi = w/\phi$, it follows that

$$vw\phi_z = \phi^2 - w(z + C),\tag{9.72}$$

which is a Riccati equation. It turns out that in order to obtain a coherent boundary layer solution at $z = 1$, we need the solution of (9.72) which tends to infinity at $z = 1$. The solution of (9.72) which does this is

$$\phi = v^{1/3} w^{2/3} \left[\frac{\text{Ai}(\zeta_1)\text{Bi}'(\zeta) - \text{Bi}(\zeta_1)\text{Ai}'(\zeta)}{\text{Bi}(\zeta_1)\text{Ai}(\zeta) - \text{Ai}(\zeta_1)\text{Bi}(\zeta)} \right],\tag{9.73}$$

where

$$\zeta = \frac{z + C}{(v^2 w)^{1/3}}, \qquad \zeta_1 = \frac{1 + C}{(v^2 w)^{1/3}}.\tag{9.74}$$

Direct consideration of (9.72) shows that there is a unique choice of $\phi_0 = \phi|_{z \to 0}$ such that $\phi \to \infty$ as $z \to 1$, providing C is not too small.

9.4.8 Boundary Layer at $z = 0$

Near $z = 0$, we put $z = \delta Z$, so that (9.69) becomes

$$2\delta\phi\phi_Z + v[\phi^2 \psi_Z]_Z = \delta^2 \phi\psi,$$
$$w\left[1 + \frac{\tilde{v}}{\delta}\psi_Z - \phi_Z\right] = \phi\psi + \frac{\varepsilon}{\delta^2}\psi_{ZZ}.\tag{9.75}$$

Correct to $O(\delta)$, we have $\phi^2 \psi_Z = \text{constant} = 0$ to match to the outer solution $\psi \sim w/\phi_0$ as $Z \to \infty$, thus $\psi = w/\phi_0 + O(\delta)$, and to leading order (9.75)$_2$ is

$$w(1 - \phi_Z) = \frac{w\phi}{\phi_0},\tag{9.76}$$

with solution

$$\phi = \phi_0\left(1 - e^{-Z/\phi_0}\right).\tag{9.77}$$

This assumes that $\phi_0 > 0$, which is certainly the case if $C > 0$. This boundary layer has been called the *compaction layer*.

9.4.9 Boundary Layer at $z = 1$

As $z \to 1$, $\phi \sim vw/(1 - z)$, $\psi \sim (1 - z)/v$, and thus we put

$$z = 1 - \delta^{1/2} Z, \qquad \phi = \Phi/\delta^{1/2}, \qquad \psi = \delta^{1/2}\Psi;\tag{9.78}$$

then (9.69) becomes

$$-2\Phi\Phi_Z + v[\phi^2\psi_Z]_Z = \delta^{3/2}\Phi\Psi,$$

$$w[1 - \tilde{v}\Psi_Z + \Phi_Z] = \Phi\Psi + \frac{\varepsilon}{\delta^{1/2}}\Psi_{ZZ}; \tag{9.79}$$

the matching conditions are

$$\Phi \sim \frac{vw}{Z}, \qquad \Psi \sim \frac{Z}{v} \quad \text{as } Z \to \infty. \tag{9.80}$$

To leading order, we have

$$-\Phi^2 + v\Phi^2\Psi_Z = 0, \tag{9.81}$$

and thus

$$\Psi = \frac{Z}{v}, \tag{9.82}$$

so that $\Psi = 0$ at $Z = 0$. To leading order, Φ then satisfies

$$w(1 + \Phi_Z) = \frac{Z\Phi}{v}, \tag{9.83}$$

and the (unique) solution which matches to the outer solution is

$$\Phi = \int_Z^\infty \exp\left[\frac{Z^2 - s^2}{2vw}\right] ds; \tag{9.84}$$

in particular, $\Phi = (vw\pi/2)^{1/2}$ on $Z = 0$, i.e.

$$\phi \approx \left(\frac{vw\pi}{2\delta}\right)^{1/2} \quad \text{on } z = 1. \tag{9.85}$$

Figure 9.15 shows a numerical solution of the equations, which clearly shows the boundary layer structure described above.

One of the more important consequences of this model and its solution is the inference that the effective pressure $\psi \propto p_s - p_l$ reaches zero at the upper boundary where refreezing occurs. Since in practice the partial melting produces a buoyant upwelling and thus a tensile stress in the mantle, this suggests that fracturing is likely to occur at this interface, affording the possibility of magma migration through the lithosphere. The fact that magma does indeed ascend through the lithosphere more or less implies that this does occur, but the formulation of the coupled fractured/porous transport is not such an easy thing to carry out. Instead, we now turn to the dynamic problem of describing how such fracture-driven transport can occur.

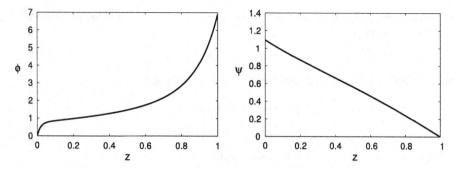

Fig. 9.15 Numerical solution for ϕ and ψ of Eqs. (9.69) with the boundary conditions (9.70), using values $\nu = 0.8$, $\tilde{\nu} = 0.12$, $w = 1$, $\delta = 0.02$, $\varepsilon = 1.2 \times 10^{-3}$. Very kindly provided by Ian Hewitt

Fig. 9.16 A linear crack in an elastic medium. A tension T acts on the medium, causing the crack to have a ·width h

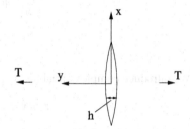

9.5 Magmafracturing in the Lithosphere

As magma oozes upwards in the asthenosphere, it ought to freeze as it reaches the cold temperature of the lithosphere. The fact that magma reaches the surface of the Earth implies that transport through the lithosphere is rapid, and the only apparent way this can be accommodated is by the process of 'magmafracturing' (by analogy to hydrofracturing). The buoyant magma flows rapidly upwards in a conduit known as a dyke, which is opened in the same way that a fracture propagates in a brittle material. The speed of ascent, of the order of metres per second, is sufficiently rapid that the lithosphere can be taken to respond elastically, and thus the width of the crack is determined in terms of the fluid pressure by quasi-static fracture mechanics (because the fluid speed is much less than the elastic wave speeds). To describe this relationship, we now discuss linear elastic fracture mechanics.

9.5.1 Fracture Mechanics

We begin by dealing fairly generally with a crack as shown in Fig. 9.16. We take axes x and y as shown, and consider the opening of a two-dimensional crack of small width h when subjected to a tension T at infinity superimposed on a lithostatic pressure $p_0 - \rho_s g x$. The fluid pressure in the crack is taken to be p. In conditions of plane strain, the displacement vector has components (u, v), and the constitutive law for linear elasticity

$$\sigma_{ij} = 2\mu\varepsilon_{ij} + \lambda\delta_{ij}\varepsilon_{kk}, \tag{9.86}$$

where λ and μ are the Lamé coefficients, takes the simple form

$$\sigma_1 = 2\mu u_x + \lambda(u_x + v_y),$$
$$\sigma_2 = 2\mu v_y + \lambda(u_x + v_y), \tag{9.87}$$
$$\tau = \mu(u_y + v_x);$$

x and y subscripts denote partial derivatives, and the three independent components of the stress tensor are $\sigma_1 \equiv \sigma_{11}, \sigma_2 \equiv \sigma_{22}, \tau \equiv \sigma_{12}$. The equations of motion are

$$\frac{\partial\sigma_1}{\partial x} + \frac{\partial\tau}{\partial y} - \rho_s g = 0,$$
$$\frac{\partial\tau}{\partial x} + \frac{\partial\sigma_2}{\partial y} = 0. \tag{9.88}$$

We introduce complex variables

$$z = x + iy, \qquad \bar{z} = x - iy, \tag{9.89}$$

and then the force balance equations can be written together as

$$\frac{\partial}{\partial z}[\sigma_1 - \sigma_2 + 2i\tau] + \frac{\partial}{\partial\bar{z}}(\sigma_1 + \sigma_2) + \frac{\partial V}{\partial\bar{z}} = 0, \tag{9.90}$$

where we define the real-valued potential

$$V = -\rho_s g(z + \bar{z}). \tag{9.91}$$

The complex displacement is

$$D = u + iv, \tag{9.92}$$

and then we find

$$\sigma_1 + \sigma_2 = 2(\lambda + \mu)(D_z + \bar{D}_{\bar{z}}),$$
$$\sigma_1 - \sigma_2 + 2i\tau = 4\mu D_{\bar{z}}. \tag{9.93}$$

From this we find that the force balance equations can be written in the form

$$\frac{\partial}{\partial z}[4\mu D_{\bar{z}}] + \frac{\partial}{\partial\bar{z}}\{2(\lambda + \mu)(D_z + \bar{D}_{\bar{z}})\} + \frac{\partial V}{\partial\bar{z}} = 0, \tag{9.94}$$

and thus there is a first integral

$$4\mu D_z + 2(\lambda + \mu)(D_z + \bar{D}_{\bar{z}}) + V = \frac{4(\lambda + 2\mu)}{\lambda + \mu}\Omega'(z), \tag{9.95}$$

where $\Omega(z)$ is an as yet arbitrary analytic function (the pre-multiplying coefficient is chosen for later algebraic simplicity).

We can solve this by firstly deriving an expression for D_z and then integrating, to find

$$2\mu D = -\frac{\mu X}{2(\lambda + 2\mu)} + (\bar{z} - z)\overline{\Omega'(z)} + \kappa \Omega(z) + \phi(\bar{z}), \qquad (9.96)$$

where we define

$$\frac{\partial X}{\partial z} = V, \qquad \kappa = \frac{\lambda + 3\mu}{\lambda + \mu}, \qquad (9.97)$$

and ϕ is an arbitrary analytic function of $\bar{z}$ (more precisely: $\phi(z)$ is an analytic function of z, but its argument in (9.96) is $\bar{z}$). See also Question 9.9. From the expressions in (9.93), we can now deduce the forms

$$\sigma_1 + \sigma_2 = 2\big(\Omega'(z) + \overline{\Omega'(z)}\big) - \frac{(\lambda + \mu)V}{\lambda + 2\mu},$$

$$\sigma_1 - \sigma_2 + 2i\tau = -\frac{\mu V}{\lambda + 2\mu} + 2\overline{\Omega'(z)} + 2(\bar{z} - z)\overline{\Omega''(z)} + 2\phi'(\bar{z}), \qquad (9.98)$$

$$\sigma_2 - i\tau = -\frac{\lambda V}{2(\lambda + 2\mu)} + \Omega'(z) - (\bar{z} - z)\overline{\Omega''(z)} - \phi'(\bar{z}),$$

where we have used the fact that we can choose

$$\frac{\partial X}{\partial \bar{z}} = V \qquad (9.99)$$

(by choosing X to be real, explicitly $X = -\frac{1}{2}\rho_s g(z + \bar{z})^2$).

So far this is all quite general. Now we consider the situation shown in Fig. 9.16. There is a crack which we denote by L on the x-axis, across which the functions Ω and ϕ will typically be discontinuous; thus L will be a branch cut for these functions.

The boundary conditions for the stresses can be expressed in the form

$$\sigma_1 + \sigma_2 \to -2p_0 + \rho_s g(z + \bar{z}),$$

$$\sigma_1 - \sigma_2 + 2i\tau \to -2T \qquad \text{as } z \to \infty, \qquad (9.100)$$

and

$$\sigma_2 - i\tau = -p,$$

$$2\mu[D]_-^+ = 2i\mu h \qquad \text{for } z \in L, \qquad (9.101)$$

where p denotes the fluid pressure in the crack, and $[D]_-^+ = D_+ - D_-$ denotes the jump in D across L, from $y = 0-$ to $y = 0+$.

Now note that, for any function $g(z)$, $g(\bar{z})_+ = g(z)_-$ and $g(\bar{z})_- = g(z)_+$. Since $\sigma_2 - i\tau$ is continuous across L, we see from (9.98) that

$$\left[\Omega'(z) - \phi'(\bar{z})\right]_-^+ = 0, \qquad (9.102)$$

and thus that $\Omega'(z) + \phi'(z)$ is an entire function. To ascertain what it is, we need the limiting behaviours of Ω' and ϕ' at ∞. These we obtain from (9.98) and (9.100); using the definition of V in (9.91), we find

$$\Omega'(z) \sim -\frac{1}{2}p_0 + \frac{\mu \rho_s g z}{2(\lambda + 2\mu)},$$

$$\phi'(\bar{z}) \sim -T + \frac{1}{2}p_0 - \frac{3\mu \rho_s g \bar{z}}{2(\lambda + 2\mu)} \quad \text{as } z \to \infty, \tag{9.103}$$

from which it follows that

$$\Omega'(z) + \phi'(z) = -T - \frac{\mu \rho_s g z}{\lambda + 2\mu}. \tag{9.104}$$

Substituting for $\phi'(\bar{z})$ in (9.98)$_3$, and letting $z \to x + i0$ (thus $\Omega'(z) \to \Omega'_+(x)$, $\Omega'(\bar{z}) \to \Omega'_-(x)$), we then find

$$\Omega'_+ + \Omega'_- = -T - \left(\frac{\lambda + \mu}{\lambda + 2\mu}\right)\rho_s g x - p \quad \text{on } L. \tag{9.105}$$

A second condition on L follows from (9.101)$_2$. Using the fact that $\Omega + \phi$ is analytic, (9.96) implies that

$$[2\mu D]^+_- = (1 + \kappa)(\Omega_+ - \Omega_-), \tag{9.106}$$

and thus (9.101)$_2$ yields

$$\Omega_+ - \Omega_- = \frac{2i\mu h}{1 + \kappa} \quad \text{on } L. \tag{9.107}$$

Together with the boundary condition (9.103)$_1$, (9.107) provides a Hilbert problem for the determination of Ω in terms of h; (9.105) then determines the crack fluid pressure p in terms of h.

The solution of (9.107) with (9.103)$_1$ is

$$\Omega(z) = \frac{1}{2\pi i} \int_L \frac{2i\mu h(s)\, ds}{(1 + \kappa)(s - z)} - \frac{1}{2}p_0 z + \frac{\mu \rho_s g z^2}{4(\lambda + 2\mu)}, \tag{9.108}$$

from which the Plemelj formulae imply

$$\Omega_+ + \Omega_- = \frac{1}{\pi i} \int_L \frac{2i\mu h(s)\, ds}{(1 + \kappa)(s - x)} - p_0 x + \frac{\mu \rho_s g x^2}{2(\lambda + 2\mu)}. \tag{9.109}$$

Differentiating this and using (9.105), we finally obtain

$$p = P - \frac{\mu}{2\pi(1 - \nu)} \int_L \frac{\partial h}{\partial s}\frac{ds}{s - x}, \tag{9.110}$$

where

$$P = p_0 - T - \rho_s g x \tag{9.111}$$

is the lithostatic compressive normal stress, and

$$\nu = \frac{\lambda}{2(\lambda + \mu)} \tag{9.112}$$

is Poisson's ratio.

9.5.2 Magma Dynamics

So far we have not specified anything about the crack L. Now we consider a situation which is appropriate to the formation of a dyke growing upwards from the asthenosphere. We imagine that the dyke originates at the asthenosphere, where we take $x = 0$, and a flux of magma $Q(t)$ is injected into the crack. We suppose, not very realistically, that the crack can be represented as having a symmetric extension below the asthenosphere, so that the crack L is the interval $(-l, l)$, where $l(t)$ is the lithospheric length of the crack, but we are concerned only with the region $x > 0$.

The crack is thin, in the sense that its width $h \ll l$, and thus the fluid flow in the crack is given by the local Poiseuille flow law

$$q = -\frac{h^3}{12\eta_m}\left(\frac{\partial p}{\partial x} + \rho_m g\right), \tag{9.113}$$

where q is the fluid flux per unit transverse width of the crack, η_m is the magma viscosity, and ρ_m is the magma density. Conservation of mass of fluid in the crack requires

$$\frac{\partial h}{\partial t} + \frac{\partial q}{\partial x} = 0. \tag{9.114}$$

The three Eqs. (9.114), (9.113) and (9.110) provide the elastohydrodynamic model for magmafracturing in the lithosphere.

9.5.3 Stress Intensity Factor

One boundary condition which we apply for (9.114) is the specified flux at the inlet:

$$q = Q(t) \quad \text{at } x = 0. \tag{9.115}$$

Another condition is necessary at the crack tip, and this is determined by a quantity called the stress intensity factor.

The stress intensity factor is associated with the stress field generated by a crack in a medium. In the simplest situation, a uniform crack overpressure σ generates a singular stress field in the solid, which has the asymptotic form

$$\sigma_{ij} = \frac{K}{\sqrt{2\pi r}} f_{ij}(\theta), \tag{9.116}$$

where the quantity K is known as the stress intensity factor, and the polar angle θ is measured from an axis along, and in the opposite direction to, the crack. For mode I cracks such as those considered here, we may define

$$K = \lim_{x \to l+} \sqrt{2\pi x} \sigma_2|_{y=0}. \tag{9.117}$$

For a crack of length $2l$, the stress intensity factor is

$$K = Y\sigma\sqrt{\pi l}, \tag{9.118}$$

where Y is an $O(1)$ numerical factor, associated with the crack geometry and the conditions of loading.

Cracks can exist as perfectly good stationary features in an elastic medium, but it is found that if the induced stress intensity factor K is large enough, then a crack will grow. There is a critical stress intensity factor K_c, such that when K reaches K_c, *dynamic* fracture occurs, and crack growth occurs at near elastic wave speeds—very rapidly—and elastic waves are generated which propagate away from the source. These are the seismic waves associated with crack propagation in earthquakes.

However, crack growth also occurs at values of $K < K_c$, albeit less rapidly. This is the phenomenon of subcritical crack propagation due to 'stress corrosion', and is familiar to us all in the cracks that migrate slowly across a pane of glass. A grander example is the rifting event which broke up Gondwanaland to form the Atlantic ocean. The crack propagated from south to north over a period of tens of millions of years.

It is found experimentally that the crack tip speed v is an increasing function of K, becoming very large as $K \to K_c$. At low values of K, it becomes exponentially small, and at higher values it reaches a plateau, before its asymptotic rise near K_c. The growth of cracks is essentially a thermodynamic process, being facilitated by the release of energy to form new crack surface and thus surface energy. At $K = K_c$, this energy is directly released from the stored elastic energy in the medium. At lower K, this is not enough, and the energy is thought to be supplied from chemical or potential energy of the fluid which migrates into the crack. It is then supposed that the crack speed is determined by the rate-limiting energy supply mechanism. At very low K, this is due to diffusion of chemical corrosive agents to the crack tip (hence the term stress corrosion), and on the plateau at higher K, it is thought to be controlled by the rate of fluid migration to the crack tip.

It is this plateau region which is relevant here, and thus we may suppose that the crack tip speed v will be controlled by the value of the stress intensity factor,

$$v = v(K), \tag{9.119}$$

where v may be a weakly increasing function of $K < K_c$. For what it is worth, values of K_c for crustal rocks have typical values $K_c \sim 10^6$ Pa m$^{1/2}$, though the extrapolation of such values to the deep lithosphere may be hazardous.

9.5.4 Non-dimensionalisation and Solution

We use (9.112) to write (9.113) in the form

$$q = -\frac{h^3}{12\eta_m}\left[-\Delta\rho_{sm}g - \frac{\partial\Pi}{\partial x}\right], \tag{9.120}$$

where

$$\Delta\rho_{sm} = \rho_s - \rho_m, \tag{9.121}$$

and

$$\Pi = \frac{\mu}{2\pi(1-v)}\oint_L \frac{\partial h}{\partial s}\frac{ds}{s-x}, \tag{9.122}$$

and then we scale the variables by writing

$$q \sim Q_0, \qquad h \sim h_0, \qquad \Pi \sim \Pi_0, \qquad x, l \sim d_L, \qquad t \sim t_0 = \frac{d_L h_0}{Q_0}, \tag{9.123}$$

where Q_0 is the order of the inlet flux, d_L is a typical lithosphere thickness, and suitable balances of the equations suggest we choose

$$h_0 = \left(\frac{12\eta_m Q_0}{\Delta\rho_{sm}g}\right)^{1/3}, \qquad \Pi_0 = \frac{\mu h_0}{(1-v)d_L}. \tag{9.124}$$

Then the dimensionless equations take the form

$$\frac{\partial h}{\partial t} + \frac{\partial q}{\partial x} = 0,$$

$$q = h^3\left[1 + \delta\frac{\partial\Pi}{\partial x}\right], \tag{9.125}$$

$$\Pi = \frac{1}{2\pi}\oint_L \frac{\partial h}{\partial s}\frac{ds}{s-x},$$

where

$$\delta = \frac{\Pi_0}{\Delta\rho_{sm}gd_L}. \tag{9.126}$$

If we take values $\rho_s = 3 \times 10^3$ kg m^{-3}, $\rho_m = 2.5 \times 10^3$ kg m^{-3}, $g = 10$ m s^{-2}, $\mu = 2 \times 10^{10}$ Pa, $v = 0.25$, $Q_0 = 1$ m^2 s^{-1}, $\eta_m = 10^2$ Pa s, $d_L = 50$ km, then we find $h_0 = 0.6$ m, $t_0 = 3 \times 10^4$ s, $\Pi_0 = 0.3 \times 10^6$ Pa, and thus $\delta \sim 10^{-3}$. The natural scale over which the singular integral term is important is of the order of 50 m, and is inconsequential over the bulk of the flow.

If we take the limit $\delta \to 0$, the solution of the model is straightforward. Suppose, for example, that the dimensionless flux $q = 1$ at the asthenosphere $x = 0$. Ignoring δ, we have $q = h^3$, and h satisfies the hyperbolic equation

$$h_t + 3h^2 h_x = 0, \tag{9.127}$$

and the solution for a crack starting at $x = 0$ at $t = 0$ is simply $h = 1$ for $0 < x < t$; the crack tip moves at unit speed to accommodate the influx of magma at the asthenosphere. In dimensional terms, this is $d_L/t_0 \sim 1.7 \, \mathrm{m\,s^{-1}}$.

This solution is invalid near the crack tip, where we must have $h \to 0$. There is thus a boundary layer near the tip in which we put

$$x = t - \sqrt{\frac{\delta}{2}} X, \qquad \Pi = \frac{\Theta}{\sqrt{2\delta}} \tag{9.128}$$

(the extra factor 2 is for cosmetic reasons), and then to leading order we find $q = h$, whence

$$\Theta_X = 1 - \frac{1}{h^2}, \qquad \Theta \approx \frac{1}{\pi} \int_0^\infty \frac{\partial h}{\partial \xi} \frac{d\xi}{\xi - X}. \tag{9.129}$$

The conditions on (9.129) are that $h \to 1$ as $X \to \infty$ and $h = 0$ at $X = 0$. More specifically, we now use the prescription for the stress intensity factor. Since the crack speed is determined by the magma flow,[1] the condition (9.119) would appear to determine K. Specifically, we use the definition of Ω in (9.108) to compute Ω' and thus $\sigma_1 + \sigma_2$ from (9.98). The local behaviour of the singular integral near the crack tip then shows that if (dimensionally) $h \sim c(l - x)^{1/2}$ near the crack tip $x = l$, then the stress is singular, as in (9.116), and the stress intensity factor is

$$K = \frac{\mu c \sqrt{\pi}}{2\sqrt{2}(1 - v)} \tag{9.130}$$

(see Question 9.9). Then we find that the required behaviour of (dimensionless) h satisfying (9.129) near $X = 0$ is that

$$h \sim 2\lambda X^{1/2}, \tag{9.131}$$

where

$$\lambda = \sqrt{\frac{2l}{\pi}} \left(\frac{K(1 - v)}{\mu h_0} \right). \tag{9.132}$$

Numerical solution of (9.129) appears to show that there is a unique solution for h, in which there is a slightly bulbous crack head, and the value of $\lambda \approx 1.3$ is determined automatically. Our interpretation of this is that this prescription is actually telling us theoretically what the appropriate value of the subcritical crack speed

---

[1] In Dave Stevenson's phrase, the tail wags the dog.

(9.119) is in this case. That is, we can determine the form of (9.119) by consideration of the elastodynamical problem. Defining the crack speed as $v = Q_0/h_0$, and
using (9.124) and (9.132), we find

$$v = \frac{\Delta\rho_{sm} g (1-v)^2 l K^2}{6\pi \lambda^2 \eta_m \mu^2}.$$ (9.133)

Using the values we introduced earlier, we find that this is

$$v = 2.2 \times 10^{-6} [l][K]^2 \text{ m s}^{-1},$$ (9.134)

where $l = [l]$ km, $K = [K]$ MPa m$^{1/2}$. With $[l] = 50$, we obtain crack speeds of
~ 1 m s^{-1} if $K \sim 10^2$ MPa m$^{1/2}$. This is a good deal larger (twenty to a hundred
times) than measured critical stress intensity factors in the crust, and requires that K_c
increases substantially with pressure, in order that this magmafracturing be aseis-
mic. This seems not unreasonable.

The elastodynamic propagation of magma through the lithosphere is thus dynamically feasible, although other questions remain, in particular the flow needs to be
sufficiently rapid that the magma does not freeze as it ascends. This is fairly simple
to evaluate. If the Péclet number of the fluid flow is large, then there will be thermal
boundary layers at the walls which will provide a much larger heat flux to the walls
than can be conducted away by the country rock; the magma will in fact melt back
the lithosphere (thus contaminating the chemistry of the host magma). For a magma
velocity $\sim v \sim 1$ m s^{-1}, crack width $h \sim 1$ m, crack length $l \sim 50$ km, and thermal
diffusivity $\kappa \sim 10^{-6}$ m^2 s^{-1}, the reduced Péclet number $Pe = \frac{vh^2}{\kappa l} \sim 20$, suggesting that even over such lengths, meltback of the channel walls is the more likely
occurrence.

The principal outstanding question is then the issue of providing a suitable inlet
boundary condition for the lithospheric fracture. Essentially, we need to glue together the porous magma transport in the asthenosphere with the fracture transport
in the lithosphere. It is not obvious how to do this. Apart from the tendency for
the magma compaction dynamics to allow fracture at the base of the lithosphere,
the magma transport solutions themselves are unstable in two dimensions to the
formation of magma channels. We might then expect a river-like system in the asthenosphere which simply continues to flow upwards through the sub-freezing lithosphere, and providing the upwards flow can be accommodated by accumulation in a
magma chamber, vented by eruptions, such a drainage system could be maintained
in a steady state. We now turn our attention to processes in magma chambers.

9.6 Crystallisation in Magma Chambers

It is generally accepted that magma initially propagates through the lithosphere by
the process of magmafracturing. The danger of freezing requires the ascent to be
rapid, and crack propagation is the only serious way for this to happen. One might
suppose that the magma would continue to ascend to the Earth's surface, but this

seems generally not to be the case; instead, magma accumulates in crustal magma chambers, typically at depths of 10–20 kilometres below the surface. The question then arises as to why this should be the case.

9.6.1 The Formation of Magma Chambers

There are various different types of magma chambers. The simplest are the dyke and the sill. The dyke we have already discussed; it is the conduit through which magma ascends through the lithosphere. It may supply a crustal magma chamber, or it may provide the vent from a magma chamber to an erupting volcano. Particularly in the latter case, the dyke may freeze after the eruption, and the solidified dyke may later be exposed at the surface by erosion.

A sill is like a dyke on its side; it is a flat tabular body which is presumably formed when an upwelling dyke reaches an unconformity, where the rock density decreases, and the propagating fracture finds it easier to propagate sideways rather than upwards. Again, such formations are commonly exhibited at the surface following erosive removal of the overlying crust.

The term laccolith refers to a magma chamber which is initiated as a sill, but in which the upwards buoyancy is sufficient to uplift the overlying crust, forming a gigantic blister. It may be that this is the principal way in which magma chambers form. It is certainly the case that upwelling magma from the deep mantle typically feeds crustal magma chambers, and these are large, long-lived bodies, which can supply volcanic eruptions for millions of years before the magma chamber finally crystallises. The dynamical processes involved in the creation of magma chambers have not apparently been studied.

An anomalous and extreme example of the magma chamber is the batholith, which refers to very large (hundreds of kilometres in horizontal dimension) intrusions of granite. Their mechanism of formation has been something of a problem to explain. The reason is that, although granites are clearly formed as igneous rocks, they are extremely silicic, and thus extremely viscous. They are too viscous to ascend via magmafracture (the velocity would be so small that they would freeze), and the other suggested ascent mechanism, as large, buoyant diapirs, may not be realistic. Another possible mechanism for the formation of granite batholiths is that they form at the base of the crust, when a mantle plume melts the crust (a process called *anatexis*), forming a hybrid, viscous magma. The consequent crystallisation forms the batholith, which is subsequently exposed at the surface after relentless erosion. Erosion of a millimetre a year allows 30 kilometres of erosion in 30 million years, which is the right kind of rate to promote the production of surface batholiths.

9.6.2 Nucleation and Crystallisation

The emplacement and subsequent crystallisation of a magma chamber is a little more complicated than the solidification of a single component material, as for example in the freezing of water. For pure liquids, we should expect a rind of crystals to grow inwards from the walls as cooling proceeds. In a multi-component liquid, a good deal more goes on, and we can use our understanding of the solidification of castings in metal alloys to explain this.

The basics may be understood by reference to a two component mixture. The thermodynamic phase diagram (like those extensively discussed in Sect. 9.3) causes the crystallisation of one substance at the expense of the other, and thus the crystallisation process releases fluid of different composition to the bulk, and thus of different density. The release of this fluid at the walls of the chamber will therefore cause compositional convection to occur. The form of this compositional convection may be compromised by the thermal convection which will occur due to cooling at the walls, and is discussed in the following Sect. 9.6.3.

For most substances, the thermal diffusivity is much larger than the compositional diffusivity, $\kappa \gg D$, and this causes crystal growth to occur in a mixed phase region, a *mush*, where solid and liquid coexist. In the laboratory, the mush is typically formed of dendritic crystals, because growth preferentially occurs on pre-existing crystal surface. In a magma chamber, growth is so slow that this may not be true.

In fact, in a casting, convective currents can allow the splintering of small crystal fragments, which can then circulate in the bulk liquid, forming seeds for the growth of so-called *equiaxed* crystals. In the final casting, one typically finds an outer dendritic or columnar rind surrounding an equiaxed core.

The principal difference between a crystallising magma chamber and a metallurgical casting lies in the chemical complexity of the magma, and a consequence of this is that the actual processes of crystal nucleation and growth are extremely slow, and in particular, the common laboratory assumption of thermodynamic equilibrium is not applicable. Crystal growth requires an undercooling.

The classical theory of non-equilibrium crystal growth is due to Avrami, and uses a concept known as the fictive volume. Suppose that the crystal nucleation rate per unit volume in a liquid is I, and the crystal surface growth rate (i.e., the surface velocity of advance) is Y, both of these being functions of temperature (and thus time, as the medium cools). A crystal nucleated at time $t = \tau$ has a volume

$$v = \frac{4\pi}{3} \left\{ \int_\tau^t Y(\theta)\, d\theta \right\}^3 \qquad (9.135)$$

at time t (assuming they are spheres: other crystal shapes will simply modify the pre-factor in (9.135)). In the time interval $(\tau, \tau + d\tau)$ there are $I(\tau)\, d\tau$ crystals generated per unit volume, and therefore the fictive crystal volume per unit volume, i.e., the fictive crystal volume fraction, ϕ', is

$$\phi' = \frac{4\pi}{3} \int_0^t I(\tau) \left\{ \int_\tau^t Y(\theta)\, d\theta \right\}^3 d\tau, \qquad (9.136)$$

whence we find

$$\frac{\partial \phi'}{\partial t} = 4\pi \int_0^t I(\tau) Y(t) \left\{ \int_\tau^t Y(\theta)\, d\theta \right\}^2 d\tau. \qquad (9.137)$$

The crux of the matter lies in the observation that this fictive volume fraction of crystals represents the quantity of crystals that would be obtained if we allow nucleation and growth of crystals within as well as outside crystals. This is clearly incorrect, but a simple adjustment allows computation of the true crystal volume fraction ϕ. Assuming the crystals are nucleated randomly in space and time, then only that proportion of the change in fictive volume $\delta\phi'$ in a small time interval δt which occurs in the intercrystal liquid will contribute to a change in the true crystal fraction. Thus, apparently,

$$\delta\phi = (1 - \phi)\,\delta\phi', \tag{9.138}$$

and therefore the evolution equation for the crystal fraction is

$$\frac{\partial\phi}{\partial t} = 4\pi(1 - \phi)Y(t)\int_0^t I(\tau)\left\{\int_\tau^t Y(\theta)\,d\theta\right\}^2 d\tau. \tag{9.139}$$

The simplest solution of this equation occurs at constant temperature and liquid composition, when it is reasonable to take I and Y as constant. With $\phi = 0$ at $t = 0$, the solution is

$$\phi = 1 - \exp\left[-\frac{\pi}{3}IY^3t^4\right]. \tag{9.140}$$

The time scale for solidification is $t \sim (\frac{1}{IY^3})^{1/4}$. The function given by (9.140) is sigmoidal, and approaches one fairly promptly at $t \approx (\frac{1.8}{IY^3})^{1/4}$. However, an inaccuracy of the model is that ϕ only tends to one as $t \to \infty$. It is fairly obvious that when ϕ approaches one, the intercrystal liquid will be confined to small tetrahedral pockets, and the specific interfacial surface area per unit volume will be $\sigma \sim (1 - \phi)^{2/3}$. Thus for $\phi \approx 1$,

$$\frac{\partial\phi}{\partial t} = \sigma Y \sim (1 - \phi)^{2/3}Y, \tag{9.141}$$

and thus

$$\phi \approx 1 - kY^3(t_c - t)^3 \tag{9.142}$$

for some finite t_c.

The difference between (9.140) and (9.142) lies in the way in which specific surface area is defined. In fact, (9.139) can be written in the form

$$\frac{\partial\phi}{\partial t} = (1 - \phi)\sigma'Y, \tag{9.143}$$

where σ' is the fictive specific surface area. Since, evidently, $\frac{\partial\phi}{\partial t} = \sigma Y$, it is clear that the critical assumption (9.138) is equivalent to

$$\sigma = (1 - \phi)\sigma', \tag{9.144}$$

and this tacitly assumes that surface area has the same distribution as volume. This may be true for small enough ϕ, but it cannot be true as $\phi \to 1$. Despite this, (9.139) is commonly used.

Undercooling The rates of nucleation I and growth Y depend on the degree of undercooling below the liquidus temperature T_L. The formation of a solid crystal of volume V is associated with a free energy change of magnitude

$$\Delta G = \sigma A - \frac{V \Delta G_{ls}}{V_m}, \tag{9.145}$$

where V_m is the liquid molar volume, ΔG_{ls} is the bulk free energy difference between solid and liquid, σ is the interfacial surface energy, and A is the crystal surface area. As with propagating fractures, there is a penalty for creating new surface energy, which is offset by the reduction in bulk free energy in forming the solid nucleus. However, since $A \propto r^2$ and $V \propto r^3$, where r is the radius of an assumed spherical nucleus, we can see that ΔG increases for small r to a maximum at $r = r_m = \frac{2\sigma V_m}{\Delta G_{ls}}$, before decreasing for larger r. There is thus an energy barrier to nucleation: nuclei larger than r_m will continue to grow, but those smaller than r_m will shrink, and the creation of sufficiently large nuclei is associated with thermal fluctuations. In keeping with the precepts of statistical mechanics, the rate at which such large fluctuations occur is of Arrhenius form, with an activation energy given by $\Delta G_m = \Delta G(r_m)$, and this suggests that the nucleation rate will be of the form

$$I = K \exp\left[-\frac{\Delta G_m}{RT}\right]. \tag{9.146}$$

The activation energy ΔG_m is itself defined by

$$\Delta G_m = \frac{16\pi \sigma^3 V_m^2}{3 \Delta G_{ls}^2}, \tag{9.147}$$

and the free energy change is related to the undercooling $\Delta T = T_L - T$. In a single component material, we have

$$\Delta G_{ls} \approx \Delta S \, \Delta T, \tag{9.148}$$

where $\Delta S \, (> 0)$ is the entropy drop on crystallisation. We thus see that we may assume that $\Delta G_m \propto \frac{1}{\Delta T^2}$. As a consequence, the nucleation rate increases rapidly with undercooling, and in practice, there is a critical undercooling ΔT_c such that nucleation is effectively absent for $\Delta T \lesssim \Delta T_c$.

At larger undercooling, other factors become important. In particular, the rate of transport of molecules (by diffusion) to the nuclear interface is itself thermally activated, and this has the effect of making the pre-factor K have an Arrhenius dependence also. If the activation energy for diffusion is ΔH, then we may write the nucleation rate in the form

$$I = K' \exp\left[-\frac{\Delta H}{RT}\right] \exp\left[-\frac{G'}{RT(\Delta T)^2}\right], \tag{9.149}$$

Fig. 9.17 Nucleation rate (arbitrary units) given by (9.149) using values $T_L = 1200$ K, $K' = 10^8$, $\frac{\Delta H}{R} = 2 \times 10^4$ K, $\frac{G'}{R} = 10^5$ K^3

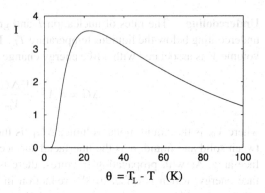

$$\theta = T_L - T \quad (K)$$

where

$$G' = \frac{16\pi\sigma^3 V_m^2}{3(\Delta S)^2}. \tag{9.150}$$

Figure 9.17 shows a typical form for the nucleation rate, which is consistent with experimentally measured rates. In particular, peak nucleation rates often occur at undercoolings of the order of hundreds of degrees Kelvin, and critical undercoolings ΔT_c can be of the order of tens of degrees. Experimental measurements normally give a quenched nucleation density of typical order $> 10^7$ cm^{-3}, indicating that nucleation is relatively easy in silicate melts. The critical undercooling is associated with the very large values of $\frac{G'}{R}$, and in fact it is appropriate to define the undercooling so that $G' = RT_L(\Delta T_c)^2$ (see Question 9.9).

Interfacial Growth   Similar sorts of ideas describe the rate of crystallisation at a pre-existing crystal interface. Essentially, crystal growth is an adsorption–desorption process, with rates of both being thermally activated. We write

$$Y = R_a - R_d, \tag{9.151}$$

where R_a and R_d are interfacial attachment and detachment rates. Attachment is thermally activated in a similar way to the transport involved in nucleation, and we write

$$R_a = K'' \exp\left[-\frac{\Delta G_{tr}}{RT}\right], \tag{9.152}$$

where ΔG_{tr} is an appropriate activation energy for diffusive transport. Most simply, R_d would be given by a similar expression, but for detachment, there is the additional energy barrier to overcome associated with the change in bulk free energy. Thus we put

$$R_d = K'' \exp\left[-\frac{\Delta G_{tr}}{RT}\right] \exp\left[-\frac{\Delta G_{ls}}{RT}\right], \tag{9.153}$$

Fig. 9.18 Growth rate (arbitrary units) given by (9.154) using values $T_L = 1200$ K, $K'' = 10^{10}$, $\frac{\Delta G_{tr}}{R} = 2 \times 10^4$ K, $\frac{\Delta S}{R} = 1$

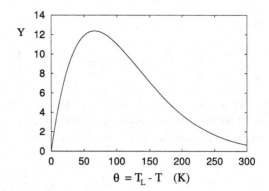

where ΔG_{ls} is given by (9.148) as before, and the pre-multiplicative constant K'' must be the same if $Y = 0$ when $T = T_L$. This leads us to the growth rate in the form

$$Y = K'' \exp\left[-\frac{\Delta G_{tr}}{RT}\right]\left(1 - \exp\left[-\frac{\Delta S \Delta T}{RT}\right]\right). \qquad (9.154)$$

The growth rate is similar to the nucleation rate, in the sense that there is a peak at a certain undercooling, but there is no critical undercooling. Figure 9.18 shows a representative illustration. As for nucleation, growth rates of silicates typically peak at undercoolings of $O(100)$ K, and the typical peak growth rates are of the order of 10^{-6}–10^{-8} m s^{-1}, with the lower values at lower liquidus temperatures. This indicates peak growth rates of the order of a metre a year for silicate melts, and suggests that crystallisation will normally occur via the slow growth of crystals in a crystal slurry.

The basic problem to study is then the crystallisation of an initial emplacement of magma, by analogy with the simplest Stefan problem of interfacial growth. We consider one-dimensional growth of a crystal mush from an initially emplaced magma in $x > 0$ next to cold country rock in $x < 0$. In the magma, the temperature will satisfy the heat conduction equation

$$\frac{\partial(\rho c_p T)}{\partial t} = \frac{\partial}{\partial x}\left(k\frac{\partial T}{\partial x}\right) + LS, \qquad (9.155)$$

where

$$\rho c_p = \rho_s c_s \phi + \rho_l c_l (1 - \phi), \qquad (9.156)$$

ρ_s and ρ_l are solid and liquid densities, c_s and c_l are solid and liquid specific heats at constant pressure,

$$k = \phi k_s + (1 - \phi)k_l \qquad (9.157)$$

is the thermal conductivity, k_s and k_l being those of solid and liquid, respectively, and L is the latent heat per unit mass. S is the rate of formation of crystal, given by

(9.139):

$$\frac{\partial \phi}{\partial t} = \frac{S}{\rho_s} = 4\pi (1 - \phi) Y(t) \int_0^t I(\tau) \left\{ \int_\tau^t Y(s) \, ds \right\}^2 d\tau. \qquad (9.158)$$

We assume the solid and liquid densities, thermal conductivities and specific heats are constant and equal, denoted ρ, k and c_p, respectively, and we define the thermal diffusivity to be $\kappa = \frac{k}{\rho c_p}$. Then we have

$$\frac{\partial T}{\partial t} = \kappa \frac{\partial^2 T}{\partial x^2} + \frac{L}{c_p} \frac{\partial \phi}{\partial t},$$

$$\frac{\partial \phi}{\partial t} = 4\pi N (1 - \phi) Z^2 Y, \qquad (9.159)$$

$$\frac{\partial Z}{\partial t} = Y,$$

where $Z = \int_0^t Y(s) \, ds$, and we have assumed that all the nucleation occurs instantly at $t = 0$, creating N nuclei per unit volume. We have

$$\phi = Z = 0 \quad \text{at } t = 0, \qquad (9.160)$$

and thus

$$1 - \phi = \exp\left[-\frac{4\pi N Z^3}{3} \right], \qquad (9.161)$$

whence the model reduces to

$$\frac{\partial T}{\partial t} = \kappa \frac{\partial^2 T}{\partial x^2} + \frac{L}{c_p} 4\pi N Z^2 Y \exp\left[-\frac{4\pi N Z^3}{3} \right],$$

$$\frac{\partial Z}{\partial t} = Y, \qquad (9.162)$$

and the growth rate Y is a given function of T. Suitable boundary and initial conditions are that

$$T = T_\infty, \qquad Z = 0 \quad \text{at } t = 0,$$

$$T = T_0 \quad \text{at } x = 0, \qquad (9.163)$$

$$T = T_\infty \quad \text{as } x \to \infty.$$

If T_L is the liquidus, we define the undercooling

$$\theta = T_L - T, \qquad (9.164)$$

and we suppose Y is a function of θ. It is convenient to non-dimensionalise the model by choosing scales

$$\theta \sim \Delta T_m, \qquad x \sim d, \qquad t \sim \frac{d^2}{\kappa} \equiv [t],$$

$$Z \sim \left(\frac{1}{4\pi N}\right)^{1/3} \equiv [Z], \qquad Y \sim \frac{[Z]}{[t]}, \tag{9.165}$$

where we suppose the peak growth rate is Y_m is at an undercooling of ΔT_m. Supposing that $T_\infty - T_0 \sim \Delta T_m$, it is also natural to scale $Y \sim Y_m$, and we can enable this by choosing the length scale d so that $[Z] = [t]Y_m$, i.e.,

$$d = \left(\frac{\kappa}{Y_m}\right)^{1/2} \frac{1}{(4\pi N)^{1/6}}. \tag{9.166}$$

The dimensionless model then becomes

$$\theta_t = \theta_{xx} - St Z^2 Y \exp\left[-\frac{1}{3}Z^3\right], \tag{9.167}$$

$$Z_t = Y,$$

with initial and boundary conditions

$$\theta = \theta_\infty, \qquad Z = 0 \quad \text{at } t = 0,$$

$$\theta = \theta_0 \quad \text{at } x = 0, \tag{9.168}$$

$$\theta = \theta_\infty \quad \text{as } x \to \infty,$$

where the Stefan number is

$$St = \frac{L}{c_p \Delta T_m} \tag{9.169}$$

and

$$\theta_0 = \frac{T_L - T_0}{\Delta T_m}, \qquad \theta_\infty = \frac{T_L - T_\infty}{\Delta T_m}. \tag{9.170}$$

We use typical values $L = 3 \times 10^5 \text{ J kg}^{-1}$, $c_p = 10^3 \text{ J kg}^{-1} \text{ K}^{-1}$, $\kappa = 10^{-6} \text{ m}^2 \text{ s}^{-1}$, $\Delta T_m = 100$ K, $N = 10^9 \text{ m}^{-3}$ (based on an eventual crystal size of 1 mm), $Y_m = 10^{-7} \text{ m s}^{-1}$, to find $d \sim 6.6$ cm, $[t] \sim 1$ hour, $St \approx 3$. The crystals grow in a fairly thin skin.

The key to understanding the solution lies in the observation that the width of the skin, $d \sim 10$ cm, is much less than the macroscopic dimension d_C of a magma chamber, so that on the large scale, the skin appears as an interface, and we might expect the usual interfacial Stefan condition to emerge in the limit $d \ll d_C$.

And indeed, this is the case. If we write (9.167) in the dimensionless form

$$\theta_t = \theta_{xx} - St\phi_t \tag{9.171}$$

(noting that $\phi = 1 - \exp(-\frac{1}{3}Z^3)$ is dimensionless), then integration on the macroscopic scale of the integral form

$$\frac{d}{dt} \int_A^B (\theta + St\,\phi)\,dx = [\theta_x]_A^B \tag{9.172}$$

from $s-$ to $s+$ (where these denote the limits as $x \to s$ from $x < s$ and $x > s$, respectively, on the macroscale, and s is the 'location' of the skin (e.g., where $\phi = \frac{1}{2}$))[2] gives the usual dimensionless Stefan condition

$$St\,\dot{s} = [\theta_x]_-^+ \tag{9.173}$$

for the solution of the macroscale temperature field. Note that this assumes that θ is continuous across the interface.

The skin acts as a 'shock' over which ϕ changes rapidly, and we thus expect that solutions of (9.167) will tend to travelling waves which accommodate this transition (and that the front speed will be given by (9.173)). We put

$$\xi = x - ct, \tag{9.174}$$

and look for solutions in which θ and Z are functions of ξ. These functions then satisfy the ordinary differential equations

$$-c\theta' = \theta'' - StZ^2 Y \exp\left[-\frac{1}{3}Z^3\right],$$

$$-cZ' = Y(\theta), \tag{9.175}$$

together with the (matching) boundary conditions

$$\theta' \to g_+, \qquad Z \to 0 \quad \text{as } \xi \to \infty,$$

$$\theta' \to g_- \quad \text{as } \xi \to -\infty, \tag{9.176}$$

where g_+ and g_- denote the limiting macroscopic values of $\theta_x|_{s+}$ and $\theta_x|_{s-}$.

For simplicity, suppose that the emplaced magma is at the liquidus, so that $\theta \to 0$ as $\xi \to \infty$, and $g_+ = 0$. Suppose also that $Y(\theta) = \theta$. Then (9.175) has a first integral

$$Z'' = -cZ' + St\left(1 - \exp\left\{-\frac{1}{3}Z^3\right\}\right), \tag{9.177}$$

using the fact that $Z \to 0$ as $\xi \to \infty$. The solutions can be analysed in the (Z, W) phase plane, where

$$Z' = W,$$

$$W' = -cW + St\left(1 - \exp\left\{-\frac{1}{3}Z^3\right\}\right). \tag{9.178}$$

---

[2]More precisely, take $A = s - \eta$, $B = s + \eta$, and then let $\eta \to 0$ with $\eta \gg d$.

Fig. 9.19 Phase portrait of (9.178) for values $St = 3$, $c = 1$

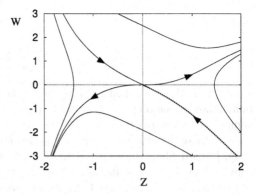

We are interested in trajectories which approach $(0, 0)$ from $Z > 0$, $W < 0$, and there is a unique such trajectory, which is the stable separatrix to the saddle point at the origin. Figure 9.19 shows the phase portrait.

It is clear that as $\xi \to -\infty$, $Z \to \infty$, and thus $W' \sim St - cW$, and W grows exponentially. This is not apparently consistent with (9.173), which requires in the present case $c = \frac{g_-}{St}$. Since $g_- = \theta' = -cZ'' = -cW'$ at $\xi = -\infty$, we see that this is only consistent with (9.178) if c is small. In fact (see Question 9.9), this is precisely the case when the travelling wave solution should be valid, since c must be small in order that it match to the outer solution near the wall of the chamber.

9.6.3 Double-Diffusive Convection

Crystals in magma chambers can hardly grow in a stagnant medium, however, since convective processes will inevitably occur. In particular, because magmas are multi-component fluids, such convection is likely to be double-diffusive, and we now discuss this subject. Double-diffusive convection refers to the buoyant motions driven by density differences of a fluid which are induced by two competing substances. Most commonly, these are temperature (T) and composition (c), and their effect on the density is represented by the relation

$$\rho = \rho_0 \big[1 - \alpha(T - T_0) + \beta c \big], \tag{9.179}$$

where conventionally we think of increasing composition causing an increased density, as is the case in the common laboratory case of salt dissolved in water.

There are now two Rayleigh numbers: first, there is the conventional thermal Rayleigh number

$$Ra = \frac{\alpha \rho_0 g \Delta T d^3}{\eta \kappa}, \tag{9.180}$$

where d is the depth of the fluid layer, ΔT the prescribed temperature excess (base over top), η the viscosity, and κ the thermal diffusivity. In addition, there is a compositional Rayleigh number

$$Rs = \frac{\beta \rho_0 g \Delta c \, d^3}{\eta \kappa}, \tag{9.181}$$

where Δc is the prescribed compositional excess. This is the conventional definition, although one might be tempted to replace κ in the denominator with D, the
compositional diffusivity.

With these definitions, positive Ra is destabilising, while positive Rs is stabilising. In particular, when $Ra > 0$ and $Rs < 0$ are numerically large enough, then
convection will occur in the usual way. There are two interesting variants which
can occur even when the fluid is statically stable (i.e., its density increases with
depth, $Ra + Rs > 0$). The first occurs for $Rs < 0$, $Ra < 0$: temperature is stabilising but composition is destabilising. In this case the static state may be unstable to
very narrow convection cells, known as fingers (often, in practice, salt fingers). This
style of convection is readily demonstrated in the laboratory, and has been observed
to occur in the oceans.[3]

In the opposite quadrant, a destabilising temperature field can render a stable
compositional field oscillatorily unstable, but this linear stability result is not often seen in practice, partly because the boundary conditions relevant to the result
are difficult to obtain, and partly because in nature, double-diffusive convection is
wildly nonlinear, and leads to other dramatic phenomena.

Perhaps the principal behavioural feature of double-diffusive convection is the
readiness with which the fluid forms separately convecting layers. Two examples
will illustrate this predilection. Suppose a stable compositional gradient has been
set up in a fluid, and it is then suddenly heated (strongly) from below. What is found
is that a layer of fluid at the base begins to convect turbulently, but not throughout the
fluid depth. In fact, the convecting fluid homogenises the density, but the convecting
layer can only extend to its level of neutral buoyancy: above this, the fluid is static.

Nevertheless, there are turbulent thermal and compositional fluxes across the interfacial boundary layer at the top of the convecting layer, and the thermal flux generates a growing thermal boundary layer which eventually becomes unstable, causing a second convecting layer to form. In this manner, a series of convecting layers
is set up, due to the fact that the thermally unstable boundary layers win because
the diffusivities satisfy $\kappa > D$ (so there is no compensating buoyancy generated by
the compositional fluxes). In the laboratory, these layers appear long-lived, being
roughly associated with the thermal diffusive time d^2/κ of the chamber. On an even
longer time scale, the convective layers merge from the base up, and the eventual
state will be one of well-mixed thermal convection, with the compositional gradients eradicated. In practice, the long transient is more interesting than the eventual
steady state.

---

[3]It is tempting to suppose that the large hexagonal basalt columns seen, for example, at the Giant's
Causeway in Northern Ireland might represent a similar phenomenon, but this appears not to be the
case; such columnar basalts are thought to arise through an instability associated with contraction-
induced fracturing.

The basic motto in this and other double-diffusive settings is simply that light fluid rises where it can. Another instance of layering exemplifies this. Again we start with a stably compositionally stratified fluid in a chamber whose walls slope inwards at the base. If the walls are cooled, a convection current (essentially like a katabatic wind) is generated which flows down the walls. Again, the flow stops when it reaches its level of neutral buoyancy, and then turns sideways into the interior fluid of the chamber. In this way, a series of (transient) horizontal convective layers are set up.

Convection in Magma Chambers   We can now paint a plausible sequence of events in a cooling magma chamber, bearing in mind that many other things are also dynamically feasible. A typical sort of shape for a magma chamber may be like a tadpole, or a balloon, with the blunt upper surface formed by doming of the overlying country rock, and the lower tail being connected to the umbilical feeder dyke. In such a chamber, the emplacement of hot magma is followed first by chilling and quenching at the margin. The cooling of the magma at the walls leads to thermal convection where cold rafts of magma detach from the sides and roof of the chamber and descend to the floor. If we suppose the feeder dyke is sealed, then there is no warming from below, and the cool magma will pond at the floor, forming a stably stratified layer, which thickens with time. The same thermal stratification is a common (and inefficient) feature of heated rooms, for example by use of a fan heater in a bathroom (although there, it is caused by ponding of hot air at the ceiling).

At a rate controlled by heat conduction (since the Stefan number is $O(1)$), crystals grow from the walls, floor and roof. In general (but not always), we may suppose that the crystals are denser than the melt, and also that light fluid is released on crystallisation. Because we may expect the crystals to grow in a slurry, it is then natural to expect sloughing off of mats of crystals from the walls and the roof, which then rain down to the floor, forming a crystal pile at the floor. Compositional convection from this pile will occur if light fluid is released, causing finger-like plumes in the overlying fluid. This compositional convection will then lead to a compositional stratification of the chamber. Evidence for such zonation can be found in relict magma chambers which are now exposed at the Earth's surface, and is indirectly evidenced by the (inversely) stratified deposits which are formed in some volcanic eruptions.

9.6.4 Layered Igneous Intrusions

An intriguing phenomenon in some fossil magma chambers is the presence of widespread layering in the rocks, at a scale from centimetres to metres. This layering can be seen through bands of differently-coloured rocks (Fig. 9.20), and there can also be layering in crystal size. The type example, for historical reasons, is the Skaergaard intrusion of East Greenland, one of a swarm of igneous intrusions associated with Tertiary volcanism and the opening of the Atlantic Ocean. It was here

Fig. 9.20 Graded layering in the Skaergaard intrusion. Photo courtesy of Kurt Hollocher

that Wager, Brown and Deer undertook their voluminous studies from the 1930s onwards which mapped the intrusion, and they also provided a tentative explanation for the observations.

It must be said (and any petrologist will do so) that the Skaergaard exhibits evidence of a bewildering variety of different dynamic features, and to attempt an explanation of all of them would be somewhat like endeavouring to explain every last eddy in a turbulent flow. This we will not (and cannot) do.

Instead we focus on some features of the large scale behaviour of the magma chamber. The chamber is divided into different zones. There is the marginal border series, associated with initial quenching and subsequent crystallisation at the sides. The main part of the chamber is divided into a number of different horizontal zones (lower zone a: LZa; lower zone b: LZb; etc.) which are distinguished by the minerals present. The same sequence of minerallisation occurs upwards as downwards, suggesting that fractional crystallisation proceeded both upwards from the floor and downwards from the roof.

The sequence of minerals present as we rise from the floor is firstly olivine and plagioclase; then augite is added; then magnetite; then pigeonite replaces olivine; and so on. These zones represent the natural process of fractional crystallisation as the residual melt alters its composition. Within the zones, we find modally graded layers. For example, a layer of LZa would have olivine at the base grading into plagioclase at the top. Sometimes these layers alternate continuously, sometimes they are intercalated by uniform gabbro layers. In thickness, they range from a scale of centimetres to metres.

The other very noticeable feature about the Skaergaard is the presence of huge blocks of anorthosite (i.e., plagioclase) within the chamber. These are thought to represent caved in (stoped) fragments of the crystallising roof of the chamber (now long eroded) which collapsed and fell to the crystal pile growing on the floor. These blocks show evidence of impact on the growing layered series. Layers have been bent, cut and warped by the blocks, and have then resumed growth about them.

Armed with the discussion of double diffusion in Sect. 9.6.3, there is an obvious possible explanation for the layering which is observed. Since there is inevitably convection, and since magmas are multi-component mixtures, double-diffusive convection must occur, and the ubiquity of layering in such convective systems then suggests itself as a cause of the layering in the rocks.

However, this idea instantly runs into problems. The situations where convective layers occur are those where there is pre-existing compositional stratification, and the layers form by a superimposed thermal convection. This is the opposite to what we expect in a magma chamber, where we suppose ponding will produce a thermal stratification, which is then subjected to compositional convection. Moreover, the layering in the Skaergaard is compositionally oscillatory. One sees, for example, olivine grading into plagioclase, then back to olivine, and so on. This is not what convective layering exhibits.

The original explanation favoured by Wager and Brown in their pioneering work was that crystal-rich turbidity currents would sweep down from the walls of the chamber, and then gravitational settling would separate the heavier olivine from the lighter plagioclase crystals. This old idea may seem implausible, but it is not without merit. While it seems unlikely that olivine and plagioclase could separate in a fairly crystal-rich slurry, it is by now well-known that differently sized particles will spontaneously separate in a granular flow, so it is at least plausible that separation could occur this way. Nor is the regular occurrence of crystal avalanches of this type so unlikely. As a crystal slurry grows and thickens, it may become gravitationally unstable, and initiate an avalanche. The resultant turbidity current can then sweep up the crystals in its path, cleaning the wall and depositing its load at the floor. Essentially the same thing happens in snow avalanches, and in submarine turbidity currents, which carve out submarine canyons in regular avalanche-induced runout events.

One of the apparent difficulties with this scenario is the question of how the growing layer can reach finite size before failing, since we envisage growth of a slurry with little intergranular contact. The apparent answer to this problem lies in the fact that silicate melts are polymerised (by chains of silica molecules), and consequently they have a yield stress, which increases with silica content. Typical sorts of values for yield stress are in the range 10^2–10^4 Pa, and these values increase with increasing crystal content. A slurry of crystals a metre thick having an excess density of, say, 500 kg m^{-3} over the bulk liquid, will exert a differential stress of 0.5×10^4 Pa. So it is entirely possible that a stationary slurry will grow and then detach at some critical thickness, and that this will happen periodically.

The yield stress of silicate magmas can also explain what has been called the plagioclase flotation problem. At least in some magmas, precipitated plagioclase crystals are lighter than their parent magma, and should thus float. It is then difficult to understand the occurrence of plagioclase crystals in the lower zone (and olivine crystals in the upper zone). But if the crystals grow in a slurry, they should remain immobile within it if $\Delta \rho g d_g < \tau_c$, where $\Delta \rho$ is the density difference between crystal and melt, d_g is the grain size, and τ_c is the yield stress. For a millimetre sized crystal, the buoyant stress is only 10 Pa, even if the density excess is 10^3 kg m^{-3}. This seems a plausible way of locking crystals in a slurry, although it will also have the effect of preventing compositional convection within the slurry.

An altogether simpler mechanism to obtain successive layers is to ascribe the cause to periodic inputs of fresh magma to the base of the chamber. This is probably not feasible for the Skaergaard, but it is a likely thing to occur, since the formation of igneous provinces and volcanic eruptions is essentially a periodic phenomenon (input of magma lowers source pressure, which lowers flow rate and causes eventual blockage of the vent, which is then subsequently re-opened by the build-up of pressure at the source).

One possible scenario is the following. Suppose a two component magma containing liquids A and B is injected into a chamber of more evolved magma. We suppose that B is heavier (more basic) and crystallises out first. When the magma is injected, it is hotter than the evolved magma, but ponds at the bottom because of its greater content of B. Being hotter, it cools by convective heat transfer to the upper layer, and as it does so, crystals of B form and will eventually sink to the floor (they may be kept in suspension for a time by the turbulent convection). The residual liquid becomes progressively enriched in A, and when its density reaches that of the liquid above, rollover occurs, and the magmas mix. If the composition reaches the eutectic point, then a different rock precipitates, containing also B. In this way, we can generate a layer of A followed by a layer of $A-B$. Subsequent injection of B-rich magma will lead to the same sequence of events, and the build-up of a periodic sequence of layers.

It is thought that this process can explain the alternating layered rocks on the Isle of Rum in the Hebrides.[4] The Rum intrusion consists of alternating layers of peridotite (essentially olivine) and allivalite (olivine-plagioclase). So in our simple model, A would be plagioclase, and B would be olivine.

Another suggested mechanism to explain the Skaergaard layering is that of oscillatory nucleation. In essence, the idea is indicated in Fig. 9.21, which shows a phase diagram (more or less the same as Fig. 9.5) of a binary mixture of two components A and B which form a eutectic (at E in the figure). Nucleation, and subsequent crystallisation, of either phase requires a finite undercooling, and so the two dashed curves below the equilibrium liquidus curves represent the *meta-liquidus* curves where nucleation and thence growth is initiated. Note that these, and the liquidus curves themselves, continue past the eutectic point. If the parental liquid temperature and composition starts out to the left of E, then cooling will cause nucleation

---

[4]The former spelling was Rhum; apparently the 'h' was added in the 1900s by the owner George Bullough, who disliked the alcoholic connotation. 'Rum' is an anglicisation of the Scots Gaelic 'Rùm', of uncertain meaning.

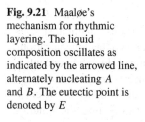

Fig. 9.21 Maaløe's mechanism for rhythmic layering. The liquid composition oscillates as indicated by the arrowed line, alternately nucleating A and B. The eutectic point is denoted by E

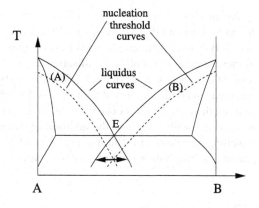

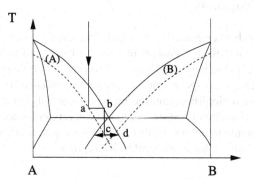

Fig. 9.22 A putative temperature-composition path for the liquid in the crystal layer. The temperature cools (slowly) until nucleation (and growth) of A begin at a. The liquid is then enriched in B until the A liquidus is reached at b, at which point growth ceases, and the liquid cools until nucleation of A recommences at c. Again enrichment of the liquid in B occurs, but this time when the A liquidus is reached at d, nucleation and growth of B has already begun, so that enrichment of A begins. Potentially, the liquid will be thus diluted of B until it reaches the B liquidus, at which point A enrichment begins again

when the temperature reaches the A meta-liquidus. The resulting growth of A moves the liquid composition to the right. Unlike the equilibrium situation, where the temperature would be constrained to move gradually down the liquidus, and would stop when it reached the eutectic point, here the temperature can move horizontally to the right through the non-equilibrium field until B starts to nucleate and grow. Then the liquid composition moves towards A and, depending on the rates, the temperature may reach the (continuation of the) B liquidus. Repeating the process leads to the oscillation shown in Fig. 9.21, and one approach to this is shown in Fig. 9.22.

While this may seem to provide a plausible oscillation mechanism, it is by no means clear that it will work in practice, when the diffusion of heat is taken into account. It appears that the necessary numerical simulation of an appropriate space-dependent crystallisation model has not yet been done, although the model is some thirty years old!

An intriguing variation on this idea uses the fact that the rate coefficients K' and K'' in (9.149) and (9.154) represent diffusion rates of the various chemical species to the crystal–liquid interface, and of course they must depend on the concentrations of these species (after all, crystallisation is simply a reaction). Thus the concentrations of chemical species in the liquid will satisfy a sequence of reaction–diffusion equations, with the kinetics of the reaction described by the dependence of the crystal growth rates on the concentrations. It is not difficult to organise the kinetics to provide oscillations, and the addition of diffusion would then naturally lead to travelling waves: layers! But it is not clear whether chemical concentrations can in practice be rate-limiting in this way (one would expect not).

9.7 Volcanic Eruptions

Not all magma chambers emplaced in the crust simply freeze. The emplacement depletes the source region, the asthenosphere, of some fraction of its partial melt, and this causes excess compaction which induces a lower melt pressure. The upwelling dyke runs out of gas, so to speak, and the chamber is inflated. But the source region will in time be replenished, and the melt pressure will build again. Thus we can expect magma chambers to be subject to the contrary influences of cooling and reinflation. When reinflation wins, further fractures are generated, and the resulting magma flow to the surface becomes a volcanic eruption.

9.7.1 Types of Eruption

There are many different types of eruption, and they differ in style because of the different types of magma which are erupted. The simplest kind of eruption is the *effusive* eruption, in which erupted lava flows in streams away from the volcanic vent. Basalts flow rapidly, but more viscous magma may be extruded very slowly, forming a lava dome.

The other types of eruption are explosive, and these range in ferocity from the relatively gentle Hawaiian and Strombolian, to the ferocious Plinian. Various other names are also used (Vulcanian, Peléean, and so on), but for our present purposes we can think of explosive eruptions as occurring in a spectrum which ranges from Strombolian to Plinian.

Essentially, the differences in eruption types are largely due to magma chemistry. The two principal constituents which affect the behaviour of the magma in an eruption are the silica content and the volatile content. Silica affects the viscosity, with more siliceous magma having higher viscosity. While basaltic magma has a typical viscosity of 10^2 Pa s, silica-rich rhyolite has a viscosity of 10^8 Pa s. Viscosity is also affected by temperature (about an order of magnitude every 100 K), by crystal content, and by the *wetness* of the magma, which is due to the other main important constituent, water. Silicate magmas are essentially solutions of (mostly metal) oxides in silica, and some of the dissolved oxides are volatile, and prone to exsolution (as gas). The principal such oxides are water, carbon dioxide and sulphur dioxide, and of these, water is usually the most important. It is the wetness of magma which causes most explosive eruptions,[5] in the following way. As magma rises in a conduit from the magma chamber, its pressure decreases. At some point, volatiles will be exsolved, in the same way that champagne effervesces when the cork is popped. Continued magma rise causes the bubbles to grow, and because of the expansion involved, there are large pressures exerted on the interstitial liquid. If the bubble volume rate of increase is sufficiently slow, then the liquid can deform to accommodate the increasing void fraction (volume fraction of bubbles) and the exit of the resulting two-phase flow at the vent leads to mildly explosive eruption.

On the other hand, if the exsolution rate is rapid, the liquid films between the bubbles may not be able to withstand the differential stresses, and may shatter. Once the integrity of the liquid is broken, the magma is essentially transformed to a gas/liquid droplet mixture which explodes from the vent. The size of the liquid droplets will depend on the degree of fragmentation, and the very smallest form the huge ash clouds which are such a dominant feature of explosive eruptions.

9.7.2 Strombolian Eruptions

Strombolian volcanoes are named for Stromboli, a volcano between Italy and Sicily, which has apparently been 'erupting' continuously for thousands of years. The eruptions consist largely of explosions, which occur fairly regularly every few minutes, and consist largely of gas. The mechanism of these eruptions can be understood by reference to gas-liquid two-phase flows.

The classic situation of industrial interest in two-phase flows is the formation of steam as water rises in an externally heated pipe. This forms the basis for studies of flow in boilers and nuclear cooling systems. As the water rises and is heated past the boiling point, bubbles of steam are nucleated and break off into the (usually turbulent) flow, forming a 'bubbly' flow. This flow régime is stable (the bubbles remain separate) until there is a transition to a régime called 'slug flow', in which the bubbles coalesce into large Taylor bubbles[6] which fill the width of the pipe, separated by large slugs of liquid, perhaps with some (small) bubbles also. This transition may be due to an instability in bubbly flow which is manifested (in steam–water flows) when the void fraction α reaches about 0.25. As α increases further, there is another transition to 'annular flow', when the liquid flows up the walls, with the gas forming a continuous (and much faster) interior core. The transition from

---

[5]This may not quite be true (at least in the way we will describe) for vulcanian eruptions, which are short term explosive events, and may be due to the build-up of (volatile) pressure in a vent which has been capped by a plug of solidified magma.

[6]After G.I. Taylor.

slug to annular flow is by way of an intermediate 'churn flow' régime, where the Taylor bubbles become irregular and intermittent.

The flow in a Strombolian vent is a two-phase flow, where the exsolution of volatiles as fluid rises is not due to external heating, but to depressurisation. The effect, however, is the same. Initially at depth the gas forms bubbles, and the flow is that of a bubbly flow. As the fluid rises, the void fraction increases, and we may expect coalescence to lead to slug flow. It is the appearance of the Taylor bubbles at the outlet of the conduit which causes the 'Strombolian burps', which are the manifestation of the eruption.

The situation is complicated by the obvious fact that if this style of eruption has been going on for millennia, then the conduit to the surface from the magma chamber has been continually open during that period. The only way this can be true when the net liquid ejection at the surface is negligible is if there is a countercurrent flow in the conduit: otherwise the liquid would simply freeze. This countercurrent flow must be due to convection of the magma in the conduit, driven by the buoyant release of volatiles. Thus we can expect the gas-rich magma to rise in the core of the conduit, while the gas-poor magma descends at the walls (or it could be the other way round).

Convection in the conduit may be periodic in time, as happens in the somewhat analogous case of a geyser, in which case the periodicity of the eruptions may be a consequence of the periodicity of the bubble rise, rather than simply being due to the inter-arrival time between bubbles.

In this scheme of things, bubbly flow leads to effusive eruptions. It is also possible to have eruptions which appear to correspond to churn flow, and these have been observed at Villarrica volcano in Chile, for example.

9.7.3 Plinian Eruptions

The formation of large Taylor bubbles from the coalescence of small bubbles involves the collapse of liquid films between bubbles. This is driven by the excess pressure in the bubbles caused by the volume expansion on exsolution. If the liquid viscosity is sufficiently low, then the liquid can drain from the films without mishap. For larger viscosities, however, this is not true, and the stress on the liquid can become so large that it ruptures. The shattering of the magma foam then causes an explosion, as the released gas erupts from the vent, carrying the shards of the liquid films with it. With the sudden release of pressure, the shards are quenched, and depending on their size, form clots of pumice and cinders, or ash. The resultant distinctive ash column is a typical product of the explosive Plinian eruption.

Pliny the Younger observed the eruption which is named for him in A. D. 79. This was the eruption of the Italian volcano Vesuvius near the present-day city of Napoli, which obliterated the Roman towns of Pompeii and Herculaneum. As is now well known, the devastation of Pompeii was caused by a dense pyroclastic flow, which rushed down the slopes of the volcano, roasting everything in its path. A pyroclastic flow is an example of a turbidity current, a turbulent gravity-driven flow which is driven by the excess density of the ash particles within it. As such, it is analogous to katabatic winds on ice slopes, snow avalanches, and submarine turbidity currents, which are instrumental in building the convoluted patterns of offshore canyons on the edges of continental shelves.

An explosive eruption column consists of a turbulent two-phase mixture of hot gas and ash. It is propelled explosively from the vent. In some circumstances, entrainment and heating of air causes a further self-driven buoyant convective uprise of the plume to the tropopause or beyond. Ash and pumice rain out from the plume, forming deposits, sometimes metres thick, over a wide surrounding area. If the ash column is too dense, or is not rising fast enough, then the fountain can collapse, and the mixture of hot gas and ash will fall back on to the slopes of the volcano, where it then forms a terrifying pyroclastic flow. The ash cloud flows downslope at speeds of tens of metres per second, devastating everything in its path. As it progresses, the heavier ash particles fall out, so that the density of the current rises. At some point, the current comes to a fairly abrupt halt; the remaining ash is deposited, and the hot gas rises harmlessly into the atmosphere.

9.8 Notes and References

The subject of magma generation, transport through the asthenosphere and lithosphere, emplacement of magma chambers, eruption and solidification, is one in which there are no final answers, and there is no one book which describes the whole sequence of events from a theoretical point of view, apart from that of Dobran (2001). The contents of his book encompass most of those of the present chapter, and provide an excellent source for both the physical ingredients of magma processes, and how to build mathematical models of them.

On the other hand, a number of textbooks describe the general phenomena, amongst which Holmes (1978) is invigorating. There are several volumes comprising edited individual chapters, in which many of the protagonist constituents are described. Among these are the volumes edited by Morgan et al. (1992) and Ryan (1990) on magma transport. The book edited by Hargraves (1980) provides a very broad synthesis of many different physical aspects of magma behaviour. Books on petrology and geochemistry are those by McBirney (1984), Hess (1989), Nockolds et al. (1978), Albarède (2003) and Mason and Moore (1982). A classic is the book by Bowen (1956). Pitcher (1997) describes the enduring complexities involved in the formation of granites.

Superplumes, Eruptions and Extinctions   It is commonly thought that the extinction of the dinosaurs at the Cretaceous–Tertiary boundary some 65 My ago was caused by the impact of a meteorite in the Yucatán peninsula in Mexico, a hypothesis first proposed by the father and son Alvarez (Alvarez et al. 1980); initially controversial, the impact hypothesis has become widely accepted (Schulte et al. 2010), though not by all: see the letters in *Science* **328**, 973–976 (2010). An alternative view, particularly advanced by Courtillot (1999) is that mass extinctions are often associated with episodes of massive extrusion of flood basalts. In particular, the formation of the Deccan Traps in India represents a stuttering outpouring of something approaching a million cubic kilometres of basaltic lava over a period of perhaps hundreds of thousands of years at about the same period, and the timing of other such flood basalt eruptions appears coeval with other extinctions in the geologic record. Courtillot thus associates extinctions, including the dinosaur extinction, with the basaltic floods; the occurrence of the meteorite is not denied, but is not considered to be a necessary cause (though it would undoubtedly help).

The gigantic eruptions themselves are thought to be associated with the existence of 'superplumes', essentially a manifestation of convection in the Earth's mantle (Schubert et al. 2001; Bercovici 2009a, 2009b). It is generally thought that 'hotspots' like Hawaii and Iceland are associated with the existence of convective upwellings ('plumes') from within the mantle, possibly originating from the core-mantle boundary, and which are consequently associated with vulcanism. In this view, the massive basaltic eruptions are associated with the sort of massive plumes which one might expect if mantle convection were episodic (as appears to be the case on Venus): see also Chap. 8. The plume hypothesis itself (Morgan 1971), seemingly an innocuous consequence of convection, is itself a matter of controversy.[7]

Magma Transport   The description of the porous flow of melt in the asthenosphere really starts with the work of Turcotte and Ahern (1978) and Ahern and Turcotte (1979), who describe the essential physics of the process, and indeed solve the problem for one-dimensional transport. While they refer to compaction, they do not model it explicitly. Compaction itself is introduced by McKenzie (1984), Scott and Stevenson (1984), and Fowler (1985b), more or less in the same way by each author, and thereafter the subject swells with the work of Hewitt, Katz, Spiegelman, Ribe, Bercovici, and many others. For no particularly good reason, much of this early literature ignored the melting term (which after all is what produces the melt), in which case the model is in some sense conservative, and one-dimensional solutions have solitary wave properties (Scott and Stevenson 1984), termed by them

---

[7]DePaolo and Manga (2003) summarise the hypothesis, and on the same page Foulger and Natland (2003) raise various objections. The root of the controversy is (as so often) a misunderstanding of how models should be applied. The objections of Foulger and Natland are that (i) there is no tomographic evidence for plume tails; (ii) hot spots are not fixed; (iii) there is no evidence for excessive temperatures at hot spots. These objections are based on a naïve understanding of how convection works. One should not expect to see plume tails tomographically, nor should hot spots be necessarily fixed, nor should there be large temperature excess in strongly variable viscosity convection. The objections are not based on data, but on preconception. The debate has become somewhat creationist in tone, with its own website (www.mantleplumes.org), and occasional inflammatory pieces, such as that of McNutt (2006), with the bye-line: "At least one chain of hot-spot volcanoes is not caused by a plume rising up from the core-mantle boundary, calling for a reexamination of the plume hypothesis." McNutt's remarks were chastised by Hofmann and Hart (2007). Just as the question for global warming sceptics is: how could CO_2 *not* cause increased temperatures? the question to be asked here is really: how could there *not* be plumes if the mantle is convecting at high Rayleigh number?

'magma solitons', or magmons. Pursuit of these mythical beasts occupied a good deal of attention for a number of years, particularly as there is a nice laboratory analogue (Scott et al. 1986), although one can show that inclusion of melting and also physically relevant boundaries suggests that in practice such waves will not be seen (Fowler 1990).

Just as for water flow under glaciers (Chap. 10) or over erodible sediment (Chap. 6), porous flow in the asthenosphere is prone to channelise (Aharonov et al. 1995; Nicolas 1986; Hewitt and Fowler 2009; Spiegelman et al. 2001; Stevenson 1989), and it seems likely that an arterial network of melt channels will be formed. The upwelling melt could freeze on to the lithosphere, form a sub-lithospheric lake, or promote fracturing and consequent continued rise through the lithosphere.

Magmafracture Fracture mechanics is really a sub-branch of the theory of (linear) elasticity. Classical books are those by Sneddon and Lowengrub (1969), Sih (1973) and England (1971). A more recent book, which also deals with time-dependent (dynamic) fracture is that by Freund (1990). Specific definitions of the stress intensity factor are given by Yang (2008) and Freund (1990). Two-dimensional crack problems often involve the methods of complex variables, for which see Carrier et al. (1966), Muskhelishvili (1953), Gakhov (1990), and also Noble (1988).

The application of magma-driven (non-equilibrium) fractures in the Earth's lithosphere was properly initiated by Spence and Turcotte (1985) (earlier work had been done by Weertman (1971)), and developed subsequently by Spence et al. (1987), and later Lister and co-workers (e.g., Lister and Kerr 1991; Richardson et al. 1996; Bolchover and Lister 1999). The second of these deals with melt channels in a viscous, porous medium, which is related to Question 9.9, while the last raises the interesting issue of side-wall solidification.

Dykes and Diapirs The issue of solidification haunts the question of transport of magma through the lithosphere. Two mechanisms are possible; the passage of magmafractures, as above, and the passage of diapirs: huge molten blobs of magma, which ascend due to their buoyancy by viscously deforming the surrounding country rock. This was suggested by Marsh (1982), following an original idea of Grout (1945). The problem has been further studied by Morris (1982), Weinberg and Podladchikov (1994) and Bittner and Schmeling (1995). There is no essential difference between the two mechanisms, they simply differ by virtue of the geometry and the deformation mechanism of the surrounding rock. Marsh suggested velocities of 10^{-9} m s^{-1} for a 3 km diapir, with a consequent Péclet number of $O(1)$.

To calculate the solidification time t_s for a diapir of diameter d, we estimate the heat flux per unit area from the diapir as $\frac{k\Delta T}{d}$, where ΔT is the temperature of the diapir above its surroundings; thus $\rho L d^3 \sim \frac{k\Delta T}{d} \cdot d^2 \cdot t_s$, where L is latent heat, whence

$$t_s \sim \frac{d^2}{\kappa} St, \qquad (9.182)$$

where κ is the thermal diffusivity, and St is the Stefan number

$$St = \frac{L}{c_p \Delta T},\tag{9.183}$$

with c_p being the specific heat. Using $L = 3 \times 10^5$ J kg^{-1}, $c_p = 10^3$ J kg^{-1} K^{-1}, $\Delta T = 10^3$ K, we have $St \sim \frac{1}{3}$, so t_s is essentially the conductive time scale. The distance which the diapir moves in this time is

$$d_s \sim Pe\,St\,d,\tag{9.184}$$

where the Péclet number is

$$Pe = \frac{Ud}{\kappa}.\tag{9.185}$$

Thus if also $Pe \sim O(1)$, then the diapir freezes.

In more detail, the diapir moves by softening a thin rind of the surrounding very viscous country rock, and moving through almost regelatively (regelation refers to the passage of a heavy object through a block of ice; cf. the discussion of glacier sliding in Chap. 10). The precepts of strongly variable viscosity (cf. Chap. 8) tell us that this rind has a thickness such that the temperature drop across it is of $O(\frac{RT_m^2}{E^*})$ (assuming $RT_m \ll E^*$), where E^* is the activation energy for creep, T_m is the freezing temperature of the diapir (or more strictly the solidus temperature of the solid/liquid interface), and R is the gas constant. For values $E^* = 500$ kJ mol^{-1}, $R = 8.3$ J mol^{-1} K^{-1}, $T_m = 1{,}200$ K,

$$\varepsilon = \frac{RT_m}{E^*} \sim 0.02\tag{9.186}$$

is indeed small, and the thickness of the rind d_r is thus

$$d_r \sim \varepsilon d\tag{9.187}$$

(there is an additional factor $\frac{T_m}{T_m - T_c}$, where T_c is the ambient temperature of the surrounding rock, but we take this to be approximately one). The velocity of ascent is thus found from balancing the viscous shear stress $\frac{\eta_r U}{d_r}$ in the rind (where η_r is the rind viscosity) with the buoyancy induced shear stress $\Delta \rho g d$, where $\Delta \rho$ is the difference in density between solid and liquid. This leads to an expression for the Péclet number which resembles a sort of Rayleigh number:

$$Pe = \frac{\varepsilon \Delta \rho g d^3}{\eta_r \kappa}.\tag{9.188}$$

Taking $d = 3$ km, $\Delta \rho = 500$ kg m^{-3}, $\eta_r = 10^{18}$ Pa s, we calculate $Pe \sim 2.7$.

In order to progress a diapir to the surface, Marsh suggested repeated injections, so that the passage way could be repeatedly warmed. This seems a little artificial, and it is not necessary for basaltic magma, where the viscosity is sufficiently low that cracks can propagate at sufficient speed that freezing does not occur.

The problem particularly arises for granitic plutons, where huge batholiths have been emplaced in the continental crust. The viscosity of rhyolite is much higher than that of basalt, so that fractures would have to be much larger, ascent times would be greater, and freezing more likely. Despite this, ascent via fracture has been suggested (Clemens and Mawer 1992; Petford et al. 1994). On the other hand, because the ascent distance d_s depends on the cube of the size, 10 km diapirs could rise 20–30 times their own size, and thus to the surface. Smaller diapirs would not do so well, though.

Against the idea of large diapir ascent may be the choice of viscosity. The value $\eta_r = 10^{18}$ Pa s is a typical asthenospheric sort of value, perhaps associated with basalts at 1,500 K. But the melting point of granites is much lower, perhaps around 1,000 K, which is suggestive of higher values of η_r. If this were the case, it would virtually rule out diapiric uprise via rind softening. An alternative possibility is that alluded to above in passing, the analogy with regelation. Regelation refers to the passage of a heavy object through a block of ice. Because increased pressure lowers the melting point of ice, a thin layer of water forms below the object and is squirted round by the pressure gradient to the rear, where it refreezes, thus enabling passage of the solid through the ice (Nye 1967). The analogy here is not quite the same, however. If a layer of hot basalt is ponded at the base of the continental crust, it will melt the crust, forming a light silicic magma which floats on top of the basalt. As the basalt cools, crystals will begin to grow at the interface with the rhyolite. As the crystals continue to grow, they will most likely form a mush or slurry, and will cause the interstitial liquid to become increasingly silicic. Eventually, the heavy mat of crystals will become unstable and sink to the floor (or this may occur continuously). Apparently this will continue until the temperature decreases to the (cotectic) point where also the rhyolite is crystallising. From this point the upper silicic layer is also being removed, and the whole system effectively moves upwards at a rate controlled by recrystallisation, and creep of the country rock need not be involved. In the process, the magma presumably becomes granitic through the incorporation of the crustal rocks. The idea is essentially due to Huppert and Sparks (1988).

Alloy Solidification   There are a number of books and proceedings volumes on solidification in multi-component systems. Amongst these, one should mention Flemings (1974), Kurz and Fisher (1998), Davis et al. (1992), Ehrhard et al. (2001) and Loper (1987), the last three of which provide multidisciplinary comparisons between different fields. Much of the interest originated from the study of the formation of metal alloy castings in metallurgy, particularly the formation of the defects known as 'freckles'. Early on it was found that an experimentally viable analogue to the industrially important alloys such as lead–tin was the aqueous ammonium chloride solution. The group led by Herbert Huppert at Cambridge has pioneered the exploration of this and other aqueous solutions in freezing conditions, and Huppert in particular, together with his colleague Steve Sparks, has uncovered, or perhaps created, a gold mine of experimentally driven insights into all sorts of geological processes in magma chambers; see, for example, Huppert (1986, 1990, 2000).

Meanwhile, Huppert's student Grae Worster (see, e.g., Worster 1997, 2000) has led the way in experimental and analytic investigations of the ammonium chloride and other aqueous crystallising solutions, more recently extending the laboratory work also to applications in sea ice growth and freezing colloidal suspensions.

Nucleation and Crystallisation   The classical theory of nucleation and crystal growth is due to Avrami (1939, 1940). Dowty (1980) gives a lucid discussion of the thermodynamics of nucleation and crystal growth rate. The theory is used in the context of magma chambers by Brandeis et al. (1984), Brandeis and Jaupart (1986), Spohn et al. (1988) and Hort and Spohn (1991). Worster et al. (1990) also use a simplified theory of non-equilibrium growth in their discussion of convective processes in magma chambers.

Double-Diffusive Convection   Salt fingers were discovered by Melvin Stern (1960), and rapidly became part of the investigative culture at Woods Hole, and in the GFD summer program there. Turner (1974) gives a nice review of the basic phenomena of double-diffusive convection. He, Huppert and Sparks are the principal advocates of the importance of fluid mechanical (including double-diffusive) processes in magma chambers, and Huppert (2000) gives a thorough and wide-ranging review of much of this work. Turner's (1973) book is the classic reference. A nice book is that edited by Brandt and Fernando (1995), which contains a number of historical recollections as well as a wide variety of applications.

Layered Igneous Intrusions   The classic work on the Skaergaard layered intrusion is the voluminous book by Wager and Brown (1968). Largely observational, but associating the layers with occasional turbidity currents, it was not until the paper by McBirney and Noyes (1979) that genuine dynamical concepts began to be applied. Noyes was a physical chemist, who brought to the problem the relatively recent (and shocking) discovery of pattern formation in the Belousov–Zhabotinskii chemical reaction, which is maintained through the interaction of nonlinear kinetics with diffusion. In their paper, they suggest that layering may occur in an analogous way to Liesegang rings, first reported in 1896 by Liesegang in photographic emulsions, where silver dichromate is precipitated in a series of bands. Similar features are seen in orbicular granites. A model for Liesegang rings was proposed by Keller and Rubinow (1981) (and is discussed further in Sect. D.11 of Appendix D), but for layered intrusions, such a model would need to be extended to multiple crystal growth, as outlined by Maaløe (1978). The book edited by Parsons (1987) contains a wealth of informative articles on the Skaergaard and other intrusions, for example in the article by Irvine (1987). Oscillatory crystallisation has also been suggested as a cause of layering by Wang and Merino (1993), but the construction of their mathematical model is not too clear.

Sparks et al. (1993) suggest an interesting mechanism, whereby the increasing crystal content of a convecting magma causes it to slow down, leading to a critical state where the crystals suddenly fall out of the solution, following which the newly particle-free magma convects more vigorously, and can again circulate the next batch of crystals. While this reasonably leads to layering, it is less obvious how chemically distinct layers can occur.

The explanation for the layering in the Rum intrusion as being due to successive injections of ultrabasic magma below a more evolved residual magma was given by Huppert and Sparks (1980). The classic description of the Rum intrusion is by Brown (1956). A more recent review is by Emeleus (1987). Note that the large scale (tens of metres) layers are themselves subject to fine-scale layering.

Volcanoes   The *Encyclopedia of volcanoes* (Sigurdsson 2000) is a voluminous compendium of knowledge written by experts. Some 1,400 large pages in length, it has the advantage of consisting of twenty page long articles, so that there is space for exposition in some depth. Each article comes with an opening glossary, and the reference lists are usefully succinct. Despite its size, it is relatively cheap. The book by Francis and Oppenheimer (2004) is more discursive and (thus) very readable, a book to read in bed, snugly situated between a popular science book and a dry monograph. The book by Sparks et al. (1997) describes volcanic eruption plumes, and relates them to mathematical models of the different flows.

Two-Phase Flows   Two-phase flows, as we have described them for flow in a volcanic vent, have mostly been studied in industrial situations; two particular applications are in boilers and nuclear cooling systems. In either case a fluid (water, for example) flows through a tube where it is externally heated sufficiently that it starts to boil. As the void fraction (i.e., gas volume fraction) increases, the flow passes through a succession of different flow régimes: bubbly, slug, churn and annular. In bubbly flow, small bubbles of gas form a *dispersed* phase in the *continuous* liquid phase. Slug flow consists of tube-wide slugs of fluid interspersed with plugs of gas. A liquid film drains at the wall as the gas bubbles ascend. This relatively regular régime becomes irregular at higher gas flow, when it is called churn flow, and at higher gas flow rates still, the gas forms a continuous core flowing through a liquid film which coats the walls; this is annular flow. The same régimes occur whether the flow is heated (e.g., steam–water) or not (e.g., air–water). Because the liquid always wets the wall, even in annular flow, boiling occurs at the internal gas–liquid interfaces, so that the liquid is mildly superheated (although nucleation will also occur at the walls, particularly in bubbly flow).

There are a number of good engineering books on two-phase flow, for example Whalley (1987), Collier and Thome (1996), Hewitt and Hall-Taylor (1970), Wallis (1969), Bergles et al. (1981) and Butterworth and Hewitt (1977). Mathematicians have tended to get hung up with details of the averaging process whereby the model equations are formulated, rather than the business of solving them. Drew and Passman (1999) gives an up-to-date account of this work. In the context of volcanic flows, two-phase flow models have been used by Melnik (2000), Starostin et al. (2005) and Bercovici and Michaut (2010), for example.

9.9 Exercises

9.1 The equations of plane strain are given by

$$\frac{\partial \sigma_1}{\partial x} + \frac{\partial \tau}{\partial y} + \rho f_1 = 0,$$

$$\frac{\partial \tau}{\partial x} + \frac{\partial \sigma_2}{\partial y} + \rho f_2 = 0,$$

where σ_1 and σ_2 are the normal stresses, τ is the shear stress, x and y subscripts denote partial derivatives, and (f_1, f_2) represents a body force per unit mass.

The elastic constitutive relations are

$$\sigma_1 = 2\mu u_x + \lambda(u_x + v_y),$$

$$\sigma_2 = 2\mu v_y + \lambda(u_x + v_y),$$

$$\tau = \mu(u_y + v_x),$$

where (u, v) are the components of displacement.

Use complex variables $z = x + iy, \bar{z} = x - iy$, to show that the force balance equations can be written in the form

$$\frac{\partial}{\partial z}[\sigma_1 - \sigma_2 + 2i\tau] + \frac{\partial}{\partial \bar{z}}(\sigma_1 + \sigma_2) + \frac{\partial V}{\partial \bar{z}} = 0,$$

where we suppose the body force can be written in terms of the potential V, thus

$$\rho f = \frac{\partial V}{\partial \bar{z}},$$

where $f = f_1 + if_2$.

The complex displacement is defined by $D = u + iv$. Show that

$$\sigma_1 + \sigma_2 = 2(\lambda + \mu)(D_z + \bar{D}_{\bar{z}}),$$

$$\sigma_1 - \sigma_2 + 2i\tau = 4\mu D_{\bar{z}},$$

and deduce that D satisfies

$$\frac{\partial}{\partial z}[4\mu D_{\bar{z}}] + \frac{\partial}{\partial \bar{z}}\{2(\lambda + \mu)(D_z + \bar{D}_{\bar{z}})\} + \frac{\partial V}{\partial \bar{z}} = 0,$$

and hence show that

$$4\mu D_z + 2(\lambda + \mu)(D_z + \bar{D}_{\bar{z}}) + V = \frac{4(\lambda + 2\mu)}{\lambda + \mu}\Omega'(z),$$

where $\Omega'(z)$ is an arbitrary (analytic) function of z.

Assuming now that V is real, show by taking the complex conjugate of this equation and eliminating $(D_z + \bar{D}_{\bar{z}})$, that

$$4\mu D_z + \frac{\mu V}{\lambda + 2\mu} = -2\overline{\Omega'(z)} + 2\kappa \Omega'(z),$$

where $\kappa = \frac{\lambda + 3\mu}{\lambda + \mu}$.

Assuming that $\frac{\partial X}{\partial z} = V$, show that this equation can be integrated to obtain

$$2\mu D = -\frac{\mu X}{2(\lambda + 2\mu)} + (\bar{z} - z)\overline{\Omega'(z)} + \kappa \Omega(z) + \phi(\bar{z}),$$

where $\phi(\bar{z})$ is a second arbitrary analytic function of its argument.

9.2 A crack L of width h containing magma in a porous medium (the asthenosphere) is described by the equations

$$\frac{\partial h}{\partial t} + \frac{\partial}{\partial x}\left[\frac{h^3}{12\eta}\left\{\Delta\rho_{sm}g + \frac{\partial \Pi}{\partial x}\right\}\right] = \frac{\mu k}{(1-v)\eta}h_{xx},$$

$$\Pi = \frac{\mu}{2\pi(1-v)}\int_L \frac{\partial h}{\partial s}\frac{ds}{s-x},$$

where x is distance upwards along the vertical crack; μ is the shear modulus, v is Poisson's ratio, k is the permeability, η is the magma viscosity, $\Delta\rho_{sm}$ the difference between solid and magma densities, and g is the acceleration due to gravity. The diffusive term represents the suction of melt into the crack from the porous medium. By choosing suitable scales for $h \sim h_0$ and $t \sim t_0$ in terms of $x \sim l$, show that the model can be written in the dimensionless form

$$\frac{\partial h}{\partial t} + \frac{\partial}{\partial x}\left[h^3\left\{1 + \delta\frac{\partial \Pi}{\partial x}\right\}\right] = h_{xx},$$

$$\Pi = \frac{1}{2\pi}\int_L \frac{\partial h}{\partial s}\frac{ds}{s-x},$$

and define the dimensionless parameter δ. Show that δ is small if $l \gg l_0$, where

$$l_0 = (12k)^{1/5}\left(\frac{\mu}{(1-v)\Delta\rho_{sm}g}\right)^{3/5},$$

and estimate l_0 if $\mu = 2 \times 10^{10}$ Pa, $\Delta\rho_{sm} = 300$ kg m^{-3}, $g = 10$ m s^{-2}, $v = \frac{1}{4}$, $k = 10^{-12}$ m^2. Deduce that on an asthenospheric length scale $l \gtrsim 10$ km, δ is indeed small.

Now suppose that a crack is nucleated at the interface $x = 0$ between the porous asthenosphere in $x < 0$ and the impermeable lithosphere in $x > 0$, where $k = 0$. Assuming $\delta = 0$, show that a similarity solution exists in $x < 0$, in which

$$h = \frac{F(\eta)}{t^{1/4}}, \qquad \eta = -\frac{x}{t^{1/2}},$$

and find the resultant equation for F.

Show from this equation that $F \to 0$ as $\eta \to \infty$, and by detailed consideration of the asymptotic form of the solution, show that in fact we require

$$\eta^{1/2}F \to 0 \quad \text{as } \eta \to \infty,$$

assuming the crack initially has finite (zero) volume.

Assuming that h is continuous at $x = 0$, show that the solution for the crack thickness in $0 < x < x_f$ is

$$h = \frac{1}{\sqrt{6}t^{1/4}}\left\{\frac{x}{\sqrt{t}} + \left(\frac{x^2}{t} + 36F_0^4\right)^{1/2}\right\}^{1/2},$$

where $F_0 = F(0)$, and show that the crack front position is

$$x_f = 2\sqrt{3}F_0^2\sqrt{t}.$$

Deduce that, if also h_x is continuous at $x = 0$, the boundary condition for F at $\eta = 0$ is

$$12F_0F_0' + 1 = 0,$$

where $F_0' = F'(0)$.

Sketch or compute the resulting form of the solution, and comment on the physical applicability of the results, in terms of magma velocity and transport time through the lithosphere. See also Fowler and Scott (1996).

9.3 A crack L is nucleated in the asthenosphere, which is treated as a viscous, porous medium of viscosity η_s and permeability k. The pore pressure in the medium is p and its far field (magmastatic) value is $p_0 - T - \rho_m gx$, where $p_0 - T - \rho_s gx = P$ is the far field (lithostatic) solid pressure, and x points vertically upwards along the crack; ρ_s and ρ_m are the solid and melt densities, respectively, while g is the acceleration due to gravity. The effective pressure $P - p$ at the crack takes a value N which is related to the rate of viscous closure w of the crack by the relation

$$N = -\frac{2\eta_s}{\pi}\int_L \frac{\partial w}{\partial s}\frac{ds}{s - x}$$

(see Ng 1998, p. 55 ff.) Assuming that $\nabla^2 p = 0$, use complex variable methods to show that the net influx of melt Ω to the crack, defined on L by

$$\Omega = \frac{2k}{\eta_m}\frac{\partial p}{\partial y}\bigg|_{0+},$$

where η_m is the melt viscosity, is given by

$$\Omega = \frac{2k}{\eta_m}\left[-2\eta_s w_{xx} + \frac{\Delta\rho_{sm}gx}{(l^2 - x^2)^{1/2}}\right],$$

where $\Delta\rho_{sm} = \rho_s - \rho_m$, and the crack is taken to be in $-l < x < l$.

Show that the second term in the above expression is absent if it is assumed that $p \to P$ in the far field.

9.4 A vertical crack L of width h is nucleated in a viscous, porous asthenosphere containing melt. The fluid flux upwards (in the x direction along the crack) is q, and conservation of melt in the crack is described by

$$\frac{\partial h}{\partial t} + \frac{\partial q}{\partial x} = \frac{m}{\rho_m} + \Omega,$$

where m is the rate of melting of the crack walls, and Ω is the supply from the porous rock via suction; ρ_m is the melt density.

The fluid flux is given by the Poiseuille law

$$q = \frac{h^3}{12\eta_m}\left\{\Delta\rho_{sm}g + \frac{\partial N}{\partial x}\right\},$$

where η_m is the melt viscosity, g is gravity, $\Delta\rho_{sm} = \rho_s - \rho_l$ is the difference in density between solid and melt, and N is the effective pressure in the crack. The wall melting, due to potential energy release, is determined from

$$mL = q\left\{\Delta\rho_{sm}g + \frac{\partial N}{\partial x}\right\},$$

where L is the latent heat.

The crack closes at a rate w, determined from the closure equation

$$N = -\frac{2\eta_s}{\pi}\int_L \frac{\partial w}{\partial s}\frac{ds}{s-x},$$

where η_s is the solid viscosity, and the crack width is described by the kinematic condition

$$\frac{\partial h}{\partial t} = \frac{m}{\rho_s} - w.$$

Finally the melt suction is given by

$$\Omega = -\frac{4k\eta_s}{\eta_m}\frac{\partial^2 w}{\partial x^2},$$

where k is the permeability.

Suppose that $x \sim l$, and that $N \ll \Delta\rho_{sm}gl$. In this case show that h satisfies the equation

$$h_t + A(h^3)_x = ABh^3 + C\frac{\partial^2}{\partial x^2}\{h_t - rABh^3\},$$

where you should define the parameters A, B, C and r, and find typical values for these assuming $\rho_s = 3 \times 10^3$ kg m^{-3}, $\rho_m = 2.5 \times 10^3$ kg m^{-3}, $g = 10$ m s^{-2}, $\eta_m = 10^2$ Pa s, $L = 3 \times 10^5$ J kg^{-1}, $k = 10^{-12}$ m^2, $\eta_s = 10^{19}$ Pa s.

By choosing $l = \sqrt{C}$ and defining suitable scales for t and h, show that the model can be written in the dimensionless form

$$h_t + (h^3)_x = h_{xxt} + \varepsilon[h^3 - r(h^3)_{xx}],$$

where

$$\varepsilon = \frac{2\Delta\rho_{sm}g}{\rho_m L}\left(\frac{k\eta_s}{\eta_m}\right)^{1/2}.$$

Show that $\varepsilon \ll 1$, and deduce that h is approximately described by the modified BBM equation[8]

$$h_t + 3h^2 h_x = h_{xxt}.$$

9.5 A pod of melt in a viscous porous medium is described by the scaled, generalised BBM equation

$$h_t + 3h^2 h_x = h_{xxt},$$

in which we assume that $h \to 0$ as $x \to \pm\infty$. Show that $\int_{-\infty}^{\infty} h\,dx = M$ is conserved, and that travelling wave solutions exist of the form $h = \phi(\eta)$ with $\eta = x - ct$, providing

$$\phi'^2 = \phi^2 - \frac{\phi^4}{2c}, \qquad \phi(\pm\infty) = 0.$$

Deduce that $\phi = \sqrt{2c}\,\mathrm{sech}\,\eta$, and thus that $c = \frac{M^2}{2\pi^2}$.

Show also that if one only requires h to be bounded at infinity, then there is a one parameter family of periodic solutions for each $c > 0$.

9.6 A magma-filled crack L of width h propagates through the lithosphere, and is described by the dimensionless equations

$$\frac{\partial h}{\partial t} + \frac{\partial}{\partial x}\left[h^3\left\{1 + \delta\frac{\partial \Pi}{\partial x}\right\}\right] = 0,$$

$$\Pi = \frac{1}{2\pi}\int_L \frac{\partial h}{\partial s}\frac{ds}{s - x},$$

where Π is the underpressure in the crack, which is assumed to lie in $-l < x < l$, and x points vertically upwards. Essentially, we view $x = 0$ as the base of the lithosphere, so that the physical crack is in $x > 0$, with its virtual extension in $x < 0$ representing conditions of symmetry about $x = 0$.

Show that exact similarity solutions can exist in terms of a similarity variable $\eta = xt^{1/3}$, and write down the equations for the similarity functions H and P corresponding to h and Π, respectively. Explain why such a similarity solution is unlikely to represent a physically appropriate solution.

---

[8] The equation is closely related to that studied by Benjamin et al. (1972) as a model for long waves in shallow water. We may infer that the present equation for h is well-posed.

Show that if δ is small and is ignored, then a similarity solution exists for input fluxes $q|_{x=0+} = h^3 = t^\nu$, and show that the appropriate similarity variable is $\eta = \frac{x}{t^\beta}$, and find β in terms of ν. Find the equation satisfied by $H(\eta)$, the appropriately rescaled definition of h. By writing $H = \eta^{1/2} G$ and $\tau = \ln \eta$, show that $G'(\tau) = F(G)$, where you should define the function F. Show that a solution is possible in which $G \sim \frac{1}{\eta^{1/2}}$ for large η, and thus we can suppose $H = 1$ at $x = 0$. In this case, show that if $0 < \beta < 1$, then $H \sim \sqrt{\frac{\eta}{3}}$ as $\eta \to \infty$, while if $\beta > 1$, $H \sim H_c + (\eta_c - \eta)^{1/2}$ as $\eta \to \eta_c$ for some finite η_c. Deduce that a physically meaningful solution of the model, in which a finite crack grows, only exists if $\nu > 0$, and give an expression for the crack front as a function of time.

9.7 The width h of a crack in an elastic medium and its underpressure Θ are determined near the crack tip by the dimensionless equations, on $[0, \infty]$,

$$\Theta_X = 1 - \frac{1}{h^2},$$

$$\Theta = \frac{1}{\pi} \int_0^\infty \frac{\partial h}{\partial \xi} \frac{d\xi}{\xi - X},$$

subject to

$$h \sim 2\lambda\sqrt{X} \quad \text{as } X \to 0,$$

$$h \to 1 \quad \text{as } X \to \infty.$$

Use complex variable methods to invert the integral equation, and by integrating the result, find an integral expression for h. Using integration by parts, derive the nonlinear integral equation for h,

$$h(X) = \frac{1}{\pi} \int_0^\infty K(x, \xi) \left(\frac{1}{h^2} - 1 \right) d\xi,$$

where

$$K(x, \xi) = (x - \xi) \ln \left| \frac{\sqrt{x} + \sqrt{\xi}}{\sqrt{x} - \sqrt{\xi}} \right| - 2\sqrt{x\xi}.$$

Deduce that if $h \sim 2\lambda\sqrt{X}$ as $X \to 0$, then

$$\lambda = \frac{2}{\pi} \int_0^\infty \sqrt{\xi} \left(\frac{1}{h^2} - 1 \right) d\xi.$$

Show also that

$$h \sim 1 + \frac{1}{2\pi X^2} \quad \text{as } X \to \infty,$$

and deduce that the fracture has a bulbous nose. Sketch the expected form of the shape (see also Spence et al. 1987).

9.8 According to Muskhelishvili (1953, pp. 73 ff.), the behaviour of the Cauchy integral

$$\Phi(z) = \frac{1}{2\pi i} \int_{-l}^{l} \frac{\phi(t)\, dt}{t - z}$$

near the end $z = l$ is given by

$$\Phi \sim -\frac{\phi^* \cosec \gamma \pi}{2i(z - l)^\gamma},$$

if

$$\phi \sim \frac{\phi^*}{(l - x)^\gamma}$$

as $x \to l$ with $0 < \gamma < 1$ (and ϕ^* is constant).

Show also that if

$$\Psi(z) = \frac{1}{2\pi i} \int_{-l}^{l} \frac{\psi(t)\, dt}{t - z},$$

and $\psi(l) = 0$, then

$$\Psi'(z) \sim \frac{1}{2\pi i} \int_{-l}^{l} \frac{\psi'(t)\, dt}{t - z} + O(1)$$

as $z \to l$.

As $x \to l+$ on $y = 0$, we have $\sigma_2 - i\tau \sim 2\Omega'$, where

$$\Omega \sim \frac{1}{2\pi i} \int_{-l}^{l} \frac{2i\mu}{1 + \kappa} \frac{h(s)\, ds}{s - z},$$

in which $\kappa = \frac{\lambda + 3\mu}{\lambda + \mu}$, and also $\nu = \frac{\lambda}{2(\lambda + \mu)}$. Show that if $h \sim c(l - x)^{1/2}$ as $x \to l-$, then

$$\sigma_2(x, 0) \sim \frac{K}{\sqrt{2\pi(x - l)}} \qquad \text{as } x \to l+,$$

where

$$K = \frac{\mu c \sqrt{\pi}}{2\sqrt{2}(1 - \nu)}.$$

9.9 The crystal nucleation rate of a liquid is given by

$$I = K' \exp\left(-\frac{\Delta H}{RT}\right) \exp\left[-\frac{G'}{RT(\Delta T)^2}\right],$$

where K' is a rate constant, G' and ΔH are activation energies, T is temperature, and $\Delta T = T_L - T$ is the undercooling below the liquidus temperature T_L. By writing

$$T = T_L - (\Delta T_c)\theta,$$

show that I can be written in the form

$$I = k' \exp\left[-\frac{\delta\theta}{\varepsilon(1-\delta\theta)}\right] \exp\left[-\frac{1}{\theta^2(1-\delta\theta)}\right],$$

provided we choose

$$\Delta T_c = \left(\frac{G'}{RT_L}\right)^{1/2},$$

and define the parameters k', δ and ε.

Use the values $T_L = 1200$ K, $K' = 10^8$, $\frac{\Delta H}{R} = 2 \times 10^4$ K, $\frac{G'}{R} = 10^5$ K to find typical values of ΔT_c, ε, δ and k', and hence show that $\delta \ll \varepsilon$, and thus that

$$I \sim k' \exp\left(-\frac{1}{\theta^2}\right) \quad \text{if } \theta \sim O(1),$$

$$I \sim k'e^{-\Theta}, \quad \text{if } \Theta \equiv \frac{\delta\theta}{\varepsilon} \sim O(1).$$

Write down a uniform approximation for I. Why is ΔT_c a measure of the critical undercooling?

9.10 The supercooling θ below the liquidus in a magma chamber satisfies the equation

$$\theta_t = \theta_{xx} - St\,\phi_t,$$

where

$$\phi = 1 - \exp\left[-\frac{1}{3}Z^3\right], \qquad Z_t = H(\theta)Y(\theta),$$

and $H(\theta)$ is the Heaviside step function. The initial and boundary conditions are

$$\theta = \theta_\infty \geq 0, \qquad Z = 0 \quad \text{at } t = 0,$$

$$\theta = \theta_0 < 0 \quad \text{at } x = 0,$$

$$\theta_x = 0 \quad \text{at } x = \frac{1}{\varepsilon}.$$

Assume that $\theta_0, \theta_\infty, St \sim O(1)$, $\varepsilon \ll 1$. By writing $x = \frac{X}{\varepsilon}$, $t = \frac{T}{\varepsilon^2}$, $Z = \frac{z}{\varepsilon^2}$, show that a suitable outer approximation for θ and z is given by

$$\theta_T = \theta_{XX}, \qquad z_T = H(\theta)Y(\theta),$$

in $0 < X < S(T)$ and $S(T) < X < 1$, where the outer solution satisfies $\theta = z = 0$ on $X = S$. Show that the outer solution for z in $X < S$ is

$$z = \int_{S^{-1}(X)}^{T} Y[\theta(X, T')]\,dT',$$

where S^{-1} is the inverse function of S. What is the outer solution for z in $X > S$?

For T close to (and greater than) $S^{-1}(X)$, show that

$$z \approx \frac{1}{2} \left[T - S^{-1}(X) \right]^2 \theta_{XX}|_{X=S-},$$

and deduce that a suitable rescaling near $X = S$ is obtained by writing

$$X = S(T) + \varepsilon \xi, \qquad \theta = \varepsilon \Theta, \qquad z = \varepsilon^2 Z,$$

and that the inner problem can be written, to leading order, as

$$0 = \Theta_{\xi\xi} + St \, \dot{S} \phi_\xi,$$
$$-\dot{S} Z_\xi = H(\Theta) y(\Theta),$$

where we write $Y(\varepsilon \Theta) = \varepsilon y(\Theta)$, supposing that $y \sim O(1)$.

Show that the matching conditions for Θ and Z are

$$\Theta \sim g_- \xi, \qquad Z \sim \frac{\xi^2}{2S^2} \Theta_{XX}|_{X=S-} \quad \text{as } \xi \to -\infty, \tag{$*$}$$
$$\Theta \sim g_+ \xi, \qquad \text{as } \xi \to \infty,$$

where $g_\pm = \theta_X|_{X=S\pm}$.

[*To compute the condition on Z, you will need to show that for $0 < T - S^{-1}(X) \ll 1$,*

$$T - S^{-1}(X) \approx -\frac{\varepsilon \xi}{\dot{S}}.]$$

Hence show that S satisfies the Stefan condition

$$St \, \dot{S} = [\theta_X]_-^+.$$

Suppose that $y(\Theta) = \Theta$. Show that Z satisfies

$$-\dot{S} Z_{\xi\xi} = g_- + St \, \dot{S} \exp\left[-\frac{1}{3} Z^3 \right],$$

and explain why, if $g_+ > 0$, appropriate boundary conditions are

$$Z = Z_\xi = 0 \quad \text{at } \xi = 0.$$

Show further that the solution for Z satisfies the second condition in $(*)$ automatically (remember that $\theta[S(T), T] \equiv 0$). What are appropriate conditions if $g_+ = 0$?

Sketch or compute the solution for Z.

9.11 Dimensionless equations describing double-diffusive convection are given by

$$\nabla.\mathbf{u} = 0,$$

$$\frac{1}{Pr}\left[\mathbf{u}_t + (\mathbf{u}.\nabla)\mathbf{u}\right] = -\nabla p + \nabla^2\mathbf{u} + Ra\,T\hat{\mathbf{k}} - Rs\,c\hat{\mathbf{k}},$$

$$T_t + \mathbf{u}.\nabla T = \nabla^2 T,$$

$$c_t + \mathbf{u}.\nabla c = \frac{1}{Le}\nabla^2 c,$$

where $\mathbf{u}$ is the fluid velocity, p is pressure, T is temperature, c is concentration, and the parameters are the Rayleigh number Ra, the solutal Rayleigh number Rs, the Prandtl number Pr and the Lewis number Le.

Assuming a two-dimensional flow, so that there is a stream function ψ satisfying

$$\mathbf{u} = (-\psi_y, \psi_x),$$

suppose that the flow is driven by prescribed temperature and salinity differences between the upper and lower surfaces of a stress-free box, so that the boundary conditions take the form

$$\psi = \nabla^2\psi = 0 \quad \text{at } z = 0, 1,$$

$$T = c = 0 \quad \text{at } z = 1,$$

$$T = c = 1 \quad \text{at } z = 0.$$

Write down the steady state in which $\mathbf{u} = 0$, and let C and θ denote perturbations to c and T, respectively. Assuming $C, \theta, \psi \ll 1$, write down the linearised equations and boundary conditions for the flow. By seeking solutions proportional to $\exp(ikx + \sigma t)\sin m\pi z$, show that

$$\left(\sigma + K^2 Pr\right)\left(\sigma + K^2\right)\left(\sigma + \frac{K^2}{Le}\right) + k^2 Pr\left[\frac{(Rs - Ra)\sigma}{K^2} + Rs - \frac{Ra}{Le}\right] = 0,$$

where

$$K^2 = k^2 + m^2\pi^2.$$

Show that this can be written in the form

$$\sigma^3 + a\sigma^2 + b\sigma + c = 0,$$

where

$$a = K^2\left(Pr + 1 + \frac{1}{Le}\right),$$

$$b = K^4\left(Pr + \frac{1}{Le} + \frac{Pr}{Le}\right) + \frac{k^2}{K^2}Pr(Rs - Ra),$$

$$c = \frac{K^6}{Le}Pr + k^2 Pr\left(Rs - \frac{Ra}{Le}\right).$$

Deduce that direct instability ($\sigma > 0$) occurs if

$$Ra - Le\, Rs > R_c = \frac{27\pi^4}{4},$$

and oscillatory instability occurs if

$$Ra > \frac{(Pr + \lambda)Rs}{1 + Pr} + \frac{(1 + \lambda)(Pr + \lambda)}{Pr} R_c,$$

where $\lambda = \frac{1}{Le}$.

Chapter 10
Glaciers and Ice Sheets

Glaciers are huge and slow moving rivers of ice which exist in various parts of the world: Alaska, the Rockies, the Alps, Spitzbergen, China, for example. They drain areas in which snow accumulates, much as rivers drain catchment areas where rain falls. Glaciers also flow in the same basic way that rivers do. Although glacier ice is solid, it can deform by the slow creep of dislocations within the lattice of ice crystals which form the fabric of the ice. Thus, glacier ice effectively behaves like a viscous material, with, however, a very large viscosity: a typical value of ice viscosity is 6 bar year (in the metre–bar–year system of units!). Since 1 bar $= 10^5$ Pa, 1 year $\approx 3 \times 10^7$ s, this is a viscosity of some 2×10^{13} Pa s, about 10^{16} times that of water. As a consequence of their enormous viscosity, glaciers move slowly—a typical velocity would be in the range 10–100 m y^{-1} (metres per year), certainly measurable but hardly dramatic. More awesome are the dimensions of glaciers. Depths of hundreds of metres are typical, widths of kilometres, lengths of tens of kilometres. Thus glaciers can have an important effect on the human environment in their vicinity. They are also indirect monitors of climate; for example, many lithographs of Swiss glaciers show that they have been receding since the nineteenth century, a phenomenon thought to be due to the termination of the 'Little Ice Age' which lasted from about 1500 to about 1900.

Where glaciers are the glacial equivalent of rivers, i.e. channelled flow, *ice sheets* are the equivalent of droplets, but altogether on a grander scale.[1] When an entire continent, or at least a substantial portion thereof, has a polar climate, then snow accumulates on the uplands, is compressed to form ice, and flows out to cover the continent, much as a drop of fluid on a table will spread under the action of gravity. However, whereas droplets can reach a steady state through the contractile effect of surface tension, this is not relevant to large ice sheets. In them, equilibrium can be maintained through a balance between accumulation in the centre and *ablation* at the margins. This can occur either through melting of the ice in the warmer climate

---

[1]Ice *caps* are smaller scale sheet flows, such as the Vatnajökull ice cap in Iceland, whose horizontal dimension is about 100 km.

A. Fowler, *Mathematical Geoscience*, Interdisciplinary Applied Mathematics 36, DOI 10.1007/978-0-85729-721-1_10, © Springer-Verlag London Limited 2011

at the ice margin, or through calving of icebergs. (Indeed, the same balance of accumulation at higher elevations with ablation at lower elevations is responsible for the normal quasi-steady profile of valley glaciers.)

There are two major ice sheets on the Earth, namely those in Antarctica and Greenland (the Arctic is an ocean, and its ice is sea ice, rarely more than three metres thick). They are on the order of thousands of kilometres in extent, and kilometres deep (up to four for Antarctica). They are thus, in fact, *shallow* flows, a fact which greatly facilitates the solution of mathematical models for their flow. Possibly more famous are the ice sheets which covered much of North America (the Laurentide ice sheet) and northern Europe (the Fennoscandian ice sheet) during the last ice age. Throughout the Pleistocene era (that is, the last two million years), there have been a succession of ice ages, each lasting a typical period of around 90,000 years, during which global ice sheet volume gradually increases, interspersed with shorter (10,000 year) *interglacials*, when the ice sheets rapidly retreat. The last ice age finished some ten thousand years ago, so that it would be no surprise if another were to start now. Perhaps the Little Ice Age was indeed the start of ice sheet build-up, only to be interrupted by the Industrial Revolution and the resultant global warming: nobody knows.

Further back in Earth's geologic history, there is evidence for dramatic, large scale glaciation in the Carboniferous (c. 300 My (million years) ago), Ordovician (c. 500 My ago), Neoproterozoic (c. 600–800 My ago) and Huronian (c. 2,500 My ago) periods. In the Neoproterozoic glaciation, it seems that the whole landmass of the Earth may have been glaciated, leading to the concept of 'snowball Earth'. It was following the shrinkage of the global ice sheet that the explosion of life on Earth started.[2]

While the motion of ice sheets and glaciers can be understood by means of viscous theory, there are some notable complications which can occur. Chief among these is that ice can reach the melting point at the glacier bed, due to frictional heating or geothermal heat input, in which case water is produced, and the ice can *slide*. Thus, unlike an ordinary viscous fluid, slip can occur at the base, and this is determined by a sliding law which relates basal shear stress τ to sliding velocity u_b and also, normally, the *effective pressure* $N = p_i - p_w$, where p_i and p_w are ice and water pressures. The determination of p_w further requires a description of the subglacial hydrology, and thus the dynamics of ice is intricately coupled to other physical processes: as we shall see, this complexity leads to some exotic phenomena.

10.1 Dynamic Phenomena

10.1.1 Waves on Glaciers

Just as on rivers, gravity waves will propagate on glaciers. Because the flow is very slow, they only propagate one way (downstream), and at speeds comparable to the surface speed (but slightly faster). These waves are known as surface waves, as they

---

[2]Snowball Earth was discussed in Chap. 2.

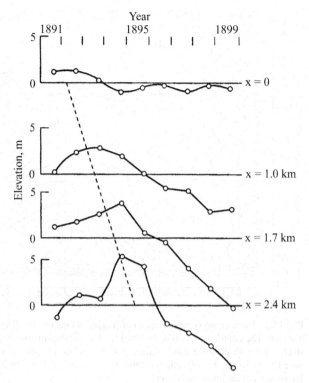

Fig. 10.1 Changes of mean surface elevation of Mer de Glace, France, along four cross-profiles over a period of 9 years. The *broken line* corresponds to a wave velocity of 800 m/a. Reproduced from Lliboutry (1958a), by permission of the International Association of Hydrological Sciences

are evidenced by undulations of the surface: an example is shown in Fig. 10.1. They are examples of *kinematic* waves, driven by the dependence of ice flux on glacier depth.

A more exotic kind of wave is the 'seasonal wave'. This has no obvious counterpart in other fluid flows. It consists of (sizeable) perturbations in the surface velocity field which propagate down-glacier at speeds in the order of 20–150 times the surface speed. There is no significant surface perturbation, and these waves must in fact be caused by variations of the basal sliding speed due to annual fluctuations in the basal water pressure. Although well known and reported at the turn of the century, little attention has been paid to these waves in recent years. Figure 10.2 shows measurements of Hodge on Nisqually Glacier which indicates the rapid passage of a velocity wave downstream.

Mention should also be made of wave ogives, although we will not deal with them here. They are bands (also known as Forbes bands) which propagate below ice-falls, and are due to the annual ablation cycle.

10.1.2 Surges

Perhaps the most spectacular form of wave motion is the glacier *surge*. Surges are large scale relaxation oscillations of the whole length of a glacier. They are roughly

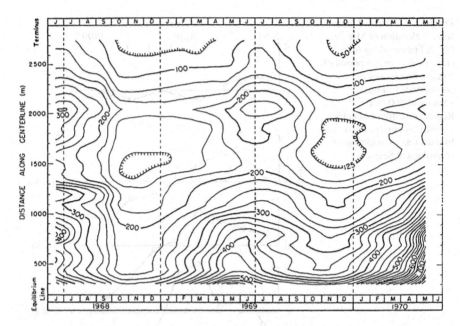

Fig. 10.2 The measured surface speed of Nisqually Glacier, Mt. Rainier, as a function of time and distance. The contour interval is $25 \ \mathrm{mm \, d^{-1}}$. The maximum and minimum speeds occur progressively later with distance down-glacier; this represents a "seasonal wave" in the ice flow. Reproduced from Hodge (1974), and reprinted from the Journal of Glaciology with permission of the International Glaciological Society

periodic, with periods of the order of 20–100 years. During a long quiescent phase, the glacier is over-extended and thin. Ice accumulation causes the glacier to thicken upstream, while the over-extended snout thins and retreats. Eventually, a critical thickness is reached, and the glacier slumps rapidly downslope again. These surges will typically last only a year or two, during which time the velocity may increase a hundred-fold. The glacier snout can then advance by several kilometres.

A typical (and much studied) example is the Variegated Glacier in Alaska. Its surge periodicity is about twenty years, while its surges last about two years. The glacier, of length twenty kilometres and depth four hundred metres, advances some six kilometres during its surge, at measured speeds of up to 65 metres per *day*. Such large velocities can only occur by basal sliding, and detailed observations during the 1982–3 surge showed that the surge was mediated by an alteration in the basal drainage system, which had the effect of raising water pressure dramatically. A dynamic model suggests, in fact, that the oscillations are caused by the competitive interaction between the basal sliding law and the hydraulics of the subglacial drainage system. When the ice is relatively thin (hence the driving shear stress is low) the drainage occurs through a network of channels incised into the ice at the glacier bed called Röthlisberger channels. These allow effective drainage at quite low water pressures (hence high effective pressures) and thus also low ice velocities. At higher driving stresses, however, an instability forces the channel system

Fig. 10.3 Variegated Glacier at the beginning of a surge, 29 August, 1964. Photograph by Austin Post, U.S. Geological Survey. From the Glacier Photograph Collection, National Snow and Ice Data Center/World Data Center for Glaciology, Boulder, Colorado

to close down, and the basal water is forced into cavities which exist between the ice and bedrock protuberances (such cavities are well known to exist). The water flow is reduced, and the sudden increase in water pressure causes a sudden increase in ice velocity—the surge. The transition front between the linked cavity drainage system and the channel system is nucleated near the maximum depth, and propagates rapidly both upstream and downstream, at (measured) speeds on the order of hundreds of metres per hour. At the end of the surge, the channel drainage system is re-established. Figures 10.3 and 10.4 show an aerial view of Variegated Glacier in pre- and post-surge states.

Our understanding of the Variegated surges relies on the concept of drainage switch between channelised flow and linked cavities, implicitly for ice flowing over (hard) bedrock. A rather different situation appears to operate in Trapridge Glacier, another well-studied surging glacier, in the Yukon. Here the glacier is cold in its interior (unlike the temperate (at the melting point) Variegated Glacier); and rests on a thick (∼6 metres) layer of *till*, sometimes more graphically called *boulder*

Fig. 10.4 Variegated Glacier at the end of a surge, 22 August, 1965. Photograph by Austin Post, U.S. Geological Survey. From the Glacier Photograph Collection, National Snow and Ice Data Center/World Data Center for Glaciology, Boulder, Colorado

clay—a non-uniform mixture of angular rock fragments in a finer-grained, clay-rich groundmass (see Fig. 10.5). Till has a bimodal grain size distribution, and is produced by the erosion of brittle underlying bedrock, and is evacuated by the slow motion of the ice downstream.

The sequence of events which appears to occur as Trapridge thickens is that, firstly, the basal ice reaches melting point (and the till thaws). When this happens, the till becomes deformable, and the basal ice can slide over the bed by riding on the deforming till. The rate at which this occurs depends on the till rheology, where opinion is currently divided as to whether a viscous or plastic rheology is the more appropriate.[3] What does seem to be clear is that the water pressure will have a major effect. Increasing saturation causes increasing water pressure, which pushes the

---

[3]Note the use of the word 'appropriate'. As a saturated, granular material, somewhat like soil, there is little argument that a plastic rheology accommodating a yield stress is the most apposite description; such a description does not in itself provide an answer to such questions as to whether till deforms with depth (i.e., shears), or whether discrete slip occurs at the ice-till interface.

Fig. 10.5 Subglacial till in a coastally exposed drumlin at Scordaun, Killough, Co. Down, Northern Ireland

sediment grains apart and allows them to move more freely, so that in effect enhanced water production causes enhanced sliding. In turn, increased sliding causes increased frictional heating, so that there is a positive feedback which potentially can cause runaway and consequent surging behaviour. Whether the effect is strong enough is not obvious, but we shall examine a simple model which suggests that it may be.

10.1.3 Ice Streams

Although ice sheets also flow under the horizontal pressure gradients induced by the glaciostatic pressures beneath their sloping surfaces, they rest on essentially un-sloping bases, and therefore have no advective component in their dynamics. Thus ice sheets do not, at least on the large scale, exhibit wave motion: the governing equations are essentially diffusive in character. On a more local scale, however, ice sheets have interesting phenomena of their own.

Principal among these may be ice streams. Ice sheets do not tend to drain uniformly to the margin from their central accumulation zones, but rather the outflows from catchment areas are concentrated into fast-moving ice streams. Examples are the Lambert Glacier in Antarctica and Jakobshavn Isbrae in Greenland, a fast-moving (more than 10 kilometres per year) outlet glacier.[4] These ice streams gain their speed by carving out deep channels through which they flow. Indeed, there is an obvious positive feedback here. The deeper an ice flow, the larger the driving basal stress, and the warmer the basal ice (due to increased frictional heat and decreased conductive heat loss), and hence the softer the ice. Both of these effects contribute to enhanced ice flow, which can explain the formation of such channels, since the erosive power of ice flow increases with the basal velocity and the basal shear stress. Indeed, flow of ice over a plane bed is subject to a lateral instability (much as overland flow of surface water is unstable to the formation of rills and gullies).

A similar kind of mechanism may operate when ice flows over deforming sediments, as in the Siple Coast of West Antarctica. Here, it is found that the flow is concentrated into five ice streams, A, B, C, D, and E, which are characterised by their heavily crevassed appearance. Ice stream B is now known as the Whillans ice stream, in memory of the glaciologist Ian Whillans. Following this, the other ice streams have also been named after individuals; specifically, A is Mercer, C is Kamb, D is Bindschadler and E is MacAyeal. The flow in these ice streams is very rapid and is due to basal sliding over the underlying sediment (except for the Kamb ice stream C, which appears to have 'switched off' several hundred years ago). Measurements on the Whillans ice stream indicate that the basal water pressure is high (within 0.4 bar of the overburden pressure), and that it is underlain by some eight metres of saturated till. A similar instability to that concerning ice flow over hard bedrock may explain the streaming nature of the flow. Where ice flow is larger, there is increased water production. *If* the drainage system is such that increased water production leads to increased water pressure (as one might expect, e.g. for a Darcy flow), then the higher water pressure decreases the viscosity of the till, and hence enhances the ice flow further. This is an instability mechanism, and the limiting factor is that when ice flow increases, there is increased heat loss from the base, which acts to limit the increase of melt rate.

---

[4]Jakobshavn has undergone a remarkable acceleration in recent years, doubling its speed from 6 to 12 kilometres a year in the ten years between 1992 and 2003.

10.1.4 Ice Shelf Instability

Where continental ice sheets are not diminished by ablation, the ice will flow to the continental margin, where it will spill into the ocean. This, for example, is the case in Antarctica, where it is so cold that ordinary mass wastage of the ice is virtually absent. As a result, *ice shelves* are formed, which are tongues of floating ice connected to the grounded ice at the *grounding line*. The grounding line is a dynamical free boundary, whose location determines the hold up of land ice, and its determination is therefore of some interest as regards sea level changes.

Over the past several decades, various arguments have been put forward to suggest that ice shelves are inherently unstable and liable to collapse. This idea was originally put forward in consideration of the West Antarctic Ice Sheet (WAIS), much of the grounded part of which lies on a submarine bed. If the WAIS were to collapse completely, global sea level would rise by some six metres, inundating many coastal cities.

The basic physical mechanism for this putative catastrophic collapse is a positive feedback between grounding line retreat and ice flow rate. Since ice shelves are not resisted at their base, they can plausibly flow more rapidly, and the consequent drawdown effect will lower ice elevation, thus allowing further grounding line retreat. The debate has been fuelled by the remarkable collapse of the Larsen B Ice Shelf on the Antarctic Peninsula in 2002, which is thought to be due to a climatic warming trend over recent decades. However, as we shall see below, it is by no means trivial to pose a theoretically coherent model for grounding line motion, and the issue of stability remains not fully resolved.

10.1.5 Tidewater Glaciers

If the position of the grounding line indicates a balance between inland ice flow and ice shelf evacuation, the actual mechanism of break up involves mass wastage by calving icebergs. Indeed, in the absence of ablation, calving is the way in which marine ice sheets (i.e., those terminating in the sea) satisfy mass balance.

Glaciers which terminate in the sea are called *tidewater glaciers*, and are susceptible to a similar kind of catastrophic retreat to that which may be important for ice shelves. They also lose mass by calving, but are distinguished from ice shelves by the fact that the ice is grounded right to the margin. Instability is promoted by the fact that the calving rate increases with depth of water. If a tidewater glacier advances (in a cold climate), it will push a ridge of moraine ahead of it, snowplough style. In a stationary state, the water depth at the calving front will then be less than it is away from the margin, because of this moraine. Then, if the glacier snout retreats due, for example, to a warming trend, the snout will suddenly find itself in deeper water. The resultant increased calving rate can then lead to a catastrophic retreat. Just such a rapid retreat was observed in the Columbia Glacier in Alaska, which retreated some 12 km in twenty years from 1982, and it seems such rapid retreat is a common behavioural feature of tidewater glaciers in conditions of warmer climates.

10.1.6 Jökulhlaups

It will be clear by now that basal water is tremendously important in determining the nature of ice flow. Equally, the basal water system can fluctuate independently of the overlying ice dynamics, most notably in the outburst floods called jökulhlaups. In Iceland, in particular, these are associated with volcanoes under ice caps, where high rates of geothermal heat flux in the confines of a caldera cause a growing subglacial lake to occur, Eventually this overflows, causing a subglacial flood which propagates down-glacier, and whose subsequent emergence at the glacier terminus releases enormous quantities of water over the southern coastal outwash plains. These floods carry enormous amounts of volcanic ash and sediments, which create vast beaches of black ash. Despite their violence, the ice flow is hardly disturbed. Jökulhlaups are essentially internal oscillations of the basal drainage system. They are initiated when the rising subglacial lake level causes leakage over a topographic rim, and the resultant water flow leads to an amplifying water flow by the following mechanism. Water flow through a channel in ice enlarges it by meltback of the walls due to frictional heating. The increased channel size allows increased flow, and thus further enlargement. The process is limited by the fact that the ice tends to close up the channel due to the excess overburden pressure over the channel water pressure, and this is accentuated when the channel is larger. In effect, the opening of the valve by the excess lake pressure is closed by the excess ice pressure. These floods occur more or less periodically, every five to ten years in the case of one of the best known, that of Grímsvötn under Vatnajökull in South-east Iceland. A theory for their formation is the subject of Chap. 11.

10.2 The Shallow Ice Approximation

10.2.1 Glaciers

We consider first the motion of a glacier in a (linear) valley. We take the x axis in the direction of the valley axis, z upwards and transverse to the mean valley slope, and y across stream. The basic equations are those of mass and momentum conservation, which for an incompressible ice flow (neglecting inertial terms) are

$$\nabla.\mathbf{u} = 0,$$
$$0 = -\nabla p + \nabla.\boldsymbol{\tau} + \rho\mathbf{g}, \tag{10.1}$$

Fig. 10.6 Typical profile of a
valley glacier

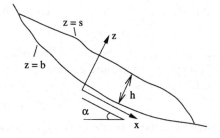

where $\mathbf{g}$ is the gravity vector, p is the pressure, and $\boldsymbol{\tau}$ is the deviatoric part of the stress tensor. These are supplemented by the energy equation, which can be written in the form

$$\rho c_p(T_t + \mathbf{u}.\nabla T) = k\nabla^2 T + \tau_{ij}\dot{\varepsilon}_{ij}, \tag{10.2}$$

where ρ is ice density, c_p is specific heat, and k is thermal conductivity. The summation convention is employed in writing the viscous dissipation term.[5] We focus for the present on the mass and momentum equations, and will deal with the energy equation later.

The stress and strain rate are related by

$$\tau_{ij} = 2\eta\,\dot{\varepsilon}_{ij}, \tag{10.3}$$

where η is the effective viscosity, and $\dot{\varepsilon}_{ij}$ is the strain rate

$$\dot{\varepsilon}_{ij} = \frac{1}{2}\left(\frac{\partial u_i}{\partial x_j} + \frac{\partial u_j}{\partial x_i}\right). \tag{10.4}$$

The most common choice of flow law is known as Glen's law, that is,

$$\dot{\varepsilon}_{ij} = A(T)\tau^{n-1}\tau_{ij}, \tag{10.5}$$

where the second stress invariant is given by $2\tau^2 = \tau_{ij}\tau_{ij}$ (using the summation convention) and $A(T)$ is a temperature-dependent rate factor which causes A to vary by about three orders of magnitude over a temperature range of 50 K: variation of A is thus significant for ice sheets (which may be subject to such a temperature range), but less so for glaciers. If we adopt the configuration shown in figure 10.6, then $\mathbf{g} = (g\sin\alpha, 0, -g\cos\alpha)$, where α is the mean valley slope downhill.

Boundary conditions for the flow are conditions of equal normal stress at the top surface $z = s(x, y, t)$; that is $\boldsymbol{\sigma}_n = -p_a\mathbf{n}$, where p_a is atmospheric pressure, or in coordinate form, $\sigma_{ij}n_j = -p_a n_i$, where $\mathbf{n} \propto (-s_x, -s_y, 1)$:

$$\begin{aligned}
(-p + \tau_{11})s_x + \tau_{12}s_y - \tau_{13} &= -p_a s_x, \\
\tau_{12}s_x + (-p + \tau_{22})s_y - \tau_{23} &= -p_a s_y, \\
\tau_{13}s_x + \tau_{23}s_y - (-p + \tau_{33}) &= p_a.
\end{aligned} \tag{10.6}$$

---

[5]The summation convention is essentially a device for omitting summation signs; it asserts that summation is implied over repeated suffixes; thus $\tau_{ij}\dot{\varepsilon}_{ij}$ means $\sum_{i,j}\tau_{ij}\dot{\varepsilon}_{ij}$.

At the base $z = b(x, y, t)$, we prescribe the velocity:

$$u = u_b, \qquad v = v_b, \qquad w = u b_x + v b_y; \tag{10.7}$$

(u_b, v_b) is the (horizontal) sliding velocity, whose form is discussed later (as are appropriate temperature conditions). Finally, the kinematic condition on the free surface $z = s$ is

$$w = s_t + u s_x + v s_y - a, \tag{10.8}$$

where a is the prescribed surface accumulation: positive for ice accumulation from snowfall, negative for ice ablation by melting.

A major simplification ensues by adopting what has been called the *shallow ice approximation*.[6] It is the lubrication theory idea that the depth $d \ll$ the glacier length l, and is adopted as follows. We scale the variables by putting

$$
\begin{aligned}
u &\sim U; & v, w &\sim \varepsilon U; \\
x &\sim l; & y, z, b, s &\sim d; & t &\sim l/U; \\
\tau_{13}, \tau_{12} &\sim [\tau]; & A &\sim [A]; & a &\sim [a]; \\
p &- p_a - (\rho g \cos \alpha)(s - z) \sim \varepsilon[\tau]; \\
\tau_{11}, \tau_{22}, &\tau_{33}, \tau_{23} \sim \varepsilon[\tau],
\end{aligned}
\tag{10.9}
$$

where

$$\varepsilon = \frac{d}{l} \tag{10.10}$$

is the aspect ratio, and we anticipate $\varepsilon \ll 1$. The choice of d and $[\tau]$ has to be determined self-consistently; we choose l from the given spatial variation of accumulation rate, and we choose U via $\varepsilon U = [a]$, which balances vertical velocity with accumulation rate. If we choose $[\tau] = \rho g d \sin \alpha$, and define

$$\mu = \varepsilon \cot \alpha, \tag{10.11}$$

then the scaled momentum equations are

$$
\begin{aligned}
\frac{\partial \tau_{12}}{\partial y} + \frac{\partial \tau_{13}}{\partial z} &= -1 + \mu \frac{\partial s}{\partial x} + \varepsilon^2 \left(\frac{\partial p}{\partial x} - \frac{\partial \tau_{11}}{\partial x} \right), \\
\frac{\partial s}{\partial y} &= \frac{\varepsilon^2}{\mu} \left[-\frac{\partial p}{\partial y} + \frac{\partial \tau_{12}}{\partial x} + \frac{\partial \tau_{22}}{\partial y} + \frac{\partial \tau_{23}}{\partial z} \right], \\
\frac{\partial p}{\partial z} &= \frac{\partial \tau_{13}}{\partial x} + \frac{\partial \tau_{23}}{\partial y} + \frac{\partial \tau_{33}}{\partial z}.
\end{aligned}
\tag{10.12}
$$

The boundary conditions (10.7) and (10.8) are unchanged in form, and the stress conditions at the top surface $z = s(x, y, t)$ (10.6) become

$$
\begin{aligned}
\varepsilon^2(-p + \tau_{11})s_x + \tau_{12}s_y - \tau_{13} &= 0, \\
\tau_{12}s_x + (-p + \tau_{22})s_y - \tau_{23} &= 0, \\
\tau_{13}s_x + \tau_{23}s_y - (-p + \tau_{33}) &= 0.
\end{aligned}
\tag{10.13}
$$

---

[6]The term arose during a discussion about glacier dynamics one tea time in the Mathematical Institute, Oxford, in 1976. It was invented in keeping with the fluid mechanical description of long waves known as shallow water theory.

Table 10.1 Typical values of constants for ice flow in glaciers and, where different, ice sheets. The activation energy for ice flow is the value appropriate between $-10°$ and $0°C$. For temperatures less than $-10°$, $E = 60$ kJ mol^{-1}. For shear-dominated flows such as ice sheets, it is the warmer value which is more relevant

Symbol	Definition	Glacier value	Ice sheet value
$[a]$	accumulation rate	$1\ \mathrm{m\,y^{-1}}$	$0.1\ \mathrm{m\,y^{-1}}$
$[A]$	flow rate constant	$0.2\ \mathrm{bar^{-n}\,y^{-1}}$	
c_p	specific heat	$2\ \mathrm{kJ\,kg^{-1}\,K^{-1}}$	
E	activation energy	$139\ \mathrm{kJ\,mole^{-1}}$	
g	gravity	$9.8\ \mathrm{m\,s^{-2}}$	
G	geothermal heat flux	$60\ \mathrm{mW\,m^{-2}}$	
k	thermal conductivity	$2.2\ \mathrm{W\,m^{-1}\,K^{-1}}$	
l	length	$10\ \mathrm{km}$	$3{,}000\ \mathrm{km}$
L	latent heat	$3.3 \times 10^5\ \mathrm{J\,kg^{-1}}$	
n	Glen exponent	3	
R	gas constant	$8.3\ \mathrm{J\,mole^{-1}\,K^{-1}}$	
$\sin \alpha$	slope	0.1	n/a
T_M	melting temperature	$273\ \mathrm{K}$	
ΔT	surface temperature deficit	$20\ \mathrm{K}$	$50\ \mathrm{K}$
ρ	ice density	$917\ \mathrm{kg\,m^{-3}}$	

To get some idea of typical magnitudes, use values $d \sim 100$ m, $l \sim 10$ km, $\tan \alpha \sim 0.1$; then $\varepsilon \sim 10^{-2}$, $\mu \sim 10^{-1}$, so that to leading order $s = s(x, t)$ and

$$\frac{\partial \tau_{12}}{\partial y} + \frac{\partial \tau_{13}}{\partial z} = -1 + \mu \frac{\partial s}{\partial x}; \qquad (10.14)$$

we retain the μ term for the moment.

The final relation to choose d (and hence also $[\tau]$) is determined by effecting a balance in the flow law. If the viscosity scale is $[\eta]$, then we choose

$$[\tau] = \frac{[\eta]U}{d}. \qquad (10.15)$$

For example, for Glen's law, we can choose $[\eta] = \frac{1}{2[A][\tau]^{n-1}}$, from which we find

$$d = \left[\frac{[a]l}{2[A](\rho g \sin \alpha)^n} \right]^{1/(n+2)}, \qquad (10.16)$$

which leads, using typical choices of the parameters given in Table 10.1, to values of d comparable to those observed ($d = 128$ m).

The two important shear stresses are given by

$$\tau_{13} = \eta \left(\frac{\partial u}{\partial z} + \varepsilon^2 \frac{\partial w}{\partial x} \right), \qquad \tau_{12} = \eta \left(\frac{\partial u}{\partial y} + \varepsilon^2 \frac{\partial v}{\partial x} \right), \qquad (10.17)$$

and the second stress invariant τ is given by

$$\tau^2 = \tau_{13}^2 + \tau_{12}^2 + \varepsilon^2\left[\frac{1}{2}(\tau_{11}^2 + \tau_{22}^2 + \tau_{33}^2) + \tau_{23}^2\right]; \qquad (10.18)$$

for Glen's flow law, the dimensionless viscosity is

$$\eta = \frac{1}{A(T)\tau^{n-1}}, \qquad (10.19)$$

where $A(T)$ is the scaled (with $[A]$) temperature-dependent rate factor. If we now put $\varepsilon = 0$ (the shallow ice approximation) we see that

$$\tau \approx \eta|\nabla u|, \qquad (10.20)$$

where $\nabla = (\partial/\partial y, \partial/\partial z)$, and for Glen's law,

$$\eta = A^{-1/n}|\nabla u|^{-(n-1)/n} \qquad (10.21)$$

(note $n = 1$ for a Newtonian flow; Glen's law usually assumes $n = 3$); the determination of velocity distribution in a cross section S of a glacier then reduces to the elliptic equation for u in S (putting $\mu = 0$ and $\varepsilon = 0$ in $(10.12)_1$):

$$\nabla.[\eta\{A, |\nabla u|\}\nabla u] = -1 \quad \text{in } S, \qquad (10.22)$$

with appropriate boundary conditions for no slip at the base being $u = 0$ on $z = b$, and the no stress condition at $z = s$ is, from $(10.13)_1$ and (10.17) with $s_y \approx 0$ and $\varepsilon = 0$, $\partial u/\partial z = 0$. The scalar s (independent of y) is determined through prescription of the downslope ice volume flux, $\int_S u\,dy\,dz = Q$, which will depend on x and t, but can be presumed to be known. In general, this problem requires numerical solution. Analytic solutions are available for constant A and a semi-circular cross section, but the free boundary choice of s cannot then be made.

Most studies of wave motion and other dynamic phenomena ignore lateral variation with y, and in this case (with $\tau_{13} = 0$ on $z = s$) (10.14) gives

$$\tau_{13} = (1 - \mu s_x)(s - z), \qquad (10.23)$$

and Glen's law is, approximately,

$$\frac{\partial u}{\partial z} = A(T)|\tau_{13}|^{n-1}\tau_{13} = A(T)|1 - \mu s_x|^{n-1}(1 - \mu s_x)(s - z)^n. \qquad (10.24)$$

If $A = 1$ is constant, then two integrations of (10.24) give the ice flux $Q = \int_b^s u\,dz$ as

$$Q = u_b H + |1 - \mu s_x|^{n-1}(1 - \mu s_x)\frac{H^{n+2}}{n+2}, \qquad (10.25)$$

where $H = s - b$ is the depth, and u_b is the sliding velocity. Integration of the mass conservation equation, together with the basal velocity condition (10.7) and the kinematic surface boundary condition (10.8), then leads to the integral conservation law

$$\frac{\partial H}{\partial t} + \frac{\partial Q}{\partial x} = a, \qquad (10.26)$$

where a is the dimensionless accumulation rate. (10.26) is an equation of convective diffusion type, with the diffusive term being that proportional to μ. For a glacier, it is reasonable to assume that $1 - \mu s_x > 0$, meaning that the ice surface is always inclined downhill, and in this case the modulus signs in (10.25) are redundant. In essence, this unidirectionality of slope is what distinguishes a glacier from an ice cap or ice sheet.

Note that if transverse variations were to be included, we should solve

$$\frac{\partial S}{\partial t} + \frac{\partial Q}{\partial x} = a, \qquad (10.27)$$

where S is the cross-sectional area, and Q would be given by $Q = \int_S u \, dS$, where u solves (10.22) in the cross section S, together with appropriate boundary conditions.

10.2.2 Ice Sheets

A model for ice sheets can be derived in much the same way—typical aspect ratios are 10^{-3}—but there is no 'downslope' gravity term $\rho g \sin\alpha$ (effectively $\alpha = 0$), and the appropriate balance determines the driving shear stress at the base in terms of the surface slope. Effectively, the advection term is lost and $\mu = 1$. Another difference is that $x \sim y \sim l$ (~ 3000 km) while $z \sim 3$ km is the only small length scale.

We will illustrate the scaling in two dimensions; the three-dimensional version is relegated to the exercises (Question 10.4). In two dimensions, we write the deviatoric stresses as

$$\tau_{11} = -\tau_{33} = \tau_1, \qquad \tau_{13} = \tau_3. \qquad (10.28)$$

Then the governing equations are

$$\begin{aligned}
u_x + w_z &= 0, \\
0 &= -p_x + \tau_{1x} + \tau_{3z}, \\
0 &= -p_z + \tau_{3x} - \tau_{1z} - \rho g, \\
\dot{\varepsilon}_{ij} &= A\tau^{n-1}\tau_{ij}.
\end{aligned} \qquad (10.29)$$

The surface stress boundary conditions are, on $z = s$,

$$\begin{aligned}
(-p + \tau_1)s_x - \tau_3 &= -p_a s_x, \\
\tau_3 s_x - (-p - \tau_1) &= p_a;
\end{aligned} \qquad (10.30)$$

at the base $z = b(x, y, t)$, we prescribe the velocity:

$$u = u_b, \qquad w = u b_x; \qquad (10.31)$$

and on $z = s$, the kinematic condition is

$$w = s_t + u s_x - a. \qquad (10.32)$$

We scale the variables by putting

$$u \sim U; \qquad w \sim [a];$$
$$x \sim l; \qquad z, b, s \sim d; \qquad t \sim l/U;$$
$$\tau_3 \sim [\tau]; \qquad A \sim [A]; \qquad a \sim [a]; \qquad (10.33)$$
$$p - p_a - \rho_i g(s - z) \sim \varepsilon[\tau];$$
$$\tau_1 \sim \varepsilon[\tau],$$

where

$$\varepsilon = \frac{d}{l}. \qquad (10.34)$$

An appropriate balance of terms is effected by choosing

$$U = 2[A][\tau]^n d = \frac{[a]}{\varepsilon}, \qquad [\tau] = \rho g d \varepsilon, \qquad (10.35)$$

and this leads to

$$d = \left(\frac{[a] l^{n+1}}{2[A](\rho g)^n} \right)^{\frac{1}{2(n+1)}}, \qquad (10.36)$$

and thus

$$\varepsilon = \left(\frac{[a]}{2[A](\rho g)^n l^{n+1}} \right)^{\frac{1}{2(n+1)}}, \qquad (10.37)$$

and the typical values of the constants in Table 10.1 do lead to a depth scale of the correct order of magnitude, $d = 3595$ m, so that $\varepsilon \sim 10^{-3}$.

The corresponding dimensionless equations are

$$u_x + w_z = 0,$$
$$0 = -s_x + \tau_{3z} + \varepsilon^2[-p_x + \tau_{1x}],$$
$$0 = -p_z + \tau_{3x} - \tau_{1z},$$
$$(u_z + \varepsilon^2 w_x) = A\tau^{n-1}\tau_3, \qquad (10.38)$$
$$2u_x = A\tau^{n-1}\tau_1,$$
$$\tau^2 = \tau_3^2 + \varepsilon^2 \tau_1^2,$$

and the boundary conditions are, on $z = s$:

$$\tau_3 + \varepsilon^2(p - \tau_1)s_x = 0,$$
$$\tau_3 s_x + p + \tau_1 = 0, \qquad (10.39)$$
$$w = s_t + us_x - a;$$

at the base $z = b(x, y, t)$:

$$u = u_b, \qquad w = ub_x. \qquad (10.40)$$

The shallow ice approximation puts $\varepsilon = 0$, and then we successively find

$$\tau_3 = -s_x(s - z), \qquad \tau = |s_x|(s - z), \qquad (10.41)$$

whence

$$p + \tau_1 = [s s_x]_x (s - z) + \frac{1}{2}\left(z^2 - s^2\right) s_{xx}, \tag{10.42}$$

and, if we assume that $A = 1$ is constant, the ice flux is

$$\int_b^s u\,dz = u_b H + \frac{H^{n+2}}{n+2}|s_x|^{n-1}(-s_x), \tag{10.43}$$

so that conservation of mass leads to

$$\frac{\partial H}{\partial t} = \frac{\partial}{\partial x}\left[\frac{H^{n+2}}{n+2}|s_x|^{n-1}s_x - H u_b\right] + a, \tag{10.44}$$

a nonlinear diffusion equation for the depth H, since $s = H + b$.

The three-dimensional version of this equation is (with $\nabla = (\partial/\partial x,\ \partial/\partial y)$)

$$H_t = \nabla \cdot \left[\left\{\frac{|\nabla s|^{n-1}H^{n+2}}{n+2}\nabla s\right\} - H\mathbf{u}_b\right] + a. \tag{10.45}$$

The term in the sliding velocity $\mathbf{u}_b$ is apparently a convective term, but in fact the sliding law usually has $\mathbf{u}_b$ in the direction of shear stress, whence $\mathbf{u}_b \propto -\nabla s$, and this term also is diffusive.

Boundary Conditions

Normally one would expect a boundary condition to be applied for (10.45) at the margin of the ice sheet, whose location itself may not be known. For an ice sheet that terminates on land, this condition would be $H = 0$ at the margin, but since the diffusion equation (10.45) is degenerate, in the sense that the diffusion coefficient vanishes where $H = 0$, no extra condition to specify the margin location is necessary, other than requiring that the ice flux also vanish where $H = 0$.

A different situation pertains for a marine ice sheet which terminates (and is grounded), let us suppose, at the edge of the continental shelf. Then the margin position is known, and the ice thickness and flux are finite. In this case, the appropriate condition is to prescribe H at the margin, on the basis that the ice (approximately) reaches flotation there.

A more representative condition for marine ice sheets occurs when the grounded ice extends into an ice shelf before the continental shelf edge is reached. Extended ice shelves occur in the Antarctic ice sheet, two notable examples being the Ronne–Filchner ice shelf and the Ross ice shelf. The grounding line where the ice changes from grounded ice to floating ice is a free boundary whose location must be determined. The appropriate boundary condition for the ice sheet at the grounding line is bound up with the mechanics of the ice shelf, whose behaviour is altogether different; the mechanics of ice shelves is studied in Sect. 10.2.6, and the problem of determining the grounding line is studied in Sect. 10.2.7.

10.2.3 Temperature Equation

Although the isothermal models are mathematically nice, they are apparently not quantitatively very realistic. For a glacier, probably the neglect of variation of the rate parameter $A(T)$ in the flow law is as important as the assumption of a two-dimensional flow, although the possible coupling of temperature to water production and basal sliding is also significant. For ice sheets, temperature variation is unquestionably significant, and cannot in practice be neglected.

Boundary Conditions

The ice temperature is governed by the energy equation (10.2), and it must be supplemented by suitable boundary conditions. At the ice surface, an appropriate boundary condition follows from consideration of energy balance, much as in Chap. 3, but for purposes of exposition, we suppose that the ice surface temperature is equal to a prescribed air temperature, thus

$$T = T_A \quad \text{at } z = s. \tag{10.46}$$

The boundary conditions at the base are more complicated. While the ice is frozen, we prescribe a geothermal heat flux G, and presume the ice is frozen to the base, so that there is no slip, thus

$$-k\frac{\partial T}{\partial n} = G, \quad T < T_M, \quad u = 0 \quad \text{at } z = b, \tag{10.47}$$

where T_M is the melting temperature, which depends weakly on pressure; $\mathbf{n}$ is the unit normal pointing upwards at the base. Classically, one supposes that when T reaches T_M, a lubricating Weertman film separates the ice from the bed, allowing slip to take place, so that we have a sliding velocity $u = u_b$, in which, for example, u_b is a function of basal shear stress τ_b. The details of the calculation of this sliding velocity are detailed in Sect. 10.3. For the moment it suffices to point out that the transition from no sliding to a full sliding velocity must occur over a narrow range of temperature near the melting point, when only a partial water film is present. In this régime, there is no net production of water at the base, the temperature is essentially at the melting point, and there is sliding, but this is less than the full sliding velocity u_b; we call this the sub-temperate regime:

$$-k\frac{\partial T}{\partial n} = G + \tau_b u, \quad T = T_M, \ 0 < u < u_b. \tag{10.48}$$

The term $\tau_b u$ represents the frictional heat delivered to the base by the work of sliding.[7]

---

[7] An alternative formulation combines the frozen and sub-temperate régimes by allowing the sliding velocity to be a function of temperature near the melting point. This may be a simpler formulation to use in constructing numerical solutions.

When the water film is completely formed, there is net water production at the base, and the sliding velocity reaches its full value; this is the temperate régime:

$$0 < -k\frac{\partial T}{\partial n} < G + \tau_b u, \quad T = T_M, \ u = u_b. \tag{10.49}$$

In all the above régimes, the ice above the bed is cold. When the heat flux $-k\frac{\partial T}{\partial n}$ reaches zero, the ice above the bed becomes temperate and moist, containing water. In this case the energy equation must be written as an equation for the enthalpy

$$h = c_p(T - T_M) + Lw, \tag{10.50}$$

where L is latent heat, and w is the mass fraction of water inclusions, and the inequalities $T \leq T_M$ in the above conditions can be replaced by the inequalities $h \leq 0$. Proper formulation of the correct thermal boundary condition when $h > 0$ now requires an appropriate formulation for the enthalpy flux q_n when $w > 0$, and this requires a description of moisture transport in the moist ice. This goes some way beyond our present concerns, and is not pursued further here.

Non-dimensionalisation

With variables scaled as in the previous section for an ice sheet, the temperature equation for an ice sheet may be written approximately as

$$T_t + \mathbf{u}.\nabla T = \frac{\alpha \tau^2}{\eta} + \beta T_{zz}, \tag{10.51}$$

where $T - T_M$ is scaled with ΔT (a typical surface temperature below melting point). The stress invariant τ is related to the horizontal velocity $\mathbf{u}_H = (u, v)$ by

$$\tau \approx \eta \left| \frac{\partial \mathbf{u}_H}{\partial z} \right| = (s - z)|\nabla s|, \tag{10.52}$$

since the horizontal stress vector $\boldsymbol{\tau} = (\tau_{13}, \tau_{23})$ satisfies

$$\boldsymbol{\tau} = \eta \frac{\partial \mathbf{u}_H}{\partial z} = -(s - z)\nabla s. \tag{10.53}$$

The parameters α and β are given by

$$\alpha = \frac{gd}{c_p \Delta T}, \quad \beta = \frac{\kappa}{d[a]}, \tag{10.54}$$

where d is the depth scale, c_p is the specific heat, g is gravity, $\kappa = \frac{k}{\rho c_p}$ is the thermal diffusivity, and $[a]$ is accumulation rate. Using the values in Table 10.1, together with $d = 3{,}500$ m, we find that typical values for an ice sheet are $\alpha \sim 0.3$, $\beta \sim 0.1$. We see that viscous heating (the α term) is liable to be significant, while thermal conduction is small or moderate.

The dimensionless forms of the temperature boundary conditions (10.46)–(10.49) take the form

$$T = T_A \quad \text{at } z = s, \tag{10.55}$$

where the scaled surface temperature T_A is negative and $O(1)$. At the base $z = b$, we have

$$-\frac{\partial T}{\partial n} = \Gamma, \quad T < 0, \; u = 0,$$

$$-\frac{\partial T}{\partial n} = \Gamma + \frac{\alpha \tau_b u}{\beta}, \quad T = 0, \; 0 < u < u_b, \tag{10.56}$$

$$0 < -\frac{\partial T}{\partial n} < \Gamma + \frac{\alpha \tau_b u}{\beta}, \quad T = 0, \; u = u_b,$$

where

$$\Gamma = \frac{Gd}{k \Delta T}. \tag{10.57}$$

For values of the parameters in Table 10.1, and with $d = 3{,}500$ m, we have $\Gamma \approx 1.9$, so that the geothermal heat flux is significant, and temperature variation is important.

The dimensional rate factor in the flow law is modelled as

$$A = A_0 \exp\left(-\frac{E}{RT}\right), \tag{10.58}$$

and in terms of the non-dimensional temperature, this can be written in the form

$$A = [A] \exp\left[\frac{E}{RT_M}\left\{1 - \left(1 + \frac{\Delta T}{T_M}T\right)^{-1}\right\}\right], \tag{10.59}$$

where

$$[A] = A_0 \exp\left(-\frac{E}{RT_M}\right). \tag{10.60}$$

Assuming $\Delta T \ll T_M$, we can write (10.58) in the approximate form

$$A \approx e^{\gamma T}, \tag{10.61}$$

where now A is the dimensionless rate factor, and

$$\gamma = \frac{E \Delta T}{R T_M^2}. \tag{10.62}$$

Using the values for the activation energy E and gas constant R in Table 10.1, we find $\gamma \approx 11.2$. Because this relatively large value occurs in an exponent, its largeness is enhanced.[8]

The temperature equation for a (two-dimensional) valley glacier is the same as (10.51), although in the definition of α, we replace d by $l \sin \alpha$. With the previous scalings, (10.52) is corrected by simply replacing $|\nabla s|$ in the last expression by $|1 - \mu s_x|$. Although the scales are different, typical values of α and β (from Table 10.1) are $\alpha \sim 0.25$, $\beta \sim 0.29$, and thus of significance. On the other hand, geothermal heat is of less importance, with a typical estimate being $\Gamma \approx 0.17$.[9]

---

[8] However, the value of E is smaller below $-10°C$, which serves to modify the severity of the temperature dependence of A at cold temperatures.

[9] This raises an interesting issue in the modelling of glaciers. So little heat is supplied geothermally that it seems difficult to raise the ice temperature much above the surface average value anywhere.

10.2.4 A Simple Non-isothermal Ice Sheet Model

The fact that the ice sheet model collapses to a simple nonlinear diffusion equation
(10.45) when the rate coefficient is independent of temperature raises the question
of whether any similar simplification can be made when temperature dependence is
taken into account. The answer to this is yes, provided some simplifying approximations are made. Before doing this, we revisit the non-dimensionalisation in (10.33).
The balance of terms in (10.35) is based on the assumption that shearing is important
throughout the thickness of the ice. However, this is not entirely accurate. The largeness of the rheological coefficient γ implies that shearing will be concentrated near
the bed, where it is warmest. This effect is amplified by the smallness of β, which
means that temperature gradients may be confined to a thin basal thermal boundary
layer, and also by the largeness of the Glen exponent n, which increases the shear
near the bed. To allow for shearing being restricted to a height of $O(\nu) \ll 1$ above
the bed, we adjust the balance in (10.35) by choosing

$$U = 2[A][\tau]^n \nu d, \tag{10.63}$$

with the other choices remaining the same. The change to the depth scale in (10.36)
is obtained by replacing $[A]$ by $\nu[A]$. We will choose ν below. Note that If $\nu = 0.1$,
then the depth scale is increased by a factor of $10^{1/8} \approx 1.33$ to 4,794 m, still appropriate for large ice sheets.

The shallow ice approximation then leads to the model

$$T_t + \mathbf{u}.\nabla T = \frac{\alpha}{\nu}\tau^{n+1}e^{\gamma T} + \beta T_{zz},$$

$$\nu\frac{\partial \mathbf{u}}{\partial z} = \tau^{n-1}\boldsymbol{\tau}e^{\gamma T},$$

$$H_t + \nabla.\left\{\int_b^s \mathbf{u}\,dz\right\} = a, \tag{10.64}$$

$$\boldsymbol{\tau} = -(s - z)\nabla s,$$

where $H = s - b$ is the depth, $\mathbf{u} = \mathbf{u}_H = (u, v)$ is the horizontal velocity, and α, β
and γ retain their earlier definitions in (10.54) and (10.62).

We now suppose that $\gamma \gg 1$ and $\beta \ll 1$. The temperature then satisfies the approximate equation

$$T_t + \mathbf{u}.\nabla T = 0, \tag{10.65}$$

where also $(10.64)_2$ implies the plug flow $\mathbf{u} = \mathbf{u}(x, y)$. This 'outer' equation is valid
away from the base, where there must be a basal boundary layer and a shear layer
in order to satisfy velocity and temperature boundary conditions.

In practice, the presence of crevasses allows meltwater and rainfall to access the glacier bed, and
the re-freezing of even a modest depth of water enormously enhances the effective heat supplied to
the bed via the latent heat released. For example, a meltwater supply of 10 cm per year to the bed
raises the effective value of Γ to 2.9.

For simplicity, let us consider the particular case of steady flow over a flat bed, so that $b = 0$, $H = s$, and time derivatives are ignored. The vertical velocity for the plug flow is

$$w \approx -z\Delta, \quad \Delta = u_x + v_y, \tag{10.66}$$

so that T satisfies

$$uT_x + vT_y - z\Delta T_z = 0. \tag{10.67}$$

In two dimensions, it is easy to solve this, because there is a stream function. For example, if $v \equiv 0$, then the general solution is $T = f(zu)$. Similarly, if $u \equiv 0$, then $T = f(zv)$. It is not difficult to see that the general solution for a two-dimensional flow is then $T = f(z|\mathbf{u}|)$.

This suggests that we seek solutions in the form

$$T = f(zU), \quad U = U(x, y). \tag{10.68}$$

Note that U is defined up to an arbitrary multiple. Substituting this into (10.67), we find that U satisfies the equation

$$\left(\frac{u}{U}\right)_x + \left(\frac{v}{U}\right)_y = 0, \tag{10.69}$$

whence there is a function χ such that

$$u = U\chi_y, \quad v = -U\chi_x. \tag{10.70}$$

Note that $(10.64)_{2,4}$ imply that

$$\mathbf{u} = -K\nabla s \tag{10.71}$$

for some scalar K, and therefore (10.70) implies that $\nabla s . \nabla \chi = 0$, and the level sets of s and χ form an orthogonal coordinate system. In particular, the coordinate $-s$ points down the steepest descent paths of the surface $z = s$, while the coordinate χ tracks anti-clockwise round the contours of s. Note that for a steady flow in which a is finite, there can be no minima of s, and the flow is everywhere downhill. In general, the contours of s are closed curves, and the coordinates $-s$, χ are like generalised polar coordinates. Complications arise if there are saddles, but these simply divide the domain into different catchments, each of which can be treated separately.

If we change to independent coordinates s and χ, then we find, after some algebra,

$$\frac{\partial}{\partial s}\left(\frac{U}{|\mathbf{u}|}\right) = -\frac{K}{|\mathbf{u}|^3}(uv_\chi - vu_\chi). \tag{10.72}$$

Let θ be the angle of the flow direction to the x axis, so that $\tan\theta = \frac{v}{u}$. Calculating the derivative of θ, we then find that Eq. (10.72) takes the form

$$\frac{\partial}{\partial s}\left(\frac{U}{|\mathbf{u}|}\right) = -\frac{1}{|\nabla s|}\frac{\partial\theta}{\partial\chi}, \tag{10.73}$$

and this is our basic equation to determine U. Note that in a two-dimensional flow where $\frac{\partial}{\partial \chi} = 0$, we regain $U = |\mathbf{u}|$ as before.

A more user-friendly version of (10.73) follows from the following geometrical considerations. From (10.70), we have $|\nabla \chi| = \frac{|\mathbf{u}|}{U}$, and since s and χ are orthogonal, it follows that

$$\frac{d\chi}{d\sigma} = \frac{|\mathbf{u}|}{U} \tag{10.74}$$

on level contours of s, where σ is arc length measured anti-clockwise. Now the curvature of a level s contour[10] is defined by

$$\kappa = \frac{d\theta}{d\sigma}. \tag{10.75}$$

Therefore $\frac{\partial \theta}{\partial \chi} = \frac{\kappa U}{|\mathbf{u}|}$; in addition distance ξ at the base below a steepest descent path on the surface satisfies $d\xi = -\frac{ds}{|\nabla s|}$, and therefore (10.73) can be written in the form

$$\frac{\partial}{\partial \xi}\left(\frac{U}{|\mathbf{u}|}\right) = \frac{U\kappa}{|\mathbf{u}|}, \tag{10.76}$$

with solution

$$U = |\mathbf{u}| \exp\left[\int^{\xi} \kappa \, d\xi\right]. \tag{10.77}$$

The constant of integration can be chosen arbitrarily.

Given a surface $z = s(x, y)$, we find the level contours and the steepest descent paths, and we compute from the geometry of the contours their curvature κ. On any steepest descent path, (10.77) then gives U, and the function f is determined by the requirement that the surface temperature

$$T_s = f(sU). \tag{10.78}$$

Along any flow path, sU is a monotone increasing function of $-s$ (see Question 10.5), and therefore (10.78) determines f uniquely. Question 10.6 shows a worked example in the particular case of a cylindrically symmetric ice sheet. Typically, with surface temperatures being lower at higher elevation, the outer temperature profiles thus constructed are inverted, i.e., they are colder at depth. There is then an inversion near the base, where a thermal boundary layer reverses the temperature gradient.

Thermal Boundary Layer Structure

We now consider Eqs. (10.64) near the base of the ice sheet. We will restrict our attention to the case where the ice is at the melting point, but where sliding is negligible. It is straightforward to deal with other cases also.

---

[10]Strictly this is twice the mean curvature, since in three dimensions the other radius of curvature of the level s contours is infinite.

Away from the base, the temperature satisfies (10.68). More specifically (see Question 10.5), we can write the outer solution in the form

$$T = f\left(\frac{z}{s}\int_s^{s_0}\frac{a}{J}\,ds\right), \tag{10.79}$$

where the integral is along a flow line $\chi =$ constant, and J is the Jacobian

$$J = -\frac{\partial(s,\chi)}{\partial(x,y)} > 0 \tag{10.80}$$

of the transformation from (x, y) to $(-s, \chi)$. Here s_0 is the surface elevation at the ridge forming the ice divide where the flow line starts.

Insofar as we might suppose the surface temperature decreases with increasing elevation, it is reasonable to suppose that f is an increasing function of its argument, and thus the temperature in (10.79) increases with height. As $z \to 0$, $T \to f(0) < 0$, and there is a temperature inversion in a thermal boundary layer through which $T \to 0$. In this layer, the plug flow velocity is unaffected, as the exponential terms are negligible. We write

$$z = \beta^{1/2}Z, \tag{10.81}$$

so that the temperature equation takes the approximate form

$$uT_x + vT_y - Z\Delta T_Z = T_{ZZ}. \tag{10.82}$$

It is convenient again to write this in terms of the coordinates s, χ, Z, when the equation becomes

$$-J[UT_s - ZU_sT_Z] = T_{ZZ}, \tag{10.83}$$

with boundary conditions

$$\begin{aligned} T &= 0 \quad \text{on } Z = 0, \\ T &\to f(0) \quad \text{as } Z \to \infty. \end{aligned} \tag{10.84}$$

We can simplify this by defining

$$X = \int_s^{s_0}\frac{ds}{J}, \qquad \Psi = ZU, \tag{10.85}$$

which reduces the problem to

$$\Psi_Z T_X - \Psi_X T_Z = T_{ZZ}. \tag{10.86}$$

A further Von Mises transformation to independent variables (X, Ψ) reduces this to a diffusion equation, for which a similarity solution is appropriate, and this is given by

$$T = f(0)\,\text{erf}\left[\frac{ZU}{2\sqrt{\Xi}}\right], \tag{10.87}$$

where

$$\Xi = \int_0^X U\,dX = \int_s^{s_0}\frac{U\,ds}{J}. \tag{10.88}$$

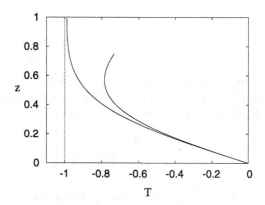

Fig. 10.7 Typical temperature profiles at $x = 0.1$ and $x = 0.5$ for the uniform approximation (10.89), for the particular assumptions in (10.90). The value of $\beta = 0.1$

A uniform approximation is then given by

$$T = f\left(\frac{z}{s}\int_s^{s_0}\frac{a}{J}\,ds\right) - f(0)\,\mathrm{erfc}\left\{\frac{zU}{2\sqrt{\beta\varXi}}\right\}. \tag{10.89}$$

To illustrate the consequent temperature profiles, we make simplifying assumptions. For a one-dimensional flow in the x-direction, we have $\chi = y$, $J = -s_x$, $U = u$. Suppose $a = 1$, $u = x$, $s = 1 - x^2$ and that the surface temperature is $T|_s = -s$. We then find that $f(x) = -(1 - x^2)$, and thence

$$T = -\left[1 - \left(\frac{zx}{1 - x^2}\right)^2\right] + \mathrm{erfc}\left[\frac{z}{\sqrt{2\beta}}\right]. \tag{10.90}$$

Figure 10.7 shows two profiles at distances of $x = 0.1$ and $x = 0.5$ along the flow line, which resemble the sorts of profiles which are typically measured (and which are also found in direct numerical simulations).

Shear Layer

Embedded within the thermal boundary layer is a thinner shear layer of thickness $O(\nu)$, to be determined. From the thermal boundary layer solution (10.87), we have $T \sim \frac{f(0)UZ}{\sqrt{\pi\varXi}}$ as $Z \to 0$; additionally we require $T \sim \frac{1}{\gamma}$ in the shear layer in order that there be non-zero shear. Thus we define

$$\theta = \gamma T, \qquad \nu = \frac{\sqrt{\beta}}{\gamma}, \tag{10.91}$$

which confirms that $\nu \ll \sqrt{\beta}$, and thus the shear layer is indeed embedded within the thermal boundary layer. Defining $z = \nu\zeta$, the shear layer equations become, to leading order,

$$\boldsymbol{\tau} = -s\nabla s,$$
$$\frac{\partial\mathbf{u}}{\partial\zeta} = \tau^{n-1}\boldsymbol{\tau}e^{\theta}, \tag{10.92}$$
$$0 = \mathcal{A}\tau^{n+1}e^{\theta} + \theta_{\zeta\zeta},$$

where

$$A = \frac{\alpha}{\sqrt{\beta}}. \tag{10.93}$$

Using the values $\alpha \sim 0.3$, $\beta \sim 0.1$, then $A \approx 1$, and it is sensible to suppose formally that $\alpha \sim \sqrt{\beta}$, so that $A \sim O(1)$. Suitable boundary conditions for the equations are then (supposing sliding is negligible)

$$\mathbf{u} = 0, \qquad \theta = 0 \quad \text{at } \zeta = 0,$$
$$\theta \sim \frac{f(0)U\zeta}{\sqrt{\pi \Xi}} \quad \text{as } \zeta \to \infty. \tag{10.94}$$

Integrating the temperature equation in the form $\theta_{\zeta\zeta} + A\boldsymbol{\tau}.\mathbf{u}_\zeta = 0$, we find that indeed $\mathbf{u} = -K\nabla s$, where

$$K = \frac{\theta_\zeta |_\infty^0}{As|\nabla s|^2}. \tag{10.95}$$

Direct integration of the temperature equation gives the solution in the form

$$\theta = -2\ln\left[B\cosh\left(\sqrt{\frac{\lambda}{2}}\frac{\zeta}{B} + C\right)\right], \tag{10.96}$$

where

$$\lambda = A\tau^{n+1}. \tag{10.97}$$

Applying the boundary conditions, we find after some algebra that, if we define $f(0) = -T_R$, then

$$K = \frac{T_R U(1 - \tanh C)}{\sqrt{\pi \Xi} As|\nabla s|^2}, \tag{10.98}$$

where

$$\cosh C = \frac{T_R U}{\sqrt{2\lambda \pi \Xi}}. \tag{10.99}$$

(10.98) defines K implicitly and non-locally along a flow line, since (10.77) implies that $U \propto K$, and thus (10.88) implies that Ξ involves an integral of K. Given s, we can determine K and hence $\mathbf{u}$, and this can be used to evolve the surface via the diffusion equation

$$s_t = \nabla.(sK\nabla s) + a. \tag{10.100}$$

This is not strictly accurate, since the (outer) temperature is assumed stationary. However, the outer temperature is only involved in the determination of the velocity field through the prescription of the ridge temperature $-T_R$. If the ridges are stationary, or if the ridge temperature is constant, then the derivation above remains appropriate; if not, T_R must be determined as well.

To illustrate how K is determined, consider the two-dimensional case, where x is in the direction of flow. We then have $\chi = y$, $J = -s_x$, $\kappa = 0$, $X = x$, $U = u$,

$\varXi = \int_0^x u\,dx$. In addition, suppose that $T_R = 1$ and that C is sufficiently large that $1 - \tanh C \approx 2e^{-2C} \approx \frac{1}{2}\operatorname{sech}^2 C$. Formally this is the case if $\mathcal{A} \ll 1$. Define

$$L = 2\sqrt{\varXi};\qquad\qquad (10.101)$$

we then find that

$$KL' = \sqrt{\pi}\,s^n|\nabla s|^{n-1}.\qquad\qquad (10.102)$$

In addition, $U = \frac{1}{2}LL' = K|\nabla s|$; eliminating K, we can derive the differential equation for L

$$LL'^2 = 2\sqrt{\pi}\,s^n|\nabla s|^n,\qquad\qquad (10.103)$$

subject to $L(0) = 0$. Solving this, we can then determine K from (10.102).

More generally, with L still defined by (10.101) and $T_R = 1$, (10.98) and (10.99) lead to

$$K = \frac{LL_X}{2W|\nabla s|} = \frac{L_X - \{L_X^2 - 2\pi\mathcal{A}s^{n+1}|\nabla s|^{n+1}\}^{1/2}}{\sqrt{\pi}\,\mathcal{A}s|\nabla s|^2},\qquad\qquad (10.104)$$

where

$$W = \exp\left\{\int_0^{\xi} \kappa\,d\xi\right\}.\qquad\qquad (10.105)$$

Solving for L_X, we find

$$L_X^2 = \frac{8\sqrt{\pi}\,W^2\tau^n}{L(4W - \sqrt{\pi}\,\mathcal{A}\tau)};\qquad\qquad (10.106)$$

both L and hence K can be determined from (10.106).

Reverting to two-dimensional flow for a moment, we might suppose that for sufficiently small x, $u \sim x$. Since with $X = x$ and $W = 1$, $u = \frac{1}{2}LL_x$, this suggests $L \sim x$, and $L_x \approx L'$ is constant. From (10.106) (with also τ small), we then have

$$L \approx \frac{2\sqrt{\pi}\,\tau^n}{L'^2}\qquad\qquad (10.107)$$

(note that this is only consistent with $L \sim x$ if $\tau \sim x^{1/n}$). From (10.104), we then have

$$K \approx \frac{\sqrt{\pi}}{L'}s^n|s'|^{n-1},\qquad\qquad (10.108)$$

and (10.100) takes the form

$$s_t = K\frac{\partial}{\partial x}\left[s^{n+1}|s_x|^{n-1}s_x\right] + a,\qquad\qquad (10.109)$$

where

$$K = \frac{\sqrt{\pi}}{L'}.\qquad\qquad (10.110)$$

One can show (cf. (10.118) below) that for small x, we indeed have $\tau \sim x^{1/n}$ as required, so that this approximation is consistent. The value of L' is determined by the boundary condition at the margin. For example, if we suppose (10.109) applies all the way to the margin at $x = 1$, where $s = 0$, then we can show for the steady state ice sheet that

$$s(0) = s_0 = \left(\frac{2n+1}{n+1}\right)^{n/(2n+1)}, \tag{10.111}$$

and that $u \approx \frac{\sqrt{\pi x}}{L's_0}$; consistency with $u \approx \frac{1}{2}LL' \approx \frac{1}{2}L'^2 x$ then implies that

$$L' = \left(\frac{2\sqrt{\pi}}{s_0}\right)^3, \qquad K = \left(\frac{\pi s_0}{2}\right)^{1/3}. \tag{10.112}$$

The success of this approach suggests an extension to the three-dimensional model (10.104). If we suppose L_X and W are approximately constant, then we find

$$L \approx \frac{8\sqrt{\pi}W^2\tau^n}{L_X^2(4W - \sqrt{\pi}A\tau)}, \qquad K \approx \frac{4\sqrt{\pi}W\tau^n}{L_X|\nabla s|(4W - \sqrt{\pi}A\tau)}, \tag{10.113}$$

and thus (10.100) takes the form

$$s_t = \nabla.[D\nabla s] + a, \tag{10.114}$$

where

$$D = \frac{\sqrt{\pi}}{L_X} \frac{s^{n+1}|\nabla s|^{n-1}}{\left\{1 - \frac{A\sqrt{\pi}s|\nabla s|}{4W}\right\}}. \tag{10.115}$$

10.2.5 Using the Equations

Nonlinear Diffusion

The derivation of (10.114) shows that for a temperature-dependent ice flow law, one reasonably obtains a nonlinear diffusion equation to govern the ice sheet evolution. Similarly, for flow over a flat base, $h = 0$, with no sliding, the isothermal ice sheet equation (10.45) is just

$$s_t = \nabla.\left[\frac{s^{n+2}|\nabla s|^{n-1}}{n+2}\nabla s\right] + a, \tag{10.116}$$

which for Glen's flow law would have $n = 3$. This is a degenerate nonlinear diffusion equation, and has singularities at ice margins ($s = 0$) or divides (where $\nabla s/|\nabla s|$ is discontinuous). In one space dimension, we have near a margin $x = x_m(t)$ where $a < 0$ (ablation), assuming zero ice flux there,

$$s \sim (a/|\dot{x}_m|)(x_m - x) \quad \text{if } \dot{x}_m < 0 \text{ (retreat)},$$
$$s \sim \left(\frac{2n+1}{n}\right)^{\frac{n}{2n+1}}[(n+1)\dot{x}_m]^{\frac{1}{2n+1}}(x_m - x)^{\frac{n}{2n+1}} \quad \text{if } \dot{x}_m > 0 \text{ (advance)}. \tag{10.117}$$

This is the common pattern for such equations: margin retreat occurs with finite slope, while for an advance, the slope must be infinite. Consequently, there is a waiting time between a retreat and a subsequent advance, while the front slope grows.

Near a divide $x = x_d$, where $s_x = 0$ and $a > 0$, s is given by

$$s \sim s_0(t) - \left(\frac{n}{n+1}\right)\left[\frac{(n+2)(a - \dot{s}_0)}{s_0^{n+2}}\right]^{1/n} |x - x_d|^{(n+1)/n}, \tag{10.118}$$

and thus the curvature is infinite. Singularities of these types need to be taken into account in devising numerical methods.

Thermal Runaway

One of the interesting possibilities of the thermomechanical coupling between flow and temperature fields is the possibility of thermal runaway, and it has even been suggested that this may provide an explanation for the surges of certain thermally regulated glaciers. The simplest model is that for a glacier, with exponential rate factor, thus

$$T_t + \mathbf{u}.\nabla T = \alpha \tau^{n+1} e^{\gamma T} + \beta T_{zz}, \tag{10.119}$$

where the stress is given by

$$\tau = s - z. \tag{10.120}$$

The simplest configuration is the parallel-sided slab in which $s = $ constant, $\mathbf{u} = (u(z), 0, 0)$, so that

$$\frac{\partial T}{\partial t} = \alpha(s - z)^{n+1} e^{\gamma T} + \beta \frac{\partial^2 T}{\partial z^2}, \tag{10.121}$$

with (say)

$$T = -1 \quad \text{on } z = s, \qquad T_z = -\Gamma \quad \text{on } z = 0. \tag{10.122}$$

For given s, (10.121) will exhibit thermal runaway for large enough α, and $T \to \infty$ in finite time. As the story goes, this leads to massive melting and enhanced sliding, thus 'explaining' surges. The matter is rather more complicated than this, however. For one thing, s would actually be determined by the criterion that, in a steady state, the flux $\int_0^s u\,dz$ is prescribed, $= B$ say, where B would be the integrated ice accumulation rate from upstream ($= \int a\,dx$).

Thus even if we accept the unrealistic parallel slab 'approximation', it would be appropriate to supplement (10.121) and (10.122) by requiring s to satisfy

$$\int_0^s u\,dz = B. \tag{10.123}$$

Since the flow law gives

$$\frac{\partial u}{\partial z} = (s - z)^n e^{\gamma T}, \tag{10.124}$$

we find, if $u = 0$ on $z = 0$, that (10.123) reduces to

$$\int_0^s (s - z)^{n+1} e^{\gamma T} \, dz = B. \tag{10.125}$$

Thermal runaway is associated with multiple steady states of (10.121), in which case we wish to solve

$$
\begin{aligned}
0 &= \alpha(s - z)^{n+1} e^{\gamma T} + \beta T_{zz}, \\
T &= -1 \quad \text{on } z = s, \\
T_z &= -\Gamma \quad \text{on } z = 0, \\
T_z &= -\left[\Gamma + \frac{\alpha B}{\beta}\right] \quad \text{on } z = s.
\end{aligned}
\tag{10.126}
$$

Putting $\xi = s - z$, we solve

$$
\begin{aligned}
T_{\xi\xi} &= -\mathcal{A}\xi^{n+1} e^{\gamma T}, \\
T - 1, \qquad T_\xi &= \Gamma + \mathcal{A}B \quad \text{on } \xi = 0,
\end{aligned}
\tag{10.127}
$$

where now

$$\mathcal{A} = \frac{\alpha}{\beta}, \tag{10.128}$$

as an initial value problem. T_ξ is monotone decreasing with increasing ξ, and thus there is a unique value of s such that $T_\xi = \Gamma$ there. It follows that there is a unique solution to the free boundary problem, and in fact it is linearly stable. It then seems that thermal runaway is unlikely to occur in practice.

A slightly different perspective may allow runaway, if we admit non-steady ice fluxes. Formally, we can derive a suitable model if $\mathcal{A} = O(1)$, $\beta \to \infty$. In this case, we can expect T to tend rapidly to equilibrium of (10.121), and then s reacts more slowly via mass conservation, thus

$$
\begin{aligned}
s_t + q_x &= a, \\
q &= \frac{1}{\mathcal{A}}[T_z]_0^s.
\end{aligned}
\tag{10.129}
$$

An x-independent version of (10.129), consistent with the previous discussion, is

$$\frac{\partial s}{\partial t} = B - q(s), \tag{10.130}$$

and this *will* allow relaxation oscillations if $q(s)$ is multivalued as a function of s—which will be the case. Surging in this sense is conceivable, but the limit $\beta \to \infty$ is clearly unrealistic, and unlikely to be attained. The earlier conclusion is the more likely.

It is also possible to study thermal runaway using the more realistic approach involving a basal shear layer, as in Sect. 10.2.4, and allowing for the separate thermal boundary conditions in (10.56). Although multiple solutions are possible, they are in reality precluded by the transition from one basal thermal régime to another as the basal ice warms. In the last thermal régime, where the basal ice becomes temperate, the dependence of the flow law on moisture content could also allow a runaway, but one which now would involve excess moisture production. Whether this can occur will depend on whether the resultant drainage to the basal stream system can be carried away subglacially, but this process requires a description of water flow within and below the glacier.

10.2.6 Ice Shelves

When an ice sheet flows to the sea, as mostly occurs in Antarctica, it starts to float at the *grounding line*, and continues to flow outwards as an *ice shelf*. The dynamics of ice shelves can be described by an approximate theory, but this is very different from that appropriate to ice sheets.

We begin with the equations in the form (10.38) and (10.39), as scaled for the ice sheet. These must be supplemented by conditions on the floating base $z = b$. To be specific, we take the level $z = 0$ to be sea level. The water depth at $z = b$ is thus $-b$, and the resulting hydrostatic pressure must balance the normal stress in the ice. In addition, there is no shear stress. The general form of the (vector) stress balance condition at an interface of this type which supports only a pressure p_i is (cf. (10.6))

$$\sigma \cdot \mathbf{n} = -p_i \mathbf{n}, \tag{10.131}$$

and in addition to this there is a kinematic boundary condition. When written in terms of the ice sheet scales, these boundary conditions become

$$-\tau_3 + \varepsilon^2(-p + \tau_1)b_x = (s + \delta b)b_x,$$
$$s = -\delta b - \varepsilon^2[\tau_3 b_x + p + \tau_1], \tag{10.132}$$
$$w = b_t + ub_x - m,$$

in which m is the bottom melting rate, and the parameter δ is given by

$$\delta = \frac{\rho_w - \rho_i}{\rho_i}, \tag{10.133}$$

where ρ_i and ρ_w are ice and water densities. The second of these conditions, the flotation condition, essentially says that 90% of the ice is below the surface, as in Archimedes' principle.

Whereas the dominant force balance in the ice sheet is between shear stress and horizontal pressure gradient, and longitudinal stresses are negligible, this is not true in the ice shelf, where the opposite is true: shear stress is small, and the primary balance is between longitudinal stress and horizontal pressure gradient. Therefore the equations must be rescaled in order to highlight this fact. The issue is complicated by the presence of two small parameters $\delta \sim 0.1$ and $\varepsilon \sim 10^{-3}$.

We suppose that the length scale for the ice shelf is $x \sim \lambda$ (relative to the horizontal ice sheet scale), and that the depth scale is $z \sim \nu$, and we anticipate that

$v \ll 1$. We then find that a suitable balance of terms reflecting the dominance of longitudinal stresses is given by writing

$$x \sim \lambda, \qquad z, b \sim v, \qquad u \sim \frac{1}{v}, \qquad w \sim \lambda,$$

$$p, \tau_1 \sim \frac{\delta v}{\varepsilon^2}, \qquad \tau_3 \sim \frac{\delta v^2}{\lambda}, \qquad \tau \sim \frac{\delta v}{\varepsilon}, \qquad s \sim \delta v. \tag{10.134}$$

The governing equations become

$$u_x + w_z = 0,$$
$$0 = -s_x + \tau_{3z} - p_x + \tau_{1x},$$
$$0 = -p_z - \tau_{1z} + \omega^2 \tau_{3x},$$
$$\left(u_z + \omega^2 w_x\right) = \omega^2 \tau^{n-1} \tau_3, \tag{10.135}$$
$$2u_x = \tau^{n-1} \tau_1,$$
$$\tau^2 = \omega^2 \tau_3^2 + \tau_1^2,$$

and the appropriate boundary conditions are, on the top surface $z = \delta s$:

$$\tau_3 + \delta(p - \tau_1)s_x = 0,$$
$$p + \tau_1 + \delta \omega^2 \tau_3 s_x = 0, \tag{10.136}$$
$$w = \lambda \delta v s_t + \delta u s_x - \lambda a;$$

and on the base $z = b$:

$$\tau_3 + (p - \tau_1)b_x = (s + b)b_x,$$
$$s + b = -\left(p + \tau_1 + \omega^2 \tau_3 b_x\right), \tag{10.137}$$
$$w = v\lambda b_t + u b_x - \lambda m;$$

in these equations,

$$\omega = \frac{v\varepsilon}{\lambda} \ll 1. \tag{10.138}$$

The length scale is as yet essentially arbitrary; observations suggest $\lambda \lesssim 1$. The parameter v is defined by the constraint that longitudinal stress balances longitudinal strain rate, and this determines

$$v = \frac{\varepsilon}{\delta}\left(\frac{\delta}{\lambda A}\right)^{1/(n+1)}, \tag{10.139}$$

where, if A varies with temperature, it is the ice upper surface (lowest) value that should be used.[11]

We let $\omega \to 0$ in these equations; it follows that $u \approx u(x, t)$, $\tau \approx |\tau_1|$, whence $\tau_1 \approx \tau_1(x, t)$; $p + \tau_1 \approx 0$, so that $\tau_{3z} \approx s_x - 2\tau_{1x}$, and thus

$$\tau_3 \approx (s_x - 2\tau_{1x})(z - \delta s) + 2\delta \tau_1 s_x. \tag{10.140}$$

---

[11]This is opposite to the situation in an ice sheet, where it is the warmest (basal) ice which is rate-controlling.

Applying the boundary conditions at $z = b$, we have

$$s = -b, \qquad 2\tau_1 b_x = (s_x - 2\tau_{1x})(b - \delta s) + 2\delta\tau_1 s_x, \tag{10.141}$$

whence, integrating, we find

$$\tau_1 = -\frac{1}{4}b, \tag{10.142}$$

and the integration constant (for $(10.141)_2$) is taken to be zero on applying an averaged force balance at the ice shelf front (see Question 10.7). Thus we finally obtain the stretching equations, noting that the ice thickness $H \approx -b$ to $O(\delta)$,

$$u_x = \frac{1}{2}\left(\frac{1}{4}H\right)^n, \qquad \nu\lambda H_t + (uH)_x = \lambda(a - m); \tag{10.143}$$

the second equation is that of mass conservation, and is derived by integrating the mass continuity equation. Note that the time scale for mass adjustment is $O(\nu\lambda) \ll 1$, so that the ice shelf responds rapidly to changes in supply. We might suppose that the choice of length scale λ would be such that $\lambda(a - m) = O(1)$, but in fact it is more likely that the extent of an ice shelf is determined by the rate of calving at the front, which is not treated here. Typical basal melt rates are comparable to accumulation rates, of the order of ten centimetres a year in some models.

Suitable initial conditions for H and u would follow from continuity of ice flux and depth across the grounding line, but the position of the grounding line $x = x_G$ is not apparently determined. Let us anticipate that suitable conditions on H and u are that

$$u \to 0, \qquad Hu \to q_I \quad \text{as } x \to x_G; \tag{10.144}$$

assuming steady conditions, it follows from (10.143) that

$$Hu = q_I + \int_{x_G}^{x} (a - m)\,dx. \tag{10.145}$$

The solution for u follows by quadrature. In the particular case that $a = m$ (and in any case as $x \to x_G$), we have $Hu = q_I$, and thus

$$u = \left\{\frac{(n+1)}{2}\left(\frac{q_I}{4}\right)^n\right\}^{1/(n+1)} (x - x_G)^{1/(n+1)}. \tag{10.146}$$

In order to find a condition for q_I and for the position of x_G, we need to consider the region near the grounding line in more detail, and this is done in the following subsection.

10.2.7 The Grounding Line

In the transition region, we need to retain terms which are of importance in both ice sheet and ice shelf approximations. This requires us to rescale the ice sheet scaled variables in the following way:

$$x - x_G \sim \gamma, \qquad z, b \sim \beta, \qquad s \sim \delta\beta, \qquad u \sim \frac{1}{\beta}, \qquad w \sim \frac{1}{\gamma},$$

$$t \sim \beta, \qquad p, \tau_1 \sim \frac{\delta\beta}{\varepsilon^2}, \qquad \tau_3 \sim \frac{\delta\beta^2}{\gamma}, \qquad \tau \sim \frac{\delta\beta}{\varepsilon}, \tag{10.147}$$

where x_G is the grounding line position; the parameters β and γ are defined by

$$\beta = \left(\frac{\varepsilon}{\delta}\right)^{\frac{n}{n+2}} \frac{1}{A^{\frac{1}{n+2}}}, \qquad \gamma = \beta\varepsilon. \tag{10.148}$$

This rescaling reintroduces the full Stokes equations. Denoting the rescaled variables (except time) by capitals, and writing

$$x - x_G(t) = \gamma X, \qquad t = \beta t^*, \tag{10.149}$$

we derive the model

$$U_X + W_Z = 0,$$
$$0 = -S_X + T_{3Z} - P_X + T_{1X},$$
$$0 = -P_Z - T_{1Z} + T_{3X},$$
$$U_Z + W_X = T^{n-1} T_3, \tag{10.150}$$
$$2U_X = T^{n-1} T_1,$$
$$T^2 = T_3^2 + T_1^2.$$

The boundary conditions are the following. On the surface $Z = \delta S$,

$$T_3 + \delta(P - T_1)S_X = 0,$$
$$P + T_1 + \delta T_3 S_X = 0, \tag{10.151}$$
$$W = \delta(\gamma S_{t^*} - \dot{x}_G S_X) + \delta U S_X - \gamma a,$$

where $\dot{x}_G = \frac{dx_G}{dt^*}$. On the base $Z = B$, when $X > 0$,

$$-T_3 + (-P + T_1)B_X = (S + B)B_X,$$
$$S + B = -[P + T_1 + T_3 B_X], \tag{10.152}$$
$$W = \gamma B_{t^*} - \dot{x}_G B_X + U B_X - \gamma m,$$

and when $X < 0$,

$$W = 0, \qquad U = 0, \tag{10.153}$$

where we assume that the sliding velocity is zero for grounded ice.[12]

---

[12]This simple assumption is not very realistic, since it is most likely that in the vicinity of the grounding line, the basal ice will be at the melting point, and the sliding velocity will be non-zero. Where ice streams go afloat, the velocity is almost entirely due to basal sliding.

To leading order, we can approximate the top surface boundary conditions as $\gamma \to 0$ and also $\delta \to 0$ by

$$T_3 = P + T_1 = W = 0 \quad \text{on } Z = 0. \tag{10.154}$$

The kinematic condition at the shelf base is approximately

$$W = -\dot{x}_G B_X + U B_X \quad \text{on } Z = B. \tag{10.155}$$

In addition, the solution must be matched to the outer (sheet and shelf) solutions. We consider first the ice sheet behaviour as $x \to x_G$. We suppose that the ice sheet is described in one dimension by (10.45), thus

$$H_t = -q_x + a, \tag{10.156}$$

where the ice flux is (in ice sheet scaled variables)

$$q = \frac{(s - b)^{n+2}(-s_x)^n}{n + 2}. \tag{10.157}$$

We can carry out a local analysis near x_G similar to those in Sect. 10.2.5. As $x \to x_G$, $s - b \to 0$ (since $s - b \sim \beta \ll 1$), but the ice flux is non-zero; in this case we find that always

$$H = s - b \sim C(x_G - x)^{\frac{n}{2(n+1)}}, \qquad q \sim q_G = \left(\frac{n}{2(n+1)}\right)^n \frac{C^{2n+2}}{n+2}. \tag{10.158}$$

When the surface slope is computed from this, we find that the requisite matching condition for the slope written in terms of the transition zone scalings is that

$$S_X \sim -\frac{nC}{2(n+1)} \frac{1}{(-\delta X)^{\frac{n+2}{2(n+1)}}} \quad \text{as } X \to -\infty. \tag{10.159}$$

Clearly the presence of the small parameter δ does not allow direct matching of the transition zone to the ice sheet.

The problem is easily resolved, however. There is a 'joining' region in which $X = \tilde{X}/\delta$, $S = \tilde{S}/\delta$, and then also $P, T_1, W \sim \delta$; the resultant set of equations is easily solved (it is a shear layer like the ice sheet), and we find

$$\tilde{S} = B_G + \left[(-B_G)^{2(n+1)/n} - \frac{2(n+1)}{n} \{(n+2)q_G\}^{1/n} \tilde{X} \right]^{n/(2n+2)}, \tag{10.160}$$

where $B = B_G$ at $x = x_G$. Expanding this as $\tilde{X} \to 0$, we find that the matching condition for S in the transition zone as $X \to -\infty$ is

$$S \sim -\Lambda X, \tag{10.161}$$

where

$$\Lambda = \frac{\{(n+2)q_G\}^{1/n}}{(-B_G)^{(n+2)/n}}. \tag{10.162}$$

A final simplification to the transition zone problem results from defining

$$\Pi = P + S; \tag{10.163}$$

to leading order in γ and δ, the transition problem is then

$$U_X + W_Z = 0,$$
$$\Pi_X = T_{3Z} + T_{1X},$$
$$\Pi_Z = -T_{1Z} + T_{3X},$$
$$U_Z + W_X = T^{n-1}T_3,$$
$$2U_X = T^{n-1}T_1,$$
$$T^2 = T_3^2 + T_1^2,$$

(10.164)

together with the boundary conditions

$$T_3 = W = 0 \quad \text{on } Z = 0,$$

(10.165)

$$B = -(\Pi + T_1 + T_3 B_X),$$
$$T_3(1 - B_X^2) = 2T_1 B_X,$$

(10.166)

$$W = (-\dot{x}_G + U)B_X \quad \text{on } Z = B, \ X > 0,$$

and

$$W = U = 0 \quad \text{on } Z = B_G, \ X < 0.$$

(10.167)

The matching conditions to the ice sheet may be summarised as

$$\Pi_X \to -\Lambda, \qquad W \to 0, \qquad T_3 \to -\Lambda Z \quad \text{as } X \to -\infty,$$

(10.168)

with the flow becoming the resultant pressure gradient driven shear flow at $-\infty$. Towards the ice shelf, a comparison of orders of magnitude shows firstly that

$$\frac{\nu}{\beta} = \left(\frac{\gamma}{\lambda}\right)^{1/(n+1)} \ll 1,$$

(10.169)

and that in the ice shelf, the transition scaled variables are

$$S, B \sim \frac{\nu}{\beta}, \qquad W \sim \frac{\gamma}{\lambda}, \qquad P, T_1 \sim \frac{\nu}{\beta}, \qquad T_3 \sim \left(\frac{\nu}{\beta}\right)^2 \frac{\gamma}{\lambda};$$

(10.170)

note also that the ice shelf time scale $\nu\lambda$ is much less than the transition zone time scale β, so that it is appropriate in the transition zone to assume that the far field ice shelf is at equilibrium, and thus described by (10.145) and (10.146). Bearing in mind (10.169), it follows from this that suitable matching conditions for the transition region are

$$T_1 \sim -\frac{1}{4}B, \qquad U \sim MX^{1/(n+1)},$$
$$W \to 0, \qquad B \sim -\frac{q_I}{U} \quad \text{as } X \to \infty,$$

(10.171)

where

$$M = \left\{\frac{(n+1)}{2}\left(\frac{q_I}{4}\right)^n\right\}^{1/(n+1)},$$

(10.172)

and the flow becomes an extensional flow as $X \to \infty$. It follows from integration of the continuity equation between B and S that the ice flux to the ice shelf, q_I, is given by

$$q_I = q_G + \dot{x}_G B_G. \tag{10.173}$$

The top surface is defined by

$$S = (\Pi + T_1)|_{Z=0}, \tag{10.174}$$

and uncouples from the rest of the problem. The extra condition on $Z = B$, $X > 0$ in (10.166) should determine B providing $\dot{x}_G$ is known. This is the basic conundrum of the grounding line determination, since there appears to be no extra condition to determine $\dot{x}_G$.

The resolution of this difficulty has not yet been finally achieved. One might wonder whether there is an extra condition hiding in the matching conditions (10.168) or (10.171), but it appears not: the conditions on T_3 and W as $X \to -\infty$ imply the pressure gradient condition, while the condition on U as $X \to \infty$ implies the other three. It seems that the answer lies in the additional posing of contact conditions. Specifically, for the solution in the transition region to have physical sense, we require that the effective normal stress downwards, $-\sigma_{33} - p_w$, be positive on the grounded base, and we require the ice/water interface to be above the submarine land surface on the floating shelf base. When written in the current scaled coordinates, these conditions become

$$\begin{aligned} B + \Pi + T_1 &> 0, & X &< 0, \\ B &> B_G, & X &> 0. \end{aligned} \tag{10.175}$$

In addition, we may add to these the condition that at the grounding line, the effective normal stress is zero, whence

$$B + \Pi + T_1 = 0 \quad \text{at } X = 0. \tag{10.176}$$

Numerical solutions appear to be consistent with the idea that, for any given $\dot{x}_G$, there is a unique value of Λ such that the contact conditions (10.175) and (10.176) are satisfied. If this is true, then (10.162) determines the ice sheet flux q_G at the grounding line as a function of x_G (through B_G) and $\dot{x}_G$, and this provides the extra condition (as well as $s \to b$ as $x \to x_G-$) for the determination of the grounding line position.

10.2.8 Marine Ice Sheet Instability

Much of the interest concerning grounding line motion concerns the possible instability of marine ice sheets. A marine ice sheet is one whose base is below sea level; the major example in the present day is the West Antarctic Ice Sheet. Marine ice sheets terminate at grounding lines, from which ice shelves protrude. Depending on the slope of the submarine surface, they can be susceptible to instability, and it

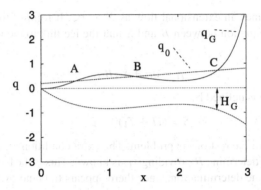

Fig. 10.8 Variation of $q_G[H_G(x)]$ and the equilibrium flux $q_0 = ax$ for the bottom depth profile H_G indicated. Equilibria occur for the points of intersection of the two flux curves, with instability occurring if $q_0' > q_G'$. Thus points A and C are stable, while B is unstable. The particular functions used are $H_G = 2x - \frac{3}{2}x^2 - \frac{1}{3}x^3$, $q_G = H^3$, and $q_0 = 0.2(1+x)$ (with the divide implicitly being at $x = -1$)

has been postulated that fluctuations in sea level, for example, might cause a catastrophic retreat of the grounding lines in West Antarctica, and consequent collapse of the ice sheet.

To understand why this might be so, consider an ice sheet governed by the mass conservation equation (10.156), and for simplicity (it does not affect the argument), take the ice flux $q = -H_x$, so that

$$H_t = H_{xx} + a, \tag{10.177}$$

with boundary conditions

$$H_x = 0 \quad \text{at } x = 0,$$
$$H = H_G(x_G), \qquad -H_x = q_G(H) \quad \text{at } x = x_G \tag{10.178}$$

(note that we retain here the finite depth of the ice sheet at the grounding line). $H_G(x)$ represents the depth of the land subsurface below sea level, and we assume that q_G is an increasing function of H, as suggested by (10.162), if Λ is constant. There is a steady solution $H = H_0(x)$; note that since the ice sheet slopes down to the ice shelf, we have $H_0'(x_G) < H_G'(x_G)$.

Consider a situation such as that shown in Fig. 10.8, in which the subsurface slopes upwards for part of the domain. In this case there can be three possible equilibria, of which the middle one is unstable. The casual argument for this is the following suggestion: if x_G advances, then the ice sheet must deliver a larger flux q_0; however, assuming $q_G'(H) > 0$, then in regions where $H_G'(x) < 0$ (i.e., the bed slopes upwards towards the grounding line), the actual flux delivered is less; consequently, the ice builds up behind the grounding line, causing its further advance.

To demonstrate this mathematically, we linearise (10.177) and (10.178) about the steady state $H = H_0(x)$, $x_G = x_G^0$, by putting $H = H_0(x) + \eta$, $x_G = x_G^0 + \gamma$; the resulting linearised system for η is (eliminating γ)

$$\eta_t = \eta_{xx},$$
$$\eta_x = 0 \quad \text{at } x = 0, \tag{10.179}$$
$$-\eta_x = K\eta \quad \text{at } x = x_G^0,$$

where

$$K = \frac{q_G'(H_G)H_G'(x_G^0) - a}{H_G'(x_G^0) - H_0'(x_G^0)}, \tag{10.180}$$

which has (stable) solutions $\eta = e^{-\lambda^2 t}\cos\lambda x$ providing $\lambda\tan\lambda x_G^0 = K$. If $K > 0$, these are the only solutions, and the steady state is stable. However, if $K < 0$, the first mode (with $\lambda x_G^0 < \frac{1}{2}\pi$) is replaced by an unstable mode $\eta = e^{\lambda^2 t}\cosh\lambda x$, where $\lambda\tanh\lambda x_G^0 = -K$. Therefore the steady state is unstable precisely if $K < 0$. Consulting (10.180), and recalling that $H_0'(x_G) < H_G'(x_G)$, it follows that the steady state is unstable if

$$\frac{dq_G}{dx_G} < a, \tag{10.181}$$

as suggested in Fig. 10.8. Question 10.8 generalises this result to the case where $q = -D(H, H_x)H_x$.

10.3 Sliding and Drainage

The sliding law relates the basal shear stress τ_b to the basal sliding velocity u_b. The classical theory, enunciated by Lliboutry, Weertman, Nye, Kamb, and others, considers ice flowing at the base of a glacier over an irregular, bumpy bedrock. The ice is lubricated at the actual interface by the mechanism of *regelation*, or melting-refreezing, which allows a thin film (microns thick) to exist at the ice-rock interface, and allows the ice to slip. The drag is then due to two processes; regelation itself, and the viscous flow of the ice over the bedrock. Regelation is dominant for small wavelength roughness, while viscous drag is dominant for large wavelengths, and early work emphasised the importance of a controlling (intermediate) wavelength (of several centimetres). More recently, the emphasis has moved away from regelation and has been put more on consideration of the viscous flow, and we do this here, assuming no normal velocity of the ice as it slides over the bed.

A suitable model for discussion is the flow of a Newtonian fluid over a rough bedrock of 'wavelength' $[x]$ and amplitude $[y]$, given by

$$y = h_D(x) \equiv [y]h\left(\frac{x}{[x]}\right), \tag{10.182}$$

where y is now the vertical coordinate.[13] The governing equations for two-dimensional flow down a slope of angle α are

$$u_x + v_y = 0,$$
$$p_x = \rho_i g \sin\alpha + \eta\nabla^2 u, \qquad (10.183)$$
$$p_y = -\rho_i g \cos\alpha + \eta\nabla^2 v,$$

where η is the viscosity. We suppose that the glacier has a depth of order d, thus providing a basal shear stress τ_b of order $[\tau]$, which drives a shear velocity of order $[u]$, and these are related by

$$[\tau] = \rho_i g d \sin\alpha = \frac{\eta[u]}{d}. \qquad (10.184)$$

The basal boundary conditions are those of no shear stress and no normal flow, and take the form

$$\sigma_{nt} = \frac{\tau_2(1 - h_D'^2) - 2\tau_1 h_D'}{1 + h_D'^2} = 0, \qquad (10.185)$$

where

$$\tau_1 = 2\eta u_x, \qquad \tau_2 = \eta(u_y + v_x), \qquad (10.186)$$

and

$$v = uh_D', \qquad (10.187)$$

both (10.185) and (10.187) being applied at $y = h_D(x)$. Note also that the normal stress is

$$-\sigma_{nn} = \frac{p(1 + h_D'^2) + \tau_1(1 - h_D'^2) + 2h_D'\tau_2}{1 + h_D'^2}. \qquad (10.188)$$

Because we describe a local flow near the base of the glacier, it is appropriate to apply matching conditions to the ice flow above. In particular, we require $\tau_2 \to \tau_b$ as y becomes large, and hence

$$u \sim u_b + \frac{\tau_b y}{\eta} \qquad (10.189)$$

far from the bed.

We non-dimensionalise the equations by scaling

$$x, y \sim [x], \qquad u, v \sim [u], \qquad \tau_b = [\tau]\tau^*, \qquad u_b = [u]u^*,$$
$$p = p_i + \frac{v[\tau]}{\sigma}P, \qquad (10.190)$$

where

$$p_i = p_a + \rho_i g(y_i - y)\cos\alpha \qquad (10.191)$$

---

[13]Because shortly we will use z for the complex variable $x + iy$.

is the ice overburden pressure, and $y = y_i \sim d$ is the ice upper surface (and taken as locally constant); this leads to the non-dimensional set

$$u_x + v_y = 0,$$
$$v P_x = \sigma^2 + \nabla^2 u, \qquad (10.192)$$
$$v P_y = \nabla^2 v,$$

subject to the boundary conditions that

$$P \to 0, \qquad u \sim u^* + \sigma \tau^* y \quad \text{as } y \to \infty, \qquad (10.193)$$

and

$$\left(1 - v^2 h'^2\right)(u_y + v_x) - 4 v h' u_x = 0, \qquad (10.194)$$
$$v = v u h',$$

on the dimensionless bed $y = vh$ (from (10.182)). The corrugation σ and the aspect ratio v are defined by

$$\sigma = \frac{[x]}{d}, \qquad v = \frac{[y]}{[x]}; \qquad (10.195)$$

v is a measure of the roughness of the bed.

We will assume that both v and σ are small. In consequence, the dimensionless basal stress τ^* in (10.193) is uncoupled from the problem; however, integration of the momentum equations over the domain yields an expression for τ^*. In dimensional terms, this relation is

$$\tau_b = \frac{1}{L} \int_0^L \sigma_{n1} \, ds, \qquad (10.196)$$

where the integral is over a length L of the base $y = h_D$, over which conditions are taken to be periodic (alternatively, the limit $L \to \infty$ may be taken).

Evidently, the velocity is uniform to leading order, and therefore we write

$$u = u^* + vU, \qquad v = vV, \qquad (10.197)$$

so that the problem reduces to

$$U_x + V_y = 0,$$
$$P_x = \frac{\sigma^2}{v} + \nabla^2 U, \qquad (10.198)$$
$$P_y = \nabla^2 V,$$

with boundary conditions

$$U \sim \frac{\sigma \tau^*}{v} y \quad \text{as } y \to \infty, \qquad (10.199)$$

and

$$\left(1 - v^2 h'^2\right)(U_y + V_x) - 4 v h' U_x = 0, \qquad (10.200)$$
$$V = (u^* + vU)h'$$

on $y = vh$. When written in dimensionless terms, the overall force balance (10.196) takes the form (with L now being dimensionless)

$$\frac{\sigma \tau^*}{v^2} = \frac{1}{L} \int_0^L \left[P(1 + v^2 h'^2) + 2(1 - v^2 h'^2) U_x + 2vh'(U_y + V_x) \right] \frac{h' \, dx}{1 + v^2 h'^2}.$$
(10.201)

It is fairly clear from (10.201) that there is a distinguished limit $\sigma \sim v^2$, which corresponds to the situation where sliding is comparable to shearing, and it is convenient to adopt this limit as an example. We introduce a stream function ψ via

$$U = \psi_y, \qquad V = -\psi_x;$$
(10.202)

then letting $v \to 0$ with $\sigma \sim v^2$, we derive the reduced model

$$P_x = \nabla^2 \psi_y,$$
$$P_y = -\nabla^2 \psi_x,$$
(10.203)

together with the boundary conditions

$$P, \psi \to 0 \quad \text{as } y \to \infty,$$
$$\psi = -u^* h(x), \qquad \psi_{yy} - \psi_{xx} = 0 \quad \text{on } y = 0.$$
(10.204)

The shear stress is determined by (10.201), whence to leading order (e.g., if h is periodic with period 2π)

$$\frac{\sigma \tau^*}{v^2} = \frac{1}{2\pi} \int_0^{2\pi} (P + 2\psi_{xy})|_{y=0} h' \, dx;$$
(10.205)

more generally a spatial average would be used. Since the expression in brackets in (10.205) is simply (minus) the normal stress, it is therefore also equal to the scaled water pressure in the lubricating film, which from (10.190) can be written in the form

$$P + 2\psi_{xy} = -N^*, \qquad N^* = \frac{\sigma(p_i - p_w)}{v[\tau]}.$$
(10.206)

The quantity N^* is the dimensionless effective pressure at the bed. We come back to this below.

A nice way to solve this problem is via complex variable theory. We define the complex variable $z = x + iy$, and note that Eqs. (10.203) are the Cauchy–Riemann equations for the analytic function $P + i\nabla^2 \psi$. Consequently, ψ satisfies the biharmonic equation, which has the general solution

$$\psi = (\bar{z} - z) f(z) - B(z) + (cc),$$
(10.207)

where f and B are analytic functions and (cc) denotes the complex conjugate, as does the overbar. The zero stress condition (10.204) requires $f = -\frac{1}{2} B'$, and also $B \to 0$ as $z \to \infty$ (with $\operatorname{Im} z > 0$), and the last condition is then

$$B + \bar{B} = u^* h \quad \text{on } \operatorname{Im} z = 0.$$
(10.208)

If h is periodic, with a Fourier series

$$h = \sum_{-\infty}^{\infty} a_k e^{ikx}, \tag{10.209}$$

then B is simply given by

$$B = u^* \sum_{1}^{\infty} a_k e^{ikz} \tag{10.210}$$

(we can assume $a_0 = 0$, i.e., the mean of h is zero). However, it is also convenient to formulate this problem as a Hilbert problem. We define $L(z) = B''(z)$, which is analytic in $\operatorname{Im} z > 0$, and then $L(z) = \overline{B''(\bar{z})}$ is analytic in $\operatorname{Im} z < 0$. From (10.207), $\nabla^2 \psi = 4\psi_{z\bar{z}} = -2(L + \bar{L})$, and therefore $P + i\nabla^2 \psi + 4iL = P + 2i(L - \bar{L})$ is analytic; since this last expression is real, it is constant and thus zero, since it tends to zero as $z \to \infty$. Applying the boundary conditions at $\operatorname{Im} z = 0$, and using the usual notation for the values on either side of the real axis, it follows that

$$L_+ + L_- = u^* h'',$$
$$L_+ - L_- = \frac{1}{2} i P, \tag{10.211}$$

which relate the values either side of $\operatorname{Im} z = 0$. From (10.207), we have $\psi_{xy} = i(\psi_{zz} - \psi_{\bar{z}\bar{z}}) = \frac{1}{2} i (z - \bar{z})(B''' - \overline{B'''})$, and thus $\psi_{xy}|_{y=0} = 0$; it follows that $P = -N^*$ on $y = 0$, and the drag (i.e., the sliding law) is then computed (for a 2π-periodic h) as

$$\frac{\sigma \tau^*}{v^2} = \frac{1}{i\pi} \int_0^{2\pi} (L_+ - L_-) h' \, dx; \tag{10.212}$$

evaluating the integral, we find

$$\frac{\sigma \tau^*}{v^2} = 4u^* \sum_{1}^{\infty} k^3 |a_k|^2. \tag{10.213}$$

For a linear model such as this, τ^* and thus τ_b is necessarily proportional to u^* and thus u_b. For Glen's flow law, the slip coefficient multiplying τ^* becomes $\frac{\sigma}{v^{n+1}}$. The problem cannot be solved exactly, but variational principles can be used to estimate a sliding law of the form

$$\tau_b \approx R u_b^{1/n}. \tag{10.214}$$

Weertman's original sliding law drew a balance between (10.214) and the linear dependence due to regelation, and the heuristic 'Weertman's law' $\tau_b \propto u_b^{1/m}$, with $m \approx \frac{1}{2}(n + 1)$ is often used.

Simplistic sliding laws such as the above have been superseded by the inclusion of cavitation. When the film pressure behind a bump decreases to a value lower than the water pressure in the local subglacial drainage system, a cavity must form,

Fig. 10.9 Stress versus
velocity for a bed of isolated
bumps. The *inset* shows the
typical form of the separated
flow on the decreasing
portion of the curve, when the
cavities reach the next bump

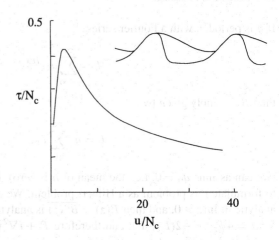

and indeed, such cavities are plentifully observed. An appropriate generalisation of
(10.211) is then

$$L_+ + L_- = u^* h'' \quad \text{in } C',$$
$$L_+ - L_- = -\frac{1}{2} i N_c \quad \text{in } C, \tag{10.215}$$

where the bed is divided into cavities (C) where P is known $(= -N_c)$, and attached
regions where h is known. One can solve this problem to find the unknown cavity
shapes, and for a bed consisting of isolated bumps, $\tau_b(u_b)$ increases monotonically
for small u_b, reaches a maximum, and then decreases for large u_b, as shown in
Fig. 10.9. The decreasing portion of the curve is unstable (increasing velocity decreases drag) and is caused by the roofs of the cavities from one bump reaching the
next bump.

From (10.206) it follows that N_c in (10.215) is proportional to the effective pressure $N = p_i - p_w$, specifically

$$N_c = \frac{\sigma N}{v[\tau]}, \tag{10.216}$$

and in fact the sliding law has the specific form $\tau_b = N f(\frac{u_b}{N})$. For a nonlinear Glen's
law, the generalisation must take the form

$$\tau_b = N f\left(\frac{u_b}{N^n}\right). \tag{10.217}$$

The reason for this is that one can scale the problem in the nonlinear case using
$p - p_i, \tau_{ij} \sim N$, $\mathbf{u} \sim A[x]N^n$, $\mathbf{x} \sim [x]$, and the consequent sliding law must be of
the form (10.217) (assuming the regelative component is small). In particular, note
that the fraction s of uncavitated bed must be a decreasing function of $\Lambda = \frac{u_b}{N^n}$.

The multivaluedness of $u_b(\tau_b)$ is very suggestive of surging—but is it realistic?
Consideration of more realistic (non-periodic) beds suggests that the multivaluedness remains so long as the peak roughness amplitude is relatively constant. However, if there are increasing large bumps—pinning points, riegels—one might expect that $f(\cdot)$ in (10.217) will be an increasing function of its argument, since when

smaller bumps start to be drowned, larger ones will take up the slack. A plausible sliding law then has $f(\xi)$ increasing as a power of ξ, whence we can obtain (for example)

$$\tau_b = c u_b^r N^s, \tag{10.218}$$

where we would expect $r, s > 0$. More specifically, (10.217) would suggest $s = 1 - rn$, and also that $r \approx \frac{1}{n}$ would be appropriate at low u_b, where cavitation is absent. When cavitation occurs, one would then expect lower stresses, so that $r < \frac{1}{n}$. There is in fact some experimental and field evidence consistent with laws of this type, with $r \approx s \approx \frac{1}{3}$, for example. More detailed theoretical studies suggest that $f(\Lambda)$ will eventually reach a maximum which is determined by the largest wavelength bumps.

An apparently altogether different situation occurs when ice slides over wet, deforming till. If the till is of thickness d_T and has (effective) viscosity η_T, then an appropriate sliding law would be

$$\tau_b = \frac{\eta_T u_b}{d_T}. \tag{10.219}$$

In fact, till is likely to have a nonlinear rheology, and also in accordance with Terzaghi's principle of soil mechanics, one would expect η_T to depend on effective pressure N. One possible rheology for till[14] gives the strain rate as

$$\dot{\varepsilon} = A_T \frac{\tau^a}{N^b}, \tag{10.220}$$

in which case the sliding law would be again of the form (10.218), with $c = (A_T d_T)^{-1/a}$, $r = 1/a$, $s = b/a$. If the till is taken to be plastic, then we would have $r = 0$, $s = 1$, corresponding to (10.220) when $a = b \gg 1$. Thus there are some good reasons to choose (10.218) as an all purpose sliding law, and this points up the necessity of a subglacial hydraulic theory to determine N.

10.3.1 Röthlisberger Channels

Subglacial water is generated both by basal melt (of significance in ice sheets) and from run-off of surface melt or rainfall through crevasses and *moulins*, which access the glacier bed. Generally the basal water pressure p_w is measured to be below the overburden ice pressure p_i, and the resulting positive effective pressure $N = p_i - p_w$ tends to cause any channels in the ice to close up (by creep of the ice). In fact, water is often seen to emerge from outlet streams which flow through large tunnels in the ice, and the theory which is thought to explain how such channels

---

[14]The choice of a suitable till rheology is problematic, since till is a granular material, and therefore has plastic behaviour, i.e., a yield stress. It is a matter of current interest whether any kind of viscous rheology is actually appropriate. Further discussion is given in the notes.

remain open asserts that the channel closure rate is balanced by melt back of the channel walls by frictional heating due to the water flow.

The classical theory of subglacial drainage is due to Röthlisberger, and is described below. Much more detail, including the effects of time dependence in the model, is provided in Chap. 11. Here we discuss only the determination of effective pressure in steady state conditions. We consider a single channel of cross-sectional area S, through which there is a water flux Q. We take Q as being determined by external factors such as surface meltwater runoff; this is appropriate for glaciers, but not for ice sheets, where Q must be determined by subglacial melting (we come back to this later). If the flow is turbulent, then the Manning law for flow in a straight conduit is

$$\rho_w g \sin\alpha - \frac{\partial p}{\partial s} = \frac{f_1 Q^2}{S^{8/3}}, \tag{10.221}$$

where ρ_w is water density, g is gravity, s is distance down channel, α is the local bed slope, and we write $p = p_w$ for water pressure; f_1 is a roughness coefficient related to the Manning friction factor.[15] If we suppose that the frictional heat dissipated by the turbulent flow is all used to melt the walls, then

$$mL = Q\left[\rho_w g \sin\alpha - \frac{\partial p}{\partial s}\right], \tag{10.222}$$

where L is the latent heat, and m is the mass of ice melted per unit length per unit time.

The last equation to relate the four variables S, Q, p and m stems from a kinematic boundary condition for the ice, and represents a balance between the rate at which the ice closes down the channel, and the rate at which melting opens it up:

$$\frac{m}{\rho_i} = K S(p_i - p)^n; \tag{10.223}$$

here m/ρ_i is the rate of enlargement due to melt back, while the term on the right hand side represents ice closure due to Glen's flow law for ice; the parameter K is proportional to the flow law parameter A.

Elimination of m and S yields a second order ordinary differential equation for the effective pressure $N = p_i - p$, which can be solved numerically. However, it is also found that typically $\partial p/\partial s \ll \rho_w g \sin\alpha$ (in fact, we expect $\partial p/\partial s \sim \rho_w g d/l$, so that in the notation of (10.11), the ratio of these terms is of $O(\mu)$); the neglect of the $\partial p/\partial s$ term in (10.221) and (10.222) is singular, and causes a boundary layer

---

[15]Retracing our steps to (4.17), we see that $f_1 = \rho_w g n'^2 G$, where the geometrical factor $G = (\frac{l^2}{S})^{2/3} = 6.57$ for a full semi-circular channel; l is the wetted perimeter.

of size $O(\mu)$ to exist near the terminus in order that p decrease to atmospheric pressure.[16] Away from the snout, then

$$S \approx \left[\frac{f_1 Q^2}{\rho_w g \sin \alpha} \right]^{3/8}, \qquad KSN^n \approx \frac{Q \rho_w g \sin \alpha}{\rho_i L}, \qquad (10.224)$$

thus

$$N \approx \beta Q^{1/4n}, \qquad (10.225)$$

where

$$\beta = \left[\frac{\rho_w g \sin^{11/8} \alpha}{\rho_i L K G^{3/8} n'^{3/4}} \right]^{1/n} \qquad (10.226)$$

is a material parameter which depends (inversely) on roughness. Taking $\rho_w = 10^3$ kg m^{-3}, $\rho_i = 917$ kg m^{-3}, $g = 9.81$ m s^{-2}, $L = 3.3 \times 10^5$ J kg^{-1}, $K = 0.5 \times 10^{-24}$ Pa^{-n} s^{-1}, $n = 3$, $\sin \alpha = 0.1$, $G = 6.57$ and $n' = 0.04$ m$^{-1/3}$ s, we find $\beta \approx 24.7$ bar (m^3 s^{-1})$^{-1/12}$, so that $N \approx 30$ bars when $Q = 10$ m^3 s^{-1}. Since $p_i = 9$ bars for a 100 metre deep glacier, it is clear that the computed N may exceed p_i. In this case, p must be atmospheric and there will be open channel flow. It is likely that seasonal variations are important in adjusting the hydraulic régime.

Arterial Drainage

A feature of the Röthlisberger system is the surprising fact that as the water pressure is increased (so N decreases), the water flux *decreases*. This is opposite to our common expectation. A consequence of this is that the channels, like Greta Garbo, want to be alone; if one puts two channels of equal size and equal effective pressures side by side, each carrying a water flux Q, then a perturbation $\Delta Q > 0$ in the flow of one channel will cause an increase in N, and thus a decrease in water pressure, relative to the other channel. Because the bed of a glacier will be leaky, this allows the now smaller channel to drain towards the bigger one, and thus the smaller one will close down. This process, the formation of larger wavelength pattern from smaller scales, is known as coarsening, and occurs commonly in systems such as granular flows, river system development and dendritic crystal growth, and is not fully understood, although the apparent mechanism may be clear (as here).

A consequence of this coarsening is that we expect a channelised system to form a branched, arborescent network, much like a subaerial river system. The difference is that tributaries oblique to the ice flow will tend to be washed away by the ice flow, so that only channels more or less parallel to the ice flow will be permanent features. Presumably, tributary flow will thus be facilitated by the presence of bedrock steps and cavities, which can shield the tributaries from the ice flow.

---

[16] At least, this would be the boundary condition if the channel were full all the way to the margin. In practice, this is not the case. Glacial streams typically emerge from a cavern which is much larger than the stream, and in this case it is appropriate to specify that the channel pressure is atmospheric where the ice pressure is positive. In any case, the Röthlisberger theory makes no sense if $p = p_i$, since then we would have $m = 0$ and thus $Q = 0$.

10.3.2 Linked Cavities

The channelised drainage system described above is not the only possibility. Since water will also collect in cavities, it is possible for drainage to occur entirely by means of the drainage between cavities. A simple way to characterise such a drainage system is via a 'shadowing function' s which is the fraction of the bed which is cavity-free. From our discussion following (10.217), s is a monotonically decreasing function of

$$\Lambda = \frac{u}{N^n}, \tag{10.227}$$

where u is the sliding velocity. If P is the normal ice stress over the cavity-free part of the bed, then a force balance over the bed suggests that

$$p_i = sP + (1 - s)p, \tag{10.228}$$

where p is the water pressure in the cavities and $p_i = N + p$ is the far field ice pressure. We imagine a system of cavities linked by Röthlisberger-type orifices.[17] If there are n_K such cavities across the width of a glacier, then the total water flow Q divides into $\frac{Q}{n_K}$ per channel, subjected to a local effective pressure $P - p = \frac{N}{s}$. Röthlisberger dynamics then dictates that the effective pressure is given by

$$\frac{N}{s(\Lambda)} = \delta N_R, \tag{10.229}$$

where N_R is given by (10.225), Λ by (10.227), and

$$\delta = \left(\frac{1}{n_K}\right)^{1/4n} < 1. \tag{10.230}$$

Linked cavity drainage thus operates at a higher pressure than a channelised drainage system. This very simple description is at best qualitatively true, but it is very powerful (and therefore tempting), as we shall see.

Stability

If there are two different styles of drainage, one may ask which will occur in practice? For the linked cavity system, the answer to this lies in the inverse to our discussion of arterial drainage above. A linked cavity system is an example of a distributed drainage system, and if we denote the corresponding effective pressure by N_K (satisfying (10.229)), then the system will be stable if $N_K'(Q) < 0$. In this case, any local enlargement of an inter-cavity passage will relax stably back to the distributed system.

It is convenient to define $L(\Lambda)$ via

$$L = \ln\left(\frac{1}{s}\right); \tag{10.231}$$

---

[17] The cavities are the veins, and the orifices are the arteries, of the subglacial plumbing system.

L is a monotonically increasing function of Λ, and for illustrative purposes we will take it, for the moment, as linear (this is inessential to the argument). Calculation of N_K' (see Question 10.10) shows that

$$-\frac{N_K'}{N_K}(n\Lambda L' - 1) = \frac{N_R'}{N_R}, \tag{10.232}$$

and thus the linked cavity system is stable if

$$\Lambda > \Lambda_c = \frac{1}{nL'}. \tag{10.233}$$

If $\Lambda < \Lambda_c$, then local perturbations will cause inter-cavity passages to grow, forming channels which will eventually coarsen to result in a single central Röthlisberger artery.

More generally, the effective pressure defined by (10.229) can be written in the form

$$n_K^{1/4n} N \exp\left[L\left(\frac{u}{N^n}\right)\right] = \beta Q^{1/4n}, \tag{10.234}$$

which yields a family of curves (depending on the number n_K of channels) which relate N to water flow Q: see Fig. 10.10. All of these curves have their turning point at $\Lambda = \Lambda_c$. To the left of these minima, $\Lambda > \Lambda_c$ and a distributed system is preferred, with n_K increasing (fine-graining) until limited by the cavity spacing. To the right of the minima, coarsening will occur until a single channel ($n_K = 1$) occurs.

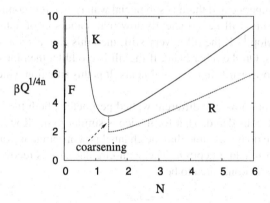

Fig. 10.10 Illustrative form of the drainage effective pressure given by (10.234) relating effective pressure N to water flow Q through a field of linked cavities. The specific functions used in the figure are $L(\Lambda) = \frac{1-e^{-k\Lambda}}{k}$, $\Lambda = \frac{u}{N^n}$, with $k = 0.2$, $u = 1$, $n = 3$. For the curve labelled K (which is the continuation of that labelled F) $n_K = 200$, while for the lower (R) curve $n_K = 1$. The three components of the curve represent F: patchy film flow; K: linked cavities; R: Röthlisberger channel. To the right of the minimum of the curve (here at $N \approx 1.41$), $\Lambda < \Lambda_c$, and the drainage channels coarsen, leading to the single Röthlisberger channel R. To the left of the minimum, distributed drainage is stable, and this takes the form of linked cavities K for $N > 0$. However, if $N \to 0$, we suppose that some of the bed remains in contact with ice, thus the shadowing function remains positive, and this allows a film flow F as the water flux decreases to zero

A question arises, what happens at very low water fluxes, when (10.234) suggests no corresponding value of N if L is linear. In reality, we expect that at very low water fluxes, water will trickle along the bed in a patchy film, while the ice is in effective contact with the larger clasts of the bed. If this is the case, then it indicates that s, and thus also L, should saturate at large Λ. The effect of this is then to cause Q given by (10.234) to reach a maximum at small N, and then decrease sharply to zero. This gives us a third branch, which we associate with film flow, when there is insufficient water flux to develop proper orifice flow between cavities.

In this view of the drainage system, there is really no difference between streams and cavities, or between linked cavities and Röthlisberger channels and patchy films; the only distinction is of one of degree. Intrinsic to our conceptual description is an assumption of bimodality of bed asperity size. The small scale granularity of the bed allows a trickling flow at small Q, while the larger bumps allow cavities and inter-cavity orifices; but while this assumption is a useful imaginative convenience, it is probably inessential.

10.3.3 Canals

A further possible type of drainage is that of a system of canals. This refers to situations where ice flows over a layer of subglacial sediments, which will commonly take the form of till, with its bimodal mixture of fine particles and coarse clasts. If the basal ice is temperate and there is subglacial water, then it is commonly thought that the permeability will be sufficiently low that some sort of subglacial stream system must develop.[18] If the till is very stiff, then this can take the form of Röthlisberger channels. On the other hand, if the till is erodible, then the channels may become incised downwards into the sediments. It is this situation which we now try and describe.

Because there are now two different wetted perimeters, both the ice and the till dynamics must be considered. That for the ice is similar to the Röthlisberger channel, except that we do not assume that the channel is semi-circular. Rather, we identify a mean width w and a depth h. The semi-circular case is recovered if $h \sim w$. We take the cross-sectional area to be

$$S = wh. \tag{10.235}$$

If we assume a Manning flow law in the canal, then (cf. (4.18))

$$\tau = \frac{\rho_w g n'^2 u^2}{R^{1/3}}, \tag{10.236}$$

---

[18]This may not always be necessary; shear and consequent fracture of the till may allow higher permeability pathways, and consequent drainage to bedrock, if there is a basal aquifer. However, this scenario seems unlikely for a deep till layer which only deforms in its uppermost part.

where u is the mean velocity, and

$$R = \frac{S}{l} \qquad (10.237)$$

is the hydraulic radius, with l being the wetted perimeter; we take

$$l \approx 2w, \qquad (10.238)$$

which will serve for both wide and semi-circular channels.

The rate of ice melting is

$$\dot{m}_i = \frac{\tau u w}{L}, \qquad (10.239)$$

while a force balance yields

$$\tau l \approx \rho_i g S S_i, \qquad (10.240)$$

where S_i is the ice surface slope (this ignores the relatively small difference between ice and water densities, and also the gradient of effective pressure). We suppose that melting balances ice closure, so that

$$\dot{m}_i = \rho_i K w^2 N^n, \qquad (10.241)$$

where K is a shape-dependent closure rate coefficient; (10.241) is appropriate for both semi-circular and wide channels. Finally, the water flux is

$$Q = Su. \qquad (10.242)$$

Counting equations, we see that only one further equation is necessary to determine N in terms of Q, and this involves a description of the sediment flow. Eliminating subsidiary variables, we find

$$w^2 = \frac{2^{4/3} \rho_w n'^2 Q^2}{\rho_i S_i h^{10/3}}, \qquad K w^2 N^n = \frac{g S_i Q}{2L}, \qquad (10.243)$$

which can be compared with (10.224). In particular, if we take $w \approx h$, then we regain the Röthlisberger relation (10.225), with

$$\beta = \left[\frac{g S_i}{2KL} \left(\frac{\rho_i S_i}{2^{4/3} \rho_w n'^2} \right)^{3/8} \right]^{1/n}, \qquad (10.244)$$

comparable to (10.226).

Now we consider the appropriate choice of depth for a canal. On the face of it, there is little difference between the combined processes of thermal erosion (melting) and ice creep, and sediment erosion and till creep. But there *is* a difference, and that lies in the rôle played by gravity. The shape of subaerial river channels is mediated by the fact that non-cohesive sediments cannot maintain a slope larger than the angle of repose, and when subject to a shear stress, the maximum slope is much less. Consequently, river beds tend to be relatively flat, and rivers are consequently wide and shallow. Therefore, if subglacial till is erodible, as we expect, the resulting channel will tend to have a depth which is not much greater than that which provides the critical stress for transporting sediment. As an approximation, we might thus take

$$\tau \approx \tau_c = \mu_c \Delta\rho_{sw} g D_s, \tag{10.245}$$

where $\mu_c \approx 0.05$ is the critical Shields stress, and D_s is a representative grain size, probably of the small size particles. In this case, the depth of the canal is given by

$$h \approx h_c = \frac{2\mu_c \Delta\rho_{sw} D_s}{\rho_i S_i}. \tag{10.246}$$

Using values $\mu_c = 0.05$, $\Delta\rho_{sw} = 1.6 \times 10^3 \text{ kg m}^{-3}$, $\rho_i = 0.917 \times 10^3 \text{ kg m}^{-3}$, $D_s \lesssim 10^{-3}$ m, $S_i = 10^{-3}$, we find $h \lesssim 20$ cm.

If we assume that the channel depth is controlled by sediment erosion as in (10.246), then the relations in (10.243) give

$$N = \frac{\gamma}{Q^{1/n}}, \tag{10.247}$$

where

$$\gamma = \left[\frac{\rho_i g S_i^2 h_c^{10/3}}{2^{7/3} K L \rho_w n'^2} \right]^{1/n}. \tag{10.248}$$

Using similar values as before with $h_c = 0.2$ m, we find $\gamma \approx 0.32$ bar $(\text{m}^3 \text{ s}^{-1})^{1/3}$.

There are two important consequences of (10.247). The first is that N decreases with Q, so that, unlike Röthlisberger channels, canals as described here will form a distributed system, just like the linked cavity system; in effect there is little difference other than a semantic one between the two systems. The second consequence is that for any reasonable values of Q, say 0.1–10 $\text{m}^3 \text{ s}^{-1}$, the effective pressure is much less than that of a channelled system, in a typical range 0.1–0.6 bars, similar to that found on the Siple Coast ice streams. The inverse dependence of N on Q also has an important dynamic effect on the ice flow, as we discuss below in Sect. 10.4.3.

An issue of concern in this description is that we have apparently ignored the details of sediment creep and canal bank erosion, despite the apparent similarity to the processes of ice creep and thermal erosion. We will have more to say on this in Sect. 10.5.2, but for the moment we simply observe that in our theoretical description, we have arbitrarily assumed that the ice surface is (relatively) flat. The rough basis for this assumption lies in our expectation that it will be appropriate if the till is much softer than the ice, but in order to quantify this, it is necessary to write down a model which allows description of both the upper ice/water interface and the lower water/sediment interface. The basis for such a model is given in Sect. 10.5.2, when we (briefly) discuss the formation of eskers.

10.3.4 Ice Streams

A modification of the discussion of ice shelves occurs when we consider an appropriate model for ice streams. Ice streams, in particular those on the Siple Coast of Antarctica, are characterised by small surface slopes and high velocities. On ice streams such as the Whillans ice stream B, the depth $d_i \sim 10^3$ m and the ice surface slope is $\sim 10^{-3}$, so that the basal shear stress is ~ 0.1 bar. If we suppose that the effective viscosity $\eta_i = \frac{1}{2A\tau^{n-1}} \approx 6$ bar y, and take the velocity scale as $U \sim 500$ m y^{-1}, then the corresponding shear stress scale $\eta_i U / d_i \sim 3$ bar: evidently motion is largely by sliding. We thus introduce a new dimensionless parameter λ, which is the ratio of the magnitude of the actual basal stress to the shear stress scale:

$$\lambda = \frac{\tau_0 d_i}{\eta_i U}, \tag{10.249}$$

where $\tau_0 = \rho_i g d_i \varepsilon$ is the basal stress scale, ε being the aspect ratio. Using the values quoted above, we may estimate $\lambda \sim 0.03$.

We follow the exposition in Sect. 10.2.2 (and its three-dimensional modification in Question 10.4), with the distinction that the shear stresses are scaled as $\tau_{13}, \tau_{23} \sim \tau_0$, while the longitudinal stresses are scaled as $p - p_a - \rho_i g(s - z)$, $\tau_{12}, \tau_{11}, \tau_{22}, \tau_{33} \sim \frac{\eta_i U}{l}$; then the scaled model (10.38) takes the form

$$u_x + v_y + w_z = 0,$$

$$\tau_{13,z} = s_x + \frac{\varepsilon^2}{\lambda}[p_x - \tau_{11,x} - \tau_{12,y}],$$

$$\tau_{23,z} = s_y + \frac{\varepsilon^2}{\lambda}[p_y - \tau_{12,x} - \tau_{22,y}],$$

$$p_z - \tau_{33,z} = \lambda(\tau_{13,x} + \tau_{23,y}),$$

$$u_z + \varepsilon^2 w_x = \lambda A \tau^{n-1} \tau_{13},$$

$$v_z + \varepsilon^2 w_y = \lambda A \tau^{n-1} \tau_{23}, \tag{10.250}$$

$$u_y + v_x = A \tau^{n-1} \tau_{12},$$

$$2u_x = A \tau^{n-1} \tau_{11},$$

$$2v_y = A \tau^{n-1} \tau_{22},$$

$$2w_z = A \tau^{n-1} \tau_{33},$$

$$\tau^2 = \tau_{13}^2 + \tau_{23}^2 + \frac{\varepsilon^2}{\lambda^2}\left[\frac{1}{2}\tau_{11}^2 + \frac{1}{2}\tau_{22}^2 + \frac{1}{2}\tau_{33}^2 + \tau_{12}^2\right].$$

This allows for a temperature-dependent rate factor A, but we will now suppose $A = 1$. The boundary conditions are, on $z = s$:

$$\tau_{13} = \frac{\varepsilon^2}{\lambda}[(-p + \tau_{11})s_x + \tau_{12}s_y],$$

$$\tau_{23} = \frac{\varepsilon^2}{\lambda}[\tau_{12}s_x + (-p + \tau_{22})s_y], \tag{10.251}$$

$$p - \tau_{33} = \lambda(-\tau_{13}s_x - \tau_{23}s_y),$$

$$w = s_t + us_x + vs_y - a,$$

while at the base $z = b(x, y, t)$:

$$(u, v) = \mathbf{u}_b, \qquad w = ub_x + vb_y. \tag{10.252}$$

The dimensionless form of the sliding law (10.218) can be written as

$$\tau_b = Ru_b^r, \tag{10.253}$$

where $u_b = |\mathbf{u}_b|$, and R is a dimensionless roughness factor which depends on effective pressure N. By choice of λ, we may suppose $R \lesssim O(1)$ in streaming flow, but we can also describe shear flow with little or no sliding by having $R \gg 1$. The corresponding vector form of (10.253) is

$$\boldsymbol{\tau}_b = Ru_b^{r-1}\mathbf{u}_b, \tag{10.254}$$

where, correct to terms of $O(\varepsilon^2)$, $\boldsymbol{\tau}_b = (\tau_{13}, \tau_{23})$.

When $\lambda \ll 1$, it is possible to use a different form of approximation (to the shallow ice approximation where $\lambda = 1$) which includes the longitudinal stress terms. This is called the membrane stress approximation. To derive it, we reconsider (10.250). Without approximation, we can integrate the vertical normal stress equation to give

$$p = -(\tau_{11} + \tau_{22}) - \lambda \left[\frac{\partial}{\partial x} \int_z^s \tau_{13}\,dz + \frac{\partial}{\partial y} \int_z^s \tau_{23}\,dz \right], \tag{10.255}$$

and substituting this into the shear stress equations, they take the form, again without approximation,

$$\tau_{13,z} = s_x - \frac{\varepsilon^2}{\lambda}(2\tau_{11,x} + \tau_{12,y} + \tau_{22,x}) - \varepsilon^2 \left(\frac{\partial^2}{\partial x^2} \int_z^s \tau_{13}\,dz + \frac{\partial^2}{\partial x \partial y} \int_z^s \tau_{23}\,dz \right),$$

$$\tau_{23,z} = s_y - \frac{\varepsilon^2}{\lambda}(\tau_{11,y} + \tau_{12,x} + 2\tau_{22,y}) - \varepsilon^2 \left(\frac{\partial^2}{\partial x \partial y} \int_z^s \tau_{13}\,dz + \frac{\partial^2}{\partial y^2} \int_z^s \tau_{23}\,dz \right). \tag{10.256}$$

The membrane stress approximation is based on the limit $\varepsilon \ll 1$, independently of the size of λ, and consists in essence of the neglect of the integral terms in (10.256), thus we have

$$\tau_{13,z} = s_x - \frac{\varepsilon^2}{\lambda}[2\tau_{11,x} + \tau_{12,y} + \tau_{22,x}],$$

$$\tau_{23,z} = s_y - \frac{\varepsilon^2}{\lambda}[\tau_{11,y} + \tau_{12,x} + 2\tau_{22,y}]. \tag{10.257}$$

If, in addition, we suppose that $\lambda \ll 1$, then it follows from (10.250)[19] that $|\frac{\partial \mathbf{u}}{\partial z}| \ll 1$, and thus $\mathbf{u} \approx \mathbf{u}_b$, and is a function of x and y. Thus, so also are the stresses τ_{11}, τ_{12} and τ_{22}, and so we can integrate (10.257) using (10.251) to obtain

$$\tau_{13} = -(s-z)s_x + \frac{\varepsilon^2}{\lambda}\left[\frac{\partial}{\partial x}\{(2\tau_{11} + \tau_{22})(s-z)\} + \frac{\partial}{\partial y}\{\tau_{12}(s-z)\} \right],$$

$$\tau_{23} = -(s-z)s_y + \frac{\varepsilon^2}{\lambda}\left[\frac{\partial}{\partial x}\{\tau_{12}(s-z)\} + \frac{\partial}{\partial y}\{(\tau_{11} + 2\tau_{22})(s-z)\} \right]. \tag{10.258}$$

---

[19] We presume that $\mathbf{u} = (u, v) \sim O(1)$.

As a matter of fact we can argue that (10.258) is still approximately true even if $\lambda \sim O(1)$, as follows. The corrective stress terms of $O(\varepsilon^2/\lambda)$ are small unless horizontal gradients of $\mathbf{u}$ are large, and the only way in which this can occur is if the sliding velocity changes rapidly in space. It follows that in regions where the corrective stresses are important, they can be accurately approximated by using the sliding velocity in computing them. There is no loss of accuracy in doing this everywhere, since the terms are in any case small when the sliding velocity is not changing rapidly. By evaluating (10.258) at the bed, we obtain a closed model for the sliding velocity, in the form

$$
\tau_1 = -H s_x + \frac{\varepsilon^2}{\lambda} \left[\frac{\partial}{\partial x} \{ (2\tau_{11} + \tau_{22}) H \} + \frac{\partial}{\partial y} \{ \tau_{12} H \} \right],
$$
$$
\tau_2 = -H s_y + \frac{\varepsilon^2}{\lambda} \left[\frac{\partial}{\partial x} \{ \tau_{12} H \} + \frac{\partial}{\partial y} \{ (\tau_{11} + 2\tau_{22}) H \} \right],
$$

(10.259)

where $H = s - b$ is the depth, and $(\tau_1, \tau_2) = \boldsymbol{\tau}_b$ is the basal shear stress, given by (10.254). This is the membrane stress approximation, in which the membrane stresses τ_{11}, τ_{12} and τ_{22} are given in terms of the sliding velocity (u, v); in the case that $\lambda \ll 1$, so that shearing is negligible, H is determined by conservation of mass in the form

$$
\frac{\partial H}{\partial t} + \frac{\partial (Hu)}{\partial x} + \frac{\partial (Hv)}{\partial y} = a,
$$

(10.260)

where a is the accumulation rate.

On the Siple Coast, fast flow alternates with inter-ice stream regions where ice flow is small, and sliding is small or negligible. In building a model for the mechanics of ice streams, it is thus advisable to allow for regions where shear flow is important. Luckily, it is easy to do this in the present context. In consideration of (10.258), we can reasonably assume that the ice surface does not change abruptly. In that case, we can remove the depth terms from inside the derivatives, and we can write (10.258) in the form

$$
(\tau_{13}, \tau_{23}) \approx (s - z)\mathbf{g},
$$

(10.261)

where

$$
\mathbf{g} = -\nabla s + \frac{\varepsilon^2}{\lambda} \mathbf{G},
$$

(10.262)

and

$$
\mathbf{G} = \left(\frac{\partial}{\partial x} \{ 2\tau_{11} + \tau_{22} \} + \frac{\partial \tau_{12}}{\partial y}, \frac{\partial \tau_{12}}{\partial x} + \frac{\partial}{\partial y} \{ \tau_{11} + 2\tau_{22} \} \right).
$$

(10.263)

In seeking a correction to (10.260) when shearing is important, we can now invert our earlier argument. The correction will only be important when sliding is small, and consequently when longitudinal stresses are small. In this case, τ does not vary rapidly, and we can take

$$
\tau = Hg;
$$

(10.264)

(10.263) can then be simplified to the form

$$\mathbf{G} \approx \frac{1}{\tau^{n-1}}(\nabla^2 \mathbf{u} + 3\nabla \Delta), \tag{10.265}$$

where Δ is the dilatation

$$\Delta = \nabla.\mathbf{u}. \tag{10.266}$$

Just as in the shallow ice approximation, we can integrate (10.261) twice to obtain the generalisation of (10.260) in the form

$$\frac{\partial H}{\partial t} + \nabla.\left\{ H\mathbf{u} + \frac{\lambda H^{n+2}}{n+2} g^{n-1}\mathbf{g} \right\} = a. \tag{10.267}$$

Together with the membrane stress approximation (10.259) and the sliding law (10.254), which can be combined to give (approximately)

$$\tau = Ru^{r-1}\mathbf{u} = H\mathbf{g}, \tag{10.268}$$

(10.267) allows for a unified description of flow in which both ice streams and non-ice stream flow are accurately described, at least for isothermal ice (i.e., constant A); $\mathbf{g}$ is defined by (10.262), and $\mathbf{G}$ by (10.263).[20]

A Simple Model of an Ice Stream

We give here a simple model of an ice stream such as those in the Siple Coast. We take axes x downstream and y cross-stream, and we suppose the velocity is purely in the x-direction, is independent of depth z, and varies only with the transverse coordinate y, thus $\mathbf{u} = (u(y), 0)$. We suppose $H = 1$ and $-\nabla s = (\alpha, 0)$, where a reasonable value of $\alpha = 0.1$. Then the basal stress is $\tau = (\tau_1, 0)$, and we have the system

$$\tau_1 = Ru^r = \alpha + \frac{\varepsilon^2}{\lambda}\frac{\partial \tau_{12}}{\partial y},$$

$$\tau^2 = \tau_1^2 + \frac{\varepsilon^2}{\lambda^2}\tau_{12}^2, \tag{10.269}$$

$$\tau_{12} = \frac{u_y}{\tau^{n-1}}.$$

For the basal sediments on the Siple coast, for example beneath the Whillans ice stream, a yield stress may be relevant, in which case we might have $r = 0$ and $R \approx 0.01$. Then we suppose that $\tau_1 \ll \frac{\varepsilon}{\lambda}\tau_{12}$, so that

$$\tau \approx \frac{\varepsilon}{\lambda}\tau_{12}. \tag{10.270}$$

---

[20]Note that τ in (10.264) is the second stress invariant, while τ in (10.268) is the basal shear stress.

It then follows that

$$\tau \approx \left(\frac{\varepsilon}{\lambda}|u_y|^{\frac{1}{n}-1}u_y\right), \qquad \tau_{12} = \left(\frac{\lambda}{\varepsilon}\right)^{1-\frac{1}{n}}|u_y|^{\frac{1}{n}-1}u_y, \tag{10.271}$$

and thus that u satisfies the equation

$$Ru^r = \alpha + \frac{\varepsilon^{1+\frac{1}{n}}}{\lambda^{\frac{1}{n}}}\frac{\partial}{\partial y}\left(|u_y|^{\frac{1}{n}-1}u_y\right). \tag{10.272}$$

We define $y = vY$, where

$$v = \frac{\varepsilon}{\lambda^{\frac{1}{n+1}}\alpha^{\frac{n}{n+1}}}, \tag{10.273}$$

and suppose that $R \ll \alpha$, connoting weak basal till. u then satisfies the equation

$$0 \approx 1 + \frac{\partial}{\partial Y}\left(|u_Y|^{\frac{1}{n}-1}u_Y\right), \tag{10.274}$$

and suitable boundary conditions are that

$$u = 0 \quad \text{on } Y = \pm L, \tag{10.275}$$

where L can be determined if we specify the normalising condition $u(0) = 1$. Adopting this, the solution is

$$u = 1 - \frac{|Y|^{n+1}}{n+1}, \tag{10.276}$$

and the dimensional ice stream width is then

$$L_{is} = 2d_i\left(\frac{n+1}{\lambda\alpha^n}\right)^{\frac{1}{n+1}}, \tag{10.277}$$

where d_i is ice depth. Taking $d_i = 1,000$ m, we can find an ice stream width of 40 km if, for example, we take $\alpha = 0.1$, $\lambda = 0.025$.

Note that in $(10.269)_1$, we have $\frac{\varepsilon}{\lambda}\tau_{12} \sim \frac{\alpha v}{\varepsilon}$, so that our earlier assumption that $\tau_1 (\sim R) \ll \frac{\varepsilon}{\lambda}\tau_{12}$ is valid if $R \ll \frac{\alpha v}{\varepsilon}$, and thus certainly if $v \gtrsim \varepsilon$, as is confirmed by (10.273).

10.4 Waves, Surges and Mega-surges

10.4.1 Waves on Glaciers

Waves on glaciers are most easily understood by considering an isothermal, two-dimensional model. We suppose the base is flat ($h = 0$), so that Eqs. (10.25) and (10.26) give

$$H_t + \left[\{1 - \mu H_x\}^n\frac{H^{n+2}}{n+2} + u_b H\right]_x = B'(x), \tag{10.278}$$

where $B'(x) = a$ is the accumulation rate, and $\mu \sim 0.1$. If we firstly put $\mu = 0$ and also $u_b = 0$, then

$$H_t + H^{n+1} H_x = B'(x), \tag{10.279}$$

which has the steady state

$$\frac{H_0^{n+2}}{n+2} = B(x). \tag{10.280}$$

With $B' > 0$ in $x < x_f$ (say) and $B' < 0$ in $x > x_f$ ($x = x_f$ is then the *firn line*), (10.280) defines a concave profile like that in Fig. 10.6. (10.279) is clearly hyperbolic, and admits wave-like disturbances which travel at a speed H^{n+1}, which is in fact $(n + 1)$ (≈ 4) times the surface speed. If we take an initial condition at $t = 0$ corresponding to a balance function $B(x) - \varepsilon D(x)$, where $\varepsilon \ll 1$, then the solution using the method of characteristics subject to an upstream boundary condition of

$$H = 0 \quad \text{at } x = 0 \tag{10.281}$$

is

$$\frac{H^{n+2}}{n+2} = B(x) - \varepsilon D(\sigma),$$
$$t = \int_\sigma^x \frac{dx'}{[(n+2)\{B(x') - \varepsilon D(\sigma)\}]^{(n+1)/(n+2)}}. \tag{10.282}$$

The characteristics of (10.279) propagate downstream and reach the snout (where $H = 0$) in finite time. (10.282) is somewhat unwieldy, and it is useful to approximate the characteristic solution for small ε. However, if we use the blunt approach, where we write $H = H_0 + \varepsilon h$, a straightforward linearisation of (10.279) shows that h grows unboundedly near the snout of the glacier. This unphysical behaviour occurs because the linearisation artificially holds the snout position fixed; mathematically, the linearisation is invalid near the snout where $H_0 = 0$ and the assumption $\varepsilon h \ll H_0$ breaks down. An apparently uniformly valid approximation can be obtained, however, by linearising the characteristics:

$$H_t + H_0^{n+1} H_x \approx B'(x). \tag{10.283}$$

For $H \approx H_0$, the general solution is

$$H = H_0(x) + \phi(\xi - t), \tag{10.284}$$

where

$$\xi = \int_0^x \frac{dx}{H_0^{n+1}(x)} \tag{10.285}$$

is a characteristic spatial coordinate (note ξ is finite at the snout). (10.284) clearly reveals the travelling wave characteristic of the solution.

Margin Response

However, although (10.284) is better than the blunt approach, it is not really good enough, as it still only defines the solution within the confines of the steady state solution domain determined by (10.285). The more methodical way to deal with the singularity of the solution at the snout is to allow margin movement by using the method of strained coordinates. That is to say, we change coordinates to

$$x = s + \varepsilon x_1(s, \tau) + \cdots, \qquad t = \tau. \tag{10.286}$$

Equation (10.279) now takes the form

$$H_\tau + H^{n+1} H_s = B' + \varepsilon \left(x_{1\tau} + H^{n+1} x_{1s} \right) H_s + \cdots, \tag{10.287}$$

and analogously to (10.282), we pose the initial condition

$$\frac{H^{n+2}}{n+2} = B(s) - \varepsilon D(s) \quad \text{at } \tau = 0. \tag{10.288}$$

In addition, we pose the boundary condition

$$H = 0 \quad \text{at } s = 0, \tag{10.289}$$

which also forces

$$x_1 = 0 \quad \text{at } s = 0. \tag{10.290}$$

We also require that x_1 is such that

$$H = 0 \quad \text{at } s = 1, \tag{10.291}$$

this being the position of the snout when $D = 0$, i.e., $B(1) = 0$.
   We put

$$H = H_0 + \varepsilon h + \cdots, \tag{10.292}$$

and hence find that

$$\frac{H_0^{n+2}}{n+2} = B(s), \tag{10.293}$$

and, using ξ defined by (10.285) (but with s as the upper limit) as the space variable, we have

$$\left(H_0^{n+1} h \right)_\tau + \left(H_0^{n+1} h \right)_\xi = (x_{1\tau} + x_{1\xi}) H_0'(\xi), \tag{10.294}$$

and the initial condition is

$$H_0^{n+1} h = -D \quad \text{at } \tau = 0. \tag{10.295}$$

The boundary condition of $h = 0$ at $s = 0$ is irrelevant here, because there is only a perturbation in the initial condition, so that for $\tau > \xi$, $h = 0$ and the steady solution is restored.
   The solution of (10.294) can be written as

$$
\begin{aligned}
H_0^{n+1} h &= -D(\xi - \tau) + U(\xi, \tau), \\
x_1 &= \int_0^\tau P(\eta + \xi - \tau, \eta) \, d\eta + x_1^0(\xi - \tau),
\end{aligned}
\tag{10.296}
$$

where U satisfies

$$U_\tau + U_\xi = P(\xi, \tau) H_0'(\xi),$$
$$U = 0 \quad \text{at } \tau = 0.$$

(10.297)

The method of strained coordinates proceeds by choosing x_1 in order that h is no more singular than H_0. Since $H_0 \sim \xi_1 - \xi$ as $\xi \to \xi_1$, where

$$\xi_1 = \int_0^1 \frac{ds}{H_0^{n+1}(s)},$$

(10.298)

we have to choose U so that the right hand side of $(10.296)_1$ is $O[(\xi_1 - \xi)^{n+2}]$ as $\xi \to \xi_1$. For $n = 3$, for example, this requires choosing the $n + 2 = 5$ conditions

$$U = D(\xi_1 - \tau), \ U_\xi = D'(\xi_1 - \tau), \ldots, U_{\xi\xi\xi\xi} = D^{\text{iv}}(\xi_1 - \tau) \quad \text{at } \xi = \xi_1. \quad (10.299)$$

As is well known, any such function will do, its importance being locally near $\xi = \xi_1$. Given U, $(10.297)_1$ defines P, and then $(10.296)_2$ defines the straining, and thus the margin position. To find U, it is convenient to solve the partial differential equation (assuming $n + 2 = 5$)

$$U_\tau = U_{10\xi},$$

(10.300)

subject to the five boundary conditions in (10.299), together with, for example,

$$U = U_\xi = \cdots = U_{4\xi} = 0 \quad \text{at } \xi = \xi_f,$$

(10.301)

where ξ_f denotes the firn line position where $H_0'(\xi_f) = 0$. The point of choosing a tenth order equation is to ensure decay away from $\xi = \xi_1$, which is cosmetically advantageous; the point of choosing ξ_f in (10.301) is to ensure that P remains bounded; again, this is largely cosmetic and one might simply replace ξ_f by ∞.

The Upstream Boundary Condition

We have blithely asserted that

$$H = 0 \quad \text{at } x = 0,$$

(10.302)

as seems fine for the diffusionless equation (10.279). Let us examine this more closely, assuming the diffusional model (10.278) with no sliding, in steady state form:

$$\frac{(1 - \mu H_x)^n H^{n+2}}{n+2} = B(x),$$

(10.303)

where we have already integrated once, applying the condition that the ice flux is zero at the glacier head $x = 0$, where $B = \int_0^x a(x') \, dx' = 0$. If we ignore μ, this seems fine, but if $\mu \neq 0$, then we can rewrite (10.303) as

$$\mu H_x = 1 - \left\{ \frac{(n+2)B}{H^{n+2}} \right\}^{1/n}.$$

(10.304)

Consideration of the direction of trajectories in the (x, H) plane shows that there is no trajectory which has $H(0) = 0$. The only alternative allowing zero flux at the head is $\mu H_x = 1$ there (physically, a horizontal surface), but then the depth is necessarily non-zero.

We have seen this problem before (see Question 4.10). Consideration of (10.304) shows that there is a unique value of $H_0 > 0$ such that if $H(0) = H_0$, then $\frac{H^{n+2}}{n+2} \sim B(x)$ for $x \gg \mu$, as is appropriate. Based on our earlier experience, we might expect that a boundary layer in which longitudinal stresses are important would provide a mechanism for the transition from $H = 0$ to $H = H_0$. This is indeed the case (see Question 10.11), but the resultant compressive boundary layer appears to be very unphysical. The theoretical description of the head of a glacier therefore remains problematical.

Shock Formation

An issue which complicates the small perturbation theory above is the possible formation of shocks. Characteristics $x(\sigma, t)$ in (10.282) intersect if $\frac{\partial x}{\partial \sigma} = 0$. Computing this we find

$$\frac{\partial x}{\partial \sigma} = \frac{H^{n+1}(x, \sigma)}{H^{n+1}(\sigma, \sigma)} - \varepsilon(n+1)D'(\sigma)H^{n+1}(x, \sigma) \int_\sigma^x \frac{dx'}{H^{2n+3}(x', \sigma)}. \qquad (10.305)$$

From this unwieldy expression, it is clear that for small ε, shocks will always form near the snout if $D' > 0$ somewhere, which is the condition for local advance of the glacier. More generally, glacier advances are associated with steep fronts, while retreats have shallower fronts, as is commonly observed.

Shocks can form away from the snout if H is increased locally (e.g., due to the surge of a tributary glacier). The rôle of the term in μ is then to diffuse such shocks. A shock at $x = x_s$ will propagate at a rate

$$\dot{x}_s = \frac{[H^{n+2}]_-^+}{(n+2)[H]_-^+}, \qquad (10.306)$$

where $[\ ]_-^+$ denotes the jump across x_s. When the shock reaches the snout, it then propagates at a speed $H_-^{n+1}/(n+2)$, which is *slower* than the surface speed.

In the neighbourhood of a shock (with $u_b = 0$), we put

$$x = x_s + \nu X, \qquad (10.307)$$

so that

$$\frac{\partial H}{\partial t} - \frac{\dot{x}_s}{\nu} \frac{\partial H}{\partial X} + \frac{1}{\nu} \left[\left\{ 1 - \frac{\mu}{\nu} H_X \right\}^n \frac{H^{n+2}}{n+2} \right]_X = B'(x_s + \nu X); \qquad (10.308)$$

if ν is small, the profile rapidly relaxes to the steady travelling wave described by

$$\dot{x}_s H_X = \left[\{1 - H_X\}^n \frac{H^{n+2}}{n+2} \right]_X, \qquad (10.309)$$

providing we choose $v = \mu$, which thus gives the width of the shock structure. (10.309) can be solved by quadrature (see Question 10.15). In practice the shock width is relatively long, so steep surface wave shocks due to this mechanism are unlikely (but they can form for other reasons, for example in surges, when longitudinal stresses become important).

Seasonal Waves

Although they constitute the more dramatic phenomenon, the seasonal wave has attracted much less attention than the surface wave, perhaps because there are less obvious comparable analogies. The surface wave is essentially the same as the surface wave in a river, while the seasonal wave bears more resemblance to a compression wave in a metal spring, even though the ice is essentially incompressible.

Apparently the waves are induced through seasonal variations in velocity, which are themselves associated with variations in meltwater supply to the glacier bed, so that a natural model for the ice flow would involve only sliding, thus (non-dimensionally)

$$H_t + (Hu)_x = a, \tag{10.310}$$

where u is the sliding velocity. If the natural time scale for glacier flow is $t_i \sim 100$ y, while the seasonal time scale is $t_s = 1$ y, then it is appropriate to rescale the time as

$$t = \varepsilon T, \quad \varepsilon = \frac{t_s}{t_i} \ll 1, \tag{10.311}$$

so that H satisfies

$$H_T = \varepsilon[a - (Hu)_x]; \tag{10.312}$$

this immediately explains why there is no significant surface perturbation during passage of the seasonal wave.

To study the velocity perturbation, we suppose that the sliding velocity depends on the basal shear stress τ (which varies little by the above discussion) and effective pressure N. If, for example, basal drainage is determined by a relation such as (10.225), then essentially $u = u(Q)$, so that waves in u are effectively waves in Q, i.e., waves in the basal hydraulic system.

Suppose that mass conservation in the hydraulic system is written non-dimensionally as

$$\phi S_T + Q_x = M, \tag{10.313}$$

where M is the basal meltwater supply rate,

$$\phi = \frac{t_h}{t_s} \tag{10.314}$$

is the ratio of the hydraulic time scale t_h to the seasonal time scale t_s, and a force balance relation such as (10.221) (cf. (10.224)) suggests $S = S(Q)$. If, for simplicity, we take $\phi S'(Q) = \kappa$ as constant, then the solution of (10.313) subject to a boundary condition of

$$Q = 0 \quad \text{at } x = 0 \tag{10.315}$$

is

$$Q = \frac{1}{\kappa}\big[J(T) - J(T - \kappa x)\big], \tag{10.316}$$

where

$$J(T) = \int_0^T M(T')\,dT'. \tag{10.317}$$

(10.316) clearly indicates the travelling wave nature of the solution.

The diagram in Fig. 10.2, sometimes called a Hodge diagram, depicts a seasonal wave through the propagation of the constant velocity contours down-glacier. The constant velocity contours are represented as functions $x(Q, t)$ (if we suppose u depends on Q). Higher Q causes higher N in Röthlisberger channels, and thus lower velocity, as seen in Fig. 10.2. A crude representation of the data is thus as a family of curves

$$x = A(Q) + X\big[T - \theta(Q)\big], \tag{10.318}$$

where A increases with Q and θ also increases with Q.

To illustrate how (10.316) mimics this, we first note that for non-negative meltwater supply rates M, J is monotonically increasing and thus invertible, whence we can write (10.316) in the form

$$x = \frac{1}{\kappa}\big[T - J^{-1}\{J(T) - \kappa Q\}\big]. \tag{10.319}$$

Suppose, for example that $M = 1 + m(T)$ where m is small and has zero mean, so that

$$J(T) = T + j(T), \quad j(T) = \int_0^T m(T')\,dT'; \tag{10.320}$$

it follows that

$$J^{-1}(u) \approx u - j(u), \tag{10.321}$$

and thus

$$x \approx Q - \frac{1}{\kappa}\big[j(T) - j(T - \kappa Q)\big]. \tag{10.322}$$

This is sufficiently similar to the putative (10.318) to suggest that this mechanism may provide an explanation for seasonal waves. According to (10.322), the dimensionless wave speed is $v_s^* \sim \frac{1}{\kappa}$, and thus the dimensional wave speed $v_s \sim \frac{l}{t_h}$. In Fig. 10.3, we see typical ice velocities of 100 m y^{-1} compared with a seasonal wave speed of some 15 km y^{-1}. Assuming a time scale of 60 y for a six kilometre long glacier, this would suggest a hydraulic time scale of about five months. On the face of it, this seems very long, but in fact the relevant hydraulic time scale should be that over which the water pressure in the cavity system can respond to changes in the channel pressure, which will be a lot longer than the adjustment time for the channel itself.

Fig. 10.11 A multivalued flux–depth relation can cause oscillatory surges

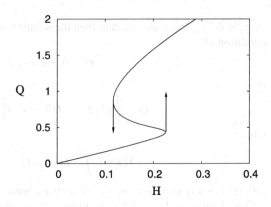

10.4.2 Surges

It has long been suggested that the fast velocities during surges could only be caused by rapid sliding. Therefore it is sufficient for our purpose to analyse the mass conservation equation in the form

$$H_t + (Hu)_x = B'(x), \qquad (10.323)$$

where u is the sliding velocity. Also, it has been thought that if the sliding velocity were a multivalued function of basal stress τ_b (i.e., $\tau_b(u)$ has a decreasing portion) then, since $\tau_b = H(1 - \mu s_x) \approx H$, this would cause the ice flux $Q = uH$ to be multivalued, as shown in Fig. 10.11. In this case we might expect relaxation oscillations to occur for values of B intermediate between the two noses of $Q(H)$. Two fundamental questions arise. Firstly, is there any genuine reason why $\tau_b(u)$ should be non-monotone, and secondly, how would such a relaxation oscillator work in the spatially dependent case? In particular, it would seem necessary to have a secondary variable, whose rapid change can facilitate the relaxation between the different solution branches (cf. Fig. 1.6 and Eqs. (1.25)).

The discussion in Sect. 10.2 suggested the possibility of non-monotone $\tau_b(u)$ for flow over a periodic bed. However, it is arguable whether real beds have this feature,[21] in which case we may suppose that τ increases with both u and N. What observations of the 1982–3 surge of Variegated Glacier showed, however, was that there is a switch in drainage pattern during its surge. There are (at least) two possible modes of drainage below a glacier. Röthlisberger channels, as described in Sect. 10.2, can form a branched arterial drainage system. In this case the value of the effective pressure at the bed N is determined by the water flow, $N = N_R$, say. Alternatively, there may be no channel system, and the water at the bed fills the cavities behind bed protuberances, and drains by a slower leakage between cavities. This is the linked cavity régime described in Sect. 10.3.2; it operates at a higher water pressure and thus lower effective pressure, N_c, than in the channel drainage.

---

[21] An exception may be the very steep 'hanging glaciers', where the periodic behaviour consists of complete detachment of the glacier snout following tensile fracture.

Fig. 10.12 N is a
multivalued function of u

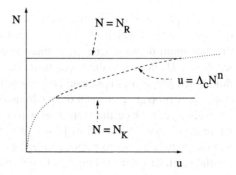

Fig. 10.13 Q is a
multivalued function of H

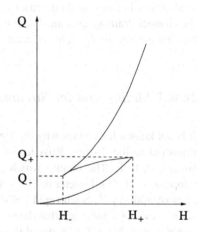

The crucial factor which enables surges to take place is the switching mechanism,
and this depends on the ice flow over the cavities.

We now combine the form of the sliding law $\tau_b = N f(u/N^n)$, as discussed in
Sect. 10.2, with a drainage system consisting either of Röthlisberger channels or
linked cavities, the choice of which depends on the value of $\Lambda = u/N^n$, with the
transition between drainage systems occurring at the critical value Λ_c. That is,

$$
\begin{aligned}
N &= N_R, \quad u/N^n < \Lambda_c; \\
N &= N_c, \quad u/N^n > \Lambda_c.
\end{aligned}
\tag{10.324}
$$

If this is written as a function $N(u)$, it is multivalued, as shown in Fig. 10.12. As a
consequence of this, the sliding law is indeed multivalued, and hence $Q(H)$ has the
form shown in Fig. 10.13.

There are two critical values of Q in Fig. 10.13, denoted Q_+, Q_-: these are the
values at the noses of the curve (where also $H = H_+, H_-$). If $B(x) < Q_+$, then
an equilibrium glacier profile exists in which $Q = B(x)$. However, if the maximum
value of B, $B_{\max}$, is greater than Q_+, then such a stable equilibrium cannot occur,
and the glacier surges.

The sequence of events in a surge is then as follows. The glacier grows from a quiescent state in which $Q < Q_+$ on the lower (slow) branch everywhere. When the maximum depth reaches H_+, there is a *reservoir* zone where $H > H_-$. The ice flux at H_+ jumps to the upper (fast) branch by switching drainage pattern, and this switch propagates upstream and downstream to where $H = H_-$. These *activation waves* propagate at rates of hundreds of metres per *hour* (and in effect have been observed). Once the activation waves have propagated to the boundaries of the reservoir zone, the ice flow is described by the fast mode on the upper branch, and the activated reservoir zone propagates rapidly downstream, possibly overriding the stagnant snout and propagating forwards as a front. In terms of Fig. 10.13, the surge terminates when H reaches H_- everywhere, and deactivation waves propagate inwards from the boundaries of the exhausted reservoir zone to re-establish the channel drainage system. There then follows another quiescent phase where the maximum value of H increases from H_- to H_+ before the next surge is initiated.

10.4.3 Sliding and Ice Streams

It is not known for certain why the ice flow on the Siple Coast of Antarctica, which flows out to the floating Ross ice shelf, segregates itself into the five distinct ice streams A to E. The picture which one has of this region is of a gently sloping (slope $\alpha \sim 10^{-3}$) kilometer thick ice sheet which flows in the ice streams at typical rates of 500 m y^{-1}. Such rapid velocity can only be due to basal sliding, and the seismic evidence indicates that the ice is underlain by several metres of wet till. One might expect that a sliding law of the form advocated previously is appropriate, that is,

$$\tau_b = c u_b^r N^s, \tag{10.325}$$

with r and s positive. The issue then arises as to how to prescribe N. Recall from Sect. 10.2 that for drainage through Röthlisberger channels, an appropriate law is $N = \beta Q_w^{1/4n}$, where Q_w is water flux. When ice flows over till, an alternative system of drainage is that of distributed 'canals' incised in the subglacial till. For such a system, an appropriate law is $N = \gamma Q_w^{-1/n}$, and the low values of effective pressure in this relation are more representative of measured basal pressures on Whillans ice stream, for example.

In this case an interesting feedback exists. In Antarctic ice streams, there is little, if any, surface melt reaching the bed, and the basal water flow is due to melting there. The quantity of meltwater produced per unit area per unit time is given by the melt velocity

$$v_m = \frac{G + \tau_b u_b - g}{\rho_w L}, \tag{10.326}$$

where ρ_w is water density, L is latent heat, G is geothermal heat flux, and g is the basal heat flux into the ice. This assumes the base is at the melting point. Thus we expect the basal water flux $Q_w \propto G + \tau_b u_b - g$, and so Q_w increases with u_b (the dependence of g on u_b is likely to be weaker—boundary layer theory would suggest $g \sim u_b^{1/2}$). If also N decreases with Q_w, then N decreases as u_b increases. But this causes further increase of u_b via the sliding law. This positive feedback can lead to a runaway phenomenon which we may call hydraulic runaway.

To get a crude idea of how this works, we denote the ice thickness as h and the surface slope as S_i. If the velocity is u, then the ice flux per unit width is

$$Q = hu; \tag{10.327}$$

the basal shear stress is

$$\tau = Rh = cu^r N^s, \tag{10.328}$$

where we define

$$R = \rho_i g S_i; \tag{10.329}$$

we suppose

$$N = \gamma Q_w^{-p}, \tag{10.330}$$

and that

$$Q_w = b[G + \tau u - g], \tag{10.331}$$

where, from (10.326), we define

$$b = \frac{l_i l_s}{\rho_w L}, \tag{10.332}$$

in which l_i is the ice flow line length scale and l_s is the stream spacing, and the heat flux to the ice is given by

$$g = au^{1/2}, \tag{10.333}$$

corresponding to a heat flux through a thermal boundary layer. Consequently

$$h = \frac{fu^r}{[G + Rhu - au^{1/2}]^m}, \tag{10.334}$$

where

$$m = ps, \qquad f = \frac{c\gamma^s}{Rb^m}. \tag{10.335}$$

It is not difficult to see from (10.334), if f is low enough (equivalently, the friction coefficient c is low enough), that u and hence the ice flux Q will be a multivalued function of h, as shown in Fig. 10.14. In fact, application of realistic parameter values suggests that such multivalued flux laws are normal. More specifically, we choose estimates for the parameters as follows. We use exponents $p = r = s = \frac{1}{3}$ and thus $m = \frac{1}{9}$, and then $c = 0.017$ bar$^{2/3}$ m$^{-1/3}$ y$^{1/3}$, based on a sliding law (10.328) with $\tau = 0.1$ bar, $N = 0.4$ bar and $u = 500$ m y^{-1}. Other parameter values are $\gamma = 0.3$ bar (m^3 s^{-1})$^{1/3}$, $S_i = 10^{-3}$, $\rho_i = 0.917 \times 10^3$ kg m^{-3}, $g = 9.8$ m s^{-2}, $G = 0.06$ W m^{-2}, $\rho_w = 10^3$ kg m^{-3}, $L = 3.3 \times 10^5$ J kg^{-1}, and in addition we

Fig. 10.14 Thermal feedback causes a multivalued ice flux. The solution of (10.334) is plotted using a value of the critical parameter $f = 70 \text{ W}^{1/9} \text{ m}^{4/9} \text{ y}^{1/3}$. Other values are as described in the text. I am indebted to Ian Hewitt for his production of this figure

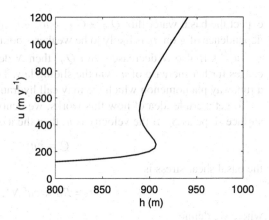

choose $l_i = 10^3$ km, $l_s = 330$ m; from these we find $R = 3 \times 10^{-7}$ W m^{-4} y, $b = 1$ J^{-1} m^5, and thus $f = 126$ W$^{1/9}$ m$^{4/9}$ y$^{1/3}$. Finally, we choose the value of a based on an assumed magnitude of $g \approx \frac{4k\Delta T}{d_i}$, where ice depth is $d_i = 10^3$ m, thermal conductivity is $k = 2.2$ W m^{-1} K^{-1}, and surface temperature below freezing is $\Delta T = 20$ K; with $u = 500$ m y^{-1}, this gives $a = 0.8 \times 10^{-2}$ W m$^{-5/2}$ y$^{1/2}$. Figure 10.14 plots velocity versus depth with these parameter values, except that we take $f = 70$ W$^{1/9}$ m$^{4/9}$ y$^{1/3}$, and Question 10.14 provides approximate analytic solutions for the different branches.

If, indeed, hydraulic feedback can cause a multivalued relationship between ice velocity and depth, what then happens in a region such as the Siple Coast of West Antarctica? We suppose that the ice flux is determined by conditions upstream, so that if the ice flux per unit width is q, and the width of the discharge region is W, then

$$Wq = B, \tag{10.336}$$

where B is the volume flux of ice discharged. If the flow law is multivalued, then there exists a range (q_-, q_+) of q such that the ice flow is unstable (see Question 10.14). If $B/W < q_-$, then a uniform slow moving ice flow is possible. Similarly, if $B/W > q_+$, a uniform fast-moving ice stream is possible. What if $q_- < B/W < q_+$? A uniform ice flow is now unstable, and we may expect a spatial instability to occur, whereby ice streams spontaneously form, as is in fact observed. Such an instability would be mediated by transitions in water pressure, since basal water will flow from fast streams at high water pressure to slower ice at low water pressure. This generates a lateral enthalpy flux, and in a steady state this can be balanced by a heat flux in the ice in the opposite direction, since cooling (g) is less effective at lower u, therefore the slow ice is warmer near the base than the ice streams.

10.4.4 Heinrich Events and the Hudson Strait Mega-surge

What if the drainage channel of an ice sheet over deforming till is relatively narrow? By analogy with the pattern formation mechanism in reaction–diffusion equations, one would expect that a multivalued flux–depth relation would not allow separate streams to form if the channel width is too small, and in this case we would expect periodic surges to occur down the channel, if the prescribed mass flux corresponds to a velocity on the unstable position of Fig. 10.14.

A situation of this type appears to have occurred during the last ice age. The Laurentide ice sheet which existed in North America drained the ice dome which lay over Hudson Bay out through the Hudson Strait, a 200 km wide trough which discharged the ice (as icebergs) into the Labrador sea and thence to the North Atlantic.

Hudson Bay is underlain by soft carbonate rocks, mudstones, which can be mobilised when wet. It has been suggested that the presence of these deformable sediments, together with the confined drainage channel, led to the occurrence of semi-periodic surges of the Hudson Strait ice stream. The evolution of events is then as follows. When ice is thin over Hudson Bay, the mudstones may be frozen at the base, there is little, if any, sliding and very little ice flow. Consequently, the ice thickens and eventually the basal ice warms. The basal muds thaw, and sliding is initiated. If the friction is sufficiently low (i.e., c and thus f is small), then the multivalued sliding law of Fig. 10.14 is appropriate, and if the accumulation rate is large enough, cyclic surging will occur. During a surge, the flow velocity increases dramatically, and there results a massive iceberg flux into the North Atlantic. On the lower branch of Fig. 10.14, water production is virtually absent, Q_w is low in (10.331) since the flow is slow and the geothermal and viscous heat at the base can be conducted away by the ice. The low value of Q_w gives high N, consistent with low u. On the upper branch, however, viscous heat dominates, and Q_w is large, N is small, also consistent with a high u.

At the end of a surge, the rapid ice drawdown causes the water production to drop, and the rapid velocities switch off. This may be associated with re-freezing of the basal mudstones.

When water saturated soils freeze, frost heave occurs by sucking up water to the freezing front via capillary action, and this excess water freezes (at least for fine grained clays and silts) in a sequence of discrete ice lenses. Heaving can occur at a typical rate of perhaps a metre per year, though less for fine grained soils, and the rate of heave is suppressed by large surface loads. Calculations suggest a surge period of perhaps a hundred years, with a drawdown of a thousand metres, and a recovery period on the order of 5,000–10,000 years. During the surge, the rapidly deforming basal muds will dilate (in the deforming horizon, likely to be only a metre or so thick). At the termination of a surge, this layer re-consolidates, and we can expect the total heave to be a certain (small) fraction of the frost penetration depth. In effect, the ice lenses freeze the muds into the ice stream, so that when the next surge phase is initiated, some of this frozen-in basal sediment will be transported downstream, and thence rafted out into the North Atlantic in iceberg discharge.

As discussed in Chap. 2, there is evidence that this rather glamorous sequence of events actually occurs. Heinrich events are layers of ice-rafted debris in deep-sea sediment cores from the North Atlantic which indicate (or are consistent with) massive iceberg discharges every 7000 years or so. In addition, oxygen isotope concentrations in ice cores from Greenland indicate that severe cooling cycles occurred during the last ice age. These cooling events may be caused by a switch-off of North Atlantic deep water (NADW) circulation—effectively switching off the convective heat transport from equatorial latitudes and thus cooling the atmosphere. It seems that sequences of these cooling cycles are terminated by Heinrich events, in the sense that following Heinrich events the climate warms dramatically, perhaps after some delay. There are two reasons why this should be so. On the one hand, the sudden reduction in ice thickness should warm the air above, and also it can be expected that a massive iceberg (and thus freshwater) flux to the North Atlantic acts as a source of thermal buoyancy, which first slows down and subsequently restarts a vigorous North Atlantic circulation. Rather than being lumbering beasts, glaciers and ice sheets show every sign of being dynamically active agents in shaping the climate and the earth's topography.

10.5 Drumlins and Eskers

There are a number of bedforms associated with the motion of ice sheets, and in this section we will discuss two of them, drumlins and eskers. Drumlins are small hills,[22] generally of oval shape, which corrugate the landscape, as shown in Figs. 10.15 and 10.16. They are formed ubiquitously under ice sheets, and take a range of shapes, depending presumably on the basal ice conditions. Ribbed moraines, also called Rogen moraines after the area in Sweden (Lake Rogen) where such features were first described, are transverse furrows like a washboard, with the undulations (presumed) perpendicular to the former ice flow. They are analogous to the transverse dunes described in Chap. 5, and as we shall see, are supposed to be formed by an analogous instability mechanism.

The three-dimensional drumlins of Fig. 10.15 may then arise through a secondary transverse instability, perhaps as some parameter associated with ice flow changes. What certainly happens under former fast-moving ice streams is that drumlins become elongated in the direction of ice flow, appearing eventually to become extremely long grooves aligned with the flow. These grooves, which can run for hundreds of kilometres, are called mega-scale glacial lineations (MSGL), and give the landscape the appearance of having been combed. Figure 10.17 shows a system of MSGL in Northern Canada.

Eskers[23] are sinuous ridges of gravel and sand, of similar dimensions to drumlins, having elevations of tens of metres. They are associated with former drainage

---

[22] The word 'drumlin' is of Irish origin, generally thought to be a diminutive of the word *druim*, meaning a hill, and thus a drumlin is a 'small hill'.

[23] The term esker is also Irish, from *eiscir*, meaning a small ridge.

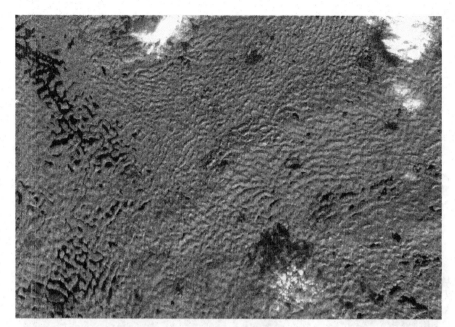

Fig. 10.15 Drumlins in Northern Ireland. Satellite view

channels under the ice, most probably Röthlisberger channels, which have become infilled with subglacial sediment. Figure 10.18 shows a satellite view of an esker system in Northern Canada. The eskers are the red lineations, and their disordered and nonlinear arrangement suggests that they may have been formed at different times, as the ice flow changes direction.

10.5.1 Drumlins

We build a theory of drumlin formation by analogy with the theory of dune formation. An ice sheet flows as indicated in Fig. 10.19 over a deformable substrate at $z = s$, where s is the elevation of the bedrock. The ice at the base is at the melting point, and there is a local drainage system for the resulting meltwater. How we treat this drainage system is key. To begin with, we suppose that the drainage system organises itself as described earlier, independently of the evolution of the bed elevation.

The surface elevation of the ice sheet is $z = z_i$, relative to a level $z = 0$ located at the elevation of the local drainage system. We suppose the bed consists of a saturated till of porosity ϕ. If the pore water pressure at the interface $z = s$ is p_w^s and the overburden normal stress there is P^s, then the corresponding pore and overburden pressures below the surface are taken to be

$$p_w = p_w^s + \rho_w g(s - z), \qquad P = P^s + \left[\rho_w \phi + \rho_s(1 - \phi)\right]g(s - z), \quad (10.337)$$

Fig. 10.16 Drumlins in the Ards Peninsula of Northern Ireland

simply through hydrostatic and lithostatic balance: ρ_w and ρ_s are the densities of water and sediment, respectively. Within the till, the effective pressure is defined as

$$p_e = P - p_w, \tag{10.338}$$

and thus

$$p_e = N + (1 - \phi)\Delta\rho_{sw}g(s - z), \tag{10.339}$$

where

$$\Delta\rho_{sw} = \rho_s - \rho_w, \tag{10.340}$$

and N is the effective pressure at the interface,

$$N = P^s - p_w^s. \tag{10.341}$$

The interfacial normal stress P^s is related to the stress in the ice by

$$P^s = -\sigma_{nn} = p_i^s - \tau_{nn}, \tag{10.342}$$

where σ_{nn} is the normal stress in the ice, τ_{nn} is the deviatoric normal stress in the ice, and p_i^s is the ice pressure at the bed. As is customary in ice sheet dynamics, we define the reduced pressure Π in the ice by

$$p_i = p_a + \rho_i g(z_i - z) + \Pi, \tag{10.343}$$

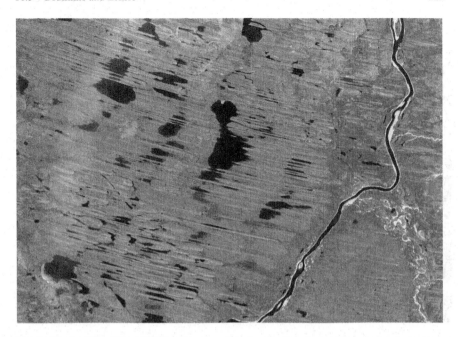

Fig. 10.17 Satellite view of MSGL in Northern Canada. The lineations are about a hundred metres in width, and of the order of a hundred kilometres in length

where p_a is atmospheric pressure, and we define the effective pressure in the drainage system as

$$N_c = p_a + \rho_i g z_i - p_c, \tag{10.344}$$

where p_c is the water pressure in the local drainage system, which we presume known. From these it follows that the effective pressure at the bed is given by

$$N = N_c + \Delta \rho_{wi} g s + \Pi - \tau_{nn}, \tag{10.345}$$

where

$$\Delta \rho_{wi} = \rho_w - \rho_i. \tag{10.346}$$

The drainage effective pressure N_c is presumed to be determined by the properties of the local hydraulic drainage system, as discussed in Sect. 10.3.

Bed Evolution

We restrict our initial presentation of the model to two dimensions (x, z), for the sake of clarity. The generalisation to three dimensions is given subsequently (see also Question 10.16). The evolution of the bed is given by the Exner equation

$$s_t + q_x = 0, \tag{10.347}$$

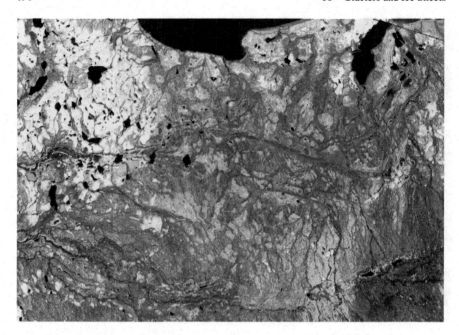

Fig. 10.18 A system of eskers in Northern Canada. This false colour satellite image shows the eskers as the criss-crossed red linear features

Fig. 10.19 System geometry

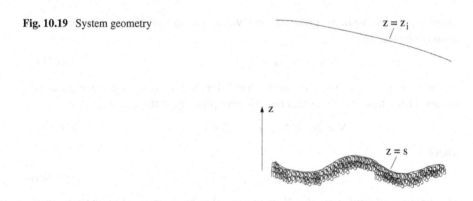

where q is the basal sediment flux. Ideally, q would be determined in the field or laboratory, but this is not very practicable. Alternatively, we might propose a sediment transport law based on a presumed rheology of the till. This also is problematic, since the determination of the rheology of granular materials is difficult and controversial. For the present purpose, we can largely avoid the issue, recognising only that there is sediment transport ($q \neq 0$), and that it is likely to depend on both the basal shear stress τ and the basal effective pressure N, thus

$$q = q(\tau, N). \tag{10.348}$$

Ice Flow

For simplicity we suppose the flow of ice is Newtonian, with a constant viscosity η. In two dimensions, the equations for the velocity (u, w) can be written in the form

$$u_x + w_z = 0,$$
$$0 = -\Pi_x - \rho_i g z_i' + \eta \nabla^2 u, \qquad (10.349)$$
$$0 = -\Pi_z + \eta \nabla^2 w,$$

where Π is the reduced pressure defined earlier, and $z_i' = \frac{\partial z_i}{\partial x}$.

Surface Boundary Conditions

The conditions which we apply at the surface are those of stress continuity and a kinematic condition, which take the form

$$\Pi - \tau_{nn} = 0, \qquad \tau_{nt} = 0, \qquad w = \frac{\partial z_i}{\partial t} + u \frac{\partial z_i}{\partial x} - a \quad \text{at } z = z_i, \qquad (10.350)$$

where τ_{nt} is the shear stress, a is the accumulation rate. We can anticipate that the horizontal length scale of interest will be that of drumlins, thus in the range 100–1000 m, and much less than the horizontal length scale appropriate to ice sheets. Therefore it seems reasonable to suppose that z_i will be almost constant, and the boundary conditions (10.350) can be approximately applied at a flat interface. Although the regional slope of the ice surface is negligible geometrically, it is necessary to retain it in the force balance equation (10.349).[24]

There is an alternative possibility for the upper boundary condition, which arises in the case that the parameter

$$\sigma = \frac{l}{d_i} \qquad (10.351)$$

is small, where l is the horizontal drumlin length scale and d_i is the ice depth scale. In this case, the flow near the base is akin to a boundary layer flow, and the appropriate condition is a matching condition to the outer ice sheet flow, which sees the base as essentially flat with small scale wrinkles. Assuming this outer ice sheet flow is a shear flow which varies on a horizontal length scale $\gg d_i$, appropriate matching conditions are

$$\Pi \to 0, \qquad \eta u_z \to \tau_b, \qquad w \to 0 \quad \text{as } z \to \infty. \qquad (10.352)$$

The quantity τ_b is the basal shear stress determined by the outer flow, and is given by

$$\tau_b = -\rho_i g z_i z_i'. \qquad (10.353)$$

---

[24] This is analogous to the Boussinesq approximation in convection: the surface slope is important in determining the driving stress, but negligible otherwise.

It is not clear which of the limits for σ is the more appropriate. Four hundred metre drumlins under eight hundred metres of ice suggest $\sigma = O(1)$, but two hundred metre drumlins under two thousand metres of ice suggest $\sigma \ll 1$. Nor is it clear whether there might be any essential difference in the resulting stability analysis. Since the limit $\sigma \ll 1$ is the simpler, we focus henceforth on that case.

Basal Boundary Conditions

To write the basal boundary conditions, we need to construct the normal and shear stress, and the tangential velocity, using the unit normal and tangent vectors. (In three dimensions, there are two tangent vectors to be used, see Question 10.16.)

In two dimensions, the normal and shear deviatoric stresses are

$$-\tau_{nn} = \frac{2\eta}{1+s_x^2}\left[\left(1-s_x^2\right)u_x + s_x(u_z + w_x)\right],$$

$$\tau = \frac{\eta}{1+s_x^2}\left[\left(1-s_x^2\right)(u_z + w_x) - 4s_x u_x\right].$$

$$(10.354)$$

We suppose that there is a sliding velocity, which we denote by U, and as for sediment transport, we suppose that this depends on the interfacial shear stress[25] τ and interfacial effective pressure N. Accounting for the tangential velocity at the bed, the sliding law then takes the form

$$\frac{u + ws_x}{(1+s_x^2)^{1/2}} = U(\tau, N).$$

$$(10.355)$$

As for the sediment transport, we avoid specification of how sliding is achieved; it might be by deformation of the underlying till, or by slip at the ice-till interface. In either case we expect dependence of U on τ and N.

The final condition at the bed is the kinematic condition,

$$w = s_t + us_x;$$

$$(10.356)$$

we ignore interfacial melting, usually of the order of millimetres per year, and negligible in this context.

Between them, Eqs. (10.345), (10.348), (10.354), (10.355) and (10.356) need to provide a total of two interfacial boundary conditions for the ice flow; the Exner equation (10.347) provides the evolution equation for s. We can take the two interfacial conditions to be the velocity conditions (10.355) and (10.356), which are given in terms of τ, N and s. Then (10.345) and (10.354)$_1$ determine N, (10.347) and (10.348) determine s, and (10.354)$_2$ determines τ. The model is therefore complete.

---

[25] Some confusion is liable to occur between the values of shear stress and other quantities at the ice-till interface, and the values of these quantities far from the interface, because it is normal to refer to the far field values as 'basal', since on the ice sheet scale, they are at the base of the large scale ice flow. We will endeavour to avoid this confusion by referring to 'far field' and 'interfacial' values (despite having defined the basal shear stress in (10.353)!).

A Reduced Model

We begin by non-dimensionalising the model. There are a number of length scales present in the equations. We define the quantities

$$z_i \sim d_i, \qquad d_D = \frac{N_c}{\Delta \rho_{iw} g}, \qquad d_T = \frac{N_c}{\Delta \rho_{sw} g (1 - \phi)}. \tag{10.357}$$

These length scales are the ice depth scale d_I, the drumlin depth scale d_D, and the till deformation depth scale d_T. To explain the significance of these scales, we resume our earlier discussion of till deformation.

Subglacial till is a granular material, consisting of rough angular fragments in a matrix of finer-grained material, the whole being water saturated when it is being deformed. In common with all granular materials, we expect that when subjected to a shear stress, it will not deform until the shear stress exceeds a critical value, called the yield stress. The reason for this is simple, insofar as we expect two solid surfaces not to permit sliding until the static coefficient of friction is exceeded. More specifically, if the normal stress between two clasts is p_e, and the shear stress is τ, then slip will occur if

$$\tau > \mu p_e, \tag{10.358}$$

where μ is the coefficient of friction. More generally, the Mohr–Coulomb yield stress τ_c in a granular material is

$$\tau_c = c + \mu p_e, \tag{10.359}$$

where c is the cohesion, often ignored as being small for subglacial till. The coefficient μ is of $O(1)$, and is related to the angle of friction ϕ_f by $\mu = \tan \phi_f$.

If we now consult (10.339) and (10.345), two observations can be made. Till deformation will cease at effective pressures larger than τ/μ. Typical basal shear stresses are likely to be in the range 0.1–1 bar, so that till will only deform at all if p_e is of this order. In particular, till deformation can only occur at values of $N_c \lesssim 1$ bar. Such low values of the effective stress have been measured under the Ross ice streams, and may be associated with a distributed, canal type of drainage. Assuming, then, that $\tau \sim N_c$, we see from (10.339) that till deformation is only viable to a depth of order d_T, as defined in (10.357). Below this depth, the effective pressure is too large to promote till deformation. This observation allows us to suggest a typical value of till transport.

The second observation is drawn from (10.345). If we anticipate that drumlins grow as a consequence of instability of a flat bed, then the effective pressure increases with drumlin elevation. When the drumlins attain an elevation of order d_D given by (10.357), the summit effective pressure will be large enough to slow down the till and thus also deformation, which presumably stunts further growth. This depth scale thus provides an estimate for the eventual height of drumlins. Choosing $N_c = 0.4$ bars gives a depth scale of 50 m, although numerical solutions generally give smaller values.

We use these ideas in choosing scales for the variables. It is first convenient to define a stream function for the flow via

$$\psi_z = u, \qquad -\psi_x = w. \tag{10.360}$$

We take the basic shear flow without bed perturbations to be

$$\psi = u_0 z + \frac{\tau_b}{2\eta} z^2, \tag{10.361}$$

and a dimensionless mean velocity $\bar{u}(t)$ is introduced as the developing bedforms alter the average sliding velocity. Specifically, we scale the model by choosing

$$z_i = d_i h, \qquad x, z \sim l, \qquad p_e, N, \Pi, \tau_{nn}, \tau \sim N_c, \qquad s \sim d_D,$$

$$\psi = u_0 \bar{u} z + \frac{\tau_b}{2\eta} z^2 + u_0 d_D \Psi, \qquad U \sim u_0, \qquad q \sim u_0 d_T, \qquad t \sim \frac{d_D l}{d_T u_0}, \tag{10.362}$$

and we scale the depth of the till by writing

$$s - z = d_T \zeta. \tag{10.363}$$

Thus the dimensionless effective pressure in the till is

$$p_e = N + \zeta, \tag{10.364}$$

and the yield criterion (10.358) becomes

$$\zeta < \frac{\tau}{\mu} - N. \tag{10.365}$$

The value of u_0 is determined by the magnitude of the sliding velocity, and the horizontal length scale is defined by balancing the stress and strain rates, thus

$$l = \left(\frac{\eta u_0 d_D}{N_c} \right)^{1/2} = \left(\frac{\eta u_0}{\Delta \rho_{iw} g} \right)^{1/2}. \tag{10.366}$$

If we choose $u_0 = 100 \text{ m y}^{-1}$ and $\eta = 6$ bar year ($\approx 2 \times 10^{13}$ Pa s), then $l = 271$ m. Other typical values, with $\rho_s = 2.5 \times 10^3 \text{ kg m}^{-3}$ and $n = 0.4$, are $d_T = 4.6$ m, and the time scale is 29 y.[26]

With this choice of scaling, the dimensionless model for the ice flow is

$$0 = -\Pi_x + \nabla^2 \Psi_z + \sigma\theta,$$
$$0 = -\Pi_z - \nabla^2 \Psi_x, \tag{10.367}$$

with far field boundary conditions (appropriate for small σ)

$$\Pi \to 0, \qquad \Psi \to 0 \quad \text{as } z \to \infty. \tag{10.368}$$

---

[26]This time scale is rather long, given recent observations of bedforms growing in a matter of years. As we shall see below, the instability does in fact occur on a much shorter time scale (and also on a shorter length scale).

The basal conditions take the form

$$-\tau_{nn} = \frac{2[(1 - v^2 s_x^2)\Psi_{zx} + v s_x(\theta + \Psi_{zz} - \Psi_{xx})]}{1 + v^2 s_x^2},$$

$$\tau = \frac{(1 - v^2 s_x^2)(\theta + \Psi_{zz} - \Psi_{xx}) - 4 v s_x \Psi_{zx}}{1 + v^2 s_x^2},$$

$$\frac{\bar{u} + v\theta z + v\Psi_z - v^2\Psi_x s_x}{(1 + v^2 s_x^2)^{1/2}} = U(\tau, N), \tag{10.369}$$

$$-\Psi_x = \alpha s_t + [\bar{u} + v\theta z + v\Psi_z]s_x,$$

$$N = 1 + s + \Pi - \tau_{nn},$$

$$s_t + q_x = 0,$$

and these are all applied at $z = vs$.

The dimensionless parameters σ, θ, v and α are defined by

$$\sigma = \frac{l}{d_i}, \qquad \theta = \frac{\tau_b}{N_c}, \qquad v = \frac{d_D}{l}, \qquad \alpha = \frac{d_T}{d_D}. \tag{10.370}$$

Supposing $l = 300$ m, $d_i = 1500$ m, thus $\tau_b = 0.15$ bar with an assumed ice surface slope of 10^{-3}, $d_D = 50$ m, $N_c = 0.4$ bar, $d_T = 5$ m, typical values are

$$\sigma \sim 0.2, \qquad \theta \sim 0.38, \qquad v \sim 0.16, \qquad \alpha \sim 0.1. \tag{10.371}$$

We now simplify the model by considering the aspect ratio $v \ll 1$. Putting $v = 0$ (and putting $\sigma = 0$ in the momentum equations), the reduced model is then

$$0 = -\Pi_x + \nabla^2\Psi_z,$$
$$0 = -\Pi_z - \nabla^2\Psi_x, \tag{10.372}$$

with matching condition (10.368), and interfacial conditions applied at $z = 0$:

$$-\tau_{nn} = 2\Psi_{zx},$$
$$\tau = \theta + \Psi_{zz} - \Psi_{xx},$$
$$\bar{u} = U(\tau, N), \tag{10.373}$$
$$-\Psi_x = \alpha s_t + \bar{u} s_x,$$
$$N = 1 + s + \Pi - \tau_{nn},$$
$$s_t + q_x = 0.$$

The stability of the uniform solution of this reduced model is studied in the following subsection.

It is straightforward to carry through this procedure in three dimensions (see also Question 10.16), and here we simply state the result. The position coordinates are now (x, y, z), with y being the transverse horizontal coordinate, and the corresponding velocity vector is $(u, v, w) = \mathbf{u}$. The reduced dimensionless model is

$$\nabla\Pi = \nabla^2\mathbf{u},$$
$$\nabla.\mathbf{u} = 0, \tag{10.374}$$

subject to

$$\Pi \to 0, \qquad u_z \to \theta, \qquad v, w \to 0 \quad \text{as } z \to \infty, \tag{10.375}$$

and

$$\tau_{nn} = 2w_z,$$
$$\tau_1 = \theta + u_z + w_x,$$
$$\tau_2 = v_z + w_y,$$
$$\tau = \left[\tau_1^2 + \tau_2^2\right]^{1/2},$$
$$\bar{u} = \frac{U(\tau, N)\tau_1}{\tau}, \tag{10.376}$$
$$0 = \frac{U(\tau, N)\tau_2}{\tau},$$
$$w = \alpha s_t + \bar{u} s_x,$$
$$N = 1 + s + \Pi - \tau_{nn},$$
$$s_t + \nabla . \mathbf{q} = 0,$$

all applied at $z = 0$. One might suppose that, since also α is quite small, it too could be neglected. As our linear stability analysis will show, this is not possible, since it provides a stabilising term at high wave number.

Ice Flow Solution

Reverting to the two-dimensional problem, the ice flow problem is linear, and can be solved conveniently using the Fourier transform

$$\hat{f}(k) = \int_{-\infty}^{\infty} f(x)e^{ikx}\,dx; \tag{10.377}$$

omitting details, we then find that

$$N = 1 + s - 2\mathcal{H}\{\alpha s_{xt} + \bar{u} s_{xx}\}, \tag{10.378}$$

where the Hilbert transform is

$$\mathcal{H}(g) = \frac{1}{\pi} \int_{-\infty}^{\infty} \frac{g(t)\,dt}{t - x}. \tag{10.379}$$

The interfacial shear stress $\tau \approx \tau_1$ can be inverted to the form

$$\tau = f(\bar{u}, N), \tag{10.380}$$

while a horizontal average of this yields the condition

$$\theta = \overline{f(\bar{u}, N)}, \tag{10.381}$$

which serves to specify the average dimensionless sliding velocity $\bar{u}$.

The sediment flux q was taken to depend on τ and N. It is the product of a dimensionless deformable till thickness A and a mean velocity V, which we suppose is

Fig. 10.20 The function
$A(N)$ given by (10.382),
where $f(\bar{u}, N) = \theta \bar{u}^a N^b$.
The parameters used are
$b = 0.6$, $\mu = 0.4$, $\theta = 0.8$,
$\bar{u} = 1$

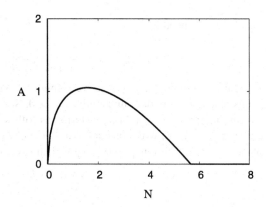

constrained by the ice velocity $\bar{u}$. The deformable depth is constrained by (10.365),
which suggests that we choose

$$A = A(N) = \left[\frac{f(\bar{u}, N)}{\mu} - N \right]_+ \tag{10.382}$$

($[x]_+ = \max(x, 0)$). Since we suppose $V \sim \bar{u}$, which itself depends on N, we may
as well take $V = 1$ and thus $q = A$. Our model for bed elevation is thus completed
by solving the Exner equation

$$\frac{\partial s}{\partial t} + \frac{\partial A(N)}{\partial x} = 0, \tag{10.383}$$

together with the normal stress condition (10.378). Note that this is a nonlinear
model for the bed elevation. A typical form of the sediment flux function $q = A$ is
shown in Fig. 10.20.

Linear Stability

We now consider the linear stability of the reduced, two-dimensional model
(10.378) and (10.383). The basic uniform state is (assuming the sliding law $\tau = \theta \bar{u}^a N^b$)

$$N = 1, \qquad s = 0. \tag{10.384}$$

For simplicity we suppose $\bar{u} \equiv 1$, which is in any case valid in the linearised theory.

We linearise about this basic state by putting $N = 1 + P$, and linearising for
small s and P. Denoting the transforms with an overhat, and using the facts that
$\widehat{f_x} = -ik\hat{f}$ and $\widehat{\mathcal{H}(g_x)} = -|k|\hat{g}$, we find that $\hat{s} \propto e^{\sigma t}$, with

$$\sigma = \frac{ikA'(1 - 2ik|k|)}{1 - 2ik|k|\alpha A'}, \tag{10.385}$$

where $A' = A'(1)$. With $\sigma = r + ikc$, this implies that the growth rate is

$$r = \frac{2k^2|k|A'(1 - \alpha A')}{1 + 4\alpha^2 A'^2 k^4}, \tag{10.386}$$

and the wave speed is

$$c = \frac{A'(1 + 4\alpha A'k^4)}{1 + 4\alpha^2 A'^2 k^4}.$$ (10.387)

Bearing in mind that α is relatively small, we see from (10.386) that the flat bed is unstable if $A' > 0$, or equivalently if $q' > 0$. Since $q = q(\tau, N)$ and $\tau = f(u, N)$, we can interpret this instability criterion as follows. If we draw the two families of curves $\tau = f(u, N)$ with constant u, and $q = q(\tau, N)$ with constant q in the (N, τ) plane, then the criterion $q' > 0$ is equivalent (assuming $q_\tau > 0$) to

$$\left. \frac{d\tau}{dN} \right|_u > \left. \frac{d\tau}{dN} \right|_q.$$ (10.388)

This criterion is easily satisfied for reasonable choices of sediment flux and sliding law. If we use the sliding law $\tau = \theta u^a N^b$, then the instability criterion is

$$\theta > \frac{\mu}{b}.$$ (10.389)

Providing the base of the ice is at the melting point, (10.389) shows that drumlins will form for large enough basal shear stress, or for low enough channel effective pressure. If we suppose that b is close to one, corresponding to a 'plastic' till, then the instability criterion is that $\tau_b \gtrsim \mu N_c$, which is simply the criterion that the till should deform. Roughly speaking, we can expect drumlinisation wherever till deforms.

The wave speed c is positive, and the growth rate has a maximum at a wave number

$$k = k_{\max} = \frac{3^{1/4}}{(2\alpha A')^{1/2}}.$$ (10.390)

If we take $\alpha = 0.1$ and $A' = \frac{1}{3}$, then $\frac{2\pi}{k_{\max}} \approx 1.23$, corresponding to a dimensional wavelength of 334 m. The corresponding growth time scale is

$$t_{\max} = \frac{1}{r} \approx \frac{1 + 4\alpha^2 A'^2 k_{\max}^2}{2k_{\max}^3 A'} \approx 0.038,$$ (10.391)

corresponding to a dimensional growth time of thirteen months. It is a hallmark of the instability that it is rapid.

Nonlinear Results

The main difficulty in computing finite amplitude solutions of the model (10.378) with (10.383) is that as the instability develops, N decreases until it inevitably reaches zero; physically, cavities form in the lee of obstacles. If s continues to denote the base of the ice flow, then while (10.378) still applies, the Exner equation (10.383) must be replaced by the cavitation condition

$$N = 0.$$ (10.392)

Fig. 10.21 Finite amplitude
till surface obtained from
solving (10.393) with a
suitable approximation for
$A(N)$ resembling Fig. 10.20.
The two surfaces are the ice
base and the base of the
deformable till layer (which
is actually $s - \alpha A$). The *thick
horizontal bands* indicate the
cavities, where $N = 0$

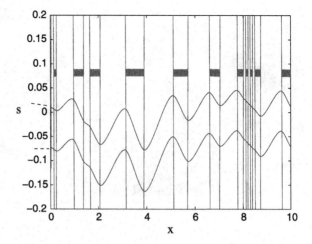

This makes the model difficult to solve numerically. One way round this is to continue to solve (10.383), but to extend the definition of A so that it is any positive value when $N = 0$. As in fact indicated in Fig. 10.20, this makes A a piecewise smooth *graph*. The combined model is thus

$$N = 1 + s - 2\mathcal{H}\{\alpha s_{xt} + \bar{u}s_{xx}\},$$
$$\frac{\partial s}{\partial t} + \frac{\partial a}{\partial x} = 0, \tag{10.393}$$

with

$$\begin{cases} a = A(N), & N > 0, \\ a > 0, & N = 0. \end{cases} \tag{10.394}$$

In practice, we approximate the graph of A by a smooth non-monotonic function. Despite this, the model is difficult to solve numerically. This is because as the oscillations grow, a transition takes place when the maximum of A is reached. When this happens, there is a rapid transition to a state in which N is piecewise constant, being positive on the upstream face of the bedforms, and (approximately) zero on the downstream cavities. At this transition, a spectral method (used because of the nice properties of the Hilbert transform in Hilbert space) generates transient high frequency components which can cause numerical breakdown. Figure 10.21 shows the result of one such calculation, in which the positions of the cavities are indicated by the horizontal bands. In this model, the drumlins reach a stationary state. In more detailed models, they form finite amplitude travelling waves, as discussed in the notes.

10.5.2 Eskers

Eskers are long, sinuous ridges of sand and gravel which, like drumlins, are associated with the existence of former ice sheets. They are thought to form through

Fig. 10.22 An esker which formed during retreat of the Stagnation Glacier, Bylot Island, Nunavut, Canada, 72°57′ 41″ N, 78°21′2″ W, in 1992. The bouldery ridge in the background is the inner face of the substantial lateral moraine surrounding the rapidly retreating glacier. The landform is actually composed mostly of glacial ice preserved by the insulating cover of about a metre of bouldery esker gravel. The height is estimated to be 8–10 m. Photograph by Christian Zdanowicz, available at www.inrs.illinois.edu/shilts

the deposition of sediments in subglacial or ice-walled pro-glacial channels, and form anastomosing patterns such as that in Fig. 10.18. On the ground, they look as shown in Fig. 10.22, although eskers which form under ice sheets are generally larger, having elevations in the range 10–50 m, and widths of 50–500 m. And they are often shrouded in trees, and only properly visible from the air. In length, they can sometimes be traced for hundreds of kilometres, although often they are segmented. Indeed, eskers are often 'beaded', either consisting of independent beads, or having oscillations in elevation along their length.

Because they consist of sorted sands and gravels, they are associated with channelised water flow, but the detailed way in which they form is not properly known. Generally, they are associated with retreating ice, and are thought to form at the margin of the ice sheet. The sediment might be deposited within the subglacial channel. Alternatively, deposits might occur pro-glacially in a stream walled by dead ice; or, if the ice terminates in a pro-glacial lake, a sub-aqueous fan may occur. In all these cases, one imagines the esker being built regressively as the ice retreats; however, little is directly known of the process, and it is not inconceivable that eskers could form wholly below the ice sheet.

Models for the construction of eskers do not yet exist, but a clue to their formulation lies in our earlier discussion of different drainage theories. In our discussion of canals (Sect. 10.3.3), we posited a drainage style in which a sediment-floored canal lay beneath an essentially flat roof. The astute reader will have been concerned as to why the ice roof should be flat, as no grounds were given for this supposition. Suppose the elevation of the ice roof above the local ice/till interface is h_i, and the depth of the stream base below it is h_s. Then the Röthlisberger channel corresponds to the assumption $h_i > 0$, $h_s = 0$, while the canal corresponds to the assumption $h_i = 0$, $h_s > 0$; an intermediate case has both being positive.

What is missing in our model is any reason for either assumption, but we can in principle supply a reason by positing a model in which both h_i and h_s are variables. We do this below, but now we can also realise that there is no reason why we cannot also have the cases $h_i > 0$, $h_s < 0$ and $h_i < 0$, $h_s > 0$; in the former case, sediment is deposited while the channel flow is maintained above it: this presumably corresponds to esker formation. In the latter case, the ice squeezes down while the stream evacuates the sediment; this corresponds to the formation of tunnel valleys.[27] Thus, at least in principle, a single model could predict all these features.

To see the structure of such a model, we generalise the discussion in Sect. 10.3.3 to allow for separate ice roof elevation h_i and sediment floor depth h_s. In its simplest form, the model is written

$$\frac{\partial(wh_i)}{\partial t} = \frac{\dot{m}_i}{\rho_i} - \frac{w^2}{\eta_i}(N + \Delta\rho_{wi}gh_i),$$
$$\frac{\partial(wh_s)}{\partial t} = \frac{\dot{m}_s}{\rho_s} - \frac{w^2}{\eta_s}(N + \Delta\rho_{sw}gh_s); \tag{10.395}$$

these represent the kinematic equations for the ice/water interface and the till/water interface, respectively, and involve a melting rate $\dot{m}_i$, erosion rate $\dot{m}_s$, and ice and till viscosities η_i and η_s.[28] The particular closure relations in (10.395) are those for a wide channel, and for geometric simplicity we suppose the wetted perimeter $l = 2w$, where w is the channel width. From (10.238), (10.239), (10.240) and (10.242), we have

$$\dot{m}_i = C_i Q, \tag{10.396}$$

where

$$C_i = \frac{\rho_i g S_i}{2L}. \tag{10.397}$$

The erosion rate of a subglacial stream is a more complex matter. For a stream with banks, we might suppose erosion of the sides by bank collapse is proportional to the stream power. However, if the till squeezes up into the channel, there are no 'banks', and the erosion rate should presumably decrease to zero, unless we imagine a stream of varying cross section, thus with erosion in the shallows, and the principal

---

[27]Tunnel valleys are large (hundreds of metres deep, kilometres wide) braided systems of former drainage channels (presumably), often infilled with sediment. They are found in Northern Germany and Denmark, for example.

[28]More exactly, they are parameters proportional to the viscosities with some suitable geometry-dependent coefficients.

downstream sediment transport in the deeper flow. For such a case, it is plausible to provide an analogous description for erosion rate,

$$\dot{m}_s = C_s Q, \tag{10.398}$$

where

$$C_s = \frac{\rho_i g S_i}{2 L_E}, \tag{10.399}$$

and L_E is a term representing latent work of erosion. Finally, the extra gravitational terms in the closure rates arise through the contribution of the respective bed elevations to the driving hydraulic closure stress.

If we take the Chézy friction law $\tau = f \rho_w u^2$, then $u = \sqrt{Ch}$, where

$$C = \frac{\rho_i g S_i}{2 f \rho_w}, \tag{10.400}$$

and the water flux is

$$Q = C^{1/2} w h^{3/2}, \tag{10.401}$$

where h is the total depth,

$$h = h_i + h_s. \tag{10.402}$$

If we suppose that sediment flux Q_s is described by a Meyer-Peter and Müller relation, then we have (cf. (6.14), or (5.5) and (5.6)),

$$Q_s = K' w (h - h_c)^{3/2}, \tag{10.403}$$

where

$$K' = \frac{\rho_i K S_i}{\Delta \rho_{sw}} \left(\frac{\rho_i g S_i}{8 \rho_w} \right)^{1/2}, \qquad h_c = \frac{2 \tau_c^* \Delta \rho_{sw} D_s}{\rho_i S_i}, \tag{10.404}$$

with $\tau_c^* \approx 0.05$, $K = 8$. The units of Q_s are m^3 s^{-1}.

In Röthlisberger channel theory, mass conservation of water determines the water volume flux Q as a function of distance downstream, and we might suppose that the equation of sediment conservation would likewise determine the sediment volume flux Q_s, both of them being increasing functions of distance downstream. This being so, the three equations in (10.395) and (10.402) serve to determine the three quantities N, h_i and h_s, with the channel width w and depth h being determined by (10.401) and (10.403). Unique values of $h > h_c$ and w exist provided

$$\frac{Q_s}{Q} < \frac{K'}{\sqrt{C}}, \tag{10.405}$$

and we suppose this to be true.

Steady solutions for h_i and N are then

$$h_i = \frac{(A_i - A_s) h^3}{\Delta \rho_{si} Q} + \frac{\Delta \rho_{sw}}{\Delta \rho_{si}} h, \tag{10.406}$$

where

$$A_i = \frac{\eta_i CC_i}{\rho_i g}, \qquad A_s = \frac{\eta_s CC_s}{\rho_s g}, \tag{10.407}$$

and

$$N = \frac{g \Delta\rho_{wi} \Delta\rho_{sw}}{\Delta\rho_{si}} \left[\frac{h^2}{\beta^2 Q} - 1 \right] h, \tag{10.408}$$

where

$$\beta = \left(\frac{\Delta\rho_{wi} \Delta\rho_{sw}}{\Delta\rho_{sw} A_i + \Delta\rho_{wi} A_s} \right)^{1/2}. \tag{10.409}$$

It is useful to write (10.406) in the form

$$h_i = \frac{B(1-\chi)h^3}{Q} + rh, \tag{10.410}$$

where

$$B = \frac{\rho_i g S_i^2 \eta_i}{4 f \rho_w \Delta\rho_{si} L}, \quad \chi = \frac{A_s}{A_i} = \frac{\rho_i L \eta_s}{\rho_s L_E \eta_i}, \quad r = \frac{\Delta\rho_{sw}}{\Delta\rho_{si}}. \tag{10.411}$$

In Fig. 10.23 we plot representative graphs of h_i versus h for values $A_i > A_s$ ($\chi < 1$) and $A_i < A_s$ ($\chi > 1$). The physically accessible space where $h > 0$ is divided into three regions. When $0 < h_i < h$, channels exist, with both the ice and the sediment being excavated. There are two particular cases: $h_i = h$ corresponds to a Röthlisberger channel, while $h_i = 0$ corresponds to a canal. If $h_i > h$, then $h_s < 0$: the sediment infiltrates the channel, causing an esker to form. If $h_i < 0$, the ice collapses, forming a tunnel valley.

As sediment flux and water discharge increase downstream, h (determined by $(1 - \frac{h_c}{h})^3 = \frac{CQ_s^2}{K^2 Q^2}$) may increase or decrease; plausibly it remains constant (if Q_s/Q

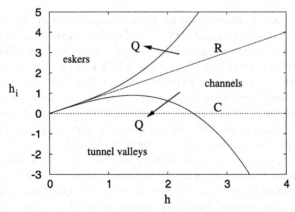

Fig. 10.23 Ice roof elevation in metres as a function of channel depth in metres, based on (10.411), using values $B = 0.8 \text{ m s}^{-1}$, $r = 0.95$, and values of $\chi = \frac{A_s}{A_i} = 0.8$ (*upper curve*) and $\chi = \frac{A_s}{A_i} = 1.2$ (*lower curve*). Channels exist for small volume and sediment fluxes, but eskers or tunnel valleys develop at larger fluxes, depending on the stiffness of the till. The lines $h_i = h$ (marked R) and $h_i = 0$ (marked C) indicate Röthlisberger channels and canals, respectively

is constant). However, as Q increases, the upper and lower curves become steeper, so that in this simple theory, eskers or tunnel valleys are promoted at larger water fluxes, and which of them occurs depends sensitively on the stiffness of the till via the definition of χ. Stiff till (high χ) promotes tunnel valley formation, while soft till (high χ) promotes esker formation. Note that, from (10.408), N increases with h, and is only positive for $h > \beta\sqrt{Q}$. We associate the lower limit with the onset of channelised flow, supposing that for lower h, a distributed film flow exists, much as discussed earlier.

The present discussion promotes a pedagogical point, which is that it may be possible to provide an understanding of eskers and tunnel valleys, as well as Röthlis-berger channels or canal, all on the basis of a self-consistent description of drainage mechanics. However, our rudimentary discussion falls a fair way short of genuine prediction. Most obviously, the ice viscosity depends on N, as does the till rheology (strongly, if it is essentially plastic), and so the critical parameter χ will depend on N and thus also h. In addition, variation with space and time is likely to be impor-tant. Furthermore, it is not immediately obvious whether the drainage characteristics of the different types of channel or canal are consistent with our earlier discussion of them.

10.6 Glaciology on Mars

We are used to the existence of glaciers and ice sheets on Earth, but ice also exists elsewhere in the solar system, and is the source of interesting and elusive phenom-ena. As an example, we consider the polar ice caps of Mars. These apparently consist largely of water ice, and that at the north pole is the larger, being comparable in size to the Greenland ice sheet. Both ice caps are covered in their respective winters by an annual layer of CO_2 frost, which sublimates in the summer, leaving the residual ice caps.

We will focus our attention on the northern ice cap, shown in Fig. 10.24, which is some 3 kilometres in depth, and 1000 km in horizontal extent. As can be seen in the picture, this ice cap is quite unlike the large ice sheets on Earth. The surface is irregular. In particular, there is a large canyon which looks as if it has been gouged from the ice surface, towards the left of the picture. This is the *Chasma Borealis*. The other pronounced feature of the ice cap consists of the stripes on the surface. These stripes are arranged in a spiral, rotating anti-clockwise, and they consist of concentrations of dust, associated with a series of troughs in the ice. Figure 10.25 shows the troughs, carved into a cross section of the ice cap.

The spiral waves suggest a formation mechanism similar to that of the Fitzhugh–Nagumo equations, which form a reaction–diffusion system of activator-inhibitor type. When the kinetics of the reaction terms are oscillatorily unstable, the addi-tion of diffusion causes the oscillations to propagate as travelling waves. The pres-ence of 'impurities' can cause these waves to propagate as spiral waves (cf. Ques-tion 1.6).

We thus set out seeking a model whose time-dependent behaviour is oscillatory. This can most easily be obtained by identifying a positive feedback in the system.

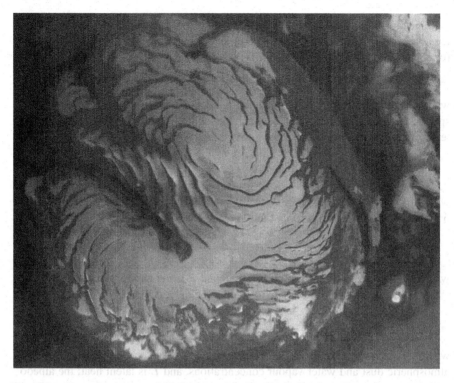

Fig. 10.24 Mars orbital camera (MOC) image of the North Polar Ice Cap of Mars. Image courtesy NASA/JPL/Malin Space Science Systems

Fig. 10.25 A cross section of the North Polar Ice Cap, showing the scarps and canyons. Redrawn from Ivanov and Muhleman (2000), reproduced with permission of Academic Press via Copyright Clearance Center. The height is relative to a plane 5 km below the mean geoid. The North Pole is near the summit

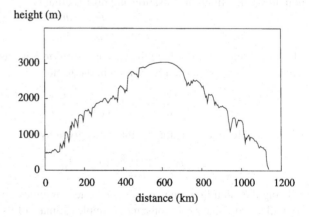

The mechanism we will use is that of dust-albedo feedback. Dust on the surface causes a reduced albedo, and thus a greater absorption of solar radiation. In turn this leads to greater sublimation of the ice, and thus increasing residual dust concentration.

Fundamental quantities in the model are thus the albedo a, the dust fraction of ice at the surface, ϕ, and the mass rate of sublimation m_i. The rate of sublimation (or condensation) is given by

$$m_i = K(p_s - p),\qquad(10.412)$$

where the rate coefficient K depends on wind speed and temperature, p is atmospheric water vapour pressure, and p_s is the saturation vapour pressures, given in terms of absolute temperature T by

$$p_s = p_s^{\mathrm{ref}} \exp\left[B\left\{ 1 - \frac{T_{\mathrm{ref}}}{T} \right\} \right],\qquad(10.413)$$

where $p_s = p_s^{\mathrm{ref}}$ at $T = T_{\mathrm{ref}}$; for water vapour, we may take $T_{\mathrm{ref}} = 273$ K at $p_s^{\mathrm{ref}} =$ 6 mbar ($=600$ Pa), the *triple point*, and the parameter B is given by

$$B = \frac{M_w L}{R T_{\mathrm{ref}}},\qquad(10.414)$$

where M_w is the molecular weight of water, and R is the gas constant.

Albedo and sublimation rate are related by the radiative energy balance law

$$I(1 - a) = \frac{\sigma T^4}{\Gamma} + m_i L,\qquad(10.415)$$

where I is received solar insolation, Γ is a greenhouse factor which may depend on atmospheric dust and water vapour concentrations, and L is latent heat; the albedo will depend on the amount of dust in the ice.

Let us denote the ice surface accumulation rate of dust as m_s, with units of mass per unit area per unit time. Then the rate of decrease of ice surface elevation due both to sublimation/condensation and dust accrual is

$$s = \frac{m_i}{\rho_i} - \frac{m_s}{\rho_s}.\qquad(10.416)$$

The basic equation describing the ice cap elevation h is then the dimensional equivalent of (10.45), which can be written in the form

$$\frac{\partial h}{\partial t} = -s + \frac{\partial}{\partial x}\left(D_i \frac{\partial h}{\partial x} \right),\qquad(10.417)$$

where the effective diffusion coefficient is given by

$$D_i = \frac{2A(\rho g)^n}{n+2} |h_x|^{n-1} h^{n+2},\qquad(10.418)$$

A being the Glen flow rate coefficient for ice, assumed constant. In our discussion we will assume that D_i is constant: a simple estimate of its appropriate size is $D_i \sim u_i l_i$, where u_i is a typical ice velocity, and l_i is the ice cap radius.[29]

---

[29] A consequence of the assumption of constant D_i will be that a finite gradient of h at the margin will imply non-zero ice flux there. This is unrealistic, and in particular, the existence of a steady state ice cap requires that the net balance be zero, i.e., $\int_0^{l_i} s\, dx = 0$.

We introduce the atmospheric water vapour concentration ρ and the atmospheric dust concentration c (both measured as mass per unit volume), noting that we may expect the greenhouse factor $\Gamma = \Gamma(\rho, c)$. The water vapour pressure is then given by the perfect gas law

$$p = \frac{\rho RT}{M_w}, \tag{10.419}$$

where R is the gas constant and M_w is the molecular weight of water. The subsidiary variables m_i, p_s, T and p are defined by (10.412), (10.413), (10.415) and (10.419). A further two relations are then necessary to determine ρ and c. These arise from the concepts of sediment transport as expounded in Chap. 5.

First, we propose two equations which describe conservation of mass of dust and water vapour in the near surface boundary layer. We will assume that a polar katabatic wind will flow downslope, transporting water vapour and dust in a thin near surface current. This current will entrain dust and water vapour from the troposphere above. If the current is of depth H, then suitable conservation laws for the vertically averaged vapour and dust concentrations ρ and c in the katabatic layer are

$$\frac{\partial(Hc)}{\partial t} + \frac{\partial(qc)}{\partial x} = E_c + \rho_s v_E - v_s c + \frac{\partial}{\partial x}\left(D_c H \frac{\partial c}{\partial x}\right),$$
$$\frac{\partial(H\rho)}{\partial t} + \frac{\partial(q\rho)}{\partial x} = E_\rho + m_i + \frac{\partial}{\partial x}\left(D_\rho H \frac{\partial \rho}{\partial x}\right), \tag{10.420}$$

where q is the katabatic wind flux, E_c and E_ρ are the entrainment rates of dust and vapour from the overlying troposphere, and D_c and D_ρ are turbulent horizontal diffusivities in the katabatic layer. Just as in Chap. 5, the term $\rho_s v_E$ represents erosion of dust from the surface, and the term $v_s c$ represents deposition of dust at the surface via settling; v_s is the settling velocity, and v_E is an erosional velocity.

The velocities v_s and v_E need to be specified, as do the amounts of frozen and unfrozen dust at the surface, and the ice dust accrual rate m_s. A reasonably general assumption about the surface is that there may be a thin surface layer of unfrozen dust which mantles the ice. If the 'depth' of this layer is F,[30] then conservation of unfrozen dust takes the form

$$\rho_s \frac{\partial F}{\partial t} = -\rho_s v_E + v_s c - m_s. \tag{10.421}$$

This layer is quite analogous to the bedload layer described in Sect. 6.4 (and F is analogous to $(1 - \phi)a$ in (6.7)).

In addition, we suppose that the ice at the surface (below the unfrozen layer) contains a volume fraction ϕ of dust. If $F \neq 0$, we can expect in general that the albedo is given by $a = a(\phi, F)$. It then remains to constitute ϕ and m_s. Quite generally, we find that for both sublimation and condensation

$$m_s = -\rho_s \phi s, \tag{10.422}$$

---

[30] More precisely, F is the volume of unfrozen dust per unit surface area.

and thus from (10.416)

$$m_i = \rho_i(1 - \phi)s. \qquad (10.423)$$

If $s > 0$,

$$\phi = \phi(x, \tau),$$

$$\int_\tau^t s(t')\,dt' = 0 \quad \text{for } s > 0. \qquad (10.424)$$

This simply states that if sublimation is occurring, the surface ice dust fraction is equal to its value at the last time the surface was exposed. Actually, (10.424) is an oversimplification, as it ignores the horizontal transport of the buried previous ice surface by ice flow. If $s < 0$, we suppose

$$\phi = \phi_s, \quad s < 0, \qquad (10.425)$$

where ϕ_s is the dust volume fraction of the unfrozen dust layer. This assumes that $F > 0$.

If $F \equiv 0$ over a time interval, then m_s is still defined by (10.422), (10.421) is irrelevant, and the ice surface dust fraction is determined by a balance between volume of ice accretion and volume of dust deposited, thus

$$\phi = \frac{cv_s - \rho_s v_E}{cv_s - \rho_s v_E - r_{si}m_i}, \qquad F \equiv 0, \qquad (10.426)$$

so long as ϕ is positive, where

$$r_{si} = \frac{\rho_s}{\rho_i}. \qquad (10.427)$$

The discussion above assumes implicitly that condensation of ice occurs directly at the surface. A different possibility is that condensation occurs as snowfall. If the snow crystals are pure, then there is no difference in the model. However, we might also suppose that ice crystals nucleate on dust particles, so that dust accumulation in the ice is partly due to this. If in fact there is no deposition of unfrozen dust (thus $F = 0$), then when $s < 0$ (it is snowing), the surface ice dust fraction will be that of the snow particles. A simple partitioning by volume fraction of dust within snow then suggests

$$\phi = \frac{c}{c + r_{si}\rho}, \quad s < 0 \quad \text{(snowfall)}. \qquad (10.428)$$

10.6.1 Non-dimensionalisation

We now proceed to make this model non-dimensional. We mostly denote appropriate scales with a subscript zero, and in particular we scale the variables as

$$\begin{aligned}
& m \sim m_0, & p, p_s \sim p_0, & \quad T \sim T_0, & \quad I \sim I_0, \\
& s \sim s_0, & h \sim h_0, & \quad t \sim t_0, & \quad x \sim l_i, & \quad \rho \sim \rho_0, \qquad (10.429) \\
& c \sim c_0, & q \sim q_0, & \quad F \sim F_0,
\end{aligned}$$

and we write

$$\frac{1-a}{1-a_0} = \alpha, \qquad K = \frac{K'\kappa}{T},$$
(10.430)

and suppose K' is constant, such that $\kappa = 1$ when there is no surface dust layer, $F = 0$: in general, we may expect κ to be a rapidly decreasing function of F; a_0 is the albedo of clean Martian ice. Balances of terms in the equations are effected by writing

$$T_0 = \left(\frac{I_0(1-a_0)}{\sigma}\right)^{1/4}, \qquad p_0 = p_s^{\text{ref}} \exp\left[B\left\{1 - \frac{T_{\text{ref}}}{T_0}\right\}\right],$$

$$m_0 = \frac{K'p_0}{T_0}, \qquad s_0 = \frac{m_0}{\rho_i}, \qquad t_0 = \frac{l_i^2}{D_i},$$
(10.431)

$$h_0 = s_0 t_0, \qquad \rho_0 = \frac{M_w p_0}{R T_0};$$

in addition, the choice of q_0 is found from a prescription for the katabatic wind (see below), while F_0 is the depth of an unfrozen dust layer over which κ decreases significantly. Of the twelve scales in (10.429), (10.431) provides definition of seven; in addition, q_0 and F_0 are determined as described above, while we suppose also that I_0 is known from the received solar radiation. This leaves us two scales (l_i and c_0) to be determined, and this will be done by prescription of two of the dimensionless parameters which emerge in the model. We also write

$$v_E = v_0 V_E,$$
(10.432)

and suppose v_0 is known.

The dimensionless version of the model can then be written in the form

$$h_t = -s + h_{xx},$$
$$\mu c_t + (qc)_x = R_c + v(\lambda V_E - c) + D c_{xx},$$
$$\mu \rho_t + (q\rho)_x = R_\rho + \gamma(1-\phi)s + D\rho_{xx},$$
$$\delta F_t = \sigma(c - \lambda V_E) + \phi s,$$
(10.433)

where

$$s = \frac{\kappa}{1-\phi}\left[\frac{1}{T}\exp\left\{\beta\left(1 - \frac{1}{T}\right)\right\} - \rho\right],$$

$$T = \Gamma\left[I\alpha - \varepsilon(1-\phi)s\right],$$
(10.434)

and the parameters are defined by

$$\mu = \frac{H l_i}{q_0 t_0}, \qquad D = \frac{D_c H}{q_0 l_i} = \frac{D_\rho H}{q_0 l_i},$$

$$\delta = \frac{F_0}{h_0}, \qquad R_c = \frac{l_i E_c}{q_0 c_0}, \qquad R_\rho = \frac{l_i E_\rho}{q_0 \rho_0},$$

$$\beta = \frac{M_w L}{R T_0}, \qquad \varepsilon = \frac{K' p_0 L}{\sigma T_0^5}, \qquad \lambda = \frac{\rho_s v_0}{v_s c_0},$$
(10.435)

$$\sigma = \frac{v_s c_0}{\rho_s s_0}, \qquad \nu = \frac{v_s l_i}{q_0}, \qquad \gamma = \frac{\rho_i s_0 l_i}{q_0 \rho_0}.$$

The issue now arises, how to choose the scales l_i and c_0. To do this, we need some understanding of how the model works, and for that, we need some idea of the size of the parameters.

We take values $I_0 = 130$ W m^{-2}, $a_0 = 0.3$, so that, with $\sigma = 5.67 \times 10^{-8}$ W m^{-2} K^{-4}, we find $T_0 = 200$ K. We use values $M_w = 18 \times 10^{-3}$ kg mole^{-1}, $L = 2.8 \times 10^6$ J kg^{-1}, $R = 8.3$ J mole^{-1} K^{-1}, $T_{\text{ref}} = 273$ K, $p_s^{\text{ref}} = 600$ Pa, from which we find $p_s^0 \approx 0.18$ Pa, and thus $\rho_0 = 2 \times 10^{-5}$ kg m^{-3}. We take $K' = 2.2 \times 10^{-5}$ m^{-1} s K, whence we find $m_0 \approx 2 \times 10^{-8}$ kg m^{-2} s^{-1}, and with $\rho_i \sim 0.9 \times 10^3$ kg m^{-3}, $s_0 \approx 2.2 \times 10^{-11}$ m s^{-1}.

We suppose that a katabatic wind of magnitude $u_w = 10$ m s^{-1} exists in a layer of depth $H = 100$ m, so that we take $q_0 = 10^3$ m^2 s^{-1}. To estimate D_i, we use (10.418) with $n = 3$ to motivate the choice $D_i \approx \bar{D} h_0^7$, where $\bar{D} = 0.4 A (\rho_i g)^3 / l_i^2$, and we use the observed value $l_i = 500$ km, and $A = 3 \times 10^{-27}$ Pa^{-3} s^{-1}, based on a basal ice temperature of 220 K, itself based on a surface ice temperature of 200 K and an areothermal heat flux of 20 mW m^{-2}. Using the definitions of the depth scale, time scale and diffusion coefficient, this leads to

$$h_0 = \left(\frac{s_0 l_i^2}{\bar{D}} \right)^{1/8} \approx 3{,}600 \text{ m}, \tag{10.436}$$

and then $t_0 \approx 1.6 \times 10^{14}$ s ≈ 5 Ma and $D_i = 1.6 \times 10^{-3}$ m^2 s^{-1} (corresponding to an ice velocity of some 0.1 m y^{-1}). We suppose that the depth scale over which a surface dust layer occludes the ice reflectivity is $F_0 = 1$ cm, and we take the turbulent diffusivity of the katabatic layer to be $D_c = D_\rho = 0.1 u_w H = 0.1 q_0$.

With all these values, we find

$$\mu \approx 0.3 \times 10^{-9}, \qquad D \approx 2 \times 10^{-5}, \qquad \delta \approx 0.3 \times 10^{-5},$$
$$\beta \approx 30.4, \qquad \varepsilon \approx 0.6 \times 10^{-3}, \qquad \gamma \approx 0.5. \tag{10.437}$$

The values of σ, λ and ν depend on what we assume about erosion and settlement of dust. We suppose that suspended dust grains have a diameter of the order of 1–2 microns. Then the Stokes settling velocity (5.8) is

$$v_s = \frac{\Delta \rho g D_s^2}{18 \eta} \approx 10^{-4} \text{ m s}^{-1}, \tag{10.438}$$

assuming $\Delta \rho = 2 \times 10^3$ kg m^{-3}, $g = 3.7$ m s^{-2}, and the atmospheric viscosity is $\eta = 10^{-5}$ Pa s; we also suppose (see (5.7) and the line after (5.9)) that $v_0 \approx 10^{-2} v_s$. If we define $A = \frac{c_0}{\rho_s}$, then we have

$$\sigma \approx 0.5 \times 10^7 A, \qquad \nu \approx 0.05, \qquad \lambda \approx \frac{10^{-2}}{A}. \tag{10.439}$$

The sizes (and signs) of the entrainment parameters R_c and R_ρ depend on what we assume about the entrainment rates E_c and E_ρ. There is little to guide us in this, except for the expectation that the numerators of R_c and R_ρ represent the total entrained dust and vapour, while the denominators represent the magnitude of the downslope fluxes; we thus expect numerators and denominators to be comparable, and this suggests that in practice R_c, $R_\rho \lesssim O(1)$.

The issue of how we choose the precise values of l_i and c_0 now arises. The fact that numerically the parameter $\gamma = O(1)$ is strongly suggestive of the idea that we choose l_i by *requiring* that $\gamma = O(1)$, and without loss of generality we may define $\gamma = 1$. The choice of c_0 depends on how we imagine the dust behaves at the ice surface. One possibility is that the dust-albedo feedback is engineered through occasional atmospheric dust storms, which affects the temperature through the dependence of Γ and α on c. In this view, $F \equiv 0$; the settlement velocity is so small that no accretion can occur, and dust is incorporated in the surface via snowfall, wherein ice accretes on dust nuclei. We then choose c_0 by requiring that

$$v = \sigma, \tag{10.440}$$

which is equivalent (with $\gamma = 1$) to choosing

$$c_0 = r_{si}\rho_0, \tag{10.441}$$

where r_{si} is given by (10.427). It then follows from (10.428) that, dimensionlessly,

$$\phi = \frac{c}{c + \rho}, \quad s < 0, \tag{10.442}$$

and in fact we will assume that (10.442) applies also for $s > 0$.

The equations for c and ρ now take the form

$$\mu c_t + (qc)_x = R_c + \phi s + Dc_{xx}, \\ \mu \rho_t + (q\rho)_x = R_\rho + (1 - \phi)s + D\rho_{xx}, \tag{10.443}$$

and can be combined if we suppose that μ, R_c, R_ρ and D are all small, and that q is constant. With the definition of ϕ in (10.442), we then have

$$\frac{dc}{d\rho} = \frac{\phi}{1 - \phi} = \frac{c}{\rho}, \tag{10.444}$$

whence $c \propto \rho$, and thus ϕ is constant. With this assumption, we can eliminate ρ from the definition of s, and the equation for c takes the form

$$\mu c_t + (qc)_x = R_c + \phi s(T, c) + Dc_{xx}, \tag{10.445}$$

where s can be expressed as

$$s = \frac{1}{(1 - \phi)T} \exp\left\{ \beta\left(1 - \frac{1}{T} \right) \right\} - \frac{c}{\phi}. \tag{10.446}$$

Because ε is small, we can take

$$T \approx I\Gamma\alpha. \tag{10.447}$$

The dimensionless incident radiation depends weakly on slope, and can be taken to be

$$I = I_0(c)(1 - mh_x), \tag{10.448}$$

where $m \approx 0.02$. The incident radiation may also depend on dustiness through the cooling effect associated with increased reflectivity in a dusty atmosphere, hence the decreasing function $I_0(c)$. In general, the scaled co-albedo α will be an increasing

Fig. 10.26 The sublimation
function s defined by
(10.446), where we take
$T = (1-\sigma c)(1-mh_x)(1+gc)$,
and use values $m = 0.02$,
$-h_x = 1$, $g = 0.05$, $\beta = 30$,
$\sigma = 0.023$, $\phi = 0.2$

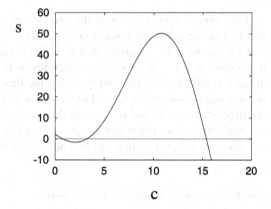

function of both ϕ and F; here we take it to be constant, $\alpha = 1$. The feedback in this version of the model thus operates through the dependence of the greenhouse factor Γ on c: Γ is an increasing function of c. Because β in (10.446) is large, s is very sensitive to c.

The 'derivation' of (10.445) is suggestive rather than rigorous, but will serve as the basis of a model for trough formation. Whether the conclusions we draw will extend to the full system, and indeed, whether the concept of dust suspension and snowfall is correct at all: these are questions which await further study.

10.6.2 Multiple Steady States

The reduced model which we now consider is that for h and c described by (10.433)$_1$ and (10.445), with s and T defined by (10.446), (10.447) and (10.448). It is clear from the definition of s in (10.446) that if T increases with c, then s can be non-monotonic. In general, s may have three zeroes as a function of c, and if we allow for the cooling effect of a dusty atmosphere at high dust concentrations, the highest zero can be quite modest, as indicated in Fig. 10.26.

The non-monotonicity of s allows the possibility of multiple steady states. The simplest way to see this is to consider (10.445) with q constant, and to ignore the very small terms in μ and D (we will reconsider their importance subsequently). With q constant, a summit condition for c must be prescribed at $x = 0$. If $s(c)$ has three zeroes as shown in Fig. 10.26, denoted c_1, c_2 and c_3, then for sufficiently small R_c, c will tend towards either the largest or smallest zero of $R_c + \phi s$. For both values, the value of $s = -R_c/\phi$ is the same and the steady profile for h is a parabola,

$$h = \frac{R_c}{2\phi}(1 - x^2),$$ (10.449)

assuming boundary conditions

$$h_x = 0 \quad \text{at } x = 0, \qquad h = 0 \quad \text{at } x = 1.$$ (10.450)

The multiplicity above depends on the choice of summit dust concentration, and this is somewhat artificial, as there is no physical reason to prescribe c at the summit. In reality, the downslope katabatic wind must be zero at the summit, so that in general q will depend on the slope $-h_x$. The simplest assumption is to take

$$q = -h_x, \tag{10.451}$$

and in this case, the equation for c is degenerate. Satisfaction of (10.445) (with $\mu = D = 0$) at $x = 0$ requires c to satisfy

$$(\phi + c)s(c) = -R_c. \tag{10.452}$$

Again, there can be three different values, and each of these leads to a genuinely different solution for h and c. In particular, if we denote a zero of (10.452) as c^*, then the steady state solution for h corresponding to $c = c^*$ is

$$h = \frac{R_c}{2(\phi + c^*)}(1 - x^2); \tag{10.453}$$

in particular, when c^* is large, the ice cap is essentially removed.

The discussion above assumes s independent of slope, i.e., $m = 0$. Non-zero values of m modify the discussion, but only quantitatively.

10.6.3 Trough Formation

The presence of multiple steady states suggests the possibility of hysteretic transitions between the lowest and highest values of c. We expect the middle steady state to be unstable. Although the steady state solutions depend on x, and in particular, $c = c(x)$ if $m \neq 0$, we will continue to refer to the steady states in terms of the (possibly three) solutions of (10.452) as c_i^*, $i = 1, 2, 3$, bearing in mind that the space-dependent solutions for c are simply the continuation to $m \neq 0$ of the constant solutions.

Suppose now that incident radiation I increases so that c_1^* and c_2^* coalesce and disappear; essentially the graph of s in Fig. 10.26 is pulled upwards. A pre-existing ice cap with $c = c_1^*$ will undergo a transition to $c = c_3^*$: the ice cap will disappear. How does this happen? The slope dependence of T and thus s means that coalescence of the roots occurs first, locally, where the slope $-h_x$ is greatest, at the margin of the ice cap. As I increases further, the point on the ice cap where coalescence occurs moves progressively back towards the summit.

In order to describe the transition, we consider the system

$$\begin{aligned} h_t &= -s + h_{xx}, \\ \mu c_t - (h_x c)_x &= R_c + \phi s + D c_{xx}, \end{aligned} \tag{10.454}$$

and to mimic the dependence of s on c and $-h_x$, we choose

$$s = f(c) + \Delta I - m h_x, \tag{10.455}$$

with

$$f(c) = A(c - c_1)(c - c_2)\left(1 - \frac{c}{c_3}\right),\qquad(10.456)$$

where we expect $c_1, c_2 \sim O(1)$, $c_3 \gg 1$, $A \gtrsim 1$. We define

$$c = c_3 C, \qquad f = A c_3^2 g, \qquad s = A c_3^2 S, \qquad t = \frac{\mu}{A c_3}\tau,$$

$$x = x_B(\tau) + \sqrt{\frac{D}{A c_3}}\,\xi, \qquad h = h_0(x) + \lambda^* H,\qquad(10.457)$$

where

$$\lambda^* = \mu\sqrt{AD}c_3^{3/2},\qquad(10.458)$$

so that

$$g = \left(C - \frac{c_1}{c_3}\right)\left(C - \frac{c_2}{c_3}\right)(1 - C) \approx C^2(1 - C),\qquad(10.459)$$

$$S = g(C) + \frac{\Delta I}{A c_3^2} - \frac{m h_0'}{A c_3^2} - \mu m H_\xi,\qquad(10.460)$$

and if we define

$$\dot{x}_B = -\frac{h_0'}{A c_3},\qquad(10.461)$$

then H and C satisfy

$$\sqrt{c_3 D A}\,H_\tau + h_0' H_\xi = -S + \frac{h_0''}{A c_3^2} + \mu\sqrt{\frac{A c_3}{D}}\,H_{\xi\xi},$$

$$C_\tau = \phi S + \frac{R_c}{A c_3^2} + \frac{h_0'' C}{A c_3} + C_{\xi\xi} + \Lambda(H_\xi C)_\xi,\qquad(10.462)$$

where

$$\Lambda = \frac{\lambda^*}{D} = \frac{\mu c_3^{3/2}\sqrt{A}}{\sqrt{D}}.\qquad(10.463)$$

Suppose firstly that $\Lambda \ll 1$. In that case, C satisfies

$$C_\tau \approx \phi g(C) + C_{\xi\xi}.\qquad(10.464)$$

For small C, $g \approx C^2$, and an initial blow-up begins to occur, in which C tends to infinity at finite time at one position. However, when $C \sim O(1)$, $g \approx C^2(1 - C)$, and C saturates at $C = 1$ and two travelling waves propagate outwards from the initial blow-up position.

The consequent perturbation to the depth is computed from $(10.462)_1$. If we assume $\sqrt{c_3 D A} \ll 1$, $\mu\sqrt{\frac{c_3 A}{D}} \ll 1$, then H is approximately given by

$$H \approx \frac{1}{h_0'}\int_\xi^\infty S\,d\xi,\qquad(10.465)$$

and this describes the front of the trough.[31] As $\xi \to -\infty$, the depth of the trough is

$$\Delta H = \frac{1}{|h_0'|} \int_{-\infty}^{\infty} S \, d\xi. \tag{10.466}$$

This is not a uniformly valid description of H, because we require $H \to 0$ as $\xi \to -\infty$. Behind the blow-up region for C, the diffusion term and time derivative of H become significant. Essentially, the trough drawdown diffuses backwards. We can recover this region by defining

$$\xi = \mu \sqrt{\frac{c_3 A}{D}} X, \qquad \tau = \mu c_3 A T, \tag{10.467}$$

and then

$$H_T - |h_0'| H_X = H_{XX} + \frac{\mu \sqrt{c_3 A}}{\sqrt{D}} \left(-S + \frac{h_0''}{A c_3^2} \right), \tag{10.468}$$

with $H = 0$ on $T = 0$, $X < 0$ and as $X \to -\infty$, and $H = -\Delta H$ at $X = 0$. At large times, the solution of this is essentially

$$H \approx -\frac{1}{2} \Delta H \operatorname{erfc} \left(-\frac{X + |h_0'| T}{2\sqrt{T}} \right), \tag{10.469}$$

and one can show that this diffusive wave travels backwards relative to x_B at the same rate that x_B travels forwards.

The consequence of all this is that local blow-up of c causes a trough to form and deepen as the region of saturated dust spreads. The trough thus formed will have an essentially stationary rear face of length $O(\mu)$, and a shallower front face of length $O(\mu \sqrt{\frac{c_3}{D}})$ (and these slopes become less severe with time).

If we take $\mu = 10^{-10}$, $D = 10^{-6}$, $c_3 = 10^4$, $A = 1$, then $\Lambda = 0.1$; uncertainty in parameter values means that in practice values of $\Lambda = O(1)$ are plausible. In this case, we cannot neglect the extra term in $(10.462)_2$. However, note that the diffusive coefficient $\mu \sqrt{\frac{A c_3}{D}} = \frac{\Lambda}{c_3}$ in $(10.462)_1$ remains small. If in addition $\sqrt{c_3 D A} \ll 1$, then it is still the case that $h_0' H_\xi \approx -g(C)$, so that the blow-up equation (10.464) is simply modified to

$$C_\tau - \frac{\Lambda}{|h_0'|} \{ C g(C) \}_\xi \approx \phi g(C) + C_{\xi\xi}, \tag{10.470}$$

and the same blow-up and formation of travelling waves occurs, modified only by the advective drift upstream.

If in addition $\sqrt{c_3 D A} \sim O(1)$, then the time derivative term in $(10.462)_1$ comes into play. By inspection, it seems that blow-up will still occur, and that there will be travelling wave solutions also in this case (see Question 10.19).

---

[31] We have applied the boundary condition that $H \to 0$ as $\xi \to \infty$, rather than $H \to 0$ as $\xi \to -\infty$; why?

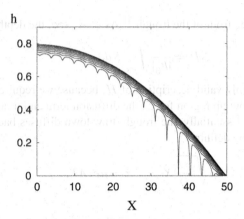

Fig. 10.27 A simulation of Eqs. (10.454), (10.455) and (10.456). The space variable has been rescaled as $x = X/L$ (largely for historical reasons, so that the trough spacing in X will be $O(1)$), and the equations are then solved in the same form, but using rescaled variables $T = L^2 t$, $S = s/L^2$, and with modified parameters $\bar{R} = R_c/L^2$, $\bar{I} = \Delta I/L^2$, $\bar{A} = A/L^2$, $\bar{m} = m/L$. The parameter values used to obtain the sequence of profiles of h above (plotted at time intervals in T of 0.2 up to a maximum of 2.6) are then $L = 50$, $\bar{m} = 0.14$, $\bar{R} = 0.001$, $\mu = 0.1$, $D = 0.002$, $\bar{I} = 0.24 + 0.02T$, $c_3 = 90$, $\bar{A} = 1$, $\phi = 0.2$, time step $\Delta T = 0.005$ and space step $\Delta X = 0.002$. The initial profile for h is $h_0 = 0.8(1 - \frac{X^2}{L^2})$, and the initial concentration profile for c is $c_0 = 1.5 - 0.005X + 0.02\sin(\frac{2\pi X}{3})$

10.6.4 Multiple Troughs

Our discussion shows that troughs can form through local blow-up of the dust concentration profile. In order to describe the Martian polar caps, we need this blow-up to occur at many different places along the surface. The simplest way in which this can occur is that as the insolation increment ΔI increases to the point where the steady states c_1 and c_2 coalesce, the resulting instability occurs at a non-zero wave number.

A straightforward local instability analysis of (10.454) and (10.455) suggests that normal modes proportional to $\exp(ikx + \sigma t)$ have slow solutions (corresponding to diffusive ice surface relaxation) $\mathrm{Re}\,\sigma \sim -k^2$, and rapid growth solutions $\mathrm{Re}\,\sigma \sim \frac{\phi f'}{\mu}$, assuming μ and D are small. A next approximation is then (if the steady dust concentration is c_0)

$$\mathrm{Re}\,\sigma \approx \phi f' + \left(\frac{\mu c_0}{\phi} - D\right)k^2 + \cdots, \tag{10.471}$$

so that we can expect growth of troughs if D is sufficiently small.[32]

Figure 10.27 shows a simulation in which troughs grow from an initial state in which a small superimposed sinusoidal variation of dust concentration is applied.

---

[32]But not if $\mu < D$, as we have suggested.

This figure is suggestive of the idea that the model has the ability to reproduce features which resemble the Martian troughs, but it is by no means clear that the simple theory suggested here is correct. Further discussion follows in the notes. In the model, trough formation occurs as the initial stages of collapse of the polar ice cap. Numerical outputs vary widely with parameter choices. In particular, it is common to find initial blow-up near the margin, leading to a large trough reminiscent of the *Chasma Borealis*.

10.7 Notes and References

The best source for general information about glaciers and ice sheets is the book by Paterson (1994). This famous book was first published in 1968, upgraded to a second edition in 1981 (but in typescript), then to an apparently terminal third edition (and in LaTeX) in 1994, and now miraculously to a fourth edition (Cuffey and Paterson 2010). Other books with a similar aim are those by Hooke (2005) and Van der Veen (1999). Books which are more concerned with observations in the field and geomorphic processes include those by Benn and Evans (1998) and Bennett and Glasser (2010), while the books by Lliboutry (1987) and Hutter (1983) are much more abstract. Lliboutry's (1964, 1965) earlier voluminous work gives useful descriptions of early work in the subject, particularly in the nineteenth century, but was unfortunately never translated from the French. There is a good deal of historical and geographical material, but the theoretical parts are inevitably dated.

From its origins as a hobby for geographers and climbers, glaciology has come to occupy centre stage in the modern preoccupation with climate, and there are many popular books detailing some of the more recent discoveries. Amongst these are the books by Imbrie and Imbrie (1979), Alley (2002) and Walker (2003): the first two by scientists, the third by a journalist, all of them entertaining.

Scaling Apart from some of the work at the end of the nineteenth century, detailed by Lliboutry (1965), the application of theoretical mechanics to problems in glaciology really begins after the second world war with the work of Nye, Weertman and Lliboutry. Egged on by the vituperative Lliboutry, the decades after the war saw enormous advances in the theoretical understanding of glacier flow. Applied mathematical principles come late to the scene, so that even something as simple as non-dimensionalisation does not happen till the mid-1970s. Possibly the first paper to do this in a formal way was that by Grigoryan et al. (1976), a paper which is not often cited because of its impenetrability, occluding the matter as it does with a heavy shroud of curvilinear coordinates. The basic lubrication approximation which describes glacier and ice sheet flow was introduced as the 'shallow ice approximation' in my thesis (Fowler 1979) and in print by Fowler and Larson (1978), and this phrase is still widely used. Fowler and Larson dealt explicitly with valley glaciers, but the same small aspect ratio approximation can be used for ice sheets (Morland and Johnson 1980; Hutter 1983; Morland 1984; Hutter et al. 1986 and Fowler 1992a, of whom we follow the latter) and ice shelves (Morland and Shoemaker 1982), although in practice similar approximations had been introduced earlier (Nye 1959; Weertman 1957b).

Waves on Glaciers   Both surface and seasonal waves were well known in the nineteenth century. An early discussion of surface waves is by Finsterwalder (1907), while Deeley and Parr (1914) provide a discussion of seasonal waves. Between the wars, glaciology enters its barren period, and the emergence from this is perhaps seeded by Gerald Seligman's foundation of the Glaciological Society (first British, later International) in 1936, but it is only after the war that some of the earlier investigations are revived. Lliboutry, in particular, was a great advocate of the expertise of the early glaciological researchers.

The modern theory of surface wave motion received its impulse from John Nye's stay at CalTech, where a youthful Gerry Whitham had recently developed the theory of kinematic waves Lighthill and Whitham (1955a, 1955b). This theory was adapted by Nye (1960, 1963) to the study of linear waves on glaciers; a nonlinear analysis is given by Fowler and Larson (1980b). A parallel development was reported by Weertman (1958). Nye's theory, based on perturbations of a parallel-sided slab, yields the unphysical singularity at the snout which was mentioned in Sect. 10.4.1.

Apart from the early work by Deeley and Parr (1914), there has been relatively little interest in seasonal waves. The main exception to this is the paper by Hodge (1974). Hewitt and Fowler (2008) provide a mathematical model which can produce certain of the observed features.

Wave ogives are lucidly discussed by Waddington (1986).

The Head and the Tail   As mentioned earlier, wave theory for glaciers is confounded by both the head of the glacier and its snout. To my knowledge, no one has paid any attention to the modelling issue with the upstream boundary condition, as discussed here in Question 10.11. The resolution of the description of the *bergschrund*[33] is likely to involve steepening bed slopes, and longitudinal stresses which reach the fracture strength of ice (around 2 bars).

The snout of a glacier is a good deal more accessible. Even so, little attention has been paid to that either. Nye and Lliboutry had a little engagement[34] about this in the 1950s (Lliboutry 1956, 1958b; Nye 1957, 1958), and Chap. 6 of my thesis (Fowler 1979) uses the method of strained coordinates to calculate the finite slope at the front. This involves inclusion of the longitudinal stresses, in much the same way as in Question 10.11. If these are ignored, then the solution of the isothermal equation has infinite snout slope if it is stationary or advancing, much as for (10.117); see also Question 10.12.

---

[33]The crevasse which marks the head of the glacier, where the ice separates from the stagnant apron of snow and ice above it.

[34]Or perhaps a spat; Nye's opening sentence in his 1958 note is the wonderful comment: "In so far as Professor Lliboutry is trying to make the theory of glacier flow more realistic one can only wish him well and hope that he is on the right track."

Boulton and Hindmarsh's (1987) seven data points of basal shear stress, effective pressure and till strain rate were presumably gleaned from observations near the glacier snout. The original data were never published, so that one can only guess how the values of shear stress were computed. In the absence of a local analysis of behaviour near the snout, such values are tantamount to guesswork.

Surges   Surging glaciers are located in various places round the world, including Alaska and Svalbard. Famously, there are no surging glaciers in the European Alps, but it is thought that there used to be at least one, Vernagtferner, in the Austrian Alps, which last surged in about 1900.[35] Early paintings, documented by Nicolussi (1990) indicate surges occurring in about 1600 and 1680, to judge from the jagged surface of the glacier in the images, and further surges occurred in 1772, 1844, and the small, perhaps final one in 1898. Apart from the last of these, the ice advanced to block the outlet stream from Hintereisferner, causing an ice-dammed lake to form, which burst through the ice, sometimes more than once, sending a flood wave down the valley to the village of Vent.

The surge on Variegated Glacier is discussed by Kamb et al. (1985), and theoretical descriptions are given by Kamb (1987) and Fowler (1987a). The present discussion is based on this latter paper, the mathematical details of which are worked out in Fowler (1989). Observations of Trapridge Glacier are described by Clarke et al. (1984) and Frappé-Sénéclauze and Clarke (2007). The issue of the Journal of Geophysical Research in which Fowler's (1987a) article appears is a collection of articles on fast glacier flow, including both ice streams, surging glaciers, and tidewater glaciers.

Streams, Shelves, Sheets   The dynamics of ice streams are reviewed by Bentley (1987), see also Engelhardt et al. (1990), while the theory of Hudson Strait megasurges is due to MacAyeal (1993). Heinrich events are discussed by Bond et al. (1992), while the discussion here is based on a paper by Fowler and Johnson (1995). The recent acceleration of Jakobshavn Isbrae in West Greenland is described by Holland et al. (2008).

The discussion of approximate temperature profiles follows that in Fowler (1992a); the profiles shown in Fig. 10.7 are reminiscent of those shown in Paterson's (1994) book, and are also similar to the computed profiles of Dahl-Jensen (1989). The concept of thermally induced instability was enunciated by Robin (1955) and taken up by Clarke et al. (1977) and Yuen and Schubert (1979), but more or less scotched by Fowler and Larson (1980a), at least in the context of two-dimensional flows; see also Fowler et al. (2009). However, Hindmarsh (2009) has shown that thermally induced instabilities do occur in three dimensions, and are capable of forming ice stream-like features.

The basic approximation for the analysis of ice shelves was done by Weertman (1957b). A general scaling analysis is given by Morland and Shoemaker (1982).

---

[35] See http://www.lrz.de/~a2901ad/webserver/webdata/vernagt/vernagt_E.html.

Typical values of sub-ice shelf melt rates are given by Holland et al. (2003), for example.

The mechanics of ice streams are thoroughly described by Van der Veen (1999). The shear stress on the Siple coast ice streams, particularly the Whillans ice stream, is small, of order 0.1–0.2 bars, but Kamb's (1991) laboratory tests indicated that the yield stress for the basal marine sediments is an order of magnitude smaller. If one supposes that the rheology of till is such that the yield stress cannot be exceeded without allowing rapid acceleration, then the presence of stable ice streams indicates that the driving stress is taken up elsewhere, most likely by lateral shear, and this is consistent with transverse velocity profiles, as shown by Van der Veen (Fig. 12.10), and as discussed in Sect. 10.3.4.

The use of longitudinal stresses in producing the membrane stress approximation is due to MacAyeal (1989). The version we present here is similar to that presented by Blatter (1995), and perhaps more in the style of Schoof and Hindmarsh (2010). Bueler and Brown (2009) present a related model, although they partition the ice velocity in an arbitrary way between shearing and sliding.

The mechanism whereby ice streams form is less clear, although some kind of spatial instability is the likely cause. As alluded to above, Hindmarsh (2009) showed, following earlier work by Payne and Dongelmans (1997), that thermal instability was a possible cause; Sayag and Tziperman (2008), following Fowler and Johnson (1996), suggested that a water-mediated feedback could also provide a mechanism.

Grounding Line   The possible collapse of the West Antarctic Ice Sheet was discussed by Hughes (1973), and Weertman (1974) gave the first theoretical discussion of grounding line stability. Subsequent authors who discuss the issue include Thomas (1979) and Hindmarsh (1993); the latter advocated a concept of neutral equilibrium for grounding line position.

The issue of the extra condition which describes the position of the grounding line is a thorny one, which is as yet not completely resolved. At a formal level, the most detailed studies are those of Wilchinsky (Chugunov and Wilchinsky 1996; Wilchinsky and Chugunov 2000, 2001), but these papers are severely impenetrable, even to initiates. Wilchinsky (2007, 2009) adds further comments to his earlier analysis. Chugunov and Wilchinsky (1996) consider the transition zone in a similar manner to that presented here. They assume Newtonian flow and a steady state, and claim to deduce the grounding line position. Two key assumptions are apparent in their reasoning. The first is the arbitrary assumption that the horizontal length scale for the ice shelf is comparable to that for the ice sheet. This allows them to deduce that (with present notation) $H_G = \beta(\varepsilon/\delta)^{1/3}$ for some $O(1)$ coefficient β (not the same β as in 10.2.7); the origin of this (correct) scale is, however, mysterious. The deduction of a numerical value of $\beta \approx 1.5$ from a numerical calculation appears to involve (in the appendix to the paper) the assumption that the bed $B(X)$ (in present notation) is smooth, i.e., $B'(0) = 0$. This assumption appears to be arbitrary, though reasonable. Wilchinsky and Chugunov (2000) extend this analysis to the junction between a rapidly moving ice stream, where shear is less important, and an ice shelf. They now state that the grounding line position is determined by the requirement of continuity of the lower ice surface at the grounding line, but they do not carry through the calculation. The scaling analysis involved is rather different than for the shear-dominated sheet/shelf transition. Finally, Wilchinsky and Chugunov (2001) extend the scaling of the 1996 paper to the nonlinear rheology of Glen's law. The flow is still steady, and it is stated that the condition $B(0+) = 0$ determines the grounding line position, and that the flux at the grounding line is, in present notation,

$$q_G = \left(\frac{\beta \varepsilon}{\delta}\right)^n H_G^{n+2}; \qquad (10.472)$$

this can be compared with (10.162). Numerical evaluation of β is again only done for the Newtonian case $n = 1$, under the additional assumptions of $B_X = B_{XX} = 0$ at $X = 0+$. Like its predecessors, this paper is hard to fathom.

More recently, the transition problem has been studied numerically by Nowicki and Wingham (2008), and it is here that the rôle of contact conditions has been emphasised. They studied the transition problem described in Sect. 10.2.7, assuming $\dot{x}_G = 0$, and for a range of incoming mass fluxes—essentially a range of values of Λ. They also allowed sliding, so that on the grounded base $X < 0$, the sliding velocity is

$$U = kT_3, \qquad (10.473)$$

which replaces the second condition in (10.153). In general, solutions are obtained for any value of Λ, but in general the (scaled) normal effective stress $B + \Pi + T_1$ on the grounded ice is singular at $X = 0$, tending to either ∞ or $-\infty$ as $X \to 0-$. In addition, one finds $B_X(0) > 0$ if $B + \Pi + T_1 \to -\infty$, and $B_X(0) < 0$ if $B + \Pi + T_1 \to \infty$. Consequently, none of these solutions are admissible. For each $k > 0$ there is precisely one value of Λ for which the contact conditions (10.175) and (10.176) are satisfied, and for this value also $B_X(0) = 0$, which can also be deduced from (10.166), which implies that

$$B + \Pi + T_1 = -\frac{2T_1 B_X^2}{1 - B_X^2} \qquad (10.474)$$

on $Z = B$, $X > 0$.

These results have not yet been extended to the non-stationary case $\dot{x}_G \neq 0$, or to the no slip case $k = 0$. The difficulty in the latter case appears to be associated with the greater numerical difficulty encountered in dealing with the more severe singularity which will occur in that case (cf. Barcilon and MacAyeal 1993). Durand et al. (2009) have used the same contact conditions in a full numerical ice sheet model in which $\dot{x}_G \neq 0$, with encouraging results.

The limit $k \to \infty$ in (10.473) corresponds to the case of sliding dominated flow as in an ice stream, and this limit has been studied directly by Schoof (2007b, 2007c) using a version of the membrane stress approximation. In order to complete his theory, he also needs an extra condition, which is taken to be that T_1 is continuous. It is not entirely obvious that this would be a consequence of the contact conditions in a suitably rescaled version of the finite k theory, although it seems likely. Schoof (2007c) is able to show directly that the Weertman slope-induced instability does indeed apply, and we have followed his presentation here.

Sliding   The theory of basal sliding over hard beds stems from Weertman (1957a) and Lliboutry (1968). Weertman presented the basic concept of the regelative lubricating film, and described in order of magnitude fashion how to obtain a sliding law. Lliboutry presents more elaborate calculations, and importantly introduces the importance of basal water. Two reviews of progress by the end of the 1970s are by Lliboutry (1979) and Weertman (1979). The linear theory is primarily due to Nye (1969, 1970) and Kamb (1970). Morland (1976a, 1976b) introduced complex variable methods, while the material presented here is based largely on Fowler (1986, 1987b). The first of these uses complex variable methods to study cavitation over simple periodic beds, and the second uses a heuristic, Lliboutry-style method to suggest a generalised Weertman model for sliding over non-periodic beds. An up to date theoretical discussion of subglacial cavitation is given by Schoof (2005), who also provides significant theoretical advances in the study of sliding over non-periodic beds, indicating in particular that Fowler's (1987b) theory is flawed, though repairable. His essential conclusion is that Iken's (1981) concept of a maximum friction (shear stress divided by normal effective stress) is valid, even for non-periodic beds, with the maximum value of the friction being set by the amplitude and slope of the largest bumps.

Weertman's original model is as follows. Consider a bed consisting of an array of (cubical) obstacles of dimension a a distance l apart, and suppose the ice flow exerts an (average) shear stress τ at the bed. The drag on each obstacle is therefore τl^2, and thus the pressure increase upstream of an obstacle is (approximately) $\tau l^2/2a^2$, while the decrease downstream is $-\tau l^2/2a^2$. The pressure difference causes a temperature difference (due to the Clapeyron effect) of

$$\delta T \approx C\tau l^2/a^2, \qquad (10.475)$$

where C is the slope of the Clapeyron curve, $-dT_m/dp = C \approx 0.0074$ K bar^{-1}; T_m is the melting temperature. Let u_R be the regelative ice velocity: then $u_R a^2$ is the regelative water flux. The latent heat required to melt this is $\rho_i L u_R a^2$, where ρ_i is ice density and L is latent heat. The heat transfer is effected through the obstacle, at a rate $(k\delta T/a)a^2 = k\delta T a$, where k is the thermal conductivity of the bedrock. Equating these suggests that

$$u_R = \left(\frac{kC}{\rho_i La}\right)\frac{\tau}{\nu^2}, \qquad (10.476)$$

where the aspect ratio $\nu = a/l$ is a measure of the roughness of the bedrock. Regelation is thus effective at *small* wavelengths.

On the other hand, let u_V be the velocity due to viscous shearing past the obstacle, with no shear stress at the bed. The differential stress generated is $\approx \tau/\nu^2$, and for a nonlinear (Glen's) flow law $\dot{\varepsilon} = A\tau^n$, the resulting strain rate is $\approx 2A(\tau/\nu^2)^n$, with $n \approx 3$. Hence we infer

$$u_V \approx 2aA\left(\tau/\nu^2\right)^n. \qquad (10.477)$$

It can be argued[36] that the stresses should be added, thus

$$\tau = v^2 \left[R_r a u + R_v (u/a)^{1/n} \right], \tag{10.478}$$

where R_r and R_v are material roughness coefficients, given approximately by

$$R_r \approx \frac{\rho_i L}{kC}, \qquad R_v \approx \left(\frac{1}{2A} \right)^{1/n}. \tag{10.479}$$

We see that motion past small obstacles occurs mainly by regelation, while motion past larger obstacles occurs largely by viscous deformation. There is a *controlling obstacle size* at which the stresses are comparable, and if we take a as this value, we obtain the Weertman sliding law

$$\tau \approx v^2 R u^{\frac{2}{n+1}}, \tag{10.480}$$

where

$$R = \left(\frac{\rho_i L}{2kCA} \right)^{1/(n+1)}. \tag{10.481}$$

Sub-Temperate Sliding   Sometimes modellers who implement sliding laws in their ice sheet computations assume that the sliding law $u = U(\tau)$ applies when the basal temperature $T = T_m$, and that $u = 0$ for $T < T_m$. This assumption is incorrect (Fowler and Larson 1980a), and it is more appropriate to allow sliding to increase continuously over a small range of temperature below the melting point, to reflect the fact that creation of a water film will occur in a patchy fashion as the melting point is approached (Hindmarsh and Le Meur 2001; Pattyn et al. 2004).

If one assumes a discontinuous sliding law, then if basal stress is continuous, one would have an inadmissible discontinuity of velocity: this was the downfall of the EISMINT ice shelf numerical modelling experiments in the 1990s. If the velocity is to be continuous, then stresses must be discontinuous and in fact singular (Hutter and Olunloyo 1980). It has indeed been suggested that such stress concentrations may have a bearing on thrust faults in glaciers (e.g., Kleman and Hättestrand 1999), but the theoretical basis for supposing they exist is dubious.

We can derive a sliding law in a Weertman-like way for basal temperatures below T_m as follows. Again we suppose that bumps of size a are spaced a distance l apart. Now we suppose that the basal temperature is at a temperature $T_b < T_m$, and we define the undercooling to be

$$\Delta T = T_m - T_b. \tag{10.482}$$

It is no longer appropriate to conceive of the water film covering the bed between the bumps, and so there is an additional component to the stress due to stick-slip friction. We will ignore this here, and suppose that as before the resistance comes primarily from the film-assisted flow over the bumps. Because the ice is below the pressure

---

[36]Weertman added the velocities instead.

melting point, there is an additional conductive heat flow away from the bumps given approximately by $\frac{k\Delta T}{l}$, and therefore (10.476) is replaced, using (10.479), by

$$R_r a u_R = \frac{\tau}{\nu^2} - \frac{\nu \Delta T}{C}, \tag{10.483}$$

and (10.478) is replaced by

$$\tau = \nu^2 \left[R_r a u + R_\nu \left(\frac{u}{a} \right)^{1/n} + \frac{\nu \Delta T}{C} \right], \tag{10.484}$$

which shows that for fixed τ, u decreases to zero continuously as ΔT increases to a temperature $\Delta T_{\max}$ given by

$$\Delta T_{\max} = \frac{C\tau}{\nu^3}. \tag{10.485}$$

For $\tau = 0.1$ bar and $\nu = 0.1$, this is ≈ 1 K.

The Rheology of Till   Although glacial geologists were aware of the widespread occurrence of subglacial drift, or subglacial till, the early theoretical studies of sliding focussed on sliding over hard beds. An abrupt shift in this view occurred on the publication of the benchmark paper by Boulton and Hindmarsh (1987), which focussed attention on the basal motion of ice due to deformation of the subglacial till. In particular, Boulton and Hindmarsh described possible viscous-type rheologies for till based on reported measurements on a subglacial till below an Icelandic glacier. Unfortunately, the original data from which the shear stresses were inferred are unavailable, and thus the experimental basis for the viscous rheology is uncertain. When laboratory measurements of subglacial till properties are made, it has been largely found that till behaves as a plastic material, having a yield stress which when reached allows indefinite strain (Kamb 1991; Hooke et al. 1997; Iverson et al. 1997; Tulaczyk et al. 2000; Rathbun et al. 2008; Altuhafi et al. 2009). This is to be expected, since till is a granular material. Ignoring cohesion, we would then have a prescription for basal shear stress in the form

$$\tau = \mu N, \tag{10.486}$$

where μ is a suitable coefficient of friction and N is effective pressure. Lliboutry suggested such a sliding law in his 1968 paper.

However, the story is more complicated than this. The rheology of a plastic material comprises the prescription of a yield stress surface (for example, the Von Mises yield stress surface $\tau_{ij}\tau_{ij} = 2\tau_c^2$) together with a flow law. The simplest such flow law allows a strain rate proportional to stress, so that the actual rheology would be that of a viscous material, where the effective viscosity is determined by the necessity to remain on the yield surface. In addition, purely geometrical considerations suggest that, in order to shear a granular material at all, a normal stress must be induced in order that the grains can move round each other. The generation of normal stresses by shear flows is a property of viscoelastic materials, and suggests that the issue of till rheology is not a simple one. The consequent dilation of the till in shear induces a reduction of pore pressure, and consequent hardening (Moore and Iverson 2002). In addition, deformation of granular materials often occurs through the formation of shear bands (Li and Richmond 1997; François et al. 2002), whose presence complicates the determination of an effective till rheology. Fowler (2003) discusses some of these issues further.

Drainage   Water is abundant under glaciers and ice sheets, and it seems usually to be the case that subglacial water cannot be evacuated through the bed, so that a subglacial hydraulic system must exist. The classical theory of drainage through channels incised upwards into the ice is due to Röthlisberger (1972), while the time-dependent development of this theory for jökulhlaups is due to Nye (1976). The ice-incised channels are called Röthlisberger, or simply R, channels, but channels cut down into underlying bedrock have been observed, and are termed Nye channels, following Nye (1973). Weertman (1972) preferred a distributed water film, although Walder (1982) showed that such a film is unstable (indeed, it is this instability which is responsible for the formation of R channels in the first place). However, the concept of a patchy film is more tenable (Alley 1989), particularly if allied to the concept that the ice-till interface can itself evolve; more on this below.

Linked cavities were first implicitly described by Lliboutry (1968), and were observed in deglaciated beds by Walder and Hallet (1979). Kamb et al. (1985) and Walder (1986) developed theoretical descriptions for the consequent hydraulic régime. While linked cavities are generally (though not necessarily, see below) associated with flow over hard beds, a similar sort of system of distributed canals was invoked by Walder and Fowler (1994) to describe channelled flow over soft till beds. For field measurements of subglacial hydrological systems, see Hodge (1974), Hubbard et al. (1995), Nienow et al. (1998) and Fudge et al. (2008). A recent review of subglacial processes of current interest is by Clarke (2005).

Drumlins   The word 'drumlin' apparently derives from the Irish, and means 'small hill'. The word appears to have first been published in the paper by Bryce (1833),[37] and is in common scientific usage by the time of Kinahan and Close

---

[37] The paper is not so easy to find. The reference in Drozdowski (1986) which most likely follows that of Menzies (1984) is marginally incorrect (it is the Journal of the Geological Society of Dublin, not of the Royal Geological Society of Dublin, and this makes a difference, since the journal subsequently changed its name to the Journal of the Royal Geological Society of Ireland). Copies of the original journal can be found in the National Library of Ireland (Kildare St., Dublin, call number IR5541g1), as well as in the library of the Royal Dublin Society in Ballsbridge. Bryce did not coin the word; he says the following: "*The gravel hills, on the other hand, have an elongated form, are generally steepest towards one side, and rise in every other direction by much more gentle acclivities. This peculiar form is so striking that the peasantry have appropriated an expressive name to such ridges ... the names* Drum *and* Drumlin (*Dorsum*) *have been applied to such hills* ...." Why the Latin word *Dorsum* (meaning back, but also ridge) is included in parentheses is not clear. Bryce's paper largely concerns the constituents of the till which constitute the drumlins of northern Ireland, from which he infers that motion was largely from the north west. He also provides what may be the first description of ribbed moraine, and deduces in effect that Belfast Lough, Lough Neagh and Lough Foyle were formed during the ice ages. Earlier uses cited in the Oxford English Dictionary are by Innes (1732) and Sinclair (1791–1799; particularly volume IX,

(1872), although the study of such bedforms was also described much earlier by Hall (1815), who was concerned with crag-and-tail features in Scotland (perhaps the best known being the Royal Mile in Edinburgh, a ridge of drift which lies in the lee of the volcanic outcrop of Edinburgh Castle). There are two interesting things about Hall's paper. First, it appears before Agassiz's glacial theory (as does Bryce's), and thus ascribes crag-and-tail features to the biblical flood. Hall and Bryce had no knowledge of ice ages. The second interesting thing is that a modern edition of Hall's biblical theory has reappeared in the flood hypothesis of John Shaw (see, e.g., Shaw 1983; Shaw et al. 1989). Shaw's ideas are largely derided, but are vigorously supported by a number of scientists.

So we need to explain Shaw's hypothesis and its reception rather carefully (see also the discussion in Sect. 11.8). Essentially, his idea is that massive subglacial meltwater floods cause the formation of drumlins, and the apparent motivation for this idea is that only the turbulent flow of water can erode such bedforms: ice is too slow. This conceit is evidently misguided, but its application requires him to produce massive subglacial floods below ice sheets. The twist in the story (see Chap. 11) is that such floods now seem likely to have occurred, but the basic difficulty with the Shaw theory remains: he needs floods to be everywhere, of incomprehensible volume, and to produce bedforms which do not actually look fluvial: a tall order.

In my view, one can be fairly circumspect about the matter. Shaw's theory, in any of its forms, is not in fact a theory: it does not provide a mechanistic process to produce the observations. A suitable point of discussion is his 1983 paper. Inspired by his mentor's monumental book masquerading as a paper (Allen 1971), Shaw provides a very persuasive analogy between some erosional marks, such as the cave scallops described by Allen, and the resulting inverted casts under ice sheets, which result as drumlins. Nothing wrong with the idea. But it is not a theory. To be a theory, it needs, for example, a predictive wavelength for scallops. Scallop formation is an interesting problem (Blumberg and Curl 1974), but is basically unsolved. One can thus criticise Shaw's ideas either on the basis that their fundamental environment is invented (massive floods), or on the basis that nothing is actually predicted. But at the same time, we have to be aware as scientists that we must try to avoid dogmatic reaction associated with paucity of imagination, because we know that this has littered our scientific history. Consider, for example, the receptions accorded to Wegener (Chap. 8) and Bretz (Chap. 11).

The development of the theory of drumlins over the nineteenth and twentieth centuries is in a similar parlous state. Although the literature describes the debate between the 'erosional' and 'depositional' theories, there is really no theory that deserves the name until Hindmarsh's landmark paper (Hindmarsh 1998), which is the first time that the word 'instability' makes an appearance, and in which an instability theory is proposed. Hindmarsh showed numerically that instability could occur,

---

pp. 131, 262–263 and volume XIX, pp. 342–344, 369), but it seems that the *dryms* of Innes on the shores of Lough Foyle in Ireland are in fact fossil dunes, while the *drums* of Sinclair near Blairgowrie in Perthshire, Scotland, are interfluves of former meltwater channels.

and essentially the same theory was solved analytically by Fowler (2000). The theory is developed further by Schoof (2007a), who revisits the same stability theory and extends it in various ways. He does, however, draw a cloud over proceedings: 'Hindmarsh and Fowler's theory does not reproduce a number of known features of drumlins'; and in his conclusions, he draws attention to certain apparent problems with the theory: the problems of three-dimensionality, the issue of stratified drumlin cores, the problem of amplitude. Gloomily, he thinks there is a 'tenuous link between the model and the origin of drumlins'. His gloom is misplaced. There is no other tenable theory in existence, and there is nothing as yet which rules it out. While the problem is certainly hard, it is likely that a clear theoretical framework will emerge over the next decade or two. In his recent book, Pelletier (2008) follows Schoof's view, and proposes a model based, bizarrely, on a compaction model for magma transport, with little connection to the physical processes involved, although he is able to produce interesting looking patterns—much like his theory of the spiral canyons on the Martian north polar ice cap (see below).

Eskers   The Irish eskers were perhaps first described scientifically by Close (1867), and later by Flint (1930). General descriptions are given by Embleton and King (1968) and Sugden and John (1976). More up to date discussions are those by Shreve (1985) and Warren and Ashley (1994). Clark and Walder (1994) noted that in the Laurentide ice sheet, the former central part, the Canadian Shield, is essentially wiped clean of sediment, which has piled up in the outer parts of the former ice sheet. And eskers are found in the Shield but not beyond it. Clark and Walder inferred the obvious conclusion that Röthlisberger channels (hence the eskers) are the drainage pattern on the (hard) Shield, while their absence on the sediment-covered margins indicates a canal-type drainage. We will come back to this observation in Chap. 11.

In our discussion of the various drainage systems which exist beneath a glacier, we have always been thinking of an isolated set of channels, or linked cavities, or canals, which somehow exist independently of the overlying ice and underlying sediments or bedrock. Finally, as we contemplate the construction of drumlins and eskers, we may come to realise that this separatist view is misguided. In our simple theory of drumlin formation, we imagine a drainage system which moves water through the landscape without interacting with it. But this is unrealistic: the development of ridges will pond water and alter drainage paths. What we need to do is to allow the drainage system to interact with the bedforms.

We might then ask ourselves, what actually is the difference between a lee-side cavity and a subglacial stream? And the answer, at least from the point of a sensible model, is none. A fully integrated model for ice, water and sediment (or rock) allows for parts of the bed where effective pressure $N > 0$ and the ice is attached, i.e., the water layer thickness $h = 0$, and parts of the bed where the ice flow is separated ($h > 0$), and where $N = 0$. In this view a cavity is the same as a stream, the precise geometrical distinction between them being simply one of degree. A model of this type has recently appeared (Fowler 2010), although its numerical solution has yet to be attempted.

Glaciology on Mars   The theory of dust–albedo feedback used in the description of the possible mechanism for the formation of the spiral troughs on the Martian polar ice caps was advanced by Howard (1978), although mathematical efforts to establish a theory had to await the models of Pelletier (2004) and Ng and Zuber (2003, 2006). Pelletier's model is essentially equivalent to the Fitzhugh–Nagumo equations, which are known to produce spiral waves, but appears to have been constructed with a view to obtaining the waves he sought. While the resulting numerical solutions which he found are suggestive, there is no coherent physical basis for the model. Ng and Zuber's model is more clearly based on Howard's idea, and uses Ivanov and Muhleman's (2000) description of radiative transport as its basis. Our description is largely based on Ng and Zuber's work, although we diverge in our development of the model and its conclusion: see Zammett and Fowler (2010).

10.8 Exercises

10.1 The downstream velocity u over the cross section S of a glacier is given by

$$\nabla.\big[\eta\{A, |\nabla u|\}\nabla u\big] = -1 \quad \text{in } S,$$

where the viscosity is given by

$$\eta = A^{-1/n}|\nabla u|^{-(n-1)/n}.$$

Assuming the rate factor $A = 1$ and a semi-circular profile for the ice cross section S. Give suitable boundary conditions for the flow, and hence derive the solution. Deduce the ice flux Q as a function of the cross-sectional area of the flow.

10.2 Use lubrication theory to derive an approximate model for two-dimensional flow of a valley glacier, assuming Glen's flow law with a rate constant independent of temperature, and no sliding at the base. Non-dimensionalise the model, and show that for typical lengths of 10 km, accumulation rates of 1 m y^{-1}, and if the rate constant in Glen's law is $0.2 \text{ bar}^{-3} \text{ y}^{-1}$ (with the Glen exponent being $n = 3$), a typical glacier depth is 100 m. Show that the dimensionless model depends on the single dimensionless parameter $\mu = d \cot \alpha / l$, where d is the depth scale, l is the length scale, and α is the valley slope. What are typical values of μ?

Show that if $\mu \ll 1$, the model takes the form of a first order hyperbolic wave equation. Write down the solution for small perturbations to the steady state, and show that the perturbations grow unboundedly near the glacier snout. Why is this? Write an alternative linearisation which allows a bounded solution to be obtained.

More generally, an exact characteristic solution of the model allows shocks to form (and thus for the glacier snout to advance). Discuss the rôle of μ in shock formation.

10.3 A glacier is subject to an accumulation rate a whose amplitude varies sinusoidally in time about a mean (space-dependent) value; specifically

$$a = a_0(x) + a_1 e^{i\omega t},$$

where a_1 is constant (the real part may be assumed). Use an appropriate linearised wave theory to determine the resultant form of the perturbed surface.
What can you say about the effect of millennial scale climate changes? About
annual balance changes?

How would you generalise your result to a general time-dependent amplitude variation?

10.4 Write down the equations governing three-dimensional flow of an ice sheet,
and show how they can be non-dimensionalised to obtain

$$H_t = \nabla.\left[\left\{\frac{|\nabla s|^{n-1} H^{n+2}}{n+2}\nabla s\right\} - H\mathbf{u}_b\right] + a,$$

assuming Glen's flow law and a temperature-independent rate coefficient.
Show that the dimensionless basal shear stress is $\boldsymbol{\tau} = -H\nabla s$.

10.5 An ice sheet in steady state has profile $z = s(x, y)$ and horizontal velocity
$\mathbf{u} = (u, v)$ independent of depth, with $\mathbf{u} = -K\nabla s$, for some scalar K. Suppose that χ is a coordinate anti-clockwise along level s contours, and that U
is a function such that $u = U\chi_y$, $v = -U\chi_x$, and which satisfies (in terms of
independent coordinates s, χ)

$$\frac{\partial}{\partial s}\left(\frac{U}{|\mathbf{u}|}\right) = -\frac{1}{|\nabla s|}\frac{\partial\theta}{\partial\chi},$$

where $\theta = \tan^{-1}(\frac{v}{u})$.

If σ measures arc length on a level s contour $C(s)$, show that

$$\frac{U\,d\chi}{|\mathbf{u}|} = d\sigma$$

on C. Show also that $\oint_{C(s)} d\theta = 2\pi$. Show that distance ξ along a flow line
satisfies $d\xi = -\frac{ds}{|\nabla s|}$, and deduce that

$$\frac{\partial}{\partial\xi}\left(\frac{U}{|\mathbf{u}|}\right) = \frac{\partial\theta}{\partial\chi}.$$

Show that integration of this equation round a closed contour $C(s)$ appears
to imply that $\frac{\partial L(s)}{\partial\xi} = 2\pi$, where $L(s) = \oint_C d\sigma$ is the circumferential length
of C. This is incorrect: why?

Show that a correct inference is that

$$-\frac{dL}{ds} = \oint_C \frac{d\theta}{|\nabla s|},$$

and show that this equation can be deduced on purely geometrical grounds.
The ice sheet profile satisfies the equation

$$\nabla.(s\mathbf{u}) = a,$$

where we suppose $a(x, y) > 0$. By using s and χ as independent coordinates, show that

$$\frac{\partial(sU)}{\partial s} = -\frac{a}{J},$$

where $J = -\frac{\partial(s,\chi)}{\partial(x,y)}$ is the Jacobian of the transformation from (x, y) to $(-s, \chi)$. Explain why $J > 0$ away from the ice sheet summit, and deduce that sU is a monotone increasing function of ξ along a flow path.

10.6 Suppose that an ice sheet has the symmetric profile $s = 1 - r^2$, where r is the polar radius from the centre. The curvature of the level s contours is thus $\kappa = \frac{1}{r}$, and the distance along a steepest descent path is r. The temperature away from the bed is given by

$$T = f(zU),$$

where

$$U = u \exp\left[\int^r \kappa \, dr\right],$$

and $u(r)$ is the radial outwards plug flow velocity of the ice; the ice depth is related to the accumulation rate $a(r)$ by

$$\frac{1}{r}\frac{\partial}{\partial r}(rsu) = a.$$

Show that $rsu = B$, where $B = \int_0^r ra\,dr$, and thus that $U = \frac{B}{s}$, and deduce that

$$T = f\left(\frac{Bz}{s}\right).$$

For the particular case where $a = 1$ and the surface temperature is $T_s = -\Gamma s$, show that the interior temperature is given by

$$T = -\Gamma\left[1 - \frac{r^2 z}{s}\right].$$

(*This temperature decreases with increasing depth, a typical result of the advection of cold inland surface ice below warmer coastal ice. Such profiles are seen in measured temperature profiles, but with an inversion near the base where the geothermal heat flux causes the basal ice to become warmer.*)

10.7 The averaged (dimensional) horizontal force balance at the calving front of an ice shelf can be written in the form

$$\int_b^s \sigma_{11}\,dz = -\int_b^0 p_w\,dz - p_a s,$$

where p_w is the (hydrostatic) water pressure, p_a is atmospheric pressure, and $z = s$ and $z = b$ are the positions of the ice top surface and bottom surface relative to sea level at $z = 0$. Show that, when written in terms of the ice sheet scales, this condition takes the form

$$\frac{(1+\delta)b^2}{2} - \frac{(s-b)^2}{2} + \varepsilon^2 \int_b^s (-p + \tau_1)\,dz = 0,$$

where $\delta = (\rho_w - \rho_i)/\rho_i$. Hence show that in terms of the ice shelf scales, the condition can be written in the form

$$b^2 - \delta s^2 + 2sb + 4\tau_1(\delta s - b) = 0,$$

assuming the approximate results $-p \approx \tau_1 = \tau_1(x, t)$. Taking $s \approx -b$, show that the vertically averaged deviatoric longitudinal stress at the calving front is

$$\bar{\tau}_1 = -\frac{1}{4}b,$$

and if this is taken as the boundary condition for the ice shelf stress τ_1, show that (by solving $(10.141)_2$)

$$\tau_1 = -\frac{1}{4}b$$

everywhere.

10.8 Suppose that

$$H_t = -q_x + a,$$

$$q = -D(H, H_x)H_x,$$

where a is constant, and the boundary conditions are

$$q = 0 \quad \text{at } x = 0,$$

$$q = q_G(x_G), \qquad H = H_G(x_G) \quad \text{at } x = x_G.$$

If the steady state solution is denoted with a suffix or superscript zero, show, by writing $H = H_0(x) + \eta$, $q = q_0(x) + Q$, and $x_G = x_G^0 + \gamma$, that the linearised system for the perturbation can be written in the form

$$\eta_t = -Q_x,$$

$$Q = -\bar{p}\eta_x - \bar{q}\eta,$$

$$Q = 0 \quad \text{at } x = 0.$$

$$Q = K\eta \quad \text{at } x = x_G^0,$$

where

$$K = \frac{q_G'(x_G^0) - q_0'(x_G^0)}{H_G'(x_G^0) - H_0'(x_G^0)}, \qquad \bar{p} = D_0 + D_{H'}H_0'', \qquad \bar{q} = D_H H_0',$$

and

$$D_{H'} = \frac{\partial D}{\partial H_x}, \qquad D_H = \frac{\partial D}{\partial H},$$

the derivatives being evaluated at the steady state. Note that

$$\bar{p} = -\frac{\partial q}{\partial H_x} > 0, \qquad \bar{q} = -\frac{\partial q}{\partial H} < 0.$$

Show that solutions exist in the form $\eta = y(x)e^{\sigma t}$, and show that the equation for y can be written in Sturm–Liouville form

$$(py')' + (s - \sigma r)y = 0,$$

where primes denote differentiation with respect to x,

$$p = \bar{p}r, \qquad r = \exp\left(\int_0^x \frac{\bar{q}\, dx}{\bar{p}}\right), \qquad s = r\bar{q}'.$$

Deduce that there exists a denumerable, decreasing sequence of eigenvalues σ, and that for the maximum of these, σ_1, the corresponding eigenfunction y_1 is of one sign (say positive).
Show that

$$\sigma_1 \int_0^{x_G^0} y_1\, dx = -K y_1(x_G^0),$$

and deduce that the steady state is unstable if and only if $K < 0$.

10.9 The drainage pressure in a subglacial channel is determined by the Röthlisberger equations

$$\rho_w g \sin\alpha_c = \frac{f_1 Q^2}{S^{8/3}},$$

$$mL = \rho_w g Q \sin\alpha_c,$$

$$\frac{m}{\rho_i} = K S N^n.$$

Explain the meaning of these equations, and use them to express the effective pressure N in terms of the water flux Q. Find a typical value of N, if $Q \sim 1\ \mathrm{m^3\ s^{-1}}$, and $\sin\alpha_c \sim 0.1$, $f_1 = f\rho_w g$, $f = 0.05\ \mathrm{m^{-2/3}\ s^2}$, $n = 3$, $L = 3.3 \times 10^5\ \mathrm{J\,kg^{-1}}$ and $K = 0.1\ \mathrm{bar^{-3}\ y^{-1}}$.

Use a stability argument to explain why Röthlisberger channels can be expected to form an arterial network.

10.10 Drainage through a linked cavity system relates the effective pressure N_K to the water flux Q by the implicit relation

$$N_K e^{S(\Lambda)} = \delta N_R(Q),$$

where $S = S'\Lambda$, (S' constant), $\Lambda = \frac{u}{N_K^n}$, $\delta < 1$ and

$$N_R(Q) = \beta Q^{1/4n},$$

with $n = 3$. Explain why this distributed system should be stable if $N_K'(Q) < 0$. Show that

$$-\frac{N_K'}{N_K}(n\Lambda S' - 1) = \frac{N_R'}{N_R},$$

and deduce that linked cavity drainage is stable for $\Lambda > \Lambda_c \equiv \frac{1}{nS'}$.

10.11 A correction for the basal shear stress near the head of a glacier which allows for longitudinal stress is

$$\tau = H(1 - H_X) + \gamma \left(H |u_X|^{\frac{1}{n}-1} u_X \right)_X,$$

where u is the velocity (assumed to be a plug flow), X is distance from the head, and γ is small. Assume that near the head of the glacier, conservation of mass takes the form

$$Hu = X,$$

and the sliding law is of the Weertman type

$$\tau = u^r,$$

where $0 < r < 1$.

We wish to apply the boundary conditions

$$H = 0 \quad \text{at } X = 0,$$

$$H \sim X^{\frac{r}{r+1}} \quad \text{as } X \to \infty.$$

Consider first an outer approximation in which the term in γ is ignored. Show that there is a unique value of $H(0) = H_0 > 0$ such that the boundary condition at ∞ can be satisfied.

Now suppose $n = 1$. Show, by writing first $X = e^{\xi}$ and then $\xi = \gamma \Xi$, that there is a boundary layer, in which H changes from 0 at $X = 0$ to H_0 as X increases, and show that $H \sim X^{H_0^2/2\gamma}$ as $X \to 0$.

Carry through the analysis when $n \neq 1$, and show that in this case $H \sim X^{(H_0^2/2\gamma)^n}$ as $X \to 0$.

Do these solutions make physical sense?

10.12 The depth H of an isothermal glacier satisfies the equation

$$H_t + \frac{\partial}{\partial x} \left[(1 - \mu H_x)^n \frac{H^{n+2}}{n+2} \right] = a,$$

and $H = 0$ at the snout $x_s(t)$. Assuming that $a_s = a(x_s) < 0$ at $x = x_s$, show, by consideration of the local behaviour of H, that if the glacier is advancing, $\dot{x}_s = v_+ > 0$, then

$$H \sim A_+(x_s - x)^{\frac{n}{n+1}},$$

and determine A_+ in terms of v_+. If the glacier is retreating, $\dot{x}_s = -v_- < 0$, show that

$$H \sim A_-(x_s - x),$$

and determine A_- in terms of v_-.

Finally show that in the steady state,

$$H \sim A_0(x_s - x)^{1/2},$$

and determine A_0.

10.13 The relation between ice volume flux and depth for a surging glacier is found
to be a multivalued function, consisting of two monotonically increasing
parts, from $(0,0)$ to (H_+, Q_+) and from (H_-, Q_-) to (∞, ∞) in (H, Q)
space, where $H_+ > H_-$ and $Q_+ > Q_-$, with a branch which joins (H_-, Q_-)
to (H_+, Q_+). Explain how such a flux law can be used to explain glacier
surges if the balance function $s(x)$ satisfies $\max s > Q_+$, and give a rough
estimate for the surge period.

What happens if $\max s < Q_-$? $\max s \in (Q_-, Q_+)$?

10.14 The depth h and velocity u of an ice sheet fan are given by the thermo-
hydraulic sliding law

$$h = \frac{f u^r}{[G + Rhu - au^{1/2}]^m},$$

where $r = \frac{1}{3}$, $m = \frac{1}{9}$, $G = 0.06\ \mathrm{W\,m^{-2}}$, $R = 3 \times 10^{-7}\ \mathrm{W\,m^{-4}}$ y, $a = 0.8 \times$
$10^{-2}\ \mathrm{W\,m^{-5/2}\ y^{1/2}}$, and $f = 126\ \mathrm{W^{1/9}\ m^{4/9}\ y^{1/3}}$. Assuming $hu \sim Q_i \approx$
$5 \times 10^5\ \mathrm{m^2\,y^{-1}}$, show how to non-dimensionalise the equation to the form

$$h = \frac{\phi u^r}{[\Gamma + hu - u^{1/2}]^m},$$

and give the definitions of the dimensionless parameters ϕ and Γ. Using the
values above, show that $\Gamma \approx 0.4$, $\phi \approx 0.77$.

Define $v = u^{1/2}$, and show that

$$L \equiv \Gamma - v + hv^2 = \left(\frac{v}{v^*}\right)^{2r/m} \equiv R,$$

where

$$v^*(h) = \left(\frac{h}{\phi}\right)^{3/2}.$$

By considering the intersections of the graphs of L and R, show that multiple
steady states are possible for sufficiently small h. Using the observation that
$\frac{2r}{m} = 6$ is large, show explicitly that if $h \lesssim \frac{1}{4\Gamma}$, then there is a solution $v \approx v^*$
for $v^* < v_-$, $v \approx v_-$ for $v^* > v_-$, and if in addition $v^* > v_+$, there are a
further two roots $v \approx v_+, v^*$, where $v_\pm$ are the two roots for v of $L = 0$.
Show also that if $h \gtrsim \frac{1}{4\Gamma}$, then there is a unique solution $v \approx v^*$.

By consideration of the graphs of $v^*(h)$ and $v_\pm(h)$ (hint: for the latter,
first draw the graph of $L = 0$ for h as a function of v), show that multiple
solutions exist for sufficiently small ϕ, and by finding when the graph of v^*
goes through the nose of the $v_\pm$ curve, show that multiple steady states exist
in the approximate range

$$\phi < \phi_c = \frac{1}{2^{8/3}\,\Gamma^{5/3}},$$

and find the value of ϕ_c.

Show that if $\phi < \phi_c$ and $hu = q$ is prescribed, there is a unique solution,
but that there is a range $q_- < q < q_+$ where such a solution is unstable (as

it lies on the negatively sloping part of the u versus h curve). What do you think happens if q lies in this intermediate range?

10.15 The depth of a glacier satisfies the equation

$$H_t + \frac{\partial}{\partial x}\left[(1 - \mu H_x)^n \frac{H^{n+2}}{n+2}\right] = B'(x),$$

where $\mu \ll 1$. Suppose first that $\mu \ll 1$, so that the diffusion term can be neglected. Write down the characteristic solution for an arbitrary initial depth profile. What is the criterion on the initial profile which determines whether shocks will form?

Now suppose $B = \frac{1}{n+2}$ is constant, so that a uniform steady state is possible. Describe the evolution of a perturbation consisting of a uniform increase in depth between $x = 0$ and $x = 1$, and draw the characteristic diagram.

Shock structure. By allowing $\mu \neq 0$, the shock structure is described by the local rescaling $x = x_s(t) + \mu X$. Derive the resulting leading order equation for H, and find a first integral satisfying the boundary conditions $H \to H_\pm$ as $X \to \pm\infty$, where $H_- > H_+$ are the values behind and ahead of the shock. Deduce that the shock speed is

$$\dot{x}_s = \frac{[H^{n+2}]_-^+}{[H]_-^+},$$

and that $\phi = H/H_+$ satisfies the equation

$$\phi_\xi = -\left[g(\phi)^{1/n} - 1\right],$$

where $\xi = X/H_+$, $\phi \to r$ as $\xi \to -\infty$, $\phi \to 1$ as $\xi \to \infty$, and

$$g(\phi) = \frac{(r^{n+2} - 1)(\phi - 1) + (r - 1)}{(r - 1)\phi^{n+2}},$$

with $r = H_-/H_+ > 1$. Show that $g(1) = g(r) = 1$, and that $g(\phi) > 1$ for $1 < \phi < r$, and deduce that a monotonic shock structure solution joining H_- to H_+ does indeed exist.

Suppose that $\delta = \Delta H/H_+$ is small, where $\Delta H = H_- - H_+$. By putting $r = 1 + \delta$ and $\phi = 1 + \delta\Phi$, show that

$$g = 1 + \frac{\delta^2(n+1)(n+2)}{2}\Phi(1 - \Phi) + \cdots,$$

and deduce that

$$\Phi_\Xi \approx -\Phi(1 - \Phi),$$

where

$$\Xi = \frac{\delta(n+1)(n+2)}{2n}\xi.$$

Deduce that the width of the shock structure is of dimensionless order

$$x - x_s \sim \frac{2n\mu H_+}{\delta(n+1)(n+2)},$$

or dimensionally

$$\frac{2n}{(n+1)(n+2)}\frac{d_+^2}{\Delta d\,\tan\alpha},$$

and that for a glacier of depth 100 m, slope $(\tan\alpha)$ 0.1, with $n = 3$, a wave of height 10 m has a shock structure of width 3 km. (*This is the monoclinal flood wave for glaciers, analogous to that for rivers discussed in Chap. 4.*)[38]

10.16　In deriving the reduced, three-dimensional model for drumlin formation, it is necessary to compute the three-dimensional components of the stress tensor at the bed. Show that the normal, x-tangential and 'y-tangential' vectors at the bed $z = s(x, y, t)$ are

$$\mathbf{n} = \frac{(-s_x, -s_y, 1)}{(1+|\nabla s|^2)^{1/2}}, \qquad \mathbf{t}_1 = \frac{(1, 0, s_x)}{(1+s_x^2)^{1/2}},$$

$$\mathbf{t}_2 = \mathbf{n}\times\mathbf{t}_1 = \frac{(-s_x s_y, 1+s_x^2, s_y)}{(1+|\nabla s|^2)^{1/2}(1+s_x^2)^{1/2}},$$

where $\nabla s = (s_x, s_y)$, and hence show that

$$-\tau_{nn} = \frac{2\eta[u_x(1-s_x^2) + v_y(1-s_y^2) + s_x(u_z + w_x) + s_y(v_z + w_y)]}{1+|\nabla s|^2}$$

$$\times \frac{[-s_x s_y(u_y + v_x)]}{1+|\nabla s|^2}.$$

Show also that the horizontal basal shear stress vector (τ_1, τ_2), where $\tau_i = \mathbf{n}.\boldsymbol{\tau}.\mathbf{t}_i$, has components

$$\tau_1 = \frac{\eta[(1-s_x^2)(u_z + w_x) - 2s_x(u_x - w_z) - s_y(u_y + v_x) - s_x s_y(v_z + w_y)]}{(1+|\nabla s|^2)^{1/2}(1+s_x^2)^{1/2}},$$

$$\tau_2 = \frac{\eta[(1+s_x^2 - s_y^2)(v_z + w_y) - 2s_y\{v_y(1+s_x^2) - w_z - s_x^2 u_x\}]}{(1+|\nabla s|^2)(1+s_x^2)^{1/2}}$$

$$\times \frac{[-s_x(u_y + v_x)(1+s_x^2 - s_y^2) - 2s_x s_y(u_z + w_x)]}{(1+|\nabla s|^2)(1+s_x^2)^{1/2}}.$$

Show also that the two x-tangential and y-tangential components of the basal velocity are

$$u_1 = \frac{u + ws_x}{(1+s_x^2)^{1/2}}, \qquad u_2 = \frac{-us_x s_y + v(1+s_x^2) + ws_y}{(1+|\nabla s|^2)^{1/2}(1+s_x^2)^{1/2}}.$$

Write down the appropriate equations for ice flow, and a suitable matching condition when $l \ll d_I$, where l is the horizontal drumlin length scale, and d_I is the ice sheet depth. By scaling the equations and assuming the aspect ratio $\nu \ll 1$, derive a reduced form of the model as in (10.374)–(10.376).

---

[38] The observation that the smallness of surface slope diffusion is offset by the smallness of surface amplitude is made, for example, by Gudmundsson (2003) (see Paragraph 16).

10.17 A model of two-dimensional ice flow over a deformable bed $z = vs$ is given by the equations

$$\Pi_x = \nabla^2 \psi_z,$$
$$\Pi_z = -\nabla^2 \psi_x,$$

with matching condition

$$\Pi \to 0, \qquad \psi_{zz} \to \theta, \qquad \psi_x \to 0 \quad \text{as } z \to \infty,$$

and at the base $z = vs$,

$$-\tau_{nn} = \frac{2[(1 - v^2 s_x^2)\psi_{zx} + vs_x(\psi_{zz} - \psi_{xx})]}{1 + v^2 s_x^2},$$

$$\tau = \frac{[(1 - v^2 s_x^2)(\psi_{zz} - \psi_{xx}) - 4vs_x\psi_{zx}]}{1 + v^2 s_x^2},$$

$$\frac{\psi_z - v\psi_x s_x}{(1 + v^2 s_x^2)^{1/2}} = U(\tau, N),$$

$$-\psi_x = \alpha v s_t + v\psi_z s_x,$$

$$N = 1 + s + \Pi - \tau_{nn},$$

$$q = q(\tau, N),$$

$$s_t + q_x = 0.$$

Assuming a basic state $\psi = \bar{u}z + \frac{1}{2}\theta z^2$, $\Pi = s = 0$, show that by putting $\psi = \bar{u}z + \frac{1}{2}\theta z^2 + \Psi$ and linearising the model, Ψ, Π and s satisfy the system

$$\Pi_x = \nabla^2 \Psi_z,$$
$$\Pi_z = -\nabla^2 \Psi_x,$$

with

$$\Pi, \Psi_{zz}, \Psi_x \to 0 \quad \text{as } z \to \infty,$$

and

$$\hat{\tau} = \Psi_{zz} - \Psi_{xx},$$

$$\Psi_z = U_\tau \hat{\tau} + U_N \hat{N},$$

$$-\Psi_x = \alpha v s_t + v\bar{u} s_x,$$

$$\hat{N} = s + \Pi + 2(\Psi_{zx} + v\theta s_x),$$

$$s_t + q_\tau \hat{\tau}_x + q_N \hat{N}_x = 0$$

at $z = 0$, where $\hat{\tau}$ and $\hat{N}$ denote the perturbations to τ and N.

Show that the solution for Ψ is of the form

$$\Psi = (a + bz)\exp[-kz + ikx + \sigma t],$$

and hence show that

$$\sigma = r - ikc,$$

where the wave speed is

$$c = \frac{R[1 + 4v^2\alpha k^3 R(\theta + k\bar{u})]}{1 + 4v^2\alpha^2 k^4 R^2},$$

where

$$R = \frac{q_N + 2k(q_N U_\tau - U_N q_\tau)}{1 + 2kU_\tau},$$

and the growth rate is

$$r = \frac{2vk^2 R[\theta + k\bar{u} - \alpha R k]}{1 + 4v^2\alpha^2 k^4 R^2}.$$

Deduce that the uniform flow is unstable if $R > 0$.

10.18 The growth rate of the instability in Question 10.17 is given by

$$r = \frac{2vk^2 R[\theta + k\bar{u} - \alpha R k]}{1 + 4v^2\alpha^2 k^4 R^2},$$

where k is the wave number,

$$R = \frac{q_N + 2k(q_N U_\tau - U_N q_\tau)}{1 + 2kU_\tau},$$

and α and v are small. Show that the maximum value of r will occur when k is large, and in this case show that we can take

$$R \approx R_\infty = q_N - \frac{U_N q_\tau}{U_\tau},$$

and hence show that the maximum of r occurs when

$$k = k_{\max} \approx \left(\frac{3}{4v^2\alpha^2 R_\infty^2}\right)^{1/4},$$

where

$$r = r_{\max} \approx \frac{3^{3/4}\bar{u}}{2^{5/2}\alpha^{3/2}(vR_\infty)^{1/2}}.$$

The uniform bed is thus unstable if $R_\infty > 0$. Suppose now that

$$q = \left[\frac{\tau}{\mu} - N\right]_+ U(\phi),$$

where $\phi = \frac{\tau}{N}$, the notation $[x]_+$ denotes $\max(x, 0)$, and $U = 0$ for $\phi < \mu$. Show that $R_\infty > 0$ when $\phi > \mu$ for any such function $U(\phi)$.

10.19 The scaled surface perturbation H and the atmospheric dust concentration C of the Martian north polar ice cap are taken to satisfy the equations

$$\alpha H_\tau - vH_\xi = -g(C),$$

$$C_\tau = \phi g(C) + C_{\xi\xi} + \Lambda(H_\xi C)_\xi,$$

where $g(C) = C^2(1 - C)$, and the positive constants α, v, ϕ and Λ are $O(1)$. If the boundary conditions are taken to be

$$H, C \to 0 \quad \text{as } \xi \to \infty, \qquad C \to 1 \quad \text{as } \xi \to -\infty,$$

show that travelling wave solutions with speed w exist if $w > w_0$, and find a (numerical) method to determine the value of w_0.

Chapter 11
Jökulhlaups

In October 1996, a volcanic eruption underneath the ice cap Vatnajökull in Iceland became an international news story. The eruption itself was spectacular enough, sending clouds of ash high into the atmosphere, but its more impressive feature was that the fissure eruption took place under about 500 metres of ice. The erupted lava melted the basal ice, causing first a huge sag in the ice surface, and eventually its complete collapse and the formation of an ice canyon, hundreds of metres deep, within the ice.

It became evident that the subglacial meltwater was flowing from the fissure towards the subglacial lake Grímsvötn, which lies in the caldera of a volcano under the ice cap. As the weeks passed, the lake level rose (as could be inferred from the uplift of the overlying ice), and eventually it reached flotation level; that is to say, the hydrostatic pressure of the lake water became sufficient to lift the overlying ice, and a *jökulhlaup* occurred.

Jökulhlaup is an Icelandic word, meaning literally a 'glacier-burst' (jökul-hlaup), and it refers to catastrophic outburst floods from glaciers which occur in various parts of the world, not only Iceland, but also Canada and the Himalayas, for example. The one which occurred in 1996 was spectacular: icequakes at Grímsvötn indicated a breach of the seal formed by the ice at the caldera rim (see Fig. 11.1), and ten hours later a huge flood emerged on the sandur plain in front of the glacier. Its peak flow was estimated as 45,000 $m^3 s^{-1}$, and the flood washed away a good part of the road bridges across the sandur—which are built to withstand such floods.

What is of interest to us about this awesome flood is that jökulhlaups occur regularly. Jökulhlaups from Grímsvötn over the last century have occurred at intervals of 5–10 years, and the same cyclicity is evident in other such floods. Nor is this evidential of cyclic volcanic activity. In fact, few of the Grímsvötn jökulhlaups are associated directly with eruptions. Rather, the release of geothermal heat from the caldera causes subglacial melting and a slow but regular rise in the lake level until a flood occurs. The water in a flood flows *underneath* the glacier, burning (somehow) its way through the ice towards the glacier terminus. The fact that the 1996 flood took 10 hours from initial rupture of the seal at the caldera rim, until its emergence at the glacier snout, indicates the travel time of the water over the intervening distance of 50 km.

A. Fowler, *Mathematical Geoscience*, Interdisciplinary Applied Mathematics 36,   741
DOI 10.1007/978-0-85729-721-1_11, © Springer-Verlag London Limited 2011

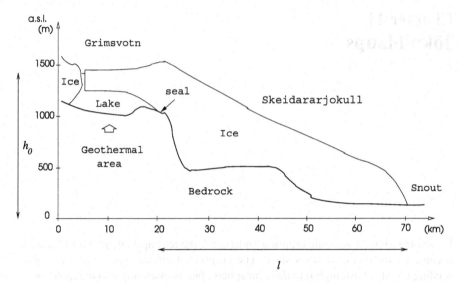

Fig. 11.1 The geometry of the lake and glacier, Grímsvötn and Skeiðarárjökull. (Figure courtesy of Felix Ng)

What we want to do in this chapter is to build a model which will help us understand how the flood can make its way under the glacier, and perhaps what causes the abatement of the flood, usually after several weeks. We might also ask what controls the recurrence time of 5–10 years, and we could consider other puzzles: in a flood, the lake is not drained completely. Nor (normally) is the flood initiated (as it was in 1996) when the lake is at flotation level; normally it is about 60 meters below this. We may not provide solutions to all these puzzles; but we will provide a mathematical framework within which one can at least consider them.

11.1 The Nye Model

The basic fact that we need to know is that in temperate glaciers (those at the melting point, for example all Alpine glaciers), surface meltwater percolates to the bed, and drains underneath the glacier in a subglacial hydraulic system which is usually thought to consist of a network of channels. At the glacier snout, one or more emergent streams often carry the discharge. The water of these pro-glacial streams is usually milky in appearance, being highly charged with glacially derived sediments, products of abrasion and comminution by the ice.

The classic Röthlisberger model of channel flow (already introduced in Chap. 10) conceptualises the subglacial streams as being cut into the ice, and existing at a pressure p which is less than the overburden pressure p_i of the ice. This is analogous to the reduced pore pressure in soils, but it occurs for a very different reason. Firstly, we expect the channel water pressure to be essentially that of the water at all parts of the bed—we visualise the boulder-strewn bed as being 'leaky', and allowing water to migrate easily; if the water pressure were not locally uniform, rapid migration would take place to ensure that it became so. But this pressure p must be less than p_i, otherwise flotation would occur, and the glacier could advance catastrophically.

Now if $p_i > p$, then the slow viscous creep of the ice will tend to cause channels to contract, just as fluid rushes into a cavity. Röthlisberger's idea was that channels could be maintained open against this contraction by having the channel walls melt—the source of the heat necessary to cause such melting being the frictional heat released by the flow of the water itself through the channel. This ingenious theory is now generally accepted as being the mechanism whereby subglacial drainage occurs, at least where there is evidence of distinct outlet streams, as in many valley glaciers.

This theory is described below. Nye used it to allow consideration of time-varying drainage, and thus developed a theoretical model for jökulhlaups, which, in spite of its simplicity, is very successful. In his model, drainage is supposed to occur through a single conduit, cut upwards from the bed into the ice, which has a semi-circular cross section of area S. The two processes which affect S are the melt rate m (mass per unit length downstream per unit time) and the viscous closure due to the creep of the surrounding ice. The equation for S is thus

$$\frac{\partial S}{\partial t} = \frac{m}{\rho_i} - K S(p_i - p)^n, \tag{11.1}$$

where ρ_i is the ice density, p is the channel pressure, and the second term represents the creep closure due to a nonlinear flow law $\dot{\varepsilon} = A\tau^n$, where $\dot{\varepsilon}$ is strain rate and τ is stress. The constant K is proportional to A, but includes also some numerical factors which arise through the exact solution of the ice creep problem. We see that the equation for S introduces further variables m and $N = p_i - p$, which is (similarly to soils) called the effective pressure.

Now suppose that x represents a downstream spatial coordinate along the axis of the channel (Nye used s, but we will reserve this for the ice surface, as in Chap. 10). We assume the channel direction is slowly varying, which avoids the necessity of including a complicated curvilinear coordinate system. We denote the volume flux of water through the channel by Q. Conservation of water mass in the channel then requires

$$\frac{\partial S}{\partial t} + \frac{\partial Q}{\partial x} = \frac{m}{\rho_w} + M, \tag{11.2}$$

where m/ρ_w represents the volumetric source due to side-wall derived melt, and M is an additional source due to tributary flow, surface meltwater supply, etc. We can consider M to be prescribed.

We also require an equation describing momentum balance for the channel water. We assume that the turbulent friction at the bed is given by a Manning correlation, and we ignore inertial effects, which is apparently equivalent to assuming that the Froude number is very small: at least during floods, it seems unlikely that this will be accurate, but in fact one can show (see Question 11.5) that in the present context, inertial terms are always small. The Manning law then gives

$$\rho_w g \sin \alpha - \frac{\partial p}{\partial x} = f \rho_w g \frac{Q|Q|}{S^{8/3}}, \tag{11.3}$$

where f is a friction factor, and α is the mean bedrock slope (of the channel). The relation of f to the Manning roughness factor n' is described below (in (11.19)).

The melt rate m is determined through an energy balance. The heat generated by friction is $Q[\rho_w g \sin \alpha - \frac{\partial p}{\partial s}]$, and this is used to control water temperature, and melt the side walls. If the average water temperature is θ_w, while that of the ice is θ_i, then an appropriate energy equation is

$$\rho_w c_w \left[S \frac{\partial \theta_w}{\partial t} + Q \frac{\partial \theta_w}{\partial x} \right] = Q \left(\rho_w g \sin \alpha - \frac{\partial p}{\partial x} \right) - m \left[L + c_w (\theta_w - \theta_i) \right]. \tag{11.4}$$

The left hand side is the material rate of change of water temperature with time, the first term on the right is the frictional source, and the second is the supply due to the enthalpy change on melting. We can consider θ_i to be known (for example, it is the pressure melting point), but θ_w must be further prescribed by a local heat transfer condition at the ice wall across the thermal boundary layer there. An empirical correlation for flow in a cylindrical tube (the Dittus–Boelter correlation) gives

$$a_{\mathrm{DB}} \left(\frac{\rho_w |Q|}{\eta_w S^{1/2}} \right)^{0.8} k(\theta_w - \theta_i) = m \left[L + c_w (\theta_w - \theta_i) \right], \tag{11.5}$$

where a_{DB} (≈ 0.2) is a constant, η_w is the viscosity of water, and k is its thermal conductivity.

Equations (11.1)–(11.5) give five equations for the five unknowns S, Q, m, p and θ_w. Certain initial and boundary conditions will also be appropriate; we will come to these in due course.

11.2 Non-dimensionalisation

Equations (11.1)–(11.5) are non-dimensionalised by writing

$$\begin{aligned} Q &= Q_0 Q^*, & S &= S_0 S^*, & p_i - p &= N_0 N^*, & m &= m_0 m^*, \\ x &= l x^*, & t &= t_0 t^*, & \theta_w &= \theta_i + \theta_0 \theta^*. \end{aligned} \tag{11.6}$$

The scales are chosen to effect the following balances in the equations. All three terms balance in (11.1), thus

$$\frac{S_0}{t_0} = \frac{m_0}{\rho_i} = K S_0 N_0^n. \tag{11.7}$$

Essentially, these give t_0 and S_0. Nothing is balanced in (11.2), but we balance the terms in (11.3) as follows. We write

$$\rho_w g \sin \alpha - \frac{\partial p}{\partial x} = \Phi + \frac{\partial (p_i - p)}{\partial x}, \tag{11.8}$$

where

$$\Phi = \rho_w g \sin \alpha - \frac{\partial p_i}{\partial x} \qquad (11.9)$$

is the hydraulic gradient (if basal water pressure equals overburden ice pressure). If $z = b$ is the altitude of the base, then $-\partial b/\partial x = \sin \alpha$, and if the ice surface is at $z = s$, then $p_i = \rho_i g(s - b)$. Hence

$$\Phi = -\frac{\partial}{\partial x}[\rho_i g s + (\rho_w - \rho_i)g b], \qquad (11.10)$$

and we write

$$\Phi = \Phi_0 \Phi^*, \qquad (11.11)$$

where

$$\Phi_0 = \frac{\rho_w g h_0}{l}, \qquad (11.12)$$

and $z = h_0$ is a typical value of the Grímsvötn lake level (see Fig. 11.1). (At the channel inlet, a value of water pressure $p = p_i$ corresponding to flotation would give $\rho_w g(h_0 - b) = \rho_i g(s - b)$, so that the hydraulic head $\rho_i g s + (\rho_w - \rho_i)g b = \rho_w g h_0$ there.) The balance of terms in (11.3) is now effected by having $\Phi \sim f \rho_w g Q^2 / S^{8/3}$, thus

$$\Phi_0 = \frac{f \rho_w g Q_0^2}{S_0^{8/3}}. \qquad (11.13)$$

In (11.4), we balance the two terms on the right, thus

$$Q_0 \Phi_0 = m_0 L, \qquad (11.14)$$

which fixes m_0, and we choose θ_0 by balancing the advective term with the source term in (11.4):

$$\frac{\rho_w c_w Q_0 \theta_0}{l} = Q_0 \Phi_0. \qquad (11.15)$$

The five relationships in (11.7), (11.13), (11.14) and (11.15) determine scales for S_0, m_0, t_0, N_0 and θ_0 in terms of Q_0, which for the moment we presume is prescribed; l is given from the geometry. It is natural to prescribe Q_0 by balancing flux with source in (11.2), i.e. $Q_0 = Ml$; indeed, this is an appropriate choice for steady subglacial drainage of Röthlisberger type, but as we shall see, it is inappropriate for the violence of jökulhlaups, where the typical flow rates are much larger.

Given Q_0, we find

$$S_0 = \left(\frac{f \rho_w g Q_0^2}{\Phi_0}\right)^{3/8}, \qquad m_0 = \frac{\Phi_0 Q_0}{L},$$
$$\theta_0 = \frac{g h_0}{c_w}, \qquad t_0 = \frac{\rho_i S_0}{m_0}, \qquad N_0 = (K t_0)^{-1/n}, \qquad (11.16)$$

and the dimensionless equations (11.1)–(11.5) become (dropping the asterisks on the variables)

Table 11.1 Physical
parameter values

Symbol	Value
c_w	$4.2 \, \text{kJ} \, \text{kg}^{-1} \, \text{K}^{-1}$
f	$0.05 \, \text{m}^{-2/3} \, \text{s}^2$
g	$9.8 \, \text{m} \, \text{s}^{-2}$
h_0	$1.5 \, \text{km}$
k	$0.56 \, \text{W} \, \text{m}^{-1} \, \text{K}^{-1}$
K	$0.5 \times 10^{-24} \, \text{Pa}^{-n} \, \text{s}^{-1}$
l	$50 \, \text{km}$
L	$3.3 \times 10^2 \, \text{kJ} \, \text{kg}^{-1}$
n	3
Q_0	$1.8 \times 10^5 \, \text{m}^3 \, \text{s}^{-1}$
η_w	$2 \times 10^{-3} \, \text{Pa} \, \text{s}$
ρ_i	$0.9 \times 10^3 \, \text{kg} \, \text{m}^{-3}$
ρ_w	$10^3 \, \text{kg} \, \text{m}^{-3}$
Φ_0	$300 \, \text{Pa} \, \text{m}^{-1}$

$$\frac{\partial S}{\partial t} = m - SN^n,$$

$$\varepsilon \frac{\partial S}{\partial t} + \frac{\partial Q}{\partial x} = \varepsilon r m + \Omega,$$

$$\Phi + \delta \frac{\partial N}{\partial x} = \frac{Q|Q|}{S^{8/3}}, \tag{11.17}$$

$$\varepsilon S \frac{\partial \theta}{\partial t} + Q \frac{\partial \theta}{\partial x} = Q \left[\Phi + \delta \frac{\partial N}{\partial x} \right] - m(1 + \varepsilon r \theta),$$

$$\theta \left(\frac{|Q|}{S^{1/2}} \right)^{0.8} = \gamma m(1 + \varepsilon r \theta),$$

and the parameters $\varepsilon, r, \Omega, \delta$ and γ are defined, after some algebra, by

$$\varepsilon = \frac{\Phi_0 l}{\rho_i L}, \qquad \delta = \frac{1}{\Phi_0 l} \left[\frac{Q_0^{1/4} \Phi_0^{11/8}}{\rho_i K L (f \rho_w g)^{3/8}} \right]^{1/n},$$

$$\gamma = \frac{\rho_w c_w}{k a_{\text{DB}} l} \left(\frac{\eta_w}{\rho_w} \right)^{0.8} Q_0^{1/2} \left(\frac{f \rho_w g}{\Phi_0} \right)^{3/20}, \tag{11.18}$$

$$r = \frac{\rho_i}{\rho_w}, \qquad \Omega = \frac{Ml}{Q_0}.$$

To estimate the sizes of the scales and parameters, we use the values in Table 11.1. The value of the closure rate coefficient K is given by $K = 2A/n^n$, and we use Paterson's (1994) recommended value of $A = 6 \times 10^{-24} \, \text{Pa}^{-n} \, \text{s}^{-1}$ at 0°C, with

$n = 3$. The definition of the friction factor f in Manning's roughness law as written here is

$$f = n'^2 \left(\frac{S}{R_H^2} \right)^{2/3}, \qquad (11.19)$$

where n' is the roughness coefficient, and R_H is the hydraulic radius ($= S/l$, where l is the wetted perimeter). For a semi-circular channel $(S/R_H^2)^{2/3} = (2(\pi + 2)^2/\pi)^{2/3} \approx 6.6$, so that if $n' = 0.09$ m$^{-1/3}$ s, then $f \sim 0.05$ m$^{-2/3}$ s^2 (and 0.01 m$^{-2/3}$ s^2 for $n' = 0.04$ m$^{-1/3}$ s). The value of Φ_0 follows from (11.13), and we have anticipated the value of Q_0 which is found below; we then find successively that

$$S_0 \sim 10^4 \text{ m}^2, \qquad m_0 \sim 163 \text{ kg m}^{-1} \text{s}^{-1}, \qquad \theta_0 \sim 3.6 \text{ K},$$
$$t_0 \sim 0.6 \times 10^5 \text{ s (0.68 day)}, \qquad N_0 \sim 3.2 \times 10^6 \text{ Pa (32 bars)}, \qquad (11.20)$$

and the dimensionless parameters are of typical sizes

$$\gamma \sim 2.5, \qquad \varepsilon \sim 0.05, \qquad r \sim 0.9, \qquad \delta \sim 0.22, \qquad \Omega \sim 0.6 \times 10^{-3}, \qquad (11.21)$$

where for Ω we assume a base flow rate of $Ml = 10^2$ m^3 s^{-1}, which represents a typical value of the discharge between jökulhlaups. We see that all the parameters are $\lesssim O(1)$, which indicates that the scales we have chosen are sensible. Next we need to find a reason why Q_0 should be chosen as large as 1.8×10^5 m^3 s^{-1}.

11.3   Boundary Conditions and Lake Refilling

The equations in (11.17) require initial conditions for S and θ, and two boundary conditions for Q or N, and one for θ. The boundary condition for θ is taken to be

$$\theta = \theta_L \quad \text{at } x = 0 \qquad (11.22)$$

(at least when $Q > 0$ at the lake). At the outlet, it seems we should prescribe

$$N = 0 \quad \text{at } x = 1, \qquad (11.23)$$

i.e., the water pressure becomes atmospheric[1] (as also does the ice pressure). At the inlet to the channel, conservation of mass requires that

$$\frac{dV}{dt} = m_L - Q(0, t), \qquad (11.24)$$

where V is the lake volume and m_L represents the geothermal melt rate. Suppose the lake level is at $z = h$, and open to the atmosphere.[2] We assume $V = V(h)$, and

---

[1] There are problems with this, however: see further discussion in the notes.

[2] At Grímsvötn, this is normally the case. Apparently the high geothermal heat levels melt ice at the caldera walls, so that water is usually present at the walls all the way to the surface. See Björnsson (1988, pp. 70–73).

in fact $V'(h) = A_L$ is the lake surface area (which may depend on h). Now the water pressure at the inlet is $\rho_w g(h - b)$, where b is the grounding line elevation of the bed, and this is equal to $p_i - N$. Therefore (if b and p_i do not vary)

$$\rho_w g \frac{dh}{dt} = -\frac{dN}{dt}(0, t), \tag{11.25}$$

and thus the boundary condition at the lake inlet is

$$-\frac{A_L}{\rho_w g} \frac{\partial N}{\partial t} = m_L - Q \quad \text{at } x = 0, \tag{11.26}$$

where $x = 0$ is taken to be the position of the lake margin. We now finally choose the scale Q_0 so that

$$Q_0 = \frac{A_L N_0}{\rho_w g t_0}, \tag{11.27}$$

where we will take A_L as constant. Using the definitions of N_0 and t_0 in (11.16), we obtain

$$Q_0 = \left[\left(\frac{A_L}{\rho_w g} \right)^n \frac{1}{K} \left\{ \left(\frac{\Phi_0}{f \rho_w g} \right)^{3/8} \frac{\Phi_0}{\rho_i L} \right\}^{n+1} \right]^{4/(3n-1)}. \tag{11.28}$$

It follows that (11.26) can be written in the dimensionless form

$$\frac{\partial N}{\partial t} = Q - \nu \quad \text{at } x = 0, \tag{11.29}$$

where

$$\nu = \frac{m_L}{Q_0}. \tag{11.30}$$

We take $A_L = 30 \text{ km}^2 = 3 \times 10^7 \text{ m}^2$, and use other values given previously; hence we compute $Q_0 \sim 1.8 \times 10^5 \text{ m}^3 \text{ s}^{-1}$, as we assumed previously.

The refilling rate m_L can be estimated from the total rate of discharge, which is about $5 \times 10^{11} \text{ kg y}^{-1}$, or $17 \text{ m}^3 \text{ s}^{-1}$. This gives a typical value of $\nu \approx 0.89 \times 10^{-4}$. We can guess that $\nu \ll 1$ controls the (large) length of the time period between jökulhlaups. Over a catchment area of perhaps $2 \times 10^8 \text{ m}^2$, this refilling rate corresponds to a melt rate of $0.8 \times 10^{-7} \text{ m s}^{-1}$. With $\rho_w = 10^3 \text{ kg m}^{-3}$ and $L = 3.3 \times 10^5 \text{ J kg}^{-1}$, the geothermal heat flux required to provide this is about 2.4 W m^{-2}; this compares with a typical (non-volcanic) value of 0.05 W m^{-2}! The heat is not, of course, delivered as a conductive heat flux, but via the upwelling of superheated geothermal fluid (essentially sulphurous groundwater).

11.4 Simplification of the Model

The parameters ε, δ and Ω are all relatively small. If we neglect them, it follows that (supposing Φ and $Q > 0$)

$$Q \approx Q(0, t), \quad \text{i.e.} \quad Q \approx Q(t),$$
$$S \approx \Phi^{-3/8} Q^{3/4}, \tag{11.31}$$

and hence that

$$m \approx \frac{\Phi^{3/20} Q^{1/2} \theta}{\gamma}, \tag{11.32}$$

and thus

$$\gamma Q^{1/2} \frac{\partial \theta}{\partial x} \approx \gamma \Phi Q^{1/2} - \theta \Phi^{3/20}. \tag{11.33}$$

We see from this that θ approaches a limiting value such that $m = \Phi Q$ over a dimensionless distance of order $\gamma Q^{1/2}$. Now since Q was scaled with a value which is somewhat larger than a typical peak discharge, it is certainly clear that between floods (when we expect $Q \sim \Omega \ll 1$) this distance will be very short; it might even be short during floods. To accommodate this suggestion, we simplify the model by assuming that $m = \Phi Q$ holds at all times, even though this may be inaccurate for a short time during maximum discharge.[3] It then follows that $N \approx N(0, t) = N(t)$, and N and Q satisfy

$$\dot{Q} = \frac{4}{3} \Phi^{11/8} Q^{5/4} - \frac{4}{3} Q N^n, \tag{11.34}$$
$$\dot{N} = Q - \nu,$$

where $\dot{N} = dN/dt$. The model thus reduces to the solution of two ordinary differential equations!

If we take Φ to be constant, Eqs. (11.34) are easily studied in the (N, Q) phase plane.[4] There is a fixed point at $Q = Q^* = \nu$, $N = N^* = \Phi^{11/8n} Q^{*1/4n}$, and if we write

$$Q = Q^* + q, \qquad N = N^* + \Pi, \tag{11.35}$$

then linearised equations for Π and q are

$$\begin{pmatrix} \dot{q} \\ \dot{\Pi} \end{pmatrix} \approx \begin{pmatrix} \frac{1}{3} N^{*n} & -\frac{4n}{3} Q^* N^{*(n-1)} \\ 1 & 0 \end{pmatrix} \begin{pmatrix} q \\ \Pi \end{pmatrix}, \tag{11.36}$$

and solutions are proportional to $e^{\lambda t}$, where

$$\lambda^2 - \frac{1}{3} N^{*n} \lambda + \frac{4n}{3} Q^* N^{*(n-1)} = 0, \tag{11.37}$$

which implies that the equilibrium is an unstable spiral (if

$$\nu > \frac{\Phi^{\frac{11(n+1)}{2(3n-1)}}}{(48n)^{\frac{4n}{3n-1}}} \approx 0.58 \times 10^{-3} \quad \text{for } n = 3; \tag{11.38}$$

---

[3]There are other considerations which suggest that a useful value of γ, at least during a violent flood, may be smaller than the value in (11.21); in this case the assumption of equilibrium of (11.33) may indeed be realistic. See also the discussion in the notes.

[4]One might wonder how the assumption that Q and N in (11.34) are functions only of t can be squared with a hydraulic potential gradient $\Phi(x)$ which depends on distance downstream. In fact, because the lake refilling condition is applied only at the channel inlet, the value of Φ in (11.34) is actually that at the inlet, i.e., $\Phi(0)$. The case where this is negative is considered below.

Fig. 11.2 Phase portrait for
N and Q, with $\nu = 0.1$ in
(11.34)

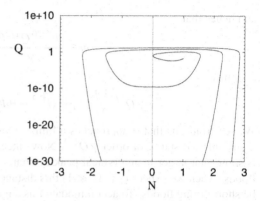

Fig. 11.3 Time series for Q
corresponding to Fig. 11.2

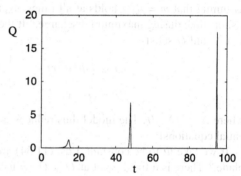

otherwise an unstable node): steady drainage from a lake is *always* unstable, in this
model.

Figure 11.2 shows a numerical solution of (11.34) in the (N, Q) phase plane.
Clearly the spiral structure continues for (N, Q) away from the unstable fixed
point. The time series corresponding to this diagram (Fig. 11.3) shows a sequence
of jökulhlaups of growing discharge, with long intervals (of $O(1/\nu)$) between the
floods. In the following section, we comment on the growing amplitude of these
oscillations. If we focus on a single flood, there is clearly no criterion to pick which
hydrograph will occur, and this is a drawback of the model. If we select initial values
and parameters to fit the rising limb of the 1972 jökulhlaup, we find (see Fig. 11.4)
that the peak discharge and decay curve do not fit the data. The resolutions of these
conundrums are given in the following sections.

11.5 Periodic Oscillations

The main problem with the result of the previous section is that there is no limit
to the growth of the jökulhlaup amplitude with time. In particular, the approximate
model allows N to become negative between floods (although Q always remains
positive). This is unphysical, for if $p > p_i$ in the channel, then in reality leakage
would occur laterally along the glacier bed, and flotation would occur.

Fig. 11.4 Hydrograph from
the 1972 jökulhlaup (*crosses*)
and a fit obtained by solving
(11.34)

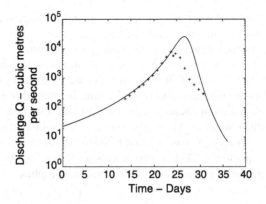

Physically, this implies that when the effective pressure N becomes zero, then the
closure equation (11.1) becomes inappropriate for determining S. We thus imagine
the following scenario. When N reaches zero, the channel can spontaneously open
(by lifting off the ice), and we might model this by imagining a very rapid 'creep'
opening of S if p approaches p_i. A suitable modification of (11.1) could then be

$$\frac{\partial S}{\partial t} = \frac{m}{\rho_i} - KSf(N), \tag{11.39}$$

where $f \approx N^n$ for $N > 0$, but $f \to -\infty$ very rapidly when $N \to 0$. The non-
dimensional version of $(11.33)_1$ is then

$$\dot{Q} = Q^{5/4} - Qf(N), \tag{11.40}$$

but now $\dot{Q} \gg 1$ if $N \approx 0$. In the limit, this simply squashes the trajectories in
Fig. 11.3 into $N > 0$, and leads to a modified phase portrait in which, however,
limit cycle periodic behaviour is still not obtained.

There are a number of other possible weaknesses in this simple discussion of
the Nye model. The difficulty associated with allowing $N = 0$ also crops up at
the glacier snout, where the imposition of $N = 0$ would cause unbounded channel
growth, according to $(11.17)_1$. Other issues which merit discussion are our neglect
of thermal advection, and the assumption of a semi-circular channel.

11.5.1 Breaking the Seal

There is, however, a far more fundamental flaw in the simple analysis given above,
and it is this: the existence of a subglacial lake in the first place implies the existence
of a local minimum in the hydraulic potential, and thus implies that the basic hydraulic gradient Φ must be negative in the vicinity of the lake. Consulting (11.17),
we see that the neglect of ε, δ and Ω then implies that there is a water divide, or *seal*,
near the lake. At the lake, $\Phi < 0$, and water flows back towards the lake, whereas

further down the ice slope, $\Phi > 0$ and drainage is towards the ice margin. This situation cannot be maintained, because then N decreases at the lake and reaches zero, at which point we suppose that a flood is initiated.

But even this description is too simplistic, at least in the case of Grímsvötn. Normally the floods from Grímsvötn are initiated when the lake level is 60–80 metres *below* flotation, and N at the inlet is inferred to be in the region of 6–8 bars. The initiation of channelised flow must therefore normally begin by a mechanism other than flotation, when there is still a hydraulic barrier to the lake at the caldera rim.

A related consideration is that mathematically, the neglect of δ in the momentum equation $(11.17)_3$ is problematical, for the following reason. Neglecting ε, integration of the mass conservation equation implies

$$Q = \Omega(x - x^*), \tag{11.41}$$

where x^* is fixed by the prescription that $\Phi(x^*) = 0$ (assuming $\delta = 0$). The momentum balance equation then gives S, so that the closure equation gives N, and the variables are independent of t. In general, the lake refilling condition is not satisfied, and this suggests that the loss of δ is a singular perturbation, which requires the consideration of a boundary layer in x.

We analyse such a boundary layer by writing $x = \delta X$, and allow for the possibility of reversed flow at the seal, whose position, however, is no longer constrained to be where $\Phi = 0$. (11.17) is written approximately as

$$\frac{\partial S}{\partial t} = \frac{|Q|^3}{S^{8/3}} - SN^n,$$

$$\frac{\partial Q}{\partial X} = \omega = \delta\Omega, \tag{11.42}$$

$$\Phi + \frac{\partial N}{\partial X} = \frac{Q|Q|}{S^{8/3}},$$

with the boundary conditions that

$$\frac{\partial N}{\partial t}(0, t) = Q(0, t) - \nu,$$

$$\frac{\partial N}{\partial X} \to 0 \quad \text{as } X \to \infty. \tag{11.43}$$

The last of these is a matching condition, which enables N to be matched to a far field slowly varying profile.

Between floods, or in conditions of normal drainage, $Q \ll 1$, and (11.43) then shows that N varies slowly, and $(11.42)_1$ implies that S relaxes to equilibrium (if $N = O(1)$) over a time scale of $O(1)$. Thus

$$S \approx \left\{|Q|^3/N^n\right\}^{3/11},$$

$$\Phi + \frac{\partial N}{\partial X} \approx \frac{N^{8n/11}}{|Q|^{2/11}} \operatorname{sgn} Q. \tag{11.44}$$

We can use this particular simplification to understand the mechanism of flood initiation. Since we suppose $\Phi < 0$ at the lake, but $\Phi > 0$ far from the lake, it is natural to choose the distinguished limit in which Φ varies over the same length scale X as is appropriate in the boundary layer. Indeed, this is the case for Grímsvötn.

Fig. 11.5 N_L versus X^*
when $\Phi = 1 - ae^{-bX}$, for
$a = 4$, $b = 2$ (strong seal) and
$a = 4$, $b = 4$ (weak seal). See
also Question 11.2

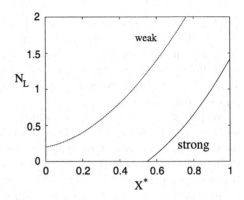

A Particular Example

In general, we need to solve the system (11.42), or (11.44), numerically. To gain
some analytic insight, it is instructive to simplify (11.44) by replacing the exponent
$8n/11$ in (11.44) by 1, and ignoring the denominator $|Q|^{2/11}$. We thus have to solve

$$\Phi + \frac{\partial N}{\partial X} = N \operatorname{sgn} Q, \tag{11.45}$$

where

$$Q = \omega(X - X^*), \tag{11.46}$$

and we require (assuming $\Phi(\infty) = 1$) that

$$
\begin{aligned}
N &\to 1 \quad \text{as } X \to \infty, \\
N &= N_L \quad \text{on } X = 0, \\
\dot{N}_L &= -(\nu + \omega X^*).
\end{aligned}
\tag{11.47}
$$

The seal position X^* is unknown, and is a function of time. It is consistent with the
slow variation of N that X^* will vary slowly with t.

One quickly finds that it is necessary that $X^* > 0$ (there is a seal) in order that N
does not grow exponentially at $+\infty$. The solution is then

$$
\begin{aligned}
N &= N_L e^{-X} - \int_0^X \Phi(\xi) e^{-(X-\xi)} \, d\xi, \quad 0 < X < X^*, \\
N &= \int_X^\infty \Phi(\xi) e^{-(\xi-X)} \, d\xi, \quad X > X^*.
\end{aligned}
\tag{11.48}
$$

Continuity of N at X^* thus requires

$$N_L(X^*) = e^{X^*} \int_0^\infty \Phi(\xi) e^{-|X^*-\xi|} \, d\xi, \tag{11.49}$$

and $(11.47)_3$ then provides an ordinary differential equation for the evolution of N_L
(or X^*): N_L decreases with time.

It is clear from our supposition that $\Phi \to 1$ as $X \to \infty$ that N_L is positive when
X^* is large; our concern is then what happens as N_L decreases in accord with

(11.47). There are two possibilities, and these are illustrated in Fig. 11.5. In the first, $N_L(0) < 0$, and thus N_L reaches zero when X^* is still positive, that is, while the seal position is still in front of the lake margin. Thus flotation occurs at the lake, and a flood ensues because of this. This appears to have been the situation in the 1996 Grímsvötn flood. We term a seal arising from a hydraulic gradient such that $N_L(0) < 0$ a *strong* seal.

In contrast, a *weak* seal is one for which $N_L(0) > 0$. In this case, X^* reaches zero while N_L is still positive, i.e., the lake level is below flotation. When the seal reaches the lake margin, the lake starts to empty, and a flood is initiated. This appears to be the normal case in Grímsvötn, where the lake level is typically about 60 metres below flotation level at flood onset, and the effective pressure is some 6 bars.

To summarise: we characterise a strong seal (in the context of (11.45)) as one with a hydraulic gradient such that

$$\int_0^\infty \Phi(\xi) e^{-\xi} d\xi < 0, \tag{11.50}$$

and a weak seal as one with

$$\int_0^\infty \Phi(\xi) e^{-\xi} d\xi > 0. \tag{11.51}$$

For a strong seal, the seal is broken when the lake level rises to flotation, but for a weak seal, the drainage divide slowly migrates backwards as the lake refills, and reaches the lake when it is still below flotation; at this point the seal is broken and the next flood is initiated. The simple formulae (11.50) and (11.51) only apply to the simplified model (11.45), but they do indicate the essential difference between a strong seal and a weak seal, which is that the strong seal has a larger negative hydraulic gradient near the lake. In the exercises, Question 11.2 calculates $N_L(X^*)$ for the representative choice

$$\Phi = 1 - ae^{-bX}, \tag{11.52}$$

from which we find that a seal is weak if

$$1 < a < a_1 = b + 1, \tag{11.53}$$

and is strong if $a > b + 1$. (This is the hydraulic gradient used in the illustrative Fig. 11.5.) Question 11.6 extends the analysis to the better approximating equation

$$\Phi + \frac{\partial N}{\partial X} = N^2 \operatorname{sgn} Q; \tag{11.54}$$

for this equation (and a hydraulic potential gradient given by (11.52)) we can show that a seal is weak if

$$a < a_2 = \left[\frac{1}{2} b j'_{(2/b),1} \right]^2, \tag{11.55}$$

where $j'_{v,k}$ is the kth zero of the Bessel function derivative $J'_v(z)$. We can use the asymptotic expansions of $j'_{v,1}$ for large and small v,

$$j'_{v,1} \approx \begin{cases} v + 0.81 v^{1/3} & \text{as } v \to \infty, \\ \sqrt{2v} & \text{as } v \to 0, \end{cases} \tag{11.56}$$

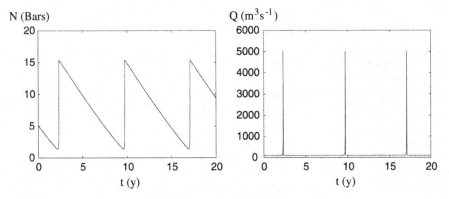

Fig. 11.6 $N_L(t)$ and the outlet discharge $Q_{\text{out}}(t) = Q_0(\Omega - \omega X^*)$ with $\omega = 0.12 \times 10^{-3}$, $\nu = 0.89 \times 10^{-4}$, $a = 2.8$, $b = 4.316$ in (11.52). To convert the results to dimensional quantities, we have used the scales $Q_0 = 0.18 \times 10^6 \ \text{m}^3 \ \text{s}^{-1}$, baseflow $\Omega Q_0 = 100 \ \text{m}^3 \ \text{s}^{-1}$, $N_0 = 32.37$ bars, $t_0 = 0.0019$ y, $\delta = 0.216$, $l = 50$ km. These are our estimated parameter values, except that the hydraulic gradient is less negative near the lake than it appears to be in reality. The (dimensionless, with δl and t_0) space and time steps used were 0.005 and 0.0005, respectively

to derive approximations for a_2 for large and small b:

$$a_2 \approx \begin{cases} [1 + 0.51 b^{2/3}]^2 & \text{as } b \to 0, \\ b & \text{as } b \to \infty. \end{cases} \tag{11.57}$$

Comparison of these approximations with the exact expression (11.55) show that the small b approximation is very accurate for $b < 1$, and quite accurate for $b < 4$; and in this range the small b approximation itself is close to $1.2 + b$; thus a_2 and a_1 are in fact quite close, and we might reasonably expect that this estimate for the occurrence of a weak seal is quite accurate also for $(11.44)_2$.

Actually, a better approximation allows for the smallness of Q by replacing $Q^{2/11}$ in $(11.44)_2$ by $\omega^{2/11}$, and then we find that that the condition for a weak seal is modified to the estimate

$$a < a_2' = \left[\frac{1}{2} \omega^{1/11} b j'_{(2/\omega^{1/11} b), 1} \right]^2 \approx 1.2 + 0.44 b \tag{11.58}$$

for $\omega = 0.12 \times 10^{-3}$ (see Question 11.6). This appears to be consistent with what is observed numerically.

Figure 11.6 shows a numerical solution of (11.42) and (11.43), using the hydraulic gradient given by (11.52) with $a = 2.8$, $b = 4.316$.[5] With these values, and the other parameters as estimated earlier, we find floods with peak discharge $5032 \ \text{m}^3 \ \text{s}^{-1}$, and periods 7.4 years, very favourably comparable to observed peak

---

[5]Reasonable estimates of Φ for Grímsvötn yield $a = 3.33$, $b = 4.316$, given our choice of Φ_0 and N_0. For these values, the seal is actually strong, but would be weak according to (11.58) if $a < 3.1$: hence our modified choice of a so that the floods are weak. A likely modification to the predicted type of seal in (11.58) arises from the effects of water temperature near the lake; this is discussed further in the notes.

Fig. 11.7 The flood hydrograph, corresponding to Fig. 11.6 (*solid curve*). Also shown is the hydrograph of the 1972 jökulhlaup (*dashed*), with the discharge scaled by 0.7, and the time scaled by 2.19. The agreement is evidently excellent. The rescaling corresponds to choosing $Q_0 = 0.26 \times 10^6 \text{ m}^3 \text{ s}^{-1}$ and $t_0 = 0.00087 \text{ y} = 0.32 \text{ d}$. See also Question 11.7

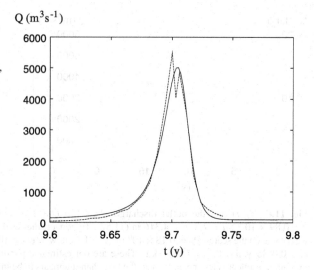

discharges and periods (but it should be mentioned that these have varied over the years). In addition, the flood duration is comparable to that generally observed (it is about twice as long), and the shape of the flood hydrograph is now very similar to that of the 1972 jökulhlaup (see Fig. 11.7). We can also see from Fig. 11.6 that floods are initiated when the lake is below flotation, with $N_L \approx 1.7$ bars, about a quarter the value observed, but this depends sensitively on the hydraulic parameters a and b. The choice of a and b is made to illustrate the pre-flotation flood initiation, but their values depend on the scales N_0 and Φ_0. In view of the simplifications made, which we discuss further below, this theory appears to give a good representation of the observations, and by tweaking the choice of parameters, it is of course possible to improve the agreement.

The 1996 Eruption

One of the surprises of the 1996 eruption was that the seal *did* fail at flotation. It had been expected to fail at the usual 60–80 m below flotation. We can guess that the reason for this was that the eruption filled the lake in a relatively short time, weeks rather than years, and therefore the equation for S in (11.42) could not be taken as being in equilibrium. For example, if the lake refilling were so rapid that $v \gg \omega$ (see Question 11.8), then N_L decreases rapidly, and it is reasonable to expect N_L to reach zero before the seal breaks, though this in fact requires a time-dependent solution of a model such as (11.42) and (11.43). In normal circumstances, slow lake refilling causes a cyclic sequence of jökulhlaups when $N_L > 0$, but if the refilling rate is dramatically increased for a short period, such as would occur following a volcanic eruption, then there is a violent flood which occurs when the lake reaches flotation.

Figure 11.8 shows a numerical experiment which illustrates this fact. The same model with the same parameters as in Fig. 11.6 is solved, but the value of v is

Fig. 11.8 Outlet discharge for a post-eruption flood; parameters as for Fig. 11.6, but $\nu = 0.56 \times 10^{-2}$ for $1 < t < 1.1$ y, corresponding to a refilling rate of 1000 m^3 s^{-1} for about a month, which resembles the conditions in 1996. The magnitude of the flood which is obtained numerically depends on the timing and magnitude of the enhanced meltwater pulse

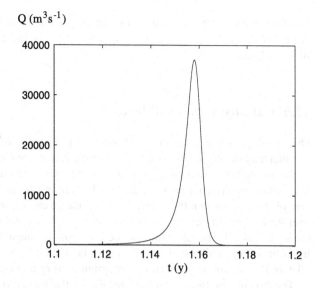

changed to be 0.56×10^{-2} between $t = 1$ y and $t = 1.1$ y. This corresponds to a flux of 10^3 m^3 s^{-1} for about a month, which corresponds to the situation after the 1996 eruption, when 3 cubic kilometres of water entered the Grímsvötn caldera in the month preceding the flood. The peak discharge is now over 35,000 m^3 s^{-1}, which compares favourably with the inferred peak of 45,000 m^3 s^{-1} (given our earlier comment about Q_0), and although the shape of the hydrograph is different (the 1996 flood lasted only a day), it is very similar to hydrographs of earlier post-volcanic floods, particularly those of 1922 and 1934, each of which lasted about ten days (0.027 y), and which had similarly shaped hydrographs, with slow rise and fast decay.

11.5.2 Wide Channels and the 1996 Eruption

In fact, the detailed behaviour of the 1996 flood was quite different from the normal, viscously controlled floods, since the water hydro-fractured its way along the base. It is possible to build an understanding of the discrepancy between the slowly rising flood hydrograph of the usual weakly sealed floods and the rapidly rising strongly sealed floods, or eruptive floods, by consideration of the way in which the flood water propagates under the glacier. When the seal is broken by overpressuring of the channel water, then the channel widens essentially as a crack does, and the channel's lateral extension must be described by hydrofracture (see also Sect. 9.5). The consequent wide channel propagates downglacier by a hydrofracturing overpressured tip, with the tail of the channel being a normal, underpressured, viscously closing wide conduit. Observations of the 1996 flood are consistent with this, since collapse of the ice post-flood indicated a wide channel near the lake of several hundred metres, and the floodwater burst out from the surface of the glacier several kilometres upstream of the snout (presumably along a thrust fault), indicating an overpressuring of several bars.

11.6 Cauldrons and Calderas

One of the plausible assumptions about our implementation of the Nye model, but one that may not always be realistic, is that the lake in the Grímsvötn caldera is open to the atmosphere. Sometimes this may be the case, and then our direct relation of lake volume to effective pressure at the lake margin is reasonable. But it is also reasonable to suppose that the overlying ice is unbroken, forming a miniature ice shelf over the caldera. In this case, we cannot simply relate water pressure to lake volume during a flood. As the lake empties, the overlying ice must deform to accommodate the volume loss, and this deformation must be related to the effective pressure in the lake. In this section we show how to compute this relationship.

The bowing of the ice surface results in the formation of a *cauldron*. Cauldrons are common where subglacial volcanoes occur, and an example of a cauldron formed in the 1996 Vatnajökull eruption is shown in Fig. 11.9. The eruption causes massive subglacial melting, and the outflow of this water (in this case, into the Grímsvötn caldera) causes the subsidence that is seen.

11.6.1 Viscous Beam Theory

We can analyse the ice cauldron deformation using a viscous analogue to the beam theory of classical elasticity. We begin by recalling the equations of two-dimensional ice motion, scaled as for ice sheets in (10.38):

$$
\begin{aligned}
u_x + w_z &= 0, \\
0 &= -s_x + \tau_{3z} + \varepsilon^2[-p_x + \tau_{1x}], \\
0 &= -p_z + \tau_{3x} - \tau_{1z}, \\
(u_z + \varepsilon^2 w_x) &= A\tau^{n-1}\tau_3, \\
2u_x &= A\tau^{n-1}\tau_1, \\
\tau^2 &= \tau_3^2 + \varepsilon^2\tau_1^2.
\end{aligned}
\tag{11.59}
$$

The ice surface is at $z = s$, and the lake roof is at $z = h$. The boundary conditions are, on $z = s$:

$$
\begin{aligned}
\tau_3 + \varepsilon^2(p - \tau_1)s_x &= 0, \\
\tau_3 s_x + p + \tau_1 &= 0, \\
w &= s_t + us_x - a,
\end{aligned}
\tag{11.60}
$$

while at the lake $z = h$,

Fig. 11.9 An ice cauldron forming during the Gjálp eruption under Vatnajökull in 1996. The cauldron is about two kilometres in diameter and 100 metres deep, and ring shear fractures can be seen, indicating yield of the ice. The subsidence rate of the ice surface was initially about 12 metres per *hour*. Photo courtesy of Magnús Tumi Gudmundsson, University of Iceland, obtained from http://www.hi.is/~mmh/gos/photos.html. For further information see Gudmundsson et al. (2004)

$$\tau_3 + \varepsilon^2(p - \tau_1)h_x = -\gamma N h_x,$$

$$\tau_3 h_x + p + \tau_1 = -\frac{\gamma N}{\varepsilon^2}, \tag{11.61}$$

$$w = h_t + u h_x - m,$$

where

$$\gamma = \frac{N_0}{\rho_i g d}, \tag{11.62}$$

d being the ice depth scale, as in Chap. 10; γ is of $O(1)$ or smaller. The terms a and m represent surface accumulation rate and basal melting rate, respectively.

We suppose in addition that pressure in the lake is hydrostatic. This condition (more precisely that there is no hydraulic gradient in the lake) can be written in terms of scaled variables as

$$\frac{\partial}{\partial x}[s + \delta_{iw}h - \gamma N] = 0, \tag{11.63}$$

where

$$\delta_{iw} = \frac{\rho_w - \rho_i}{\rho_i}, \tag{11.64}$$

and the lake effective pressure is defined (as for a channel) as the cryostatic overburden ice pressure minus the water pressure. The definition of δ_{iw} is identical to that of δ in Chap. 10, but the symbol δ in this chapter has been reserved (see (11.18)).

The final equation is the lake refilling equation, which balances the net outflow from the lake with its rate of loss of volume. In dimensional terms, this can be written in the form

$$Q_+ - Q_- = -\int_{A_L} w\, dS, \tag{11.65}$$

where A_L is the lake area. Its dimensionless form is given later.

Normally, we would rescale the equations into a form appropriate for an ice shelf, but when rapid subsidence occurs, this is inappropriate. Rapid deflation of an ice cauldron requires a positive effective pressure in the lake in order to suck the ice down; for an ice shelf (see Sect. 10.2.6), we would suppose that a flotation condition applies.

The rescaling we choose balances different terms in the equations, and is analogous to that used when scaling the equations of elasticity for a beam. We rescale the variables as follows:

$$p, \tau_1 \sim \Lambda, \qquad \tau \sim \varepsilon \Lambda, \qquad \tau_3 \sim \frac{\varepsilon^2 \Lambda}{\lambda},$$

$$w \sim W, \qquad u \sim \frac{\varepsilon^2 W}{\lambda}, \qquad x \sim \lambda, \qquad t \sim \frac{\mu}{W}. \tag{11.66}$$

The length scale λ is prescribed (from the lake geometry), but Λ, μ and W must be chosen appropriately. We expect $\Lambda \gg 1$, and that $W \gg 1$ is determined by the rate of lake deflation. We take the deflation rate scale to be w_s, so that (compare (10.33))

$$W = \frac{w_s}{[a]}, \tag{11.67}$$

and from (11.65), we define w_s (and thus W) by

$$Q_0 = A_L w_s. \tag{11.68}$$

The parameter μ is chosen so that the new time scale is that of the channel flow, t_0 in (11.6); this leads to the definition

$$\mu = \frac{w_s t_0}{d}. \tag{11.69}$$

With these rescalings, the equations take the form

$$v^2 u_x + w_z = 0,$$
$$0 = -s_x + \varepsilon^2 \Lambda[\tau_{3z} - p_x + \tau_{1x}],$$
$$0 = -p_z - \tau_{1z} + v^2 \tau_{3x},$$
$$u_z + w_x = v^2 \left(\frac{\varepsilon^{n-1} \Lambda^n}{v^2 W}\right) A\tau^{n-1} \tau_3, \tag{11.70}$$
$$2u_x = \left(\frac{\varepsilon^{n-1} \Lambda^n}{v^2 W}\right) A\tau^{n-1} \tau_1,$$
$$\tau^2 = v^2 \tau_3^2 + \tau_1^2,$$

where

$$v = \frac{\varepsilon}{\lambda}, \tag{11.71}$$

and this will be small.

The boundary conditions become, on the surface $z = s$,

$$\tau_3 + (p - \tau_1)s_x = 0,$$
$$v^2 \tau_3 s_x + p + \tau_1 = 0, \tag{11.72}$$
$$s_t = \mu \left(w - v^2 u s_x + \frac{a}{W} \right),$$

and on the lake $z = h$,

$$\tau_3 + (p - \tau_1)h_x = -\frac{\gamma}{\varepsilon^2 \Lambda} N h_x,$$
$$v^2 \tau_3 h_x + p + \tau_1 = -\frac{\gamma}{\varepsilon^2 \Lambda} N, \tag{11.73}$$
$$h_t = \mu \left(w - v^2 u h_x + \frac{m}{W} \right),$$

and the lake hydrostatic condition is still (11.63). The lake refilling equation (11.65) becomes in dimensionless terms

$$Q_+ - Q_- = - \int_{x_-}^{x_+} w \, dx, \tag{11.74}$$

assuming a two-dimensional geometry; x_- and x_+ are the upstream and downstream positions of the lake margins.

Equations (11.70) describe the deformation of the viscous beam, with the boundary conditions in (11.72) and (11.73) sufficing also to determine the evolution of the ice surface s and the lake roof h, providing the effective pressure N is known: this is determined by the buoyancy condition (11.63). The equations and boundary conditions include six parameters v, ε, Λ, W, γ and μ; ε is the ice sheet aspect ratio, defined in (10.37), v is the beam aspect ratio, defined in (11.71), γ was defined in (11.62), μ in (11.69), and W in (11.67); this leaves the parameter Λ yet to be chosen.

For Grímsvötn, the lake diameter is of order 5 km, while the Vatnajökull ice cap is of depth *ca.* 500 m, and is some 100 km in extent. Thus $\varepsilon \sim 0.005$ and $\lambda \sim 0.05$; hence $v \sim 0.1$ in this case, and generally we assume it to be small. Since v is the aspect ratio for the ice 'beam', the beam equation follows from the assumption that it is small.

The parameter Λ is determined by requesting a balance in the constitutive law for longitudinal stress, thus we choose

$$\frac{\varepsilon^{n-1} \Lambda^n \bar{A}}{v^2 W} = 1; \tag{11.75}$$

$\bar{A}$ is the relevant value of the flow rate coefficient A, and is explicitly included because the rate controlling value of A is the smallest, i.e., that at the surface, which in an ice sheet may be three orders of magnitude smaller than that at the base. We also define a parameter

$$\beta = \frac{1}{\varepsilon^2 \Lambda} = \left(\frac{\bar{A}}{\varepsilon^{n+1} \nu^2 W} \right)^{1/n}. \tag{11.76}$$

When written in terms of local dimensional quantities, this is

$$\beta = \frac{\rho_i g d}{\tau_A}, \tag{11.77}$$

where the beam stress scale is

$$\tau_A = \left(\frac{\nu^2 w_s}{d A_s} \right)^{1/n}, \tag{11.78}$$

with A_s being the surface flow law coefficient (thus $\bar{A} = A_s/[A]$, cf. (10.33)) and w_s the surface deformation rate scale. For a subsidence rate appropriate to Fig. 11.9 of 12 metres an hour, $w_s \approx 10^5 \, \mathrm{m \, y^{-1}}$, and with $A_s = 10^{-2} \, \mathrm{bar^{-3} \, y^{-1}}$, $d = 500$ m and $\nu = 0.25$, this gives $\tau_A \approx 11$ bars and $\beta \approx 4$. In other circumstances, the beam stress is lower, and the value of β may be large.

From the equations and boundary conditions, we see that $p + \tau_1 \sim \nu^2$, and therefore we define

$$p + \tau_1 = -\nu^2 \sigma_3. \tag{11.79}$$

Using (11.79), (11.75) and (11.76), the rescaled equations become

$$\nu^2 u_x + w_z = 0,$$
$$p + \tau_1 = -\nu^2 \sigma_3,$$
$$0 = -\beta s_x + \tau_{3z} + 2\tau_{1x} + \nu^2 \sigma_{3x},$$
$$0 = \sigma_{3z} + \tau_{3x}, \tag{11.80}$$
$$u_z + w_x = \nu^2 A \tau^{n-1} \tau_3,$$
$$2u_x = A \tau^{n-1} \tau_1,$$
$$\tau^2 = \nu^2 \tau_3^2 + \tau_1^2,$$

and the boundary conditions become, on $z = s$,

$$\tau_3 - \left(2\tau_1 + \nu^2 \sigma_3 \right) s_x = 0,$$
$$\tau_3 s_x - \sigma_3 = 0, \tag{11.81}$$
$$s_t \approx \mu \left(w - \nu^2 u s_x \right),$$

and on the lake $z = h$,

$$\tau_3 - \left(2\tau_1 + \nu^2 \sigma_3 \right) h_x = -\gamma \beta N h_x,$$
$$\tau_3 h_x - \sigma_3 = -\frac{\gamma \beta N}{\nu^2}, \tag{11.82}$$
$$h_t = \mu \left(w - \nu^2 u h_x + m^* \right),$$
$$\frac{\partial}{\partial x} [s + \delta_{iw} h - \gamma N] = 0.$$

taking $W \gg 1$, but retaining the melt term $m^* = m/W$ since it may be the increase in basal melt rate which causes cauldron formation.

While all the parameters in the equations have been defined, the choice of the volume flux scale Q_0, just as in the Nye model, is still open. It is natural to balance terms in the hydrostatic equation $(11.82)_4$; bearing in mind $(11.81)_3$, this suggests $\mu \sim \gamma$, and in fact our earlier choice in (11.27) corresponds to the choice

$$\mu = \frac{\gamma}{1 + \delta_{iw}}, \tag{11.83}$$

and we will see that it remains appropriate to define μ in this way for closed subsurface lakes. The three other parameters involved in finding an approximate solution to this set of equations are γ, β and ν. We have indicated that $\gamma \lesssim 1$, $\nu \ll 1$ and $\beta \gtrsim 1$. We therefore begin by seeking approximate solutions for $\nu \ll 1$, and then subsequently considering possible choices for γ, β and μ. The limit $\nu \to 0$ is the classical approximation associated with beam theory.

With $\nu \ll 1$, we have $w \approx w(x, t)$, $\tau \approx |\tau_1|$ and $u_z + w_x \approx 0$, whence

$$2u_x \approx A|\tau_1|^{n-1}\tau_1,$$
$$u \approx V - zw_x, \tag{11.84}$$

where $V(x, t)$ is to be determined. The coefficient A has now been rescaled with $\bar{A}$, if this is not equal to one. From (11.84), we find

$$\tau_1 = \left(\frac{2}{A}\right)^{1/n} \frac{V_x - zw_{xx}}{|V_x - zw_{xx}|^{(n-1)/n}}. \tag{11.85}$$

We now define the three quantities M, the bending moment, S, the shear force, and T, the tension, as

$$M = -\int_h^s 2z\tau_1\,dz, \qquad S = \int_h^s \tau_3\,dz, \qquad T = \int_h^s \tau_1\,dz. \tag{11.86}$$

We also define the secondary quantities U, the uplift force, and L, the lifting torque, to be

$$U = \int_h^s \sigma_3\,dz, \qquad L = \int_h^s z\sigma_3\,dz. \tag{11.87}$$

Integrating the force balance equations and applying the boundary conditions on $z = s$ and $z = h$, we then find that

$$2\frac{\partial T}{\partial x} + \nu^2 \frac{\partial U}{\partial x} = -\gamma\beta Nh_x + \beta(s - h)s_x,$$
$$\frac{\partial M}{\partial x} - \nu^2 \frac{\partial L}{\partial x} + S = \gamma\beta Nhh_x - \frac{1}{2}\beta(s^2 - h^2)s_x, \tag{11.88}$$
$$\frac{\partial S}{\partial x} = \frac{\gamma\beta N}{\nu^2}.$$

Subject to suitable boundary conditions at the margins $x_\pm$, these equations will enable us to provide a closed solution which determines the surface depression rate $-w$ in terms of the underlying effective pressure N.

Some care needs to be taken with this discussion. The rescaling of the stresses in (11.66) is such that they may be large. Comparison with the original choice of scales in (10.9) shows that the greatest stress is the longitudinal stress, which is of order $\tau_A = \frac{\rho_i g d}{\beta}$ as given by (11.78), and this can be some way larger than typical stress magnitudes in glaciers. Since the yield strength of ice (before it fractures) is generally thought to be of the order of bars, it is clear that the stresses in the viscous ice beam may exceed this. In this case, we can expect the ice to fracture, so that its effective rheology becomes, for example, plastic. Indeed, we see extensive fracturing in Fig. 11.9.

Marginal Boundary Conditions

We forgo discussion of such plastic subsidence, and assume that the ice retains its integrity. In order to motivate plausible boundary conditions, we consider the lake margins $x_\pm$ to be fixed, so that

$$w = 0 \quad \text{at } x = x_\pm. \tag{11.89}$$

In addition, we compute typical viscosity scales η_I for the ice sheet ice, and η_L for the overlake ice. These are taken to be

$$\eta_I = \frac{1}{2[A][\tau]^{n-1}}, \qquad \eta_L = \frac{1}{2A_s \tau_A^{n-1}}, \tag{11.90}$$

where the ice sheet scales are those defined in (10.33). We take $[A] = 0.2 \, \text{bar}^{-3} \, \text{y}^{-1}$, $[\tau] = 0.2 \, \text{bar}$, which yields $\eta_I \approx 60 \, \text{bar y}$, while for the overlake ice, we take $A_s = 10^{-2} \, \text{bar}^{-3} \, \text{y}^{-1}$, $\tau_A = 11 \, \text{bars}$, yielding $\eta_s = 0.4 \, \text{bar y}$. On this basis, we see that the overlake ice can be much less stiff than the ice sheet ice, and then it seems appropriate to pose conditions of no horizontal velocity at the margins, thus

$$V = w_x = 0 \quad \text{at } x = x_\pm. \tag{11.91}$$

Because (11.88) are vertically averaged equations, we might expect not to be able to satisfy both conditions in (11.91), but simply an integrated condition of no horizontal mass flux at the margins. Using (11.84)$_2$, this condition can be written in the form

$$V = \frac{1}{2}(s + h)w_x \quad \text{at } x = x_\pm. \tag{11.92}$$

In fact, we shall see that both conditions in (11.91) can be satisfied.

The boundary conditions (11.89) and (11.92) are applicable for soft lake ice which may be appropriate for Vatnajökull, but in other circumstances, the lake ice can be hard. This is probably the case in Antarctica, where the surface value of A_s will be much lower, and the beam stress τ_A is also lower. In this case it is appropriate to apply conditions of zero longitudinal stress at the margins, i.e., $\tau_1 = 0$, and if we suppose that this condition can be applied in a vertically integrated sense, then

$$M = T = 0 \quad \text{at } x = x_\pm. \tag{11.93}$$

Summary

From the rescaled equations of motion (11.80), together with the upper and lower boundary conditions (11.81) and (11.82), we have derived vertically integrated equations (11.88), together with suitable lateral boundary conditions, either (11.89) and (11.92) for soft lake ice (Vatnajökull), or (11.93) for hard lake ice (Antarctica). We now need to use the beam approximation based on $v \ll 1$ to derive the viscous beam equation.

The Case $n = 1$

To illustrate the derivation of the beam equation, we take $n = 1$. We then have, from (11.85),

$$\tau_1 = \frac{2}{A}\{V_x - z w_{xx}\}, \tag{11.94}$$

and thus

$$M = a_m V_x + b_m w_{xx}, \\ T = a_T V_x + b_T w_{xx}, \tag{11.95}$$

where

$$a_m = -\frac{2}{A}(s^2 - h^2), \qquad b_m = \frac{4}{3A}(s^3 - h^3), \\ a_T = \frac{2}{A}(s - h), \qquad b_T = -\frac{1}{A}(s^2 - h^2), \tag{11.96}$$

and we take A to be constant. Thus V and w satisfy the equations

$$\frac{\partial^2}{\partial x^2}[a_m V_x + b_m w_{xx}] = -\frac{\gamma \beta N}{v^2} + \frac{\partial}{\partial x}\left[\gamma \beta N h h_x - \frac{1}{2}\beta(s^2 - h^2)s_x\right], \\ 2\frac{\partial}{\partial x}[a_T V_x + b_T w_{xx}] = -\gamma \beta N h_x - \frac{1}{2}\beta(s^2 - h^2)s_x, \tag{11.97}$$

together with the approximate kinematic conditions (if $m^* = 0$)

$$s_t \approx h_t \approx \mu w, \tag{11.98}$$

whence the depth $H = s - h$ is a function only of space: for simplicity we take it to be a constant. The consequent hydrostatic condition is, using (11.82)$_3$ and (11.83),

$$s_x = \mu N_x. \tag{11.99}$$

Integrating this, we have

$$s = s_+ + \mu\{N - N_+(t)\}, \tag{11.100}$$

where $s = s_+$ (constant, since $w = 0$ there) and $N = N_+$ at $x = x_+$. Differentiating, we finally have

$$N_t = \dot{N}_- + w. \tag{11.101}$$

11.6.2 The Beam Boundary Layer

To complete the prescription for N, we need to find w in terms of N by solving the beam equations (11.97). The product $\gamma\beta = N_0/\tau_A$ is the ratio of the channel effective pressure scale to the beam stress, and is generally of $O(1)$ or larger. It is therefore reasonable to suppose that $\gamma\beta \gg \nu^2$. In this case, the solution becomes of singular perturbation type, and supports boundary layers near the lake margins.

Note first that since $s_x \sim \mu \sim \gamma$, the term $\frac{\gamma\beta N}{\nu^2}$ is the largest of those on the right hand side of (11.97). A leading order approximation (the outer solution) to (11.97)$_1$ is then simply

$$N \approx 0, \tag{11.102}$$

whence also

$$s_x \approx 0, \tag{11.103}$$

and thus

$$a_T V_x + b_T w_{xx} \approx T, \tag{11.104}$$

where $T(t)$ is the tension. From this we have

$$V_x = \frac{T}{a_T} + \frac{1}{2}(s + h)w_{xx}. \tag{11.105}$$

Since $s_x \approx h_x \approx 0$, then a_T is approximately constant, and so

$$V \approx \frac{Tx}{a_T} + \frac{1}{2}(s + h)w_x + V_0 \approx \frac{Tx}{a_T} + V_0 \tag{11.106}$$

since $w \approx \dot{s}/\mu$ is independent of x. It follows from (11.101) that in fact

$$w \approx -\dot{N}_+, \tag{11.107}$$

and this together with (11.74) implies that the boundary condition for the channel is

$$\dot{N}_+ = \frac{Q_+ - Q_-}{x_+ - x_-}, \tag{11.108}$$

which is essentially the same as we assumed earlier (cf. (11.29)). This relation is essentially the punch line of our analysis, at least for describing subglacial floods. The subsidence of the surface requires further discussion.

The outer solution for w does not satisfy the boundary conditions at the margins, so that a boundary layer is needed near both margins. Both are similar, and we treat that at x_+. The choice of boundary layer scale depends on the size of the parameters. We define

$$\sigma = \frac{\nu^{1/2}}{(\gamma\beta)^{1/4}}, \qquad \omega = \nu\sqrt{\gamma\beta}, \tag{11.109}$$

and we will assume, as is likely, that $\omega \lesssim 1$. The boundary layer variables X and v are defined by

$$x = x_+ - \sigma X, \qquad V = \frac{v}{\sigma}, \tag{11.110}$$

and Eqs. (11.97), (11.98) and (11.99) become

$$\frac{\partial^2}{\partial X^2}[-a_m v_X + b_m w_{XX}] = -N + \omega \frac{\partial}{\partial X}\left[Nhh_X - \frac{1}{2\gamma}(s^2 - h^2)s_X\right],$$

$$2[-a_T v_X + b_T w_{XX}] = -\omega\left[Nh_X + \frac{1}{2\gamma}(s^2 - h^2)s_X\right], \qquad (11.111)$$

$$s_t = h_t = \mu w, \qquad s_X = \mu N_X.$$

Hard Lake Ice Boundary Conditions

For simplicity, suppose that $\omega \ll 1$, and consider first the ice sheet type boundary conditions for hard overlake ice given by (11.93); these require

$$a_m v_X + b_m w_{XX} = a_T v_X + b_T w_{XX} = w = 0 \quad \text{at } X = 0, \qquad (11.112)$$

i.e.,

$$v_X = w_{XX} = w = 0 \quad \text{at } X = 0, \qquad (11.113)$$

and the matching conditions to the outer solution are

$$v \to \sigma\left(\frac{Tx_+}{a_T} + V_0\right), \qquad w \to -\dot{N}_+ \quad \text{as } X \to \infty. \qquad (11.114)$$

Neglecting terms of $O(\omega)$, we find that

$$v_X = -\frac{1}{2}(s + h)w_{XX},$$
$$D_1 w_{XXXX} = -N, \qquad (11.115)$$

where the flexural viscosity D_1 is given by

$$D_1 = \frac{H^3}{3A}, \qquad H = s - h; \qquad (11.116)$$

the second equation in (11.115) is the viscous beam equation.

(11.100) is still the integral of the hydrostatic equation, and thus $s_t = \mu w = \mu(N_t - \dot{N}_+)$, whence

$$w = N_t - \dot{N}_+; \qquad (11.117)$$

differentiating (11.115)$_2$, we derive the viscous beam equation for w in the form

$$D_1 w_{XXXXt} = -w - \dot{N}_+, \qquad (11.118)$$

with the boundary conditions for w in (11.113) and (11.114). Exactly the same boundary layer description applies at the left margin x_-.

Soft Lake Ice Boundary Conditions

Suppose instead that we take the boundary conditions (11.91) which are appropriate for soft overlake ice, such as might be the case in the Vatnajökull cauldrons. Neglecting $O(\omega)$, a first integral of $(11.111)_2$ is

$$-a_T v_X + b_T w_{XX} = T_+, \tag{11.119}$$

where $T_+(t)$ is an integration function. Using the definitions of a_T and b_T and the kinematic conditions, this can be integrated subject to $v = 0$ at $X = 0$ to obtain

$$v = -\frac{1}{2}(s+h)w_X + \int_0^X \frac{s_X s_{Xt}}{\mu}\, dX + T_+ X. \tag{11.120}$$

Applying the matching conditions at $X \to \infty$, we find $T_+ = 0$ and

$$\sigma\left(\frac{Tx_+}{a_T} + V_0\right) = \frac{d}{dt}\int_0^\infty \frac{s_X^2}{2\mu}\, dX. \tag{11.121}$$

An entirely similar analysis at the other margin x_- yields the comparable condition

$$\sigma\left(\frac{Tx_-}{a_T} + V_0\right) = -\frac{d}{dt}\int_0^\infty \frac{s_X^2}{2\mu}\, dX, \tag{11.122}$$

whence it follows (since each boundary solution is the same) that

$$\frac{\sigma(x_+ - x_-)T}{a_T} = \frac{d}{dt}\int_0^\infty \frac{s_X^2}{\mu}\, dX, \tag{11.123}$$

and this determines the tension T. The beam equation (11.118) for w is derived the same way, the only difference being that the boundary conditions are

$$w = w_X = 0 \quad \text{at } X = 0, \qquad w \to -\dot{N}_+ \quad \text{as } X \to \infty. \tag{11.124}$$

A Uniform Approximation for $n \neq 1$

A uniform approximation to the solution can evidently be made by retaining only the term in N in (11.97), or equivalently by formally assuming that $\gamma\beta \sim v^2$. We make this formal assumption, and additionally allow $n \neq 1$, thus we revert to the equations in the form (11.88).

For the case of hard overlake ice, we have $T = 0$ everywhere, and thus (11.85) and (11.86) imply by inspection that

$$V_x = \frac{1}{2}(s+h)w_{xx}, \tag{11.125}$$

if we take A to be constant. We then calculate

$$M = \frac{D_n w_{xx}}{|w_{xx}|^{(n-1)/n}}, \tag{11.126}$$

where

$$D_n = \frac{nH^{(2n+1)/n}}{(2n+1)A^{1/n}},$$ (11.127)

and the beam equation takes the form

$$\frac{\partial^2}{\partial x^2}\left\{\frac{D_n w_{xx}}{|w_{xx}|^{(n-1)/n}}\right\} = -\frac{\gamma\beta N}{v^2},$$ (11.128)

subject to

$$w = w_{xx} = 0 \quad \text{at } x = x_\pm,$$ (11.129)

and N is given by (11.101).

The case of soft overlake ice is somewhat impassable: the beam equation for $n = 1$ becomes non-local. If μ is small, then so is T, and the derivation for hard ice works in the same way, leading to the same approximate equation (11.128).

11.6.3 Similarity Solutions

We would like to solve the beam equation (11.128) together with the buoyancy condition (11.101), in order to trace the subsidence of a cauldron. A simple way to do this is motivated by Nye's pioneering paper of 1976 on jökulhlaups, where he showed that the rising limb of the 1972 Grímsvötn flood hydrograph could be well fitted by a power law in the form

$$Q \propto \frac{1}{(\bar{t}-t)^4}, \qquad S \propto \frac{1}{(\bar{t}-t)^3},$$ (11.130)

which arises from the flood model (11.42) on neglecting the closure term and taking $Q \sim S^{4/3}$. Since the lake refilling equation implies $\dot{N}_+ \approx Q$, this approximation also implies $N_+ \propto \frac{1}{(\bar{t}-t)^3}$, and this observation suggests how a useful similarity solution to the present beam boundary layer model can be obtained, if we suppose that the rising stage of a flood can be described by a water flux varying as in (11.130).

We consider the nonlinear model (11.128) and (11.101) together with the boundary conditions in (11.129). The boundary layer model near $x = x_+$ for this equation is obtained by writing

$$x = x_+ - \sigma_n X, \qquad \sigma_n = \left(\frac{D_n v^2}{\gamma\beta}\right)^{n/[2(n+1)]}.$$ (11.131)

In addition we define

$$w = -\dot{N}_+ + \Upsilon.$$ (11.132)

The bending moment is thus

$$M = \frac{D_n \Upsilon_{XX}}{\sigma_n^{2/n}|\Upsilon_{XX}|^{(n-1)/n}},$$ (11.133)

and Υ and N satisfy the boundary layer equations

$$\left[\frac{\Upsilon_{XX}}{|\Upsilon_{XX}|^{(n-1)/n}}\right]_{XX} = -N,$$

$$N_t = \Upsilon,$$
(11.134)

with boundary conditions

$$N = N_+, \qquad \Upsilon = \dot{N}_+, \qquad \Upsilon_{XX} = 0 \quad \text{on } X = 0,$$

$$\Upsilon \to 0 \quad \text{as } X \to \infty.$$
(11.135)

Suppose now that

$$N_+ = \frac{c}{(\bar{t} - t)^\alpha}, \qquad t < \bar{t}.$$
(11.136)

We seek a similarity solution in the form

$$N = N_+ \psi(\eta), \qquad \Upsilon = \dot{N}_+ \phi(\eta), \qquad \eta = \frac{mX}{(\bar{t} - t)^b};$$
(11.137)

substituting these forms into the boundary layer equations, we find that we require

$$b = \frac{(n-1)\alpha - 1}{2(n+1)},$$
(11.138)

and then ϕ and ψ satisfy the boundary value problem

$$\left[\frac{\phi''}{|\phi''|^{(n-1)/n}}\right]'' = -\Gamma \psi,$$

$$\varepsilon \eta \psi' + \psi = \phi,$$
(11.139)

where

$$\varepsilon = \frac{(n-1)\alpha - 1}{2(n+1)\alpha}, \qquad \Gamma = \left[\frac{c^{n-1}}{\alpha m^{2(n+1)}}\right]^{1/n};$$
(11.140)

Γ can be chosen arbitrarily through the choice of m. To be precise, we choose $\Gamma = 1$, so that

$$m^{2(n+1)} = \frac{c^{n-1}}{\alpha};$$
(11.141)

it follows from this that

$$\eta = \left(\frac{N_+^n}{\dot{N}_+}\right)^{\frac{1}{2(n+1)}} X.$$
(11.142)

The boundary conditions for ϕ and ψ are

$$\phi = 1, \qquad \psi = 1, \qquad \phi'' = 0 \quad \text{on } \eta = 0,$$

$$\phi \to 0 \quad \text{as } \eta \to \infty.$$
(11.143)

To solve these equations, it is convenient to define the subsidiary function χ via

$$\chi = \frac{\phi''}{|\phi''|^{(n-1)/n}},$$
(11.144)

Fig. 11.10 Solution of (11.146) and (11.147) for ϕ, with $n = 1$ (as in (11.148)) and $n = 3$

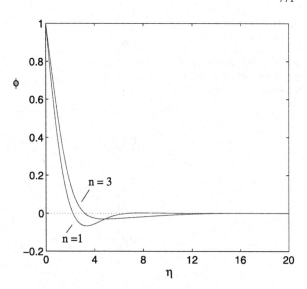

so that

$$\phi'' = \chi|\chi|^{n-1}, \qquad \chi'' = -\phi. \tag{11.145}$$

We note that if $n = 3$ and $\alpha = 3$, the parameter $\varepsilon \approx 0.2$ and is small. This suggests a perturbation solution for the equations. Although apparently a singular perturbation, we see from the boundary conditions that the perturbation is actually regular; the neglect of the term in ε suppresses a singular solution $\psi \propto \eta^{-1/\varepsilon}$ which is precluded by the boundary condition for ψ. At leading order $\psi = \phi$, and the problem thus reduces to

$$\begin{aligned} \phi'' &= \chi|\chi|^{n-1}, \\ \chi'' &= -\phi, \end{aligned} \tag{11.146}$$

with

$$\begin{aligned} \phi &= 1, \quad \chi = 0 \quad \text{on } \eta = 0, \\ \phi &\to 0, \quad \chi \to 0 \quad \text{as } \eta \to \infty. \end{aligned} \tag{11.147}$$

When $n = 1$, the problem can be solved analytically, and we obtain

$$\phi = e^{-\eta/\sqrt{2}} \cos(\eta/\sqrt{2}), \qquad \chi = e^{-\eta/\sqrt{2}} \sin(\eta/\sqrt{2}) \tag{11.148}$$

(see also Question 11.10). For $n > 1$, a numerical solution is necessary. Figures 11.10 and 11.11 show the solution of (11.146) with (11.147) for $n = 1$, as well as a numerical solution for $n = 3$.

To find the subsidence profile, we solve (11.98), using the definition of w in (11.132) and (11.137), thus

$$\frac{\partial s}{\partial t} = -\mu \dot{N}_+ \big[1 - \phi(\eta) \big]. \tag{11.149}$$

Figure 11.12 shows a simulation of this model.

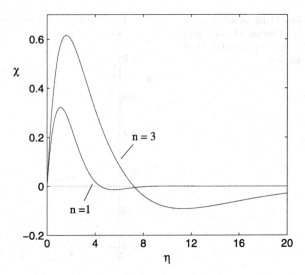

Fig. 11.11 Solution of (11.146) and (11.147) for χ, with $n = 1$ (as in (11.148)) and $n = 3$. When $n = 3$, the maximum value of $\chi = 0.615$ at $\eta = 1.6$

It is of interest to calculate the maximum stress and bending moment in the beam. From (11.85) and (11.125), the maximum stress is at the surface and base of the ice, and is given by

$$|\tau_1|_{\max} = \left| \frac{H w_{xx}}{A} \right|^{1/n}. \tag{11.150}$$

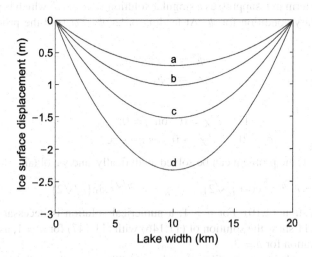

Fig. 11.12 Solution of (11.149) (starting at $t = 0$ days) at times (a) $t = 350$ days, (b) $t = 371$ days, (c) $t = 382$ days, (d) $t = 393$ days using (11.136) for N_+, with $\bar{i} = 1$, $c = 2.5 \times 10^{-3}$, $\alpha = 3$, $\gamma \beta / \nu^2 = 15.8$. The dimensional blow-up time is 416 days, and the units of vertical subsidence are metres, corresponding to a choice of $\mu d = N_0 / \rho_i g = 3$ m, and thus $N_0 = 0.27$ bars. The results are plotted as if for a lake of width 20 km, in which case if $d = 4$ km, then $\nu = 0.2$ and thus $\gamma \beta = 0.63$, corresponding to a value of $\tau_A = 0.43$ bars

After some algebra, we find that the maximum stress is

$$|\tau_1|_{\max} = \left(\frac{H}{A}\right)^{1/n}\left(\frac{\gamma\beta}{v^2 D_n}N_+\dot{N}_+\right)^{1/(n+1)}|\chi|, \qquad (11.151)$$

and the bending stress M in (11.133) is given by

$$M = \left[\frac{\gamma\beta}{v^2}D_n^n N_+\dot{N}_+\right]^{1/(n+1)}\chi. \qquad (11.152)$$

When $n = 1$, the maximum stress and bending stress are at $\eta = 1.11$, where $\chi = 0.32$. For $n = 3$, the maximum dimensionless bending stress is $\chi = 0.62$ at $\eta = 1.6$.

Ring Fractures

Since (via (11.66), (11.76) and (11.77)) the longitudinal stress is scaled with τ_A, we see from (11.151) that when $n = 3$, the maximum stress in the overlake ice has a dimensional value of

$$\tau_{\max} \approx 0.76\left(\frac{\gamma\beta}{v^2}N_+\dot{N}_+\right)^{1/4}\tau_A, \qquad (11.153)$$

where we take $H = A = 1$ and use the definition of D_n in (11.127) and the maximum of $\chi \approx 0.62$; the maximum occurs at a distance from the margin of

$$\Delta x = 1.6\left(\frac{\dot{N}_+}{N_+^3}\right)^{1/8}\frac{v^{1/2}l_w}{(\gamma\beta)^{1/4}}, \qquad (11.154)$$

where l_w is the lake width. If $\tau_{\max}$ reaches the yield stress τ_c of ice, then the ice will fracture, forming a crevasse, and the overlake ice will reset itself to the application of effective boundary conditions at the position of this ring fracture. Thereafter, continued rise of effective pressure will allow a new ring fracture to occur in-lake of the old fracture, and in this way a sequence of such fractures may form, with a spacing indicated by (11.154), as seen in Fig. 11.9. If we equate $\tau_{\max} = \tau_c$, and use (11.153) to eliminate $\dot{N}_+$, we find that the fracture spacing should be

$$\Delta x \approx 1.84\left(\frac{\tau_c}{\tau_A}\right)^{1/2}\left(\frac{v^2}{\gamma\beta}\right)^{3/8}\frac{l_w}{N_+^{1/2}}. \qquad (11.155)$$

Estimates of this are typically less than the lake width, but do not appear consistent with the small scale (tens of metres) cracks visible in Fig. 11.9, for which some further development of the theory would appear to be necessary, for example associated with the finite strain of the surface ice.

11.7 Floods from Ice Sheets

We have mostly focussed our attention on subglacial floods from beneath glaciers or small ice caps. There is no intrinsic reason why floods should not occur from lakes beneath ice sheets. It is now known that there are many lakes beneath the Antarctic ice sheet, for example, and it seems reasonable to suppose that these might also drain semi-periodically. There is in fact now a good deal of evidence for this, and there is plenty of evidence of floods from ice sheets during the last ice ages. We now gather together some of this story.

Badlands   According to Webster's dictionary, badlands are regions marked by intricate erosional sculpturing, scanty vegetation, and fantastically formed hills. In eastern Washington State in the U.S.A., the *Channelled Scablands* are an example of such a landscape. They were formed as a result of massive floods from glacial Lake Missoula, a pro-glacial lake which formed south of the Cordilleran Ice Sheet (the western part of the Laurentide Ice Sheet). It is thought that the drainage of Missoula was blocked by a lobe of the ice sheet, and that the resulting massive build-up of the lake to a volume in excess of 2,000 cubic kilometres, led to a sequence of floods (at least 40, probably more, at intervals of years to decades) of enormous magnitude (estimates for peak discharge are in the region of 10^7 m^3 s^{-1}); these floods caused the massive erosion which formed the scablands. Flow speeds of the order of 25 m s^{-1} and flow depths of up to 300 m caused the erosion of channels into the solid basaltic rock, and the formation of gigantic forms of lateral bars and ripples.

The 8,200 Year B.P. Cooling Event   We have already discussed the sudden climatic cooling event at 8,200 years B.P. (before present, the 'present' being taken as 1950). It is thought that this is due to the catastrophic flooding of glacial Lake Agassiz, which formed south of the dwindling Laurentide ice sheet, in the vicinity of Hudson Bay, as the ice melted. The meltwater builds up behind the ice, but is blocked from escaping to the south. Eventually, a flood is initiated, probably under the ice sheet, causing massive influx of fresh water to the Hudson Strait, and thence to the North Atlantic, where, as discussed in Chap. 2, it can temporarily shut down the oceanic thermohaline circulation, and thus cause a sudden cooling in the northern hemisphere.

Floods from the Antarctic   Similar scabland landscapes have been reported on high terrain in Antarctica, which indicate that massive floods occurred there also. It is thought that these floods are subglacial (because they are at such high elevation), and this suggests that massive sub-Antarctic jökulhlaups have occurred in the past. The likely candidate for the source of such a massive flood is Lake Vostok, containing some 5,400 cubic kilometres of water, and situated under the central part of East Antarctica. There is in fact little to distinguish sub-ice sheet floods mathematically from sub-ice cap floods, beyond the different scales. If there is net drainage towards the lake, and this is not taken up by basal freeze-on to the base of the ice, then inevitably, it seems, a flood will occur. Modelling of the filling of Lake Vostok, for example, suggests that floods with peak discharges of the order of 10^5 m^3 s^{-1} lasting for a year can occur, with a period of the order of 40,000 years.

Much smaller floods are known to occur (from other lakes) at present, and appear to constitute the natural way in which drainage takes place beneath the ice sheet. Satellite imagery has revealed relatively rapid ice surface collapses (of order of metres in a year), which are presumably due to one subglacial lake discharging into another. We can imagine drainage under the ice sheet as consisting of a pseudo-porous flow effected by short term drainage events between the lakes, which act as the pores of the medium.

A question which arises is whether the Nye–Röthlisberger theory can predict these small amplitude fluctuations. Returning to the theory, we see that (11.27) implies that, since the total discharge $\sim Q_0 t_0$ is of the order of $A_L h_d$, where h_d is the drawdown depth, we have

$$h_d \sim \frac{N_0}{\rho_w g}. \tag{11.156}$$

This suggests that if the drawdown is, as frequently observed, of the order of a metre, then $N_0 \sim \rho_w g h_d \sim 0.1$ m. A drawdown of 4 metres is consistent with $N_0 \sim 0.4$ m, similar to values inferred on the Whillans ice stream B.

In turn, our discussion of drainage mechanics in Chap. 10 suggests that such low effective pressures are consistent with a canal-type drainage over sediments. Two questions now arise: first, can our previous theory reproduce such differently scaled floods, and second, is such a theory consistent with canal-style drainage?

To answer the first, we consider the dependence of the effective pressure scale N_0 on the volume flux scale Q_0. From (11.16), we can deduce

$$N_0 = \left[\frac{\Phi_0^{11/8}}{\rho_i K L (f \rho_w g)^{3/8}} \right]^{1/n} Q_0^{1/4n} \tag{11.157}$$

(cf. (10.226)). If a 20 km by 20 km lake deflates by 4 m in a year, we can estimate the volume flux as 50 m^3 s^{-1}. Even if this is two orders of magnitude less than the floods found in Fig. 11.6, it changes N_0 by less than a factor of two. Although there are many parameters in (11.157), few of them are adjustable. The simplest way to reduce N_0 in (11.157) from 32 bars (see (11.20)) to < 1 bar is to increase the closure coefficient K by (say) four orders of magnitude. In terms of viscosity, this means replacing an ice viscosity of 2×10^{13} Pa s by one of 2×10^9 Pa s, coincidentally similar to early estimates of till 'viscosity'. Recalling that Q_0 is calculated independently via (11.28), we see that increasing K by 10^4 does indeed reduce Q_0 by 10^2, and thus reduces N_0 by $10^{3/2} \approx 31.6$ to a value ≈ 1 bar. The time scale also becomes a little longer, although not apparently as much as one would like. Nevertheless, if one simply takes the Grímsvötn model and increases K by 10^4, together with a suitable refilling rate to give a period of ten years, we obtain the fit shown in Fig. 11.13 to data measured on the Lambert Glacier in Antarctica, where there is an apparent flood every ten years or so.

Thus it seems we can replicate these short term, small amplitude floods; but there are issues in the data in Fig. 11.13 which give cause for concern. The recovery phase shows that surface elevation accelerates as it rises, completely at odds with the Nye theory. And, while the abrupt phase of the flood is well represented by the model, the slow down of the subsidence towards the end is also problematic.

In fact, one should have misgivings about applying the Nye theory as it stands. Apparently, (11.157) indicates the usual Röthlisberger balance which presumably

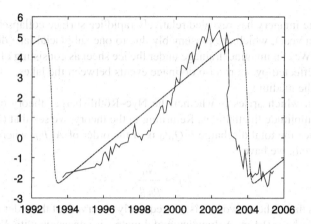

Fig. 11.13 The Nye theory for surface elevation (in metres) (computed from the effective pressure variation), assuming geometric parameters appropriate to the Lambert Glacier. The *solid line* indicates surface elevation, from ERS satellite altimeter data provided by Andy Shepherd, while the *dashed line* comes from the solution of the Nye model. The principal changes in the parameters are that we take $K = 0.5 \times 10^{-20}$ Pa^{-3} s^{-1}, which enables the relatively small amplitude change, and the lake refilling rate is taken to be 7 m^3 s^{-1}, in order to obtain a period of ten years

is inappropriate for drainage through sediments, where the equilibrium of $(11.17)_1$ ought to correspond to N being a decreasing function of Q. What is needed is a theory of floods through sediment-floored canals, but that has yet to be constructed.

Dansgaard–Oeschger Events   Dansgaard–Oeschger events were described in Sect. 2.5.5. They are rapid changes of northern hemisphere climate (by five to ten degrees Celsius) which occurred on a time scale of decades during the last ice age, and which occur semi-periodically, with a rough period of some 1,500 years. In Sect. 2.5.5, we raised the idea that sub-Laurentide jökulhlaups might provide a mechanism for the freshwater release, but postponed a detailed discussion. We provide some further discussion now.

The main problem with sub-Laurentide lake floods is the problem of where do you store the water. Since we know that there were massive pro-glacial lakes (Glacial Lake Agassiz, Glacial Lake Missoula, for example), we can certainly contemplate the existence of massive sub-ice sheet lakes. But why would they form? Lake Vostok in Antarctica lies in a deep tectonic basin, but such tectonic features are largely absent under the Laurentide Ice Sheet.

A clue may lie in the contemplation of Fig. 11.14, which shows the association of eskers with the area of exposed bedrock (shield), suggesting that drainage beyond this region might have taken the form of a distributed, canal-like system at a lower effective pressure. If that were the case, then the subglacial streams making their way from the central ice domes would encounter a transition at the edge of the bedrock where they would face a virtual escarpment. The hydraulic head (cf. (11.8)) of the subglacial water is

$$\phi = -N + \rho_i g s + \Delta \rho_{wi} g b, \qquad (11.158)$$

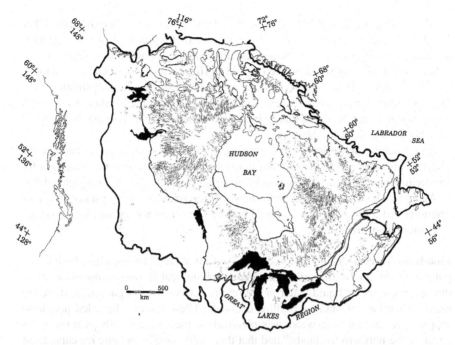

Fig. 11.14 The Laurentide Ice Sheet during the last glaciation (*heavy line*). The *inner solid curve* marks the boundary of the mostly exposed crystalline bedrock. Note the gap to the upper right of the Great Lakes, where the St. Lawrence ice stream flowed into the Gulf of St. Lawrence. The *short fine lines*, mostly within the bedrock region, represent eskers. The *black regions* represent major lakes, ranging westwards from the Great Lakes through Lake Winnipeg, Great Slave Lake, and Great Bear Lake. Figure copyright Geological Society of America, reproduced from Clark and Walder (1994), Fig. 5, and kindly provided by Peter Clark

so, presuming the ice surface s is continuous, a jump *down* in effective pressure of ΔN corresponds to a jump *up* in bed elevation b of

$$\Delta b = \frac{\Delta N}{\Delta \rho_{wi} g}. \tag{11.159}$$

If the effective pressure were 10 bars on the bedrock side and 1 bar on the sediment side, then in effect the streams would encounter a barrier of elevation ≈ 1100 m. This is a fairly significant barrier, requiring an ice surface drop of 100 m to overcome it. If the transition takes place over 100 km, then the ice surface slope can probably compensate, but if it occurs more rapidly, then water may accumulate at the base of the virtual scarp, causing a subglacial lake to form. This is a runaway process, because the formation of the lake drops the basal shear stress to zero, thus tending to flatten the ice surface and removing its slope.

This raises the possibility that the present lakes surrounding the Canadian Shield are the remnant of former subglacial lakes. The lakes might have been strung together like a necklace, allowing for circumferential drainage between them, ending with discharge down the St. Lawrence spillway.

There are a number of effects of such putative lakes on the ice dynamics. The ice above a lake becomes an ice shelf, controlled by longitudinal stresses. There is a grounding line at each margin, and the upstream grounding line, at least, has an enhancement of shear stresses (see (10.133), (10.147) and (10.148)) by a factor of $(\frac{\delta}{\varepsilon})^{2/(n+2)} \approx 6.3$. This suggests that the ice will dig itself a hole upstream of the lake, providing for the continued existence of the lake after the ice sheet has gone. It seems possible that the Great Lakes, indeed all of these shield-marginal lakes, could have been constructed in this way.

The dynamics at the downstream grounding line are less clear, partly because the effective pressure is low. Indeed, the transition to low N heralds the condition for ice stream dynamics, and it may thus be no coincidence that several subglacial lakes in Antarctica have been found at the heads of ice streams. The presence of a lake promotes faster flow, and this may well provide the necessary seeding for ice stream formation downstream.

Floods on Mars   As we discussed in Chap. 10, there is ice on Mars, both in the polar ice caps, and in the soil. Many geomorphological features show evidence of the past presence of liquid water or ice on the surface, and in particular, there are many indications of massive outflow channels. These features have led people to suppose that at one time water was plentiful on the surface, perhaps forming an ocean in the northern lowlands,[6] and that there may have been large ice caps, from beneath which powerful jökulhlaups emerged to carve the Martian surface features.[7]

The question arises, how could this come about? One attractive idea views the hydrological cycle on Mars in the following way. Just as on Earth, planetary outgassing from volcanoes produces water vapour and carbon dioxide. And just as on Earth, volcanoes erupt periodically. The larger the volcano, the larger the eruption and the longer the period between eruptions. The largest eruptions on Earth (the basaltic flood eruptions, described in Sect. 9.1) are thought to be associated with the arrival at the crust of giant thermal plumes in the Earth's mantle (see Chap. 8), associated with time-periodicity of mantle convection. On Mars, which hosts the largest volcano in the Solar System, Olympus Mons, it is reasonable to suppose that extremely massive eruptions may occur at intervals of perhaps tens of millions of years. The resulting greenhouse-induced heating of permafrost causes massive outflows and the collapse of the source terrain: essentially a hydraulic volcano. These floods flow into the northern lowlands, forming an ocean, whose evaporation and subsequent precipitation forms large ice sheets in the southern highlands, whence form the glacially sculpted features which are observed. The outgassing produces enough greenhouse gas to warm the atmosphere (allowing water to form), and also increases atmospheric pressure above the triple point pressure so that water is stable. As the volcanic activity subsides, the atmosphere cools, and the climate reverts to its present cold, dry interglacial type. As with much else concerning Mars and

---

[6]The hypsometry of Mars is odd: the southern hemisphere is elevated, and the northern hemisphere is relatively flat, and much lower.

[7]The study of Martian landforms is called areomorphology.

the other planets, data are scarce, hypotheses are cheap, and speculation is rife. But then, it is the imaginative aspect of science that makes it so much fun.

11.8 Notes and References

Much of the early literature on Icelandic jökulhlaups is in Icelandic (!), for example in the journal *Jökull*; a more accessible introduction to the subject are the papers by Björnsson (1974, 1992), and in particular the same author's book (Björnsson 1988) gives a thorough review, and is an essential classic. The theoretical literature is somewhat sparse, but shows signs of maturation with the increasing interest in palaeo-floods on Earth and Mars. The seminal paper is by Nye (1976), who presented the basic theory. Variants on the model are discussed by Spring and Hutter (1981) and Clarke (1982); in particular, they emphasise the importance of the temperature equation. Clarke (2003) critically reviews flood modelling, and in particular discusses and recalibrates the flow resistance of the channel. There is also a brief discussion in the book by Paterson (1994).

The recent 1996 eruption and its aftermath was described at the time on a number of web pages; for example see http://www.spri.cam.ac.uk/jok/jok.htm. A description is given in the paper by Gudmundsson et al. (1997). A recent review which gives a very complete historical account of the subject, and which also emphasises the differences between the Nye-type melt-opening jökulhlaup and the Gjálp-type hydrofracturing jökulhlaup, is that by Roberts (2005).

Nye's Model   Nye's (1976) model is a fairly astonishing *tour de force*. In particular, his derivation of the energy equation (11.4) is succinct and to the point. It is, however, a physicist's derivation (there is work done here and here, energy gained there, etc.), and it is not something that in the nature of things satisfies a mathematician, who wants to see a formal derivation. Spring and Hutter (1982) provided such a derivation, and this exercise is repeated in Appendix E, which attempts to draw a middle line between Nye's direct approach, and the very abstract approach of Spring and Hutter. Although Eq. (11.4) appears 'obvious', this appendix suggests that it is far from being so.

Wide Channels and Other Problems   Despite its quite astonishing success, the Nye model (and our solution of it) has a number of difficulties associated with it. The first of these is the assumption of a semi-circular channel. With an assumption of easy lateral slip of ice at the bed, the resulting closure can be calculated from that for a circular channel. But why should the channel be semi-circular? This assumption seems completely arbitrary, yet the resulting theory appears to work exceptionally well.[8] Particularly when one considers the peak size of S (≈ 600 m^2 for Fig. 11.6,

---

[8]At least during floods. In normal circumstances, wide channels may be preferred (Hooke et al. 1990). See also the discussion on hybrid channels in the notes for Chap. 10.

and $\approx 2{,}900$ m^2 for Fig. 11.8), which imply maximum channel radii in the region
of 20 m and 43 m, respectively, it is pertinent to wonder whether the assumption of
semi-circular cross section is likely to be correct.

We can give some plausibility arguments to suggest why a circular shape is
adopted. For 'weak' floods in which the channel effective pressure is always positive, channel closure is that of a more viscous 'fluid' pushing back a less viscous
one. For this situation, viscous fingering does not occur, and a circularly deforming
boundary is stable. Further arguments in this direction are that, since the ice pressure is larger at the glacier bed, the contracting effective stress is larger there: this
mitigates against wider channels, while taller channels are not favoured because of
the consequent increased lateral melting.

None of these arguments work for 'strong' (flotation driven) floods, which as we
have discussed, propagate as viscous fractures at the tip, and in which the channels
must therefore be wide (Jóhannesson 2002a, 2002b).

Temperature   The assumption of equilibrium temperature (and thus the removal
of θ from the Nye model) seems rather unfortunate, and it would seem that at least it
would affect the flood dynamics in a quantitative manner. Let us therefore reconsider
the energy equation. Neglecting ε in (11.17), we have

$$Q\frac{\partial\theta}{\partial x} = Q\hat{\phi} - m, \tag{11.160}$$

where we write the hydraulic gradient as

$$\hat{\phi} = \Phi + \frac{\partial N}{\partial X} \tag{11.161}$$

(here $x = \delta X$). Since

$$Q|Q| = S^{8/3}\hat{\phi}, \qquad m \approx \frac{\theta}{\gamma}\left(\frac{|Q|}{S^{1/2}}\right)^{0.8}, \tag{11.162}$$

the temperature equation simplifies to

$$\frac{\partial\theta}{\partial x} = \hat{\phi} - \frac{\theta}{\lambda}, \tag{11.163}$$

where

$$\lambda = \frac{\gamma|Q|^{0.5}\,\mathrm{sgn}\,Q}{|\hat{\phi}|^{0.15}}. \tag{11.164}$$

Our simplification of the Nye model to the form (11.42) was based on the neglect
of the advective term in (11.163). As we stated, between floods this seems entirely
reasonable, since in equilibrium $\theta \approx \lambda\hat{\phi} \ll 1$. During floods, if we take $\hat{\phi}$ to be
constant, we see that the advective term $\theta_x = \exp(-x/\lambda)$, and in fact this is small
away from the lake, since for weak floods, we have computed the peak $Q \sim 10^{-2}$,
thus $\lambda \sim 0.1$, and $\theta_x \sim 10^{-4}$. However, the neglect of the advective term is inaccurate near the lake where $x = \delta X$, and in this region the equilibrium assumption
leads to the approximation for the advective term

$$\theta_x \approx \frac{\gamma\sqrt{Q}}{\delta}(\hat{\phi}^{0.85})', \tag{11.165}$$

where the prime denotes differentiation of $\hat{\Phi}^{0.85}$ with respect to X: γ and δ are defined in (11.18), with typical values given in (11.21). This term is of order $\gamma\sqrt{Q}/\delta$, and becomes significant at peak discharge. Its effect on the melt rate m which appears in the channel closure equation (11.17)$_1$ is to replace $m = \hat{\Phi}Q$ by

$$m \approx Q[\hat{\Phi} - \theta_x] \qquad (11.166)$$

(see (11.160)). Inspection of our numerical results shows that $S_X < 0$ at the lake at peak discharge, and thus (since $\hat{\Phi} = Q^2/S^{8/3}$), $\hat{\Phi}' > 0$, and thus $\theta_x > 0$. The effect of including temperature adjustment should therefore be to reduce the peak discharge.

This assumes the lake temperature is essentially equal to the freezing temperature, i.e., $\theta = 0$ at $x = 0$. If the lake is superheated, as might be the case following an eruption, then we expect $\theta_x < 0$, and the peak discharge will be enhanced. As we have seen, the floods may be rather different in this case. In particular, we associate warm water temperature with rapid lake filling, and a likely strong flood propagating down glacier as a wide fracture channel. Now consideration of the empirical heat transfer relation (11.5) shows that the term raised to the power 0.8 is essentially the Reynolds number for a semi-circular channel. For a wide channel, the relevant length scale is the channel depth, and this implies that to use (11.5) for a wide channel, we should multiply the constant a_{DB} by $(h/w)^{0.4}$, where h is channel depth and w is channel width. For a wide channel, this cause an increase in γ, which enhances the importance of the thermal advective term.

However, our problem is really the opposite of this. According to (11.163) and (11.164), the exit temperature of the water ought to be $\theta \sim \gamma\sqrt{Q}$. For a peak discharge of $Q = 0.033$ (corresponding to 6,000 m^3 s^{-1}) and with $\gamma \sim 2.5$ as in (11.21), we have $\theta \sim 0.45$ corresponding to an exit temperature of 1.6°, using the scale for θ_0 in (11.20). This is about thirty times higher than is observed (Clarke 2003), suggesting that heat transfer should be much more efficient than that given by the Dittus–Boelter relation assumed by Nye (1976). Two possible reasons are that the measurements used in establishing the Dittus–Boelter relation were done at Reynolds numbers between 10^4 and 10^5, which is two or three orders of magnitude less than our situation. It may simply be that the heat transfer parameterisation is not very accurate. Possibly more likely is that other physical processes contribute to a larger effective heat transfer; for example, mechanical erosion of ice by the turbulent, sediment-laden flow. In either event, it seems there are good reasons to suppose that a practical value of γ may be much smaller than given in (11.21).

It has to be said in any case that inclusion of the temperature equation in the form (11.163) is not a straightforward addition to the numerical problem of solving the Nye equations. The reason for this is that when $X^* > 0$, the θ equation (11.163) must be solved in a direction away from the seal: that is, we prescribe (or in fact require) $\theta = 0$ at $X = X^*$, and must solve for θ by stepping backwards into $X < X^*$ and forwards into $X > X^*$. This is not an insurmountable problem, but it is at least awkward, because of the singularity at $X = X^*$. A reasonable alternative would be to ignore advection if $X^* < 0$, and only solve (11.163) if $X^* > 0$. If the lake temperature is positive, this gives a discontinuity in θ at flood initiation, and this would cause further numerical awkwardness.

Snout Closure and Open Channel Flow   If we reconsider the channel closure
equation

$$\frac{\partial S}{\partial t} = \frac{|Q|^3}{S^{8/3}} - S|N|^{n-1}N, \tag{11.167}$$

it is clear that the boundary condition $N = 0$ at the snout $x = 1$ is problematic, since
it predicts indefinite opening of the channel. The problem is that the closure rate
term is based on Nye's (1953) calculation of closure of a cylindrical borehole in an
infinite medium, which becomes increasingly irrelevant at the snout. It is simple to
modify Nye's analysis to consider the effect of a stress-free outer boundary at radius
R_f, and the effect of this is to modify the closure term in (11.167) so that the closure
equation becomes

$$\frac{\partial S}{\partial t} = \frac{|Q|^3}{S^{8/3}} - \frac{S|N|^{n-1}N}{\left\{1 - \left(\frac{S}{S_f}\right)^{1/n}\right\}^n}, \tag{11.168}$$

where $S_f = \pi R_f^2$. This limits the growth of S because $S \leq S_f$ everywhere; (11.168)
applies if $S < S_f$, and is replaced by the condition $S = S_f$ if $S \geq S_f$. In practice we
might take S_f proportional to ice depth (perhaps to some power).

A further complication is that in practice the channel may reach atmospheric
pressure at some distance up stream from the snout. In this case, the location of the
position where $N = 0$ is a free boundary, with the extra condition being that the flux
becomes equal to the open channel value.

Inertia Terms   One of the subtleties that Spring and Hutter introduced was the
inertial acceleration terms in the water momentum equation. Later, Clarke (2003)
found a way to solve this formulation of the problem numerically. The inertial terms,
when scaled, are multiplied by a form of Froude number squared, but one computed
with the ice depth, not the channel depth (see Question 11.5). In general, this term
is small and can be safely neglected. Inclusion of inertia terms simply introduces
complication without any advantage.

Channel Roughness   In his seminal paper, Nye (1976) used a Manning roughness
of $n' = 0.12$ m$^{-1/3}$ s in order to fit the rising limb of the 1972 Grímsvötn hydrograph. This represents a very rough channel, and Clarke (2003) thought that such a
value was too high. Nye only used this to fit the rising limb, and did not otherwise
solve his model. In fact, it is perfectly possible to fit the whole 1972 hydrograph
using lower choices of roughness; for example, Fowler (2009) fitted the 1972 peak
discharge and duration using values $A_L = 10$ km^2 (cf. Björnsson 1992, Fig. 2)[9]
and $n' = 0.04$ m$^{-1/3}$ s: see also Question 11.7. On the other hand, we might well
suppose that a sediment-laden torrent at Reynolds number 10^8 might well be very
rough.

---

[9]Our previous choice of $A_L = 30$ km^2 (after (11.30)) corresponded to the maximum area before
1940, since when the lake area has declined; 10 km^2 is approximately the minimum lake area in
1972. A_L is an approximately linearly increasing function of lake level, and hence a decreasing
function (see (11.25)) of N_0.

The Viscous Beam Thin films abound in applied mathematics, having applications in, for example, glass blowing, foam drainage, and coating flows, and in the first two of these, where there is no shear stress applied at the upper and lower surfaces, the problem is essentially the same as that describing ice shelves, or where there is a load, ice cauldrons. An entry into this literature is through the papers by Howell (1996) and Teichman and Mahadevan (2003), for example.

Floods from Ice Sheets The description of the Channelled Scablands of Eastern Washington State and their origin by massive floodwaters is due to Bretz (1923, 1969). Initially the floods were associated with glacial meltwater, and later with the ice-dammed glacial Lake Missoula. The sequential nature of the Missoula floods is described by Waitt (1984), for example.

The 8,200 year cold event is discussed by Alley et al. (1997), and a theoretical discussion is given by Clarke et al. (2004). Commentary on this paper by Sharpe (2005) and the authors' reply (Clarke et al. 2005) focusses on the related work by John Shaw and his co-workers (e.g., Shaw 1983; Shaw et al. 1989). Shaw's central thesis is that drumlins (see Chap. 10) were formed subglacially by flowing meltwater. The analogy with fluvial erosional forms then dictates high Reynolds number flows, and consequently that such erosion could only have occurred in huge floods (so the argument goes). Shaw and his followers have been somewhat messianic in their pursuit of this thesis, but it has to be said that many of the central pillars of the argument are coming to be accepted: the existence of subglacial lakes, the existence of pro- and subglacial floods. Like a Shakespearean tragedy, the single flaw in the argument stems from the unnecessary assumption that because drumlins look 'like' fluvial erosion forms, they must *be* such forms. Most scientists are put off the Shaw theory because of the apparently unrealistic constraints which the theory seeks to impose: vast lakes, enormous floods, beyond the ability of a reputable theory to explain as being physically possible.

Sub-Antarctic lakes are described by Siegert et al. (1996), for example, who give an inventory of such lakes. A more recent review is by Siegert (2005). The largest and best known is Lake Vostok (Siegert et al. 2001), which has an approximate volume of 5,000 km^3. Lake Vostok may 'drain' by basal freeze on, which currently appears to remove the lake water. For the sorts of subglacial floods described by Denton and Sugden (2005) (see also Sugden and Denton 2004[10]), some such huge body of water must drain to the coast, although whether this is possible for Vostok is not known. Goodwin (1988) describes a jökulhlaup of six months duration observed near the coast in East Antarctica. More recently, Wingham et al. (2006) have observed small scale floods of a subglacial lake in the Adventure Trough of East Antarctica. Erlingsson (2006) raises the possibility of a Vostok jökulhlaup.

Dansgaard–Oeschger events were already discussed in Sect. 2.5.5. Ganopolski and Rahmstorf (2001) provide the model result that freshwater flux oscillations of magnitude 0.1 Sv (10^5 m^3 s^{-1}) can cause switching of the circulatory state. Similar

---

[10]Note: in Fig. 1 of this paper, the two marks of 130° E should both read 163° E (David Sugden, private communication).

results are given by Stocker and Wright (1991). Alley et al. (2001) and Ganopolski
and Rahmstorf (2002) suggest that a very weak periodicity may be amplified to
produce the 1,500 year cycle by means of stochastic resonance.

Mars   An elegant, if slightly manic, discussion of recent theories about Mars is
the semi-popular book by Kargel (2004). The list of contents is off-putting, but the
material is stimulating, and there are many excellent photographs from the various Mars orbiting spacecraft. The literature on Mars is mostly based around these
images and their interpretation. An early paper on massive floods is by Baker and
Milton (1974), and a more recent review is that by Baker (2001). Two contrasting
views (CO_2 or water?) are described by Hoffman (2000) and Coleman (2003), and
the possibility of jökulhalups on Mars, possibly associated with sub-ice volcanoes,
is considered by Chapman et al. (2003).

11.9 Exercises

11.1 In the absence of lake refilling at the inlet, a suitable boundary condition for
(11.17) at $s = 0$ is $Q = 0$. Assuming Q_0 is chosen so that $\Omega = 1$ (why?), find
typical values of the dimensionless parameters, and hence derive an approximate ordinary differential equation for N describing steady state drainage.
Show further that, if δ is small, a further simplification is possible, and in this
case derive an approximate *drainage law* in the form $N = cQ^\nu$, and give explicit expressions and typical values for c and ν. The approximation $\delta \to 0$ is
a *singular perturbation*; where does if fail, why, and what is the resolution?

11.2 Suppose the hydraulic gradient near $X = 0$ is given by

$$\Phi = 1 - ae^{-bX},$$

and N and Q are determined on $0 < X < \infty$ by

$$\Phi + \frac{\partial N}{\partial X} = N \operatorname{sgn} Q,$$

$$Q = \omega(X - X^*),$$

with

$$N \to 1 \quad \text{as } X \to \infty,$$

$$N = N_L \quad \text{on } X = 0,$$

$$\dot{N}_L = -(\nu + \omega X^*),$$

where ω and ν are small. Show that

$$N_L(0) = 2e^{X^*} - \left[1 + \frac{a}{b-1}\right] + \frac{2a}{b^2 - 1}e^{-(b-1)X^*},$$

and deduce that the seal is weak (seal breaking occurs when the lake level is
below flotation: $N_L(0) > 0$) if $a < b + 1$.

11.3 Use the relations

$$S_0 = \left(\frac{f\rho_w g Q_0^2}{\Phi_0}\right)^{3/8}, \qquad m_0 = \frac{\Phi_0 Q_0}{L},$$

$$\Phi_0 = \frac{\rho_w g h_0}{l}, \qquad t_0 = \frac{\rho_i S_0}{m_0}, \qquad N_0 = (K t_0)^{-1/n},$$

to find explicit relations for the scales S_0, m_0, etc. in terms of Q_0, and hence show that the parameters

$$\varepsilon = \frac{S_0 l}{t_0 Q_0}, \qquad \delta = \frac{N_0}{\Phi_0 l}$$

are given by

$$\varepsilon = \frac{\rho_w g h_0}{\rho_i L}, \qquad \delta = \frac{1}{\rho_w g h_0}\left[\frac{Q_0^{1/4} \Phi_0^{11/8}}{\rho_i K L (f\rho_w g)^{3/8}}\right]^{1/n}.$$

11.4 Use the definitions

$$Q_0 = \frac{A_L N_0}{\rho_w g t_0}, \qquad S_0 = \left(\frac{f\rho_w g Q_0^2}{\Phi_0}\right)^{3/8}, \qquad m_0 = \frac{\Phi_0 Q_0}{L},$$

$$\Phi_0 = \frac{\rho_w g h_0}{l}, \qquad t_0 = \frac{\rho_i S_0}{m_0}, \qquad N_0 = (K t_0)^{-1/n},$$

to estimate values for the peak discharge $Q_{\max} \sim Q_0$ and the total discharge $V_{\max} \sim Q_0 t_0$ (assuming A_L is constant). By varying in turn A_L, K, f and Φ_0, show that

$$Q_{\max} \propto V_{\max}^b,$$

and compare the values of b with the Clague–Mathews result $b = \frac{2}{3}$, or that of Björnsson, $b = 1.84$.

[*The Clague and Mathews (1973) peak discharge/volume relationship is a venerable curiosity of subglacial hydrology, and appears to have no simple explanation. Other studies include those by Björnsson (1992), Walder and COsta (1996), and Ng and Björnsson (2003), the last of which provides a detailed analytic investigation.*]

11.5 The momentum equation in the model for subglacial drainage is modified to include inertial acceleration terms, thus

$$\rho_w (u_t + u u_x) = \Phi + \frac{\partial N}{\partial x} - \frac{f\rho_w g Q|Q|}{S^{8/3}},$$

where $\Phi = \frac{\rho_w g d_i}{l}$ is the basic hydraulic gradient, d_i is ice depth, l is glacier length, N is the effective pressure, Q is volume flux, S is cross-sectional area, and $u = \frac{Q}{S}$ is the mean velocity. By scaling the variables as

$$\Phi = \Phi_0 \Phi^*, \qquad N \sim N_0, \qquad x \sim l, \qquad Q \sim Q_0,$$

$$S \sim \left(\frac{f\rho_w g Q_0^2}{\Phi_0}\right)^{3/8}, \qquad t \sim \frac{\rho_i L S_0}{\Phi_0 Q_0},$$

show that the dimensionless form of this equation can be written as

$$F^2(\varepsilon u_t + u u_x) = \Phi^* + \delta \frac{\partial N}{\partial x} - \frac{Q|Q|}{S^{8/3}},$$

where

$$\delta = \frac{N_0}{\rho_w g d_i}, \qquad \varepsilon = \frac{\rho_w g d_i}{\rho_i L}, \qquad F^2 = \frac{u_0^2}{g d_i},$$

and $u_0 = \frac{Q_0}{S_0}$. Show that, if $Q_0 = 10^5$ m^3 s^{-1}, $S_0 = 10^4$ m^2, $g = 10$ m s^{-2} and $d_i = 500$ m, then $F^2 \sim 0.02$, and explain why more generally the acceleration terms can be neglected. What complications do they cause if they are included?

11.6 In an approximate model for the channel effective pressure between floods, N satisfies the Riccati equation

$$\Phi(X) + \frac{\partial N}{\partial X} = N^2 \operatorname{sgn}(X - X^*),$$

with boundary conditions

$$N \to 1 \quad \text{as } X \to \infty,$$
$$N = N_L \quad \text{at } X = 0,$$

where we assume $\Phi(\infty) = 1$.

Assume that

$$\Phi = 1 - a e^{-bX},$$

and use appropriate substitutions in $X \lessgtr X^*$ of the form $N = w'/cw$ to reduce the model to a pair of linear equations for w. Hence show that for $X > X^*$,

$$N = -\frac{v'}{v}, \quad v = J_\nu\left(\lambda e^{-X/\nu}\right),$$

where

$$\lambda = \frac{2\sqrt{a}}{b}, \qquad \nu = \frac{2}{b},$$

and for $X < X^*$,

$$N = \frac{w'}{w}, \quad w = J_{i\nu}\left(i\lambda e^{-X/\nu}\right) + \alpha Y_{i\nu}\left(i\lambda e^{-X/\nu}\right),$$

where α must be chosen so that N is continuous at X^*.

Hence show that

$$N_L(0) = \frac{\lambda J_\nu'(\lambda)}{\nu J_\nu(\lambda)},$$

and deduce that the seal is weak if $\frac{J_\nu'(\lambda)}{J_\nu(\lambda)} > 0$, and thus if

$$a < a_2 = \left[\frac{1}{2} b j'_{(2/b),1}\right]^2,$$

where $j'_{\nu,1}$ is the first zero of $J_\nu'(z)$.

Use tables to evaluate $j'_{\nu,1}$ for $\nu = \frac{1}{2}, 1, 2, 3, 4, 5$, and compare the consequent values of a_2 for $b < 4$ with those obtained from the asymptotic result $j'_{\nu,1} \approx \nu + 0.81\nu^{1/3}$ when $\nu \gg 1$, and deduce that the small b approximation is accurate in this range. By plotting the results numerically, show that a useful approximation to a_2 is then

$$a_2 \approx 1.2 + b,$$

for $b < 4$.

An even better approximate model is to take

$$\Phi + \frac{\partial N}{\partial X} = \frac{N^2}{\sigma^2} \operatorname{sgn}(X - X^*),$$

where $\sigma = \omega^{1/11} \approx 0.44$ (since $Q \sim \omega$) for $\omega = 1.2 \times 10^{-4}$. In this case, modify the analysis above to show that a seal is weak if

$$a < a'_2 = \left[\frac{1}{2}\sigma b j'_{(2/\sigma b),1}\right]^2,$$

and deduce that

$$a'_2 \approx 1.2 + 0.44b$$

for $b < 9$.

11.7 The dimensionless, reduced Nye model for jökulhlaups depends on dimensionless parameters $\omega = \delta\Omega$ and ν defined by

$$\omega = \frac{M}{\Phi_0 Q_0}\left[\frac{\Phi_0^{11/8} Q_0^{1/4}}{\rho_i K L (f\rho_w g)^{3/8}}\right]^{1/n}, \qquad \nu = \frac{m_L}{Q_0},$$

and the dimensional results depend on the volume flux scale Q_0 and time scale t_0 defined by

$$t_0 = \frac{\rho_i L}{\Phi_0} \frac{(f\rho_w g)^{3/8}}{Q_0^{1/4}},$$

$$Q_0 = \left[\left(\frac{A L}{\rho_w g}\right)^n \frac{1}{K}\left\{\left(\frac{\Phi_0}{f\rho_w g}\right)^{3/8} \frac{\Phi_0}{\rho_i L}\right\}^{n+1}\right]^{4/(3n-1)}.$$

The periodic solutions therefore have dimensional period

$$P = \frac{t_0}{\nu} P^*,$$

and the floods have peak discharge

$$Q_{\max} = Q_0 Q^*$$

and duration

$$t_F = t_0 t^*.$$

In general, the starred quantities are functions of ω and v, but let us assume they depend only on the quantity

$$\alpha = \frac{\omega^{4n/(4n-1)}}{v} = \frac{1}{m_L}\left(\frac{M}{\Phi_0 m_L}\right)^{4n/(4n-1)}\left[\frac{\Phi_0^{11/8}}{\rho_i K L(f\rho_w g)^{3/8}}\right]^{4/(4n-1)}.$$

Suppose that a particular numerical solution, with particular choices for the scales and parameters (and in particular $f = f_n$, $A_L = A_n$ and $m_L = m_n$), produces floods of (dimensional) duration t_n, peak discharge Q_n and with period P_n, whereas the actual observed floods have corresponding quantities t_a, Q_a and P_a. Show how, by choosing new values $f = f_a$, $A_L = A_a$ and $m_L = m_a$, the numerical solution can be made to fit the data, and give explicit expressions for the ratios $\frac{f_a}{f_n}$, $\frac{A_a}{A_n}$ and $\frac{m_a}{m_n}$.

11.8 The Nye model for jökulhlaups can be reduced to the system

$$\frac{\partial S}{\partial t} = \frac{|Q|^3}{S^{8/3}} - S|N|^{n-1}N,$$

$$Q = \omega(X - X^*),$$

$$\Phi + \frac{\partial N}{\partial X} = \frac{Q|Q|}{S^{8/3}},$$

with the boundary conditions that

$$\frac{\partial N}{\partial t}(0, t) = Q(0, t) - v,$$

$$\frac{\partial N}{\partial X} \to 0 \quad \text{as } X \to \infty,$$

assuming that the effect of thermal advection is ignored.

Suppose that $\omega \sim v \ll 1$. Show that between floods, the variables can be rescaled (explaining how) so that the equations take the form

$$\beta\frac{\partial s}{\partial T} = \frac{|q|^3}{s^{8/3}} - |\Pi|^{n-1}\Pi,$$

$$q = \xi - \xi^*,$$

$$\Phi + \Pi_\xi = \frac{|q|q}{s^{8/3}},$$

subject to

$$\Pi_T = -1 + \alpha q \quad \text{on } \xi = 0,$$

$$\Pi_\xi \to 0 \quad \text{as } \xi \to \infty,$$

where

$$\alpha = \frac{\omega^{4n/(4n-1)}}{v}, \qquad \beta = \frac{v}{\omega^{(n+1)/(4n-1)}},$$

and show that for values of ω, $v \sim 10^{-3}$, $\beta \ll 1$ but $\alpha = O(1)$ (e.g., if $n = 3$).

Show that if β is put to zero, then the Nye model between floods (when $\xi^* > 0$ and $\Pi > 0$) can be solved by computing the solution to the system

$$\frac{d}{d\xi}\left[|\xi - \xi^*|^{1/4n} R\right] = R^{8n/11} \operatorname{sgn}(\xi - \xi^*) - \Phi(\xi),$$

subject to

$$R \to \Phi_\infty^{11/8n} \quad \text{as } \xi \to \infty$$

and

$$\frac{d}{dT}\left[\xi^{*(1/4n)} R\,|_{\xi=0}\right] = -1 - \alpha\xi^*.$$

Consider and explain the difficulties in solving this problem numerically. Explain why this approximation can break down if $v \gg \omega$.

11.9 The vertically integrated dimensionless viscous ice beam equations are given by

$$2\frac{\partial T}{\partial x} + v^2\frac{\partial U}{\partial x} = \gamma\beta Nh_x + \beta(s - h)s_x,$$

$$\frac{\partial M}{\partial x} - v^2\frac{\partial L}{\partial x} + S = -\gamma\beta Nhh_x - \frac{1}{2}\beta(s^2 - h^2)s_x,$$

$$\frac{\partial S}{\partial x} = -\frac{\gamma\beta N}{v^2},$$

where M, S, T, U and L are defined by

$$M = -\int_h^s 2z\tau_1\,dz, \qquad S = \int_h^s \tau_3\,dz, \qquad T = \int_h^s \tau_1\,dz,$$

$$U = \int_h^s \sigma_3\,dz, \qquad L = \int_h^s z\sigma_3\,dz,$$

$s(x, t)$ is the ice surface elevation, $h(x, t)$ is the lake roof, and $p + \tau_1 = -v^2\sigma_3$.

Show that if

$$-p + \tau_1 = \beta(s - z), \qquad \tau_3 = 0$$

on $x = x_\pm$, which denote the subglacial lake margins, then

$$2T + v^2 U = \frac{1}{2}\beta(s - h)^2,$$

$$-M + v^2 L = \frac{1}{6}\beta(s - h)^2(s + 2h),$$

$$S = 0,$$

at $x = x_\pm$. Show that

$$\int_{x_-}^{x_+} N\,d\xi = 0,$$

and, assuming that $s|_{x_+} = s|_{x_-}$ and $h|_{x_+} = h|_{x_-}$,

$$\frac{\gamma\beta}{\nu^2} \int_{x_-}^{x_+} \xi N \, d\xi = \int_{x_-}^{x_+} \left[2T + \nu^2 U - \frac{1}{2}\beta(s-h)^2 \right] h_\xi \, d\xi,$$

and deduce that if ν is sufficiently small, then

$$\int_{x_-}^{x_+} \xi N \, d\xi \approx 0.$$

Interpret these results in terms of overall force and torque on the ice beam.

11.10 The subsidence rate $w(X, t)$ of an ice cauldron on an ice sheet is governed by the beam equation

$$w_{XXXX} = -N,$$

where the effective load $N(X, t)$ satisfies

$$N_t = \dot{N}_+ + w,$$

and $N_+(t)$ is a function of t ($\dot{N}_+$ being its derivative). The model is to be solved on $0 < X < \infty$, and we prescribe

$$w = w_{XX} = 0, \qquad N = N_+, \quad \text{on } X = 0,$$

$$w \to -\dot{N}_+ \quad \text{as } X \to \infty.$$

(i) By writing $w = -\dot{N}_+ + W$ and integrating repeatedly, show that

$$W = \int_X^\infty \frac{1}{6}(\xi - X)^3 N(\xi, t) \, d\xi,$$

and deduce that

$$\dot{N}_+ = \int_0^\infty \frac{1}{6}\xi^3 N(\xi, t) \, d\xi.$$

Hence show that N satisfies the integro-differential equation

$$N_t = \dot{N}_+ + \int_0^\infty G(X, \xi) N(\xi, t) \, d\xi,$$

and give the definition of G.

Show that $\int_0^\infty \xi N(\xi, t) \, d\xi = 0$, and deduce that G can be written in the symmetric form

$$G(X, \xi) = \begin{cases} -\frac{1}{2}\xi^2 X - \frac{1}{6}X^3, & \xi > X, \\ -\frac{1}{2}X^2\xi - \frac{1}{6}\xi^3, & \xi < X. \end{cases}$$

(ii) Show directly from the governing equations that if

$$N_+ = \frac{c}{(t_0 - t)^\alpha},$$

there is a similarity solution of the form

$$N = N_+ \psi(\eta), \qquad W = \dot{N}_+ \phi(\eta),$$

where

$$\eta = mX(t_0 - t)^\beta,$$

and β should be determined. Show that, by choosing the value of m suitably, the equation for ϕ can be written in the form (after eliminating ψ)

$$\varepsilon\eta\phi^v - \phi^{iv} - 4\phi = 0,$$

where the Roman numeral superscripts indicate the number of derivatives. Give the value of ε, and write down suitable boundary conditions for ϕ.

Appendix A
The Schwarzschild–Milne Integral Equation

The exact solution of (2.15)–(2.17) is obtained as follows. We define, as before, the local average intensity

$$J(\tau) = \frac{1}{2}\int_{-1}^{1} I(\tau, \mu)\, d\mu, \tag{A.1}$$

and the formal solution of (2.15) is

$$I = \begin{cases} \int_{\tau}^{\infty} e^{-(t-\tau)/\mu} J(t)\frac{dt}{\mu}, & \mu > 0, \\ \int_{0}^{\tau} e^{-(t-\tau)/\mu} J(t)\frac{dt}{(-\mu)}, & \mu < 0, \end{cases} \tag{A.2}$$

providing J does not grow exponentially as $\tau \to \infty$ (specifically, $J = o(e^{\tau})$). Substituting this expression back into (A.1), we find, after some algebra, that J satisfies the Schwarzschild–Milne integral equation

$$J(\tau) = \frac{1}{2}\int_{0}^{\infty} E_1\big(|t - \tau|\big) J(t)\, dt, \tag{A.3}$$

and the flux conservation law (2.17) can be written in the form

$$\Phi = 2\pi\left[\int_{\tau}^{\infty} J(t) E_2(t - \tau)\, dt - \int_{0}^{\tau} J(t) E_2(\tau - t)\, dt\right]. \tag{A.4}$$

The exponential integrals E_1 and E_2 are defined by

$$E_2(y) = y\int_{y}^{\infty} \frac{e^{-s}}{s^2}\, ds, \qquad E_1(y) = \int_{y}^{\infty} \frac{e^{-s}}{s}\, ds; \tag{A.5}$$

(A.4) acts as a normaliser for the linear equation (A.3).

Equation (A.3) is amenable to treatment by the Wiener–Hopf technique. It defines J for $\tau > 0$, and we extend the definition of J so that

$$J = 0, \quad \tau < 0, \tag{A.6}$$

A. Fowler, *Mathematical Geoscience*, Interdisciplinary Applied Mathematics 36, DOI 10.1007/978-0-85729-721-1, © Springer-Verlag London Limited 2011

and we define a function $h(\tau)$, $h = 0$ for $\tau > 0$, so that

$$J(\tau) = \frac{1}{2} \int_{-\infty}^{\infty} E_1\big(|t - \tau|\big) J(t)\, dt + h(\tau), \tag{A.7}$$

for all values of τ. Write $K(t) = \frac{1}{2} E_1(|t|)$, so that, if we take Fourier transforms of (A.7), we get

$$\hat{J}_+ = \hat{K} \hat{J}_+ + \hat{h}_-, \tag{A.8}$$

where $\hat{J}_+(z)$ is the transform of J and the $+$ indicates that $\hat{J}_+(z)$ is analytic in an upper half plane (since $J = 0$ for $\tau < 0$). Since $J = o(e^\tau)$ as $\tau \to \infty$, this is at least $\operatorname{Im} z > 1$. Similarly $\hat{h}_-$ is analytic in a lower half-plane.

The solution of (A.8) is now effected through the splitting of $(1 - \hat{K})$ into factors analytic in upper and lower half planes, and this can be done by solution of an appropriate Hilbert problem. The transform $\hat{K}$ is defined as

$$\hat{K}(z) = \int_{-\infty}^{\infty} K(s) e^{isz}\, ds, \tag{A.9}$$

and we find that

$$\hat{K} = \frac{1}{2iz} \ln\left(\frac{1 + iz}{1 - iz}\right) = \frac{1}{z} \tan^{-1} z. \tag{A.10}$$

We will now strengthen our assumption on J so that J does not grow exponentially as $\tau \to \infty$, i.e., $J = o(e^{\alpha \tau})$ for any $\alpha > 0$; then $\hat{J}_+$ is analytic in $\operatorname{Im} z > 0$. Our aim now is to find a function G analytic in $\operatorname{Im} z \lessgtr 0$ such that $G_+/G_- = 1 - \hat{K}$ on $\mathbf{R}$, and this is done by solving the Hilbert problem $\ln G_+ - \ln G_- = \ln(1 - \hat{K})$. To do this we wish to have $1 - \hat{K} \neq 0$, in order that $\ln(1 - \hat{K})$ be Hölder continuous. On the other hand we want $\ln\{1 - \hat{K}(t)\} \to 0$ as $t \in \mathbf{R} \to \pm\infty$. These concerns motivate the modification of $1 - \hat{K}(t)$ by a factor $(t^2 + 1)/t^2$, since $1 - \hat{K} = O(t^2)$ as $t \to 0$ (and is non-zero for $t \neq 0$), so that we seek a function G such that

$$\frac{G_+(t)}{G_-(t)} = \left(\frac{t^2 + 1}{t^2}\right)\left[1 - \frac{1}{2it} \ln\left(\frac{1 + it}{1 - it}\right)\right], \tag{A.11}$$

for $t \in \mathbf{R}$. Clearly G is only determined up to a multiplicative analytic function, and to be specific we will suppose $G_\pm \to 1$ as $z \to \infty$. We take the branches of $\ln(1 \pm it)$ to be such that $\ln 1 = 0$. The solution of (A.11) is

$$G(z) = \exp\left[\frac{1}{2\pi i} \int_{-\infty}^{\infty} \ln\left\{\left(\frac{t^2 + 1}{t^2}\right)\left(1 - \frac{1}{t} \tan^{-1} t\right)\right\} \frac{dt}{t - z}\right], \tag{A.12}$$

and with this definition of $G(z)$ (and thus $G_+(t)$ and $G_-(t)$), Eq. (A.8) for $\hat{J}_+$ can be written in the form, for $t \in \mathbf{R}$,

$$\frac{z^2}{z + i} G_+ \hat{J}_+ = (z - i) \hat{h}_- G_-. \tag{A.13}$$

Fig. A.1 Inversion contour
for (A.16)

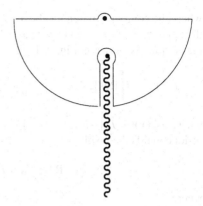

Clearly the left hand side defines the limit on $\mathrm{Im}\, z = 0+$ of a function analytic in the upper half plane $\mathrm{Im}\, z > 0$, while the right hand side is the limit on $\mathrm{Im}\, z = 0-$ of a function analytic in $\mathrm{Im}\, z < 0$ (since (A.7) implies that h grows no faster than $J(-\tau)$). We infer that each function can be analytically continued into its opposite half plane, thus defining an entire function $E(z)$, so that

$$\hat{J}_+(z) = \frac{(z+i)E(z)}{z^2 G_+(z)}. \tag{A.14}$$

The definition of $\hat{J}_+$ as a Fourier transform requires $\hat{J}_+ \to 0$ as $z \to \infty$, while also $G_+ \to 1$ as $z \to \infty$. It follows that $\hat{J}_+ \sim E/z$, which requires that $E = ic$ is constant, i.e.,

$$\hat{J}_+ = \frac{ic(z+i)}{z^2 G_+(z)}, \tag{A.15}$$

and the constant c is determined by the normalising condition (A.4). (The factor i is inserted for later convenience.)

Some information on the structure of $\hat{J}_+$ can be gleaned from (A.11). Evidently G_+ can be extended to $\mathrm{Im}\, z < 0$, and G_- to $\mathrm{Im}\, z > 0$ by the reciprocal relationship

$$\frac{G_+(z)}{G_-(z)} = \left(\frac{z^2+1}{z^2}\right)\left[1 - \frac{1}{2iz}\ln\left(\frac{1+iz}{1-iz}\right)\right]. \tag{A.16}$$

Care needs to be used in interpreting (A.16). If $\mathrm{Im}\, z < 0$, then (A.16) provides an analytic continuation for G_+ there, which shows that the continuation of G_+ to $\mathrm{Im}\, z < 0$ (very definitely *not* equal to G_-) has a logarithmic branch point at $z = -i$. Similarly G_-, extended to $\mathrm{Im}\, z > 0$, has a logarithmic branch point at $z = +i$. Therefore $\hat{J}_+$, extended via (A.15) to $\mathrm{Im}\, z < 0$, has a double pole at $z = 0$ (as $G_+(0) = \frac{1}{\sqrt{3}} \neq 0$) and a branch cut which we may take from $-i$ to $-i\infty$.

The inverse transform of (A.15) is

$$J(\tau) = \frac{1}{2\pi}\int_{-\infty}^{\infty} \hat{J}_+(z)e^{-iz\tau}\, dz, \tag{A.17}$$

where the contour is indented above the origin. If $\tau < 0$, we complete the contour in the upper half plane, whence we have $J = 0$ (as we assumed). If $\tau > 0$, we complete the contour as shown in Fig. A.1. The result of this is that

$$J(\tau) = -i\left[\text{Res}\{\hat{J}_+e^{-iz\tau}\}\big|_{z=0} + \frac{1}{2\pi}\int_0^\infty e^{-\tau(1+x)}[\hat{J}_+^+ - \hat{J}_+^-]\,dx\right], \quad (A.18)$$

where $\hat{J}_+^+(x) = \hat{J}_+[-i + xe^{-i\pi/2}]$, $\hat{J}_+^-(x) = \hat{J}_+[-i + xe^{3i\pi/2}]$. Calculation of the residue yields the result

$$\text{Res}\big|_{z=0} = ic\sqrt{3}(1 + \tau - j), \quad (A.19)$$

where

$$j = \frac{1}{\pi}\int_0^\infty\left[\frac{1}{(1 - t^{-1}\tan^{-1}t)} - 1 - \frac{3}{t^2}\right]\frac{dt}{1+t^2}. \quad (A.20)$$

We use (A.16) to substitute for G_+ in (A.15), and then we find

$$\hat{J}_+^\pm(x) = \frac{-c}{(2+x)G_-[-i(1+x)]l_\pm(x)}, \quad (A.21)$$

where

$$l_\pm(x) = 1 - \frac{1}{2(2+x)}\left[\ln\left(\frac{2+x}{x}\right) \pm i\pi\right]. \quad (A.22)$$

It follows that

$$\hat{J}_+^+ - \hat{J}_+^- = \frac{i\pi c}{g_-(x)\left[\{2 + x - \frac{1}{2}\ln(\frac{2+x}{x})\}^2 + \frac{\pi^2}{4}\right]}, \quad (A.23)$$

where $g_-(x) = G_-(-i - ix)$, and from (A.12), we find

$$g_-(x) = \exp\left[-\frac{(1+x)}{2\pi}\int_{-\infty}^\infty\ln\left[\left(\frac{t^2+1}{t^2}\right)\left(1 - \frac{1}{t}\tan^{-1}t\right)\right]\frac{dt}{\{t^2 + (1+x)^2\}}\right]. \quad (A.24)$$

Finally, therefore, $J = cJ_0(\tau)$, where

$$J_0(\tau) = \sqrt{3}(1 + \tau - j) + \frac{\pi}{2}e^{-\tau}\int_0^\infty\frac{e^{-x\tau}\,dx}{g_-(x)\left[\{2 + x - \frac{1}{2}\ln(\frac{2+x}{x})\}^2 + \frac{\pi^2}{4}\right]}. \quad (A.25)$$

Evidently $J \approx c\sqrt{3}(1 - j + \tau) + o(e^{-\tau})$ as $\tau \to \infty$, which confirms the assumption of non-exponential growth.

It only remains to compute c (which is evidently real, hence the choice of constant ic in (A.15)), and there seems no obvious short cut other than laborious substitution of the expression (A.25) for J into (A.4), which can be written in the form

$$c = \frac{\Phi}{2\pi\int_0^\infty J_0(t)H(\tau - t)\,dt}, \quad (A.26)$$

where

$$H(\theta) = \begin{cases} E_2(-\theta), & \theta < 0, \\ -E_2(\theta), & \theta > 0. \end{cases} \tag{A.27}$$

A.1 Exercises

A.1 What is wrong with the following argument? To determine c in (A.26), write (A.4) in the form (since $J = 0$ for $\tau < 0$)

$$\Phi = 2\pi \int_{-\infty}^{\infty} J(t) H(\tau - t) \, dt,$$

where

$$H(\theta) = \begin{cases} E_2(-\theta), & \theta < 0, \\ -E_2(\theta), & \theta > 0. \end{cases}$$

A Fourier transform yields, via the convolution theorem,

$$\frac{\Phi}{2\pi i z} = \hat{J}_+(z) \hat{H}(z),$$

where

$$\hat{H}(z) = -2i \int_0^{\infty} E_2(\theta) \sin z\theta \, d\theta.$$

Show that

$$-\int_0^{\infty} E_2(\theta) e^{iz\theta} \, d\theta = \frac{\ln(1 - iz) + iz}{z^2},$$

so that

$$\frac{\Phi}{2\pi i z} = 2i \hat{J}_+ \frac{\left[2iz - \ln\left(\frac{1+iz}{1-iz}\right) \right]}{z^2}.$$

Since also

$$\hat{J}_+ = \frac{ic(z + i)}{z^2 G_+(z)},$$

this implies

$$G_+(z) = \frac{A(z + i)\left[1 - \frac{1}{2iz} \ln\left(\frac{1+iz}{1-iz}\right) \right]}{z^2},$$

where $A = \frac{8\pi c}{\Phi}$; but this is not analytic in Im $z > 0$.

Appendix B
Turbulent Flow

Shear flows become turbulent if the Reynolds number Re is sufficiently large. Usually, this means $Re \sim 10^3$. For flow in a cylindrical pipe, the Reynolds number is conventionally chosen to be

$$Re = \frac{Ud}{\nu}, \tag{B.1}$$

where U is the mean velocity, d is the pipe diameter, and ν is the kinematic viscosity. With this definition, the onset of turbulence occurs at $Re = 2,300$, although the details of the transition process are complicated (Fowler and Howell 2003), and occur over a range of Reynolds number.

Most obviously, one might suppose that turbulence arises because of an instability of the uniform (laminar) flow, and for half a century this motivated the study of the famous Orr–Sommerfeld equation (one version of which is studied in Appendix C), which describes normal modes of the linearised Navier–Stokes equations describing perturbations about a steady uniform flow. Commonly such studies are done in two dimensions, for example for plane Poiseuille flow, when the Reynolds number is defined in terms of the maximum (centre-line) speed of the laminar flow and the half-width. This leads to a definition which is $\frac{3}{4}$ of that which would arise using the mean velocity and width. For plane Poiseuille flow, it is found that the steady flow is linearly unstable if $Re > 5,772$; on the other hand, turbulence sets in at $Re \approx 1,000$ (Orszag and Patera 1983). For pipe flow, the flow is linearly stable at all Reynolds numbers, although the decay rate of disturbances tends to zero as $Re \to \infty$.

It appears that the transition to turbulence is only vaguely related to the stability of the uniform state. The story is most simply told in the plane Poiseuille case. The instability at $Re = Re_c = 5,772$ is subcritical, and an (unstable) branch of finite amplitude stationary solutions bifurcates for $Re < Re_c$, and exists down to about $Re = 2,900$ before bending back on to a higher amplitude stable branch. Crucially, the (two-dimensional) stability or instability occurs on a long viscous time scale. However, these stationary solutions are subject to a three-dimensional instability which occurs on the fast convective time scale, and it is this which appears to cause the transition. Its occurrence at $Re \approx 1,000$ is associated with the fact that while the two-dimensional equilibria no longer exist there, two-dimensional disturbances will still decay on the slow viscous time scale, thus allowing the rapid three-dimensional growth. Essentially the same story occurs in pipe flow, although there it seems that $Re_c = \infty$. Numerical experiments have also found unstable travelling wave structures, now in the form of arrays of longitudinal vortices, and transition is associated with their existence (Eckhardt et al. 2007).

Since in fact, turbulence is an irregular, chaotic motion, it seems most likely that its occurrence is associated with the occurrence of a homoclinic bifurcation (Sparrow 1982), which not only produces the strange turbulent motion, but also the various travelling wave structures that can be found.

B.1 The Reynolds Equation

The actual calculation of turbulent flows is usually done following Reynolds's (1895) formulation of averaged equations. We write the Navier–Stokes equations for an incompressible flow in the form

$$\frac{\partial u_i}{\partial x_i} = 0,$$

$$\rho \frac{\partial u_i}{\partial t} + \rho \frac{\partial}{\partial x_j}(u_i u_j) = -\frac{\partial p}{\partial x_i} + \mu \nabla^2 u_i,$$

(B.2)

where suffixes i represent the components, and the summation convention is used (i.e., summation over repeated suffixes is implied). If we denote time averages by an overbar, and fluctuations by a prime, thus

$$u_i = \bar{u}_i + u_i',$$ (B.3)

then averaging of (B.2) yields

$$\frac{\partial \bar{u}_i}{\partial x_i} = 0,$$

$$\rho \frac{\partial}{\partial x_j}(\bar{u}_i \bar{u}_j) + \frac{\partial}{\partial x_j}(\overline{\rho u_i' u_j'}) = -\frac{\partial \bar{p}}{\partial x_i} + \mu \nabla^2 \bar{u}_i.$$

(B.4)

The second of these can be written in the form

$$(\bar{\mathbf{u}}.\nabla)\bar{\mathbf{u}} = -\nabla p + \nabla.\{\tau + \tau^T\},$$ (B.5)

where

$$\tau_{ij} = 2\mu \dot{\bar{\varepsilon}}_{ij}, \quad \dot{\bar{\varepsilon}}_{ij} = \frac{1}{2}\left(\frac{\partial \bar{u}_i}{\partial x_j} + \frac{\partial \bar{u}_j}{\partial x_i}\right)$$ (B.6)

is the ordinary molecular mean stress, and

$$\tau_{ij}^T = -\overline{\rho u_i' u_j'}$$ (B.7)

is called the Reynolds stress. The essential problem in describing fully turbulent flows is to close the averaged model by prescribing the Reynolds stress.

B.2 Eddy Viscosity

The simplest way to close the Reynolds equation is to suppose that

$$\tau_{ij}^T = 2\mu_T \dot{\bar{\varepsilon}}_{ij}, \tag{B.8}$$

by analogy to (B.6). The coefficient μ_T is called the eddy viscosity. This itself can be prescribed in various ways, but the simplest is to take it as constant. For example, in a channel flow we might take

$$\mu_T = \rho \varepsilon_T \bar{u} d, \tag{B.9}$$

where d is the depth and $\bar{u}$ the mean velocity. More generally, one allows μ_T to vary with distance from bounding walls, as described below.

Measurements in turbulent wall-bounded flows lead to the definition of a friction factor f through the wall stress

$$\tau_w = f \rho \bar{u}^2. \tag{B.10}$$

Here, $\bar{u}$ is the mean velocity, and the friction factor $f = \frac{1}{8}\lambda$ in Schlichting's (1979) notation. For an open channel flow, (B.9) is consistent with (B.10) if $\varepsilon_T = \frac{1}{3}f$. Typical values for f are small, for example Blasius's law in smooth-walled pipe flows has

$$f \approx \frac{0.04}{Re^{1/4}} \tag{B.11}$$

for Reynolds numbers in the range 10^4–10^5, and thus $f \sim 0.004$ and $\varepsilon_T \sim 0.001$. Roughness of the wall gives correspondingly larger values of f and ε_T. Notice that ε_T^{-1} is the Reynolds number based on the eddy viscosity, and is relatively large, reflecting the well-known fact that the turbulent eddies disturbing the mean flow are of relatively small amplitude. A more realistic form for the eddy viscosity uses Prandtl's mixing length theory, which is motivated by observations that the mean velocity profile is approximately logarithmic. The following discussion is based on that of Schlichting (1979).

The friction velocity is defined as

$$u_* = \sqrt{\frac{\tau_w}{\rho}} \tag{B.12}$$

(note that $u_* \ll \bar{u}$ since generally $f \ll 1$), thus

$$f = \left(\frac{u_*}{\bar{u}}\right)^2. \tag{B.13}$$

For a one-dimensional shear flow, with coordinate z normal to the wall (at $z = 0$), Prandtl's mixing length hypothesis is

$$\tau = \rho l^2 \left| \frac{\partial u}{\partial z} \right| \frac{\partial u}{\partial z},$$ (B.14)

where τ is the shear stress, l is the mixing length, and u the velocity; Prandtl further suggests

$$l = \kappa z,$$ (B.15)

with κ a constant. If we suppose $\tau = \tau_w = \text{constant}$, then

$$u_* = \kappa z \frac{\partial u}{\partial z},$$ (B.16)

thus

$$\frac{u}{u_*} = C + \frac{1}{\kappa} \ln\left(\frac{u_* z}{\nu} \right),$$ (B.17)

which is the famous universal logarithmic velocity profile. See also Question 5.11 and the discussion on turbulent flow and eddy viscosity in the notes in Sect. 5.9 for Chap. 5.

B.3 Pipe Flow

We now consider the case of flow in a pipe of radius a, and suppose that (B.17) applies, where z is radial distance inwards from the wall. If u_m is the maximum velocity at $z = a$, then (B.17) implies

$$u_m - u = \frac{u_*}{\kappa} \ln\left(\frac{a}{z} \right),$$ (B.18)

and the mean velocity $\bar{u} = \frac{2}{a^2} \int_0^a (a - z) u \, dz$ satisfies

$$u_m - \bar{u} = \frac{3 u_*}{2\kappa}.$$ (B.19)

In addition, comparison of (B.17) and (B.18) implies

$$u_m = \frac{u_*}{\kappa} \ln\left(\frac{a u_*}{\nu} \right) + u_* C.$$ (B.20)

Using (B.19) and (B.13), and defining the Reynolds number

$$Re = \frac{\bar{u} d}{\nu},$$ (B.21)

where the pipe diameter $d = 2a$, we find

$$\frac{1}{\sqrt{f}} = \frac{1}{\kappa} \ln\left[Re\sqrt{f}\right] + C - \frac{3}{2\kappa} - \frac{1}{\kappa} \ln 2. \tag{B.22}$$

Extensive measurements indicate that this formula is very successful in predicting $f(Re)$ assuming $\kappa = 0.4$, $C = 5.5$. The principal assumption involved is that of an eddy viscosity

$$\nu_T = \kappa^2 z^2 \left|\frac{\partial u}{\partial z}\right|. \tag{B.23}$$

B.4 Extension to Rivers

The above results are easily extended to a river of depth d. Suppose now that

$$\tau = \tau_w \left(1 - \frac{z}{d}\right) = \rho \kappa^2 z^2 u'^2, \tag{B.24}$$

where $u' = \partial u / \partial z$. Integrating, we find, with $u = u_m$ at $z = d$,

$$u_m - u = \frac{u_*}{\kappa} \int_{z/d}^{1} (1-\xi)^{1/2} \frac{d\xi}{\xi} = 2\frac{u_*}{\kappa}\left[\ln \cot \frac{1}{2}\alpha - \cos\alpha\right], \tag{B.25}$$

where $\alpha = \sin^{-1}\sqrt{\frac{z}{d}}$. With the mean flow $\bar{u} = \frac{1}{d}\int_0^d u\,dz$, we find

$$u_m - \bar{u} = \frac{2u_*}{3\kappa}, \tag{B.26}$$

while comparison of (B.25) as $z \to 0$ with (B.17) yields

$$\frac{u_m}{u_*} = C - \frac{2}{\kappa} + \frac{1}{\kappa} \ln\left(\frac{4u_* d}{\nu}\right), \tag{B.27}$$

and elimination of u_m between (B.26) and (B.27) gives, with $Re = \bar{u}d/\nu$,

$$\frac{1}{\sqrt{f}} = \frac{1}{\kappa} \ln\left[Re\sqrt{f}\right] + C - \frac{8}{3\kappa} + \frac{1}{\kappa} \ln 2, \tag{B.28}$$

essentially the same result as (B.22).

B.5 Manning's Law

It is of interest to compare the laboratory born flow law (B.28) with a flow law such as that of Manning. Manning's law is

$$\bar{u} = \frac{R^{2/3} S^{1/2}}{n}, \tag{B.29}$$

where R is the hydraulic radius and S is the downstream slope. For a wide river, we take $R = d$ and $\tau_w = \rho g d S$. We thus have

$$\bar{u} R = \nu Re, \qquad f \bar{u}^2 = g R S, \tag{B.30}$$

from which we find

$$\bar{u} = \left(\frac{g S \nu Re}{f} \right)^{1/3}, \qquad R = \left(\frac{\nu^2 Re^2 f}{g S} \right)^{1/3}, \tag{B.31}$$

and Manning's law (B.29) can be written in the form

$$f = \left[\frac{g S^{1/10} n^{9/5}}{\nu^{1/5}} \right] Re^{-1/5}, \tag{B.32}$$

broadly comparable to (B.28). (As mentioned above, the often used Blasius relation (B.11) approximating (B.28) has $f \propto Re^{-1/4}$.)

B.6 Entry Length

It is well-known that the development of laminar pipe Poiseuille flow from a plug entry flow occurs over an extended distance (the entry length) which scales as $d Re$. The entry length scale is determined by the diffusion of vorticity through laminar boundary layers into the core potential flow. If we scale up this process to rivers, with $d = 1$ m, $Re = 10^6$, it would suggest entry lengths of 1000 km! In reality, however, such boundary layers would be turbulent, and a better notion of entry length would be d/ε_T, perhaps 100 m; and in fact sinuous channels and bed roughness will ensure that river flow will always be fully turbulent.

However, the entry length concept provides a framework within which one can pose Kennedy's (1963) potential flow model for dune formation (see Chap. 5), even if in practice it is not realistic. Further, if one adopts a constant eddy viscosity model of turbulent flow, then the small value of ε_T is consistent with an inviscid outer solution away from the boundary, even if the assumption of a shear free velocity is not. On the other hand, it is conceivable that in laboratory experiments, the outer inviscid flow might indeed be a plug flow if the entry conditions are smooth.

B.7 Sediment Deposition

Suppose now that a suspended sediment concentration $c(z)$ is maintained in a turbulent flow by the action of an eddy viscosity. The units of c are taken to be mass per unit volume of the stream. In equilibrium, we have a balance between the upward turbulent flux and the downward velocity, which we take as v_s:

$$-v_T \frac{\partial c}{\partial z} = v_s c. \tag{B.33}$$

We suppose Reynolds' analogy that the eddy momentum diffusivity is equal to the eddy sediment diffusivity, and between (B.23) and (B.24), we have

$$v_T = \kappa u_* z \left(1 - \frac{z}{d}\right)^{1/2}.$$ (B.34)

Solving this gives

$$c = c_s \left(\frac{z}{d}\right)^Z \exp\left[-Z \int_{z/d}^1 \frac{d\xi}{\xi}\left[\frac{1}{(1-\xi)^{1/2}} - 1\right]\right],$$ (B.35)

where Z is the Rouse number,

$$Z = \frac{v_s}{\kappa u_*}.$$ (B.36)

Unfortunately, this gives $c = 0$ at $z = 0$ and thus zero deposition there! This is due to the artificial singularity in u as $z \to 0$, and an artificial escape from this quandary is to evaluate c at a small distance above the bed. As a simple alternative we suppose v_T is constant, given by (B.9) for example. Then

$$c = c_0 \exp\left[-\frac{v_s z}{v_T}\right],$$ (B.37)

and the mean concentration is

$$\bar{c} = \frac{c_0}{R}(1 - e^{-R}),$$ (B.38)

where

$$R = \frac{v_s d}{v_T}.$$ (B.39)

If we use (B.9) and (B.13), then

$$R = \frac{v_s}{\varepsilon_T \bar{u}} = \frac{\kappa \sqrt{f}}{\varepsilon_T} Z.$$ (B.40)

The sediment deposition rate is, from (B.33) and cf. (5.10),

$$\rho_s v_D = c_0 v_s = \bar{c} v_s D,$$ (B.41)

where (B.38) implies

$$D(R) = \frac{R}{1 - e^{-R}}.$$ (B.42)

Other expressions involving $v_T(z)$ give similar expressions which increase with R (or Z) (Einstein 1950).

Appendix C
Asymptotic Solution of the Orr–Sommerfeld Equation

In this appendix we provide an asymptotic solution of the Orr–Sommerfeld equation describing rapid shear flow over a slightly wavy boundary. The description is based on the asymptotic theory described by Drazin and Reid (1981), which itself describes a body of research stemming from original investigations by Heisenberg and Tollmien. The theory is, however, rather difficult to follow, and is gone through in detail here for that reason.

The Orr–Sommerfeld equation is

$$ik\left[U\left(\Psi'' - k^2\Psi\right) - U''\Psi\right] = \frac{1}{R}\left[\Psi^{iv} - 2k^2\Psi'' + k^4\Psi\right], \tag{C.1}$$

and describes the z-dependent amplitude of a horizontal Fourier mode (of zero wave speed) of wave number k. $U(z)$ is the basic horizontal velocity profile. The boundary conditions we impose are those corresponding to no slip at the perturbed boundary and free slip at the top surface:

$$\begin{aligned} \Psi = 0, \qquad \Psi' = 1 \quad \text{at } z = 0, \\ \Psi = 0, \qquad \Psi'' = 0 \quad \text{at } z = 1. \end{aligned} \tag{C.2}$$

We seek asymptotic solutions for $R \gg 1$. Accordingly, there is an outer solution

$$\Psi \sim \Lambda\left[\Psi_0 + \frac{1}{R}\Psi_1 + \cdots\right], \tag{C.3}$$

where Λ is a scaling parameter to be chosen so that $\Psi_0 = O(1)$. The equation for Ψ_0 is the inviscid (Rayleigh) equation

$$U\left(\Psi_0'' - k^2\Psi_0\right) - U''\Psi_0 = 0, \tag{C.4}$$

and we might expect to satisfy the boundary conditions on the free surface $z = 1$. In fact, we see that specification of $\Psi_0 = 0$ on $z = 1$ automatically implies that $\Psi_0'' = 0$ there. The outer solution is written in terms of two independent Frobenius series of

(C.4), expanded about $z = 0$. Assuming $U(0) = 0$, $U'(0) = U'_0 \neq 0$, we have these two solutions given by

$$\psi_1 = z P_1(z),$$

$$\psi_2 = P_2(z) + \frac{U''_0}{U'_0} \psi_1 \ln z, \tag{C.5}$$

where

$$P_1 = 1 + \frac{U''_0}{2U'_0} z + \frac{1}{6} \left(\frac{U'''_0}{U'_0} + k^2 \right) z^2 + \cdots,$$

$$P_2 = 1 + \left(\frac{U'''_0}{2U'_0} - \frac{U''^2_0}{U'^2_0} + \frac{1}{2} k^2 \right) z^2 + \cdots, \tag{C.6}$$

and the functions P_1 and P_2 are easily found numerically (Drazin and Reid 1981, pp. 137–138).

We denote

$$P_1(1) \equiv P_{11}, \qquad P_2(1) \equiv P_{21}; \tag{C.7}$$

then the outer solution at leading order is

$$\Psi \sim \Lambda \left[P_{21} \psi_1 - P_{11} \psi_2 + O\left(R^{-1} \right) \right]. \tag{C.8}$$

Evidently, this does not satisfy the boundary conditions at $z = 0$, and we anticipate a boundary layer of thickness $\varepsilon \ll 1$ (to be chosen), in which the neglected terms become important. We define

$$z = \varepsilon \zeta, \tag{C.9}$$

and expand (C.8) in terms of ζ. The result is that

$$\Psi \sim \Lambda \left[-P_{11} + \varepsilon \zeta \left\{ P_{21} - P_{11} \frac{U''_0}{U'_0} \ln(\varepsilon \zeta) \right\} + \cdots \right], \tag{C.10}$$

and Van Dyke's (1975) matching principle indicates that we may need two terms of the inner expansion to match to this.

In the boundary layer, it is appropriate to choose

$$\varepsilon = \frac{1}{(ikRU'_0)^{1/3}}, \tag{C.11}$$

with the phase of ε (ph ε) defined as $-\pi/6$ (we suppose $U'_0 > 0$ and $k > 0$). In this case $R^{-1} \sim \varepsilon^3$, and the second term in the outer solution is of relative order ε^3. We then write

$$\Psi \sim \Lambda [\chi_0 + \varepsilon \chi_1 + \cdots], \tag{C.12}$$

Fig. C.1 Contours for the
Airy integral (C.15)

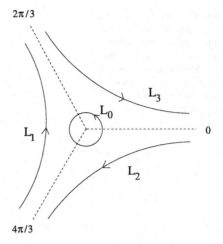

and the equations for χ_0 and χ_1 are

$$LD^2\chi_0 = 0,$$

$$LD^2\chi_1 = \frac{\zeta^2 U_0''}{2U_0'}\chi_0'' - \frac{U_0''}{U_0'}\chi_0, \tag{C.13}$$

where the operators L and D are defined by

$$D = \frac{d}{d\zeta}, \qquad L = D^2 - \zeta. \tag{C.14}$$

Reid (1972), see also Drazin and Reid (1981, pp. 465 ff.) shows how to solve these equations in terms of a class of generalised Airy functions.
   We begin by defining the functions

$$A_p^{(L)}(\zeta) = \frac{1}{2\pi i}\int_L t^{-p}e^{\zeta t - \frac{1}{3}t^3}\,dt, \tag{C.15}$$

where L is one of the contours shown in Fig. C.1, and p is an integer. We denote the function defined via the contour L_k as $A_p^{(k)}$. (Drazin and Reid's notation is different; they write $A_p^{(k)}(\zeta)$ as $A_k(\zeta, p)$.) These functions are analytic, and satisfy the third order differential equation

$$(LD + p - 1)A_p = 0. \tag{C.16}$$

The functions $A_p^{(1)}, A_p^{(2)}, A_p^{(3)}$ are independent, and by contraction of $L_1 \cup L_2 \cup L_3$, we see that

$$A_p^{(1)} + A_p^{(2)} + A_p^{(3)} = A_p^{(0)} = -B_p(\zeta), \tag{C.17}$$

Fig. C.2 The Stokes sectors T_i (bounded by the Stokes lines) and the anti-Stokes sectors S_i (bounded by the anti-Stokes lines) for (C.15). The signs in the sectors indicate the sign of $\arg \frac{2}{3} z^{3/2}$ as $z \to \infty$

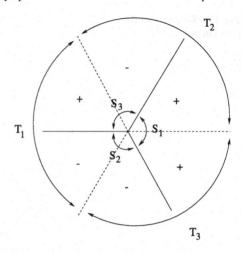

where B_p is a polynomial in ζ for integral p, in particular $B_p = 0$ for $p \le 0$, and

$$B_1(\zeta) = 1, \qquad B_2(\zeta) = \zeta, \qquad B_3(\zeta) = \frac{1}{2}\zeta^2. \tag{C.18}$$

The functions A_p satisfy the equations

$$LD^2 A_{p+1} = -(p-1)A_p,$$
$$DA_p = A_{p-1}, \tag{C.19}$$
$$\zeta A_p = pA_{p+1} + A_{p-2},$$

the last of these following from the first two together with (C.16). In particular, $LA_0 = 0$ and $A_0^{(k)}$ are the Airy functions; for example, $A_0^{(1)}(\zeta) = \mathrm{Ai}\,(\zeta)$. We also have the rotation formulae

$$A_p^{(2)}(\zeta) = e^{-2(p-1)\pi i/3} A_p^{(1)}\left(\zeta e^{2\pi i/3}\right),$$
$$A_p^{(3)}(\zeta) = e^{2(p-1)\pi i/3} A_p^{(1)}\left(\zeta e^{-2\pi i/3}\right). \tag{C.20}$$

It is clear from (C.19) that the solution for χ_0 in (C.13) is of the form

$$\chi_0 = \chi_{00} + \chi_{01}\zeta + \alpha_0 A_2^{(1)}(\zeta) + \beta_0 A_2^{(3)}(\zeta). \tag{C.21}$$

(Although $A_2^{(2)}$ is another possible solution, it is not independent because of (C.17), and because $B_2(\zeta) = \zeta$.)

Drazin and Reid (1981) give the asymptotic behaviour as $\zeta \to \infty$ of the functions $A_p^{(k)}$, based on the method of steepest descents and the rotation formulae (C.20). The Stokes sectors T_i are delimited by Stokes lines at $\arg \zeta = 0, 2\pi/3, 4\pi/3$, and within these, the anti-Stokes lines are $\arg \zeta = \pi/3, \pi, 5\pi/3$ (see Fig. C.2). Note that we seek the behaviour of $A_2^{(k)}$ as $\zeta \to \infty$ along $\arg \zeta = \pi/6$ (since $\zeta = (ikRU_0')^{1/3}z$), which lies in the sector S_1: $-\frac{\pi}{3} < \arg \zeta < \frac{\pi}{3}$, in which the functions A_+ and A_- defined by Drazin and Reid (p. 463, Eq. (A12)) respectively grow and decay exponentially. From their Eq. (A14), we then see that $A_p^{(1)} \to 0$ as $\zeta \to \infty e^{i\pi/6}$, while $A_p^{(3)}$ grows exponentially. Therefore $\beta_0 = 0$ in (C.21).

Next we turn to the solution for χ_1. From (C.13), we have, using $D^2 A_2 = A_0$,

$$LD^2\chi_1 = \frac{U_0''}{U_0'}\left[\frac{1}{2}\alpha_0\zeta^2 A_0^{(1)} - \chi_{00} - \chi_{01}\zeta - \alpha A_2^{(1)}\right]. \tag{C.22}$$

The solution to this equation is (using (C.18))

$$\chi_1 = \chi_{10} + \chi_{11}\zeta + \alpha_1 A_2^{(1)}(\zeta) + \frac{U_0''}{U_0'}\left[\alpha_0\left\{-\frac{1}{2}A_0^{(1)} + \frac{1}{10}A_{-3}^{(1)} + A_3^{(1)}\right\}\right.$$
$$\left. + \frac{1}{2}\chi_{01}\zeta^2 - \chi_{00}\phi\right], \tag{C.23}$$

where we use $LD^2 B_3 = -B_2$ and again suppress $A_2^{(3)}(\zeta)$, and ϕ is a particular solution to

$$LD^2\phi = B_1. \tag{C.24}$$

For matching purposes, ϕ must not grow exponentially at $\infty e^{\pi i/6}$.

The use of the relation $LD^2 B_{p+1} = -(p-1)B_p$ does not help here, because if $p = 1$, then $LD^2 B_2 = 0$. To find a solution, we now define the further generalised Airy functions

$$A_{pq}^{(k)}(\zeta) = \frac{1}{2\pi i}\int_{L_k} t^{-p}(\ln t)^q e^{\zeta t - \frac{1}{3}t^3}\,dt, \tag{C.25}$$

where $\arg t \in (0, 2\pi)$. (Drazin and Reid write $A_{pq}^{(k)}(\zeta)$ as $A_k(\zeta, p, q)$.) We also define the loop integrals

$$B_{pq}^{(k)}(\zeta) = \frac{1}{2\pi i}\int_{\infty e^{2(k-1)i\pi/3}}^{(0+)} t^{-p}(\ln t)^q e^{\zeta t - \frac{1}{3}t^3}\,dt, \tag{C.26}$$

where the loop contours in (C.26) are defined by Erdélyi et al. (1953, p. 13), and used by Olver (1974) and Reid (1972). The notation $\int_a^{(0+)}$ denotes an integral over a contour which is a loop beginning and ending at the point a, and which encloses the origin (and encircles it counterclockwise). For the integrands with branch points as in (C.26), these are thus the keyhole contours $\hat{L}_k$ as indicated in Fig. C.3.

It is straightforward to derive analogues of (C.19) (which apply to any of the contours L_k or $\hat{L}_k$), and these are (for A_{pq} or B_{pq})

$$DA_{pq} = A_{p-1,q},$$
$$(LD + p - 1)A_{pq} = qA_{p,q-1}, \tag{C.27}$$
$$LD^2 A_{p+1,q} = -(p-1)A_{pq} + qA_{p,q-1},$$

Fig. C.3 Two of the three
loop contours for (C.26), $\hat{L}_1$
and $\hat{L}_2$

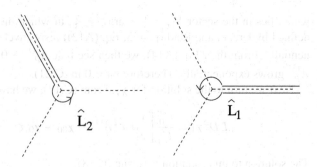

and in particular we see that

$$LD^2 A_{21} = A_1, \tag{C.28}$$

since it is clear that $A_{p0} = A_p$ for any p. Incidentally, note that when $q = 0$, the integrands of (C.26) do not have a branch point, and therefore the loop contours $\hat{L}_k$ are all equivalent to L_0, so that $B_{p0}^{(k)} = B_p$, and in particular

$$LD^2 B_{21}^{(k)} = B_1 \tag{C.29}$$

for each contour $\hat{L}_k$. Consulting (C.24), we see that any of $B_{21}^{(k)}$ is a particular solution for ϕ in (C.23), but we require one which does not grow exponentially. It is clear, since $LD^2 A_2^{(k)} = 0$, that the difference between the various $B_{21}^{(k)}$ for different k will be a sum of multiples of $A_2^{(k)}$, and this is explicitly provided by the connection formulae of Drazin and Reid (p. 475, Eq. (A43)):

$$\begin{aligned}
B_{21}^{(2)} - B_{21}^{(3)} &= 2\pi i A_2^{(1)}, \\
B_{21}^{(1)} - B_{21}^{(2)} &= 2\pi i A_2^{(3)}.
\end{aligned} \tag{C.30}$$

The object now is to find an appropriate solution of (C.29) which does not grow exponentially as $\zeta \to \infty e^{\pi i/6}$, and for this we need to know the asymptotic behaviour of one of the $B_{21}^{(k)}$. At this point we diverge from the discussion by Drazin and Reid (pages 178, 474). We consider explicitly the contour integral over $\hat{L}_2$:

$$B_{p1}^{(2)} = \frac{1}{2\pi i} \int_{\infty e^{2\pi i/3}}^{(0+)} t^{-p} \ln t \, e^{\zeta t - \frac{1}{3}t^3} \, dt. \tag{C.31}$$

In choosing the contour, we anticipate that we will require $\mathrm{Re}(\zeta t) < 0$, and to be specific, we define $\arg t \in (-\frac{4\pi}{3}, \frac{2\pi}{3})$ in (C.31). We have, successively,

$$B_{p1}^{(2)} = -\frac{\partial}{\partial p} \left[\frac{1}{2\pi i} \int_{\infty e^{2\pi i/3}}^{(0+)} t^{-p} e^{\zeta t - \frac{1}{3}t^3} \, dt \right] \tag{C.32}$$

and thus (put $t^3 = 3u$)

$$B_{p1}^{(2)} = -\frac{\partial}{\partial p}\left[\frac{1}{2\pi i}\int_{\infty e^{2\pi i/3}}^{(0+)}\sum_{n=0}^{\infty}\frac{\left(-\frac{1}{3}\right)^n}{n!}t^{3n-p}e^{\zeta t}\,dt\right]; \tag{C.33}$$

the method of proof of Watson's lemma then implies

$$B_{p1}^{(2)} \sim -\frac{\partial}{\partial p}\left[\sum_{n=0}^{\infty}\frac{\left(-\frac{1}{3}\right)^n}{n!}\frac{1}{2\pi i}\int_{\infty e^{2\pi i/3}}^{(0+)}t^{3n-p}e^{\zeta t}\,dt\right], \tag{C.34}$$

provided $\mathrm{Re}(\zeta t) < 0$.

Equation (6), page 14, of Erdélyi et al. (1953) gives

$$\frac{1}{2\pi i}\int_{\infty e^{i\delta}}^{(0+)}t^{-s}e^{-tX}\,dt = \frac{(Xe^{-i\pi})^{s-1}}{\Gamma(s)} \tag{C.35}$$

for any value of s, where, if $\arg t \in (\delta, 2\pi + \delta)$, then $-(\frac{1}{2}\pi + \delta) < \arg X < \frac{1}{2}\pi - \delta$.

In the present case, $\arg\zeta = \frac{\pi}{6}$, so that if we define $\delta = -\frac{4\pi}{3}$, (and note that $\infty e^{-4\pi i/3} = \infty e^{2\pi i/3}$), $X = \zeta e^{i\pi}$, then $\arg X = \frac{7\pi}{6}$ and lies between $-\frac{\pi}{2} - \delta = \frac{5\pi}{6}$ and $\frac{\pi}{2} - \delta = \frac{11\pi}{6}$. We thus have, for $\arg\zeta = \frac{\pi}{6}$,

$$\frac{1}{2\pi i}\int_{\infty e^{2\pi i/3}}^{(0+)}t^{-s}e^{t\zeta}\,dt = \frac{\zeta^{s-1}}{\Gamma(s)}, \tag{C.36}$$

and hence (C.34) gives, with $s = p - 3n$,

$$B_{p1}^{(2)}(\zeta) \sim -\frac{\partial}{\partial p}\left[\sum_{n=0}^{\infty}\frac{\left(-\frac{1}{3}\right)^n}{n!}\frac{\zeta^{p-3n-1}}{\Gamma(p-3n)}\right]. \tag{C.37}$$

Carrying out the differentiation,

$$B_{p1}^{(2)}(\zeta) \sim \sum_{n=0}^{\infty}\frac{\left(-\frac{1}{3}\right)^n}{n!}\left\{-\frac{\zeta^{p-3n-1}\ln\zeta}{\Gamma(p-3n)} + \zeta^{p-3n-1}\frac{\Gamma'(p-3n)}{\Gamma^2(p-3n)}\right\}. \tag{C.38}$$

Finally we put $p = 2$. Noting that Γ'/Γ^2 is finite and $1/\Gamma(r) = 0$ for non-positive integers r, we have

$$B_{21}^{(2)}(\zeta) \sim -\zeta\ln\zeta + \psi(2)\zeta + O(\zeta^{-2}), \tag{C.39}$$

for $\zeta \to \infty$ with $-\frac{\pi}{6} < \arg\zeta < \frac{5\pi}{6}$, and in particular when $\arg\zeta = \frac{\pi}{6}$; $\psi = \Gamma'/\Gamma$ is the digamma function.

We may now finally define a particular solution for ϕ in (C.24) to be (cf. (C.29))

$$\phi = B_{21}^{(2)}(\zeta). \tag{C.40}$$

Before we complete the solution by matching to the outer solution, we compare
(C.40) with results of Drazin and Reid (page 178). They choose (Eq. (27.49)) $\phi_{DR} = B_{21}^{(3)}$, and match in the sector $-\pi < \arg \zeta < \frac{1}{3}\pi$, where their Eq. (27.50) implies

$$\phi_{DR} \sim -\zeta[\ln \zeta - 2\pi i] + \psi(2)\zeta. \tag{C.41}$$

The connection formula (C.30)$_1$ implies that ϕ_{DR} and ϕ have the same asymptotic behaviour, since $A_2^{(1)}$ is exponentially small for $-\frac{\pi}{3} < \arg \zeta < \frac{\pi}{3}$ (Drazin and Reid, Eq. (A36), page 473). The only distinction between (C.39) and (C.41) is thus in the phase of $\ln \zeta$. (Note that the error term in Eq. (27.50) of Drazin and Reid should read $O(\xi^{-2})$.)

In fact, neither Drazin and Reid (nor Reid 1972) are specific about the phase either of t or of ζ in the definition of the loop integrals $B_{pq}^{(k)}$, although earlier (page 468) they suppose $-\frac{4}{3}\pi < \arg \zeta < \frac{2}{3}\pi$. If we define the modified loop integral

$$\hat{B}_{21}^{(2)}(\zeta) = \frac{1}{2\pi i} \int_{[\infty e^{2\pi i/3}, \infty e^{8\pi i/3}]}^{(0+)} t^{-2} \ln t \, e^{\zeta t - \frac{1}{3}t^3} \, dt, \tag{C.42}$$

just as in (C.31), but with $\arg t \in (\frac{2\pi}{3}, \frac{8\pi}{3})$, then we see immediately that (since $B_{20}^{(k)}(\zeta) = B_2(\zeta) = \zeta$)

$$\hat{B}_{21}^{(2)}(\zeta) = B_{21}^{(2)}(\zeta) + 2\pi i \zeta, \tag{C.43}$$

which allows consistency between (C.39) and (C.41) if $\phi_{DR} = \hat{B}_{21}^{(2)}$ or, equivalently, $\hat{B}_{21}^{(3)}$. We thus consider the discrepancy between the two accounts to be due to the choice by Reid (1972) of a different phase of ζ in applying Erdélyi et al.'s formula.

C.1 Matching

To summarize thus far, we have an outer solution (C.8):

$$\Psi \sim \Lambda[P_{21}\Psi_1(z) - P_{11}\psi_2(z) + O(\varepsilon^3)], \tag{C.44}$$

where, as $z = \varepsilon\zeta \to 0$,

$$\Psi \sim \Lambda\left[-P_{11} + \varepsilon\zeta\left\{P_{21} - P_{11}\frac{U_0''}{U_0'}\ln\varepsilon\right\} - \varepsilon P_{11}\frac{U_0''}{U_0'}\zeta\ln\zeta + \cdots\right]. \tag{C.45}$$

The inner solution is, from (C.12), (C.21) with $\beta_0 = 0$, (C.23) and (C.40),

$$\Psi \sim \Lambda\left[\left\{\chi_{00} + \chi_{01}\zeta + \alpha_0 A_2^{(1)}(\zeta)\right\}\right.$$
$$+ \varepsilon\left\{\chi_{10} + \chi_{11}\zeta + \alpha_1 A_2^{(1)}(\zeta) + \frac{U_0''}{U_0'}\left[\alpha_0\left\{-\frac{1}{2}A_0^{(1)} + \frac{1}{10}A_{-3}^{(1)} + A_3^{(1)}\right\}\right.\right.$$
$$\left.\left.\left. + \frac{1}{2}\chi_{01}\zeta^2 - \chi_{00}B_{21}^{(2)}(\zeta)\right]\right\} + \cdots\right], \tag{C.46}$$

which must satisfy the boundary conditions (from (C.2)) $\Psi = 0$, $d\Psi/d\zeta = \varepsilon$ on $\zeta = 0$. To accommodate these, we choose

$$\Lambda = \varepsilon \Lambda_1 + \varepsilon^2 \Lambda_2 + \cdots, \tag{C.47}$$

and thus specify (using the fact that $DA_p = A_{p-1}$, $DB_{pq} = B_{p-1,q}$)

$$\chi_{00} + \alpha_0 A_2^{(1)}(0) = 0,$$

$$\chi_{01} + \alpha_0 A_1^{(1)}(0) = 1/\Lambda_1,$$

$$\chi_{10} + \alpha_1 A_2^{(1)}(0) + \frac{U_0''}{U_0'}\left[\alpha_0\left\{-\frac{1}{2}A_0^{(1)}(0) + \frac{1}{10}A_{-3}^{(1)}(0) + A_3^{(1)}(0)\right\}\right.$$

$$\left. - \chi_{00}B_{21}^{(2)}(0)\right] = 0, \tag{C.48}$$

$$\chi_{11} + \alpha_1 A_1^{(1)}(0) + \frac{U_0''}{U_0'}\left[\alpha_0\left\{-\frac{1}{2}A_{-1}^{(1)}(0) + \frac{1}{10}A_{-4}^{(1)}(0) + A_2^{(1)}(0)\right\}\right.$$

$$\left. - \chi_{00}B_{11}^{(2)}(0)\right] = -\Lambda_2/\Lambda_1^2.$$

It remains to choose $\alpha_0, \alpha_1, \Lambda_1, \Lambda_2$, and these must follow from matching (C.45) and (C.46). For large ζ, (C.46) is

$$\Psi \sim \Lambda\left[\chi_{00} + \chi_{01}\zeta + \varepsilon\left\{\chi_{10} + \chi_{11}\zeta\right.\right.$$

$$\left.\left. + \frac{U_0''}{U_0'}\left[\frac{1}{2}\chi_{01}\zeta^2 - \chi_{00}\{-\zeta\ln\zeta + \psi(2)\zeta\}\right]\right\} + \cdots\right]. \tag{C.49}$$

Matching thus requires (we telescope the terms in $\ln\varepsilon$)

$$\chi_{00} = -P_{11},$$

$$\chi_{01} = 0,$$

$$\chi_{10} = 0, \tag{C.50}$$

$$\chi_{11} = P_{21} - \frac{P_{11}U_0''}{U_0'}\ln\varepsilon - \chi_{00}\psi(2)\frac{U_0''}{U_0'}.$$

The eight equations in (C.48) and (C.50) determine the unknowns $\alpha_0, \alpha_1, \Lambda_1, \Lambda_2$, $\chi_{00}, \chi_{01}, \chi_{10}$ and χ_{11}. In particular, we want to calculate $\frac{d^2\Psi}{dz^2}|_{z=0}$. At leading order, this is (with $\varepsilon = (ikRU_0')^{-1/3}$)

$$\left.\frac{d^2\Psi}{dz^2}\right|_{z=0} \sim (ikRU_0')^{1/3}\Lambda_1\alpha_0 A_0^{(1)}(0), \tag{C.51}$$

so it suffices to determine Λ_1 and α_0. We have $\chi_{00} = -P_{11}$ which is known by solving the Rayleigh equation, and $\chi_{01} = 0$. Therefore

$$\alpha_0 = \frac{P_{11}}{A_2^{(1)}(0)}, \qquad \Lambda_1 = \frac{A_2^{(1)}(0)}{P_{11}A_1^{(1)}(0)}. \qquad (C.52)$$

Notice that calculation of other coefficients requires the knowledge of $B_{21}^{(2)}(0)$ and $B_{11}^{(2)}(0)$. In view of our circumspection concerning $B_{pq}^{(2)}$, we would need to be suspicious of the definitions given by Drazin and Reid (Eq. (A39), page 474). The values of $A_p^{(1)}(0)$ are given by Drazin and Reid (page 468, Eq. (A11)), in particular,

$$A_1^{(1)}(0) = -\frac{1}{3}, \qquad A_2^{(1)}(0) = \frac{1}{3^{4/3}\Gamma\left(\frac{4}{3}\right)}. \qquad (C.53)$$

Note that $\alpha_0\Lambda_1 = 1/A_1^{(1)}(0) = -3$, and $A_0^{(1)}(0) = \text{Ai}(0) = \frac{1}{3^{2/3}\Gamma\left(\frac{2}{3}\right)} \approx 0.355$, thus

$$\left.\frac{d^2\Psi}{dz^2}\right|_0 \sim -3(ikRU_0')^{1/3}\text{Ai}(0) \approx -1.06(ikRU_0')^{1/3}. \qquad (C.54)$$

Note that this result (see comment after (C.11)) applies for $k > 0$ (and $U_0' > 0$). For $k < 0$, we use the fact that Ψ is the Fourier transform of a real function, and hence

$$\Psi(z, -k) = \overline{\Psi(z, k)}. \qquad (C.55)$$

Appendix D
Melting, Dissolution, and Phase Changes

The study of phase change and chemical reactions involves from the outset the magical art of thermodynamics. I have yet to meet an applied mathematician who claims to understand thermodynamics, and the interface of the subject with fluid dynamics raises serious fundamental issues. These we skirt, providing instead a cookbook of recipes. The initial material can be found in Batchelor (1967), while its extension to phase change and reaction involves (geo)chemical thermodynamics, as expounded by Kern and Weisbrod (1967) and Nordstrom and Munoz (1994), for example.

D.1 Thermodynamics of Pure substances

The state of a pure material is described by two independent quantities, such as temperature and pressure. Any other property of the material is then in principle a function of these two. Among such properties we have the volume, V; the internal energy, E; and a number of thermodynamic variables: the entropy S, the enthalpy H, the Helmholtz free energy F, and the Gibbs free energy G.

We distinguish between *intensive* and *extensive* variables. Intensive variables are those which describe properties of the material; they are local. Pressure and temperature are examples of intensive variables. Extensive variables are those which depend on the amount of material; volume is one such variable. Typically, extensive variables are simply intensive variables multiplied by the amount of substance, measured in *moles*.[1] If n moles of a substance have extensive variables V, H, S, E, F and G (all capitals), then the corresponding intensive variables are the specific volume $v = V/n$, and the specific enthalpy, entropy, internal energy, Gibbs free en-

---

[1] A mole of a substance is a fixed number (Avogadro's number, $\approx 6 \times 10^{23}$) of molecules (or atoms, as appropriate) of it. The weight of one mole in grams is called the (gram) molecular weight. The molecular weight of compound substances is easily found. For example, carbon (C) has a molecular weight of 12, while oxygen (O_2) has a molecular weight of 32; thus the molecular weight of CO_2 is 44, and we can write $M_{CO_2} = 44 \times 10^{-3} \text{ kg mole}^{-1}$.

A. Fowler, *Mathematical Geoscience*, Interdisciplinary Applied Mathematics 36,
DOI 10.1007/978-0-85729-721-1, © Springer-Verlag London Limited 2011

ergy and Helmholtz free energy are defined similarly (and may be denoted as lower case variables). In addition, the material density ρ is equal to $1/v$.

Definitions of H, F and G are

$$H = E + pV,$$
$$F = E - TS, \tag{D.1}$$
$$G = H - TS.$$

Two further relations are then necessary to determine E and S. An equation of conservation of energy (discussed in Sect. D.2) determines E, and the entropy S is determined via the differential relation

$$T\,dS = dE + p\,dV. \tag{D.2}$$

It will be convenient sometimes to work with the intensive forms of the variables, thus division of (D.2) yields

$$T\,ds = de + p\,dv. \tag{D.3}$$

From this latter relation we have the expressions

$$\left(\frac{\partial e}{\partial v}\right)_s = -p, \qquad \left(\frac{\partial e}{\partial s}\right)_v = T, \tag{D.4}$$

and if we now form the mixed second derivative $\frac{\partial^2 e}{\partial s\,\partial v}$ in two ways, we derive the relation

$$\left(\frac{\partial p}{\partial s}\right)_v = -\left(\frac{\partial T}{\partial v}\right)_s, \tag{D.5}$$

which is one of the four Maxwell relations. The others are derived in a similar way by considering mixed partial derivatives of h, f and g, yielding

$$\left(\frac{\partial v}{\partial s}\right)_p = \left(\frac{\partial T}{\partial p}\right)_s,$$

$$\left(\frac{\partial v}{\partial T}\right)_p = -\left(\frac{\partial s}{\partial p}\right)_T, \tag{D.6}$$

$$\left(\frac{\partial p}{\partial T}\right)_v = \left(\frac{\partial s}{\partial v}\right)_T.$$

Four partial derivatives are associated with specifically named quantities, which can be measured. These are the coefficient of thermal expansion

$$\beta = \frac{1}{v}\left(\frac{\partial v}{\partial T}\right)_p, \tag{D.7}$$

the coefficient of isothermal compressibility

$$\xi = -\frac{1}{v}\left(\frac{\partial v}{\partial p}\right)_T, \tag{D.8}$$

the specific heat at constant pressure,

$$c_p = T\left(\frac{\partial s}{\partial T}\right)_p, \tag{D.9}$$

and the specific heat at constant volume,

$$c_v = T\left(\frac{\partial s}{\partial T}\right)_v. \tag{D.10}$$

With these definitions, we can write

$$T\,ds = de + p\,dv = c_p\,dT - \beta v T\,dp, \tag{D.11}$$

which is useful in writing the energy equation, as we will now see.

D.2 The Energy Equation

The basic equations of conservation of mass, momentum and energy for a fluid with density ρ, velocity $\mathbf{u}$ and internal energy e are

$$\frac{\partial \rho}{\partial t} + \nabla.(\rho\mathbf{u}) = 0,$$

$$\frac{\partial \rho u_i}{\partial t} + \nabla.(\rho u_i \mathbf{u}) = \nabla.\sigma_i + \rho f_i, \tag{D.12}$$

$$\frac{\partial}{\partial t}\left[\frac{1}{2}\rho u^2 + \rho e + \rho\chi\right] + \nabla.\left[\left\{\frac{1}{2}\rho u^2 + \rho e + \rho\chi\right\}\mathbf{u}\right] = \nabla.(\sigma_i u_i) - \nabla.\mathbf{q},$$

where $\sigma_i = \sigma_{ij}\mathbf{e}_j$, $\mathbf{q}$ is the heat flux, and the conservative body force $\mathbf{f}$ is defined by

$$\mathbf{f} = -\nabla\chi, \tag{D.13}$$

where χ is the potential. Algebraic manipulation of the energy equation using the other two leads to the energy equation in the form

$$\rho\frac{de}{dt} = \sigma_{ij}\dot{\varepsilon}_{ij} - \nabla.\mathbf{q}, \tag{D.14}$$

where

$$\dot{\varepsilon}_{ij} = \frac{1}{2}\left(\frac{\partial u_i}{\partial x_j} + \frac{\partial u_j}{\partial x_i}\right) \tag{D.15}$$

is the strain rate. We can write

$$\sigma_{ij}\dot{\varepsilon}_{ij} = -p\nabla.\mathbf{u} + \tau_{ij}\dot{\varepsilon}_{ij}, \tag{D.16}$$

where τ_{ij} is the deviatoric stress tensor, and using the conservation of mass equation
$(D.12)_1$, we find

$$\rho\left[\frac{de}{dt} + p\frac{dv}{dt}\right] = \tau_{ij}\dot{\varepsilon}_{ij} - \nabla.\mathbf{q} \equiv R. \tag{D.17}$$

The right hand side R of this equation consists of the viscous dissipation and the
heat transport. Using (D.11), this leads to

$$\rho T\frac{ds}{dt} = R. \tag{D.18}$$

Using the relation in (D.11), the energy equation can also be written in the form

$$\rho c_p \frac{dT}{dt} - \beta T\frac{dp}{dt} = R, \tag{D.19}$$

and using the definition of (specific) enthalpy, it takes the form

$$\rho\frac{dh}{dt} - \frac{dp}{dt} = R. \tag{D.20}$$

These different forms are variously of use depending on the material properties.
In particular, for a perfect gas one can show (see Question D.12) that

$$dh = c_p\,dT, \qquad de = c_v\,dT. \tag{D.21}$$

The second of these also applies to an incompressible fluid.

D.3 Phase Change: Clapeyron Equation

The use of the free energies G (Gibbs free energy) and F (Helmholtz free energy)
is that they describe thermodynamic equilibrium conditions. Specifically, they take
minimum (and thus stationary) values at equilibrium. The difference between them
resides in the external conditions. At constant temperature and pressure, the Gibbs
free energy is a minimum, while at constant temperature and volume, the Helmholtz
free energy is a minimum. Of course, we are *never* really interested in systems
which are at equilibrium. Implicitly, thermodynamics is useful because we typically
assume that in systems away from equilibrium (pretty much everything), there is
a rapid relaxation of some parts of the system towards equilibrium. For example,
it is common to assume that in melting or freezing, the solid–liquid interface is at
the melting point. This is often a good assumption, but not always. One needs to be
aware that in practice, we assume thermodynamic relations in a quasi-equilibrium

manner. If there is a gradient in the Gibbs free energy, then transport will occur to try to minimise the free energy. A gradient in temperature causes heat transport; a gradient in pressure causes fluid flow. A gradient in chemical potential (discussed in Sect. D.4) causes Fickian diffusion.

A simple use of the Gibbs free energy is in determining the Clapeyron relation, which relates melting temperature (or any phase change temperature) to pressure. The Gibbs free energy is $G = H - TS$, and using (D.2), we find (for intensive variables)

$$dg = v\, dp - s\, dT. \tag{D.22}$$

Suppose now that we have a phase boundary between, say, solid and liquid (of the same material), denoted by subscripts s and l. At the phase boundary, equilibrium dictates that $g_s = g_l$, where these are the free energies in the solid and liquid phase. Inequality would cause transport, as we have said. Suppose the melting temperature is T_M, and the system moves to a different temperature and pressure. At the new equilibrium, the perturbations to the free energies must be equal, thus $\Delta g_s = \Delta g_l$, and thus

$$v_s\, \Delta p - s_s\, \Delta T = v_l\, \Delta p - s_l\, \Delta T, \tag{D.23}$$

whence

$$\frac{\Delta T_M}{\Delta p} = \frac{\Delta v}{\Delta s}, \tag{D.24}$$

where

$$\Delta v = v_l - v_s \tag{D.25}$$

is the change of specific volume on melting, and

$$\Delta s = s_l - s_s \tag{D.26}$$

is the change of specific entropy on melting. We define the latent heat to be

$$L = T_M\, \Delta s, \tag{D.27}$$

so that (D.24) takes the form of the Clapeyron equation,

$$\frac{L\, \Delta T_M}{T_M} = \left(\frac{1}{\rho_l} - \frac{1}{\rho_s}\right) \Delta p. \tag{D.28}$$

This relation, or its differential equivalent, describes the form of the phase transition curves which, for ice-water-water vapour, have been drawn in Fig. 2.7.

D.4 Phase Change in Multi-component Materials

Now we consider materials, such as alloys or aqueous solutions, which contain more than one substance. In a sense, we have already introduced this by considering two different phases of a pure material. If we suppose that we have n_i moles of substance i (these are thus extensive variables), then each substance has its own Gibbs free energy, and these contribute additively to the total free energy. The free energy of each phase is called its *chemical potential*, and the chemical potential μ_i of phase i is defined more precisely by asserting that the total Gibbs free energy satisfies

$$dG = V\,dp - S\,dT + \sum_i \mu_i\,dn_i, \tag{D.29}$$

thus

$$\mu_i = \frac{\partial G}{\partial n_i}, \tag{D.30}$$

where the derivative is evidently at constant temperature and pressure. The chemical potential is thus an intensive variable. Suppose we have a solid in equilibrium with a liquid. Since the differential increments in (D.29) are all independent, we can imagine a change of solid i to liquid i, such that $dn_i^L = -dn_i^S$. The consequent change in Gibbs free energy is $(\mu_i^L - \mu_i^S)\,dn_i^L$, and in equilibrium this must be zero. Thus we must have

$$\mu_i^L = \mu_i^S \tag{D.31}$$

at equilibrium, in each component. Just as heat flows down a temperature gradient, so substance is transported down a chemical potential gradient.

For a perfect gas, the specific Gibbs free energy $g(T, p)$ satisfies

$$\left.\frac{\partial g}{\partial p}\right|_T = v = \frac{RT}{p} \tag{D.32}$$

(since $G = ng$ and $pV = nRT$, for n moles of the gas), and thus

$$g = g^0 + RT \ln p. \tag{D.33}$$

In a mixture of gases, the partial pressure of each component gas is that pressure it would have if the other gases were removed. Dalton's law says that the partial pressures are additive, so that their sum is the total pressure of the gas mixture. If we suppose in a mixture that the analogue of (D.33) holds for partial energies and pressures, i.e.,

$$g_i = g_i^0 + RT \ln p_i, \tag{D.34}$$

then since $p_i V = n_i RT$ and g_i is the chemical potential of gas i, we can write

$$\mu_i = \mu_i^0 + RT \ln c_i, \tag{D.35}$$

where c_i is the molar fraction of phase i ($= \frac{n_i}{\sum_i n_i}$). This relation more generally characterises an *ideal* mixture, whether it be of gases, liquids or solids.

Now let us consider an interface (we will think of it as a solid-liquid interface) between the melt and solid of a two component mixture containing substances A

Fig. D.1 The double tangent
construction for c_S and c_L.
The curves are the graphs of
the functions g_S and g_L
defined by (D.38), in which
we define (the units are
arbitrary)
$\mu_A^0(L) = \mu_B^0(L) = RT$,
$\mu_A^0(S) = 1$, $\mu_B^0(S) = 4$. The
figure shows the construction
at $RT = 2.5$. c denotes the
concentration as mole
fraction of A

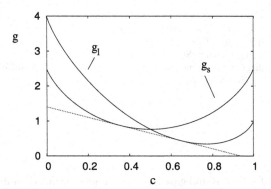

Fig. D.2 Typical phase
equilibrium for an ideal
solution. The same formulae
are used as in constructing
Fig. D.1, with the range
corresponding to $1 \le RT \le 4$

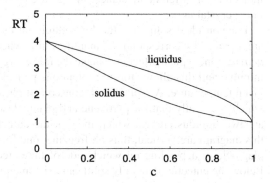

and B. We will suppose the mixture is ideal. At the interface, the chemical potentials
of each component must be equal, thus

$$\mu_A^L = \mu_A^S, \qquad \mu_B^L = \mu_B^S, \qquad \text{(D.36)}$$

and these will determine the interfacial concentrations as functions of temperature.
To be specific, let c denote the molar fraction of component A, so that $1 - c$ is the
molar fraction of B. Then the bulk Gibbs free energies (one in each phase) are

$$g = \mu_A c + \mu_B(1 - c), \qquad \text{(D.37)}$$

and for an ideal solution, we have

$$g = \mu_B^0(1 - c) + \mu_A^0 c + RT\left[c \ln c + (1 - c) \ln(1 - c)\right]. \qquad \text{(D.38)}$$

The two functions g_S and g_L are thus convex upwards functions, and the criterion
for equilibrium as in (D.36) is obtained by drawing a common tangent to g_S and g_L,
as indicated in Fig. D.1, and done in Question D.3; this gives the solid and liquid
concentrations in equilibrium for a particular temperature; as the temperature varies,
we obtain the typical phase diagram shown in Fig. D.2.

Although our discussion is motivated by gases, the concept of an ideal solution
applies equally to liquids and solids. Indeed, Fig. 9.4 shows a phase diagram essentially the same as that in Fig. D.2, for the solid solution of albite and anorthite. As

Fig. D.3 A typical phase
diagram for a mixture
(pyroxene–plagioclase) with
a eutectic point. Such
diagrams are common for
aqueous solutions

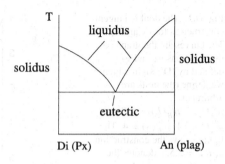

for liquids and gases, ideal solutions occur when there is no penalty for introducing
molecules of different substance. In the case of solids, this means replacing atoms
in the crystal lattice.

For non-ideal solutions, the logarithmic terms such as $\ln c$ in the free energy
are replaced by corresponding quantities $\ln a$, where a is a function of c called the
activity. One typical effect is to make the free energy curves g_S and g_L have multiple
minima, and this allows for more than one pair of liquidus and solidus values at a
given temperature. A typical such consequent phase diagram is shown in Fig. D.3,
which is actually that for pyroxene and plagioclase shown in Fig. 9.12. Here there
are two liquidus curves, which meet at the eutectic point. The solidus curves in
this diagram are vertical, thus on freezing, one forms either pure pyroxene or pure
plagioclase, depending on which side of the eutectic the liquid composition lies.
Below the eutectic point only solid can exist in equilibrium.

D.5 Melting and Freezing

In discussing phase change, we have mostly referred to melting and freezing. In
terms of pure materials, there is no distinction to be made between this, boiling and
condensation (of liquid and gas), and sublimation and condensation (of solid and
gas). A point we will now make is that there is similarly no distinction between the
different corresponding situations which refer to multi-component phase change.
The melting and freezing of an alloy is familiar in industrial contexts (in forming
solid castings) as well as the environment. The simplest example is the case of an
iceberg, consisting of fresh water ice in equilibrium with a slightly salty ocean. Icebergs are of course not formed by freezing the ocean (but sea ice is), but the principle
will serve. Freezing of salty sea water occurs on a diagram similar to Fig. D.3; for a
sufficiently dilute solution, freezing forms more or less pure water ice, with the salt
being rejected into the water. We routinely refer to this as freezing.

D.6 Precipitation and Dissolution

Suppose, however, that we take a salty solution at high temperature. Better, think
of sugar dissolved in water (or tea) at high temperatures. The solubility is greater at

higher temperatures, and if we cool the tea (a lot), eventually the sugar will come out of solution; it precipitates, while at high temperature it dissolves. We do not normally think of this as melting and freezing, but the process is exactly the same. The only difference to the iceberg is that we are on the other side of the eutectic. Now, when we take our saline solution at high salt concentration and high temperature, and lower the temperature, we reach a liquidus on the other side of the eutectic to that of the iceberg; solid salt is frozen (but we say it is precipitated), and the remnant water becomes purer. Or, if we pour salt into water when we cook, it dissolves as we heat the water; we aid the dissolution by stirring, which increases the available surface area for dissolution. We do not think that the salt is melting; but it is. There is no distinction between the processes of melting and freezing of alloys and precipitation and dissolution of solutes.

D.7 Evaporation and Boiling

Surely, however, evaporation and boiling are not the same at all? Evaporation occurs continually at temperatures below the boiling point: we sweat; boiling occurs at a fixed temperature. For water, boiling occurs at 100°C at sea level. But evaporation occurs from oceans at their much lower temperatures. Certainly, on the top of Mount Everest, boiling temperature is reduced, but this is because the pressure is lower, and occurs through the Clapeyron effect.

So then, what is evaporation? The saturation vapour pressure of water vapour, p_{sv}, is a function of temperature, given by the solution of (2.56), and it increases to a pressure of one bar (sea level atmospheric pressure) at a temperature of 100°C, where boiling occurs spontaneously.

It is all, in fact, the same story. The ocean, let us say, is pure water (ignore salt). The atmosphere is a two component mixture (let us say) of water vapour and air; it is alloy. If we take a hot atmosphere and reduce its temperature, condensation occurs at a temperature which depends on atmospheric composition. The molar fraction of water vapour in the atmosphere is just p_v/p_a, the vapour pressure divided by the atmospheric pressure. On what would be the liquidus (but now must be the vaporus[2]), the vapour pressure has its saturation value, the molar fraction of water vapour is $p_{sv}/p_a = c_{sv}$, and the saturation temperature T_s is a function of c_{sv}. What has boiling to do with this? Not much! Evaporation *is* boiling. What we normally call boiling refers to the position of the vaporus when $c_{sv} = 1$, i.e. $p_{sv} = p_a$. For given atmospheric pressure, we cannot raise the liquid temperature beyond the vaporus temperature at vapour concentration of one. If we change atmospheric pressure, then this temperature will change. Yes, because of Clapeyron, but also because pressure dictates concentration. Gases are different because the amount of gas depends on pressure. For liquids and solids, this is mostly not the case.

---

[2]Solidus is a perfectly good Latin word, but liquidus is not; vaporus is invented here.

D.8 Chemical Reactions

Surely chemical reactions are different? So it would appear. If we pour vinegar (acetic acid) into a kettle furred up with limescale (calcium carbonate), the limescale will dissolve, or react, forming carbon dioxide in the process. In a coal fire, the carbon in the coal reacts with oxygen, forming carbon dioxide. There is no equilibrium surface or phase diagram here, surely?

But in fact the difference is only one of degree. When a salt M dissolves in water to the point of saturation, the equilibrium that results is a consequence of a simple reversible reaction

$$M^S \underset{k_P}{\overset{k_D}{\rightleftharpoons}} M^L,$$ (D.39)

where k_D is the rate of dissolution and k_P is the rate of precipitation. The fact that there is an equilibrium is a consequence of the reversibility. The only effective difference between this and a chemical reaction is that the examples cited above are almost irreversible. If we burn coal in a sealed environment, the carbon reacts with the oxygen to form a mixed atmosphere of O_2 with CO_2, just as when we evaporate water vapour in air. If the reaction is reversible, then an equilibrium will be obtained. In practice (in this example) the backward reaction rate is negligible, and so the equilibrium which obtains occurs when the coal is (almost) entirely used up. Chemical reaction is thus the process describing the evolution towards thermodynamic equilibrium.

D.9 Surface Energy

Interfaces between two materials, be they both fluids, fluid and solid, or any other such combination, carry a surface energy per unit area, denoted γ. The existence of a surface energy causes a pressure jump across the interface, and the requirement of force balance (Newton's third law) on the massless interface means that the interface appears to carry a tension, the surface tension. To see how the surface energy induces this pressure jump, we consider equilibrium of a system containing an interface. For example, we may think of a box containing fluid with a gas bubble in it. To change the surface area of the interface, we may alter the external pressure, and thus the equilibrium is that associated with constant volume and temperature, for which the relevant minimum is obtained by the Helmholtz free energy F. The basic recipe for an increment of F for each phase follows from (D.1) and (D.2), and is

$$dF = -p\,dV - S\,dT;$$ (D.40)

when the surface area of a phase interface has a surface energy per unit area γ, then a change in surface area dA causes an additional contribution $\gamma\,dA$, which must also be included. Suppose the two sides of the interface are denoted by subscripts $-$ and $+$, and have corresponding pressures p_- and p_+. For an isothermal change at constant total volume, $dV_- = -dV_+$, and thus the increment of the total Helmholtz free energy of the system is

$$dF = -p_- dV_- - p_+ dV_+ + \gamma \, dA = -(p_- - p_+) dV_- + \gamma \, dA = 0, \quad \text{(D.41)}$$

and thus

$$p_- - p_+ = \gamma \frac{\partial A}{\partial V_-}. \quad \text{(D.42)}$$

This determines the pressure jump at the interface. It is a result of differential geometry that $\frac{\partial A}{\partial V_-} = 2\kappa$, where κ is the mean curvature of the surface (the average of the two principal curvatures); for example the mean curvature κ of a spherical surface measured from the side on which the centre of the sphere lies is just $1/R$, where R is the sphere radius.

D.9.1 The Gibbs–Thomson Effect

The curvature of an interface also has an effect on the melting temperature, and this is known as the Gibbs–Thomson effect. For this we may go back to the Clapeyron type argument and specific Gibbs free energy of each phase (i.e., their chemical potentials). Denoting these as before as g_s and g_l, but now allowing solid and liquid pressures to change independently, we have

$$v_s \Delta p_s - s_s \Delta T = v_l \Delta p_l - s_l \Delta T, \quad \text{(D.43)}$$

and with $L = T_M \Delta s$ being the latent heat, we have the generalised Clapeyron relation

$$\frac{L \Delta T_M}{T_M} = \left(\frac{1}{\rho_l} - \frac{1}{\rho_s} \right) \Delta p_l - \frac{(p_s - p_l)}{\rho_s}, \quad \text{(D.44)}$$

in which the first term on the right is the Clapeyron effect of changing pressure, and the second is the Gibbs–Thomson effect, which describes change of melting temperature with surface curvature, since $p_s - p_l = 2\gamma\kappa$, with the curvature measured from the solid side of the interface.

D.10 Pre-melting

It is commonly the case that a solid will maintain a thin liquid film of its melt at an interface with, for example, a quartz grain, even at temperatures below the freezing point. This phenomenon is known as 'pre-melting' (Dash et al. 2006; Wettlaufer and Worster 2006), and is associated with an excess free energy manifested by very thin films due to a variety of intermolecular forces, for example Van der Waals forces. The scale on which these forces act is measured in molecular diameters, and so the film thicknesses over which these free energy effects are important are of the order of nanometres. Just as for surface energy, pre-melting causes an excess pressure, called the *disjoining pressure*, to occur in the film, and it causes a displacement of the freezing temperature. A particular geophysical problem in which this disjoining pressure is important is in the phenomenon of frost heave (Rempel et al. 2004), wherein freezing soil is uplifted, causing the heave which can be very damaging to roads and structures. The force generated in frost heave can be very large, of the order of bars, and this force is due to the disjoining pressure in the thin water films which separate the ice from the soil grains.[3]

To understand the dynamic effects, we consider a thin film of thickness h separating an ice surface from a foreign solid surface. In the absence of the film, the ice-solid interface has a surface energy which we denote by γ_{si}, while the interposition of a liquid film creates two new surfaces, of interfacial energies γ_{sw} (solid-water) and γ_{iw} (ice-water). In addition, the liquid film has a Gibbs free energy per unit area of the form

$$G = \rho_l \mu_l h + \Phi(h), \tag{D.45}$$

where μ_l is the chemical potential energy of the bulk liquid, and Φ is the free energy associated with intermolecular forces. In particular, we suppose

$$\Phi(0) = \gamma_{si}, \qquad \Phi(\infty) = \gamma_{sw} + \gamma_{iw}; \tag{D.46}$$

the liquid film is energetically preferred if $\Delta\gamma < 0$, where

$$\Delta\gamma = \gamma_{sw} + \gamma_{iw} - \gamma_{si}, \tag{D.47}$$

and it is in this case that a positive disjoining pressure occurs. We write

$$\Phi = \gamma_{si} + \Delta\gamma \phi(h), \tag{D.48}$$

where ϕ increases monotonically from zero at $h = 0$ to one at $h = \infty$. For example, Van der Waals forces lead to a form for ϕ of

$$\phi = \left[1 - \frac{\sigma^2}{h^2} \right]_+, \tag{D.49}$$

where the constant σ is of the order of a molecular diameter. Clearly, if $\Delta\gamma < 0$, then Φ is a monotonically decreasing function of h, while the bulk free energy is an increasing function, and thus a minimum of G in (D.45) will be obtained when h is finite, if $|\Delta\gamma|$ is sufficiently large. This causes the wetting film.

---

[3]This is perhaps an inverted way of looking at it. Heaving requires the maintenance of the film between ice and soil grains; as long as the film is maintained, heave will occur. The presence of a large overburden pressure will eventually suppress heave, but the necessary pressures are large.

D.10.1 Disjoining Pressure

We consider the Helmholtz free energy of a film of thickness h. Following a small perturbation to the film thickness,

$$dF = -p_w \, dV_w - p_i \, dV_i - S \, dT + A \, d\Phi, \tag{D.50}$$

where A is surface area. We have $dV_w = A \, dh$; for an isothermal change at constant volume $dV_w = -dV_i = A \, dh$, and therefore

$$p_i - p_w = -\Phi'(h) = -\Delta\gamma\phi'(h); \tag{D.51}$$

this is the disjoining pressure. For (D.49), this leads to

$$p_i - p_w = -\frac{\mathcal{A}}{6\pi h^3}, \tag{D.52}$$

where $\mathcal{A}$ is the Hamaker constant

$$\mathcal{A} = 12\pi\sigma^2\Delta\gamma. \tag{D.53}$$

D.10.2 Freezing Point Depression

Finally we consider the effect of a thin film on the freezing point. This simply follows from (D.44), which we write in the form

$$\frac{L\Delta T}{T_M} = (v_w - v_i)\Delta p_w - v_i\Delta(p_i - p_w), \tag{D.54}$$

and thus, from (D.51), (ignoring liquid pressure variations)

$$\frac{L(T - T_M)}{T_M} \approx \frac{\Delta\gamma \, \phi'(h)}{\rho_i}. \tag{D.55}$$

For $\Delta\gamma < 0$, this represents the freezing point depression due to pre-melting only; the Clapeyron and Gibbs–Thomson effects can be added to the right hand side. Because $\phi' \propto \frac{1}{h^3}$, these thin films can be maintained to temperatures quite a way below the normal freezing point.

D.11 Liesegang Rings

As discussed in Chap. 9, Liesegang rings can form when crystals are precipitated in a dilute solution. Liesegang himself put some silver nitrate on a gel containing potassium dichromate, and the resulting silver dichromate crystals precipitate in bands. In this section, we describe a model due to Keller and Rubinow (1981) which aims to explain the phenomenon, based on the earlier ideas of Ostwald.

Keller and Rubinow consider the reaction scheme

$$A + B \underset{k_-}{\overset{k_+}{\rightleftharpoons}} C \overset{p}{\to} D, \tag{D.56}$$

in which A would represent the silver nitrate seed crystal, B would be the dilute dichromate solution, C is the reaction product silver dichromate, and D is the solid precipitate. In one dimension, a suitable set of equations is

$$\begin{aligned} a_t &= D_A a_{xx} - r, \\ b_t &= D_B b_{xx} - r, \\ c_t &= D_C c_{xx} + r - p, \\ d_t &= p, \end{aligned} \tag{D.57}$$

where r is the reaction rate and p is the precipitation rate, given respectively by

$$r = k_+ ab - k_- c, \tag{D.58}$$

and

$$p = \begin{cases} q(c - c_s) & \text{if } c \geq c_n \ (> c_s) \text{ or } d > 0, \\ 0 & \text{if } c < c_n \text{ and } d = 0, \end{cases} \tag{D.59}$$

where c_s is the saturation concentration of C and c_n is the required supersaturation for nucleation. Let us suppose that $D_B = D_C$, and the reaction is very fast, so that $r \approx 0$. Then

$$c \approx Kab, \tag{D.60}$$

where

$$K = \frac{k_+}{k_-}. \tag{D.61}$$

Suitable initial conditions are

$$a = 0, \qquad b = b_0, \qquad c = d = 0 \quad \text{at } t = 0, \tag{D.62}$$

and suitable boundary conditions are

$$a = a_0, \qquad b_x = c_x = 0 \quad \text{at } x = 0. \tag{D.63}$$

Adding the equations for b and c, and defining $B = b + c$, we obtain

$$B_t = D_B B_{xx} - p, \tag{D.64}$$

and in addition (D.60) implies

$$c = AB, \tag{D.65}$$

where

$$A = \frac{Ka}{1 + Ka}. \tag{D.66}$$

Keller and Rubinow assume that the reaction term r can be neglected in the equation for a, essentially on the basis that if $b_0 \ll a_0$ (the dichromate is very dilute), then very little A is removed in forming the product. In this case A simply diffuses away from the seed crystal, providing an expression for a as

$$a = a_0 \operatorname{erfc}\left(\frac{x}{2\sqrt{D_A t}}\right). \tag{D.67}$$

It is convenient to scale the equations, and we therefore choose the scales

$$d, c, B \sim b_0, \qquad p \sim q b_0, \qquad t \sim \frac{1}{q}, \qquad x \sim \sqrt{\frac{D_B}{q}}; \tag{D.68}$$

then the dimensionless model is

$$\begin{aligned} B_t &= B_{xx} - p, \\ d_t &= p, \end{aligned} \tag{D.69}$$

where

$$p = \begin{cases} AB - A_s & \text{if } AB \geq A_n \ (> A_s) \text{ or } d > 0, \\ 0 & \text{if } AB < A_n \text{ and } d = 0, \end{cases} \tag{D.70}$$

where we define

$$c_s = b_0 A_s, \qquad c_n = b_0 A_n. \tag{D.71}$$

The function A is given by

$$A = \frac{\kappa \operatorname{erfc} \theta}{1 + \kappa \operatorname{erfc} \theta}, \qquad \theta = \frac{\beta x}{2\sqrt{t}}, \tag{D.72}$$

in which

$$\kappa = K a_0, \qquad \beta = \sqrt{\frac{D_B}{D_A}}. \tag{D.73}$$

Note that A is a monotonically $O(1)$ decreasing function of θ, which tends to zero at infinity. The initial and boundary conditions are

$$\begin{aligned} B = 1, \quad p = 0 \quad &\text{at } t = 0; \\ B_x = 0 \quad &\text{at } x = 0, \\ B \to 1 \quad &\text{as } x \to \infty. \end{aligned} \tag{D.74}$$

It should be noted that since the time scale is that of precipitation (and thus quite fast in the laboratory), we can expect the length and time scales to be small, so that large space and time solutions of this model are of interest.

D.11.1   Central Precipitation

The maximum value of $A = \frac{\kappa}{1+\kappa}$ is at $\theta = 0$, and thus precipitation will begin at $x = 0$ providing $\frac{\kappa}{1+\kappa} > A_n$; we presume this to be the case. Keller and Rubinow give an ingenious (but heuristic) approximate solution for their model, which we now emulate. Initially, there is a central precipitating region $0 < x < R(t)$, where $p > 0$, and $p = 0$ outside this. First, they suppose that A is slowly varying in space, and that R is slowly varying in time, so that a quasi-static solution is appropriate. Since B is continuous at R, then $AB = A_n$ there, and this solution is

$$AB = A_s + \frac{(A_n - A_s)\cosh(\sqrt{A}x)}{\cosh(\sqrt{A}R)}. \tag{D.75}$$

For $x > R$, a stationary solution is not possible, but for slowly varying R,

$$B = 1 + \left(\frac{A_n}{A} - 1\right)\operatorname{erfc}\left\{\frac{x - R}{2\sqrt{t}}\right\}. \tag{D.76}$$

Equating the derivatives B_x at $R\pm$, we find that R is determined by the relation

$$\sqrt{A} - \frac{A_n}{\sqrt{A}} = (A_n - A_s)\sqrt{\pi t}\tanh(\sqrt{A}R), \tag{D.77}$$

in which $A(\theta)$ is given by (D.72), with

$$\theta = \frac{\beta R}{2\sqrt{t}}. \tag{D.78}$$

To solve this, we define

$$u = \sqrt{A}R, \tag{D.79}$$

and then (D.77) can be written in the form

$$u \tanh u = \frac{2\theta}{\beta\sqrt{\pi}}\left\{\frac{A(\theta) - A_n}{A_n - A_s}\right\}. \tag{D.80}$$

The right hand side is a unimodal (one-humped) function of θ, while $u\tanh u$ is an increasing function of u. Therefore $u(\theta)$ is a positive function in the range $0 < \theta < \theta_n$, where $A(\theta_n) = A_n$. Consulting (D.77), we see that initially $A = A_n$ and thereafter increases with t. Therefore, initially $\theta = \theta_n$ and decreases with increasing t. Since A is increasing as is R, u must increase, but it cannot do so indefinitely, because of the maximum value of $u(\theta)$. In consequence, there is a finite time t^* when R reaches a maximum R^*, and the solution cannot be continued beyond this time.

Keller and Rubinow go on to suggest that a sequence of precipitation bands will subsequently form, and they analyse these based on the same approximating solutions. The question arises, whether there is any rational basis for supposing that their approximation method is valid.

The two principal assumptions in the solution method are that A is slowly varying in space for $x < R$, and that R is slowly varying in time. The first of these requires θ defined by (D.78) to be small, and since A ranges from

$$A_0 = \frac{\kappa}{1+\kappa} \tag{D.81}$$

to A_n at $x = R$, this requires

$$\delta = \frac{\kappa}{1+\kappa} - A_n \ll 1. \tag{D.82}$$

The assumption that R is slowly varying, i.e., that the time derivative in (D.69)$_1$ can be ignored, requires $t \gg x^2 \sim R^2$, and thus, from (D.78), $\theta \ll \beta$. Assuming $\beta \sim O(1)$, as seems likely, this condition is included by (D.82).

We write

$$\theta = \delta\Theta, \tag{D.83}$$

and then (D.80) can be approximated by

$$u^2 \approx \frac{2\delta^2\Theta(1 - a'\Theta)}{\beta\sqrt{\pi}(A_n - A_s)}, \tag{D.84}$$

where

$$a' = \frac{2\kappa}{\sqrt{\pi}(1+\kappa)^2}. \tag{D.85}$$

From (D.78) and (D.79), we then find

$$R \approx \frac{\delta\kappa\sqrt{\pi t}}{A_0\{\beta A_0 + (A_n - A_s)\pi\kappa t\}}, \tag{D.86}$$

and R reaches its maximum

$$R^* = \frac{\delta}{2A_0}\left\{\frac{\kappa}{\beta A_0(A_n - A_s)}\right\}^{1/2} \tag{D.87}$$

at time

$$t^* = \frac{\beta A_0}{\pi\kappa(A_n - A_s)}. \tag{D.88}$$

These results provide a basis for a direct asymptotic approach, based, for example, on the limit $\delta \ll 1$, with the other parameters being taken as $O(1)$.

D.12 Exercises

D.1 The density ρ, velocity $\mathbf{u}$ and internal energy e of a fluid are given by the conservation laws

$$\frac{\partial \rho}{\partial t} + \nabla.(\rho \mathbf{u}) = 0,$$

$$\frac{\partial \rho u_i}{\partial t} + \nabla.(\rho u_i \mathbf{u}) = \nabla.\boldsymbol{\sigma}_i + \rho f_i,$$

$$\frac{\partial}{\partial t}\left[\frac{1}{2}\rho u^2 + \rho e + \rho \chi\right] + \nabla.\left[\left\{\frac{1}{2}\rho u^2 + \rho e + \rho \chi\right\}\mathbf{u}\right] = \nabla.(\sigma_i u_i) - \nabla.\mathbf{q},$$

where $\boldsymbol{\sigma}_i = \sigma_{ij}\mathbf{e}_j$, $\mathbf{q}$ is the heat flux, and the conservative body force $\mathbf{f}$ is defined by

$$\mathbf{f} = -\nabla\chi,$$

where χ is the potential.

Show that the momentum equation can be written in the form

$$\rho\left[\frac{\partial u_i}{\partial t} + \mathbf{u}.\nabla u_i\right] = \frac{\partial \sigma_{ij}}{\partial x_j} + \rho f_i,$$

and that the energy equation can be reduced to

$$\rho\frac{de}{dt} = \sigma_{ij}\dot{\varepsilon}_{ij} - \nabla.\mathbf{q}.$$

D.2 The perfect gas law may be written in the form

$$v = \frac{RT}{Mp},$$

where R is the gas constant, and M is the molecular weight. Show that $\beta = \frac{1}{T}$, and deduce that for a perfect gas,

$$dh = c_p\,dT,$$

where h is specific enthalpy.

Use the relation

$$de = T\,ds - p\,dv$$

and the definition of the specific heat at constant volume,

$$c_v = T\left(\frac{\partial s}{\partial T}\right)_v,$$

to show that

$$de = c_v\, dT - p\, dv + T\left(\frac{\partial p}{\partial T}\right)_v dv$$

[*hint*: use the Maxwell relations]. Hence show that, for a perfect gas,

$$de = c_v\, dT.$$

D.3 The functions $g_S(c)$ and $g_L(c)$ are defined by

$$g = Ac + B(1 - c),$$

for coefficients A_S and B_S, and A_L and B_L, respectively, and these are defined by

$$A = a + RT \ln c,$$
$$B = b + RT \ln(1 - c),$$

with similar subscripting S, L of the coefficients a and b.

Show that the conditions $A_L = A_S$ and $B_L = B_S$ are solved by values c_L, c_S which satisfy

$$g_S'(c_S) = g_L'(c_L) = \frac{g(c_S) - g(c_L)}{c_S - c_L}.$$

Appendix E
Averaged Equations in Two Phase Flow

E.1 Discontinuities and Jump Conditions

Suppose we have a conservation law of the form

$$\frac{\partial \phi}{\partial t} + \nabla . \mathbf{F} = 0, \tag{E.1}$$

which is derived from the integral conservation law

$$\frac{d}{dt} \int_V \phi \, dV = -\int_{\partial V} \mathbf{F}.\mathbf{n} \, dS. \tag{E.2}$$

From first principles we can derive the jump condition across surfaces where ϕ and $\mathbf{F}$ are discontinuous:

$$[\phi]_-^+ = V_n [F_n]_-^+, \tag{E.3}$$

where $+$ and $-$ refer to the values either side of the surface of discontinuity, and $\mathbf{n}$ is the unit normal at this surface (pointing either way); V_n is the speed of the surface in the direction of the normal, and $F_n = \mathbf{F}.\mathbf{n}$. In the common case of a fluid in motion, where the conservation law takes the form

$$\frac{\partial \phi}{\partial t} + \nabla . (\phi \mathbf{u}) = \nabla . \mathbf{J}, \tag{E.4}$$

the corresponding jump condition is

$$\left[\phi(u_n - V_n) - J_n \right]_-^+ = 0. \tag{E.5}$$

The basic equations of conservation of mass, momentum and energy for a fluid with density ρ, velocity $\mathbf{u}$ and internal energy e were given in (D.12), and are repeated here:

$$\frac{\partial \rho}{\partial t} + \nabla . (\rho \mathbf{u}) = 0,$$

A. Fowler, *Mathematical Geoscience*, Interdisciplinary Applied Mathematics 36, DOI 10.1007/978-0-85729-721-1, © Springer-Verlag London Limited 2011

$$\frac{\partial \rho u_i}{\partial t} + \nabla.(\rho u_i \mathbf{u}) = \nabla.\sigma_i + \rho f_i, \qquad (E.6)$$

$$\frac{\partial}{\partial t}\left[\frac{1}{2}\rho u^2 + \rho e + \rho\chi\right] + \nabla.\left[\left\{\frac{1}{2}\rho u^2 + \rho e + \rho\chi\right\}\mathbf{u}\right] = \nabla.(\sigma_i u_i) - \nabla.\mathbf{q};$$

in the last equation, χ is the potential energy. The corresponding jump conditions are

$$\left[\rho(u_n - V_n)\right]_-^+ = 0,$$

$$\left[\rho u_i(u_n - V_n) - \sigma_{in}\right]_-^+ = 0, \qquad (E.7)$$

$$\left[\left\{\frac{1}{2}\rho u^2 + \rho e + \rho\chi\right\}(u_n - V_n)\right]_-^+ = [\sigma.\mathbf{u}.\mathbf{n} - q_n]_-^+.$$

Note that these jump conditions are implied automatically by the integral forms of the conservation laws, assuming there is no production at the surface (e. g., of energy by a surface reaction). Therefore the integral forms can be applied directly to find the total mass, momentum and energy conservation laws for a two phase flow in which the density and energy in particular may be discontinuous.

Let us define the interfacial source term

$$\Gamma = \Gamma_- = -\left[\rho(u_n - V_n)\right]_-, \qquad (E.8)$$

where we define the unit normal $\mathbf{n}_-$ here to be pointing from the $-$ phase towards the $+$ phase. If we suppose that there is no slip across an interface, $[\mathbf{u}.\mathbf{t}]_-^+ = 0$, where $\mathbf{t}$ is any tangent vector at the interface, then the momentum jump condition $(E.7)_2$ implies

$$[\sigma_{nt}]_-^+ = 0, \qquad [\sigma_{nn}]_-^+ = -\Gamma[u_n]_-^+, \qquad (E.9)$$

and the energy jump condition becomes

$$\Gamma\left[\frac{1}{2}u^2 + e\right]_-^+ + [\sigma_{nn}u_n - q_n]_-^+ = 0, \qquad (E.10)$$

since we take the potential energy χ to be continuous.

E.2   Averaging Methods

Next, we consider the derivation of averaged equations for two-phase flows. This is a subject which has been the subject of a number of different investigations, see for example Ishii (1975) and Drew and Passman (1999), and also the thorough overview by Drew and Wood (1985). Averaging proceeds as in the derivation of averaged equations for turbulent flows (see Sect. B.1), but the choice of average is not clear cut. A local space average seems the most obvious choice, but only for homogeneous flows. A local time average is a better choice, but in fact preference is usually given to the ensemble average over a number of realisations of the flow. For stationary flows, this is likely to be equivalent to a local time average.

Further complication arises since often one is concerned with axial flows in a pipe (for example in a volcanic vent), where a cross-sectional average is appropriate either as well as, or instead of a local time average. There seem to be few examples where two-phase models in two or three dimensions are proposed.

There are various different ways to derive averaged equations. We follow Drew and Wood (see also Fowler 1997) in using an indicator function X_k which is equal to one in phase k ($k = 1, 2$) and zero otherwise. We denote averages by overbars, and the averaged equations are obtained by multiplying the point forms of the governing equations by X_k and then averaging. This procedure introduces derivatives of the piecewise continuous X_k, and these must be interpreted using generalised functions. To see how this works, consider a general conservation law of the form

$$\frac{\partial}{\partial t}(\rho \psi) + \mathbf{V}.(\rho \psi \mathbf{u}) = -\mathbf{V}.\mathbf{J} + \rho f, \tag{E.11}$$

where ψ is the conserved quantity (per unit mass), $\mathbf{u}$ is the fluid velocity, $\mathbf{J}$ is the flux, and f is a volumetric source. Multiplying by X_k and averaging yields the exact equation

$$\frac{\partial}{\partial t}(\overline{X_k \rho \psi}) + \mathbf{V}.[\overline{X_k \rho \psi \mathbf{u}}]$$

$$= -\mathbf{V}.[\overline{X_k \mathbf{J}}] + \overline{X_k \rho f}$$

$$+ \overline{\rho \psi \left\{ \frac{\partial X_k}{\partial t} + \mathbf{u}_i.\mathbf{V} X_k \right\}} + \overline{\{\rho \psi(\mathbf{u} - \mathbf{u}_i) + \mathbf{J}\}.\mathbf{V} X_k}, \tag{E.12}$$

where $\mathbf{u}_i$ is the velocity of the interface between the phases, and we assume that $\overline{\mathbf{V} f} = \mathbf{V} \bar{f}$, $\overline{\partial f/\partial t} = \partial \bar{f}/\partial t$, which will be the case for sufficiently well-behaved f. Derivatives of X_k are interpreted as generalised functions. Thus, for example, $\mathbf{j}.\mathbf{V} X_k$ is defined, for any smooth test function ϕ which vanishes at infinity, through the identity

$$\int_V \phi \mathbf{j}.\mathbf{V} X_k \, dV = -\int_V X_k \mathbf{V}.(\phi \mathbf{j}) \, dV = -\int_{V_k} \mathbf{V}.(\phi \mathbf{j}) \, dV = -\int_{S_k} \phi j_n \, dS, \tag{E.13}$$

where j_n is the normal component of $\mathbf{j}$ at the interface, pointing *away* from phase k. This suggests that $\overline{\mathbf{j}.\mathbf{V} X_k}$ can be identified with the specific surface average of $-\mathbf{j}.\mathbf{n}$, which is consistent with the fact that $\mathbf{V} X_k$ is essentially a delta function centred on the interface.

To interpret the interfacial advective derivative of X_k, we have, for test functions $\phi(\mathbf{x}, t)$ which vanish both at $\mathbf{x} \to \infty$ and $t \to \pm\infty$,

$$\int\int \phi \left[\frac{\partial X_k}{\partial t} + \mathbf{u}_i . \nabla X_k \right] dV \, dt$$

$$= -\int\int X_k \left[\frac{\partial \phi}{\partial t} + \mathbf{u}_i . \nabla \phi \right] dV \, dt$$

$$= -\int_{-\infty}^{\infty} \int_{V_k(t)} \left[\frac{\partial \phi}{\partial t} + \mathbf{u}_i . \nabla \phi \right] dV \, dt$$

$$= -\int_{-\infty}^{\infty} \frac{d}{dt} \int_{V_k(t)} \phi \, dV \, dt = -\left[\int_{V_k(t)} \phi \, dV \right]_{-\infty}^{\infty} = 0. \qquad (E.14)$$

The averaged form of (E.12) is now derived in terms of the averaged volume fraction α_k, density ρ_k, velocity $\mathbf{u}_k$, species ψ_k, flux $\mathbf{J}_k$, and source f_k, defined as

$$\alpha_k = \overline{X_k}, \qquad \alpha_k \rho_k = \overline{X_k \rho}, \qquad \alpha_k \rho_k \mathbf{u}_k = \overline{X_k \rho \mathbf{u}},$$

$$\alpha_k \rho_k \psi_k = \overline{X_k \rho \psi}, \qquad \alpha_k \mathbf{J}_k = \overline{X_k \mathbf{J}}, \qquad \alpha_k \rho_k f_k = \overline{X_k \rho f}, \qquad (E.15)$$

and the conservation law (E.12) then takes the form

$$\frac{\partial}{\partial t}(\alpha_k \rho_k \psi_k) + \nabla . \{\alpha_k \rho_k \psi_k (\mathbf{u}_k + \mathbf{U}_k^\psi)\}$$

$$= -\nabla . (\alpha_k \mathbf{J}_k) + \alpha_k \rho_k f_k + \overline{\{\rho\psi(\mathbf{u} - \mathbf{u}_i) + \mathbf{J}\} . \nabla X_k}, \qquad (E.16)$$

where the profile velocity $\mathbf{U}_k^\psi$ is defined by

$$\alpha_k \rho_k \psi_k \mathbf{U}_k^\psi = \overline{X_k \rho \psi \mathbf{u}} - \alpha_k \rho_k \psi_k \mathbf{u}_k. \qquad (E.17)$$

In one-dimensional flows, $\mathbf{U}_k^\psi = (D_k^\psi - 1)\mathbf{u}_k$, and D_k^ψ is called a profile coefficient. Apart from this, the last term in (E.16), representing interfacial transfer of ψ, must be constituted.

E.3 Mass and Momentum Equations

Conservation of mass is determined from (E.11) by putting

$$\psi = 1, \qquad \mathbf{J} = \mathbf{0}, \qquad f = 0. \qquad (E.18)$$

The corresponding equations for each phase are, from (E.16), with $\psi_k = 1$, $\mathbf{J}_k = \mathbf{0}$ and $\mathbf{U}_k^1 = \mathbf{0}$,

$$\frac{\partial}{\partial t}(\alpha_k \rho_k) + \nabla . [\alpha_k \rho_k \mathbf{u}_k] = \Gamma_k, \qquad (E.19)$$

where

$$\Gamma_k = \overline{\rho(\mathbf{u} - \mathbf{u}_i).\nabla X_k}, \tag{E.20}$$

and Γ represents a mass source due to phase change (without which $\mathbf{u} = \mathbf{u}_i$ at the interface).

Next, consider momentum conservation. With appropriate interpretation of tensor notation, we put

$$\psi = \mathbf{u}, \qquad \mathbf{J} = -\mathbf{T} \equiv p\mathbf{I} - \tau, \qquad f = \mathbf{g}, \tag{E.21}$$

where p is the pressure, τ is the deviatoric stress tensor, and $\mathbf{g}$ is gravity. In addition, we write

$$\overline{X_k \rho \mathbf{u} \mathbf{u}} = \alpha_k \rho_k \mathbf{u}_k \mathbf{u}_k - \alpha_k \mathbf{T}'_k; \tag{E.22}$$

the second term can be interpreted as (minus) the Reynolds stress (cf. (B.7)). The momentum equation can thus be written as

$$\frac{\partial}{\partial t}(\alpha_k \rho_k \mathbf{u}_k) + \nabla.[\alpha_k \rho_k \mathbf{u}_k \mathbf{u}_k] = \nabla.\left[\alpha_k(\mathbf{T}_k + \mathbf{T}'_k)\right] + \alpha_k \rho_k \mathbf{g} + \mathbf{M}_k + \mathbf{u}^m_{ki} \Gamma_k, \tag{E.23}$$

where

$$\alpha_k \mathbf{T}_k = \overline{X_k(-p\mathbf{I} + \tau)}, \qquad \mathbf{M}_k = \overline{(p\mathbf{I} - \tau).\nabla X_k},$$

$$\mathbf{u}^m_{ki} = \frac{\overline{\rho \mathbf{u}(\mathbf{u} - \mathbf{u}_i).\nabla X_k}}{\overline{\rho(\mathbf{u} - \mathbf{u}_i).\nabla X_k}}. \tag{E.24}$$

We define the average pressure and deviatoric stress in phase k to be

$$p_k = \frac{\overline{X_k p}}{\alpha_k}, \qquad \tau_k = \frac{\overline{X_k \tau}}{\alpha_k}. \tag{E.25}$$

It is conventional to separate the local interfacial stresses from those due to large scale variations in α_k by writing the interfacial momentum source as

$$\mathbf{M}_k = p_{ki}\nabla \alpha_k + \mathbf{M}'_k, \tag{E.26}$$

where

$$\mathbf{M}'_k = \overline{(p - p_{ki})\nabla X_k} - \overline{\tau.\nabla X_k}, \tag{E.27}$$

p_{ki} is the average interfacial pressure in phase k, and we use the fact that $\overline{\nabla X_k} = \nabla \alpha_k$. Thus the momentum equation can be written as

$$\frac{\partial}{\partial t}(\alpha_k \rho_k \mathbf{u}_k) + \nabla.(\alpha_k \rho_k \mathbf{u}_k \mathbf{u}_k) = -\alpha_k \nabla p_k + (p_k - p_{ki})\nabla \alpha_k + \nabla.[\alpha_k \tau_k]$$

$$+ \nabla.[\alpha_k \mathbf{T}'_k] + \alpha_k \rho_k \mathbf{g} + \mathbf{M}'_k + \mathbf{u}^m_{ki} \Gamma_k. \tag{E.28}$$

Often we may ignore the Reynolds stresses as well as the macroscopic viscous stresses, and if we ignore surface energy effects, we may take $p_k = p_{ki}$. The term $\mathbf{M}'_k$ is the interfacial force, and is generally much larger than the other stress terms. In this case, the momentum equation becomes

$$\frac{\partial}{\partial t}(\alpha_k \rho_k \mathbf{u}_k) + \nabla.(\alpha_k \rho_k \mathbf{u}_k \mathbf{u}_k) = -\alpha_k \nabla p_k + \alpha_k \rho_k \mathbf{g} + \mathbf{M}'_k + \mathbf{u}^m_{ki} \Gamma_k. \qquad (E.29)$$

The interfacial force $\mathbf{M}'_k$ includes the important interfacial drag, as well as other forces, in particular the virtual mass force. Interfacial drag is due to friction at the interface, while virtual mass terms are associated with relative acceleration. There are various other forces which are sometimes included, also (see Drew and Wood 1985). The momentum source from phase change $\mathbf{u}^m_{ki} \Gamma_k$ is often ignored. In conditions of slow flow, constitution of the interfacial drag as a term proportional to the velocity difference between the phases leads to Darcy's law.

E.4   Energy Equation

The point form of the energy equation is given in (E.6) or (D.12); we use the form of (D.14), specifically

$$\frac{\partial}{\partial t}(\rho e) + \nabla.(\rho e \mathbf{u}) = -\nabla.\mathbf{q} + \mathbf{T}:\nabla \mathbf{u}. \qquad (E.30)$$

To derive the averaged version, we put $\psi = e$, $\mathbf{J} = \mathbf{q}$, $\rho f = \mathbf{T}:\nabla \mathbf{u}$ in (E.12). By analogy with (E.22), we define the turbulent heat transport $\mathbf{q}'_k$ via

$$\overline{X_k \rho e \mathbf{u}} = \alpha_k \rho_k e_k \mathbf{u}_k + \alpha_k \mathbf{q}'_k; \qquad (E.31)$$

we then obtain (cf. (E.16)) the averaged energy equation

$$\frac{\partial}{\partial t}(\alpha_k \rho_k e_k) + \nabla.\{\alpha_k \rho_k e_k \mathbf{u}_k\} = -\nabla.\{\alpha_k(\mathbf{q}_k + \mathbf{q}'_k)\} + \alpha_k D_k + e_{ki} \Gamma_k + E_k, \qquad (E.32)$$

where

$$D_k = \frac{\overline{X_k \mathbf{T}:\nabla \mathbf{u}}}{\alpha_k}, \qquad e_{ki} = \frac{\overline{\rho e(\mathbf{u} - \mathbf{u}_i).\nabla X_k}}{\overline{\rho(\mathbf{u} - \mathbf{u}_i).\nabla X_k}}, \qquad E_k = \overline{\mathbf{q}.\nabla X_k}, \qquad (E.33)$$

and are respectively the average viscous dissipation, the interfacial internal energy transfer, and the interfacial heat transfer. The first two of these are generally negligible, while the third is usually large, at least if the two phases have different average temperatures. It is because of this that typically temperature does not vary locally, so that it suffices to consider total energy conservation. To see why this should be, we need to consider the averaged jump conditions between the phases.

E.5 Jump Conditions

The jump conditions for the point forms of the conservation laws in (E.7) have their counterpart in the averaged equations. For the general conservation law (E.11), the corresponding jump condition at an interface is

$$[-\{\rho\psi(\mathbf{u}-\mathbf{u}_i)+\mathbf{J}\}.\mathbf{n}]_-^+ = m_\psi, \tag{E.34}$$

where $\mathbf{n}=\mathbf{n}_-$ points from $-$ to $+$, and m_ψ represents a surface production term, which is normally zero. From (E.13), we can identify

$$\langle -\{\rho\psi(\mathbf{u}-\mathbf{u}_i)+\mathbf{J}\}.\mathbf{n}_k\rangle|_{\partial V_k} = \overline{\{\rho\psi(\mathbf{u}-\mathbf{u}_i)+\mathbf{J}\}.\nabla X_k}, \tag{E.35}$$

where $\mathbf{n}_k$ points out of phase k, and the angle brackets denote a specific surface average (i.e., a surface integral over the interface divided by the volume); thus (with no surface source term) the jump conditions for the averaged equations take the form

$$\sum_k \overline{\{\rho\psi(\mathbf{u}-\mathbf{u}_i)+\mathbf{J}\}.\nabla X_k} = 0, \tag{E.36}$$

bearing in mind that $\mathbf{n}_1 = -\mathbf{n}_2$.

Mass and momentum jump conditions are quite straightforward. Consulting (E.18) and (E.20), we have

$$\sum_k \Gamma_k = 0; \tag{E.37}$$

consulting (E.21) and (E.24), we have

$$\sum_k (\mathbf{M}_k + \mathbf{u}_{ki}^m \Gamma_k) = \mathbf{0}. \tag{E.38}$$

Energy is slightly more opaque, since we have to go back to the conservation form of the equation in (E.6) to derive the appropriate jump condition. This takes the form

$$\sum_k \left\{ \frac{1}{2}(u_{ki}^e)^2 \Gamma_k + e_{ki}\Gamma_k + E_k + W_k \right\} = 0, \tag{E.39}$$

where the extra terms not defined in (E.33) are given by

$$(u_{ki}^e)^2 = \frac{\overline{\rho u^2(\mathbf{u}-\mathbf{u}_i).\nabla X_k}}{\overline{\rho(\mathbf{u}-\mathbf{u}_i).\nabla X_k}}, \qquad W_k = -\overline{\mathbf{T}.\mathbf{u}.\nabla X_k}, \tag{E.40}$$

representing the interfacial kinetic energy transport and the interfacial work.

E.5.1 Practical Approximations

Generally speaking, the interfacial momentum flux $\mathbf{u}_{ki}^m \Gamma_k$ can be neglected, so that (E.38) reduces to the force balance

$$\sum_k \mathbf{M}_k \approx \mathbf{0}. \tag{E.41}$$

The interfacial kinetic energy and interfacial work terms in (E.40) are generally small, and additionally we suppose $e_{ki} \approx e_k$, so that (E.39) becomes the Stefan condition

$$\sum_k (e_k \Gamma_k + E_k) = 0. \tag{E.42}$$

We can normally also neglect the dissipation term in (E.32). If we suppose that the interfacial transport terms E_k, typically proportional to the difference in temperature between the phases, are large, then the conclusion is that the temperatures must be equal, and a single equation for the temperature then follows from adding the energy equations for the two phases. Adopting the jump condition (E.42), this leads to

$$\frac{\partial}{\partial t} \left\{ \sum_k \alpha_k \rho_k e_k \right\} + \nabla \cdot \left\{ \sum_k \alpha_k \rho_k e_k \mathbf{u}_k \right\} = -\nabla \cdot \left\{ \sum_k \alpha_k (\mathbf{q}_k + \mathbf{q}_k') \right\}. \tag{E.43}$$

Generally, we are only concerned with energy conservation when there is phase change, i.e., $\Gamma_k \neq 0$. In this case, the assumption of local thermodynamic equilibrium prescribes the local temperature as the freezing or boiling temperature as appropriate. Thus the energy equation does not in fact determine the temperature, but serves to determine the mass source due to phase change, Γ_k. To see how this happens, we need to relate the internal energies e_k to temperature T.

E.5.2 Thermodynamics

Quite generally, (D.3), (D.9) and (D.10) imply that the enthalpy and internal energy satisfy

$$\left. \frac{\partial h}{\partial T} \right|_p = c_p, \qquad \left. \frac{\partial e}{\partial T} \right|_v = c_v, \tag{E.44}$$

and one usually takes

$$h = c_p T, \qquad e = c_v T. \tag{E.45}$$

Commonly one rewrites the energy equation in terms of the enthalpy, since in phase change problems the latent heat is defined (at a fixed pressure and temperature) by

$$L = T_M \Delta s \approx \Delta e + p \Delta v = \Delta h. \tag{E.46}$$

Using (E.19) and (E.37), we can write the energy equation in the form

$$\sum_k \left[\left\{ \Gamma_k + \alpha_k \rho_k \frac{d}{dt_k} \right\} \left(h_k - \frac{p_k}{\rho_k} \right) \right] = \nabla.[\overline{K} \nabla T], \qquad \text{(E.47)}$$

where $\overline{K}$ is the phase-averaged thermal conductivity, including both molecular and turbulent conductivities,

$$\overline{K} = \sum_k \alpha_k K_k, \qquad \text{(E.48)}$$

and

$$\frac{d}{dt_k} = \frac{\partial}{\partial t} + \mathbf{u}_k.\nabla. \qquad \text{(E.49)}$$

For example, consider a vapour–liquid flow, with $h_g - h_l = L$. Denoting specific heats as c_{pk}, we can write (E.47) in the form

$$\Gamma_g L + \sum_k \left[\alpha_k \rho_k c_{pk} \frac{dT_M}{dt_k} - \left\{ \frac{\partial}{\partial t}(\alpha_k p_k) + \nabla.(\alpha_k p_k \mathbf{u}_k) \right\} \right] = \nabla.[\overline{K} \nabla T_M], \quad \text{(E.50)}$$

and this determines the mass source term Γ_g (which is positive for boiling, and negative for condensation).

E.6 Nye's Energy Equation in a Subglacial Channel

A particular variant of the procedures outlined above is Nye's derivation of the energy equation governing water flow in a sub-glacial channel. In particular, the variables and thus also the equations are cross-sectionally averaged. Nye (1976) provided his Eq. (11.4) with the minimum of fuss. Let us now try and derive this equation using the principles enunciated above. The equation is

$$\rho_w c_w \left[S \frac{\partial \theta_w}{\partial t} + Q \frac{\partial \theta_w}{\partial x} \right] = Q \left(\rho_w g \sin \alpha - \frac{\partial p}{\partial x} \right) - m[L + c_w(\theta_w - \theta_i)], \quad \text{(E.51)}$$

in which θ_w is the water temperature, S is the channel cross-sectional area, $Q = Su$ is the volume flux, p is the channel pressure, and θ_i is the surrounding ice temperature, taken as constant. See Chap. 11 for further details.

To derive this in detail, we need to derive also the appropriate forms of the conservation of mass and momentum equations in the channel. To begin with, we note the general relation

$$\frac{d}{dt} \int_{V(t)} L \, dV = \int_{V(t)} \frac{\partial L}{\partial t} \, dV + \int_{\partial V} L V_n \, dS, \qquad \text{(E.52)}$$

where V_n is the normal velocity of the moving boundary ∂V of the time dependent volume V. This applies whether or not the volume V is a material volume. If it is, then $V_n = u_n$, the fluid normal velocity. For a subglacial channel, this is not the case. We can then relate the rate of change of the integral of L over $V(t)$ to that over the material volume which is instantaneously coincident with V:

$$\frac{d}{dt}\int_{V(t)} L\,dV = \frac{d}{dt}\int_{V_m(t)} L\,dV - \int_{\partial V} L(u_n - V_n)\,dS, \tag{E.53}$$

where V_m is the corresponding material volume.

The equation of conservation of mass follows from putting $L = \rho\ (= \rho_w)$. The mass $\int_{V_m}\rho\,dV$ is conserved, and mass conservation takes the form

$$\frac{\partial}{\partial t}\int_{V(t)}\rho\,dV + \int_{\partial V_\parallel}\rho u_n\,dS = -\int_{\partial V_\perp}\rho(u_n - V_n)\,dS, \tag{E.54}$$

where we take the volume V to be the cross section of the channel times a small (fixed) increment δx in the downstream direction, $\partial V_\parallel$ denotes the end faces of the volume (on which $V_n = 0$), and $\partial V_\perp$ denotes the ice-water interface. Dividing by δx and letting $\delta x \to 0$, we obtain conservation of mass in the form

$$\frac{\partial}{\partial t}(\rho S) + \frac{\partial}{\partial x}(\rho S u) = m, \tag{E.55}$$

where u is the average velocity and ρ is the average density over the cross section, and

$$m = \int_{\partial S_-}\Gamma\,ds = -\int_{\partial S_-}[\rho(u_n - V_n)]_-\,ds; \tag{E.56}$$

∂S_- is the perimeter of the cross section S, taken on the inside.

This same procedure allows us to form averaged momentum and energy equations. The basic momentum equation in integral form is

$$\frac{d}{dt}\int_{V_m(t)}\rho u_i\,dV = \int_{\partial V}\boldsymbol{\sigma}_i.\mathbf{n}\,dS + \int_V \rho f_i\,dV, \tag{E.57}$$

and performing the same reduction as above leads to

$$\frac{\partial}{\partial t}\int_V \rho u_i\,dV + \int_{\partial V_\parallel}\rho u_i u_n\,dS$$
$$= -\int_{\partial V_\perp}\rho u_i(u_n - V_n)\,dS - \int_{\partial V}pn_i\,dS + \int_{\partial V}\boldsymbol{\tau}_i.\mathbf{n}\,dS + \int_V \rho f_i\,dV, \tag{E.58}$$

where we write $\boldsymbol{\sigma} = -p\boldsymbol{\delta} + \boldsymbol{\tau}$, $\boldsymbol{\delta}$ being the unit tensor and $\boldsymbol{\tau}$ being the deviatoric stress tensor. We now use the divergence theorem on the pressure term to write this as

$$\frac{\partial}{\partial t}\int_V \rho u_i\,dV + \int_{\partial V_\parallel}\rho u_i u_n\,dS$$
$$= -\int_{\partial V_\perp}\rho u_i(u_n - V_n)\,dS - \int_V \frac{\partial}{\partial x_i}(p + \rho\chi)\,dV + \int_{\partial V}\boldsymbol{\tau}_i.\mathbf{n}\,dS, \tag{E.59}$$

in which we suppose that ρ is constant; χ is the gravitational potential energy. Taking $i = 1$ (the x direction) and averaging, we obtain the momentum equation in the form

$$\frac{\partial}{\partial t}(\rho Su) + \frac{\partial}{\partial x}(\rho Su^2) = -S\frac{\partial}{\partial x}(p + \rho\chi) - \tau_w l, \qquad (E.60)$$

where l is the wetted perimeter, τ_w is the wall stress. Importantly, no slip at the wall implies $u_1 = 0$ on $\partial V_\perp$ (if we assume downstream ice velocity is negligible). We have neglected deviatoric longitudinal stress on the ends of ∂V. Note that it is important to convert the surface integral in pressure in (E.58) to the volume integral in (E.59) before deriving (E.60) (otherwise we would be tempted to put the S coefficient of the pressure term inside the derivative). Specifically, (E.59) takes the form

$$\frac{\partial}{\partial t}(\rho u S \delta x) + \cdots = -\frac{\partial}{\partial x}(p + \rho\chi)S\,\delta x + \cdots = -S\,\delta(p + \rho\chi) + \cdots, \qquad (E.61)$$

and on dividing by δx, we obtain (E.60). (This is analogous to the absorption of the term $p_k \nabla \alpha_k$ into the interfacial term in (E.26).)

Note also that taking $i = 3$ (the z direction) gives us the hydrostatic condition (if we neglect deviatoric normal stress)

$$\frac{\partial}{\partial z}(p + \rho\chi) = 0. \qquad (E.62)$$

The energy equation is derived in a similar way. We take the integral form of the third equation in (E.6), and apply the same procedure as above. This leads us to

$$\frac{\partial}{\partial t}\int_V \rho\left[e + \frac{1}{2}u^2 + \chi\right]dV + \int_{\partial V_\parallel} \rho\left[e + \frac{1}{2}u^2 + \chi\right]u_n\,dS$$

$$= -\int_{\partial V_\perp} \rho\left[e + \frac{1}{2}u^2 + \chi\right](u_n - V_n)\,dS + \int_{\partial V}(\sigma_{ij}u_i n_j - q_n)\,dS. \qquad (E.63)$$

We split the stress tensor up as before and conflate the pressure and potential energy term. Averaging the consequent result, putting $u_1 = 0$ on $\partial V_\perp$, assuming $\rho_t = 0$, and neglecting deviatoric longitudinal stress, then leads (after a good deal of manipulation) to the averaged energy equation

$$\rho S\left[\frac{\partial}{\partial t}\left(e + \frac{1}{2}u^2\right) + u\frac{\partial}{\partial x}\left(e + \frac{1}{2}u^2\right)\right] = -\int_{\partial S} q_n\,ds - Su\frac{\partial}{\partial x}(p + \rho\chi) - \frac{1}{2}mu^2. \qquad (E.64)$$

This is essentially Nye's equation (E.51), if we neglect the kinetic energy terms proportional to $\frac{1}{2}u^2$, and put $e = c_w\theta_w$ and $\chi = g(z\cos\alpha - x\sin\alpha)$. The details of the algebraic manipulation form the substance of Question E.1.

To complete the derivation of (E.51), we need to constitute the heat flux term $\int_{\partial S} q_n\,ds$. In Nye's equation, this is given by

$$\int_{\partial S} q_n\,ds = m[L + c_w(\theta_w - \theta_i)]. \qquad (E.65)$$

To derive this, we go back to the jump conditions (E.9) and (E.10). First we note that the right hand side of (E.65) is the jump in enthalpy $-[h]_-^+ = h_w - h_i$, where the enthalpy is $h = e + \frac{p}{\rho}$. Next we assume that the ice is at the melting point, so that θ_i is constant and there is no heat flux from the ice to the interface. Therefore the heat flux in (E.65) is $-\int_{\partial S}[q_n]_-^+ ds$, and thus (E.65) will follow from the result that

$$\Gamma[h]_-^+ = [q]_-^+, \tag{E.66}$$

where Γ is given by (E.8). The jump in enthalpy is related to the jump in internal energy by the relation

$$[h]_-^+ = [e]_-^+ - p\Delta v, \tag{E.67}$$

where $v = 1/\rho$ is the specific volume, and we define the change of volume Δv on melting as

$$\Delta v = -[v]_-^+ = \frac{1}{\rho_w} - \frac{1}{\rho_i}. \tag{E.68}$$

From (E.8), we derive

$$\Gamma \Delta v = [u_n]_-^+, \tag{E.69}$$

and thus from (E.9), we have

$$\sigma_{nn}^+ = -p - \Gamma(u_n^+ - u_n^-) = -p - \Gamma^2 \Delta v. \tag{E.70}$$

Hence we obtain

$$[\sigma_{nn}u_n]_-^+ = -p\Gamma \Delta v - \Gamma^2 \Delta v u_n^+, \tag{E.71}$$

and thus (E.10) implies, using (E.67),

$$\Gamma[h]_-^+ = [q_n]_-^+ + \Gamma^2 \Delta v u_n^+, \tag{E.72}$$

where we take $[\frac{1}{2}u^2]_-^+ = 0$ assuming no slip at the interface; (E.66) and then also (E.65) follow on neglecting the term $\Gamma^2 \Delta v u_n^+$, which is comparable to the kinetic energy of the ice.

E.7 Exercises

E.1 Consider the energy equation in the form of (E.63):

$$\frac{\partial}{\partial t} \int_V \rho \left[e + \frac{1}{2}u^2 + \chi \right] dV + \int_{\partial V_{\parallel}} \rho \left[e + \frac{1}{2}u^2 + \chi \right] u_n \, dS$$

$$= -\int_{\partial V_{\perp}} \rho \left[e + \frac{1}{2}u^2 + \chi \right] (u_n - V_n) \, dS + \int_{\partial V} (\sigma_{ij}u_in_j - q_n) \, dS,$$

where the volume V is a short cylindrical segment of length δx, with the ice/water interface being denoted as $\partial V_\perp$ with normal in the $(\mathbf{e}_2, \mathbf{e}_3)$ plane, and the ends being denoted as $\partial V_\|$, with normal in the $\mathbf{e}_1$ direction. By using the relationship that

$$\frac{d}{dt}\int_{V(t)} L\,dV = \int_{V(t)} \frac{\partial L}{\partial t}\,dV + \int_{\partial V} L V_n\,dS,$$

where V_n is the normal velocity of ∂V, show that the energy equation can be written in the form

$$\frac{\partial}{\partial t}\int_V \rho\left[e + \frac{1}{2}u^2\right]dV + \int_{\partial V_\|} \rho\left[e + \frac{1}{2}u^2\right]u_n\,dS + \int_V \rho_t \chi\,dV$$

$$= -\int_{\partial V_\perp} \rho(u_n - V_n)\left(e + \frac{1}{2}u^2\right)dS - \int_{\partial V_\perp} q_n\,dS$$

$$- \int_{\partial V_\perp} (p + \rho\chi)u_n\,dS - \int_{\partial V_\|} (p + \rho\chi)u_n\,dS,$$

where we take $\partial V_\|$ to be fixed in space, write $\sigma_{ij} = -p\delta_{ij} + \tau_{ij}$, and assume that deviatoric longitudinal stress and longitudinal heat flux are negligible, and that $u_1 = 0$ on $\partial V_\perp$. What does the term $\int_V \rho_t \chi\,dV$ represent physically?

Assuming now that ρ is constant, show that the averaged energy equation can be written as

$$\frac{\partial}{\partial t}\left[\rho S e + \frac{1}{2}\rho S u^2\right] + \frac{\partial}{\partial x}\left[\rho S e u + \frac{1}{2}\rho S u^3\right]$$

$$= me - \frac{\partial}{\partial x}[(p + \rho\chi)Su] - \int_{\partial S} (p + \rho\chi)u_n\,ds - \int_{\partial S} q_n\,ds,$$

where S is the cross-sectional area, and

$$m = -\int_{\partial S} \rho[u_n - V_n]\,ds.$$

Using this last equation, show that

$$\frac{m}{\rho} = -\int_{\partial S} u_n\,ds + S_t,$$

and by assuming that $p + \rho\chi$ is constant (why?) on ∂S, show that

$$-\int_{\partial S} (p + \rho\chi)u_n\,ds = (p + \rho\chi)(Su)_x,$$

where you should assume conservation of mass in the form

$$(\rho S)_t + (\rho S u)_x = m.$$

Using conservation of mass again, deduce from the above that the energy equation can finally be written in the form

$$\rho S\left[\frac{\partial}{\partial t}\left(e+\frac{1}{2}u^2\right)+u\frac{\partial}{\partial x}\left(e+\frac{1}{2}u^2\right)\right]=-\int_{\partial S}q_n\,ds-Su\frac{\partial}{\partial x}(p+\rho\chi)-\frac{1}{2}mu^2.$$

References

Aagaard P, Helgeson H (1983) Activity/composition relations among silicates and aqueous solutions: II. Chemical and thermodynamic consequences of ideal mixing of atoms on homological sites in montmorillonites, illites, and mixed-layer clays. Clays Clay Miner 31:207–217

Abbott MR, Lighthill MJ (1956) A theory of the propagation of bores in channels and rivers. Math Proc Camb Philos Soc 52:344–362

Abramowitz M, Stegun I (1964) Handbook of mathematical functions. Dover, New York

Agustin L et al (EPICA community) (2004) Eight glacial cycles from an Antarctic ice core. Nature 429:623–628

Aharonov E, Whitehead JA, Kelemen PB, Spiegelman M (1995) Channeling instability of upwelling melt in the mantle. J Geophys Res 100:20433–20450

Ahern JL, Turcotte DL (1979) Magma migration beneath an ocean ridge. Earth Planet Sci Lett 45:115–122

Ahnert F (1996) Introduction to geomorphology. Arnold, London

Albarède F (2003) Geochemistry: an introduction. Cambridge University Press, Cambridge

Allen JRL (1971) Transverse erosional marks of mud and rock: their physical basis and geological significance. Sediment Geol 5:167–385

Allen JRL (1985) Principles of physical sedimentology. Chapman and Hall, London

Alley RB (1989) Water-pressure coupling of sliding and bed deformation: I. Water system. J Glaciol 35:108–118

Alley RB (2002) The two-mile time machine: ice cores, abrupt climate change, and our future. Princeton University Press, Princeton

Alley RB, Mayewski PA, Sowers T, Stuiver M, Taylor KC, Clark PU (1997) Holocene climatic instability: a prominent, widespread event 8200 years ago. Geology 25:483–486

Alley RB, Anandakrishnan S, Jung P (2001) Stochastic resonance in the North Atlantic. Paleoceanography 16(2):190–198

Altuhafi FN, Baudet BA, Sammonds P (2009) On the time-dependent behaviour of glacial sediments. Quat Sci Rev 28:693–707

Alvarez LW, Alvarez W, Asaro F, Michel HV (1980) Extraterrestrial cause for the Cretaceous-Tertiary extinction. Science 208:1095–1108

Anderson DL (2007) New theory of the Earth. Cambridge University Press, Cambridge

Andrews DG (2000) An introduction to atmospheric physics. Cambridge University Press, Cambridge

Angevine CL, Turcotte DL (1983) Porosity reduction by pressure solution: a theoretical model for quartz arenites. Geol Soc Amer Bull 94:1129–1134

Aris R (1956) On the dispersion of a solute in a fluid flowing through a tube. Proc R Soc A 235:67–78

Aris R (1975) Mathematical theory of diffusion and reaction in permeable catalysts. Two volumes. Oxford University Press, Oxford

Arrhenius S (1896) On the influence of carbonic acid in the air upon the temperature of the ground. Philos Mag 41:237–275

Athy LF (1930) Density, porosity, and compaction of sedimentary rocks. Am Assoc Pet Geol Bull 14:1–22

Avrami M (1939) Kinetics of phase change. I. General theory. J Chem Phys 7:1103–1112

Avrami M (1940) Kinetics of phase change. II. Transformation–time relations for random distribution of nuclei. J Chem Phys 8:212–224

Audet DM, Fowler AC (1992) A mathematical model for compaction in sedimentary basins. Geophys J Int 110:577–590

Bader FG (1978) Analysis of double-substrate limited growth. Biotechnol Bioeng 20:183–202

Bagnold RA (1936) The movement of desert sand. Proc R Soc Lond A 157:594–620

Bagnold RA (1941) The physics of blown sand and desert dunes. Methuen, London

Baker VR (2001) Water and the martian landscape. Nature 412:228–236

Baker VR, Milton DJ (1974) Erosion by catastrophic floods on Mars and Earth. Icarus 23:27–41

Baldwin P (1985) Zeros of generalized Airy functions. Mathematika 32:104–117

Balmforth NJ, Mandre S (2004) Dynamics of roll waves. J Fluid Mech 514:1–33

Balmforth NJ, Provenzale A, Whitehead JA (2001) The language of pattern and form. In: Balmforth NJ, Provenzale A (eds) Geomorphological fluid mechanics. Springer, Berlin, pp 3–33

Barcilon V, MacAyeal DR (1993) Steady flow of a viscous ice stream across a no-slip/free-slip transition at the bed. J Glaciol 39(131):167–185

Barnard JA, Bradley JN (1985) Flame and combustion, 2nd edn. Chapman and Hall, London

Barry RG, Chorley RJ (1998) Atmosphere, weather and climate, 7th edn. Routledge, London

Batchelor GK (1967) An introduction to fluid dynamics. Cambridge University Press, Cambridge

Baú D, Gambolati G, Teatini P (2000) Residual land subsidence near abandoned gas fields raises concern over Northern Adriatic coastland. Eos 81(22):245–249

Bear J (1972) Dynamics of fluids in porous media. Elsevier, Amsterdam (Dover reprint, 1988)

Bear J, Bachmat Y (1990) Introduction to modelling of transport phenomena in porous media. Kluwer, Dordrecht

Bear J, Verruijt A (1987) Modelling groundwater flow and pollution. Reidel, Dordrecht

Bebernes J, Eberly D (1989) Mathematical problems from combustion theory. Springer, New York

Beeftink HH, van der Heijden RTJM, Heijnen JJ (1990) Maintenance requirements: energy supply from simultaneous endogenous respiration and substrate consumption. FEMS Microbiol Lett 73:203–209

Bender CM, Orszag SA (1978) Advanced mathematical methods for scientists and engineers. McGraw-Hill, New York

Benjamin TB (1959) Shearing flow over a wavy boundary. J Fluid Mech 6:161–205

Benjamin TB, Bona JL, Mahony JJ (1972) Model equations for long waves in nonlinear dispersive systems. Philos Trans R Soc Lond A 272:47–78

Benn DA, Evans DJA (1998) Glaciers and glaciation. Edward Arnold, London

Bennett MR, Glasser NF (2010) Glacial geology: ice sheets and landforms, 2nd edn. Wiley, London

Bensoussan A, Lions JL, Papanicolaou G (1978) Asymptotic analysis for periodic structures. North-Holland, Amsterdam

Bentley CR (1987) Antarctic ice streams: a review. J Geophys Res 92:8843–8858

Bercovici D (1993) A simple model of plate generation from mantle flow. Geophys J Int 114:635–650

Bercovici D (2003) The generation of plate tectonics from mantle convection. Earth Planet Sci Lett 205:107–121

Bercovici D (ed) (2009a) Mantle dynamics. Treatise on geophysics, vol 7. Elsevier, Amsterdam

Bercovici D (2009b) Mantle dynamics past, present and future: an introduction and overview. In: Bercovici D (ed) Mantle dynamics. Treatise on geophysics, vol 7. Elsevier, Amsterdam, pp 1–30

Bercovici D, Michaut C (2010) Two-phase dynamics of volcanic eruptions: compaction, compression and the conditions for choking. Geophys J Int 182:843–864

Bergles AE, Collier JG, Delhaye JM, Hewitt GF, Mayinger F (1981) Two-phase flow and heat transfer in the power and process industries. Hemisphere, McGraw-Hill, New York

Berner RA, Lasaga AC, Garrels RM (1983) The carbonate–silicate geochemical cycle and its effect on atmospheric carbon dioxide over the past 100 million years. Am J Sci 283:641–683

Bigg G (2003) The oceans and climate, 2nd edn. Cambridge University Press, Cambridge

Birchwood RA, Turcotte DL (1994) A unified approach to geopressuring, low-permeability zone formation, and secondary porosity generation in sedimentary basins. J Geophys Res 99:20051–20058

Bittner D, Schmeling H (1995) Numerical modelling of melting processes and induced diapirism in the lower crust. Geophys J Int 123:59–70

Björnsson H (1974) Explanation of jökulhlaups from Grímsvötn, Vatnajökull, Iceland. Jökull 24:1–26

Björnsson H (1988) Hydrology of ice caps in volcanic regions. Societas Scientarium Islandica, University of Iceland, Reykjavik

Björnsson H (1992) Jökulhlaups in Iceland: prediction, characteristics and simulation. Ann Glaciol 16:95–106

Blatter H (1995) Velocity and stress fields in grounded glaciers: a simple algorithm for including deviatoric stress gradients. J Glaciol 41:333–344

Blumberg PN, Curl RL (1974) Experimental and theoretical studies of dissolution roughness. J Fluid Mech 65:735–751

Bolchover P, Lister JR (1999) The effect of solidification on fluid-driven fracture, with application to bladed dykes. Proc R Soc Lond A 455:2389–2409

Bolshakov VA (2003) Modern climatic data for the Pleistocene: implications for a new concept of the orbital theory of paleoclimate. Russ J Earth Sci 5:125–143

Bond G et al (1992) Evidence for massive discharges of icebergs into the North Atlantic ocean during the last glacial period. Nature 360:245–249

Bond GC, Showers W, Cheseby M, Lotti R, Almasi P, de Menocal P, Priore P, Cullen H, Hajdas I, Bonani G (1997) A pervasive millennial-scale cycle in North Atlantic Holocene and glacial climates. Science 278:1257–1266

Bond GC, Showers W, Elliott M, Evans M, Lotti R, Hajdas I, Bonani G, Johnson S (1999) The North Atlantic's 1–2 kyr climate rhythm: relation to Heinrich events, Dansgaard/Oeschger cycles and Little Ice Age. In: Clark PU, Webb RS, Keigwin LD (eds) Mechanisms of global climate change. Geophys monogr, vol 112. AGU, Washington, pp 35–58

Boulton GS, Hindmarsh RCA (1987) Sediment deformation beneath glaciers: rheology and geological consequences. J Geophys Res 92:9059–9082

Bowden FP, Yoffe YD (1985) Initiation of growth of explosion in liquids and solids. Cambridge University Press, Cambridge

Bowen NL (1956) The evolution of the igneous rocks. Dover, New York

Brandeis G, Jaupart C (1986) On the interaction between convection and crystallisation in cooling magma chambers. Earth Planet Sci Lett 77:345–361

Brandeis G, Jaupart C, Allègre CJ (1984) Nucleation, crystal growth and the thermal regime of cooling magmas. J Geophys Res 89:10161–10177

Brandt A, Fernando HJS (eds) (1995) Double-diffusive convection. AGU, Washington

Brenner H (1980) A general theory of Taylor dispersion phenomena. Phys Chem Hydrodyn 1:91–123

Bretz JH (1923) The channeled scablands of the Columbia Plateau. J Geol 31:617–649

Bretz JH (1969) The Lake Missoula floods and the Channeled Scabland. J Geol 77:505–543

Broecker WS (1991) The great ocean conveyor. Oceanography 4:79–89

Broecker WS, Bond G, Klas M (1990) A salt oscillator in the glacial Atlantic? I: the concept. Paleoceanography 5:469–477

Brown GM (1956) The layered ultrabasic rocks of Rhum, Inner Hebrides. Philos Trans R Soc A 240:1–53

Bryce J (1833) On the evidences of diluvial action in the north of Ireland. J Geol Soc Dublin 1:34–44

Buckmaster JD, Ludford GSS (1982) Theory of laminar flames. Cambridge University Press, Cambridge

Budyko MI (1969) The effect of solar radiation variations on the climate of the Earth. Tellus 21:611–619

Bueler E, Brown J (2009) Shallow shelf approximation as a "sliding law" in a thermomechanically coupled ice sheet model. J Geophys Res 114:F03008. doi:10.1029/2008JF001179

Burgers JM (1948) A mathematical model illustrating the theory of turbulence. Adv Appl Mech 1:171–199

Busse FH (1985) Transition to turbulence in Rayleigh–Bénard convection. In: Swinney HL, Gollub JP (eds) Hydrodynamic instabilities and the transition to turbulence, 2nd edn. Topics in applied physics, vol 45. Springer, Berlin

Butterworth D, Hewitt GF (1977) Two-phase flow and heat transfer. Oxford University Press, Oxford

Cadbury D (2000) The dinosaur hunters. Fourth estate, London

Calov R, Ganopolski A, Petoukhov V, Claussen M, Greve R (2002) Large-scale instabilities of the Laurentide ice sheet simulated in a fully coupled climate-system model. Geophys Res Lett 29(24):69. doi:10.1029/2002GL016078

Carrier GF, Krook M, Pearson CE (1966) Functions of a complex variable. McGraw-Hill, New York

Cathles LM (1975) The viscosity of the Earth's mantle. Princeton University Press, Princeton

Chandler MA, Sohl LE (2000) Climate forcings and the initiation of low-latitude ice sheets during the Neoproterozoic Varanger glacial interval. J Geophys Res D 105(10):20737–20756

Chandrasekhar S (1960) Radiative transfer. Dover, New York

Chandrasekhar S (1981) Hydrodynamic and hydromagnetic stability. Dover, New York

Chanson H (2005) Physical modelling of the flow field in an undular tidal bore. J Hydraul Res 43:234–244

Chanson H (2009) Current knowledge in hydraulic jumps and related phenomena A survey of experimental results. Eur J Mech B, Fluids 28:191–210

Chapman MG, Gudmundsson MT, Russell AJ, Hare TM (2003) Possible Juventae Chasma sub-ice volcanic eruptions and Maja Valles ice outburst floods, Mars: implications of Mars Global Surveyor crater densities, geomorphology, and topography. J Geophys Res 108(E10):5113. doi:10.1029/2002JE002009

Chapman S (1930) A theory of upper atmospheric ozone. Mem R Meteorol Soc 3:103–125

Charru F, Hinch EJ (2006) Ripple formation on a particle bed sheared by a viscous liquid. Part 1. Steady flow. J Fluid Mech 550:111–121

Chorley RJ (ed) (1969) Introduction to physical hydrology. Methuen, London

Chow VT (1959) Open-channel hydraulics. McGraw-Hill, New York

Christensen UR (1984a) Heat transport by variable viscosity convection and implications for the Earth's thermal evolution. Phys Earth Planet Inter 35:264–282

Christensen UR (1984b) Convection with pressure- and temperature-dependent non-Newtonian rheology. Geophys J R Astron Soc 77:343–384

Christensen UR (1985) Heat transport by variable viscosity convection. II: pressure influence, non-Newtonian rheology and decaying heat sources. Phys Earth Planet Inter 37:183–205

Christensen U, Harder H (1991) Three-dimensional convection with variable viscosity. Geophys J Int 104:213–226

Christensen UR, Yuen DA (1984) The interaction of a subducting lithospheric slab with a chemical or phase-boundary. J Geophys Res 89:4389–4402

Christensen UR, Yuen DA (1985) Layered convection induced by phase transitions. J Geophys Res 90:10291–10300

Chugunov VA, Wilchinsky AV (1996) Modelling of a marine glacier and ice-sheet–ice-shelf transition zone based on asymptotic analysis. Ann Glaciol 23:59–67

Clague JJ, Mathews WH (1973) The magnitude of jökulhlaups. J Glaciol 12:501–504

Clark PU, Walder JS (1994) Subglacial drainage, eskers, and deforming beds beneath the Laurentide and Eurasian ice sheets. Geol Soc Amer Bull 106:304–314

Clarke GKC (1982) Glacier outburst floods from 'Hazard Lake', Yukon Territory, and the problem of flood magnitude prediction. J Glaciol 28:3–21

Clarke GKC (2003) Hydraulics of subglacial outburst floods: new insights from the Spring-Hutter formulation. J Glaciol 49:299–313

Clarke GKC (2005) Subglacial processes. Annu Rev Earth Planet Sci 33:247–276

Clarke GKC, Nitsan U, Paterson WSB (1977) Strain heating and creep instability in glaciers and ice sheets. Rev Geophys Space Phys 15:235–247

Clarke GKC, Collins SG, Thompson DE (1984) Flow, thermal structure, and subglacial conditions of a surge-type glacier. Can J Earth Sci 21:232–240

Clarke GKC, Leverington DW, Teller JT, Dyke AS (2004) Paleohydraulics of the last outburst flood from glacial Lake Agassiz and the 8200 BP cold event. Quat Sci Rev 23:389–407

Clarke GKC, Leverington DW, Teller JT, Dyke AS, Marshall SJ (2005) Fresh arguments against the Shaw megaflood hypothesis. A reply to comments by David Sharpe on "Paleohydraulics of the last outburst flood from glacial Lake Agassiz and the 8200 BP cold event". Quat Sci Rev 24:1533–1541

Clemens JD, Mawer CK (1992) Granitic magma transport by fracture propagation. Tectonophysics 204:339–360

Close MH (1867) Notes on the general glaciation of Ireland. J R Geol Soc Irel 1:207–242

Cocks D (2005) Mathematical modelling of dune formation. DPhil thesis, Oxford University. http://eprints.maths.ox.ac.uk

Cogan NG, Keener JP (2004) The role of the biofilm matrix in structural development. Math Med Biol 21:147–166

Coleman NM (2003) Aqueous flows carved the outflow channels on Mars. J Geophys Res 108(E5):5039

Collier JG, Thome JR (1996) Convective boiling and condensation, 3rd edn. Clarendon, Oxford

Colling A (ed) (2001) Ocean circulation, 2nd edn. Butterworth–Heinemann, Oxford

Colombini M (2004) Revisiting the linear theory of sand dune formation. J Fluid Mech 502:1–16

Courtillot V (1999) Evolutionary catastrophes: the science of mass extinction (transl J McClinton). Cambridge University Press, Cambridge

Crowley TJ, Baum SK (1993) Effect of decreased solar luminosity on late Precambrian ice extent. J Geophys Res 98(D9):16723–16732

Cuffey K, Paterson WSB (2010) The physics of glaciers, 4th edn. Elsevier, Amsterdam

Cushman JH (ed) (1990) Dynamics of fluids in hierarchical porous media. Academic Press, London

Dahl-Jensen D (1989) Steady thermomechanical flow along two-dimensional flow lines in large grounded ice sheets. J Geophys Res 94:10335–10362

Dash JG, Rempel AW, Wettlaufer JS (2006) The physics of premelted ice and its geophysical consequences. Rev Mod Phys 78:695–741

Davies GF (1999) Dynamic Earth: plates, plumes and mantle convection. Cambridge University Press, Cambridge

Davies JH, Davies DR (2010) Earth's surface heat flux. Solid Earth 1:5–24

Davis SH, Huppert HE, Müller U, Worster MG (eds) (1992) Interactive dynamics of convection and solidification. Kluwer, Dordrecht

Davis WM (1899) The geographical cycle. Geogr J 14:481–504

DePaolo DJ, Manga M (2003) Deep origin of hotspots—the mantle plume model. Science 300:920–921

Deeley RM, Parr PH (1914) On the Hintereis glacier. Philos Mag 27(6):153–176

Defant A (1958) Ebb and flow: the tides of Earth, air and water. University of Michigan Press, Ann Arbor

Denton GH, Sugden DE (2005) Meltwater features that suggest Miocene ice-sheet overriding of the Transantarctic Mountains in Victoria Land, Antarctica. Geogr Ann 87A:67–85

Dewynne JN, Fowler AC, Hagan PS (1993) Multiple reaction fronts in the oxidation-reduction of iron-rich uranium ores. SIAM J Appl Math 53:971–989

Dobran F (2001) Volcanic processes: mechanisms in material transport. Kluwer, New York

Dockery J, Klapper I (2001) Finger formation in biofilm layers. SIAM J Appl Math 62:853–869

Dold JW (1985) Analysis of the early stage of thermal runaway. Q J Mech Appl Math 38:361–387

Dowty E (1980) Crystal growth and nucleation theory and the numerical simulation of igneous crystallisation. In: Hargraves RB (ed) Physics of magmatic processes. Princeton University Press, Princeton, pp 419–485

Drazin PG, Johnson RS (1989) Solitons: an introduction. Cambridge University Press, Cambridge

Drazin PG, Reid WH (1981) Hydrodynamic stability. Cambridge University Press, Cambridge

Dressler RF (1949) Mathematical solution of the problem of roll waves in inclined open channels. Commun Pure Appl Math 2:149–194

Drew DA, Passman SL (1999) Theory of multicomponent fluids. Springer, New York

Drew DA, Wood RT (1985) Overview and taxonomy of models and methods for workshop on two-phase flow fundamentals. National Bureau of Standards, Gaithersburg

Drozdowski E (1986) An international drumlin biography. Boreas 15:310

Dullien FAL (1979) Porous media: fluid transport and pore structure. Academic Press, New York

Durán O, Herrmann HJ (2006) Vegetation against dune mobility. Phys Rev Lett 97:188001

Durand G, Gagliardini O, Zwinger T, Le Meur E, Hindmarsh RCA (2009) Full Stokes modeling of marine ice sheets: influence of the grid size. Ann Glaciol 50(52):109–114

Eberl D, Hower J (1976) Kinetics of illite formation. Geol Soc Amer Bull 87:1326–1330

Eberl HJ, Parker DF, Van Loosdrecht MCM (2001) A new deterministic spatio-temporal continuum model for biofilm development. Comput Math Methods Med 3:161–175

Eckhardt B, Schneider TM, Hof B, Westerweel J (2007) Turbulent transition in pipe flow. Annu Rev Fluid Mech 39:447–68

Edelstein-Keshet L (2005) Mathematical models in biology. SIAM, Philadelphia

Ehrhard P, Riley DS, Steen PH (eds) (2001) Interactive dynamics of convection and solidification. Kluwer, Dordrecht

Einstein HA (1950) The bedload function for bedload transportation in open channel flows. Tech Bull No 1026, USDA, Soil Conservation Service, pp 1–71

Embleton C, King CAM (1968) Glacial and periglacial geomorphology. Edward Arnold, London

Emeleus CH (1987) The Rhum layered complex, Inner Hebrides, Scotland. In: Parsons I (ed) Origins of igneous layering. NATO ASI series C, vol 196. Reidel, Dordrecht, pp 263–286

Emerson SR, Hedges JI (2008) Chemical oceanography and the marine carbon cycle. Cambridge University Press, Cambridge

Engelhardt H, Humphrey N, Kamb B, Fahnestock M (1990) Physical conditions at the base of a fast moving Antarctic ice stream. Science 248:57–59

Engelund F (1970) Instability of erodible beds. J Fluid Mech 42:225–244

Engelund F, Fredsøe J (1982) Sediment ripples and dunes. Annu Rev Fluid Mech 14:13–37

England AH (1971) Complex variable methods in elasticity. Wiley-Interscience, London

Erdélyi A, Magnus W, Oberhettinger F, Tricomi FG (1953) Higher transcendental functions, vol 1. McGraw-Hill, New York

Erlingsson U (2006) Lake Vostok behaves like a 'captured lake' and may be near to creating an Antarctic jökulhlaup. Geogr Ann 88A:1–7

Evatt G, Fowler AC, Clark CD, Hulton N (2006) Subglacial floods beneath ice sheets. Philos Trans R Soc 364:1769–1794

Finsterwalder S (1907) Die Theorie der Gletscherschwankungen. Z Gletschkd 2:81–103

Fisher RA (1937) The wave of advance of advantageous genes. Ann Eugen 7:353–369

Flemings MC (1974) Solidification processing. McGraw-Hill, New York

Flint RF (1930) The origin of the Irish "eskers". Geogr Rev 20:615–630

Foulger GR, Natland JH (2003) Is "hotspot" volcanism a consequence of plate tectonics? Science 300:921–922

Fowkes ND, Mahony JJ (1994) An introduction to mathematical modelling. Wiley, Chichester

Fowler AC (1979) Glacier dynamics. DPhil thesis, University of Oxford

Fowler AC (1983) On the thermal state of the earth's mantle. J Geophys 53:42–51

Fowler AC (1985a) Fast thermoviscous convection. Stud Appl Math 72:189–219

Fowler AC (1985b) A mathematical model of magma transport in the asthenosphere. Geophys Astrophys Fluid Dyn 33:63–96

Fowler AC (1986) A sliding law for glaciers of constant viscosity in the presence of subglacial cavitation. Proc R Soc Lond A 407:147–170

Fowler AC (1987a) A theory of glacier surges. J Geophys Res 92:9111–9120

Fowler AC (1987b) Sliding with cavity formation. J Glaciol 33:255–267

Fowler AC (1989) A mathematical analysis of glacier surges. SIAM J Appl Math 49:246–262

Fowler AC (1990) A compaction model for melt transport in the Earth's asthenosphere. Part II: applications. In: Ryan MP (ed) Magma transport and storage. Wiley, Chichester, pp 15–32

Fowler AC (1992a) Modelling ice sheet dynamics. Geophys Astrophys Fluid Dyn 63:29–65

Fowler AC (1992b) Convection and chaos. In: Yuen DA (ed) Chaotic processes in the geological sciences. Springer, New York, pp 43–69

Fowler AC (1993a) Towards a description of convection with temperature and pressure dependent viscosity. Stud Appl Math 88:113–139

Fowler AC (1993b) Boundary layer theory and subduction. J Geophys Res 98:21997–22005

Fowler AC (1997) Mathematical models in the applied sciences. Cambridge University Press, Cambridge

Fowler AC (2000) An instability mechanism for drumlin formation. In: Maltman A, Hambrey MJ, Hubbard B (eds) Deformation of glacial materials. Spec pub geol soc, vol 176. Geological Society, London, pp 307–319

Fowler AC (2003) On the rheology of till. Ann Glaciol 37:55–59

Fowler AC (2009) Dynamics of subglacial floods. Proc R Soc A 465:1809–1828. doi:10.1098/rspa.2008.0488

Fowler AC (2010) The formation of subglacial streams and mega-scale glacial lineations. Proc R Soc Lond A 466:3181–3201. doi:10.1098/rspa.2010.0009

Fowler AC, Howell PD (2003) Intermittency in the transition to turbulence. SIAM J Appl Math 63:1184–1207

Fowler AC, Johnson C (1995) Hydraulic runaway: a mechanism for thermally regulated surges of ice sheets. J Glaciol 41:554–561

Fowler AC, Johnson C (1996) Ice sheet surging and ice stream formation. Ann Glaciol 23:68–73

Fowler AC, Larson DA (1978) On the flow of polythermal glaciers. I. Model and preliminary analysis. Proc R Soc Lond A 363:217–242

Fowler AC, Larson DA (1980a) The uniqueness of steady state flows of glaciers and ice sheets. Geophys J R Astron Soc 63:333–345

Fowler AC, Larson DA (1980b) On the flow of polythermal glaciers II. Surface wave analysis. Proc R Soc Lond A 370:155–171

Fowler AC, McGuinness MJ (1982) A description of the Lorenz attractor at high Prandtl number. Physica D 5:149–182

Fowler AC, O'Brien SBG (1996) A mechanism for episodic subduction on Venus. J Geophys Res 101:4755–4763

Fowler AC, O'Brien SBG (2003) Lithospheric failure on Venus. Proc R Soc Lond A 459:2663–2704

Fowler AC, Schiavi E (1998) A theory of ice sheet surges. J Glaciol 44:104–118

Fowler AC, Scott DR (1996) Hydraulic crack propagation in a porous medium. Geophys J Int 127:595–604

Fowler AC, Yang X-S (1998) Fast and slow compaction in sedimentary basins. SIAM J Appl Math 59:365–385

Fowler AC, Yang X-S (1999) Pressure solution and viscous compaction in sedimentary basins. J Geophys Res 104:12989–12997

Fowler AC, Yang X-S (2003) Dissolution/precipitation mechanisms for diagenesis in sedimentary basins. J Geophys Res 108(B10):2509. doi:10.1029/2002JB002269

Fowler AC, Kopteva N, Oakley C (2007) The formation of river channels. SIAM J Appl Math 67:1016–1040

Fowler AC, Toja R, Vázquez C (2009) Temperature dependent shear flow and the absence of thermal runaway in valley glaciers. Proc R Soc Lond A 466:363–382

Fowler AC, McGuinness MJ, Ellis AS (2011) On an evolution equation for sand dunes. SIAM J Appl Math, submitted

Francis P, Oppenheimer C (2004) Volcanoes, 2nd edn. Oxford University Press, Oxford

François B, Lacombe F, Herrmann HJ (2002) Finite width of shear zones. Phys Rev E 65:031311

Frappé-Sénéclauze T-P, Clarke GKC (2007) Slow surge of Trapridge Glacier, Yukon Territory, Canada. J Geophys Res 112:F03S32. doi:10.1029/2006JF000607

Fredsøe J (1974) On the development of dunes in erodible channels. J Fluid Mech 64:1–16

Freed RL, Peacor DR (1989) Geopressured shale and sealing effect of smectite to illite transition. Am Assoc Pet Geol Bull 73:1223–1232

Freeze RA, Cherry JA (1979) Groundwater. Prentice-Hall, London

French RH (1994) Open-channel hydraulics. McGraw-Hill, New York

Freund LB (1990) Dynamic fracture mechanics. Cambridge University Press, Cambridge

Fricker HA, Scambos T, Bindschadler R, Padman L (2007) An active subglacial water system in West Antarctica mapped from space. Science 315:1544–1548

Fudge TJ, Humphrey NF, Harper JT, Pfeffer WT (2008) Diurnal fluctuations in borehole water levels: configuration of the drainage system beneath Bench Glacier, Alaska, USA. J Glaciol 54:297–306

Gakhov FD (1990) Boundary value problems. Dover, New York

Ganopolski A, Rahmstorf S (2001) Rapid changes of glacial climate simulated in a coupled climate model. Nature 409:153–158

Ganopolski A, Rahmstorf S (2002) Abrupt glacial climate changes due to stochastic resonance. Phys Rev Lett 88(3):038501. doi:10.1103/PhysRevLett.88.038501

García M, Parker G (1991) Entrainment of bed sediment into suspension. J Hydraul Eng 117:414–435

Ghil M, Childress S (1987) Topics in geophysical fluid dynamics. Springer, Berlin

Gill AE (1982) Atmosphere-ocean dynamics. Academic Press, San Diego

Glassman I (1987) Combustion, 2nd edn. Academic Press, Orlando

Goodwin ID (1988) The nature and origin of a jökulhlaup near Casey Station, Antarctica. J Glaciol 34:95–101

Goudie A (1993) The nature of the environment, 3rd edn. Blackwell, Oxford

Gradshteyn IS, Ryzhik IM (1980) Table of integrals, series and products, Corrected and enlarged edition. Academic Press, New York

Gray P, Scott SK (1990) Chemical oscillators and instabilities: non-linear chemical kinetics. Clarendon, Oxford

Grigoryan SS, Krass MS, Shumskiy PA (1976) Mathematical model of a three-dimensional non-isothermal glacier. J Glaciol 17:401–417

Grindrod P (1991) Patterns and waves. Oxford University Press, Oxford

Grout FF (1945) Scale models of structures related to batholiths. Am J Sci 243-A:260–284. (Daly volume)

Gudmundsson GH (2003) Transmission of basal variability to a glacier surface. J Geophys Res 108(B5):2253. doi:10.1029/2002JB002107

Gudmundsson MT, Sigmundsson F, Björnsson H (1997) Ice-volcano interaction of the 1996 Gjálp subglacial eruption, Vatnajökull, Iceland. Nature 389:954–957

Gudmundsson MT, Sigmundsson F, Björnsson H, Högnadóttir T (2004) The 1996 eruption at Gjálp, Vatnajökull ice cap, Iceland: efficiency of heat transfer, ice deformation and subglacial water pressure. Bull Volcanol 66:46–65

Haberman R (1998) Mathematical models. Society for Industrial and Applied Mathematics, Philadelphia

Hack JT (1957) Studies of longitudinal profiles in Virginia and Maryland. USGS Prof Paper, 294-B

Hagan PS, Polizzotti RS, Luckman G (1986) Internal oxidation of binary alloys. SIAM J Appl Math 45:956–971

Hall J (1815) On the revolutions of the Earth's surface. Trans R Soc Edinb 7:169–184

Hargraves RB (ed) (1980) Physics of magmatic processes. Princeton University Press, Princeton

Harland WB (1964) Critical evidence for a great infra-Cambrian glaciation. Geol Rundsch 54:45–61

Harland WB (2007) Origins and assessment of snowball Earth hypotheses. Geol Mag 144:633–642

Heinrich H (1988) Origin and consequences of cyclic ice rafting in the Northeast Atlantic Ocean during the past 130000 years. Quat Res 29:142–152

Hershenov J (1976) Solutions of the differential equation $u''' + \lambda^2 z u' + (\alpha - 1)\lambda^2 u = 0$. Stud Appl Math 55:301–314

Hess PC (1989) Origins of igneous rocks. Harvard University Press, Cambridge

Hewitt GF, Hall-Taylor NS (1970) Annular two-phase flow. Pergamon, Oxford

Hewitt IJ, Fowler AC (2008) Seasonal waves on glaciers. Hydrol Process 22:3919–3930

Hewitt IJ, Fowler AC (2009) Melt channelization in ascending mantle. J Geophys Res 114:B06210. doi:10.1029/2008JB006185

Hinch EJ (1991) Perturbation methods. Cambridge University Press, Cambridge

Hindmarsh RCA (1993) Qualitative dynamics of marine ice sheets. In: Peltier WR (ed) Ice in the climate system. Springer, Berlin, pp 67–99

Hindmarsh RCA (1998) The stability of a viscous till sheet coupled with ice flow, considered at wavelengths less than the ice thickness. J Glaciol 44:285–292

Hindmarsh RCA (2009) Consistent generation of ice-streams via thermo-viscous instabilities modulated by membrane stresses. Geophys Res Lett 36:L06502

Hindmarsh RCA, Le Meur E (2001) Dynamical processes involved in the retreat of marine ice sheets. J Glaciol 47:271–282

Hodge SM (1974) Variations in the sliding of a temperate glacier. J Glaciol 13:349–369

Hofmann AW, Hart SR (2007) Another nail in which coffin? Science 315:39–40

Hoffman N (2000) White Mars: a new model for Mars' surface and atmosphere based on CO_2. Icarus 146:326–342

Hoffman PF, Kaufman AJ, Halverson GP, Schrag DP (1998) A Neoproterozoic snowball Earth. Science 281:1342–1346

Holland DM, Jacobs SS, Jenkins A (2003) Modelling the ocean circulation beneath the Ross Ice Shelf. Antarct Sci 15:13–23

Holland DM, Thomas RH, de Young B, Ribergaard MH (2008) Acceleration of Jakobshavn Isbrae triggered by warm subsurface ocean waters. Nat Geosci 1:659–664

Holmes A (1978) Principles of physical geology, 3rd edn, revised by Doris Holmes. Wiley, New York

Holmes MH (1995) Introduction to perturbation theory. Springer, New York

Holmes MH (2009) Introduction to the foundations of applied mathematics. Springer, Dordrecht

Holton JR (2004) An introduction to dynamic meteorology, 4th edn. Elsevier, Burlington

Hooke RLeB (2005) Principles of glacier mechanics, 2nd edn. Cambridge University Press, Cambridge

Hooke RLeB, Laumann T, Kohler J (1990) Subglacial water pressures and the shape of subglacial conduits. J Glaciol 36:67–71

Hooke RLeB, Hanson B, Iverson NR, Jansson P, Fischer UH (1997) Rheology of till beneath Storglaciären, Sweden. J Glaciol 43(143):172–179

Hoppensteadt F (1975) Mathematical theories of populations: demographics, genetics and epidemics. Society for Industrial and Applied Mathematics, Philadelphia

Hort M, Spohn T (1991) Numerical simulation of the crystallization of multicomponents in thin dikes or sills. 2. Effects of heterocatalytic nucleation and composition. J Geophys Res 96:485–499

Horton RE (1945) Erosional development of streams and their drainage basins; hydrophysical approach to quantitative morphology. Bull Geol Soc Am 56:275–370

Houghton JT (2002) The physics of atmospheres, 3rd edn. Cambridge University Press, Cambridge

Houghton JT, Meira Filho LG, Callander BA, Harris N, Kattenberg A, Maskell K (eds) (1996) Climate change 1995: the science of climate change. Cambridge University Press, Cambridge

Howard AD (1978) Origin of the stepped topography on the Martian poles. Icarus 34:581–599

Howard AD (1994) A detachment-limited model of drainage basin evolution. Water Resour Res 30:2261–2285

Howard LN (1966) Convection at high Rayleigh number. In: Görtler H (ed) Proc 11th int cong appl mech. Springer, Berlin, pp 1109–1115

Howard LN, Kopell N (1977) Slowly varying waves and shock structures in reaction-diffusion equations. Stud Appl Math 56:95–145

Howell PD (1996) Models for thin viscous sheets. Eur J Appl Math 7:321–343

Howison SD (2005) Practical applied mathematics: modelling, analysis, approximation. Cambridge University Press, Cambridge

Hubbard BP, Sharp MJ, Willis IC, Nielsen MK, Smart CC (1995) Borehole water-level variations and the structure of the subglacial hydrological system of Haut Glacier d'Arolla, Valais, Switzerland. J Glaciol 41:572–583

Hughes TJ (1973) Is the West Antarctic ice sheet disintegrating? J Geophys Res 78:7884–7910

Hunt JCR, Leibovich S, Richards KJ (1988) Turbulent shear flows over low hills. Q J R Meteorol Soc 114:1435–1470

Hunt JM (1990) Generation and migration of petroleum from abnormally pressured fluid compartments. Am Assoc Pet Geol Bull 74:1–12

Huppert HE (1986) The intrusion of fluid mechanics into geology. J Fluid Mech 173:557–594

Huppert HE (1990) The fluid mechanics of solidification. J Fluid Mech 212:209–240

Huppert HE (2000) Geological fluid mechanics. In: Batchelor GK, Moffatt HK, Worster MG (eds) Perspectives in fluid dynamics. Cambridge University Press, Cambridge, pp 447–506

Huppert HE, Sparks RSJ (1980) The fluid dynamics of a basaltic magma chamber replenished by influx of hot, dense, ultrabasic magma. Contrib Mineral Petrol 75:279–289

Huppert HE, Sparks RSJ (1988) The fluid dynamics of crustal melting by injection of basaltic sills. Trans R Soc Edinb 79:237–243

Hutter K (1983) Theoretical glaciology. Reidel, Dordrecht

Hutter K, Olunloyo VOS (1980) On the distribution of stress and velocity in an ice strip, which is partly sliding over and partly adhering to its bed, by using a Newtonian viscous approximation. Proc R Soc Lond A 373:385–403

Hutter K, Yakowitz S, Szidarovsky F (1986) A numerical study of plane ice sheet flow. J Glaciol 32:139–160

Hüttmann A, Wilson RD, Thornton SF, Lerner DN (2003) Natural attenuation of ammonium at a former coal carbonisation plant (Mansfield, UK): conceptual model for biodegradation processes. In: Consoil 2003, Gent, Conference Proceedings CD, pp 1542–1547

Hyde WT, Crowley TJ, Baum SK, Peltier WR (2000) Neoproterozoic 'snowball Earth' simulations with a coupled climate/ice sheet model. Nature 405:425–429

Iken A (1981) The effect of subglacial water pressure on the sliding velocity of a glacier in an idealized numerical model. J Glaciol 27:407–422

Imbrie J, Imbrie KP (1979) Ice ages; solving the mystery. Harvard University Press, Cambridge

Innes R (1732) Miscellaneous letters on several subjects in philosophy and astronomy, I, p 4. S Birt, London

Irvine TN (1987) Layering and related structures in the Duke Island and Skaergaard intrusions: similarities, differences, and origins. In: Parsons I (ed) Origins of igneous layering. NATO ASI series C, vol 196. Reidel, Dordrecht, pp 185–245

Ishii M (1975) Thermo-fluid dynamic theory of two-phase flow. Eyrolles, Paris

Ivanov AB, Muhleman DO (2000) The role of sublimation for the formation of the Northern ice cap: results from the Mars Orbiter Laser Altimeter. Icarus 144:436–448

Iverson NR, Baker RW, Hooyer TS (1997) A ring-shear device for the study of till deformation: tests on tills with contrasting clay contents. Quat Sci Rev 16(9):1057–1066

Izumi N, Parker G (1995) Inception of channelization and drainage basin formation: upstream-driven theory. J Fluid Mech 283:341–363

Izumi N, Parker G (2000) Linear stability analysis of channel inception: downstream-driven theory. J Fluid Mech 419:239–262

Jackson PS, Hunt JCR (1975) Turbulent wind flow over a low hill. Q J R Meteorol Soc 101:929–955

Jarvis GT, McKenzie D (1982) Mantle convection as a boundary layer phenomenon. Geophys J R Astron Soc 68:389–427

Jaupart C, Labrosse S, Mareschal J-C (2009) Temperatures, heat and energy in the mantle of the Earth. In: Bercovici D (ed) Mantle dynamics. Treatise on geophysics, vol 7. Elsevier, Amsterdam, pp 253–303

Jeffrey A (2004) Handbook of mathematical formulas and integrals, 3rd edn. Elsevier, Amsterdam

Jeffreys H (1925) The flow of water in an inclined channel of rectangular section. Philos Mag 49:793–807

Jeffreys H, Jeffreys B (1953) Methods of mathematical physics. Cambridge University Press, Cambridge

Jimenez J, Zufiria JA (1987) A boundary layer analysis of Rayleigh-Bénard convection at large Rayleigh number. J Fluid Mech 178:53–71

Jóhannesson T (2002a) The initiation of the 1996 jökulhlaup from Lake Grímsvötn, Iceland. In: Snorrason Á, Finnsdóttir HP, Moss ME (eds) The extremes of the extremes: extraordinary floods. IASH publ, vol 271, pp 57–64

Jóhannesson T (2002b) Propagation of a subglacial flood wave during the initiation of a jökulhlaup. Hydrol Sci J 47:417–434

Johnsen SJ, Clausen HB, Dansgaard W, Fuhrer K, Gundestrup N, Hammer CU, Iversen P, Jouzel J, Stauffer B, Steffensen JP (1992) Irregular glacial interstadials recorded in a new Greenland ice core. Nature 359:311–313

Jones M (1994) Mechanical principles of sediment deformation. In: Maltman A (ed) The geological deformation of sediments. Chapman and Hall, London, pp 37–71

Julien PY (1995) Erosion and sedimentation. Cambridge University Press, Cambridge

Kalnay E (2003) Atmospheric modeling, data assimilation and predictability. Cambridge University Press, Cambridge

Kamb B (1987) Glacier surge mechanism based on linked cavity configuration of the basal water conduit system. J Geophys Res 92:9083–9100

Kamb B (1991) Rheological nonlinearity and flow instability in the deforming bed mechanism of ice stream motion. J Geophys Res 96(B10):16585–16595

Kamb B, Raymond CF, Harrison WD, Engelhardt H, Echelmeyer KA, Humphrey N, Brugman MM, Pfeffer T (1985) Glacier surge mechanism: 1982–1983 surge of Variegated Glacier, Alaska. Science 227:469–479

Kamb WB (1970) Sliding motion of glaciers: theory and observation. Rev Geophys Space Phys 8:673–728

Kardar M, Parisi G, Zhang YC (1986) Dynamic scaling of growing interfaces. Phys Rev Lett 56:889–892

Kargel JS (2004) Mars—a warmer, wetter planet. Springer, Berlin

Kasting JF (1989) Long-term stability of the Earth's climate. Palaeogeogr Palaeoclimatol Palaeoecol 75:83–95

Kasting JF, Ackermann TP (1986) Climatic consequences of very high carbon dioxide levels in Earth's early atmosphere. Science 234:1383–1385

Kaye GWC, Laby TH (1960) Physical and chemical constants, 12th edn. Longman, Harlow

Keener JP (1980) Waves in excitable media. SIAM J Appl Math 39:528–548

Keener JP (1986) A geometrical theory for spiral waves in excitable media. SIAM J Appl Math 46:1039–1056

Keller JB, Rubinow SI (1981) Recurrent precipitation and Liesegang rings. J Chem Phys 74:5000–5007

Kennedy JF (1963) The mechanics of dunes and anti-dunes in erodible-bed channels. J Fluid Mech 16:521–544

Kern R, Weisbrod A (1967) Thermodynamics for geologists. Translation from French by D McKie. Freeman Cooper and Co, San Francisco

Kevorkian J, Cole JD (1981) Perturbation methods in applied mathematics. Springer, Berlin

Kinahan GH, Close MH (1872) The general glaciation of Iar-Connaught and its neighbourhood, in the counties of Galway and Mayo. Hodges, Foster and Co, Dublin

Kleman J, Hättestrand C (1999) Frozen-bed Fennoscandian and Laurentide ice sheets during the Last Glacial Maximum. Nature 402:63–66

Knighton D (1998) Fluvial forms and processes: a new perspective. Arnold, London

Knittle E, Jeanloz R (1991) Earth's core–mantle boundary: results of experiments at high pressures and temperatures. Science 251:1438–1443

Koestler A (1964) The sleepwalkers. Penguin Books, London

Köhler P, Fischer H (2006) Simulating low frequency changes in atmospheric CO_2 during the last 740000 years. Clim Past 2:57–78

Köhler P, Fischer H, Munhoven G, Zeebe RE (2005) Quantitative interpretation of atmospheric carbon records over the last glacial termination. Glob Biogeochem Cycles 19:GB4020. doi:10.1029/2004GB002345

Kopell N, Howard LN (1973) Plane wave solutions to reaction-diffusion equations. Stud Appl Math 42:291–328

Korteweg DJ, de Vries G (1895) On the change of form of long waves advancing in a rectangular canal, and on a new type of long stationary waves. Philos Mag Ser 5 39:422–443

Kramer S, Marder M (1992) Evolution of river networks. Phys Rev Lett 68:205–208

Krauskopf KB, Bird DK (1995) Introduction to geochemistry. McGraw-Hill, New York

Kroy K, Sauermann G, Herrmann HJ (2002a) Minimal model for sand dunes. Phys Rev Lett 88:054301

Kroy K, Sauermann G, Herrmann HJ (2002b) Minimal model for aeolian sand dunes. Phys Rev E 66:031302

Kurz W, Fisher DJ (1998) Fundamentals of solidification, 4th edn. Trans Tech, Zurich

Lakin WD, Ng BS, Reid WH (1978) Approximations to the eigenvalue relation for the Orr-Sommerfeld problem. Philos Trans R Soc 289:347–371

Lamb, Sir Horace (1945) Hydrodynamics, 6th edn. Dover reprint of the 1932 sixth edition. Dover, New York

Lambe TW, Whitman RV (1979) Soil mechanics, SI version. Wiley, New York

Lang C, Leuenberger M, Schwander J, Johnsen S (1999) 16°C rapid temperature variation in central Greenland 70000 years ago. Science 286:934–937

Lasaga AC, Berner RA, Garrels RM (1985) An improved geochemical model of atmospheric CO_2 fluctuations over the past 100 million years. In: Sundquist ET, Broecker WS (eds) The carbon cycle and atmospheric CO_2: natural variations Archaen to present. AGU, Washington, pp 397–411

Le Grand HE (1988) Drifting continents and shifting theories. Cambridge University Press, Cambridge

Leuenberger MC, Lang C, Schwander J (1999) Delta ^{15}N measurements as a calibration tool for the paleothermometer and gas–ice age differences: a case study for the 8200 BP event on GRIP ice. J Geophys Res 104(D18):22163–22170

Li M, Richmond O (1997) Intrinsic instability and non-uniformity of plastic deformation. Int J Plast 13:765–784

Lighthill MJ, Whitham GB (1955a) On kinematic waves. I. Flood movement in long rivers. Proc R Soc Lond A 229:281–316

Lighthill MJ, Whitham GB (1955b) On kinematic waves. II. A theory of traffic flow on long, crowded roads. Proc R Soc Lond A 229:317–345

Lin CC, Segel LA (1974) Mathematics applied to deterministic problems in the natural sciences. MacMillan, New York

Liñán A, Williams FA (1993) Fundamental aspects of combustion. Oxford University Press, Oxford

Liou KN (2002) An introduction to atmospheric radiation, 2nd edn. Academic Press, San Diego

Lister JR, Kerr RC (1991) Fluid-mechanical models of crack propagation and their application to magma transport in dykes. J Geophys Res 96:10049–10077

Lliboutry LA (1956) La mécanique des glaciers en particulier au voisinage de leur front. Ann Geophys 12:245–276

Lliboutry LA (1958a) La dynamique de la Mer de Glace et la vague de 1891–95 d'après les mésures de Joseph Vallot. In: Physics of the movement of ice (Chamonix symposium). IAHS publ, vol 47. IAHS Press, Wallingford, pp 125–138. Available to download at http://www.iahs.info/redbooks/047.htm

Lliboutry LA (1958b) Glacier mechanics in the perfect plasticity theory. J Glaciol 3:162–169

Lliboutry LA (1964) Traité de glaciologie, vol I. Glace, neige, hydrologie nivale. Masson, Paris

Lliboutry LA (1965) Traité de glaciologie, vol II. Glaciers, variations du climat, sols gelés. Masson, Paris

Lliboutry LA (1968) General theory of subglacial cavitation and sliding of temperate glaciers. J Glaciol 7:21–58

Lliboutry LA (1979) Local friction laws for glaciers: a critical review and new openings. J Glaciol 23:67–95

Lliboutry LA (1987) Very slow flows of solids. Basics of modeling in geodynamics and glaciology. Martinus Nijhoff, Dordrecht

Loewenherz DS (1991) Stability and the initiation of channelized surface drainage: a reassessment of the short wavelength limit. J Geophys Res 96:8453–8464

Loewenherz-Lawrence DS (1994) Hydrodynamic description for advective sediment transport processes and rill initiation. Water Resour Res 30:3203–3212

Loper DE (ed) (1987) Structure and dynamics of partially solidified systems. Martinus Nijhoff, Dordrecht

Lorenz EN (1963) Deterministic non-periodic flow. J Atmos Sci 20:130–141

Lynch DK (1982) Tidal bores. Sci Am 247:131–143

Lynch P (2006) The emergence of numerical weather prediction: Richardson's dream. Cambridge University Press, Cambridge

Maaløe S (1978) The origin of rhythmic layering. Mineral Mag 42:337–345

MacAyeal DR (1989) Large-scale ice flow over a viscous basal sediment. J Geophys Res 94:4071–4087

MacAyeal DR (1993) Binge/purge oscillations of the Laurentide ice sheet as a cause of the North Atlantic's Heinrich events. Paleoceanography 8:775–784

Malkus WVR, Veronis G (1958) Finite amplitude cellular convection. J Fluid Mech 4:225–260

Manabe S, Stouffer RJ (1995) Simulation of abrupt climate change induced by freshwater input to the North Atlantic Ocean. Nature 378:165–167

Marsh BD (1982) On the mechanics of igneous diapirism, stoping, and zone melting. Am J Sci 282:808–855

Marsh SP, Glicksman ME (1996) Overview of geometric effects on coarsening of mushy zones. Metall Mater Trans 27A:557–567

Marshall HG, Walker JCG, Kuhn WR (1988) Long term climate change and the geochemical cycle of carbon. J Geophys Res 93:791–801

Mason B, Moore CB (1982) Principles of geochemistry, 4th edn. Wiley, Chichester

Massey BS (1986) Measures in science and engineering. Ellis Horwood, Chichester

Matson LE (2007) The Malkus–Lorenz water wheel revisited. Am J Phys 75:1114–1122

Matuszkiewicz A, Flamand JC, Bouré JA (1987) The bubble-slug flow pattern transition and instabilities of void-fraction waves. Int J Multiph Flow 13:199–217

Mayer KU, Benner SG, Frind EO, Thornton SF, Lerner DN (2001) Reactive transport modeling of processes controlling the distribution and natural attenuation of phenolic compounds in a deep sandstone aquifer. J Contam Hydrol 53:341–368

McBirney AR (1984) Igneous petrology. Freeman Cooper and Co, San Francisco

McBirney AR, Noyes RM (1979) Crystallisation and layering of the Skaergaard intrusion. J Pet 20:487–554

McCoy RM (2006) Ending in ice: the revolutionary idea and tragic expedition of Alfred Wegener. Oxford University Press, Oxford

McKenzie DP (1984) The generation and compaction of partially molten rock. J Pet 25:713–765

McNutt MK (2006) Another nail in the plume coffin? Science 313:1394

Meinhardt H (1982) Models of biological pattern formation. Academic Press, New York

Meinhardt H (1995) The algorithmic beauty of sea shells. Springer, Berlin

Melnik O (2000) Dynamics of two-phase conduit flow of high-viscosity gas-saturated magma: large variations of sustained explosive eruption intensity. Bull Volcanol 62:153–170

Menzies J (1984) Drumlins: a bibliography. Geo Books, Norwich

Meyer-Peter E, Müller R (1948) Formulas for bed-load transport. In: Proc int assoc hydraul res, 3rd annual conference, Stockholm, pp 39–64

Miller RN (2007) Numerical modelling of ocean circulation. Cambridge University Press, Cambridge

Millero FJ (1995) Thermodynamics of the carbon dioxide system in the oceans. Geochim Cosmochim Acta 59:661–677

Monod J (1949) The growth of bacterial cultures. Annu Rev Microbiol 3:371–394

Moore DR, Weiss NO (1973) Two-dimensional Rayleigh-Bénard convection. J Fluid Mech 58:289–312

Moore PL, Iverson NR (2002) Slow episodic shear of granular materials regulated by dilatant strengthening. Geology 30:843–846

Moresi L-N, Solomatov VS (1995) Numerical investigation of 2D convection with extremely large viscosity variations. Phys Fluids 7:2154–2162

Moresi L-N, Solomatov VS (1998) Mantle convection with a brittle lithosphere: thoughts on the global tectonic styles of the Earth and Venus. Geophys J Int 133:669–682

Morgan JP, Blackman DK, Sinton JM (eds) (1992) Mantle flow and melt generation at mid-ocean ridges. Geophysical monograph, vol 71. AGU, Washington

Morgan WJ (1971) Convection plumes in the lower mantle. Nature 230:42–43

Morland LW (1976a) Glacier sliding down an inclined wavy bed. J Glaciol 17:447–462

Morland LW (1976b) Glacier sliding down an inclined wavy bed with friction. J Glaciol 17:463–477

Morland LW (1984) Thermo-mechanical balances of ice sheet flow. Geophys Astrophys Fluid Dyn 29:237–266

Morland LW, Johnson IR (1980) Steady motion of ice sheets. J Glaciol 25:229–246

Morland LW, Shoemaker EM (1982) Ice shelf balances. Cold Reg Sci Technol 5:235–251

Morris S (1982) The effects of strongly temperature-dependent viscosity on slow flow past a hot sphere. J Fluid Mech 124:1–26

Morris S, Canright D (1984) A boundary layer analysis of Bénard convection in a fluid of strongly temperature-dependent viscosity. Phys Earth Planet Inter 29:320–329

Munhoven G, François LM (1996) Glacial–interglacial variability of atmospheric CO_2 due to changing continental silicate rock weathering: a model study. J Geophys Res 101:21423–21437

Murray JD (2002) Mathematical biology, 2 volumes. Springer, Berlin

Muskhelishvili NI (1953) Singular integral equations (Translation edited by JRM Radok). Noordhoff, Groningen

Nataf H-C, Richter FM (1982) Convection experiments in fluids with highly temperature-dependent viscosity and the thermal evolution of the planets. Phys Earth Planet Inter 29:320–329

Nayfeh AH (1973) Perturbation methods. Wiley-Interscience, New York

Needham DJ, Merkin JH (1984) On roll waves down an open inclined channel. Proc R Soc Lond A 394:259–278

Newell AC (1985) Solitons in mathematics and physics. Society for Industrial and Applied Mathematics, Philadelphia

Newell AC, Whitehead JA (1969) Finite bandwidth, finite amplitude convection. J Fluid Mech 38:279–303

Ng FSL (1998) Mathematical modelling of subglacial drainage and erosion. DPhil thesis, Oxford University

Ng F, Björnsson H (2003) On the Clague-Mathews relation for jökulhlaups. J Glaciol 49:161–172

Ng FSL, Zuber MT (2003) Albedo feedback in the patterning mechanisms of Martian polar caps. In: 3rd international conference on Mars polar science and exploration, abstract #8061, Lunar and Planetary Science Institute, Houston (CD-ROM)

Ng FSL, Zuber MT (2006) Patterning instability on the Mars polar ice caps. J Geophys Res 111:E02005. doi:10.1029/2005JE002533

Nicolas A (1986) A melt extraction model based on structural studies in mantle peridotites. J Pet 27:999–1022

Nicolussi K (1990) Bilddokumente zur Geschichte des Vernagtferners im 17. Jahrhundert. Zeit Gletschkd Glazialgeol 26(2):97–119

Nienow P, Sharp M, Willis I (1998) Seasonal changes in the morphology of the subglacial drainage system, Haut Glacier d'Arolla. Switz Earth Surf Process Landf 23:825–843

Noble B (1988) Methods based on the Wiener–Hopf technique, 2nd (unaltered) edn. Chelsea, New York

Nockolds SR, O'B Knox RW, Chinner GA (1978) Petrology for students. Cambridge University Press, Cambridge

Nordstrom DK, Munoz JL (1994) Geochemical thermodynamics, 2nd edn. Blackwell Scientific Publications, Cambridge

North GR (1975a) Analytical solution to a simple climate model with diffusive heat transport. J Atmos Sci 32:1301–1307

North GR (1975b) Theory of energy-balance climate models. J Atmos Sci 32:2033–2043

North GR, Mengel JG, Short DA (1983) Simple energy balance model resolving the season and continents: applications to astronomical theory of ice ages. J Geophys Res 88:6576–6586

Nowicki SMJ, Wingham DJ (2008) Conditions for a steady ice sheet–ice shelf junction. Earth Planet Sci Lett 265:246–255

Nye JF (1953) The flow law of ice from measurements in glacier tunnels, laboratory experiments and the Jungfraufirn borehole experiment. Proc R Soc Lond A 219:477–489

Nye JF (1957) Glacier mechanics; comments on Professor L Lliboutry's paper. J Glaciol 3:91–93

Nye JF (1958) Comments on Professor Lliboutry's paper. J Glaciol 3:170–172

Nye JF (1959) The motion of ice sheets and glaciers. J Glaciol 3:493–507

Nye JF (1960) The response of glaciers and ice sheets to seasonal and climatic changes. Proc R Soc Lond A 256:559–584

Nye JF (1963) The response of a glacier to changes in the rate of nourishment and wastage. Proc R Soc Lond A 275:87–112

Nye JF (1967) Theory of regelation. Philos Mag Ser 8 16(144):1249–1266

Nye JF (1969) A calculation on the sliding of ice over a wavy surface using a Newtonian viscous approximation. Proc R Soc Lond A 311:445–477

Nye JF (1970) Glacier sliding without cavitation in a linear viscous approximation. Proc R Soc Lond A 315:381–403

Nye JF (1973) Water at the bed of a glacier. IASH Publ 95:189–194

Nye JF (1976) Water flow in glaciers: jökulhlaups, tunnels, and veins. J Glaciol 17:181–207

Ockendon H, Ockendon JR (2004) Waves and compressible flow. Springer, New York

Olbers D (2001) A gallery of simple models from climate physics. Prog Probab 49:3–63

Olson P, Corcos GM (1980) A boundary layer model for mantle convection with surface plates. Geophys J R Astron Soc 62:195–219

Olver FWJ (1974) Asymptotics and special functions. Academic Press, New York

O'Malley K, Fitt AD, Jones TV, Ockendon JR, Wilmott P (1991) Models for high-Reynolds-number flow down a step. J Fluid Mech 222:139–155

Oreskes N (1999) The rejection of continental drift. Oxford University Press, New York

Orme AR (2007) The rise and fall of the Davisian cycle of erosion: prelude, fugue, coda, and sequel. Phys Geogr 28:474–506

Orszag SA, Patera AT (1983) Secondary instability of wall-bounded shear flows. J Fluid Mech 128:347–385

Ortoleva P (1994) Geochemical self-organisation. Oxford University Press, Oxford

Parker G (1975) Sediment inertia as a cause of river antidunes. J Hydraul Div ASCE 101:211–221

Parker G (1978) Self-formed straight rivers with equilibrium banks and mobile bed. Part 1. The sand-silt river. J Fluid Mech 89:109–125

Parker G (2004) 1D sediment transport morphodynamics with applications to rivers and turbidity currents. http://vtchl.uiuc.edu/people/parkerg/morphodynamics_e-book.htm

Parsons B, Sclater JG (1977) An analysis of the variation of ocean floor depth and heat flow with age. J Geophys Res 82:803–827

Parsons DR, Walker IJ, Wiggs GFS (2004) Numerical modelling of flow structures over idealized transverse aeolian dunes of varying geometry. Geomorphology 59:149–164

Parsons I (ed) (1987) Origins of igneous layering. NATO ASI series C, vol 196. Reidel, Dordrecht

Parteli EJR, Durán O, Herrmann HJ (2007) Minimal size of a barchan dune. Phys Rev E 75:011301

Parteli EJR, Durán O, Tsoar H, Schwämmle V, Herrmann HJ (2009) Dune formation under bi-modal winds. Proc Natl Acad Sci 106:22085–22089

Paterson WSB (1994) The physics of glaciers, 3rd edn. Pergamon, Oxford

Pattyn F, de Smedt B, Souchez R (2004) Influence of subglacial Vostok lake on the regional ice dynamics of the Antarctic ice sheet: a model study. J Glaciol 50:583–589

Payne AJ, Dongelmans PW (1997) Self-organization in the thermomechanical flow of ice sheets. J Geophys Res 102:12219–12233

Pearson JRA (1958) On convection cells induced by surface tension. J Fluid Mech 4:489–500

Pedlosky J (1987) Geophysical fluid dynamics, 2nd edn. Springer, Berlin

Pelletier JD (2004) How do spiral troughs form on Mars? Geology 32(4):365–367

Pelletier JD (2008) Quantitative modeling of Earth surface processes. Cambridge University Press, Cambridge

Peregrine DH (1966) Calculations of the development of an undular bore. J Fluid Mech 25:321–330

Petford N, Lister JR, Kerr RC (1994) The ascent of felsic magmas in dykes. Lithos 32:161–168

Petit JR, Jouzel J, Raynaud D, Barkov NI, Barnola J-M, Basile I, Bender M, Chappellaz J, Davis M, Delaygue G, Delmotte M, Kotlyakov VM, Legrand M, Lipenkov VY, Lorius C, Pépin L, Ritz C, Saltzman E, Stievenard M (1999) Climate and atmospheric history of the past 420000 years from the Vostok ice core, Antarctica. Nature 399:429–436

Picioreanu C, van Loosdrecht MCM, Heijnen JJ (1998) Mathematical modeling of biofilm structure with a hybrid differential-discrete cellular automaton approach. Biotechnol Bioeng 58:101–116

Pierrehumbert RT (2004) High atmospheric carbon dioxide necessary for the termination of global glaciation. Nature 429:646–648

Pillow AF (1952) The free convection cell in two dimensions. Dept of Supply, Aeronautical Re-search Laboratories, Report A.79 [Dept of Mathematics, University of Queensland, St Lucia, Queensland 4067, Australia]

Pitcher WS (1997) The nature and origin of granite, 2nd edn. Chapman and Hall, London

Polubarinova-Kochina PYa (1962) Theory of ground water movement. Princeton University Press, Princeton

Price M (1985) Introducing groundwater. George Allen and Unwin, London

Prosperetti A, Satrape JV (1990) Stability of two-phase flow models. In: Joseph DD, Schaeffer DG (eds) Two-phase flow models and waves. Springer, New York, pp 98–117

Pugh DT (1987) Tides, surges and mean sea-level. Wiley, Chichester

Pye K, Tsoar H (1990) Aeolian sand and sand dunes. Unwin Hyman, London

Quareni F, Yuen DA, Sewell G, Christensen UR (1985) High Rayleigh number convection with strongly variable viscosity: a comparison between mean field and two-dimensional solutions. J Geophys Res 90:12633–12644

Rahmstorf S (1995) Bifurcations of the Atlantic thermohaline circulation in response to changes in the hydrological cycle. Nature 378:145–149

Rahmstorf S (2002) Ocean circulation and climate during the past 120000 years. Nature 419:207–214

Rathbun AP, Marone C, Alley RB, Anandakrishnan S (2008) Laboratory study of the frictional rheology of sheared till. J Geophys Res 113:F02020. doi:10.1029/2007JF000815

Rayleigh, Lord (1908) Note on tidal bores. Proc R Soc Lond A 81:448–449

Rayleigh, Lord (1916) On convective currents in a horizontal layer of fluid when the higher tem-perature is on the under side. Philos Mag 32:529–546

Reese CC, Solomatov VS (2002) Mean field heat transfer scaling for non-Newtonian stagnant lid convection. J Non-Newton Fluid Mech 107:39–49

Reese CC, Solomatov VS, Moresi L-N (1999) Non-Newtonian stagnant lid convection and mag-matic resurfacing on Venus. Icarus 139:67–80

Reid WH (1972) Composite approximations to the solutions of the Orr-Sommerfeld equation. Stud Appl Math 51:341–368

Rempel AW, Wettlaufer JS, Worster MG (2004) Premelting dynamics in a continuum model of frost heave. J Fluid Mech 498:227–244

Reynolds AJ (1965) Waves on the erodible bed of an open channel. J Fluid Mech 22:113–133

Reynolds O (1895) On the dynamical theory of incompressible viscous fluids and the determination of the criterion. Philos Trans R Soc Lond A 186:123–164

Ribe NM (2009) Analytical approaches to mantle dynamics. In: Bercovici D (ed) Mantle dynamics. Treatise on geophysics, vol 7. Elsevier, Amsterdam, pp 167–226

Richards K (1982) Rivers: form and process in alluvial channels. Methuen, London

Richards KJ (1980) The formation of ripples and dunes on an erodible bed. J Fluid Mech 99:597–618

Richardson CN, Lister JR, McKenzie D (1996) Melt conduits in a viscous porous matrix. J Geophys Res 101:20423–20432

Rittmann BE, McCarty PL (1980) Model of steady-state-biofilm kinetics. Biotechnol Bioeng 22:2343–2357

Roberts GO (1977) Fast viscous convection. Geophys Astrophys Fluid Dyn 8:197–233

Roberts GO (1979) Fast viscous Bénard convection. Geophys Astrophys Fluid Dyn 12:235–272

Roberts MJ (2005) Jökulhlaups: a reassessment of floodwater flow through glaciers. Rev Geophys 43:RG1002

Robin G de Q (1955) Ice movement and temperature distribution in glaciers and ice sheets. J Glaciol 2:523–532

Robinson JR (1967) Finite amplitude convection cells. J Fluid Mech 30:577–600

Rodríguez-Iturbe I, Rinaldo A (1997) Fractal river basins. Cambridge University Press, Cambridge

Röthlisberger H (1972) Water pressure in intra- and subglacial channels. J Glaciol 11:177–203

Rowbotham F (1970) The Severn bore, 2nd edn. David and Charles, Newton Abbot

Rubinstein J, Mauri R (1986) Dispersion and convection in periodic porous media. SIAM J Appl Math 46:1018–1023

Ruddiman WF (2001) Earth's climate: past and future. Freeman, New York

Ryan MP (ed) (1990) Magma transport and storage. Wiley, Chichester

Saffman PG (1959) A theory of dispersion in a porous medium. J Fluid Mech 6:321–349

Sahimi M (1995) Flow and transport in porous media and fractured rock. VCH, Weinheim

Samarskii AA, Galaktionov VA, Kurdyumov SP, Mikhailov AP (1995) Blow-up in quasilinear parabolic equations. de Gruyter expositions in mathematics, vol 19. de Gruyter, Berlin

Sanchez-Palencia E (1983) Homogenization method for the study of composite media. Springer, Berlin

Sass BM, Rosenberg PE, Kittrick JA (1987) The stability of illite/smectite during diagenesis: an experimental study. Geochim Cosmochim Acta 51:2103–2115

Sauermann G, Kroy K, Herrmann HJ (2001) Continuum saltation model for sand dunes. Phys Rev E 64:031305

Sayag R, Tziperman E (2008) Spontaneous generation of pure ice streams via flow instability: role of longitudinal shear stresses and subglacial till. J Geophys Res 113:B05411. doi:10.1029/2007JB005228

Schlichting H (1979) Boundary layer theory. McGraw-Hill, New York

Schmidt MW, Vautravers MJ, Spero HJ (2006) Rapid subtropical North Atlantic salinity oscillations across Dansgaard–Oeschger cycles. Nature 443:561–564

Schoof C (2005) The effect of cavitation on glacier sliding. Proc R Soc Lond A 461:609–627

Schoof C (2007a) Pressure-dependent viscosity and interfacial instability in coupled ice-sediment flow. J Fluid Mech 570:227–252

Schoof C (2007b) Marine ice-sheet dynamics. Part 1. The case of rapid sliding. J Fluid Mech 573:27–55

Schoof C (2007c) Ice sheet grounding line dynamics: steady states, stability and hysteresis. J Geophys Res 112:F03S28. doi:10.1029/2006JF000664

Schoof C, Hindmarsh RCA (2010) Thin-film flows with wall slip: an asymptotic analysis of higher order glacier flow models. Q J Mech Appl Math 63:73–114

Schubert G, Turcotte DL, Olson P (2001) Mantle convection in the Earth and planets. Cambridge University Press, Cambridge

Schulte P et al (2010) The Chicxulub asteroid impact and mass extinction at the Cretaceous-Paleogene boundary. Science 327:1214–1218

Schwämmle V, Herrmann H (2004) Modelling transverse dunes. Earth Surf Process Landf 29:769–784

Scott DR, Stevenson DJ (1984) Magma solitons. Geophys Res Lett 11:1161–1164

Scott DR, Stevenson DJ, Whitehead JA (1986) Observations of solitary waves in a viscously deformable pipe. Nature 319:759–761

Segel LA (1969) Distant side-walls cause slow amplitude modulation of cellular convection. J Fluid Mech 38:203–224

Selby MJ (1993) Hillslope materials and processes, 2nd edn. Oxford University Press, Oxford

Sellers WD (1969) A climate model based on the energy balance of the Earth-atmosphere system. J Appl Meteorol 8:392–400

Sellmeijer JB, Koenders MA (1991) A mathematical model for piping. Appl Math Model 15:646–651

Severinghaus JP, Brook EJ (1999) Abrupt climate change at the end of the last glacial period inferred from trapped air in polar ice. Science 286:930–934

Shapiro MA, Keyser DA (1990) Fronts, jet streams and the tropopause. In: Newton CW, Holopainen EO (eds) Extratropical cyclones. The Erik Palmén memorial volume. Amer Met Soc, Boston, pp 167–191

Sharpe D (2005) Comments on: "Paleohydraulics of the last outburst flood from glacial Lake Agassiz and the 8200 BP cold event" by Clarke et al. [Quat Sci Rev 23:389–407 (2004)]. Quat Sci Revs 24:1529–1532

Shaw J (1983) Drumlin formation related to inverted meltwater erosional marks. J Glaciol 29:461–479

Shaw J, Kvill D, Rains B (1989) Drumlins and catastrophic subglacial floods. Sediment Geol 62:177–202

Shields A (1936) Anwendung der Ähnlichkeits mechanik und der Turbulenzforschung auf die Geschiebebewegung. Mitteilung der Preussischen Versuchanstalt für Wasserbau und Schiffbau, Heft 26, Berlin

Shreve RL (1985) Esker characteristics in terms of glacier physics, Katahdin esker system, Maine. Geol Soc Amer Bull 96:639–646

Siegert MJ (2005) Lakes beneath the ice sheet: the occurrence, analysis, and future exploration of Lake Vostok and other Antarctic subglacial lakes. Annu Rev Earth Planet Sci 33:215–245

Siegert MJ, Dowdeswell JA, Gorman MR, McIntyre NF (1996) An inventory of Antarctic subglacial lakes. Antarct Sci 8:281–286

Siegert MJ, Ellis-Evans JC, Tranter M, Mayer C, Petit J-R, Salamatin A, Priscu JC (2001) Physical, chemical and biological processes in Lake Vostok and other Antarctic subglacial lakes. Nature 414:603–608

Sigurdsson H (ed) (2000) Encyclopedia of volcanoes. Academic Press, San Diego

Sih GC (ed) (1973) Methods of analysis and solutions of crack problems. Noordhoff, Leyden

Sinclair, Sir John (ed) (1791–1799) The statistical account of Scotland: drawn up from the communications of the ministers of the different parishes, 21 vols. William Creech, Edinburgh

Smith JD (1970) Stability of a sand bed subjected to a shear flow of low Froude number. J Geophys Res 75:5928–5940

Smith JD, McLean SR (1977) Spatially averaged flow over a wavy surface. J Geophys Res 83:1735–1745

Smith JE (1971) The dynamics of shale compaction and evolution in pore-fluid pressures. Math Geol 3:239–263

Smith TR (2010) A theory for the emergence of channelized drainage. J Geophys Res 115:F02023. doi:10.1029/2008JF001114

Smith TR, Bretherton FP (1972) Stability and the conservation of mass in drainage basin evolution. Water Resour Res 8:1506–1529

Smith TR, Birnir B, Merchant GE (1997a) Towards an elementary theory of drainage basin evolution: I. The theoretical basis. Comput Geosci 23:811–822

Smith TR, Birnir B, Merchant GE (1997b) Towards an elementary theory of drainage basin evolution: II. A computational evaluation. Comput Geosci 23:823–849

Sneddon IN, Lowengrub M (1969) Crack problems in the classical theory of elasticity. Wiley, New York

Solomatov VS (1996) Stagnant lid convection on Venus. J Geophys Res 101:4737–4753

Solomon S (1999) Stratospheric ozone depletion: a review of concepts and history. Rev Geophys 37:275–316

Sparks RSJ, Huppert HE, Koyaguchi T, Hallworth MA (1993) Origin of modal and rhythmic igneous layering by sedimentation in a convecting magma chamber. Nature 361:246–249

Sparks RSJ, Bursik MI, Carey SN, Gilbert JS, Glaze LS, Sigurdsson H, Woods AW (1997) Volcanic plumes. Wiley, Chichester

Sparrow C (1982) The Lorenz equations: bifurcations, chaos, and strange attractors. Springer, New York

Sparrow EM, Cess RD (1978) Radiation heat transfer. Hemisphere, Belmont

Spence DA, Turcotte DL (1985) Magma driven propagation of cracks. J Geophys Res 90:575–580

Spence DA, Sharp PW, Turcotte DL (1987) Buoyancy-driven crack propagation: a mechanism for magma migration. J Fluid Mech 174:135–153

Spiegelman M, Kelemen PB, Aharonov E (2001) Causes and consequences of flow organization during melt transport: the reaction infiltration instability in compactible media. J Geophys Res 106:2061–2078

Spohn T, Hort M, Fischer H (1988) Numerical simulation of the crystallization of multicomponent melts in thin dikes or sills. 1. The liquidus phase. J Geophys Res 93:4880–4894

Spring U, Hutter K (1981) Numerical studies of jökulhlaups. Cold Reg Sci Technol 4:227–244

Spring U, Hutter K (1982) Conduit flow of a fluid through its solid phase and its application to intraglacial channel flow. Int J Eng Sci 20:327–363

Starostin AB, Barmin AA, Melnik OE (2005) A transient model for explosive and phreatomagmatic eruptions. J Volcanol Geotherm Res 143:133–151

Stern ME (1960) The 'salt fountain' and thermohaline convection. Tellus 12:172–175

Stevenson DJ (1989) Spontaneous small-scale melt segregation in partial melts undergoing deformation. Geophys Res Lett 16:1067–1070

Stocker TF, Johnsen SJ (2003) A minimum thermodynamic model for the bipolar seesaw. Paleoceanography 18:1087. doi:10.1029/2003PA000920

Stocker TF, Wright DG (1991) Rapid transitions of the ocean's deep circulation induced by changes in surface water fluxes. Nature 351:729–732

Stoker JJ (1957) Water waves: the mathematical theory with applications. Interscience, New York

Stommel H (1961) Thermohaline convection with two stable régimes of flow. Tellus 13:224–230

Strahler AN (1952) Hypsometric (area altitude) analysis of erosional topography. Geol Soc Amer Bull 63:1117–1142

Strutt JW (Lord Rayleigh) (1871) On the light from the sky, its polarization and colour. Phil Mag 41:107–120

Su MD, Xu X, Zhu JL, Hon YC (2001) Numerical simulation of tidal bore in Hangzhou Gulf and Qiantangjiang. Int J Numer Methods Fluids 36:205–247

Sugden D, Denton G (2004) Cenozoic landscape evolution of the Convoy Range to Mackay Glacier area, Transantarctic Mountains: onshore to offshore synthesis. Geol Soc Am Bull 116:840–857

Sugden DE, John BS (1976) Glaciers and landscape: a geomorphological approach. Edward Arnold, London

Sumer BM, Bakioglu M (1984) On the formation of ripples on an erodible bed. J Fluid Mech 144:177–190

Sykes RI (1980) An asymptotic theory of incompressible turbulent boundary-layer flow over a small hump. J Fluid Mech 101:647–670

Tackley P (1998) Self-consistent generation of tectonic plates in three-dimensional mantle convection. Earth Planet Sci Lett 157:9–22

Tackley P (2000a) Self-consistent generation of tectonic plates in time-dependent, three-dimensional mantle convection simulations. 1. Pseudoplastic yielding. Geochem Geophys Geosyst 1(8):1021. doi:10.1029/2000GC000036

Tackley P (2000b) Self-consistent generation of tectonic plates in time-dependent, three-dimensional mantle convection simulations. 2. Strain weakening and asthenosphere. Geochem Geophys Geosyst 1(8):1026. doi:10.1029/2000GC000043

Tackley PJ (2009) Mantle geochemical dynamics. In: Bercovici D (ed) Mantle dynamics. Treatise on geophysics, vol 7. Elsevier, Amsterdam, pp 437–505

Takahashi T, Sutherland SC, Sweeney C, Poisson A, Metzl N, Tilbrook B, Bates N, Wanninkhof R, Feely RA, Sabine C, Olafsson J, Nojiri Y (2002) Global sea–air CO_2 flux based on climatological surface ocean pCO_2, and seasonal biological and temperature effects. Deep-Sea Res II 49:1601–1622

Tayler AB (1986) Mathematical models in applied mechanics. Clarendon, Oxford

Taylor GI (1953) Dispersion of soluble matter in a solvent flowing slowly through a tube. Proc R Soc Lond A 219:186–203

Taylor KC, Lamorey GW, Doyle GA, Alley RB, Grootes PM, Mayewski PA, White JWC, Barlow LK (1993) The 'flickering switch' of late Pleistocene climate change. Nature 361:432–436

Taylor KC, Mayewski PA, Alley RB, Brook EJ, Gow AJ, Grootes PM, Meese DA, Saltzman ES, Severinghaus JP, Twickler MS, White JWC, Whitlow S, Zielinski GA (1997) The Holocene–Younger Dryas transition recorded at Summit, Greenland. Science 278:825–827

Teichman J, Mahadevan L (2003) The viscous catenary. J Fluid Mech 478:71–80

Thomas GE, Stamnes K (1999) Radiative transfer in the atmosphere and ocean. Cambridge University Press, Cambridge

Thomas RH (1979) The dynamics of marine ice sheets. J Glaciol 24:167–177

Toggweiler JR, Russell JL, Carson SR (2006) Mid-latitude westerlies, atmospheric CO_2, and climate change during the ice ages. Paleoceanography 21:PA2005. doi:10.1029/2005PA001154

Tricker RAR (1965) Bores, breakers, waves and wakes. Elsevier, New York

Tucker GE, Slingerland RL (1994) Erosional dynamics, flexural isostasy, and long-lived escarpments: a numerical modeling study. J Geophys Res 99:12229–12243

Tulaczyk SM, Kamb B, Engelhardt HF (2000) Basal mechanics of Ice Stream B, West Antarctica. I. Till mechanics. J Geophys Res 105(B1):463–481

Turcotte DL (1992) Fractals and chaos in geology and geophysics. Cambridge University Press, Cambridge

Turcotte DL, Ahern JL (1978) A porous flow model for magma migration in the asthenosphere. J Geophys Res 83:767–772

Turcotte DL, Oxburgh ER (1967) Finite amplitude convection cells and continental drift. J Fluid Mech 28:29–42

Turner JS (1973) Buoyancy effects in fluids. Cambridge University Press, Cambridge

Turner JS (1974) Double-diffusive phenomena. Annu Rev Fluid Mech 6:37–54

Vallis G (2006) Atmospheric and oceanic fluid dynamics. Cambridge University Press, Cambridge

Van der Veen CJ (1999) Fundamentals of glacier dynamics. Balkema, Rotterdam

Van Dyke MD (1975) Perturbation methods in fluid mechanics. Parabolic Press, Stanford

Van Rijn LC (1984) Sediment transport. Part II. Suspended load transport. J Hydraul Eng 110:1613–1641

Vosper SB, Mobbs SD, Gardiner BA (2002) Measurements of the near-surface flow over a hill. Q J Meteorol Soc 128:2257–2280

Waddington ED (1986) Wave ogives. J Glaciol 32:325–334

Wager LR, Brown GM (1968) Layered igneous rocks. Oliver and Boyd, Edinburgh

Waitt RB Jr (1984) Periodic jökulhlaups from Pleistocene Glacial Lake Missoula—new evidence from varved sediment in Northern Idaho and Washington. Quat Res 22:46–58

Walder JS (1982) Stability of sheet flow of water beneath temperate glaciers and implications for glacier surging. J Glaciol 28:273–293

Walder JS (1986) Hydraulics of subglacial cavities. J Glaciol 32:439–446

Walder JS, Costa JE (1996) Outburst floods from glacier-dammed lakes: the effect of mode of lake drainage on flood magnitude. Earth Surf Proc Landf 21:701–723

Walder JS, Fowler A (1994) Channelised subglacial drainage over a deformable bed. J Glaciol 40:3–15

Walder J, Hallet B (1979) Geometry of former subglacial water channels and cavIties. J Glaciol
    23:335–346
Walker JCG, Hays PB, Kasting JF (1981) A negative feedback mechanism for the long-term stabi-
    lization of Earth's surface temperature. J Geophys Res 86(C10):9776–9782
Walker G (2003) Snowball Earth: the story of the great global catastrophe that spawned life as we
    know it. Bloomsbury, London
Wallis GB (1969) One-dimensional two-phase flow. McGraw-Hill, New York
Wang Y, Merino E (1993) Oscillatory magma crystallisation by feedback between the concentra-
    tions of the reactant species and mineral growth rates. J Pet 34:369–382
Wanner O, Eberl H, Morgenroth E, Noguera DR, Picioreanu C, Rittmann BE, van Loosdrecht
    MCM (2006) Mathematical modeling of biofilms. Report of the IWA Biofilm Modeling Task
    Group, Scientific and Technical Report No 18, IWA Publishing, London
Ward RC, Robinson M (2000) Principles of hydrology, 4th edn. McGraw-Hill, New York
Warren WP, Ashley GM (1994) Origins of the ice-contact stratified ridges (eskers) of Ireland.
    J Sediment Res A 64:433–449
Watson GN (1944) A treatise on the theory of Bessel functions, 2nd edn. Cambridge University
    Press, Cambridge
Wealthall GP, Thornton SF, Lerner DN (2001) Natural attenuation of MTBE in a dual porosity
    aquifer. In: 6th international conference on in situ and on site bioremediation, San Diego, pp 59–
    66
Weertman J (1957a) On the sliding of glaciers. J Glaciol 3:33–38
Weertman J (1957b) Deformation of floating ice shelves. J Glaciol 3:39–42
Weertman J (1958) Travelling waves on glaciers. In: IUGG symposium, Chamonix. Int assoc hy-
    drol sci publ, vol 47, pp 162–168
Weertman J (1971) Velocity at which liquid-filled cracks move in the Earth's crust or in glaciers.
    J Geophys Res 76:8544–8553
Weertman J (1972) General theory of water flow at the base of a glacier or ice sheet. Rev Geophys
    Space Phys 10:287–333
Weertman J (1974) Stability of the junction of an ice sheet and an ice shelf. J Glaciol 13:3–11
Weertman J (1979) The unsolved general glacier sliding problem. J Glaciol 23:97–115
Wegener A (1966) The origin of continents and oceans, 4th edn, transl J Biram. Dover, New York
Weinberg RF, Podladchikov YY (1994) Diapiric ascent of magmas through power-law crust and
    mantle. J Geophys Res 99:9543–9559
Weng WS, Hunt JCR, Carruthers DJ, Warren A, Wiggs GFS, Livingstone I, Castro I (1991) Air
    flow and sand transport over sand-dunes. Acta Mech, Suppl 2:1–22
Wesseling P (1969) Laminar convection cells at high Rayleigh number. J Fluid Mech 36:625–637
Wettlaufer JS, Worster MG (2006) Premelting dynamics. Annu Rev Fluid Mech 38:427–452
Whalley PB (1987) Boiling, condensation, and gas-liquid flow. Clarendon, Oxford
Whitham GB (1974) Linear and nonlinear waves. Wiley, New York
Wilchinsky AV (2007) The effect of bottom boundary conditions in the ice-sheet to ice-shelf tran-
    sition zone problem. J Glaciol 53:363–367
Wilchinsky AV (2009) Linear stability analysis of an ice sheet interacting with the ocean. J Glaciol
    55:13–20
Wilchinsky AV, Chugunov VA (2000) Ice-stream–ice-shelf transition: theoretical analysis of two-
    dimensional flow. Ann Glaciol 30:153–162
Wilchinsky AV, Chugunov VA (2001) Modelling ice flow in various glacier zones. J Appl Math
    Mech 65:479–493. In Russian: Prikl Mat Mekh 65:495–510
Willett SD, Brandon MT (2002) On steady states in mountain belts. Geology 30:175–178
Willgoose G (2005) Mathematical modeling of whole landscape evolution. Annu Rev Earth Sci
    33:443–459
Willgoose G, Bras RL, Rodríguez-Iturbe I (1991) A coupled channel network growth and hillslope
    evolution model: I. Theory. Water Resour Res 27:1671–1684
Williams FA (1985) Combustion theory, 2nd edn. Benjamin/Cummings, Menlo Park
Winchester S (2001) The map that changed the world. Viking, Penguin Books, London

Wingham DJ, Siegert MJ, Shepherd A, Muir AS (2006) Rapid discharge connects Antarctic sub-glacial lakes. Nature 440:1033–1037

Winstanley H (2001) The formation of river networks. MSc dissertation, Oxford University

Wolanski E, Williams D, Spagnol S, Chanson H (2004) Undular tidal bore dynamics in the Daly Estuary, Northern Australia. Estuar Coast Shelf Sci 60:629–636

Worster MG (1997) Convection in mushy layers. Annu Rev Fluid Mech 29:91–122

Worster MG (2000) Solidification of fluids. In: Batchelor GK, Moffatt HK, Worster MG (eds) Perspectives in fluid dynamics. Cambridge University Press, Cambridge, pp 393–446

Worster MG, Huppert HE, Sparks RSJ (1990) Convection and crystallization in magma cooled from above. Earth Planet Sci Lett 101:78–89

Yang X-S (2000) Nonlinear viscoelastic compaction in sedimentary basins. Nonlinear Process Geophys 7:1–7

Yang X-S (2008) Mathematical modelling for earth scientists. Dunedin Academic Press, Edinburgh

Yu J, Kevorkian J (1992) Nonlinear evolution of small disturbances into roll waves in an inclined open channel. J Fluid Mech 243:575–594

Yuen DA, Schubert G (1979) The role of shear heating in the dynamics of large ice masses. J Glaciol 24:195–212

Yuen DA, Maruyama S, Karato S-I, Windley BF (eds) (2007) Superplumes: beyond plate tectonics. Springer, Dordrecht

Zammett RJ, Fowler AC (2010) The morphology of the Martian ice caps: a mathematical model of ice-dust kinetics. SIAM J Appl Math 70:2409–2433

Zeebe RE, Wolf-Gladrow D (2001) CO_2 in seawater: equilibrium, kinetics, isotopes. Elsevier, Amsterdam

Index

Printed by Printforce, United Kingdom